Laser Remote Sensing

OPTICAL ENGINEERING

Founding Editor
Brian J. Thompson
University of Rochester
Rochester, New York

Laser Remote Sensing

edited by
Takashi Fujii
Central Research Institute of Electrical Power Industry
Tokyo, Japan

Tetsuo Fukuchi
Central Research Institute of Electrical Power Industry
Tokyo, Japan

CRC Press
Taylor & Francis Group
Boca Raton London New York

CRC Press is an imprint of the
Taylor & Francis Group, an **informa** business

CRC Press
Taylor & Francis Group
6000 Broken Sound Parkway NW, Suite 300
Boca Raton, FL 33487-2742

First issued in paperback 2019

© 2005 by Taylor & Francis Group, LLC
CRC Press is an imprint of Taylor & Francis Group, an Informa business

No claim to original U.S. Government works

ISBN-13: 978-0-8247-4256-0 (hbk)
ISBN-13: 978-0-367-39252-9 (pbk)

Visit the Taylor & Francis Web site at
http://www.taylorandfrancis.com

and the CRC Press Web site at
http://www.crcpress.com

Preface

With growing interest in the earth's atmospheric environment, measurement of aerosols, water vapor, clouds, winds, trace constituents, and temperature have increasingly become important for understanding the complex mechanisms that govern the atmosphere. Range-resolved measurements, which are frequently necessary for comparison with atmospheric model studies, can be most conveniently and efficiently performed by a laser remote sensor (lidar), which can, in principle, provide four-dimensional (space and time) maps of these quantities.

The last definitive book on lidar, *Laser Remote Sensing* by R. M. Measures, was published in 1984. Although the fundamentals and theory of operation remain since unchanged, numerous new technologies have emerged in lasers and optoelectronics, some of which have been successfully applied to lidar and have made large contributions to the development of novel lidar systems. Most notably, the wide range of solid-state laser sources that have recently become available has extended the probing wavelength range into the infrared and also made possible the development of more compact systems suitable for airborne or spaceborne platforms. The recent progress in ultrafast laser technology also offers an interesting new tool for atmospheric probing. Although information on the recent progress in lidar technology and atmospheric applications can be found in numerous journal articles and conference proceedings, there has been no work summarizing the recent advances and achievements in a comprehensive format.

The aim of this book is to provide an up-to-date, comprehensive review of lidar, focusing mainly on applications to current topics in atmospheric science. The scope of the book includes laser remote sensing of the atmosphere, including measurement of aerosols, water vapor, clouds, winds, trace constituents, and temperature. In addition, other interesting applications such as vegetation monitoring and altimetry are also covered. The lidar systems described herein include ground-based (fixed or mobile), airborne, and spaceborne (satellite based) systems. Emphasis is placed on instrumentation and measurement techniques, to enable the reader to understand what kind of lidar system is necessary for a particular application.

The individual chapters are self-contained and written by authors who are outstanding experts in their fields. Chapter 1 is an introductory chapter presenting brief basics of lidar as well as a summary of each chapter. Chapter 2 presents the newest lidar technology, using an ultrafast laser. Chapter 3 presents Mie lidar measurements of aerosols and clouds in the troposphere. Chapter 4 presents trace gas species measurements by different absorption lidars in the troposphere as well as the pump and probe OH lidar. Chapter 5 presents the measurements of temperature, wind, and constituent structures such as metal layers and polar mesospheric clouds in the middle and upper atmosphere by resonance fluorescence lidar and Rayleigh lidar. Chapter 6 presents fluorescence spectroscopy and imaging of lidar targets such as marine, terrestrial vegetation, and facades of historical buildings. Chapter 7 presents wind measurements by coherent and direct detection Doppler wind lidars. Chapters 8 and 9 present various lidar measurements from airborne and space-based platforms, respectively. Selected figures in color representing a wealth of information that can be obtained by lidar are collected in a color insert.

The book is intended for scientists, researchers, and students who are interested in the atmospheric environment and wish to learn about the measurement capabilities of state-of-the-art lidar systems. However, because the book only briefly covers the fundamentals, the reader is expected to be familiar with lidar theory and operation, as described in Measures' book.

We would like to thank each of the authors of the chapters for their outstanding contributions to this book. We also thank the staff of CRC Press, Taylor & Francis Group, Inc., for their collaboration and patience during the production of this book.

Takashi Fujii
Tetsuo Fukuchi
Tokyo, Japan

Contributors

Farzin Amzajerdian
Lidar Applications Group
NASA Langley Research Center
Hampton, Virginia

R. Bourayou
Lidar Applications Group
Institut für
Experimentalphysik
  Freie Universität Berlin
Berlin, Germany

Edward V. Browell
Lidar Applications Group
NASA Langley Research Center
Hampton, Virginia

Bertrand Calpini
Air Pollution Laboratory, Swiss
Federal Institute of Technology
Switzerland
Presently at
Measurement Teet Department
Aerological Station Payerne
Payerne, Switzerland

Xinzhao Chu
Department of Electrical and
Computer Engineering
University of Illinois at
  Urbana-Champaign
Urbana, Illinois

S. Frey
Lidar Applications Group
Laboratoire de Spectrométrie
  Ionique et Moléculaire
Université Claude Bernard
Villeurbanne Cedex, France

Phillip Gatt
Coherent Technologies Inc.
Louisville, Connecticut

William B. Grant
Lidar Applications Group
NASA Langley Research Center
Hampton, Virginia
Presently at
Sunlight, Nutrition and Health
Research Center
SUNARC
San Francisco, California

Sammy W. Henderson
Coherent Technologies Inc.
Louisville, California

R. Milton Huffaker
Coherent Technologies Inc.
Louisville, Connecticut

Syed Ismail
Lidar Applications Group
Atmospheric Sciences
  Research
NASA Langley Research Center
Hampton, Virginia

Jerome Kasparian
Laboratoire de Spectrométrie
  Ionique et Moléculaire
Université Claude Bernard
Villeurbanne Cedex, France

Michael J. Kavaya
NASA Langley Research Center
Hampton, Virginia

J. C. Luderer
Lidar Applications Group
Institut für
  Experimentalphysik
Freie Universität Berlin
Berlin, Germany

G. Méjean
Lidar Applications Group
Laboratoire de Spectrométrie
  Ionique et Moléculaire
Université Claude Bernard
Villeurbanne Cedex, France

George C. Papen
Department of Electrical and
  Computer Engineering
University of Illinois at Urbana-
  Champaign
Urbana, IL
Presently at
Department of Electrical and
  Computer Engineering
University of California
  at San Diego
La Jolla, California

David Rees
Hovemere Ltd
Arctic House
Kent, United Kingdom

M. Rodriguez
Institut für
  Experimentalphysik
Freie Universität Berlin
Berlin, Germany

E. Salmon
Laboratoire de Spectrométrie
  Ionique et Moléculaire
Université Claude Bernard
Villeurbanne Cedex, France

Valentin Simeonov
Air Pollution Laboratory, Swiss
  Federal Institute of Technology
Lausanne, Switzerland

Upendra N. Singh
NASA Langley Research Center
Hampton, Virginia

Sune Svanberg
Department of Physics
Lund Institute of Technology
Lund, Sweden

Nobuo Tekeuchi
Center for Environmental
  Remote Sensing
Cheba University
Cheba, Japan

Claus Weitkamp
Institut für Küstenforschung /
  Physikalische und Chemische
  Analytik
Geesthacht, Germany

H. Wille
Institut für
  Experimentalphysik, Freie
Universität Berlin
Berlin, Germany

David M. Winker
NASA Langley Research Center
Hampton, Virginia

J.-P. Wolf
Laboratoire de Spectrométrie
  Ionique et Moléculaire
Université Claude Bernard
Villeurbanne Cedex, France

L. Wöste
Institut für
  Experimentalphysik
Freie Universität Berlin
Berlin, Germany

J. Yu
Laboratoire de Spectrométrie
  Ionique et Moléculaire
Université Claude Bernard
Villeurbanne Cedex, France

Contents

1

Lidar: Introduction

CLAUS WEITKAMP

Institut für Küstenforschung,
GKSS-Forschungszentrum Geesthacht GmbH,
Geesthacht, Germany

1. INTRODUCTION

1.1. From Visual Perception to Lidar

Sensual perception of remote objects by processing of radiative stimuli is a capability widely encountered in animals. Humans perceive only *passively*, that is, see and hear objects that either radiate themselves or scatter and retransmit radiation from an external source. Some animals can locate objects *actively* as well, using optical or acoustical radiation they generate on purpose. Several species of fish literally carry lanterns. Horses and elephants are known to utilize active acoustic means for orientation in complete darkness. Bats use it to locate and catch prey.

Man has been using illumination with visible light from artificial sources for active optical detection of objects. Distance is inferred stereoscopically, that is, from the slightly different images obtained at the viewing angles of the two eyes, by mental construction, from two two-dimensional images, of a three-dimensional geometric relationship between different parts of the scene, or, for more remote objects, from the decrease of visual contrast.

Except for stereoscopic viewing, which fails at longer distance, these methods yield relative values only. Distance can be determined in a quantitative way by measuring the transit time of radiation from the source to the object and back. Systems that rely on this principle require a pulsed or modulated source and a detection system with adequate time resolution, approximately a million times better in the optical case than in the acoustical case. Man's senses do not nearly meet this requirement. Therefore, the technique became available to us only after the advent of suitable microwave, light, and sound sources and time-resolving detection systems.

Depending on whether sound, radiowaves, or light is used, these systems are called SONAR (*SO*und *NA*vigation *R*anging) or SODAR (*SO*und *D*etection *A*nd *R*anging), RADAR (*RA*diowave *D*etection *A*nd *R*anging), or LIDAR (*LI*ght *D*etection *A*nd *R*anging). Sonar works under water, the remaining techniques in the atmosphere. Lidar uses not just visible wavelengths ($400\,nm < \lambda < 700\,nm$), but also ultraviolet ($225\,nm < \lambda < 400\,nm$) and infrared radiation ($0.7\,\mu m < \lambda < 12\,\mu m$).

All these techniques are based on the same simple principle. A short pulse of radiation is transmitted into water or air, and the backscattered radiation is detected and analyzed. Clearly, radiation scattered from an object at a closer distance comes back sooner than that from an object at a longer distance. Scattering occurs not only on solid objects but also from the molecules and particulate matter in air and water. The return signal will, therefore, not be of the same length as the transmitted pulse, but extended in time, with a huge, but short peak from a solid object (if there is any) sitting on a much weaker, but temporally extended signal from air or water.

1.2. What This Book Does Not Consider

Some techniques such as most variants of sonar and navigation radar as is used on airplanes and boats only look for the large peak and determine the distance and direction to the corresponding object, which may be the ground, a building, a mountain, iceberg or other obstacle, a satellite or balloon being tracked, or another boat or airplane. These techniques are not considered in the present context, even if the technology is very similar and many other aspects are important for lidar as well. We exclude, in particular, a technique that has been known as *ladar*. Ladar can be described as an imaging system similar to a lidar that rapidly scans small solid angles at elevations close to the horizon in search for solid objects such as buildings, vehicles, artillery; infrared wavelengths are often used to better penetrate haze and fog and to reduce the risk of the ladar's own discovery. We also exclude all kinds of range-finding equipment and technology except cloud-base sensing devices. We exclude short-distance optical remote sensing used in industrial, security, and medical applications. We also exclude laser art and laser shows and entertainment.

Despite its title, this book is meant to be on lidar, not on optical remote sensing in general. We exclude all passive optical techniques including FTIR, or Fourier-transform infrared spectrometry. We also exclude optical remote sensing that provides no depth resolution or that is not truly monostatic. If, for example, the transmitter and receiver of a system that otherwise appear very similar to those of a lidar are located at different positions, with the receiver scanning the beam along its path, a system that could work with a continuous wave (cw) instead of a pulsed source, we do not consider that a lidar. We further exclude all types of DOAS, or differential optical absorption spectrometers, that is, systems that are monostatic except for a distant retroreflector and allow the determination of the average, not depth-resolved, concentration between the locations of the system and the reflector. Some DOAS systems use a multitude of reflectors at several points along one line, thus providing average concentrations out to different distances. By consecutive subtraction of the total load between the system and one reflector from the measured total load between the system and the subsequent reflector, some kind of depth resolution can also be obtained (Platt, 1978). On the other hand, there are lidars that, under unfavorable conditions, have to use such long integration depths that the resulting number of effective depth bins gets comparable to that of a well-tuned multimirror DOAS. So, the limit between lidar and nonlidar methods is not always easy to define. Also, the measurement of algae in a shallow layer of water or on the surface of a building or monument (as presented, e.g., in Chapter 6) is on the borderline because depth information is unimportant in these measurements. It is true that the systems used are genuine lidars; however, no depth profiles are generated, the distance information is only used to unambiguously identify the object under investigation, and range gating (or time gating, which is the same) is used to reduce the intensity of ambient background light or to gain additional information on the nature of the object by determination of its fluorescence lifetime. Other examples not commensurate with the strict definition of lidar as given above are the laser altimetry subsections of Chapters 8 and 9 which, however, because the instruments used have all characteristics of lidar systems and usually profiles of atmospheric parameters are measured simultaneously, have been included in the present treatment.

1.3. How It All Began

Lidar is not a new technique. Apparently, the first device was successfully operated soon after the secrecy on radar had been lifted at the end of the Second World War (Jones, 1949). A high-voltage spark between aluminum electrodes was used as a source, two searchlight mirrors were the transmitter and receiver optics, and a photoelectrical cell the detector. The system was successfully used to measure cloud-base heights up to 5.5 km in bright daylight. The term "lidar" was coined several years later (Middleton and Spilhaus, 1953) as a mere analog to the better-known radar, without expressly telling what it could be the acronym of.

Although photomultiplier tubes were already available at that time, sparks and flashlamps were not the ideal sources for applications that require, in addition to small divergence and short pulse duration, a spectrally narrow beam as well. If, as it had jokingly been put, the invention of the laser in 1960

(Maiman, 1960) was "a solution looking for a problem," then lidar was clearly a problem.

1.4. Lidar Literature and Information Dissemination

Lidar had a relatively slow start. It took 8 years before (in 1968) the first major conference that was exclusively devoted to the new technique could be held (ILRC, 1968). Another 8 years later, the first book appeared (Hinkley, 1976). After another 7 and 8 years, two more books were published (Killinger and Mooradian, 1983; Measures, 1984). Although thousands of articles have appeared in archival journals, these three were the only major publications in book form. All three, and particularly the monograph by Measures (1984), have remained the efficient shortcut for everyone who wants to familiarize himself with the lidar technique. A short treatise that concentrates on the basic principles (Weitkamp, 1990) and a larger book on the remote determination of pollutants (Klein and Werner, 1993) were never translated to English and have thus not seen widespread distribution. International Laser Radar Conferences, which were held in irregular, 1- to 2-year intervals between 1968 and 1982, are now convened regularly every 2 years; their proceedings also provide an excellent overview over lidar work of the preceding 2 years (ILRC, 1982–2004). Some of them (ILRC, 1996, 2000) contain only selected papers that went through a critical evaluation and editing phase, making these proceedings somewhat incomplete, but high-standard books of reference. An important regional series of conferences are the Japanese Laser Sensing Symposia held for the 20th time in 1999 and grown to such international importance that their title is now "International Laser Sensing Symposium" (ILSS) (ILSS'99, 1999). Within the lidar community subgroups have developed, some of which stage their own workshops and conferences that are also held more or less regularly. One of them is the coherent lidar group that holds meetings every second year and had its 12th session in Bar Harbor, Maine, in 2003 (CLRC, 2003). Another is the group on multiple-scattering lidar experiments (MUSCLE), which also meets in different countries approximately every year and a half. The latest MUSCLE meetings were held in 2002 in Oberpfaffenhofen, Germany (MUSCLE, 2002), in 2004 in St. Petersburg, Russia (MUSCLE, 2004), and will be followed in 2005 by a conference in Quebec, Canada. A series of conferences on "Optical Remote Sensing of the Atmosphere," better known as the "Lake Tahoe Conferences" named after the place where the first couple of meetings were held, with approximately half its contributions devoted to lidar, was started in 1985 and continued in 18-month intervals until 1993 (ORSA, 1990, 1991, 1993). Another series of meetings is especially devoted to lower tropospheric profiling, its needs, and technologies (ISTP, 1988–2003). Finally, large geoscience conferences such as the International Geoscience and Remote Sensing Symposium (IGARSS) and the Conferences of the Society of Photo-Electrical Instrumentation Engineers (SPIE) as well as the annual meetings of the American and European Geophysical Societies (AGS, EGS) usually contain one or several sessions that are expressly or from the nature of their subject dominated by contributions about lidar.

1.5. What a Lidar Is

A lidar consists essentially of five subsystems: (1) a transmitter, in all practical cases a laser, mostly, but not necessarily pulsed; (2) a transmitter optics; (3) a receiver optics; (4) a detector; and (5) an electronic system for data acquisition, processing, evaluation, display, and storage (Figure 1.1). Most lidars have additional components that differ considerably according to type and purpose of the lidar. These can be one or more additional lasers; one or more additional receivers; optical fibers in the transmitter or receiver optics; passive wavelength-selective devices such as absorption cells, filters, monochromators, and spectrometers; active devices for wavelength selection and adjustment; polarizing and polarization-sensitive components; some kind of beam-steering gear; manually operated or automated alignment aids; one-way or interactive diagnostic systems for long-term unattended operation; protective shutters; and some kind of shelter or housing often called "platform" with an infrastructure that provides

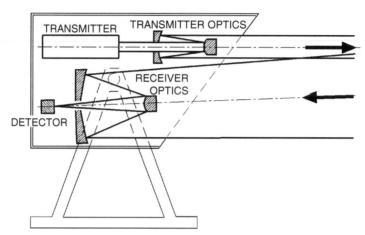

Figure 1.1   Essential optical components of a lidar system.

power, temperature control, operator accommodation, displacement possibilities, orientational stability, telemetry, etc., as the case may be. Lidars range in size from a shoebox to a 40-ft container, and beyond.

Lidars are by definition monostatic, that is, the transmitter and the receiver are at the same location. They may be coaxial, with the axes of the transmitted beam and receiver field of view coinciding, or side by side or biaxial, with the two axes parallel or near-parallel, but not identical.

The transmitter optics of a lidar serves the dual purpose of expanding the laser-beam diameter in order to reduce the area density of the laser pulse energy and of reducing the divergence of the laser beam. The receiver optics collects the backscattered light and focuses it onto the detector. Suitably programmed, the associated electronics takes the signals from the detector, extracts and conditions the relevant information, processes the signals to convert them to the physical quantities of interest, and displays, recombines, and compresses the data for long-term storage.

1.6.   The Lidar Return Signal and Lidar Equation

If the speed of light is denoted by c, then the delay t between the transmitted and backscattered pulses from an object at distance x is given by

$$t = 2\frac{x}{c} \tag{1.1}$$

if only the direct path is considered, that is, multiple scattering is excluded for the time being. Equation (1.1) relates the return time with the distance of the scatterer. Time and distance can thus be, and will be used synonymously in this book. Differentiated, Equation (1.1) also shows that the smallest discernable depth interval

$$\Delta x \geq \frac{c}{2}\Delta t \tag{1.2}$$

and, thus, depth resolution is limited by the laser pulse length, detection system time constant, or digitizer or photon-counting time-bin width, whichever is the longest. Clearly, the delay between successive pulses must be longer than $2/c$ times the distance from which no return signal can be detected any more. This is usually quite a bit longer than the lidar range, or maximum distance out of which meaningful data can be collected.

The profile of the lidar signal, that is, the power P registered by a lidar detector from distance x if light of wavelength λ_0 is transmitted from the system and light of wavelength λ is backscattered, depends on the properties of the lidar and the properties of the atmosphere. The atmosphere can scatter, absorb, and depolarize laser light, shift its wavelength, and broaden its spectral distribution. The parameters that relate the measured lidar signal $P(\lambda, x)$ with the properties of the atmosphere are the extinction coefficients $\alpha(\lambda, x)$ and $\alpha(\lambda_0, x)$ and the scattering coefficient $\beta(\theta, \lambda_0, \lambda, x)$. Because lidars only sense radiation that is scattered back in the direction toward the system, only scattering angles $\theta = \pi$ are relevant. We call the scattering coefficient in the backward direction, $\beta(\pi, \lambda_0, \lambda, x) \equiv \beta(\lambda_0, \lambda, x)$, the *backscatter coefficient*. The α describes by what fraction dI/I the intensity I of a laser beam is attenuated when passing a thin layer dx of the atmosphere located around distance x, assuming the validity of the Lambert–Beer law. The dimension of the attenuation coefficients is consequently an inverse unit of length such as m^{-1}. The backscatter coefficient $\beta(\lambda_0, \lambda, x)$ describes what fraction dI of the incident radiation I is shifted (if $\lambda \neq \lambda_0$) and reemitted from a layer of thickness dx at distance x into a small reception angle that is defined by A/x^2, when A is the receiver area. A suitable unit of β is m^{-1} sr^{-1}. In elastic backscatter lidar, only light is registered for which $\lambda = \lambda_0$. In this case, the backscatter coefficient is usually written as $\beta(\lambda_0, \lambda_0, x) \equiv \beta(\lambda, x)$ or simply $\beta(x)$.

The calculation of the detector output or lidar signal can be carried out rigorously, although hardly ever in closed form, if the spectral, temporal, and spatial properties of the laser light and the optical properties of the lidar receiver are to be taken into account in full detail. Unless chirped beams are used (which were hard to avoid in the early, ruby-laser-dominated times of lidar), the spectral and spatial–temporal properties can be treated separately. The ways the atmosphere interacts with the spectral properties of the laser light differ very much for the different types of lidar and will, therefore, be treated in the chapters that follow. The behavior of the laser beam as to its temporal and geometric properties, however, is very similar for all types of lidar. Therefore, a short description of the different steps necessary for the calculation of the profile $P(\lambda, x)$ is given here. The quantitative derivation and the mathematical prescriptions to be followed for the calculation of the detector response $P(\lambda_0, \lambda, x)$ are given in detail, e.g., in Harms et al. (1978).

The detector response is obtained by imaging the radiation backscattered from a plane perpendicular to the lidar beam at distance x from the lidar onto the detector. In doing so, two things must be noticed. The first is that the detector is usually positioned close to or in the focal plane of the receiver system. The image plane of an object at finite distance x, however, is not identical with the focal plane of the imaging system. Its distance from the main mirror is longer than the focal distance *and different for each distance x*. Light from a point in the atmosphere that would be concentrated in a point in the image plane is thus distributed over a whole area, the circle of confusion, in the focal plane. The second thing to consider is the temporal distribution of the laser power at distance x. If the rise time of the pulse is t_1 and the fall time is t_2 ($t_1, t_2 > 0$), so that the pulse duration is $\tau = t_1 + t_2$, the light that reaches the detector at time t does not originate from distance $x = ct/2$ alone, but from the whole interval $(t - t_1)c/2 < x < (t + t_2)c/2$. The temporal distribution of backscattered radiation on the detector is thus obtained from the image of the convolution, not simply of the product of the irradiance and the backscatter coefficient, with the temporal distribution of power in the laser pulse.

One usually starts with the assumption that the transmitted beam is Gaussian, that is, it comes from a laser in TEM_{00} mode, and that its diameter increases with distance. Then, the irradiance or distribution of laser power in a plane perpendicular to the beam axis at distance x is calculated, multiplied with the backscatter coefficient $\beta(\lambda_0, \lambda, x)$, and convolved with the temporal power distribution function of the laser pulse.

The result does not yet take into account the fact that, on its way from the lidar transmitter to the object plane, the lidar beam is attenuated by the factor

$$\exp\left[-\int_0^x \alpha(\lambda_0, \xi)\,d\xi\right] \tag{1.3}$$

and that the fraction of light that is scattered back toward the receiver is attenuated by another factor

$$\exp\left[-\int_0^x \alpha(\lambda, \xi)\,d\xi\right] \tag{1.4}$$

It is common practice to modify the result of the described procedure in such a way that the convolution is eliminated by the assumption of a laser pulse short in relation to both the depth over which changes in β are noticeable (which is usually justified) and the length of the time bins of the digitizing equipment (which is not always the case), and to bury the cumbersome integrations that involve the geometry parameters in a quantity called laser-beam detector-field-of-view overlap function $O(x)$, with ($0 \leq O \leq 1$).

With the above simplifications and the introduction of an additional efficiency factor η that considers losses by mirrors and filters, detector efficiency, etc. and which may be wave-length-dependent, but does not depend on distance, we get the commonly used lidar equation

$$P(x, \lambda_0, \lambda) = \frac{c\tau}{2} \bar{P} \frac{A\eta(\lambda)O(x)}{x^2}$$

$$\times \beta(x, \lambda_0, \lambda) e^{\int_0^x -\alpha(\lambda_0,\xi)d\xi} e^{\int_0^x -\alpha(\lambda,\xi)d\xi}$$

(1.5)

Here, $\tau\bar{P}$ is the product of laser pulse duration τ and average power $\bar{P}$ is the laser pulse energy, and A is the receiver area; these as well as η and O are parameters of the system alone. α and β are the extinction and backscatter coefficients of the atmosphere; the exponentials at the right are called atmospheric extinction at the laser wavelength λ_0 and return wavelength λ, respectively.

Each of the following chapters will start from Equation (1.5) or an equation similar to it. Notation will sometimes be different, some of the quantities A, η, or O will be assumed to depend on more or on fewer parameters, or may be combined with one another or with c, τ, and $\bar{P}$. For certain variants of the lidar technique, it is more practical to use the product $x^2P(x, \lambda_0, \lambda)$ instead of $P(x, \lambda_0, \lambda)$ itself. The very principle of any kind of lidar whatsoever is to extract α and β from $P(x, \lambda_0, \lambda)$ and the atmospheric quantity of interest from α and β. In which way α and β depend on the different properties of state of the atmosphere depends on the processes considered and on the wavelength(s) involved. Each of the following chapters will describe the relation of α and β with the quantity or quantities that are to be determined. An overview is found on pages 11 to 145 of the book by Measures (1984), but there are much more detailed treatments of just about every possible interaction of light with the atmosphere and its constituents.

1.7. Atmospheric Parameters that Can Be Measured

We consider the atmosphere as being made up of gases that can be molecular, atomic, or ionic in nature, and cloud and aerosol particles of solid, liquid, or mixed-phase aggregation. Gases are normally sufficiently well characterized by their volume (molecule or atom number or mass) density and its spatial and temporal distributions. For particles mass or number densities are more difficult to determine, so often their contributions to α and β, the wavelength dependency of these contributions, and their share in the degree to which polarized radiation is depolarized upon backscattering are measured. Sometimes microscopic quantities such as mean square radius and other shape parameters can be inferred; chemical composition is more difficult to assess. The atmosphere is further characterized by quantities like pressure, temperature, turbulence, and wind. All these quantities are amenable to measurement by lidar. From α and β, secondary data can be determined such as horizontal, vertical, and slant visual range.

More highly integrated atmospheric data like cloud cover and cloud albedo, surface and subsurface properties of the ocean and of inland waters, and surface properties of the solid Earth will not be treated in the present book.

1.8. Interaction Processes Used

The terminology of interactions that light undergoes with constituents of the atmosphere has grown historically. Therefore, some of the definitions are not very systematic, particularly with respect to the different variants of the lidar technique. We do not give a comprehensive treatment here, although we try to define most of the different technical terms as used in the remainder of this book. In doing so, we partly follow the recent treatise by She (2001) and partly use older classifications (Young, 1982). Although both historically correct and logically simple, She's terminology is only now beginning to be generally adopted in the lidar community. We do not, in particular, define "elastic scattering" as a variant of scattering in which there occurs no change of wavelength, only that shift is too small to be noticeable with the equipment available at the times of Lord Rayleigh's experiments. We define Rayleigh scattering as all elastic scattering that is not Mie scattering, i.e., as the sum of Cabannes and S- and S'-branch purely rotational Raman scattering, with Cabannes scattering being the sum of coherent elastic scattering and Q-branch (i.e., degenerate) rotational (incoherent) Raman scattering.

In Figure 1.2, an attempt is made to visualize the different processes in graphic form. Both the Stokes and anti-Stokes branches from the diatomic, homonuclear molecules that constitute most of the air are of S type. This is because of the convention to define the change ΔJ in rotational quantum number J, which determines O ($\Delta J = -2$), Q ($\Delta J = 0$), and S branches ($\Delta J = +2$) not as the difference of J numbers of the final and initial, but of the higher and lower of the two states involved; this difference cannot get nonnegative if no vibrational energy is exchanged. In Figure 1.2, the relation of the intensities of the different types of scattering with the on- and off-diagonal elements of the molecular polarizability tensor, a and γ, is also indicated. However, there are other dependencies such as those on the wavelength of the primary radiation and on Boltzmann factors. For a detailed treatment, the reader is referred to the book by Long (2002).

All light elastically backscattered in the atmosphere is broadened, or individual photons are shifted, by pressure and

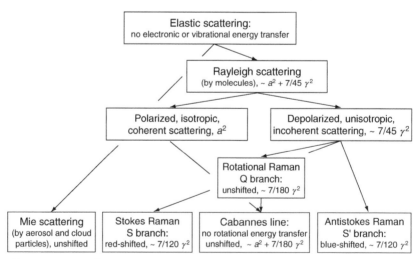

Figure 1.2   Elastic scattering nomenclature.

temperature effects and also by collective motion of the air molecules. This broadening amounts, for ambient conditions, to about 3 GHz (FWHM) at visible wavelengths and applies to the Cabannes lines and each one of the rotational Raman lines alike. It is approximately 100 times narrower for the Mie line.

Unless otherwise stated, we use in this book the following definitions:

Absorption	The removal of light from a beam without reemission.
Fluorescence	The removal of light from a beam and subsequent delayed reemission, usually in all directions, at the primary or another (usually longer) wavelength.
Scattering	The change of direction of a photon or a beam of light.
Extinction	The removal of light from a beam by the sum of absorption, scattering, and fluorescence.
Backscattering	Scattering by 180° with respect to the incoming photon, or beam.
Elastic scattering	Scattering with no apparent change of wavelength: sum of Mie and Rayleigh scattering.
Mie scattering	Scattering from cloud droplets and aerosol particles. Although Mie's theory (Mie, 1908) is valid for spherical particles only, we sloppily use the term for scattering by nonspherical particles as well.
Rayleigh scattering	Elastic scattering from atmospheric molecules: sum of Cabannes and rotational-Raman S and S' branch scattering.
Cabannes scattering	Sum of coherent, isotropic, polarized scattering as described by the trace elements of the polarizability tensor

	of a diatomic molecule and rotational-Raman Q branch, that is, the degenerate, incoherent, unisotropic, depolarized scattering described by the off-trace elements of the polarizability tensor of a diatomic molecule.
Rotational Raman scattering	Term confined to the rotational Raman S (Stokes) and S' (anti-Stokes) branches.
Prompt inelastic scattering	(Rotation–vibration) Raman scattering.
Delayed inelastic scattering	Fluorescence.
Brillouin scattering	Scattering from phonons, mainly in condensed matter and usually negligible in the atmosphere, resulting in a small distortion of the Rayleigh backscatter peak.
Resonant absorption, resonant scattering by molecules, atoms, and ions	Absorption or scattering of light by a molecule, atom, or ion with internal transitions matching or approximately matching the energy (and other properties) of the photon. These processes are characterized by large interaction cross-sections.
Depolarization	The partial backscattering of linearly polarized light in the plane of polarization perpendicular to that of the transmitted beam.

1.9. Lidar Systematics

There are several ways of classifying lidar systems: by the physical effect utilized, by purpose, by wavelength (IR, VIS, UV), by technology (cw/pulsed, coaxial/side-by-side, analog/photon-counting), by type of platform, etc. Lidars can also be grouped according to the way the backscatter signal profiles have to be treated to obtain the target quantities, or quantities of interest: differentiating lidars need little or no information in addition to the measured signals, but are very sensitive to noise and to differences in the receiving channels if more than one channel is used, an example being double-(wavelength- as well as range) differential DAS lidar. Proportional lidars such as water-vapor rotation–vibration Raman lidar are much less sensitive to noise, but need an absolute calibration of the relative sensitivities of the water-vapor and reference (usually nitrogen) channels. Integrating lidars such as Rayleigh lidar for the determination of the temperature in the upper stratosphere are very insensitive to noise but need a boundary value with which to start the integration procedure.

In Figure 1.3, an overview is given of the lidar schemes described in this volume. It shows which processes are exploited in which systems, and what atmospheric parameters can be determined. As can be guessed from Figure 1.3, the terminology is neither very strict nor very logical. For example, Raman lidars are usually rotation–vibration Raman systems unable to separate pure rotational Raman radiation from the rest of the Rayleigh return signal, but do have an elastic channel in which the totality of the elastic signal (Rayleigh + Mie) is registered. In case they transmit more than one wavelength, they can work very well as differential-absorption lidars if use is made of two of the elastic return signals. With one wavelength only, they can still work as DIALs if the primary signal is combined with one of the Raman wavelengths. Using two Raman wavelengths can be even more advantageous, allowing the elimination of uncertainties due to the presence of aerosols. This means that the subjects of the different chapters are largely, but not totally mutually exclusive.

1.10. Lidars Considered in This Book

Except for Chapter 2, which anticipates recent advances, this book follows approximately the historical sequence in which the different ideas developed and, thus, the different variants

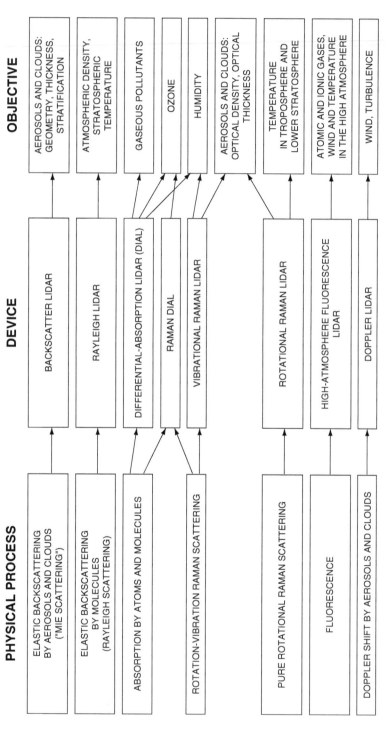

Figure 1.3 Physical processes, lidar devices, and atmospheric quantities determined with the systems described in this book.

of the technique entered the scene. Today, work is being done on all of these methods, although not with the same effort and intensity. Very few techniques have been omitted. All of those either could not prove viability at least on an experimental scale, require unproportionally complicated equipment, have not found widespread application because of limited perform-ance, or simply do not qualify as genuine lidars as defined earlier in this chapter. An exception is Raman lidar. Although a powerful technique, Raman lidar has not been included in the present book for lack of space and also because several other publications on this subject are in preparation.

The book is arranged in such a way that each chapter is self-sufficient in that there is no need for the reader to start from the beginning if he or she is interested in one particular kind of lidar. Instead, the basic facts are laid down anew at the beginning of each chapter, even if this results in a certain duplication of material. The layout of chapters is usually such that the physical process or processes on which the technique is based are briefly reviewed, then the theory of the measurement is developed, followed by typical examples of measurements, and ending with a summary of the merits and potential, and also of the shortcomings and limitations of the method. Each chapter also contains its own bibliography, resulting again in some duplication of certain frequently cited references.

Chapter 2: Femtosecond White-Light Lidar

Chapter 2 of this book, as already mentioned, does not follow the general arrangement that is more or less in the same sequence as the historical development of lidar systems. It is, on the contrary, a description of the most recent, most advanced, and, to many lidar scientists, most unexpected development, femtosecond white-light lidar. Except in the very early days when no other powerful sources of light were available, lidar had always been operated with colored, more or less narrow-bandwidth optical radiation, and pulse dur-ations were in the nanosecond to microsecond range, more than six orders of magnitude longer than the femtosecond

(1 fsec $= 10^{-15}$ sec) pulses used here. As a pulse energy of 1 mJ and a pulse duration of 1 fsec translate into an average optical power of 1 TW, "terawatt lidar" is an alternative expression for the new technique.

Femtosecond white-light lidar is possible because media such as the atmosphere behave entirely differently when power densities get so high that interactions cease to be linear. More power then not merely produces more of the well-known, low-power-density effect, that is, not *more scattering, more absorption*, etc., but qualitatively new effects begin to occur. One important nonlinear effect is self-focusing of the laser beam through radial changes of the air's refractive index, which is stronger on the beam axis than in the beam center, leading to a kind of nonmaterial lens called "Kerr lens." Another is multiphoton ionization and plasma generation in the focus of these Kerr lenses. This, in turn, causes defocusing and the generation of white light even from initially monochromatic laser beams by a process called self-phase modulation. Finally, unlike conventional laser beams with angles of divergence that, in a scintillating atmosphere, cannot be made much smaller than 100 μrad (i.e., 100 mm at a distance of 1 km), TW pulses can be made to exhibit atmospheric filamentation and, thus, provide a means to deliver high laser intensities to remote locations.

The possibilities of TW white-light lidar include hitherto difficult or unsuccessful ventures such as atmospheric trace-molecule, trace-ion, and trace-radical measurements, aerosol remote chemical analysis, and direct aerosol-particle sizing.

When this chapter was selected to be placed right after the Introduction, it was to show the extent to which lidar continues to be a challenge for physicists, optical engineers, meteorologists, atmospheric chemists, and scientists of many other disciplines. It was also to document the importance the editors attribute to this new technique. Those readers who feel themselves a bit at a loss when trying to follow all the ideas of Chapter 2 are advised to skip it and to proceed to Chapter 3, giving Chapter 2 another try toward the end when the simpler variants of the lidar technique have been well understood. In this sense what is now Chapter 2 could as well have been placed at the end of the book as a brilliant illustration of the outlook and the concluding remarks.

Chapter 3: Elastic Lidar Measurement of the Troposphere

Historically, Mie lidar was the first practical application of the lidar technique. It is, therefore, presented in Chapter 3. Although conceptually simple if only geometric properties of the atmosphere are required (such as the bottom of an overhead cloud or the thickness of a smokestack plume or a patch of fog), it requires complicated data evaluation procedures if the microscopic properties of the particles that make up the cloud, fog, haze, or smoke are to be determined. The reason for this is the fact that, at one or several given wavelengths λ, *one* vector of measured quantities, viz., $P(x, \lambda)$ is available from the measurement, but that *two* vectors of unknowns, $\alpha(x, \lambda)$ and $\beta(x, \lambda)$, must be determined from these. This requires either additional measurements or additional assumptions. Because of the high temporal and spatial variability of the atmosphere and because aerosol *must* be measured *in situ* if the important properties such as particle number density, particle size, and particle shape are to be conserved, assumptions rather than measurements are the preferred source of additional information. Mie lidar, thus, turns out to present one of the greatest challenges in lidar at the present time.

More than 99.9% of all aerosols and virtually all anthropogenic aerosol are found in the troposphere, that is, in that part of the atmosphere in which there occurs weather. Its upper boundary, the tropopause or the altitude at which temperature has a local minimum, ranges from about 6 km in the polar winter to 17 km at equatorial latitude.

Chapter 3 presents, in Section 3.1, the theoretical foundations of present state-of-the-art techniques to extract profiles of α and β from the lidar return signals, both for the case of negligible absorption by gases and sufficiently high absorption by gas molecules to require special attention. A lidar apparatus suitable for that kind of measurements is then described. Information on the thermodynamic phase (solid or liquid) and shape (spherical or nonspherical) of aerosol particles can be obtained if, for a linearly polarized laser beam, the ratio of depolarized to polarized *single-scattering* return signal is available, but multiple scattering also depolarizes the light even when occurring on perfectly spherical particles. In Section 3.2, the problem is treated for two cases: clouds and mineral dust. Mie lidars, because of their conceptual simplicity, lend themselves more easily than other types of lidar to fully automated, unattended operation. Section 3.3 describes several types of lidar that work without a human operator. In Section 3.4, multiwavelength aerosol lidar is addressed, with special emphasis on systems with two and four wavelengths. Finally, Section 3.5 is devoted to aerosol lidar network measurements and systems with special emphasis on the European Aerosol Lidar Network (EARLINET), the Micro-Pulse Lidar Net (MPLNet), and the Asian Dust Net (ADNet).

Chapter 4: Trace Gas Species Detection in the Lower Atmosphere by Lidar

Next to Mie (or aerosol) lidar, differential-absorption lidar has probably been devoted most of the total lidar activity until now. Differential-absorption lidar requires the use of at least two wavelengths even in its very simplest implementation. It requires the possibility to select or, even better, tune the wavelengths to be used in such a way that two closely spaced wavelengths are available that differ considerably in absorption cross-section of the gas of interest. It requires pulses that, except for wavelength, resemble each other very closely in stability and in spatial and temporal behavior. The experimental effort is thus much higher than it is for Mie lidar, and differential-absorption lidars require great experience in operation. Chapter 4 deals with this variant of lidar.

Unlike other types of lidar, differential-absorption lidar allows to write down the unknown quantity, the concentration of the gas of interest, in closed form. The only quantities that need to be known are the absorption cross-sections of the gas at the two wavelengths used. Clearly, these wavelengths must remain stable during the experiment. In Section

4.1, the inversion of the lidar equations is carried out to yield the expression for the unknown gas concentration. The next three sections cover applications in the ultraviolet (defined here as ranging from 200 to 450 nm), the mid-infrared (1 to 4 μm), and the far infrared (9 to 11 μm). For the gases amenable to measurement in these three ranges of wavelengths examples are given of ozone, nitrogen oxides, sulfur dioxide, chlorine, aromatic hydrocarbons, and mercury in the UV, hydrogen chloride in the near IR, and sulfur hexafluoride, ammonia, and ozone in the far IR. In another section, a measurement of the hydroxyl radical (OH) is presented, an atmospheric species of extremely low concentration of approximately 10^{-13}, or only 10^6 radicals per cubic centimeter, but of enormous importance as an "atmospheric detergent" because of its high reactivity and speedy destruction of many noxious pollutants. The technique does not precisely measure the concentration of the radical, but deduces concentrations from the dynamics of the radical in the atmosphere. OH is created with a powerful pulsed UV laser, and the subsequent degradation of the radical is measured using its fluorescence signal. Not yet matured into a lidar in the strict definition of the word as outlined above, the technique still deserves to be described in this chapter because of its extraordinary success in the determination of an import atmospheric constituent present only in ultratrace concentrations.

Chapter 5: Resonance Fluorescence Lidar for Measurements of the Middle and Upper Atmosphere

Aerosols are not confined to the troposphere. Volcanic emissions sometimes occur with so much vertical thrust that they penetrate the tropopause and extend far into the stratosphere where they are converted to (usually liquid) volcanic stratospheric aerosol that can have massive effects on the insolation of the Earth's cloud system down to the ground. Polar stratospheric clouds (PSCs) have been known to form above the Antarctic and Arctic where they interact strongly with the stratospheric ozone layer, particularly in spring. Higher up, in the mesosphere, and less frequently observed, other objects such as polar mesospheric clouds (PMCs) and noctilucent clouds (NLCs), believed to be the PMCs' "ragged edges," can be studied. A particularly interesting phenomenon is the occurrence of metallic species in atomic and ionic forms around the mesopause at approximately 90 km height. Although present in ultratrace concentrations only, these can be measured with lidar to high sensitivity and accuracy. They not only present great interest in themselves, but can be used as a kind of natural tracer for the determination of wind and temperature and for the study of the thermal structure as well as of dynamic processes such as tides and gravity waves, heat, momentum, and sodium flux, and also chemical processes in the higher layers of the atmosphere. Finally, meteor trails are an important application of high-atmosphere lidar measurements. The influx rate, chemical composition, size distribution, approach velocity, and ablation rate can all be studied with lidar, showing interesting and partially unexplained phenomena.

The process of choice for the detection of these metallic species is resonance fluorescence. Chapter 5 is, therefore, devoted to resonance fluorescence lidar. It is by far the most comprehensive treatment of this compendium. Not only is the lidar equation and, in particular, its fluorescence variant developed in great detail, the basic processes used for the determination of the different properties of state are also presented. At least one example is given for each type of relevant apparatus. The results are put into their historical context and discussed in meticulous detail. It can safely be stated that a reader able to memorize the essentials of Chapter 5 can consider himself or herself as extremely well informed about the actual state of middle and upper atmosphere lidars without recurring to any further references.

Chapter 6: Fluorescence Spectroscopy and Imaging of Lidar Targets

Resonance fluorescence is particularly efficient in the middle and upper atmosphere where quenching processes, that is, radiationless transitions of excited atoms back to their ground state, do not play a dominant role. In the troposphere, these quenching effects, along with the relatively much lower abundance of alkali, earth-alkali, and other metal atoms, render narrow-bandwidth, high-sensitivity fluorescence lidar unpractical. Even mercury, with its two giant resonance lines in the near-UV that make it fluoresce extremely intensely, is more easily measured with differential-absorption than resonance-fluorescence lidar (Edner et al., 1989).

This does not mean that resonance fluorescence lidar is limited to the higher layers of the atmosphere. It is true that signals from naturally fluorescing airborne substances are normally too weak to be detected below the atmospheric signal background. If artificial fluorescent tracers, usually released as monodisperse aerosols, are used, then the location, speed, and turbulent diffusion of air parcels and with it all essential parameters of cloud and plume diffusion can be measured with little experimental effort. According to Kyle et al. (1982), a concentration of 2.2×10^{-7} g of fire orange dye per cubic meter of air is sufficient for measurements out to 1 km distance, over a moonlit landscape. Uthe et al. (1985) used the same dye to pursue an artificial cloud to a distance of several hundred kilometers with airborne fluorescence lidar; however, they needed 50 kg of tracer material.

These are not the applications with which Chapter 6 of this book is concerned. Instead, a target at an arbitrary, but fixed distance is illuminated with the lidar beam. The receiver, unlike receivers in conventional lidars, images the illuminated object in the plane perpendicular to the optical axis (as does a ladar) and carries out a spectral analysis of the return light for each pixel of the image, or object. The time information is used to reduce the effect of stray light and to unambiguously identify the object, but not for any further information on depth. Unlike isolated atoms or ions, the fluorescing substances in the targets have broad absorption bands and broad features in the fluorescence spectra as well, but which show sufficient specificity for the extraction of the relevant information.

The most commonly used lasers are pulsed excimer and frequency-tripled and frequency-quadrupled Nd:YAG lasers, but nitrogen lasers can also be used if power requirements are moderate. Optical parametric oscillators and other tunable sources such as dye lasers can yield additional information from the differences in fluorescence spectra obtained at different excitation wavelengths. On the receiver side, the devices have at least two, but usually several wavelength channels, up to the whole dipersive power of a spectrometer. Ratioing the signals in the fluorescence channels with the elastic return signal makes the system relatively immune to laser-pulse power fluctuations, variations in overlap integral, and other system parameter variations. For aquatic targets, the liquid–water Raman return signal is often used as the reference.

Applications include the measurement of gelbstoff or dissolved organic matter, suspended organic matter, algae, and oil slicks. For the identification of the latter, the method is particularly specific. Not only can weathered oil and sunk oil, that is, oil that had been on the surface for some time and that sank into the water, be distinguished, the method also allows to determine the origin of the oil.

Vegetation monitoring allows one to distinguish different types of crops, trees, and again algae, and to measure their biochemical activity. This latter capability is made use of in the investigation of historical monuments. The presence of organisms that deteriorate the stone, so-called biodeteriogens, can be mapped in this way; the areas treated with chemical agents to stop algae growth can also be clearly recognized. As a side effect, but nonetheless highly welcome, the spectra yield the different kinds of stone from which the building was erected.

Chapter 7: Wind Lidar

The speed and direction with which its constituents move in Earth's coordinate frame is probably the most important information on the state of the atmosphere. Depending on the size of the moving objects, we use different terms for the phenomenon. The uncoordinated movement of individual atoms, molecules, clusters, or aerosol particles defines what we call temperature. The collective movement of objects in the size range from meters to hundreds of meters with frequent changes in both speed and direction is called turbulence. The collective movement of objects on a scale of tens of meters to tens of kilometers and fairly slow changes of speed and direction is called wind.

From the wind a number of other phenomena are easily deduced. These include large-scale circulations like polar vortices, stratospheric jets, gravity waves, and the transport of dust, aerosols, moisture, and gases like ozone with atmospheric air masses. Wind is, thus, one of the key parameters in atmospheric research including atmospheric chemistry, aerosol physics, and weather and climate research. At least equally important is the knowledge of the wind field, both directly and indirectly, not just for air traffic, but for many operations on the solid Earth and on oceans and estuarine waters as well. Examples are hurricane and tornado warnings for inhabited areas, storm warnings for boats, the prediction of tide heights, storm tides, surges, and many others. It is obvious that ground-level wind measurements alone are not adequate for these purposes.

Chapter 7 is, therefore, devoted to wind lidar. The physical effects used are elastic backscattering of the transmitted laser radiation and the shift of the laser wavelength by the Doppler effect: particles moving toward the lidar increase the apparent frequency causing a shift to the blue, and vice versa. Only the line-of-sight, or longitudinal, component of the three-dimensional wind vector can thus be determined in a single measurement.

Because typical wind speeds are so small in comparison with the speed of light, the wavelength shifts are extremely small. This requires special analysis techniques for the extraction of the wavelength shift from the backscattered lidar signal, which are not necessary for most other types of lidar. Two principles have been used for this purpose, coherent detection and incoherent or direct detection. They differ in many respects. Most sections of Chapter 7 are, therefore, divided into a "coherent" and a "direct" subsection. In the remaining sections coherent Doppler lidar clearly dominates. The reason for this is the fact that coherent Doppler lidar has a longer history, is better developed, is in much wider practical use, and provided virtually all of the important results.

Chapter 7 is made up of seven sections: an introduction, some general background information about the two detection techniques, a more detailed section on qualitative aspects of the operation of wind lidars, a mathematical treatment of the physics involved, a description of instruments for both types of technique, a section on measurements with results that could not have been obtained with other than lidar methods, and a brief summary. After Chapter 5, Chapter 7 is the most comprehensive and longest chapter of the book. This may reflect the importance that is today attributed to wind measurements on all scales — local, regional, global — and from all kinds of platforms — stationary, mobile, ground-based, shipborne, airborne, and from space.

Chapter 8: Airborne Lidar Systems

The merit of ground-based lidars is often said to have been the addition of the third dimension to measurements that had previously been limited to the Earth's surface or to points easily accessible. Seldom heard, but no less important, is the addition of the fourth dimension, time: Ground-based lidars provide, indeed, continuously in time, height (z) profiles of virtually all atmospheric parameters. Their limitation resides in the fact that there is only one point in the $x-y$ (or $\theta-\varphi$) plane: clearly, a ground-based lidar can at one moment in time take measurements from one point on the Earth's surface only. A lidar can, quasi-simultaneously, that is, within times during which the essential features to be measured do not undergo dramatic changes, cover a large part of the surface or even the whole globe if it can be built into an aircraft or spacecraft. This has indeed been done.

The advantage of continental or global coverage of the Earth's atmosphere with lidars is not for free. The price to pay is an increase in sophistication and thus in development and manufacturing cost by an order of magnitude each when taking a lidar from the laboratory to, say, a seagoing boat, from boat to airplane, and from airplane to space. Although the principle of operation is the same, satellite lidars bear little resemblance to laboratory versions of systems that serve the same purpose. Requirements for airborne and spaceborne lidars encompass so many parameters that the last two chapters of the book, Chapters 8 and 9, are devoted to these two fields.

In Chapter 8, which deals with airborne lidar systems, these advantages and challenges that vary greatly with purpose are described in detail. The main part of the chapter is devoted to the discussion of these questions, and also to the achievements obtained and to noteworthy results measured with these systems. Applications have been numerous: aerosols, clouds, wind, PSCs, ozone, water vapor, metals in the ionosphere, density, temperature, and others. Not strictly in the lidar mode, these systems have also been used for the determination of a large number of hydrospheric and land-use parameters such as oil slicks, phytoplankton, water depth, and turbidity and, on land, the type of land use, health state of crops and forests, tree-stem diameters, above-ground biomass and biomass volume per acre, leaf–area index, etc.

Chapter 9: Space-Based Lidar

Spaceborne lidars must meet even more stringent requirements as to the size, weight, power consumption, and, most important, remote-control capabilities and long-term unattended operation. Chapter 9 represents in a way the continuation of the previous chapter. Following a short introduction, Section 9.2 is again devoted to the general technological problems with lasers, telescopes, and detectors. Sections 9.3 through 9.6 then describe, for cloud and aerosols, altimetry, wind, and differential absorption, respectively, the specific technical challenges, the corresponding programs, the solutions found, and, finally, to the extent that sufficient progress has already been made, the results available so far. Spaceborne lidar is today that field of lidar in which activities are most intense and progress is fastest. Although there remain enough deficiencies to be overcome and problems to be solved, it can be expected that spaceborne lidar will in the foreseeable future provide, in real or near-real time, all data whose unavailability is to date a frequent excuse for the failure of systems designed to predict the behavior of the Earth's atmosphere. It is to be hoped that the actual tendency is not continued to provide funds for the missions proper, but not for the subsequent evaluation, reduction, visualization, archiving, and retrievable storage of the data as well as the necessary support for those who need the data for applications. This phantom is unfortunately present today, throwing its shade on the brilliant scene as which space lidar proponents like to make the near-future situation of high-temporal-resolution, high-spatial resolution, instant-availability, global-coverage atmospheric data appear.

1.11. Lidar Guidelines

Several lidar techniques have to date reached a high degree of maturity. Like every measurement device, lidar systems need the possibility to have their results certified according to the criteria of present-day quality assurance. The German Commission on Air Pollution Prevention (KRdL) of the Association of German Engineers (VDI) and German Institute of Standardization (DIN) established a working group whose task was to prepare a set of recommendations for the use and operation of lidar systems. Three of these guidelines are now available in print, the guidelines for differential-absorption lidar (VDI, 1999), Doppler wind lidar (VDI, 2001), and visibility lidar (VDI, 2004).

The guidelines describe the actual state of the art of the respective lidar technique. Their prime intended purpose is to serve as a decision-making aid in the preparatory stages of legislation and application of legal regulations and ordinances. Important purposes are also to:

- Provide and unify definitions for lidar performance data.
- Help lidar users avoid severe mistakes in lidar operation and data reduction.

- Provide the foundations for lidar quality assurance.
- Help render results traceable to fundamental constants.
- Help prospective users decide, for a given problem, whether to use lidar or some conventional technique.
- Help prospective lidar users compare different instruments and choose a suitable technique.
- Help inexperienced lidar users understand operational characteristics.
- Enable industry and authorities to replace slow, expensive, error-prone, or otherwise inappropriate techniques with better-suited alternatives.
- Help designers define performance aims.

Lidar guidelines have a number of peculiarities for the following reasons. Because of the relative complexity of lidars, few systems have been built more than once. Existing systems show great variability in operation mode and system parameters. Actual definitions of performance characteristics and other technical terms as used for *in situ* systems are not applicable for lidar: technical terms such as onset distance, range, limit of detection, and limit of quantification need a more complex, but nevertheless precise, definition. The latter three depend so strongly on system and on weather parameters that a set of parameters was fixed for which they are determined in order to make systems comparable. These data are defined as the agreed range, agreed limit of detection, and agreed limit of quantification. Finally, the calibration procedures which differ greatly from the calibration of local measurement devices are addressed.

Target variables of the guidelines that have been completed are the concentrations of trace gases commonly considered as pollutants, the wind vector, and visibility or visual range. A distinction is made in all three guidelines between measured variables (usually the return signals or the beat signal), primary target variables (concentration, wind vector, some well-defined kind of visual range), and secondary target variables such as average concentration data, concentration gradients, concentration trends, stratification phenomena, shear wind situations, aircraft wake vortices, updraft and downdraft regions, target recognition probability, etc.

Algorithms turned out to be amenable to standardization up to a certain point. An example is the effect of aerosol data on gas concentrations determined with differential-absorption lidar. No standardization has been possible so far, however, for the procedures to determine the relevant aerosol data, essentially the aerosol extinction and backscatter coefficients and their derivatives with respect to distance and wavelength. Nevertheless do these guidelines contain a valuable set of definitions and technical rules. They may thus constitute a first step toward the establishment of similar rules for the operation of lidar systems in other countries and, in the long run, on the regional level by organizations such as the European Committee for Standardization (CEN) and on a global scale by the International Organization for Standardization (ISO).

REFERENCES

CLRC (2003): 12th Coherent Laser Radar Conference. Bar Harbor, Maine, U.S.A., June 15–21, 2003.

Edner, H., G.W. Faris, A. Sunesson, and S. Svanberg (1989): Atmospheric atomic mercury monitoring using differential absorption lidar techniques. *Applied Optics* **28**, 921–930.

Harms, J., W. Lahmann, and C. Weitkamp (1978): Geometrical compression of lidar return signals. *Applied Optics* **17**, 1131–1135.

Hinkley, E.D., ed. (1976): *Laser Monitoring of the Atmosphere.* Springer, Berlin, 380 pp.

ILRC (1968): International Laser Radar Conference. Boulder, CO; no proceedings found.

ILRC (1982): Eleventh International Laser Radar Conference. Madison, Wisconsin, USA, June 21–25 1982. NASA Conference Publication 2228.

ILRC (1984): 12 Conférence Internationale Laser Radar/12 International Laser Radar Conference. Résumés des Communications/ Abstracts of papers, Aix-en-Provence, France, August 13–17, 1984, 451 pp.

ILRC (1986): Thirteenth International Laser Radar Conference. Abstracts of papers presented at the conference held in Toronto,

Ontario, Canada, August 11–16, 1986. NASA Conference Publication 2431, 321 pp.

ILRC (1988): 14 International Laser Radar Conference/14 Conferenza Internazionale Laser Radar. Conference abstracts, Innichen — San Candido, Italy, June 20–23, 1988, 512 pp.

ILRC (1990): Fifteenth International Laser Radar Conference. Abstracts of papers, Tomsk, USSR, July 23–27, 1990. Institute of Atmospheric Optics, Tomsk, USSR, Part I xvii + 404 pp., Part II xvii + 430 pp.

ILRC (1992): M.P. McCormick, ed.: Sixteenth International Laser Radar Conference. Abstracts of papers presented at a conference and held in Cambridge, Massachusetts, July 20–24, 1992. NASA Conference Publication 3158. Part 1 xxx pp. + pp. 1–380, Part 2 xxx pp. + pp. 381–732.

ILRC (1994): 17th International Laser Radar Conference. Abstracts of papers, Sendai, Japan, July 25–29, 1994. Sendai International Center, Sendai, Japan, xix + 592 pp.

ILRC (1996): A. Ansmann, R. Neuber, P. Rairoux, U. Wandinger, eds.: *Advances in Atmospheric Remote Sensing with Lidar.* Selected papers of the 18th International Laser Radar Conference (ILRC), Berlin, July 22–26, 1996. Springer, Berlin, xxi + 590 pp.

ILRC (1998): U.N. Singh, S. Ismail, G.K. Schwemmer, eds.: Nineteenth International Laser Radar Conference, Annapolis, MD, July 6–10, 1998. National Aeronautics and Space Administration, Langley Research Center, Hampton, VA. Abstracts of papers, NASA/CP-1998-207671/PT1, xxvi pp. + pp. 1–552 + pp. A1–A17; NASA/CP-1998-207671/PT2, xxvi pp. + pp. 553–986 + pp. A1–A17.

ILRC (2000): A. Dabas, C. Loth, J. Pelon, eds.: *Advances in Laser Remote Sensing.* Selected papers presented at the 20th International Laser Radar Conference (ILRC), Vichy, France, July 10–14, 2000. École Polytechnique, Palaiseau, France, xxvii + 491 pp.

ILRC (2002): L.R. Bissonnette, G. Roy, G. Vallée, eds.: *Lidar Remote Sensing in Atmosphere and Earth Sciences.* Reviewed and revised papers presented at the 21st International Laser Radar Conference (ILRC21), Québec, Canada, July 8–12, 2002. Defence R&D Canada Valcartier, Val-Bélair, QC, Canada, Part I xxvi pp. + pp. 1–416 + pp. A1–A9, Part II xxvi pp. + pp. 417–864 + pp. A1–A9.

ILRC (2004): G. Pappalardo, A. Amodeo, eds.: Reviewed and revised papers presented at the 22nd International Laser Radar

Conference (ILRC 2004), Matera, Italy, July 12–16, 2004. European Space Agency ESA/ESTEC, Noordwijk, The Netherlands, report SP-561. Volume I xxxii pp. + pp. 1–552, Volume II xxiv pp. + pp. 553–1043.

ILSS'99 (1999): International Laser Sensing Symposium/20th Japanese Laser Sensing Symposium. Abstract of papers, Fukui, Japan, September 6–8, 1999. Laser Radar Society of Japan, x + 312 pp.

ISTP (1988): Lower Tropospheric Profiling: Needs and Technologies. Boulder, CO, May 31–June 3, 1988, Extended abstracts. American Meteorological Society, xxiv + 260 pp. + pp. A1–A10.

ISTP (1991): Lower Tropospheric Profiling: Needs and Technologies. Boulder, CO, September 10–13, 1991, Extended abstracts. National Center for Atmospheric Research, Boulder, CO, xx + 214 pp.

ISTP (1994): Third International Symposium on Lower Tropospheric Profiling: Needs and Technologies. Hamburg, Germany, August 30–September 3, 1994. Extended abstracts. Max-Planck-Gesellschaft zur Förderung der Wissenschaften, Volume 1 xxii pp. + pp. 1–200; Volume 2 xxii pp. + pp. 201–462.

ISTP (1998): Fourth International Symposium on Tropospheric Profiling. Snowmass, CO, September 1998. Book of abstracts, two volumes.

ISTP (2000): Fifth International Symposium on Lower Tropospheric Profiling: Needs and Technology. Adelaide, Australia, December 4–8, 2000.

ISTP (2003): Sixth International Symposium on Tropospheric Profiling: Needs and Technologies. Leipzig, Germany, September 14–20, 2003, 528 + xxvi pp. See http://istp2003.tropos.de:8085.

Jones, F.E. (1949): Radar as an aid to the study of the atmosphere. *Journal of the Royal Aeronautical Society* **53**, 433–448.

Killinger, D.K., and A. Mooradian, eds. (1983): *Optical and Laser Remote Sensing*. Springer, Berlin, 383 pp.

Klein, V., and C. Werner (1993): *Fernmessung von Luftverunreinigungen*. Springer, Berlin, 254 pp.

Kyle, T.G., S. Barr, and W.E. Clements (1982): Fluorescent particle lidar. *Applied Optics* **21**, 14–15.

Long, D.A. (2002): *The Raman Effect*. Wiley, Chichester, UK, 597 pp.

Maiman, T.H. (1960): Stimulated optical radiation in ruby. *Nature* **187**, 493–494.

Measures, R.M. (1984): *Laser Remote Sensing*. Wiley-Interscience, New York, 510 pp.

Middleton, W.E.K., and A.F. Spilhaus (1953): *Meteorological Instruments*. Third Edition. University of Toronto Press, Toronto, p. 207.

Mie, G. (1908): Beiträge zur Physik trüber Medien, speziell kolloidaler Metallösungen. *Annalen der Physik, 4. Folge* **25**, 376–445.

MUSCLE (2002): C. Werner, U.G. Oppel, T. Rother, eds.: 12th International Workshop on Multiple Scattering Lidar Experiments. Oberpfaffenhofen, Germany, September 10–12, 2002. SPIE, The International Society for Optical Engineering, SPIE Volume 5059, 234 pp.

MUSCLE (2004): Multiple Scattering Lidar Experiments — MUSCLE XIII. St. Petersburg, Russia, June 28–July 1, 2004. In print.

ORSA (1990): Optical Remote Sensing of the Atmosphere. Summaries of papers presented at the Optical Remote Sensing of the Atmosphere Topical Meeting. Incline Village, NV, February 12–15, 1990. Technical Digest Volume 4. Optical Society of America, Washington, DC, xviii + 650 pp.

ORSA (1991): Optical Remote Sensing of the Atmosphere. Summaries of papers presented at the Optical Remote Sensing of the Atmosphere Topical Meeting. Williamsburg, VA, November 18–21, 1991. Technical Digest Volume 18. Optical Society of America, Washington, DC, xvi + 332 pp.

ORSA (1993): Optical Remote Sensing of the Atmosphere. Summaries of papers presented at the Optical Remote Sensing of the Atmosphere Topical Meeting. Salt Lake City, UT, March 8–12, 1993. Technical Digest Volume 5. Optical Society of America, Washington, DC, xvi + 468 pp.

Platt, U. (1978): Dry deposition of SO_2. *Atmospheric Environment* **12**, 363–367.

She, C.-Y. (2001): Spectral structure of laser light scattering revisited: bandwidths of nonresonant scattering lidars. *Applied Optics* **40**, 4875–4884.

Uthe, E.E., W. Viezee, B.M. Morley, J.K.S. Ching (1985): Airborne lidar tracking of fluorescent tracers for atmospheric transport and diffusion studies. *Bulletin of the American Meteorological Society* **66**, 1255–1262.

VDI (1999): *VDI 4210 Part 1 Remote Sensing — Atmospheric Measurements with LIDAR — Measuring Gaseous Air Pollution with DAS LIDAR*. German/English, June 1999, Beuth Verlag, Berlin, 47 pp.

VDI (2001): *VDI 3786 Part 14 Environmental Meteorology — Ground-Based Remote Sensing of the Wind Vector — Doppler*

Wind LIDAR. German/English, December 2001, Beuth Verlag, Berlin, 47 pp.

VDI (2004): *VDI 3786 Part 15 Environmental Meteorology — Ground-Based Remote Sensing of Visual Range — Visual Range Lidar*. German/English, August 2004, Beuth-Verlag, Berlin, 32 pp.

Weitkamp, C. (1990): Lidar. In: B. Ruck, ed.: *Lasermethoden in der Strömungsmeßtechnik*. AT-Fachverlag, Stuttgart, pp. 151–208.

Young, A.T. (1982): Rayleigh scattering. *Physics Today* **35**(1), 42–48.

2

Femtosecond White-Light Lidar

**J. KASPARIAN[a], R. BOURAYOU[b], S. FREY[a], J. C. LUDERER[b],
G. MÉJEAN[a], M. RODRIGUEZ[b], E. SALMON[a], H. WILLE[b],
J. YU[a], J.-P. WOLF[a], and L. WÖSTE[b]**

[a]Teramobile Project, Laboratoire de Spectrométrie
Ionique et Moléculaire, UMR CNRS 5579, Université Claude
Bernard Lyon 1, Villeurbanne Cedex, France
[b]Teramobile Project, Institut für Experimentalphysik,
Freie Universität Berlin, Berlin, Germany

2.1. INTRODUCTION

While light detection and ranging (lidar) allows the determination of three-dimensional distributions of atmospheric trace gases as well as information about the abundance, size, and phase of the aerosol, however, it is generally restricted to the detection of only one gaseous substance at a time, and does not allow the remote determination of aerosol chemical composition, as would, for example, be required to identify bioagents. On the contrary, long-path optical absorption methods like Fourier-transform infrared spectroscopy (FTIR) or differential optical absorption spectroscopy (DOAS) simultaneously yield precise concentration data of a large group of atmospheric constituents from the absorption of light from a broadband source, either natural (usually the Sun or the Moon) or artificial (lamp), along its path across the atmosphere. But they do not give access to the three-dimensional mappings allowed by lidar.

The evaluation of climatological developments like global warming, depletion of ozone in the stratosphere and its increase in the troposphere, as well as the requirement for more precise weather predictions define a clear need for better performing and more versatile remote sensing apparatus, which would combine the range resolution of lidar with the multicomponent analysis capability of DOAS or FTIR. This would require the generation of a remote white-light atmospheric lamp, free from electric cables, which could be placed arbitrarily in the atmosphere. For this purpose, there have been attempts[1] to generate a laser-induced plasma focus at remote distances by launching a high-power femtosecond (fsec) laser (100 fsec, 3 TW) into the atmosphere. The ultrashort pulses were also adequately chirped to compensate group-velocity dispersion in the air (see Section 2.4). This assured that the pulses are temporally focused at a predetermined distance, where the initial transform-limited fsec-pulses are reestablished. When coincidence of this temporal focus with the geometrical one is achieved, the required intensity for creating a plasma spot is reached. However, instead of the expected plasma focus, an extended white-light channel is observed. Unlike the fundamental laser wavelength (~800 nm), this channel is clearly visible even by the naked eye (as shown in Figure 2.1).[2] The spectrum of the emitted white light covers the entire visible range from the ultraviolet (UV) to the infrared (IR), and its signals could be detected from altitudes beyond 10 km. Similar white-light filaments were also observed by Mourou and Korn in the laboratory, over distances of some meters.[3]

Figure 2.1 (Color figure follows page 398). Terawatt white-light beam in the atmosphere. Photograph taken at the Jena University. The upper edge of the planetary boundary layer (~1.5 km) is visible via enhanced aerosol scattering (Teramobile©).

The attractive perspectives for applications of white light in the field of optical remote sensing (fsec-lidar) provided the motivation for the *Teramobile* project.[4,5] Besides other atmospheric applications such as lightning control,[6] this French–German collaborative project aims at developing a new lidar technique that allows multipollutant detection and composition analysis of the aerosols, as well as remote sensing of the water-vapor saturation in view of a better forecast of rain, hail, or snow.

2.2.   THE TERAMOBILE SYSTEM[4]

The basic setup of our fsec-lidar experiment is depicted in Figure 2.2.[5] It shows a fsec-laser system that is based on the well-known chirped pulse amplification (CPA) technique,[7] with a titanium–sapphire oscillator and an Nd:YAG pumped titanium–sapphire amplification chain including a regenerative amplifier and two four-pass amplifiers. It provides 350 mJ pulses with 70 fsec duration, resulting in a peak power of 5 TW at a wavelength of about 800 nm and with a repetition rate of 10 Hz. The classical compressor setup has been improved to allow easy precompensation of the group-velocity dispersion (GVD) in air (see Section 2.4). Together with an adjustable focus, this feature allows the generation of filaments at a predetermined distance away from the sending optics.

The backward-emitted white light is collected on a 40 cm receiving telescope, which can work both on- or off-axis with respect to the emitted laser beam. The lidar detection chain consists of a high-resolution spectrometer with three integrated gratings and four possible detectors, allowing the simultaneous temporal and spectral analysis of the return signal in a wavelength range between 190 nm and 2.5 μm. This range can be extended in the IR up to 5 μm by using interference filters and an LN$_2$-cooled InSb photodiode. In addition, depolarization measurements of the return signal can be performed to characterize the aerosol phase.

Mobility is a key feature of the Teramobile laser system. Therefore, the experiment is hosted in a self-contained mobile

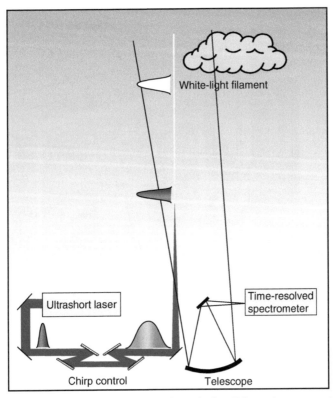

Figure 2.2 Setup of the fsec-white-light lidar. A terawatt fsec-pulse, with a negative chirp to compensate the group-velocity dispersion, is send out to the atmosphere and produces white light because of the high intensity in self-guided filaments. The backscattered light is then collected by a telescope and analyzed with high temporal and spectral resolutions. (From Kasparian J. et al., *Science*, 301, 61, 2003. With permission.)

unit shown in Figure 2.3,[4] which contains the laser. Its integration in the reduced space of the mobile laboratory, embedded in a standard 20 ft freight container, required a particularly compact design. Further, the laboratory hosts the lidar emission and detection devices and working space for operators. The container is equipped with the necessary power and cooling supplies, temperature stabilization, vibration control, and an additional standard standalone lidar to

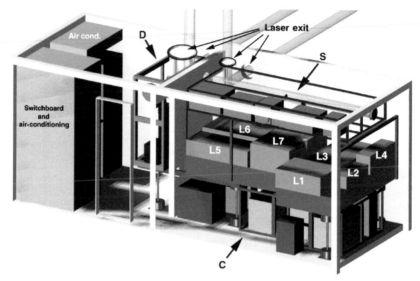

Figure 2.3 Three-dimensional view inside the Teramobile container. Laser room: CPA laser chain (L1–L7), power units and cooling system (C), sending telescope (S). Control and detection room: detection system (D), computer and electronics, with desk, electrical switchboard and air conditioning (hidden). The laser beam is send through a large mirror telescope to one of four possible exits (horizontal or vertical, on-axis with or beside the detection path). (From Wille et al., *Eur. Phys. J. AP*, 20, 183, 2002. With permission.)

ensure eye safety. Therefore, the system is capable of operating at virtually any location, including urban areas or industrial sites, and even abandoned airports or astronomical observatories useful for long-distance propagation studies.

2.3. NONLINEAR PROPAGATION OF TW PULSES

High-power laser pulses propagating in transparent media undergo nonlinear propagation. Nonlinear self-action leads to strong evolutions of the spatial (self-focusing,[8] self-guiding,[9] self-reflection[10]), spectral (four-wave mixing,[11] self-phase modulation[12–14]), and temporal (self-steepening,[15–18] pulse splitting[19]) characteristics of the pulse. The propagation medium is also strongly affected. It undergoes multiphoton ionization,[20–24] resulting in the formation of an ionized plasma channel.

These phenomena have been extensively studied in condensed matter since the early 1970s, from the theoretical as well as the experimental points of view. However, it was only in 1985 that the development of the CPA technique[7] permitted the production of ultrafast laser pulses reaching intensities as high as 10^{20} W/cm^2, and hence the observation of highly nonlinear propagation even in only slightly nonlinear dilute media such as atmospheric pressure gases. We will focus on nonlinear propagation in air resulting in coherent white-light generation and filamentation when adequate high-power fsec-laser pulses propagate across transparent media. The strongly nonlinear optical phenomenon results from a critical balance between self-focusing Kerr lenses and defocusing plasma lenses, which are formed by a nonuniform, intensity-dependent refractive index across the laser beam profile.[9,25–27]

2.3.1. Kerr Self-Focusing

For high luminous intensity I, the refractive index n of air is modified by the Kerr effect[15]:

$$n(I) = n_0 + n_2 I \tag{2.1}$$

where n_2 is the nonlinear refractive index of air ($n_2 = 3 \times 10^{-19}$ cm^2/W).[28] Since the intensity profile of a laser beam is usually rather bell-shaped than uniform, the refractive index increase is larger in the center of the beam than on the edge (Figure 2.4A). This induces a radial refractive index gradient equivalent to a lens (called "Kerr lens"), with a focal length depending on the intensity I. The beam is focused by this lens, which leads to an intensity increase, resulting in the formation of a shorter focal length lens and so on until the beam collapses on its axis.

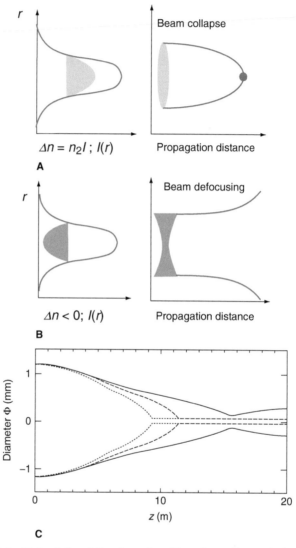

Figure 2.4 Principle of the focusing Kerr lens (A) and the defocus-ing plasma lens (B). A balance of both effects results in self-guided high-intensity (10^{13} to 10^{14} W/cm^2) filaments with diameters in the range of 100 μm. (C) Theoretical calculation of the propagation of fsec-pulses. The curves show the evolution of the beam diameter as a function of the propagation distance, considering strong (continuous lines), weak (dashed lines), and no (dotted lines) retarded Kerr effect. (Panel C from Chiron et al., *Eur Phys. J. D*, 6, 383, 1999. With permission.)

However, self-focusing can overcome the natural diffraction of the beam only if the overall beam power is above a critical power P_{crit} given by

$$P_{crit} = I\pi w^2 = \frac{\pi}{k^2 n_2} = \frac{\lambda^2}{4\pi n_2} \qquad (2.2)$$

where w is the beam waist. It should be pointed out that this is a critical *power* rather than a critical intensity. For a titanium–sapphire laser ($\lambda = 800$ nm) in air, $P_{crit} \sim 3$ GW. Conversely, the distance at which the beam is focused is related to the successive focal lengths $f(I)$, and is a function of the initial *intensity*.

2.3.2. Multiphoton Ionization and Plasma Generation

If the laser pulse intensity reaches 10^{13} to 10^{14} W/cm^2, higher-order nonlinear processes such as multiphoton ionization (MPI) occur. At 800 nm, eight to ten photons are needed to ionize N_2 and O_2 molecules, respectively, and produce an ionized plasma. The ionization process can involve tunneling as well, because of the very high electric field carried by the laser pulse. However, following Keldysh's theory,[29,30] MPI dominates for intensities lower than 10^{14} W/cm^2. In contrast to longer pulses, fsec-pulses combine a high MPI efficiency due to their very high intensity, but with a limited overall energy so that the generated electron densities (10^{16} to 10^{17} cm^{-3}) are far from saturation. Moreover, since the laser pulse duration at play $\tau_L = 100$ fsec is much shorter than the electron–ion collision time in air (which is in the order of $\tau_{ei} \approx 1.5$ to 2 psec), avalanche ionization of the electrons accelerated in the intense laser field cannot occur. This prevents a subsequent optical breakdown that would lead to a high electron density. Finally, losses by inverse Bremsstrahlung are therefore negligible, which allows long-distance pulse propagation.

However, the electron density ρ, directly coupled to the radial intensity profile of the laser beam, creates a negative refractive index gradient. This gradient acts as a negative lens, which tends to defocus the laser beam (as schematically shown in Figure 2.4B).

2.3.3. Filamentation of High-Power Laser Beams

Based on their effect described above, both Kerr self-focusing and plasma defocusing should prevent long-distance propagation of high-power laser beams. However, a remarkable situation occurs in air, where both effects exactly compensate and give rise to a self-guided quasisolitonic[31] propagation. The laser beam is first self-focused by Kerr effect (Figure 2.4A); this focusing then increases the beam intensity and generates a plasma by MPI, which, in turn, defocuses the beam (Figure 2.4B). The intensity then decreases as the beam broadens and plasma generation stops, which allows Kerr refocusing to take over again. This dynamic balance between Kerr effect and plasma generation leads to the formation of stable narrow structures called filaments (Figure 2.4C).[39] Light filaments were first described in condensed media.[12,32,33] Subsequently, they were observed in air by Braun et al.,[9] who discovered that mirrors could be damaged by high-power ultrashort laser pulses even at a large distance away from the laser source. These light filaments have remarkable properties; in particular, they can propagate over tens or even hundreds of meters,[34] although their diameter is only 100 to 200 μm, and have almost constant values of intensity[35,36] (typically 10^{14} W/cm^2), energy (some millijoules), diameter, and electron density (typically 10^{16} cm^{-3}).[37,38]

The beam propagation is governed by Maxwell's equation:

$$\nabla^2 E - \frac{1}{c^2} \frac{\partial^2 E}{\partial t^2} = \mu_0 \sigma \frac{\partial E}{\partial t} + \mu_0 \frac{\partial^2 P}{\partial t^2} \tag{2.3}$$

where σ is the conductivity and accounts for losses, and P is the polarization of the medium. In contrast with the linear wave propagation equation, P now contains a self-induced nonlinear contribution corresponding to Kerr focusing and plasma generation:

$$P = P_L + P_{NL} = \varepsilon_0(\chi_L + \chi_{NL})E \tag{2.4}$$

where χ_L and χ_{NL} are the linear and nonlinear susceptibilities, respectively. A radially symmetric pulse propagating along the z-axis in a reference frame moving at the group velocity v_g leads to the following nonlinear Schrödinger equation (NLSE)[39]:

$$\nabla_\perp^2 \varepsilon + 2i\left(k\frac{\partial \varepsilon}{\partial z}\right) + 2k^2 n_2 |\varepsilon|^2 \varepsilon - k^2 \frac{\rho}{\rho_c}\varepsilon = 0 \tag{2.5}$$

where $\varepsilon = \varepsilon(r, z, t)$ is the pulse envelope of the electric field and ρ_c the critical electron density (1.8×10^{21} cm^{-3} at 800 nm). ε is assumed to vary slowly compared to the carrier oscillation and to have a smooth radial decrease. In this first-order treatment, GVD and losses due to multiphoton and plasma absorption are neglected ($\sigma = 0$). In Equation (2.5), the Laplacian operator accounts for wave diffraction in the transverse plane, while the two last terms represent the nonlinear contributions of Kerr focusing and plasma defocusing (notice the opposite signs). The electronic density $\rho(r, z, t)$ is computed using the rate equation (2.6) in a self-consistent way with (2.5):

$$\frac{\partial \rho}{\partial t} - \gamma |\varepsilon|^{2\alpha}(\rho_n - \rho) = 0 \tag{2.6}$$

where ρ_n is the neutral molecular concentration in air, γ the MPI efficiency, and α the number of photons needed to ionize an air molecule (typically $\alpha = 10$).[36]

Many theoretical studies have been carried out to simulate the nonlinear propagation of high-power laser beams, both in the single as well as in the multiple filamentation regimes.

Numerically solving the NLSE[39–41] equation leads to the evolution of the pulse intensity $I = |\varepsilon|^2$ as a function of propagation distance. Initial Kerr lens self-focusing and subsequent stabilization by the MPI-generated plasma are well reproduced by these simulations. However, a numerical instability related to the high nonlinearity of the NLSE leads

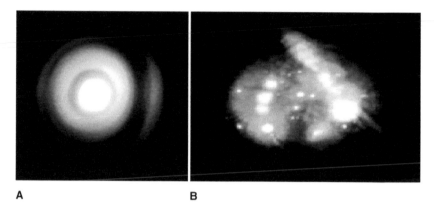

A   B

Figure 2.5   (Color figure follows page 398). (A) Filament image on
a screen (laser energy 8 mJ, pulse duration 120 fsec) showing conical
emission. (B) Multiple filamentation in a TW beam (350 mJ, 80 fsec).

to unreasonable computation times, which prevent simulations over long distances.[42]

At laser powers much above P_{crit}, the incident beam
breaks up into several localized filaments.[42–44] The intensity
in each filament is indeed clamped at 10^{14} to 10^{13} W/cm^2,
corresponding to a few millijoules,[35] while the number of
filaments is roughly proportional to the input power.
Figure 2.5 shows a cross-section of laser beams undergoing
filamentation (A, 5 mJ) and multifilamentation (B, 400 mJ).
The stability of this quasisolitonic structure is remarkable:
filaments, although only 100 μm in diameter, have been observed to propagate over 200 m,[34] and recent vertical experiments provided evidence of filamentation at kilometer-range
altitudes.[45]

2.3.4.   White-Light Generation and Self-Phase
Modulation

The spectral content of the emitted light is of particular importance for lidar applications. Nonlinear propagation of high-
intensity laser pulses not only provides self-guiding of the
light but also an extraordinary broad continuum spanning
from the UV to the IR. This supercontinuum is generated by

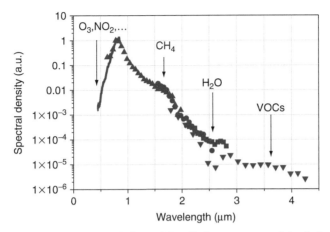

Figure 2.6 Spectrum of the white light measured in laboratory experiments. (From Kasparian J. et al., *Opt. Lett.*, 25, 1397, 2000. With permission.)

self-phase modulation as the high-intensity pulse propagates. As depicted above, the Kerr effect leads, because of the spatial intensity gradient, to self-focusing of the laser beam. However, the intensity also varies with time, and the instantaneous refractive index of the air is modified as

$$n(t) = n_0 + n_2 I(t) \tag{2.7}$$

The spectrum of the emitted white light strongly depends on the initial chirp settings; it can cover — as shown in the measurement depicted in Figure 2.6[46] — a wide range between 230 nm[47] and 4.5 μm.[46]

2.4. ATMOSPHERIC FILAMENTATION EXPERIMENTS

The delivery of high laser intensities at high altitudes is very attractive for atmospheric research because it allows the exploitation of nonlinear optical processes like white-light generation, multiphoton-induced fluorescence, or harmonic generation for diagnostic purposes. Since filamentation counteracts diffraction over long distances, it opens an opportunity to deliver high laser intensities over long distances and at high altitudes.

However, when propagating ultrashort pulses over a long distance in air, GVD has to be taken into account, since these pulses intrinsically have a broad spectral bandwidth. For example, Fourier-limited 100 fsec pulses have a bandwidth of typically 16 nm. In normally dispersive air the longer wavelength ("red") components of the laser spectrum propagate faster than the shorter wavelength ("blue") ones, resulting in an increase of the pulse duration along its propagation path in air. This temporal broadening amounts to 1 psec per kilometer propagation for a 100 fsec pulse, resulting in a tenfold decrease of the peak power. This effect, however, can be turned into an advantage by sending out adequately chirped laser pulses, in which the blue component of the laser spectrum precedes its red component. This is easily achieved by a realignment of the fsec-laser compressor. As a result, the negatively chirped pulse shortens temporally during its propagation and its intensity increases until the temporal focus is reached and the conditions for filamentation are reached.

The chirp-based control of the emitted white-light supercontinuum has been well demonstrated by using the high-resolution imaging mode of the 2-m telescope of the Thüringer Landessternwarte Tautenburg in Germany.[5,45] For these experiments, the Teramobile laser was placed next to the telescope. The laser was launched into the atmosphere and the backscattered light was imaged through the telescope. Figure 2.7A shows a typical image obtained at the fundamental of the laser pulse ($\lambda = 800$ nm), over an altitude range from 3 to 25 km. On this picture, two multiple scattering halos on aerosol layers were observed at altitudes of 4 and 9 km, respectively. In the blue–green band (400 to 500 nm) of the white-light supercontinuum, the corresponding images — as shown in Figure 2.7B and C — strongly depend on the initial chirp settings of the laser pulse. White-light signal can only be observed for an adequate GVD precompensation.

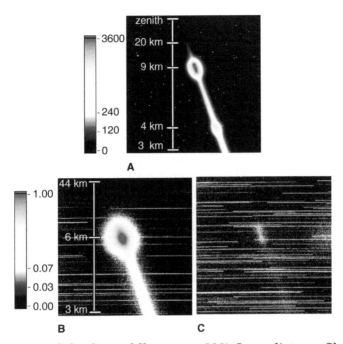

Figure 2.7 (Color figure follows page 398). Long-distance filamentation and control of nonlinear optical processes in the atmosphere.[5] Pictures of the fsec-laser beam propagating vertically, imaged by the CCD camera of the 2-m telescope at the Thüringer Landessternwarte Tautenburg. (A) Fundamental wavelength, visible up to 25 km through two aerosol layers appearing as multiple-scattering halos. (B and C) Blue light (390 to 490 nm) generated by two 600 fsec pulses of same initial peak power, with negative (GVD precompensating) and positive chirps. The stripes across the images are due to the star motion during the acquisition time of several minutes. Note that the white-light generation requires GVD precompensation. (Derived from Kasparian J. et al., *Science*, 301, 61, 2003. With permission.)

2.5. LIDAR REMOTE SENSING OF ATMOSPHERIC TRACES

The principle of standard lidar systems is based on the effect of optical scattering of emitted light on atmospheric constituents. The Rayleigh-, Raman-, fluorescence-, or Mie-scatter process returns a small portion of the emitted light back to the observer. This necessarily leads to an unfavorable $1/R^2$ — dependency of the received light, where R is the distance from the scatter location to the observer. When spectrally dispersed, the signal at the receiver is usually too weak, as arc-lamp-based lidar experiments have shown in the past. Since the spectrum of the emitted white light — as shown in Figure 2.6 — also covers a wide spectral range, the same question with regard to the unfavorable $1/R^2$-dependency must be raised for the fsec-lidar measurements as well. Only a small amount of white-light return-signals could therefore be expected normally, which would prevent an identification of atmospheric trace substances, requiring highly resolved absorption analyses over extended observation lengths. An important experiment in this regard was performed by Yu et al.,[48] who measured, with the experimental setup shown in Figure 2.8A, the angular emissivity of fsec-white-light filaments. The result — as plotted in Figure 2.8B — shows a surprisingly strong backward component of the emitted white light. This unexpected behavior makes fsec-filaments in particular promising for lidar applications because it opens the perspective to establish a directional, white-light source in the atmosphere radiating predominantly in the backward direction, where the receiver is.

The basic setup of our fsec-lidar — as depicted in Figure 2.2 — shows the fsec-laser pulse, which, after passing a negative chirp generator, is vertically launched into the atmosphere. As a result, filaments are generated at a predetermined distance. The backward-emitted white light is then collected in a telescope and focused across a spectrometer on a detector, which allows temporally resolved observations. Figure 2.9 shows three transients of spectrally filtered return signals recorded in the time window after the laser was fired.[5] Two of them were recorded — as an indicator for white light — at $\lambda = 300\,nm$ and $\lambda = 600\,nm$, while the third one was recorded at the third harmonic (TH) wavelength at $\lambda = 267\,nm$. The horizontal axis of these signals corresponds to the altitude from where it was backscattered. While the signal in the visible range is still intense over 5 km altitude,

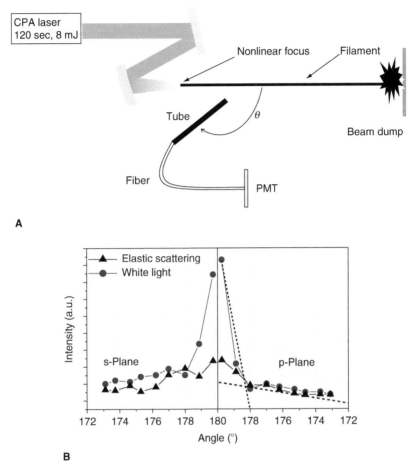

Figure 2.8 Self-reflection of the white-light supercontinuum: (A) experimental setup; (B) results around backscattering. (Derived from Yu J. et al., *Opt. Lett.*, 26, 533, 2001. With permission.)

the UV signals are more attenuated due to a more efficient Rayleigh scattering, as well as absorption by the high tropospheric ozone concentrations at the TH wavelength. Hence, they vanish around 1.5 km.

The unique capability of white-light lidar to perform high-resolution measurements over a wide spectral range is clearly illustrated in Figure 2.10A.[5] The measurement was

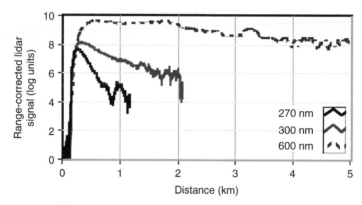

Figure 2.9 Vertical white-light lidar profile at three wavelengths: 270 nm (third harmonic), 300 nm, and 600 nm (white light). (Derived from Kasparian J. et al., *Science*, 301, 61, 2003. With permission.)

recorded from an altitude of 4.5 km; it presents the highly resolved optical fingerprints of the atmosphere along this absorption path. The richness of the atmospheric absorption lines appearing represents, for example, extended ro-vibrational progressions of highly forbidden transitions of the O_2 A-band ($\sigma \approx 10^{-25}$ to 10^{-24} cm^2) and H_2O overtone bands. Figure 2.10B shows that simulations based on the HITRAN database[49] of the water-vapor lines yield an almost perfect agreement with the measured result, except for weak lines missing in the database since they cannot be measured on laboratory-scale absorption paths. The observable resolution also permits the retrieval of temperature profiles, while Doppler-shift measurements might allow for wind measurements. The results show a quality of data that is already now comparable with results obtained from dedicated DIAL systems.

2.6. AEROSOLS

The strong Mie-scatter behavior of the atmospheric aerosol allows an easy lidar observation. The set of measurable properties, however, is restricted to abundance, size, and refractive index of the aerosol, even when strong a priori assumptions about the particle size distribution are made and multispectral

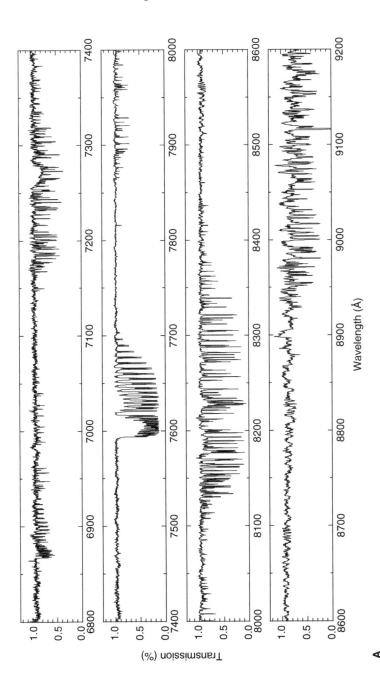

Figure 2.10 High-resolution spectrum of atmospheric absorption measured with the fsec-white-light lidar from an altitude of 4.5 km: (A) broad spectral range acquired in a single lidar acquisition (6000 pulses), showing several oxygen and water-vapor bands[5]; (B) section of the same spectrum with a fit based on HITRAN, to retrieve the averaged humidity. (Derived from Kasparian J. et al., *Science*, 301, 61, 2003. With permission.)

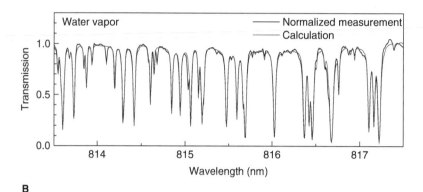

Figure 2.10   (*Continued*)

lidar techniques are employed.[50] This situation can certainly be further improved by exploiting the spectral information which is available with a white-light lidar.

No remote sensing method, however, allows yet the determination of the chemical composition of aerosols. Here, the high peak power of the fsec-pulses offers the perspective to generate nonlinear effects directly inside the aerosol, from which such information can be retrieved. Recently, it has been shown that fsec-laser light can be focused onto a very bright hot spot inside the microcavity of a water droplet, yielding a very large cross-section for nonlinear processes like multiphoton-excited fluorescence (MPEF).[51,52] Due to the reciprocity principle, this fluorescence light, which initially is emitted isotropically from the hot spot, is then refocused into the backward direction by the droplet itself, favoring its lidar detection.

We have recently demonstrated the capability of the MPEF process for the remote identification of clean and contaminated water particles.[5,53] In this experiment, we produced a controlled distribution of water droplets of about 1.5 μm in size in an open cloud chamber, which was placed 50 m away from the Teramobile system. On demand, the droplets were doped with a biosimulant, namely riboflavin, which exhibits a strong fluorescence emission around 540 nm. With adequate peak power at 800 nm (400 GW), a significant two-photon excited fluorescence signal emerged from the doped droplets allowing to distinguish them — as shown in Figure 2.11[5] — unambiguously from the clean water cloud. To our knowledge, this measurement represents the first remote composition analysis of an aerosol, in this case a biosimulant.

The production of charges inside the filament provides the perspective to determine directly if the atmosphere is supersaturated. Supersaturation refers to a metastable state of the atmosphere, where water condensation does not occur, and hence no clouds form, in spite of favorable thermodynamical conditions (humidity, temperature, and pressure). Supersaturation is due to the lack of sufficient condensation nuclei on which water vapor can nucleate. Those germs can be either natural, like aerosols or dust, or artificial, for example, heavy molecules like AgI, a substance commonly used for cloud injection experiments. It is also known that electric charges can act as nucleation germs.[54] We have recently demonstrated that the charge density in laser-induced plasma filaments is sufficient to achieve condensation. Launching the fsec-laser beam into a fog chamber where supersaturation was maintained, we observed water droplets nucleating around the plasma filament. The resulting cloud line strongly scatters

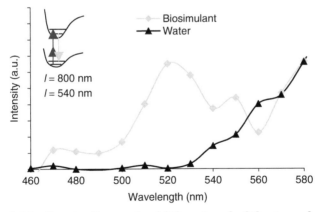

Figure 2.11 Spectrally resolved lidar signal of the two-photon excited fluorescence from a biosimulant (riboflavin-doped water droplets) at 50 m distance.[5] The signal from natural aerosols (water) is given for reference. (From Kasparian J. et al., *Science*, 301, 61, 2003. With permission.)

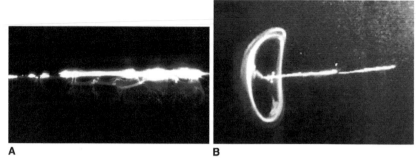

A   B

Figure 2.12 Laser-induced nucleation of water droplets in a cloud chamber.[5] The nucleated droplets appear as a clear line, since they scatter subsequent laser pulses. On picture B, the shockwave produced by the expansion of the laser-generated ionized filament is visible, as a ring of higher aerosol density. (From Kasparian J. et al., *Science*, 301, 61, 2003. With permission.)

subsequent pulses, appearing as a bright line on photographs (as shown in Figure 2.12[5]). This experiment demonstrates the possibility to create laser-induced nucleation germs. Aiming the laser at a particular direction in the sky, and adjusting the distance where the ionized filaments are produced, it should therefore be possible to probe whether nucleation occurs, and hence whether the atmosphere is supersaturated. This observable is of great importance with regard to the prediction of rain, hail, or snow.

2.7. CONCLUSION

In the last few years, the knowledge about the propagation of ultrashort laser pulses in air and the relating nonlinear optical effects has progressed immensely. In particular, the discovery of extended fsec-laser induced white-light filaments, which exhibit exciting new physical phenomena, makes them very interesting for applications in the field of atmospheric remote sensing. Several encouraging experiments have raised hopes for a remote multicomponent diagnosis of the atmosphere by means of fsec-white-light lidar, the remote composition analysis of the aerosol, and the remote detection of atmospheric supersaturation. The development of these applications will further be backed by the expected progresses in laser ultrafast technology, such as direct diode pumping, miniaturization or the development of pulse shaping techniques, which will improve strongly their versatility, reliability, and ease of operation, and hence make them highly valuable for atmospheric applications.

ACKNOWLEDGMENTS

This work was performed in the framework of the Teramobile project, jointly funded by the Centre National de la Recherche Scientifique (France) and the Deutsche Forschungsgemeinschaft (Germany). We acknowledge the valuable collaborations with the team at the Thüringer Landessternwarte Tautenburg. The strong support by our technical staffs in the universities of Berlin, Jena, and Lyon, especially M. Barbaire, M. Kerleroux, M. Néri, W. Ziegler, and M. Kregielski was of particular value to us. The Teramobile web site is www. teramobile.org.

REFERENCES

1. P. Rairoux, H. Schillinger, S. Niedermeier, M. Rodriguez, F. Ronneberger, R. Sauerbrey, B. Stein, D. Waite, C. Wedekind, H. Wille, L. Wöste, *Applied Physics B* **71**, 573 (2000).
2. L. Wöste, C. Wedekind, H. Wille, P. Rairoux, B. Stein, S. Nikolov, Chr. Werner, S. Niedermeier, H. Schillinger, R. Sauerbrey, *Laser und Optoelektronik* **29**(5), 51 (1997).
3. A. Braun, G. Korn, X. Liu, D. Du, J. Squier, G. Mourou, *Optics Letters* **20**, 73 (1995).
4. H. Wille, M. Rodriguez, J. Kasparian, D. Mondelain, J. Yu, A. Mysyrowicz, R. Sauerbrey, J.-P. Wolf, L. Wöste, *European Physical Journal — Applied Physics* **20**, 183 (2002).
5. J. Kasparian, M. Rodriguez, G. Méjean, J. Yu, E. Salmon, H. Wille, R. Bourayou, S. Frey, Y.-B. André, A. Mysyrowicz, R. Sauerbrey, J.-P. Wolf, L. Wöste, White light filaments: sources for long-range atmospheric analysis, *Science* **301**, 61 (2003).

6. M. Rodriguez, R. Sauerbrey, H. Wille, L. Wöste, T. Fujii, Y.-B. André, A. Mysyrowicz, L. Klingbeil, K. Rethmeier, W. Kalkner, J. Kasparian, E. Salmon, J. Yu, J.-P. Wolf, *Optics Letters* **27**, 772 (2002).

7. D. Strickland, G. Mourou, *Optics Communications* **56**, 219 (1985).

8. D. Strickland, P.B. Corkum, *Journal of the Optical Society of America B* **11**, 492 (1994).

9. A. Braun, G. Korn, X. Liu, D. Du, J. Squier, G. Mourou, *Optics Letters* **20**, 73 (1995).

10. L. Roso-Franco, *Physical Review Letters* **55**, 2149 (1995).

11. R.R. Alfano, S.L. Shapiro, *Physical Review Letters* **24**, 584 (1970).

12. R.R. Alfano, S.L. Shapiro, *Physical Review Letters* **24**, 592 (1970).

13. R.R. Alfano, S.L. Shapiro, *Physical Review Letters* **24**, 1217 (1970).

14. A. Brodeur, S.L. Chin, *Journal of the Optical Society of America B* **16**, 637 (1999).

15. Y. Shen, *The Principles of Nonlinear Optics*, John Wiley & Sons, New York (1984).

16. G. Yang, Y. Shen, *Optics Letters* **9**, 510 (1984).

17. Y.S. Kivshar, B.A. Malomed, *Review of Modern Physics* **61**, 763 (1989).

18. A.L. Gaeta, *Physical Review Letters* **84**, 3582 (2000).

19. J.K. Ranka, R.W. Schirmer, A.L. Gaeta, *Physical Review Letters* **77**, 3783 (1996).

20. A. Proulx, A. Talebpour, S. Petit, S.L. Chin, *Optics Communications* **174**, 305 (2000).

21. S. Tzortzakis, M.A. Franco, Y.-B. André, A. Chiron, B. Lamouroux, B.S. Prade, A. Mysyrowicz, *Physical Review E* **60**, R3505 (1999).

22. S. Tzortzakis, B. Prade, M. Franco, A. Mysyrowicz, *Optics Communications* **181**, 123 (2000).

23. H. Schillinger, R. Sauerbrey, *Applied Physics B* **68**, 753 (1999).

24. J. Schwarz, P. Rambo, J.C. Diels, *Applied Physics B* **72**, 343 (2001).

25. E.T.J. Nibbering, P.F. Curley, G. Grillon, B.S. Prade, M.A. Franco, F. Salin, A. Mysyrowicz, *Optics Letters* **21**, 62 (1996).

26. A. Brodeur, C.Y. Chien, F.A. Ilkov, S.L. Chin, O.G. Kosareva, V.P. Kandidov, *Optics Letters* **22**, 304 (1997).

27. M. Mlejnek, E.M. Wright, J.V. Moloney, *Optics Letters* **23**, 382 (1988).
28. X.M. Zhao, P. Rambo, J.-C. Diels, *QELS'95* **16**, 178 (1995).
29. L.V. Keldysh, *Soviet Physics JETP* **20**, 1307 (1965).
30. M. Geissler, G. Tempea, A. Scrinzi, M. Schnürer, F. Krausz, T. Barbec, *Physical Review Letters* **83**, 2930 (1999).
31. L. Bergé, A. Couairon, *Physical Review Letters* **86**, 1003 (2001).
32. R.G. Brewer, J.R. Lifsitz, E. Garmire, R.Y. Chiao, C.H. Townes, *Physical Review* **166**, 326 (1968).
33. C.A. Sacchi, C.H. Townes, J.R. Lifsitz, *Physical Review* **174**, 439 (1968).
34. B. La Fontaine, F. Vidal, Z. Jiang, C.Y. Chien, D. Comtois, A. Desparois, T.W. Johnston, J.-C. Kieffer, H. Pépin, H.P. Mercure, *Physics of Plasmas* **6**, 1615 (1999).
35. A. Becker, N. Aközbek, K. Vijayalakshmi, E. Oral, C.M. Bowden, S.L. Chin, *Applied Physics B* **73**, 287 (2001).
36. J. Kasparian, R. Sauerbrey, S.L. Chin, *Applied Physics B* **71**, 877 (2000).
37. H. Schillinger, R. Sauerbrey, *Applied Physics B* **68**, 753 (1999).
38. Tzortzakis, M.A. Franco, Y.-B. André, A. Chiron, B. Lamouroux, B.S. Prade, A. Mysyrowicz, *Physical Review E* **60**, R3505 (1999).
39. A. Chiron, B. Lamouroux, R. Lange, J.-F. Ripoche, M. Franco, B. Prade, G. Bonnaud, G. Riazuelo, A. Mysyrowicz, *The European Physical Journal D* **6**, 383 (1999).
40. J.K. Ranka, A.L. Gaeta, *Optics Letters* **23**, 534 (1998).
41. M. Mlejnek, E.M. Wright, J.V. Moloney, *Optics Letters* **23**, 382 (1998).
42. L. Bergé, S. Skupin, F. Lederer, G. Méjean, J. Yu, J. Kasparian, E. Salmon, J.P. Wolf, M. Rodriguez, L. Wöste, R. Bourayou, R. Sauerbrey, Multiple filamentation of TW laser pulses in air, *Physical Review Letters* 92, 225002 (2004).
43. M. Mlejnek, M. Kolesik, J.V. Moloney, E.M. Wright, *Physical Review Letters* **83**, 2938 (1999).
44. K. Germaschewski, R. Grauer, L. Bergé, V.K. Mezentsev, J. Juul Rasmussen, *Physica D* **151**, 175 (2001).
45. M. Rodriguez, R. Bourayou, G. Méjean, J. Kasparian, J. Yu, E. Salmon, A. Scholz, B. Stecklum, J. Eislöffel, U. Laux, A.P. Hatzes, R. Sauerbrey, L. Wöste, J.-P. Wolf, Kilometer-range non-linear propagation of femtosecond laser pulses, Physical Review E 69, 036607 (2004).

46. J. Kasparian, R. Sauerbrey, D. Mondelain, S. Niedermeier, J. Yu, J.-P. Wolf, Y.-B. André, M. Franco, B. Prade, A. Mysyrowicz, S. Tzortzakis, M. Rodriguez, H. Wille, L. Wöste, *Optics Letters* **25**, 1397 (2000).
47. N. Aközbek, A. Iwasaki, A. Becker, M. Scalora, S.L. Chin, C.M. Bowden, *Physical Review Letters* **89**, 143901 (2002).
48. J. Yu, D. Mondelain, G. Ange, R. Volk, S. Niedermeier, J.-P. Wolf, J. Kasparian, R. Sauerbrey, *Optics Letters* **26**, 533 (2001).
49. L.S. Rothman, A. Barbe, D.C. Benner, L.R. Brown, C. Camy-Peyret, M.R. Carleer, K. Chance, C. Clerbaux, V. Dana, V.M. Devi, A. Fayt, J.-M. Flaud, R.R. Gamache, A. Goldman, D. Jacquemart, K.W. Jucks, W.J. Lafferty, J.-Y. Mandin, S.T. Massie, V. Nemtchinov, D.A. Newnham, A. Perrin, C.P. Rinsland, J. Schroeder, K.M. Smith, M.A.H. Smith, K. Tang, R.A. Toth, J. Vander Auwera, P. Varanasi, K. Yoshino, *Journal of Quantitative Spectroscopy and Radiation Transfer* **82,** 5 (2003).
50. B. Stein, C. Wedekind, H. Wille, F. Immler, M. Müller, L. Wöste, M. del Guasta, M. Morandi, L. Stefanutti, A. Antonelli, P. Agostini, V. Rizi, G. Readelli, V. Mitev, R. Matthey, R. Kivi, E. Kyrö, *Journal of Geophysical Research* **104**, S23983 (1999).
51. S.C. Hill, V. Boutou, J. Yu, S. Ramstein, J.-P. Wolf, Y.-L. Pan, S. Holler, R.K. Chang, *Physical Review Letters* **85**, 54 (2000).
52. V. Boutou, C. Favre, S.C. Hill, Y.L. Pan, R.K. Chang, J.P. Wolf, *Applied Physics B* **75**, 145 (2002).
53. G. Méjean, J. Kasparian, J. Yu, S. Frey, E. Salmon, J.-P. Wolf, Remote detection and identification of biological aerosols using a femtosecond terawatt lidar system, *Applied Physics B* **78**, 535–537 (2004).
54. K.J. Oh, G.T. Gao, X.C. Zeng, *Physical Review Letters* **86**, 5080 (2001).

3

Elastic Lidar Measurement of the Troposphere

NOBUO TAKEUCHI

Center for Environmental Remote Sensing,
Cheba University,
Japan

An elastic lidar is a lidar system that is based on the elastic scattering, is well developed, and is a very popular lidar system. The definition of elastic scattering is "scattering with no apparent wavelength change and the sum of Mie scattering and Rayleigh scattering," as explained in Chapter 1. In a typical case, it uses a solid-state laser as a light source. The elastic scattering is unable to identify the gas species. However, it is suitable for the detection of particles (aerosol) and cloud, and in the limiting case of no aerosol, air molecules as the limit of small particles. A relatively compact lidar system is constructed for monitoring the elastic scattering. Particle and cloud are good tracers for the detection of the atmospheric boundary phenomena, and monitoring a cloud and long-range transportation of pollutant and dust in the free troposphere.

In this chapter, an outline of the troposphere by lidar monitoring, the lidar equation and its analytical solution, optical properties of aerosol and air molecules, the lidar system and examples of monitoring, monitoring of optical properties of aerosols, and lidar network monitoring are described.

3.1. OUTLINE OF THE TROPOSPHERE BY LIDAR MONITORING

The troposphere extends from the ground surface to 11–13 km in altitude in the middle latitude, to 17 km in the tropics, and to 7 km in the Polar Regions. The influence of Earth's surface reaches the top of the troposphere, that is, the tropopause. The principal constituents of the dry atmosphere of the Earth are nitrogen (78% in volume) and oxygen (21%). They absorb radiation slightly. The atmospheric gases in the remaining 1% are argon (0.9%), carbon dioxide (0.03%), and trace amounts of hydrogen, ozone, methane, carbon monoxide, helium, neon, krypton, and xenon. In addition to these, varying amounts of water vapor exist depending on the situation; typically, the maximum amount is up to 4%. Some of the minor and trace components, such as water vapor, carbon dioxide, and ozone, absorb radiation and change the climate. An elastic lidar, in the case of negligible aerosol, can observe a weak backward scattering from air molecules, which are used for monitoring of pressure. The direct monitoring of atmospheric species becomes possible by differential absorption lidar (DIAL) method, which is the combination of elastic scattering and absorption process and explained in another chapter in this book. The aerosol shows strong backscattering interaction, which is strong enough to monitor the atmospheric structure using it as an indicator.

The aerosol is generated naturally and anthropogenically. It is classified into three categories according to size. The one that is larger than 2 μm in diameter is called coarse particle, which is usually a natural source such as dust and sea salt. The one smaller than 2 μm in diameter is called fine particle, which is further divided into two classes depending on the range: smaller than 0.1 μm (nucleation mode or Aitken particle) and larger than 0.1 μm (accumulation mode). Most of all anthropogenic sources of aerosol are from industrial emission and transportation exhaust, such as nitrogen dioxide, sulfur dioxide, and soot. Industrial emission gas is transformed to ammonium nitrate and ammonium sulfate through gas-to-particle conversion. They become Aitken particles and then coagulate to a fine particle. A low volatile organic compound is usually a natural source from vegetation and is a composition of a fine particle. The aerosol is removed from the atmosphere by wet deposition (rain out through a cloud or wash out through a precipitation) or dry deposition. However, its lifetime is around several days to a week and it is a major constituent of the troposphere, especially in urban areas. The concentration and the spatial and temporal distributions of aerosol are important factors for meteorological prediction, global radiation budget, and air pollution.

The water-vapor content in air considerably varies depending on the temperature. The saturable water-vapor content of air varies from 190 parts per million (ppm) at $-40°C$ to 42,000 ppm (4.2%) at 30°C. The direct detection of water vapor is necessary to use the DIAL method or Raman scattering lidar, but the water vapor affects the condensation of aerosol and generation of cloud (the cloud is generated when the water-vapor pressure exceeds the saturation value). The phase of cloud is liquid or solid depending on the surrounding temperature. Lidar can detect a cloud base height and in the case of a thin cloud it gives the vertical profile of cloud concentration. However, when the attenuation is too large, it gives the concentration profile until the limit of attenuation, which is usually between 2 and 3 in the optical depth. A cloud has an effective radius of 2 to 30 μm, which is larger than the wavelength, so that a lidar is not suitable to derive the size distribution of a cloud. An ice crystal cloud shows a large depolarization effect, so that depolarization monitoring can be used to distinguish the ice from the water droplet in a cloud.

The wavelength of the lidar is selected in a high transmission region of light in the atmosphere, which is shown in Figure 3.1. For trace gas monitoring, a spectroscopic technique, such as specific absorption by gas, is used. This is applied in DIAL. In this chapter, the description is limited to elastic (Mie or Rayleigh) scattering (Born and Wolf, 1950; van der Hulst, 1957).

3.2. LIDAR EQUATION AND ANALYTICAL SOLUTION

The principle of lidar is described by lidar equation, Equation (3.1), when the multiple scattering is ignored:

$$P_r(R) = K\beta(R) \exp\left[-2\int_0^R \alpha(r)dr\right]/R^2 \qquad (3.1)$$

where $P_r(R)$ is the backscattering signal power, K the system constant, $\beta(R)$ the backscattering coefficient, and $\alpha(r)$ the extinction coefficient, where $\alpha(r) = \alpha_s(r) + \alpha_a(r)$ ($\alpha_s(r)$ and $\alpha_a(r)$ are scattering and absorption coefficients of a particle, respectively) (Hinkley, 1976; Measures, 1984). The system constant K consists of the product of the laser emitting power P_0, optical efficiency of the receiving optics η, receiving area A_r, and the half of the pulse spatial width l ($= c\tau/2$, where c is the light velocity and τ is the pulse temporal width). In Equation (3.1), the quantities of $\beta(R)$ and $\alpha(r)$ are two unknown quantities. Usually, both $\beta(R)$ and $\alpha(r)$ consist of contributions from both air molecules and particulates. The lidar equation has the analytical solution when we have a simple relation between $\beta(R)$ and $\alpha(r)$. In this section, we mention two kinds of retrieval methods that give analytical solutions. As to the independent methods to determine $\alpha(r)$ and $\beta(R)$, some are described in Section 3.4.

3.2.1. One-Component Case — Klett Method

In the case of dense aerosol concentration, the molecular component of the atmosphere can be ignored and the result of extinction coefficient α is obtained under the assumption of $\beta = \alpha^k/S_1$ as

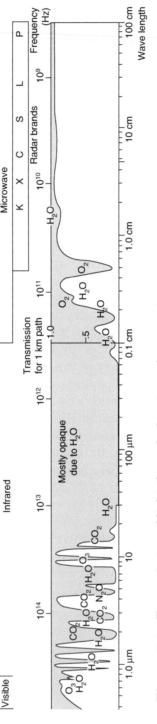

Figure 3.1 Transmission of light through the atmosphere.

$$\alpha(R) = \frac{X(R)}{(X(R_f)/\alpha(R_f)) - 2/k \int_{R_0}^{R} X(r)dr} \tag{3.2}$$

where $X(R) = P(R)R^2$, and R_0 is a boundary condition at a near end (Davis, 1969). However, this solution is unstable when the value of the second term in the denominator approaches the first term. Klett (1981) set the boundary condition at a far end R_f and obtained a stable solution:

$$\alpha(R) = \frac{X(R)}{(X(R_f)/\alpha(R_f)) + 2/k \int_{R}^{R_f} X(r)dr} \tag{3.3}$$

Here, k is a constant with value 0.7 to 1.3, typically 1 (in the following we use $k = 1$). In this case, the error of the extinction coefficient from the real value due to the boundary condition may give the deviation of the following (Hughes et al., 1985):

$$\frac{\delta\alpha(R)}{\alpha(R)} = \exp[-2\tau(R, R_f)]\frac{\delta\alpha(R_f)}{\alpha(R_f)} \tag{3.4}$$

This equation shows that the relative error of extinction coefficient becomes smaller in proportion to $\exp[-2\tau(R, R_f)]$, where $\tau(R, R_f)$ is the optical thickness from R_f to R, which is given as $\int_{R}^{R_f} \alpha(r)dr$.

However, the boundary condition can be taken at the near distance R_0. Then, the extinction coefficient is expressed in Equation (3.2) and the error becomes

$$\frac{\delta\alpha(R)}{\alpha(R)} = \exp[2\tau(R_0, R)]\frac{\delta\alpha(R_0)}{\alpha(R_0)} \tag{3.5}$$

which means the error increases with optical thickness $\tau(R_0, R)$. From this comparison, the Klett method, which puts the boundary condition at a far end, provides a much smaller error and a stable solution.

This situation is shown in Figure 3.2. In Figure 3.2(a), the lidar signal is shown with the noise. Then, the derived extinction profile is shown in the case of backward retrieval with the boundary condition at $R = R_f$ (6 km in this case) for various amounts of error. In this case, 100% error at $R = R_f$ gives the error of 30% at $R = 0$. However, 30% error at $R = 0$

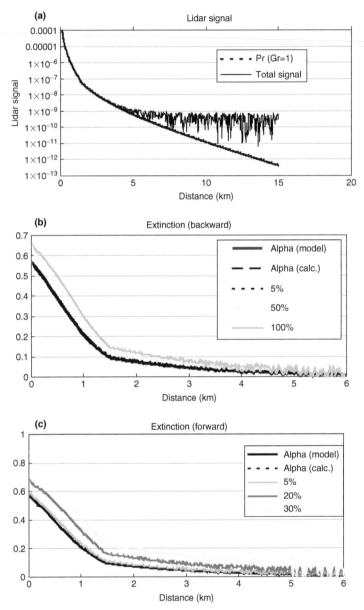

Figure 3.2 Simulation of retrieval by forward and backward (Klett) methods: (a) model lidar signal for overlapping function $G(r) = 1$, without noise (Pr) and with noise (total signal); (b) extinction coefficient retrieved by backward method with 0%, 5%, 50% and 100% boundary condition error; (c) extinction coefficient retrieved by forward method with 0%, 5%, 20% and 30% boundary condition error.

gives more than 100% error at $R = 6\,\mathrm{km}$, and when the error exceeds 50%, the solution diverges and oscillates. Thus, it is seen that the backward retrieval method is stable and gives small errors.

3.2.2.   Two-Component Case — Fernald Methods

When the aerosol concentration is thin and the molecular component cannot be ignored, two components (molecule and aerosol) have to be considered, and the extinction and backscattering coefficients are the sum of two components:

$$\alpha = \alpha_1 + \alpha_2$$
$$\beta = \beta_1 + \beta_2 \qquad (3.6)$$
$$\alpha_1(R) = S_1(R)\beta_1(R)$$

and

$$\alpha_2(R) = S_2\beta_2(R)$$

Here, the suffix 1 shows the aerosol component and suffix 2 shows the molecular component. Then, the equation corresponding to Equation (3.2) takes the form

$$\alpha_1(R) = -\frac{S_1(R)}{S_2}\alpha_2(R) + \frac{S_1(R)X(R)\exp I(R)}{\frac{X(R_{\mathrm{f}})}{(\alpha_1(R_{\mathrm{f}})/S_1(R_{\mathrm{c}}))+(\alpha_2(R_{\mathrm{f}})/S_2)}} + J(R)$$

$$(3.7)$$

$$I(R) = 2\int_R^{R_{\mathrm{f}}}\left[\frac{S_1(r)}{S_2} - 1\right]\alpha_2(r)\mathrm{d}r,$$

$$J(R) = 2\int_R^{R_{\mathrm{f}}} S_1(r)X(r)\exp I(r)\mathrm{d}r$$

The error of the extinction coefficient corresponding to Equations (3.4) and (3.5) takes the form (Hughes et al., 1985)

$$\frac{\delta\alpha_1(R)}{\alpha_1(R)} = \exp\left[-2\tau_1(R,R_{\mathrm{f}}) - 2\frac{S_1}{S_2}\tau_2(R,R_{\mathrm{f}})\right]\frac{B(R_{\mathrm{f}})}{B(R)}\frac{\delta\alpha_1(R_{\mathrm{f}})}{\alpha_1(R_{\mathrm{f}})}$$

$$(3.8)$$

and

$$\frac{\delta\alpha_1(R)}{\alpha_1(R)} = \exp\left[2\tau_1(R,R_0) + 2\frac{S_1}{S_2}\tau_2(R,R_0)\right]\frac{B(R_0)}{B(R)}\frac{\delta\alpha_1(R_0)}{\alpha_1(R_0)}$$

$$(3.9)$$

where $B(R) = \beta_1/(\beta_1 + \beta_2)$. The quantity τ_1 is the optical thickness for aerosol and τ_2 is that for the molecular species. So the error becomes smaller by selecting the boundary condition at the far end R_f.

In order to obtain the quantitative amount of extinction coefficient, the choice of the value of S_1 is significant. Usually, the extinction coefficient becomes larger with S_1. If the optical thickness is determined separately with the sun-photometer or with other instrument, the S_1 value is determined in order to match the optical thickness of both measurements (τ-matching method). Or simultaneous monitoring of Mie and Raman scatterings (Ansmann et al., 1995) or high spectral resolution lidar (Shipley et al., 1983) may determine α and β, independently, and thus the value of S_1.

3.3. LIDAR SYSTEM AND EXAMPLE OF MONITORING

An elastic lidar is classified into two types, monostatic and bistatic, by the relative location of a transmitter and a receiver. In the latter case, a transmitter and a receiver are separated more than several tens of meters and a continuous wave laser can be used as a light source (Devara and Raj, 1991), and the vertical distribution is measured by changing the looking angle. A CCD sensor is used as the detector without scanning for monitoring the vertical profile (Lin et al., 1999). However, a bistatic lidar is a special case and in the following, a monostatic lidar, which installs both a transmitter and a receiver at the same location, is discussed.

Lidars are also classified by platform. The range of detection is usually several tens of kilometers at most. In order to cover a wider range, a mobile lidar is used: vehicle, airplane, satellite-borne lidar systems have been developed.

Airborne and satellite-borne systems are explained in Chapters 8 and 9 in this book. Another method for covering a wide range is a network observation by multiple lidar systems. This is mentioned in Section 3.5.

As examples of elastic lidars, single-wavelength unattended lidar, multiwavelength lidar, polarization diversity lidar, and high-spectrum-resolution lidar are treated in this chapter.

3.3.1. Characteristics of Performance

The performance of an elastic lidar is characterized by the detection range satisfying an appropriate signal-to-noise ratio (SNR). The SNR is described in the case of the Mie/Rayleigh scattering lidar as

$$\mathrm{SNR} = \frac{E[P_\mathrm{S}]}{\sigma[P_\mathrm{S}]} = \frac{\sqrt{M}\sqrt{\eta_\mathrm{Q}\tau_\mathrm{S}/h\nu}P_\mathrm{S}}{\sqrt{\mu}\sqrt{P_\mathrm{S}+P_\mathrm{b}+P_\mathrm{D}}} \tag{3.10}$$

where P_S is the lidar signal power, P_b the background radiation power, P_D the power equivalent to the dark current noise, $E[x]$ the expectation value of x, $\sigma[x]$ the standard deviation, M an integration time, η_Q the quantum efficiency, τ_S the sampling time, and μ an excess noise factor.

Usually, the detection limit is defined as SNR = 1; however, the practical detection range is obtained as the distance where the SNR is 3 (error, defined as the reciprocal of SNR, is 33%).

3.3.2. Single-Channel Lidar with Scanning Mechanism

A single-channel lidar is the simplest case of elastic scattering lidars. A vertical-looking lidar generates a time series of the vertical profiles of aerosol and cloud concentrations. The indication where the abscissa shows the elapsed time, the ordinate, the height, and the intensity, the concentration, is called time–height indicator. In order to monitor the spatial profiles of aerosol and cloud, the monitoring direction is scanned vertically and horizontally. The data of vertical scanning are called range–height indicator (RHI), and the data of the horizontal scanning are called plan-position indicator.

The time series of a three-dimensional scan by lidar was done by the Wisconsin University Lidar group (Piironen and Eloranta, 1995) and is called volume imaging lidar (VIL). A VIL scans vertically (RHI) at a definite azimuthal angle and shifts the azimuthal angle with a small step and repeats the RHI measurement. The collected data are constructed three-dimensionally in a computer. One of the RHI measurements in VIL is shown in Figure 3.3. The VIL monitored convective boundary layer mean depths, cloud base altitudes, cloud top altitudes, cloud coverages, and cloud shadows. From the reliability and the compactness points of view, an Nd:YAG (Nd-doped yttrium aluminum garnet) solid-state laser's second harmonics (532 nm) and third harmonics (355 nm) are predominantly used as a light source of a lidar system. The details of laser can be found elsewhere (e.g., Weber, 2000). In the case of VIL, the fundamental frequency of the Nd:YAG laser (1064 nm) is used.

3.3.3. Multiwavelength Lidar

As an example of an elastic lidar, a four-wavelength lidar system (Takeuchi et al., 1997) and its specification are shown in Figure 3.4 and Table 3.1.

In this system, two Nd:YAG lasers are used as the basic excitation source. One is used for providing the fundamental (1064 nm), second harmonic (532 nm), and third harmonic (355 nm) frequencies. The other one is used for pumping Ti:sapphire laser at 756 nm wavelength by the second harmonic of an Nd:YAG laser. The four-wavelength laser lights are combined into the same optical path using dichroic mirrors and are emitted from the backside of the secondary mirror along the optical axis of a Newtonian telescope, which receives the backscattered signal. The received signal is split into each wavelength by another set of dichroic mirrors and then detected by photomultiplier tubes (PMTs). An example of the four-wavelength data analyzed is displayed in

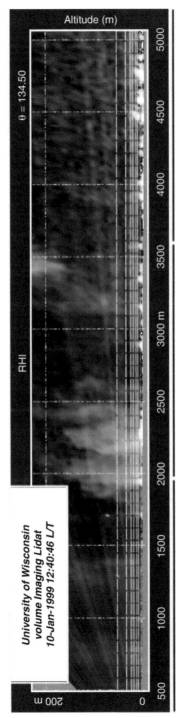

Figure 3.3 An example of volume imaging lidar data. (From Piironen A.K. and Eloranta E.W., *J. Geophys. Res.*, 100, 25567, 1995. With permission.)

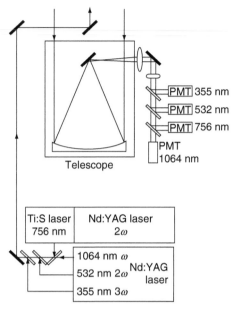

Figure 3.4 Block diagram of four-wavelength lidar.

Table 3.1 Specification of Four-Wavelength Lidar

Laser	Fundamental, SH, TH of an Nd:YAG laser, Ti:Al$_2$O$_3$ laser pumped by an Nd: YAG laser (SH)			
Wavelength	1064 nm	756 nm	532 nm	355 nm
Laser	Nd:YAG	Ti:Al$_2$O$_3$	Nd:YAG (SH)	Nd:YAG (TH)
Repetition rate	10 Hz			
Output stability	<5%			
Pulse energy	500 mJ	120 mJ	300 mJ	150 mJ
Beam divergence	<5 mrad			
Telescope	Newtonian, vertical looking			
Diameter	0.8 m			
Efficiency	0.3			
Detector	PMT	PMT	PMT	PMT
Quantum efficiency	0.001	0.084	0.2	0.3
Filter width	3.9 nm	1.9 nm	1.9 nm	1.9 nm
Sky radiance (W/m^2/sr/nm)	0.0234	0.072	0.142	0.098

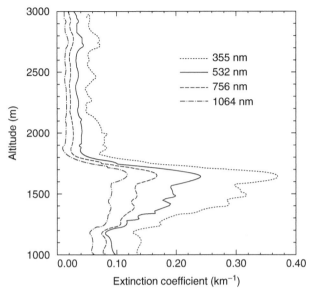

Figure 3.5 An example of four-wavelength lidar data (Yabuki et al. 2000).

Figure 3.5. Here, the signal is averaged for 3 min (1800 shots) and retrieved by the Fernald method (Equation (3.7)).

3.3.4. Polarization Diversity Lidar

In Mie scattering, the backscattering from a spherical particle does not change the polarization; the nonspherical particle changes the polarization. So if we monitor the polarization status of the scattered light, information on the shape of the particle can be obtained. However, multiple scattering of the spherical particle also changes the polarization. So the cloud will give the depolarization even if a droplet consists of water (spherical). But the ice crystal cloud (nonspherical) will have higher depolarization δ, which is usually defined by $\delta = I_\perp / I_\parallel$, where $I_\perp$ is the intensity of perpendicular component and $I_\parallel$ is the intensity of the parallel component to the incident light. Typical examples of depolarization studies are the mineral dust aerosol (usually desert origin) and cloud particles.

The lidar that has two channels to detect different polarizations is called a polarization diversity lidar (Sassen et al., 1992; Sassen, 1994). The Mie scattering theory shows that optically homogeneous and spherically symmetrical scatterers do not produce depolarization in the exact backward single scattering, whereas crystalline particles, such as ice crystals, generate depolarization as a result of the internal reflections responsible for much of the backscattering, and mineral dust gives large deporalization due to its nonsphericity. In Figure 3.6, a polarization diversity lidar system in Utah University is shown with the specifications in Table 3.2. In this case, two wavelengths of the Nd:YAG laser (1064 and 532 nm) are used for monitoring; four channels are used for detecting polarizations both parallel and perpendicular to the transmitting polarization. The wavelength of 532 nm is detected by PMTs and the wavelength of 1064 nm is detected by avalanche photodiode (APD). Two sets of dual 100 MHz/8-bit transient digitizers were used for digitizing the analog signal. An example of cirrus lidar signal returns with simultaneous four-channel observation is shown in Figure 3.7. Ten shots averaging yield of 1 Hz temporal resolution with 30 m spatial resolution is shown. However, Sassen points out the necessity of higher resolution, both temporally and spatially, to study the microphysical process of clouds.

3.3.4.1. Monitoring of Mineral Dust

As mentioned in the previous section, in the troposphere, mineral dust particles (mainly form the desert dust) give a large depolarization ratio. The extinction-to-backscatter ratio S_R is different for spherical, S_s, and nonspherical particles, S_{ns} (Mishchenko and Hovenier, 1995). The ratio S_{ns}/S_s becomes over 1.5 with a maximum of 3 from a model consideration. Without multiple scattering, the depolarization ratio δ ($=P_\perp/P_\parallel$) reaches more than 0.4 (Barnaba and Gobbi, 2001). As the depolarization ratio δ of mineral dust is large compared to other aerosol particles, δ is used to discriminate desert

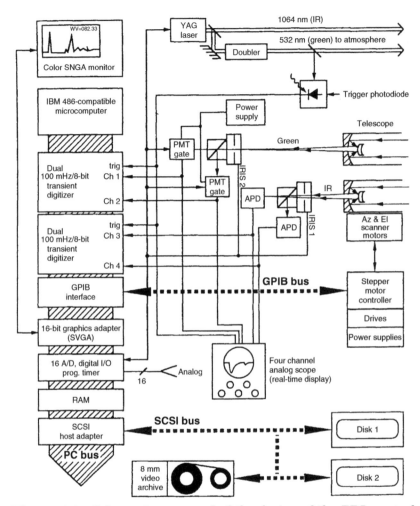

Figure 3.6 Schematic portrayal of the design of the PDL control and data acquisition system. (From Sassen K., *Proc. IEEE*, 82, 1907, 1994. With permission.)

origin dust from other aerosol particles. The Sahara desert sand is flown to Mediterranean Sea and South Europe, further to North Europe, and some crosses the Atlantic Ocean to South America. The identification is usually checked by the depolarization ratio. In Asia, the desert sand from Gobi or

Table 3.2 Specifications of The University of Utah Two-Color
Polarization Diversity Lidar (PDL) System

Operational Wavelength (Nd:YAG)	$0.532 + 1.064\,\mu m$
Peak energy	0.45 J for each color
Maximum PRF	10 Hz
Pulse width	9 nsec
Laser beam divergence	0.45 mr
Receiver FOV	0.2–3.8 mr
Receiver diameter	30 cm (two telescopes)
Detector — visible	Two gated PMTs
Detector — IR	Two APDs
Maximum scan rate	5°/s

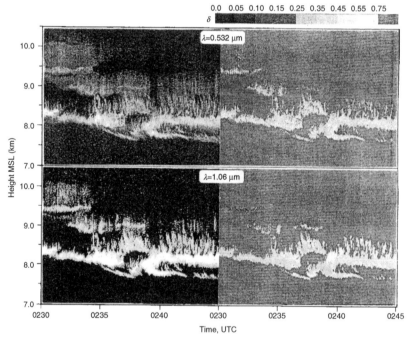

Figure 3.7 An example of simultaneous four-channel data products derived from vertically pointing PDL measurements of layered
cirrus clouds collected on April 9, 1993. In the left are shown range-
squared corrected, returned laser signals in arbitrary units based on
a logarithmic gray scale. In the right are linear depolarization ratio
displays derived from the ratio of the perpendicular-to-parallel laser
returns.

Takhramakan desert is called Asian dust (Kosa), which is significant from March to May. It is observed by the lidar network system in Asia (Murayama et al., 2001; AD-NET).

Some examples of Asian dust observed in AD-NET are shown in Figure 3.8. In Figure 3.8(a), the range-corrected backscattering intensity during April 15 to 25, 1998, in Tsukuba, is shown. In Figure 3.8(b) and (c), simultaneous monitoring of scattering ratio and depolarization ratio at 532 nm at Nagoya, Tokyo, and Tsukuba is shown. In this case, Asian dust was transported by a downward moving low pressure and the altitude of Asian dust was below 3 km. However, there was a thin dust layer around 6 to 8 km altitude, which has a large depolarization ratio. This thin layer is ascribed by the Asian dust component and it may act as the cloud condensation nuclei (CCN) of cirrus cloud. Another typical example of mineral dust is Saharan desert dust, which is transported to north and west. In the EU NAUL II campaign in spring 1999, Saharan dust was monitored by a lidar at Crete. The dust continued for 8 days and the extinction coefficient profile was monitored by a lidar. The contour of the extinction profile is shown in Figure 3.9. The maximum altitude of dust reached up to 10 km (Gobbi et al., 2000).

3.3.4.2. Cloud Measurement

Not only does cloud thermodynamic phase discrimination gives depolarization, but the experiment shows a considerable amount of depolarization ratio δ. Cirrus and winter mountain storm clouds are typical examples of ice crystals. Unfortunately, the exact solution of light scattering by hexagonal ice crystal cannot be obtained. In a water droplet cloud, the effect of multiple scattering increases with the field of view of the detection and the penetration depth into the cloud. Sassen et al. (1992) simulated the contribution of ice crystals in a cloud by numerical estimation. The depolarization ratio increases with the nonsphericity of the particle and multiple scattering. The existence of ice crystal of five particles in litter (with 1 to 5 μm in radius) in a water droplet cloud ($200 \, cm^{-3}$) gives a significant change of the extinction

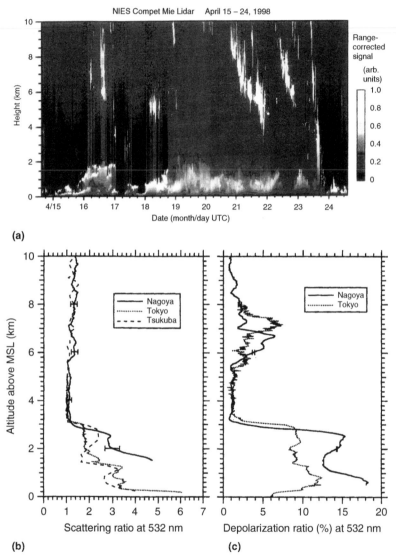

(a)

(b)   (c)

Figure 3.8 (a) Lidar observation of Asian dust. Time-to-height indication (THI) of the range-corrected backscattering intensity by the NIES lidar in Tsukuba, during April 15 to 25, 1998. (b) Simultaneous lidar observation of scattering ratio at 532 nm, at Nagoya, Tokyo, and Tsukuba around 12:30 UTC April 20, 1998. (c) Depolarization ratio at 532 nm corresponding to (b). (From Murayama T. et al., *J. Geophys. Res.*, 106, 18345, 2001. With permission.)

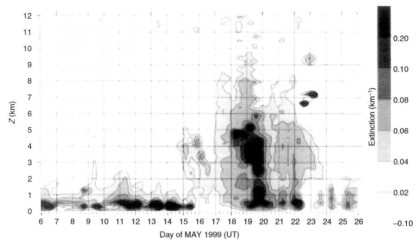

Figure 3.9 Contour plot of aerosol extinction (km^{-1}) estimated by lidar observation during PAUL II campaign (spring 1999) at Crete island (35N–23E). Saharan dust event continues for 8 days from May 15 to 23. The altitude of dust event reaches 10 km. (From Gobbi G.P. et al., *Atmos. Environ.*, 34, 5119, 2000. With permission.)

and depolarization; however, the separation with the multiple scattering effect is difficult. The dependence of depolarization with penetration depth is shown in Figure 3.10. It shows the depolarization ratio increases over 0.5 in the penetration of 140 m, which corresponds to the extinction of roughly 5.

3.3.5. Unattended Lidar

3.3.5.1. Micro-Pulse Lidar

Mie scattering shows the vertical profile of the extinction coefficient, which gives information of the aerosol and the cloud structure. Although a thick cloud hinders light transmission, if optical thickness is small it gives the structure of the cloud. As mentioned previously, the depolarization ratio measurement gives some hint of the ice crystal. In the cloud-free case, the concentration of the aerosol is obtained by extinction profiles. If the conversion coefficient of aerosol mass concentration from the extinction coefficient is known, then

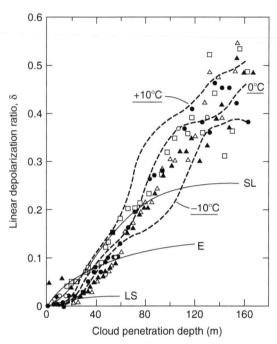

Figure 3.10 Comparison of PDL data with several depolarization simulations from Sassen K. et al. (*Appl. Opt.*, 31, 2914, 1992), LS (Liou K.-N. and Schotland R.M., *Atmos. Sci.*, 28, 772, 1971), E (Eloranta E.W., PhD dissertation, University of Wisconsin, 1972), and SL (Sun Y.Y. and Li Z.-P., *Appl. Opt.*, 28, 3633, 1989. With permission.)

the mass concentration vertical profile is obtained. However, it is a function of composition, refractive index, size distribution, and relative humidity, and it is very difficult to obtain it only from the lidar data. However, the extinction profile is a good indicator of the aerosol concentration. The autonomous monitoring of aerosol is strongly desired, and the micro-pulse lidar (MPL) (Spinhirne, 1993; Welton et al., 2001) is one of a typical instrument for automatic operation. It has the following features:

1. The use of diode pumped solid-state laser (the second harmonics of LD-pumped Nd:YLF).

2. High repetition frequency and low pulse power per pulse (2500 Hz and 10 μJ/pulse typically).
3. Eye safe level of the transmitting laser (large expansion by 20 cm telescope).
4. Photon counting detection.
5. The use of the same telescope for transmission and receiving (less possibility of alignment deterioration).
6. No warming up time.
7. Compact system to carry by one person.

However, it is necessary to prevent the strong reflecting intensity from reaching the detector.

Recently, in order to prevent the extra reflection from the secondary mirror, a doughnut shape of the transmitting beam is considered by combination of two cone-shaped prism (Shiina et al., 2002).

MPL is a conventional time-gated, incoherent detection lidar approach that is able to profile all significant atmospheric cloud and aerosol structures with a compact, fully eye-safe instrument. Eye safety allows for full-time, long-term unattended operation, and is achieved by transmitting low-power pulses through an expanded beam (10 μJ with a 0.2 m aperture and 50 μrad beam divergence), with a high-pulse repetition frequency (routinely 2.5 kHz) much higher than in ordinary lidar systems. Signal acquisition is handled via photon counting, which gives a relatively high accuracy, and is a problem-free method of handling low-level signal compared to analog detection. Its block diagram is shown in Figure 3.11 and specifications are summarized in Table 3.3. As shown in Figure 3.11, an MPL has a configuration of a transceiver type. A Cassegrainian telescope is used both for transmitting light and receiving a scattered signal. Then the alignment is free from the relative configuration of the transmitting optics and the receiving optics. The eye-safety level of the transmitting signal is roughly 10 μJ/pulse when the beam is expanded to 20 cm in diameter. An MPL has the ability of reaching approximately 10 km for an aerosol target and 20 km for a cloud.

A prototype of the MPL instrument was developed at NASA-Goddard Space Flight Center (GSFC) as a result of

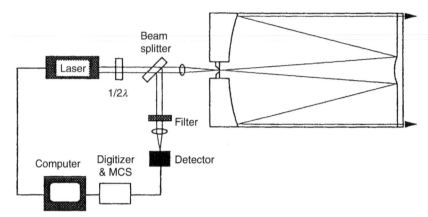

Figure 3.11   Block diagram of micro-pulse lidar. (From Spinhirne J.D., *IEEE Trans. Geosci. Remote Sensing*, 31, 48, 1993. With permission.)

research on efficient lidars for space-borne applications. The initial system was first used in the southwest Pacific PROBE study (1993). In October of that year, an MPL began taking full-time measurements for the Atmospheric Radiation Measurement (ARM) program at a Cloud and Radiation Testbed (CART) site in north-central Oklahoma. An example of long-term monitoring using an MPL at the CART site is shown in Figure 3.12. Additional systems have been installed at ARM tropical western Pacific Atmospheric Radiation and Cloud Stations (ARCS) at Papua New Guinea (September 1996) and Manus Island (November 1998), and the North Slope,

Table 3.3   Specification of MPL

Laser	LD Pumped Nd:YLF (SH)
Wavelength	523 nm
Pulse energy	5–10 µJ
Repetition freq.	2500 Hz
Receiver	
Telescope	Cassegrainian
Aperture	20 cmφ
Detection	
Photon counting	Si-APD (SPCU-AQ)

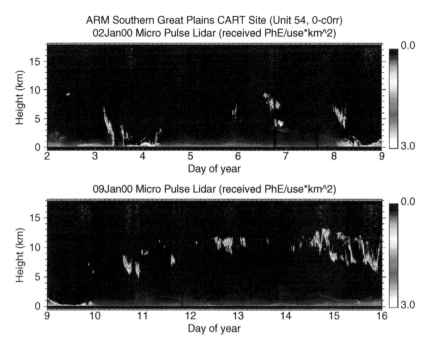

Figure 3.12 MPL image data for a week at the ARM CART site (MPL-NET).

AK CART (March 1998). Also, some are used in Arctic and Antarctic monitoring (NPRI) and in Thailand. An example of continuous 1-month observation is shown in Figure 3.13.

3.3.5.2. Other Systems

Similar unattended systems were also developed by other groups (Grund and Sandberg, 1996; Goldsmith et al., 1998; Pinandito et al., 2000; Sugimoto et al., 2000a,b; Eloranta and Ponsardin, 2001) Among them, diode pumped solid-state laser (DPSS) is a main trend of the light source. If the lifetime of LD becomes practically infinitive, the LD pump solid-state laser will be used as an ideal compact laser source. However, a flash-lamp pumped solid-state laser (Nd:YAG) is also a practical powerful light source with no replacement over a half year. This kind of system is used in lidar network monitoring of Asian area by NIES (AD-NET) (Murayama et al., 2001).

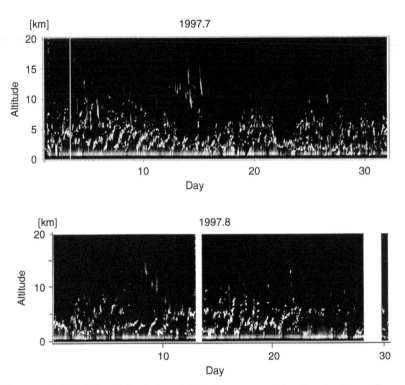

Figure 3.13 Monthly data at Sri Samrong, Thailand. Quasidiurnal variation of cloud motion is observed in a rainy season of tropical region (Takeuchi et al., 2001).

The continuous monitoring of aerosol is desired to monitor the vertical structure of the atmosphere. The aerosol is a good indicator of the atmospheric condition. So Mie lidar monitoring of aerosols is a powerful tool for the monitoring of the boundary layer, the mixing layer, the planetary boundary layer, and their vertical motions. Also, if aerosol is free (over 30 km), temperature is monitored by a Rayleigh lidar (Chanin and Hauchecorne, 1984). Aerosol is used as an indicator for atmospheric stability, upwind motion, and downwind motion (Lagroses et al., 2004).

3.3.6. Combination with Radar — Cloud Measurement

Cloud, as well as aerosol, is a common target of Mie scattering lidar from the beginning of lidar development. A cloud is usually optically thick and it is well observed by radar technology. Geometrical properties of clouds, such as fractional cover and cloud boundaries, are fundamental to radiative transfer study. Microphysical parameters, such as phase (ice or water), particle size, and water content are also significant. Even using only a lidar, Pal and Carswell (1973, 1976, 1978, 1985) were the first to study the polarization features of a cloud using a ruby laser. The anisotropy of polarization is explained with the multiple scattering of a spherical particle. Platt (1979) studied the radiation properties of a cloud with lidar and radiometer. The Experimental Cloud Lidar Pilot Study (ECLIPS) (Platt et al., 1994) was carried out in 1989 and 1991 to obtain statistics on cloud base height, extinction, optical depth, cloud brokenness, and surface fluxes. Cloud base and top altitudes were measured by a threshold method, and cloud properties of emittance and optical depth were derived from lidar and radiometer (Platt, 1979). The collected data are archived in NASA Langley Research Center. Sassen (1991) gave an overview of polarization on clouds. The explanation of polarization lidar was given in Section 3.3.4.

As the lidar can measure the cloud optical thickness up to 3, at most 4, the combination of lidar and microwave radar can give more general information of the cloud. Weitkamp et al. (1999) measured clouds simultaneously with lidars (277 nm hydrogen-shifted Raman lidar (HSRL) pumped with KrF eximer laser and 720 to 780 nm $Ti:Al_2O_3$ tunable laser) and a 95 GHz radar. In most of the cases, they obtained similar profiles. However, the profiles were different in some cases. In Figure 3.14(a) and (b), lidar and radar monitoring, respectively, of stratus cloud around 1 km in altitude and altocumulus cloud around 2.6 km are shown. In Figure 3.14(a), a higher-altitude cloud is not seen by the strong attenuation of

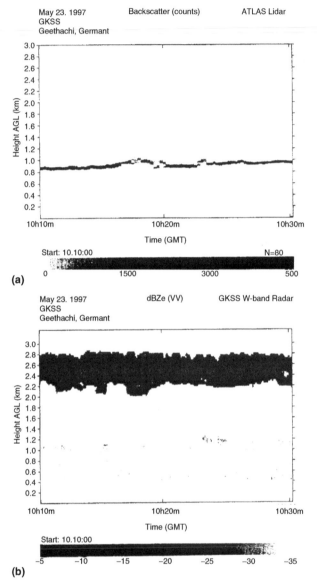

Figure 3.14   (a) Lidar observation of a cloud at 277 nm on May 23, 1997. (b) Simultaneous radar observation. (From Weitkamp C. et al., *Phys. Chem. Earth*, 24, 163, 1999. With permission.)

the 277 nm wavelength due to tropospheric ozone. In Figure 3.14(b), the 1 km stratus cloud is scarcely perceptible due to the small diameter. However, the higher-altitude cloud is clearly seen. Occasionally, the high-altitude cloud can be seen by a lidar in the visible wavelength (720 to 780 nm), but not by a radar.

Donovan et al. (2001) mentioned that the complementary nature of lidar and radar provides a more complete description of cloud boundaries than either instrument alone. In a cloud, multiple scattering is important due to dense concentration. The particle size of a cloud is larger than 1 μm, and the combination of lidar and radar is used for studying the effective size of the cloud.

The effect of clouds and aerosols on regional and global climate is of great importance. NASA pursues a project climate and radiation science program and field studies incorporating airborne remote sensing and in situ measurements of clouds and aerosols. These projects involve coordination of ground-based and satellite measurements with the airborne observations. The Cloud Physics Lidar (CPL; McGill et al., 2002) is an airborne lidar system designed specifically for studying clouds and aerosols using the ER-2 High Altitude Aircraft (which typically flies at 20 km). The CPL flies on the ER-2 along with other instruments. A window at the bottom of the superpod allows the instrument to look directly at the nadir. The CPL provides the following data products:

1. Cloud profiling with 30 m vertical and 200 m horizontal resolution at 1064, 532, and 355 nm, providing cloud location and internal backscatter structure.
2. Aerosol, boundary layer, and smoke plume profiling at all three wavelengths.
3. Depolarization ratio to determine the phase (e.g., ice or water) of clouds using the 1064 nm output.
4. Direct determination of the optical depth of cirrus clouds (up to ~OD 3) using the 355 nm output.

3.4. MONITORING OF AEROSOL OPTICAL PROPERTIES

In a case of unattended lidar systems, monitoring is usually done to find the optical properties of aerosols in the troposphere, such as the vertical profile of the extinction coefficient. The most important optical properties are particle size distribution, and complex the refractive index. If we assume a spherical shape of particle, Mie scattering theory (van der Hulst, 1957) provides the phase function, the single-scattering albedo, the asymmetric parameter, and the S_1 parameter (extinction-to-backscattering coefficient ratio: lidar ratio). A single-scattering albedo is defined by the ratio of scattering to total extinction. The deviation from unity shows the absorption by particles and it is an important parameter for earth radiation balance. The S_1 parameter is important for the derivation of the correct extinction value from the lidar measurement.

3.4.1. Model of Aerosol Size Distribution

The fundamental properties of aerosols relating to lidar monitoring are described by Measures (1984) and Hinkley (1976). The optical properties of aerosols are determined by size distribution, refractive index, and shape. The typical size distributions are log normal distribution, power law (Junge distribution), and modified gamma distribution:

3.4.1.1. Log Normal Distribution

Aerosol size distribution is usually given by the sum of three log normal distribution modes:

$$\frac{dN}{d \ln r_p} = \sum_{1}^{3} \frac{N_i}{(2\pi)^{1/2} \sigma_{gi}} \exp \left(\frac{(\ln r_p - \ln \bar{r}_{pi})^2}{2\sigma_{gi}^2} \right) \quad (3.11)$$

Here, N is the total particle number, r_p the particle radius, N_i the total number of mode i, $\bar{r}_{pi}$ the median radius for mode

i, σ_{gi} the width of mode i, and the total particle number is $N = \Sigma N_i$. Usually, three modes are selected for the construction of one kind of aerosol size distribution; one mode from each size range. The range of diameter is classified into three ranges: nucleation mode for the radius less than $0.1\,\mu$m, accumulation mode for $0.1\,\mu$m $<$ radius $<$ $1\,\mu$m, and coarse mode for radius larger than $1\,\mu$m. Sometimes, two modes are selected from the accumulation mode. The nucleation mode does not affect the optical properties. So the accumulation mode and coarse mode are important for optical properties such as scattering and absorption.

3.4.1.2. The Power Law Distribution

The wavelength dependence of the extinction coefficient α, or its integral optical thickness τ, is often depicted by empirical Angstrom's law, which is expressed by

$$\tau_i = \tau_0 \left(\frac{\lambda_i}{\lambda_0}\right)^{-a} \tag{3.12}$$

Then the exponent a, which is called Angstrom exponent, is related to the exponent ν of the size distribution in the power law (Junge distribution) by $\nu = a + 2$:

$$n(\log r_{\mathrm{p}}) = \frac{\mathrm{d}N}{\mathrm{d}\ln r_p} = \begin{cases} 0 & r_p < 0.02\,\mu\text{m} \\ const & 0.02 < r_{\mathrm{p}} < 0.1\,\mu\text{m} \\ C/(r_{\mathrm{p}})^{\nu} & 0.1 < r_{\mathrm{p}} < 10\,\mu\text{m} \\ 0 & r_{\mathrm{p}} > 10\,\mu\text{m} \end{cases} \tag{3.13}$$

where C and ν are constants. The value $\nu = 3$ corresponds to a constant volume per unit size logarithmic interval in the Junge volume distribution ($\mathrm{d}V/\mathrm{d}\ln r_{\mathrm{p}} = (4\pi/3))\, r_{\mathrm{p}}^3 \mathrm{d}N/\mathrm{d}\ln r_{\mathrm{p}}$.

When a is >1, the smaller particle is dominant; when a is <1, the large particle is dominant. The quantity a is obtained as $a = \ln(\tau_2/\tau_1)/\ln(\lambda_1/\lambda_2)$ from the lidar monitored data. So the power law distribution is used for convenience of the feature of the particle size distribution.

3.4.1.3. Modified Gamma Distribution

Another popular size distribution of aerosol is a modified Gamma function, which is given as

$$n(r_p) = Ar_p^b \exp(-Br_p^c) \tag{3.14}$$

where A, b, B, and c are all positive parameters. This function provides a very flexible form; however, its use is often cumbersome. The total aerosol number $N = \int n(r_p)dr_p$ is equal to

$$N = \frac{AB^{-(b+1)/c}}{c}\Gamma\left(\frac{b+1}{c}\right)$$

where Γ is the gamma function. The maximum of the distribution occurs at radius $r_p = (b/Bc)^{1/c}/2$.

The size distribution and refractive index is roughly determined by the chemical composition. It depends on the origin of the particulate. Hess et al. (1998) provide a software OPAC that gives the optical properties of various aerosols. There aerosols are classified into 10 categories: water insoluble, water soluble, soot, sea salt (accumulation mode), sea salt (coarse mode), mineral (nucleation mode), mineral (accumulation mode), mineral (coarse mode), mineral-transported, and sulfate droplet. For each mode, the mean radius in number, also in volume, both limits of the size, the width of the spread of the distribution, density, and mass per particle are given. From these data the optical properties of the aerosol can be calculated.

From the origin of aerosols, it is sometimes convenient to specify six elements as an indication of different aerosol components. They are organic carbon (OC), elementary carbon (EC), sea salt, ammonium nitrate (NH_4NO_3), ammonium sulfate (($NH_4)_2SO_4$), and soil (Kaneyasu et al., 1995). They constructed aerosol types such as urban, marine, rural, remote continental, free troposphere, polar, and desert (Jaenicke, 1993).

The aerosol size distribution is determined from in situ monitoring. One example of size distribution in an urban area is shown in Figure 3.15 (Takeuchi, 1999).

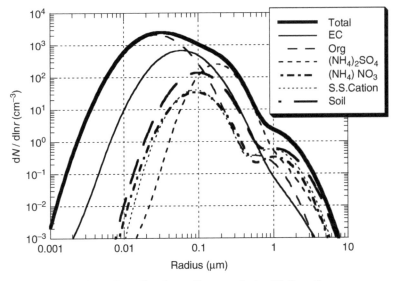

Figure 3.15 Size distribution of aerosols in Chiba, Japan.

3.4.2. Optical Properties of Aerosol and Air Molecule

The characteristics of the lidar equation are described by extinction coefficient, backscattering coefficient, and its ratio. They are characterized by size distribution function and complex refractive index.

The extinction coefficient and the backscattering coefficient are written (van de Hulst, 1957) as

$$\alpha(\lambda) = \int n(r)\pi r_p^2 Q_{ext}(\lambda, r_p, m) dr_p,$$

$$Q_{ext}(\lambda, r_p, m) = \sigma_{ext}(\lambda, r_p, m)/\pi r_p^2$$

(3.15)

$$\beta(\lambda) = \int n(r_p)\pi a^2 \left[\frac{dQ_{scat}(\lambda, r_p, m)}{d\Omega}\right]_{\Omega=\pi} dr_p,$$

$$Q_{scat}(\lambda, r_p, m) = \sigma_{scat}(\lambda, r_p, m)/\pi r_p^2$$

(3.16)

where $n(r_p)$ is the particle size distribution function as the function of particle radius r_p. The extinction efficiency

$Q_{ext}(\lambda, r_p, m)$ and scattering efficiency $Q_{scat}(\lambda, r_p, m)$ are given by extinction cross-section $\sigma_{ext}(\lambda, r_p, m)$ and scattering cross-section $\sigma_{scat}(\lambda, r_p, m)$ normalized by the geometrical cross-section of the particle πr_p^2. When a spherical particle is assumed, the calculation is done by Mie theory. The refractive index m is a complex number, which is determined by chemical composition.

The aerosol is usually classified into two groups by size: coarse particle (larger than $2\,\mu m$ in diameter) and fine particle (smaller than $2\,\mu m$). The former is usually of natural origin and consists of sea salt, soil, and dust. The latter is of natural or anthropogenic origin. The fine particle is further classified into two, larger and smaller around $0.1\,\mu m$. The smaller type is called Aitken particle or nucleation mode. It grows with coagulation or by absorbing ambient gases. The larger type is called accumulation mode. However, the accumulation mode usually does not grow to the size of a coarse particle by crossing the $2\,\mu m$ limit.

Total scattering efficiency Q_s approaches 2 at the large limit of $x = kr_p$ ($k = 2\pi/\lambda$). It becomes smaller than 0.1 for $x < 1$. So Mie scattering becomes negligible for r_p (particle radius), being smaller than $\lambda/2\pi$. Here the contribution from air molecules is much larger than from aerosols. This region is called Rayleigh scattering and the scattering intensity is inversely proportional to the quartet of wavelength.

In the case of light transmitting in the atmosphere, the transmission of the wavelength shorter than 300 nm is absorbed by ozone and oxygen (as shown in Figure 3.1). So the minimum wavelength for lidar is around 300 nm. (The wavelength region shorter than 300 nm is of solar blind, where the lidar can be operated in a shorter detection range without being affected by daytime solar radiation.) A particle smaller than $0.05\,\mu m$ in radius shows negligible scattering during lidar monitoring. Therefore, the particle smaller than $0.05\,\mu m$ does not contribute to the lidar scattering signal. The fine particles are emitted from combustion, burning smoke, fog, and industrial emission, and their size increases with the condensation and coagulation. They then become the accumulation mode type and are detectable.

3.4.3. Monitoring by High-Spectral-Resolution Lidar

Mie–Rayleigh scattering is called as elastic scattering, where the scattering wavelength does not change. However, when the spectral width of the emitting laser is narrow enough, the scattering light has the Doppler broadening width. As the particle has the Maxwellian distribution, the concentration of particles is described as

$$n(v) = \sqrt{\frac{mv^2}{2kT}}\exp\left(-\frac{mv^2}{2kT}\right)$$

The Doppler shift $\delta\nu$ is given by $2\nu/c\nu$. The air molecule has lighter weight than an aerosol particle and has a wider spectral width of about 400 to 600 MHz at 300 K. Compared to the air molecule, the Doppler broadening of an aerosol is negligible, and is given by the broadening of the emitting laser light itself. So if we have two channels to discriminate the narrow and the broad spectra, we can determine the intensity of molecular and aerosol scattering. Sroga et al. (1983) proposed a high-finesse, multietalon system to discriminate two channels (the estimation is given by Shipley et al., 1983). Later they used an atomic blocking filter (Shimizu et al., 1983) for monitoring of cirrus cloud optical properties (Eloranta and Pirronen, 1996). The block diagram is shown in Figure 3.16. The pre-etalon, which consists of a narrow gap etalon and an interference optical filter, limits the background radiation. The light that passes through the pre-etalons is split into two beams. One detects the total of Rayleigh (molecular) and Mie (particle) scattering components and the other blocks the narrow width particle component. From the assumption of the spectral profile of molecular and particle scatterings, the intensities of the both scatterings are independently calculated.

If both components are derived independently, then the number of photons for Rayleigh (molecular) scattering $N_m(R)$ and that for Mie (particle) scattering $N_p(R)$ are expressed as

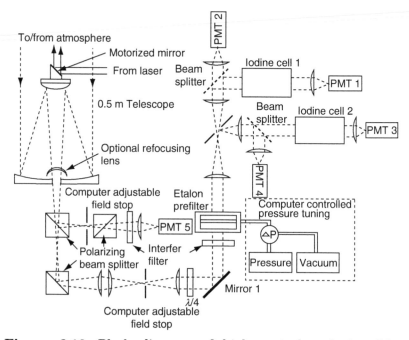

Figure 3.16 Block diagram of high-spectral-resolution lidar. (From Eloranta E.W. and Pirronen P., in: *Advances in Atmospheric Remote Sensing with Lidar*, A. Ansmann et al., eds., Springer, Berlin, 1996. With permission.)

$$N_p(R) = \frac{\lambda_0\, E_0\, Ar\delta t}{2h}\, \frac{\alpha_p(R)}{R^2}\, \Phi_p(\pi, R)\, \exp\left[-2\int_0^R \alpha_t(R')dR'\right]$$

(3.17)

$$N_m(R) = \frac{\lambda_0\, E_0\, Ar\delta t}{2h}\, \frac{\alpha_m(R)}{R^2}\, \Phi_m(\pi)\, \exp\left[-2\int_0^R \alpha_t(R')dR'\right]$$

(3.18)

Here $N_p(R)$ is the receiving photon number for particle scattering, $N_m(R)$ the receiving photon number for molecular scattering, λ_0 the incident laser wavelength, E_0 the emitting laser pulse energy, Ar the aperture area of a receiver, δt the sampling time, $\alpha_p(R)$ the extinction coefficient of the particle, $\alpha_m(R)$ the extinction coefficient of the molecule, $\alpha_t(R)$ the total extinction coefficient, $\Phi_p(\pi,R)$ the phase function of particle for backward scattering at the range of R, and $\Phi_m(\pi)$ the backward scattering phase function of molecule $(= 3/8\pi)$.

Since the transmission of the laser light is common for both scatterings, if we know both $N_p(R)$ and $N_m(R)$, then the ratio of particle to molecular backscattering $B(R) = N_p(R)/N_m(R)$ is obtained as

$$B(R) = \frac{N_p(R)}{N_m(R)} = \frac{\alpha_p(R)}{\alpha_m(R)} \frac{\Phi_p(\pi, R)}{\Phi_m(\pi)} \tag{3.19}$$

and $\alpha_m(R)$ is proportional to the pressure and can be theoretically calculated if the absolute pressure is known. Thus, the backscattering coefficient of the particle can be calculated as

$$\alpha_p(R)\Phi_p(\pi, R) = B(R) \cdot \alpha_m(R) \cdot \Phi_m(\pi) \tag{3.20}$$

($\alpha_m(R)$ is directly proportional to the atmospheric density). Equation (3.18) can be solved for the total optical thickness over a layer in the atmosphere between R_1 and R_2:

$$\tau(R_2) - \tau(R_1) = \int_{R_1}^{R_2} \sigma_t(r)\mathrm{d}r = \frac{1}{2}\ln\left[\frac{\alpha_m(R_2)}{\alpha_m(R_1)}\frac{R_1^2 N_m(R_1)}{R_2^2 N_m(R_2)}\right] \tag{3.21}$$

The total extinction coefficient $\alpha_r(R)$ can be computed from the range derivative of the optical thickness:

$$\overline{\alpha_t(R)} = \frac{\partial \tau(R)}{\partial R} \approx \frac{\tau(R_2) - \tau(R_1)}{R_2 - R_1} \tag{3.22}$$

If the single-scattering albedo of the particles ω_0 is known or the extinction coefficient can be assumed to be dominated by scattering $(\sigma_{scat} = \sigma_t)$, Equations (3.20) and (3.21) can be solved for the average backscatter phase function, which is the reciprocal of the extinction-to-backscattering coefficient ratio for particles, over the layer from range R_1 to range R_2:

$$\overline{\Phi_p(\pi, R)} = \frac{\overline{B(R)\alpha_m(R)}3/8\pi}{\overline{\omega_0\alpha_t(R)} - \overline{\alpha_m(R)}} \tag{3.23}$$

Then, the extinction coefficient $\overline{\alpha_p(R)} = \omega_0\overline{\alpha_t(R)} - \overline{\alpha_m(R)}$ and the backscattering coefficient $\overline{\alpha_p(R)} \cdot \overline{\Phi_p(\pi, \overline{R})}$ for the particles can been obtained.

Another method of HSRL is the use of an atomic or molecular absorption filter. The atomic line absorption was supposed to be used for a narrow spectrum-width blocking filter (Shimizu et al., 1983). In a lidar system, the received signal is divided into two channels. One channel detects a signal for the whole spectrum and the other passes a signal for the wing parts of spectrum (molecular scattering), absorbing the central part of the aerosol's scattering by a narrow-width atomic filter. Then the difference shows the scattering of the aerosol. A block diagram already shown in Figure 3.16 uses this design. Thus, the HRSL can separate the backscattering coefficient of a molecule and an aerosol, which gives the lidar backscattering ratio.

3.4.4.   Derivation Using a Raman–Mie Lidar

Independent derivation of extinction and backscattering coefficients is also possible using a Raman–Mie lidar, when the air pressure is assumed to be known (Ansmann et al., 1992). From the Raman signal of air molecule (nitrogen or oxygen), the total extinction coefficient is obtained just like in Equation (3.25):

$$N_L(R) = \frac{\lambda_0\, E_0\, Ar\delta t}{2h} \frac{[\beta_{L,a}(R) + \beta_{L,m}(R)]}{R^2}$$
$$\times \exp\left[-2\int_0^R [\alpha_{L,a}(r) + \alpha_{L,m}(r)]dr\right] \tag{3.24}$$

$$N_R(R) = \frac{\lambda_R\, E_0\, Ar\delta t}{2h} \frac{n_R(R)}{R^2} \frac{d\sigma_R(\pi)}{d\Omega}$$
$$\times \exp\left[-\int_0^R [\alpha_{L,a}(r) + \alpha_{L,m}(r) + \alpha_{R,a}(r) + \alpha_{R,m}(r)]dr\right]$$
$$\tag{3.25}$$

Here, suffixes L and R mean the elastic scattering (laser frequency) and Raman scattering (Raman-shifted frequency), respectively. The Raman scattering coefficient is proportional to the atmospheric density (in this case, nitrogen or oxygen is used as the Raman material), $n_R(R)$. So if the density is known and the backscattering coefficient is given, the aerosol extinction coefficient $\alpha_{L,a}(R)$ at laser frequency is obtained as

$$\alpha_{L,a}(R) = \frac{d/dR[\ln(n_R(R)/N_R(R)R^2)] - \alpha_{L,m}(R) - \alpha_{R,m}(R)}{1 + (\lambda_L/\lambda_R)^k}$$

(3.26)

where we assume Angstrom law for the wavelength dependence of the aerosol extinction, that is, the particle extinction is proportional to λ^{-k}. For an aerosol particle and a water droplet with diameter comparable to the measurement wavelength, $k = 1$ is appropriate; for material larger than the laser wavelength, $k = 0$ is approximated.

For obtaining $\beta_{L,a}$,

$$\frac{N_L(R)N_R(R_0)}{N_L(R_0)N_R(R)}$$

is considered:

$$\beta_{L,a}(R) = -\beta_{L,m}(R) + [\beta_{L,a}(R_0) + \beta_{L,m}(R_0)] \times \frac{N_L(R)N_R(R_0)}{N_L(R_0)N_R(R)}$$

$$\times \frac{\exp\left[-\int_{R_0}^R [\alpha_{R,a}(R) + \alpha_{R,m}(R)]dR\right]}{\exp\left[-\int_{R_0}^R [\alpha_{L,a}(R) + \alpha_{L,m}(R)]dR\right]} \times \frac{n_R(R)}{n_R(R_0)}$$

(3.27)

The reference altitude is chosen so that $\beta_{L,m}(R_0) \gg \beta_{L,a}(R_0)$ and $\beta_{L,a}(R_0) + \beta_{L,m}(R_0) \approx \beta_{L,m}(R_0)$. These clear air conditions are satisfied in the upper atmosphere. The extinction to backscattering coefficient is obtained as $S_{L,a} = \alpha_{L,a}/\beta_{L,a}$. Usually, the inelastic scattering is weak, so the daylight measurement is difficult due to the strong background radiation noise.

3.4.5. Monitoring by Multiwavelength Lidar

3.4.5.1. Two-Wavelength Lidar Method

In the previous section, the optical properties of aerosol, such as extinction coefficient and backscattering coefficient, determined independently from HSRL or Raman-elastic lidar data alone were described. However, HSRL is very sophisticated and the Raman lidar signal is not strong enough for obtaining good results. So if the properties of the aerosol are determined only from the elastic lidar signal, then it is very convenient. Potter (1987) proposed a method to determine the extinction profile from two-wavelength measurement with the assumption $\alpha_s(R) = k\alpha_L (R)$, where $\alpha_i(R)$ ($i = $ s or L) is the extinction coefficient of the single component and s and L are wavelengths at which the extinction coefficient is smaller and larger, respectively. Then, Ackermann (1998, 1999) extended the inversion algorithm to a two-component atmosphere, and obtained the analytical expression of one component (Ackermann, 1997). Although Kunz (1999) claimed that Potter's (1987) and Ackermann's (1999) work do not yield unique solutions, Gimmestad (2001) showed that there is a unique solution as proposed by Potter and Ackermann. In the following, a brief explanation for two-wavelength lidar analysis is given.

In two-wavelength lidar, the one-component lidar signals are written as

$$X_L(R) = K_L \beta_L(R) \exp\left[-2\int_0^R \alpha_L(r)dr\right] \tag{3.28}$$

$$X_S(R) = K_S \beta_S(R) \exp\left[-2\int_0^R \alpha_S(r)dr\right] \tag{3.29}$$

where $X(R)$ is the background-subtracted and range-corrected lidar signal, K includes all instrument constants, $\beta(R)$ is the volume backscatter coefficient, $\alpha(R)$ is the extinction coefficient, and the subscripts L and S refer to the two wavelengths, where the extinction is larger and smaller, respectively.

Here, K_L and K_S are not known, molecular scattering and molecular extinction can be ignored (so we can treat it as one component), and the following three relations are assumed to be valid for all ranges in an interval from R_0 to R_f:

$$\alpha_S(R) = k\alpha_L(R) \tag{3.30}$$

$$\beta_S(R) = \alpha_S(R)/S_S \tag{3.31}$$

$$\beta_L(R) = \alpha_L(R)/S_L \tag{3.32}$$

The constants k, S_S, and S_L are not known.

Equations (3.28) and (3.29) have a similar solution as Equation (3.3). However, here we use transmission $T(R_0, R_F)$ as the boundary condition. Then the solutions are described as

$$\alpha_L = \frac{X_L(R)}{\dfrac{2\int_{R_0}^{R_F} X_L(r)dr}{T_L(R_0,\,R_F)^{-2}-1} + 2\int_R^{R_F} X_L(r)dr} \tag{3.33}$$

$$\alpha_S = \frac{X_S(R)}{\dfrac{2\int_{R_0}^{R_F} X_S(r)dr}{T_S(R_0,\,R_F)^{-2}-1} + 2\int_R^{R_F} X_S(r)dr} \tag{3.34}$$

where the boundary conditions $T_L(R_0, R_F)$ and $T_S(R_0, R_F)$ are the atmospheric transmittances over the range R_0–R_F, defined by

$$T_L(R_0, R_F) = \exp\left[-\int_{R_0}^{R_F} \alpha_L(r)dr\right] \tag{3.35}$$

$$T_S(R_0, R_F) = \exp\left[-\int_{R_0}^{R_F} \alpha_S(r)dr\right] \tag{3.36}$$

Then we consider the double ratio of T_L and T_S at two different ranges R_1 and R_2 of lidar signals X_L and X_S:

$$\frac{X_L(R_1)X_S(R_2)}{X_S(R_1)X_L(R_2)} = \frac{\exp\left[-2\int_{R_1}^{R_2}\alpha_S(r)dr\right]}{\exp\left[-2\int_{R_1}^{R_2}\alpha_L(r)dr\right]}$$

$$= \frac{T_S^2(R_1, R_2)}{T_L^2(R_1, R_2)} \tag{3.37}$$

Then from

$$\frac{X_L(R_i)X_S(R_j)}{X_S(R_i)X_L(R_j)} = \exp\left[-2(k-1)\int_{R_i}^{R_j}\alpha_L(r)dr\right] \qquad (3.38)$$

the uniqueness of k and $\alpha_L(R)$ is validated.

From (3.38), k is determined as (Potter, 1987)

$$k_{i,j} = \frac{1}{2}\left\{2 - \ln\left[\frac{X_L(R_i)X_S(R_j)}{X_S(R_i)X_L(R_j)}\right] \Big/ \int_{R_i}^{R_2}\alpha_L(r)dr\right\} \qquad (3.39)$$

In this case, $k_{i,j}$ is assumed to be constant independent of R_i and Rj. So changing the initial condition $T_L(R_0, R_F)$ and moving R_j, if $k_{i,j}$ does not change, i.e., is constant, then the real value has been obtained, without additional information.

Ackermann (1997) extended Potter's method to include both aerosol and molecular scattering.

In the above discussion, the extinction coefficient can be derived only from lidar data. When the chemical composition and the size distribution of the aerosol are known, the optical properties can be estimated using Mie scattering theory for the case of a spherical particle. Del Guasta and Martini (2000) proposed the derivation of mass concentration from the two-wavelength monitoring method and demonstrated the urban aerosol monitoring using 532 and 1064 nm lidars. In this treatment, the effect of uncertainty in the width of particle size distribution on mass concentration derivation from lidar data is considered in the case of a monomodal, log normal size distribution. The mass-to-backscatter ratio and its uncertainty were computed for six accumulation-mode aerosol models as the function of backscatter angstrom coefficient and relative humidity. A mass-to-backscatter ratio uncertainty of less than $\pm 30\%$ was obtained for all six models. The aerosol mass concentration was derived by fitting lidar data at 532 and 1064 nm with a monomodal distribution of urban aerosols of finite distribution width σ_g (see Equation (3.11)) of log 1.4 to log 2.0. An example of the measurement is shown in Figure 3.17.

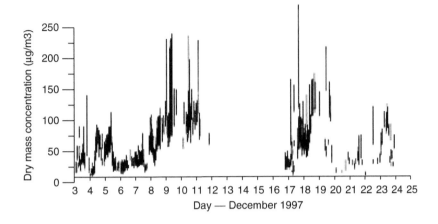

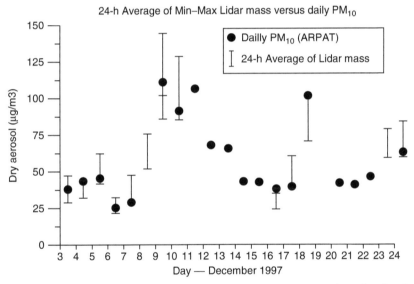

Figure 3.17 (Upper) Dry min.–max. mass concentration obtained from 532 to 1064 nm lidar data with 5 min resolution. (Lower) Twenty-four-hour average of lidar derived, dry aerosol, min.–max. mass concentrations compared with daily PM_{10} data. (From Del Guasta M. and Martini S., *J. Aerosol Sci.*, 31, 1469, 2000. With permission.)

3.4.5.2. Four-Wavelength Lidar Method

In the previous section, proportionality of extinction coefficients between different wavelengths is assumed and the vertical profile of the extinction coefficient could be obtained. For a two-component atmosphere, inversion of lidar data requires the extinction-to-backscattering ratio (S_1 parameter) for each wavelength to determine aerosol extinction profiles. For usually encountered conditions (particle size distributions and refractive indices) of tropospheric aerosols, the Mie theory indicates that values of S_1 parameter can vary in a wide range of 10 to 100 sr. This S_1 problem has so far hindered general application of multiwavelength lidars to aerosol studies.

Another approach based on a lookup table (LUT) calculated by the Mie theory has been proposed (Yabuki et al., 2003). A set of aerosol optical parameters (wavelength dependence of extinction coefficient, scattering coefficient, and S_1 parameter) is computed by assuming appropriate aerosol size distributions and complex refractive indices. The wavelength dependence of the observed lidar signals is compared with that of the LUT, leading to the determination of aerosol properties at each altitude.

If only the single scattering is considered, the two-component lidar equation (3.1) is described as

$$P(R) = K \frac{G(R)}{R^2} [\beta_1(R) + \beta_2(R)]$$
$$\times \exp\left[-2 \int_{R_0}^{R_f} \alpha_1(r) dr - 2 \int_{R_0}^{R_f} \alpha_2(r) dr \right] \qquad (3.40)$$

Here K denotes a system constant, $G(z)$ the overlapping function, β_1 and β_2 the backscattering coefficients, and α_1 and α_2 the extinction coefficients. The subscripts 1 and 2 refer to the aerosol and molecular constants, respectively. By setting $\alpha_1 = S_1 \beta_1$, $\alpha_2 = S_2 \beta_2$ ($S_2 = 8.52$ sr), and $X(R) = P(R)R^2$, $\alpha_1(R)$ is obtained by the Fernald method (1984). An LUT is constructed based on Mie-scattering theory. Here the mixture of urban and maritime aerosols is considered. The parameters used are summarized as follows: size distribution $s(u)$, 11 models ($u = 0$ for the urban to $u = 10$ for the maritime as two

extreme cases, and nine logarithmically interpolating intermediate cases); complex refractive index is taken between
1.40 and 1.60 with a step of 0.01 in the real part ($m'(j_1)$,
$j_1 = 0, 1, \ldots, 20$) and between 0.0 and 0.03 with a step of 1.0
$\times 10^{-4}$ in the imaginary part ($m''(j_2)$, $j_2 = 0, 1, \ldots, 300$); four
wavelength ($\lambda(i)$, $i = 1$ to 4 corresponding to 355, 532, 756,
1064 nm). The extinction coefficient α_1 and the backscattering
coefficient β_1 are calculated for each wavelength λ as $\alpha_1 = \int_0^\infty$
$n(r_p)\alpha_{\text{ext}}(r_p, \lambda)dr_p$ and $\beta_1 = \int_0^\infty n(r_p) (d\sigma_{\text{scat}} (r_p, \lambda)/d\Omega)_{\theta = \pi} dr_p$,
respectively. Here $n(r_p)$, α_{ext}, and $(d\sigma_{\text{scat}}/d\Omega)_{\theta = \pi}$ are the size
distribution, the extinction coefficient, and the backscattering
cross-section, respectively. S_1 is obtained as the ratio of α_1 and
β_1. LUT data consist of a set of α_1 and S_1 parameters for the
four wavelengths calculated with the same set of the size
distribution and the refractive index. Figure 3.18(a) shows
particle size distribution with urban and maritime as the
two extreme ends, and logarithmically uniformly separated.
In Figure 3.18(b), the wavelength dependence of the extinction coefficient for various size distributions is shown. The
dependence is found to be large for the urban model ($u = 0$),
and it becomes smaller as the distribution comes closer to the
maritime model ($u = 10$). Figure 3.19 shows the wavelength
dependence of the S_1 parameter for (a) $m'' = 0.0$, (b) $m'' = 0.01$,
and (c) $m'' = 0.03$. For the urban model, the S_1 parameter

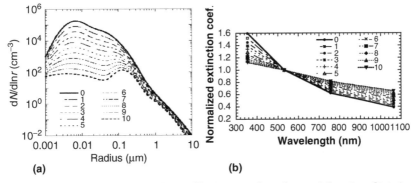

Figure 3.18 (a) Log-uniformally interpolated particle size distribution, (0) urban, (10) maritime. (b) The wavelength dependence of
the extinction coefficient for various size distributions.

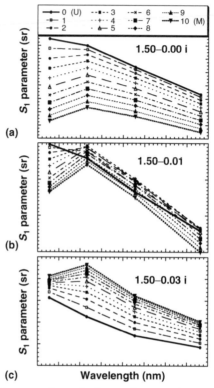

Figure 3.19 The wavelength dependence of the S_1 parameter for (a) $m'' = 0.0$, (b) $m'' = 0.01$, and (c) $m'' = 0.03$.

decreases with increasing wavelength, while for the maritime model the parameter exhibits a maximum at 532 nm. The increase in the imaginary part infers the increase of absorption in the aerosol scattering process. Generally, the increase in absorption results in the increase of the S_1 parameter: this behavior is more pronounced for the maritime model than for the urban model, leading to the S_1 size inversion between the urban and maritime models. In the following, $e_1(\lambda(i), m'(j_1), m''(j_2), s(u))$ is used for the extinction coefficient for the LUT, which corresponds to the extinction coefficient for the relative size distribution profile.

An LUT of $e_1(\lambda(i))$ and $S_1(\lambda(i))$ is constructed based on the Mie-scattering theory as the function of $m'(j_1)$, $m''(j_2)$, and $s(u)$.

The LUT method determines the vertical profile of the extinction coefficient, the S_1 parameter, the refractive index, the size distribution, and the signal power. There the altitude R is discretely specified as $R(q)$ $(q=1,2,\ldots,Q)$ with $R_1=R(1)$ being the lowest altitude in the analysis, which is taken as an altitude where the overlapping function $G(R)$ becomes unity. The most probable profile is determined from the following conditions:

1. $\alpha_1(R_1)=\alpha_1(R_2)$ and $S_1(R_1)=S_1(R_2)$ for all wavelengths. (The lowest two levels have the same α_1 and S_1.)
2. The ratio of the observation signal and that of the theoretical signal for successive layers are considered. The theoretical signal is estimated from the S_1 value of the LUT. Both ratios are assumed to be coincident.
3. The most probable profile is obtained when the wavelength dependence of the extinction coefficient of the observed signal ratio matches the theoretically obtained ratio using the LUT.

The details of the calculation process are as follows. We define a function Dif, the difference between the calculated ratio of the signal intensity at altitude R_{q-1} and R_q, and the observed ratio of the signal intensity at R_{q-1} and R_q:

$$Dif = \frac{P_{\text{cal.}}(R(q),\lambda(i),m'(j_1),m''(j_2),s(u))}{P_{\text{cal.}}(R(q-1),\lambda(i),m'(j_1'),m''(j_2'),s(u'))}$$
$$- \frac{P_{\text{obs.}}(R(q),\lambda(i))}{P_{\text{obs.}}(R(q-1),\lambda(i))} \tag{3.41}$$

where

$$P_{\text{cal.}}(R(q),\lambda(i),m'(j_1),m''(j_2),s(u))$$
$$= \frac{K}{R(q)^2}\left[\frac{\alpha_1(\lambda(i),m'(j_1),m''(j_2),s(u))}{S_1(\lambda(i),m'(j_1),m''(j_2),s(u))}+\frac{\alpha_2(R(q),\lambda(i))}{S_2}\right]$$
$$\times \exp\left[-2\int_0^{R(q)}\alpha_1(R'\lambda(i),m'(j_1),m''(j_2),s(u))\mathrm{d}R'\right.$$
$$\left.-2\int_0^{R(q)}\alpha_2(R',\lambda(i))\mathrm{d}R'\right] \tag{3.42}$$

1. The derivation of the extinction coefficient from the least-square sum of *Dif* for all wavelengths.
2. Comparison of wavelength dependence between the extinction profile derived by step (1) and that of the LUT.
3. The determination of the most probable profile for each altitude.

For the set of $m'(J_1)$, $m''(J_2)$, and $s(U)$, the most probable profile is obtained:

$$\alpha_1^*(R(q), \lambda(i)) = \alpha_1(\lambda(i), m'(J_1), m''(J_2), s(U))$$

$$S_1^*(R(q), \lambda(i)) = S_1(\lambda(i), m'(J_1), m''(J_2), s(U))$$

$$m'^*(R(q)) = m'(J_1)$$

$$m''^*(R(q)) = m''(J_2)$$

$$s^*(R(q_2)) = s(U)$$

The actual size distribution is obtained from

$$n^*(r, R(q)) = n_s(r) \frac{\alpha_1^*(R(q), \lambda(j))}{\int_0^\infty n_s(r)\alpha_{ext}(r, \lambda(j))dr} \tag{3.43}$$

where $n_s(r)$ is the normalized size distribution for $s(u)$.

The LUT method was applied to the actual lidar data obtained with a multiwavelength lidar at Chiba University with a vertical resolution of 15 m. The vertical profiles of the extinction coefficient, the S_1 parameter, and the refractive index are shown in Figure 3.20(a), Figure 3.21(a), and Figure 3.22, respectively. In Figure 3.22, the dotted line shows the original value of the refractive index and the solid line shows the running mean with 75 m resolution. Recalculated profiles of the extinction coefficient and the S_1 parameter are plotted in Figure 3.20(b) and Figure 3.21(b), respectively, after using the running mean of the complex refractive index and size distribution function with 75 m range resolution.

For all wavelengths, the S_1 values are relatively large in the mixed layer; the values decrease with altitude in the free troposphere. It is also seen that the differences of S_1 among

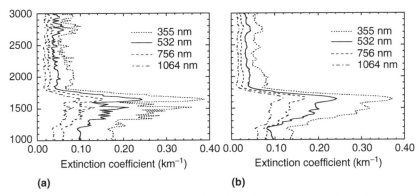

(a)   **(b)**

Figure 3.20 The vertical profile of extinction coefficient for each wavelength by the LUT method. The data were taken on October 21, 1998 (JST 13:30): (a) original solution, (b) recalculated profile after the complex refractive index and size distribution in (a) are smoothed by the running mean with 75 m resolution.

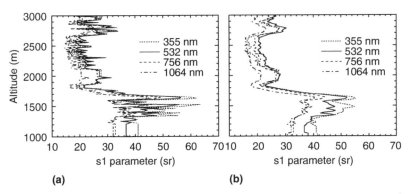

(a)   **(b)**

Figure 3.21 The vertical profile of the S_1 parameter for each wavelength by the LUT method: (a) original solution, (b) recalculated profile after the complex refractive index and size distribution in (a) are smoothed by the running mean with 75 m resolution.

wavelengths tend to decrease with the altitude. The S_1 parameter for 532 nm varies in a range of 20 to 60 sr, which is reasonable compared with those reported previously (Takamura et al., 1994; Müller et al., 1998; Doherty et al., 1999).

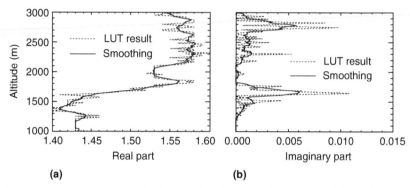

Figure 3.22 The vertical profile of refractive index: (a) real part, (b) imaginary part for each wavelength by the LUT method.

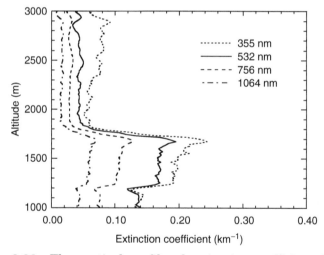

Figure 3.23 The vertical profile of extinction coefficient derived by the Fernald method, assuming refractive index of 1.5–0.01i and urban model.

In Figure 3.23, the result retrieved by the Fernald method is shown. The S_1 parameters derived from the Mie theory with a refractive index of 1.50 to 0.01i, a value typical for the urban aerosol model, are used (S_1 are 49.8, 47.9, 43.3, and 37.9 for 355, 532, 756, and 1064 nm, respectively). It is seen that the extinction profiles between 1100 and 1500 m are different in Figure 3.23 and Figure 3.20. Although the far-end boundary condition is selected in the conventional method, however, in the LUT method, the boundary condition starts from the near distance, where the effect of detector noise is small. So, in this simulation, it is found that when a noise of 250% is added to the input extinction coefficient of two wavelengths at an intermediate altitude, the resulting errors in S_1 parameters for the four wavelengths are less than 20% for higher altitudes.

3.5. LIDAR NETWORK MONITORING

The coverage area that one lidar station can monitor is limited to several tens of kilometers at the most. So in order to monitor an operationally wide area from the ground, network monitoring is necessary. In this section, some lidar network systems are introduced.

3.5.1. Earlinet

A European Aerosol Research Lidar Network (EARLINET) (Bosenberg et al., 2001) was established with the support of the European Commission to establish an Aerosol Climatology in European Continent. EARLINET will establish a quantitative comprehensive statistical database of both horizontal and vertical distribution of aerosols on a continental scale using a network of advanced laser remote-sensing stations distributed over most of Europe. The goal is to provide aerosol data for unbiased sampling, selected important processes, and airmass history, together with comprehensive analyses of these data.

The objectives will be reached by implementing a network of 21 stations (Figure 3.24) using advanced quantitative laser remote sensing (multispectral backscatter lidar mostly combined with Raman lidar) to measure directly the vertical distribution of aerosol. These stations are distributed over most of Europe and will be operated by researchers who

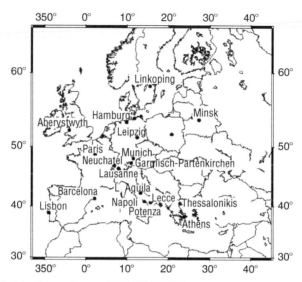

Figure 3.24 Locations of EARLINET sites.

have long experience both in the development of these systems and their application in different areas of atmospheric
research. For the operation of the network only existing instrumentation will be used, with the exception of very few
necessary upgrades.

A major part of the measurements will be performed on a
fixed schedule to provide an unbiased statistically significant
dataset. Additional measurements will be performed to specifically address important processes that are localized either
in space or time. All lidar measurements will be supported by
a suite of more conventional observations. Back-trajectories
derived from operational weather prediction models will be
used to characterize the history of the observed air parcels,
accounting explicitly for the vertical distribution.

The data will be collected in a central database, made
accessible to all internal and possible external partners, and
will be analyzed with respect to the different objectives by
specially formed task groups. The work will proceed mainly
in three steps: a short instrument preparation phase, a long
data-collection period, accompanied by combined analysis of
data. Special care will be taken to assure data quality, including intercomparisons at instrument and evaluation levels.

3.5.2. MPL-Net

MPL-Net (http://mplnet.gsfc.nasa.gov/) is a worldwide network of micropulse lidar (MPL) systems. MPL-Net is run by NASA Goddard Space Flight Center (GSFC). The MPL is a single-channel (523 nm), autonomous, eye-safe lidar system originally developed at GSFC and is commercially available. The MPL is used to determine the vertical structure of clouds and aerosols. The MPL data are analyzed to produce optical properties such as extinction and optical depth profiles of the clouds and aerosols. The primary goal of MPL-Net is to provide long-term datasets of cloud and aerosol vertical distributions at key sites around the world (Figure 3.25). The long-term datasets will be used to validate and help improving global and regional climate models, and also serve as ground-truth sites for NASA/EOS satellite programs such as the Geoscience Laser Altimeter System (GLAS) on the ICESat spacecraft launched in January 2003.

Figure 3.25 Locations of MPL-NET sites.

MPL-Net consists of NASA-operated sites, incorporated sites from the ARM MPL network, and sites privately operated by researchers from around the world. Also, all MPL-Net sites are colocated with AERONET sunphotometers. Instrument calibrations and data processing for all sites are accomplished using techniques developed by NASA. In addition to the long-term sites, MPL-Net provides support for field experiments each year using MPL systems reserved for field use (land- and ship-based deployments possible).

3.5.3. AD-Net

Asian Dust Network (AD-Net; http://info.nies.go.jp:8094/kosapub/) is an international virtual community constituted in February 2001, based on a 4 year of spring campaign (from March 1 to May 31) for the rapid communication through the internet on Asian dust events since 1997. The first campaign was mainly led by lidar groups and limited in Japan. Hence, the former activity was tentatively named as Lidar Network Observation of Kosa in Japan (LINK-J, Kosa is a Japanese word that means "yellow sand").

Since 1998, the exchange of Asian dust information with lidar and other surface observations in East Asia countries (China, Korea, and Japan) started. In April 1998, a well-known huge Asian dust event occurred and the coordinated network worked effectively (Murayama et al., 2001). In these unusual events, a large amount of Asian dust was transported to North America beyond the Pacific (Husar et al., 2001). Since 1999, the community was expanded and included various fields of atmospheric sciences, for example, lidar, radiation, and other ground chemical and physical measurements, transport model, and satellite remote-sensing groups. Some results were shown in Section 3.3.4.1. The map of AD-NET is shown in Figure 3.26.

In the above, typical lidar networks were introduced. In addition, a German network system is also operated (Eixmann et al., 2002), and three lidar stations in Jakarta are also constructing a network observation for covering Jakarta city (Pinandito et al., 2000).

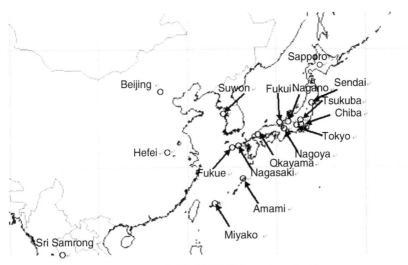

Figure 3.26 Locations of AD-NET lidar sites in Japan.

3.6. CONCLUSIONS AND FUTURE TREND

In this chapter, the recent results of tropospheric lidar monitoring are introduced. With the development of electronics and information technology, unattended lidar systems will be introduced for global and local environmental monitoring. One of the targets of lidar technology is a "turn-key" device similar to home electronics, such as TV, washing machine, and electrical range. An elastic lidar is the most basic, technically complete, and nearest to automatic operation. The function of the elastic lidar is the monitoring of spatially and temporally distributed profiles of atmospheric particles and clouds in the troposphere, especially in the boundary layer, and the stratosphere. With the progress of lasers, especially DPSS (diode pumped solid state) laser, and electronics, a compact, robust, long lifetime, and portable system will be realized in the near future.

REFERENCES

Ackermann, A. (1997) Two-wavelength lidar inversion algorithm for a two-component atmosphere, *Appl. Opt.* **36**, 5134–5143.

Ackermann, J. (1998) Two-wavelength lidar inversion algorithm for a two-component atmosphere with variable extinction-to-backscatter ratio, *Appl. Opt.* **37**, 3164–3171.

Ackermann, J. (1999) Analytical solution of the two-frequency lidar inversion technique, *Appl. Opt.* **38**, 7414–7418.

AD-NET; http://info.nies.go.jp:8094/kosapub/

Ansmann, A., U. Wanginger, M. Riebesell, C. Weitkamp and W. Michaelis (1992) Independent measurement of extinction and backscatter profiles in cirrus clouds by using a combined Raman elastic-backscatter lidar, *Appl. Opt.* **33**, 7113–7131.

Barnaba, F. and G.P. Gobbi (2001) Lidar estimation of tropospheric aerosol extinction, surface area, and volume: maritime and desert-dust cases, *J. Geophys. Res.* **106**, 3005–3018.

Born, M. and E. Wolf (1950) *Principle of Optics*, Pergamon Press, Oxford.

Bosenberg, J., A. Ansmann et al. (2001) EARLINET: European Aerosol Research Lidar Network. In: *Advances in Laser Remote Sensing*, A. Dabas, C. Loth, J. Pelon, eds., pp. 155–159, E. poly, Cedex, France.

Chanin, M.L. and A. Hauchecorne (1984) Lidar studies of temperature and density using Rayleigh scattering, *MAP Handbook* **13**, 87–98.

Davis, P.A. (1969) The analysis of lidar signatures of cirrus clouds. *Appl. Opt.* **8**, 2099–2102.

Doherty, S.J., T.L. Anderson and R.J. Charlson (1999) Measurement of the lidar ratio for atmospheric aerosols with a 180° backscatter nephelometer, *Appl. Opt.* **38**, 1823–1832.

Del Guasta, M. and S. Martini (2000) On the retrieval of urban aerosol mass concentration by a 532 and 1064 nm lidar, *J. Aerosol Sci.* **31**, 1469–1488.

Devara, P.C.S. and P.E. Raj (1991) Study of atmospheric aerosols in a terrain-induced nocturnal boundary layer using bistatic lidar, *Atmos. Env. A* **25**, 655–660.

Donovan, D.P., A.C.A.P. van Lammeren, R.J. Hogan, H.W.J. Russchenberg, A. Apituley, P. Francis, J. Testud, J. Pelon, M. Quante and J. Goddard (2001) Cloud effective particle size and water content profile retrievals using combined lidar and radar observations, 2, Comparison with IR radiometer and in situ measurements of ice clouds, *J. Geophys. Res.* **106**, 27449–27464.

Eixmann, R., C. Böckmann, B. Fay, V. Matthias, I. Mattis, D. Müller, S. Kreipl, J. Schneider and A. Stohl (2002) Tropospheric aerosol layers after a cold front passage in January 2000 as observed at several stations of the German Lidar Network, *Atmos. Res.* **63**, 39–58.

Eloranta, E.W. (1972) Calculation of doubly scattered lidar returns, PhD dissertation (University of Wisconsin).

Eloranta, E.W. and P. Pirronen (1996) Measurements of cirrus cloud optical properties and particle size with the University of Wisconsin high spectral resolution lidar. In: *Advances in Atmospheric Remote Sensing with Lidar*, A. Ansman et al., eds., pp. 83–86, Springer, Berlin.

Eloranta, E.W. and P. Ponsardin (2001) A high spectral resolution lidar designed for unattended operation in the Arctic. *Optical Society of America Topical Meeting on Optical Remote Sensing of the Atmosphere*, February 5–8, Coeur d'Alene, Idaho.

Gimmestad, G.G. (2001) Comment on two-wavelength lidar inversion techniques, *Appl. Opt.* **40**, 2004–2009.

Gobbi, G.P., F. Barnaba, R. Giorgi and A. Santacasa (2000) Altitude-resolved properties of a Saharan dust event over the Mediterranean, *Atmos. Environ.* **34**, 5119–5127.

Goldsmith, J.E.M., F.H. Blair, S.E. Bisson and D.D. Turner (1998) Turn-key Raman lidar for profiling atmospheric water vapor, clouds, and aerosols, *Appl. Opt.* **37**, 4979–4990.

Grund, C.J. and S.P. Sandberg (1996) Depolarization and backscatter lidar for unattended operation. *Proceedings, 18th International Laser Radar Conference*, July 22–26, 1996, Berlin, Germany, A. Ansmann et al., eds., pp. 3–6, Springer, Berlin.

Hess, M., P. Koepke and I. Schult (1998) Optical properties of aerosols and clouds: the software package OPAC, *Bull. Am. Met. Soc.* **79**, 831–844.

Hinkley, E.D., ed. (1976) *Laser Monitoring of the Atmosphere*, Topics in Applied Physics, vol. 14, Springer, New York.

Hughes, H.G., J.A. Fergnson and D.H. Stephans (1985) Sensitivity of a lidar inversion algorithm to parameters relating atmospheric backscatter and extinction, *Appl. Opt.* **24**, 1609–1613.

Husar, R.B., D.M. Tratt, B.A. Schichtel, S.R. Falke, F. Li, D. Jaffe, S. Gasso, T. Gill, N.S. Laulainen, F. Lu, M.C. Reheis, Y. Chun, D. Westphal, B.N. Holben, C. Gueymard, I. McKendry, N. Kuring, G.C. Feldman, C. McClain, R.J. Frouin, J. Merrill, D. DuBois, F. Vignola, T. Murayama, S. Nickovic, W.E. Wilson, K. Sassen,

N. Sugimoto and W.C. Malm (2001) The Asian dust events of April 1998, *J. Geophys. Res.* **106**, 18317–18330.

Jaenicke, R. (1993) Tropospheric aerosols. In: *Aerosol–Cloud–Climate Interactions*, P.V. Hobbs, ed., pp. 1–31, Academic Press, San Diego.

Kaneyasu, N., S. Ohta and N. Murao (1995) Seasonal variation in the chemical composition of atmospheric aerosols and gaseous species in Sapporo, Japan, *Atmos. Environ.* **29**(13), 1559–1568.

Klett, J.D. (1981) Stable analytical inversion solution for processing lidar returns, *Appl. Opt.* **20**, 211–220.

Kunz, G.J. (1999) Two-wavelength lidar inversion algorithm, *Appl. Opt.* **38**, 1015–1020.

Lagrosas, N., Y. Yoshii, H. Kuze, N. Takeuchi, S. Naito, A. Sone, and H. Kan (2004), Observation of boundary layer aerosols using a continuously operated, portable lidar system, Atmospheric Env. **38**, 3885–3892.

Lin, J., H. Mishima, Y. Kubota, F. Kobayashi, T. Kawahara, Y. Saito, A. Nomura, K. Yamaguchi and K. Morikawa (1999) Bistatic imaging lidar measurements in the lower atmosphere, *Rev. Laser Eng.* **27**, 827–834.

Liou, K.-N. and R.M. Schotland (1971) Multiple backscattering and depolarization from water clouds for a pulsed lidar system, *Atmos. Sci.* **28**, 772–784.

McGill, M., D. Hlavka, W. Hart, V.S. Scott, J. Spinhirne and B. Schmid (2002) Cloud physics lidar: instrument description and initial measurement results, *Appl. Opt.* **41**(18), 3725–3734.

Measures, R.M. (1984) *Laser Remote Sensing*, John Wiley & Sons, New York.

Mishchenko, M.I. and J.W. Hovenier (1995) Depolarization of light backscattered by randomly oriented nonspherical particles, *Opt. Lett.* **20**, 1356–1358.

MPL-NET; http://mplnet.gsfc.nasa.gov/

Müller, D., U. Wandinger, D. Althausen, I. Mattis and A. Ansmann (1998) Retrieval of physical particle properties from lidar observations of extinction and backscatter at multiple wavelengths, *Appl. Opt.* **37**, 2260–2263.

Murayama, T., N. Sugimoto, I. Uno, K. Kinoshita, K. Aoki, N. Hagiwara, Z. Liu, I. Matsui, T. Sakai, T. Shibata, K. Arao, B-J. Shon, J-G. Won, S-C. Yoon, T. Li, J. Zhou, H. Hu, M. Abo, K. Iokibe, R. Koga and Y. Iwasaka (2001) Ground-based network observation of Asian dust events of April 1998 in East Asia, *J. Geophys. Res.* **106**, 18345–18360.

Pal, S.R. and A.I. Carswell (1973) Polarization properties of lidar backscattering from clouds, *Appl. Opt.* **12**, 1530–1535.

Pal, S.R. and A.I. Carswell (1976) Multiple scattering in atmospheric clouds: lidar observations, *Appl. Opt.* **15**, 1990–1995.

Pal, S.R. and A.I. Carswell (1978) Polarization properties of lidar scattering from clouds at 347 nm and 694 nm, *Appl. Opt.* **17**, 2321–2328.

Pal, S.R. and A.I. Carswell (1985) Polarization anisotropy in lidar multiple scattering from atmospheric clouds, *Appl. Opt.* **24**, 3464–3471.

Pinandito, M., I. Rosananto, I. Hidayat, S. Sugondo, S. Asiati, A. Pranowo, I. Matsui and N. Sugimoto (2000) Mie scattering lidar observation of aerosol vertical profiles in Jakarta, Indonesia, *Environ. Sci.* **13**(2), 205–216.

Piironen, A.K. and E.W. Eloranta (1995) Convective boundary layer mean depths, cloud base altitudes, cloud top altitudes, cloud coverages, and cloud shadows obtained from volume imaging lidar data, *J. Geophys. Res.* **100**(D12), 25569–25576.

Platt, C.M.R. (1979) Remote sensing of high clouds: I. Calculation of visible and infrared optical properties from lidar and radiometer measurements, *J. Appl. Meteorol.* **18**, 1130–1143.

Platt, C.M.R., S.A. Young et al. (1994) The Experimental Cloud Lidar Pilot Study (ECLIPS) for cloud-radiation research, *Bull. Am. Meteorol. Soc.* **75**, 1635–1654.

Potter, J.F. (1987) Two-frequency lidar inversion technique, *Appl. Opt.* **26**, 1250–1256.

Sassen, K. (1991) The polarization lidar technique for cloud research: a review and current assessment, *Bull. Am. Meteorol. Soc.* **72**(12), 1848–1866.

Sassen, K. (1994) Advances in polarization diversity lidar for cloud remote sensing, *Proc. IEEE* **82**, 1907–1914.

Sassen, K., H. Zhao and G.C. Dodd (1992) Simulated polarization diversity lidar returns from water and precipitating mixed phase clouds, *Appl. Opt.* **31**, 2914–2923.

Shiina, T., E. Minami, M. Ito, Y. Okamura (2002) Optical Circulator for an in-line-type compact lidar, *Appl. Opt.* **41**(19) 3900–3905.

Shimizu, H., S.A. Lee and C.Y. She (1983) High spectral resolution lidar system with atomic blocking filters for measuring atmospheric parameters, *Appl. Opt.* **22**(9), 1373–1381.

Shipley, S.T., D.H. Tracy, E.W. Eloranta, J.T. Trauger, J.T Sroga, F.L. Roesler and J.A. Weinman (1983) High spectral resolution lidar to measure optical scattering properties of atmospheric aerosols. 1: theory and instrumentation, *Appl. Opt.* **22**(23), 3716–3723.

Spinhirne, J.D. (1993) Micro pulse lidar, *IEEE Trans. Geosci. Remote Sensing* **31**, 48–55.

Sroga, J.T., E.W. Eloranta, S.T. Shipley, F.L. Roesler and P.J. Tryon (1983) High spectral resolution lidar to measure optical scattering properties of atmospheric aerosols. 2: calibration and data analysis, *Appl. Opt.* **22**(23), 3725–3732.

Sugimoto, N., I. Matsui, A. Shimizu, M. Pinandito and S. Sugondo (2000a) Climatological characteristics of cloud distribution and planetary boundary layer structure in Jakarta, Indonesia revealed by lidar observation, *Geophys. Res. Lett.* **27**, 2909–2912.

Sugimoto, N., I. Matsui, Z. Liu, A. Shimizu, I. Tamamushi and K. Asai (2000b) Observation of aerosols and clouds using a two-wavelength polarization lidar during the Nauru99 experiment, *J. Mar. Meteorol. Soc.* **76**, 93–98.

Sun, Y.-Y. and Z.-P. Li (1989) Depolarization of polarized light caused by high altitude clouds. 2: depolarization of lidar induced by water clouds, *Appl. Opt.* **28**, 3633–3638.

Takamura, T., Y. Sasano and T. Hayasaka (1994) Tropospheric aerosol optical properties derived from lidar, sun photometer, and optical particle counter measurements, *Appl. Opt.* **33**, 7132–7140.

Takeuchi, N., H. Kuze, Y. Sakurada, T. Takamura, S. Murata, K. Abe and S. Moody (1997) Construction of a multi-wavelength lidar system for satellite data atmospheric correction. In: *Advances in Atmospheric Remote Sensing with Lidar*, A. Ansmann et al., eds., pp. 71–74, Springer, Berlin.

Takeuchi, N. (1999) Aerosol characterization for lidar monitoring in Asia. *Proc. CLEO/'99, Pacific RIM (ThESS2-1)*, pp. 1048–1049.

Takeuchi, N., H. Kuze, T. Takamura, T. Nakajima, S. Baimoung (2000), Continuous cloud observation in tropics by Micro-Pulse-Lidar, Proc. of 20th Int. Laser Radar Conf. Vichy, France.

van der Hulst, H.C. (1957) *Light Scattering by Small Particles*, Dover Publications, Inc., New York.

Weber, M.J. (2000) *Handbook of Lasers*, CRC Press, Boca Raton, FL.

Weitkamp, C., H. Flint, W. Lahmann, F. Theopold, O. Danne, M. Quante and E. Raschke (1999) Simultaneous radar and lidar cloud measurements at Geesthacht (53.5 N, 10.5 E), *Phys. Chem. Earth* **24**, 163–166.

Welton, E.J., J.R. Campbell, T.A. Berkoff, J.D. Spinhirne, S. Tsay and B. Holben (2001) First Annual Report: The Micro-pulse Lidar Worldwide Observational Network, Project Report.

Yabuki, M., H. Kinjo, H. Kuze and N. Takeuchi (2000) Derivation of aerosol optical properties from four-wavelength lidar observations, *Proc. SPIE* **4153**, 132–142.

Yabuki, M., H. Kinjo, H. Kuze and N. Takeuchi (2003) Determination of vertical distribution of aerosol optical parameters by use of multi-wavelength lidar data, *Jpn. J. Appl. Phys.* **42**, 1–9.

4

Trace Gas Species Detection in the Lower Atmosphere by Lidar: From Remote Sensing of Atmospheric Pollutants to Possible Air Pollution Abatement Strategies

BERTRAND CALPINI and VALENTIN SIMEONOV

Air Pollution Laboratory, Swiss Federal
Institute of Technology, Lausanne, Switzerland

4.1.   INTRODUCTION

Trace gases are atmospheric constituents that occur in relatively small and sometimes highly variable concentrations. Most of them are of natural or anthropogenic origin, but some like chlorofluorocarbons are only generated by human activities. Some trace gas concentrations have changed rapidly during the last three centuries and human activities account for most of these changes. As some of these gases may play a crucial role in the earth's radiative balance and atmospheric chemistry and be important pollutants, they are the subject of intensive scientific study. Once emitted into the atmosphere they can be transformed by a chain of complex chemical reactions. Air quality models are used as a unique method for the simulation of trace gas concentrations in time and space, thus yielding a powerful tool for the achievement of optimal pollution abatement strategies.

A large number of methods are used to measure trace gas concentrations. Among the different chemical and physical methods used for trace gas monitoring, optical methods have in recent years attracted a great deal of attention. Lidar methods, and more specifically the differential absorption lidar (DIAL) method, give a unique possibility to measure three-dimensional (3D) trace gas concentrations with a temporal and spatial resolution of direct interest for the applied models.

This chapter is dedicated to DIAL methods for monitoring trace gases in the planetary boundary layer (PBL). The basic principles of lidar operation with emphasis on the DIAL method are outlined together with information about the experimental setup, the lasers relevant to pollution monitoring, the detection system, and error sources. This is followed by an overview of different trace gas measurement obtained by DIAL. Both ultraviolet (UV) and infrared (IR) systems are considered with emphasis on ozone.

Studies of tropospheric ozone concentrations and variability are motivated by the impact of ozone on human health and vegetation as well as its significance as a greenhouse gas. Hence, 3D measurements of ozone are a focal point in atmospheric measurements. Therefore a more detailed description of an ozone DIAL setup is given in the following text. Lidar data analysis is also outlined while underlining error analysis. Further information on lidar principles can be found in Ref. [1].

In the last part of this chapter, the use of lidar measurements in air quality modeling is presented. Eulerian mesoscale air pollution models are introduced and some case studies are presented, in particular a field study in the area of Grenoble, France, 1999. This overview is concluded by a description of the novel pump and probe OH lidar.

4.2. DIFFERENTIAL ABSORPTION LIDAR EQUATION

The basic lidar equation formally relates the power of the laser light originally transmitted to the atmosphere $P_L(\lambda)$, to the power of the backscattered light $P(\lambda, R)$ received at the system from a distance R,

$$P(\lambda, R) = C(\lambda, R)P_{\mathrm{L}}(\lambda)\frac{A}{R^2}\Delta R\beta(\lambda, R)$$

$$\exp\left(-2\int_0^R [\alpha(\lambda, R) + \sigma N(R)]\mathrm{d}r\right) \tag{4.1}$$

where $C(\lambda, R)$ is a parameter characterizing the total efficiency of the lidar transmission receiving and detection parts, A is the effective receiver area, ΔR is the range resolution of the lidar, which depends both on the laser pulse-duration and the detection resolution time, and $\beta(\lambda, R)$ is the volume backscattering coefficient of the atmosphere. In this particular version of the lidar equation, we intentionally separate the extinction term under the integral into two parts. The first component is the volume atmospheric extinction, $\alpha(\lambda, R)$ that takes into account the extinction due to light scattering from particles and molecules, and the second component, σN, accounts for the absorption by gas with an absorption cross section σ and volume concentration $N(R)$. Note that Equation (4.1) already contains information about the concentration (i.e., $N(R)$) of the absorbing gas. However, to derive the concentration one needs to know precisely the instrument parameters (i.e., $C(\lambda, R)$, $P_{\mathrm{L}}(\lambda)$, A) and the atmospheric variables ($\beta(\lambda, R)$ and $\alpha(\lambda, R)$).

The instrumental and atmospheric parameters in Equation (4.1) which we need to know a priori can be eliminated or reduced if two or more lidar signals with wavelengths that are absorbed differently by the gas of interest are used. In the instance where two wavelengths are used, the wavelength, which is more strongly absorbed is usually denoted as λ_{ON} and the other wavelength as λ_{OFF}. The gas number density is derived from the λ_{ON} and λ_{OFF} lidar equations as,

$$N_X(R) = \underbrace{\frac{1}{2\Delta\sigma}\frac{\mathrm{d}}{\mathrm{d}R}\ln\frac{P(R, \lambda_{\mathrm{OFF}})}{P(R, \lambda_{\mathrm{ON}})}}_{\mathrm{A}} - \underbrace{\frac{1}{2\Delta\sigma}\frac{\mathrm{d}}{\mathrm{d}R}\ln\frac{\beta(R, \lambda_{\mathrm{OFF}})}{\beta(R, \lambda_{\mathrm{ON}})}}_{\mathrm{B}}$$

$$- \underbrace{\frac{\alpha(R, \lambda_{\mathrm{OFF}}) - \alpha(R, \lambda_{\mathrm{ON}})}{\Delta\sigma}}_{\mathrm{C}} \tag{4.2}$$

known as the "DIAL equation". Here $\Delta \sigma = \sigma(\lambda_{ON}) - \sigma(\lambda_{OFF})$ is the so-called differential absorption cross section of the molecular species of interest. The sensitivity and lower detection limit of a lidar for a specific compound are defined by $\Delta \sigma$.

The elimination of the instrument parameters is only possible if the ratio of the lidar constants and the transmitted powers at the two wavelengths are range-independent, i.e.,

$$\frac{d}{dR} \frac{C(\lambda_{ON})}{C(\lambda_{OFF})} = 0 \quad \text{and} \quad \frac{d}{dR} \frac{P(\lambda_{ON})}{P(\lambda_{OFF})} = 0 \tag{4.3}$$

This can be achieved by carefully designing the lidar.

For most molecular species such as H_2O, SO_2, NO_2, and NO the wavelength separation between the ON and OFF wavelengths can be smaller than $1\,cm^{-1}$. If this is the case, the differences in the scattering properties of the atmosphere (term B) and the differential extinction due to aerosol and interfering gases (term C) can be neglected and the gas number density is given simply by term A in Equation (4.2). For the particular species mentioned above great care should be taken to avoid systematic errors or random deviations due to large laser linewidth or wavelength instability of the lidar transmitter. This problem is not essential for species like ozone or chlorine that has broad absorption features. However, to obtain significant differential cross sections, wider ($\Delta \lambda \cong 5 - 20\,nm$) wavelength separation is needed and terms B and C in the DIAL equation can no longer be neglected. This point is addressed in more detail in the part dedicated to ozone DIAL systems.

4.3. THE DETECTION OF TRACE GAS SPECIES BY DIAL

The DIAL technique is a range resolved technique allowing the detection of various atmospheric trace species. To do this, the pollutant's absorption spectrum must be known and the corresponding laser sources must be available. For a number of the species of interest, this spectral information is summarized in Table 4.1. A comprehensive list of atmospheric and trace gases has been compiled by the Geophysics Laboratory,

Table 4.1   Absorption Cross-Section of Pollutant Gases at
Wavelengths Suitable for DIAL Measurement in the Troposphere

Molecule	Typical Wavelength	Laser (Examples)	Absorption Cross Section $(10^{-18}\,cm^2)$
Nitric oxide, NO	226.8 nm	Dye	4.6
Benzene, C_6H_6	252.9 nm	Dye	5.7
Mercury, Hg	253.65 nm	Dye	3.3×10^4
Toluene, C_7H_8	266.9 nm	Dye, Ti:Sa	2.8
Ozone, O_3	266.0 nm	Dye, Nd:YAG-IV	9.49
	289.0 nm	Raman D_2	1.59
	299.0 nm	Raman H_2	0.42
Formaldehyde, CH_2O	286.5 nm	Dye, Ti:Sa	0.068
Sulfur dioxide, SO_2	300.0 nm	Dye, Ti:Sa	1.3
Chlorine, Cl_2	330.0 nm	Dye, XeCl + Raman cell	0.26
Nitrous acid, HONO	354.0 nm	Dye, Ti:Sa	0.50
Nitrogen dioxide, NO_2	448.1 nm	Dye	0.69
Methane, CH_4	3.270 μm	OPO	2.0
	3.391 μm		0.6
Propane, C_3H_8	3.391 μm	OPO	0.8
Hydrogen chloride, HCl	3.636 μm	DF, OPO	0.20
Methane, CH_4	3.715 μm	DF, OPO	0.002
Sulfur dioxide, SO_2	3.984 μm	DF, OPO	0.42
Carbon monoxide, CO	4.709 μm	CO_2	2.8
	4.776 μm		0.8
Nitric oxide, NO	5.215 μm	CO	0.67
	5.263 μm	CO_2	0.6
Propylene, C_3H_6	6.069 μm	CO	0.09
1,3-Butadiene, C_4H_6	6.215 μm	CO	0.27
Nitrogen dioxide, NO_2	6.229 μm	CO	2.68
Sulfur dioxide, SO_2	9.024 μm	CO_2	0.25
Freon-11, CCl_3F	9.261 μm	CO_2	1.09
Ozone, O_3	9.505 μm	CO_2	0.45
	9.508 μm		0.9
Fluorocarbon-113, $C_2Cl_3F_3$	9.604 μm	CO_2	0.77
Benzene, C_6H_6	9.621 μm	CO_2	0.07
MMH, $CH_3N_2H_3$	10.182 μm	CO_2	0.06
Ethyl mercaptan, C_2H_5SH	10.208 μm	CO_2	0.02
Chloroprene, C_4H_5Cl	10.261 μm	CO_2	0.34
Monochloroethane, C_2H_5Cl	10.275 μm	CO_2	0.12
Ammonia, NH_3	10.333 μm	CO_2	1.0
Ethylene, C_2H_4	10.533 μm	CO_2	1.19
Sulfur hexafluoride, SF_6	10.551 μm	CO_2	30.3

Table 4.1 Absorption Cross-Section of Pollutant Gases at Wavelengths Suitable for DIAL Measurement in the Troposphere — (Continued)

Molecule	Typical Wavelength	Laser (Examples)	Absorption Cross Section $(10^{-18} \, cm^2)$
Trichloroethylene, C_2HCl_3	10.591 μm	CO_2	0.49
1,2-Dichloroethane, $C_2H_4Cl_2$	10.591 μm	CO_2	0.02
Hydrazine, N_2H_4	10.612 μm	CO_2	0.18
Vinyl chloride, C_2H_3Cl	10.612 μm	CO_2	0.33
UDMH, $(CH_3)_2N_2H_2$	10.696 μm	CO_2	0.08
Fluorocarbon-12, CCl_2F_2	10.719 μm	CO_2	1.33
Perchloroethylene, C_2Cl_4	10.742 μm	CO_2	0.18
1-Butene, C_4H_8	10.787 μm	CO_2	0.13
Perchloroethylene, C_2Cl_4	10.834 μm	CO_2	1.14

U.S. Air Force Systems Command [2], which covers the spectral range from 0 to $23,000 \, cm^{-1}$ (435 nm). A compilation of the IR absorption spectra is also available [3] as DIAL is possible in the UV, the visible, and the IR.

Molecular absorption features can vary greatly and thus be more or less optimal for DIAL experiments, as will be discussed in the following examples in the UV spectral region. Some trace gasses such as NO, NO_2, or SO_2 are characterized by very narrow lines, for example, the full width at half maximum (FWHM) of NO lines around 226 nm is only 0.01 nm. Toluene and benzene, on the other hand, show broader lines, of the order of 1 nm FWHM. The absorption features of O_3 and Cl_2 are characterized by very broad structures (ca. 50 nm FWHM) without any strong absorbing narrow spectral lines in the region of interest, that is, around 250 and 350 nm.

The accuracy of the DIAL measurement is directly dependent on the accuracy of the differential absorption cross section. The absorption lines in the visible and near-UV can, in principle, be measured accurately in the laboratory, and can generally be transposed to the atmospheric conditions without problematic adjustments. In some cases, corrections must be made for changes in temperature and pressure, as the pressure and the temperature of the lower troposphere are generally higher than those of the higher altitudes. These necessary corrections can often be estimated from laboratory measurements.

Due to the low signal to noise ratio (SNR) of individual DIAL signals, a certain amount of signal averaging is necessary. Further analysis of the DIAL lidar signals consists of (1) determining the offset and identifying the range, (2) calculating a preliminary concentration profile using the lidar equation, often called "uncorrected concentration" profile, and (3) taking into account and correcting for the wavelength dependence of possible interferences, in particular the interferences due to aerosols.

Signal averaging over consecutive signals is required for all DIAL measurements because smooth representative signals are needed to yield species concentration, which is obtained from the signal derivative as shown in Equation (4.2). Spatial averaging, over neighboring time (range) bins is traditionally performed with either a sliding average, or a polynomial or exponential fitting technique [4]. Although more sophisticated smoothing techniques have been proposed by some authors [5–8], in our experience the improvement they bring about is in most cases insignificant. It has been shown in the literature that the smoothing procedure is critical only in the case of low SNR and when the concentration of the species to be measured varies strongly with the distance [4].

The offset of the signal, which has its origin in the solar background, the detector dark current, and the analog-to-digital pedestal level, can then be subtracted. The range at which the species can be measured is determined, at the near end, by finding the distance of full overlap between the laser beam and the telescope's field of view, and at the far end by finding the distance that corresponds to the lowest acceptable SNR.

At this point, an "uncorrected" concentration profile is determined using Equation (4.2). For species with smooth optical absorption features (O_3, Cl_2) that vary slowly with the wavelength, a large separation between the ON and OFF wavelengths is necessary, so that a correction which takes into account all wavelength-dependent parameters may be applied. It includes extinction and backscattering for both molecular and aerosol components, and absorption by gases with overlapping lines. Occasionally, a careful selection of the ON and OFF DIAL wavelengths can lead to the cancellation or the significant decrease of the hindering effect of interfering species. A typical example is the interference of SO_2 in the case of DIAL ozone measurement.

The correction of the molecular extinction and backscattering is performed either by climatological models, or by independent measurement (e.g., balloon borne). The correction of the aerosol interference is particularly difficult because modeling is impossible in polluted environment, and the lidar signal itself is not sufficient to unambiguously retrieve the aerosol microphysical parameters. Strong aerosol loading and gradients are one of the characteristics of polluted air masses, and, therefore, represent a major source of uncertainty in DIAL ozone measurements in the PBL. Several aerosol correction schemes are outlined in the following text.

The algorithm proposed in Ref. [9] is the most widely used aerosol correction scheme. It makes use of the Fernald–Klett method at less absorbed wavelengths (λ_{OFF}) to deduce the correction for the differential backscatter that is most significant in inhomogeneous aerosol layers. It has the drawback of requiring three parameters that must be estimated "a priori": the extinction-to-backscatter ratio or the so-called "lidar ratio" profile, the backscattering coefficient β_a^{ref} at a reference altitude for λ_{OFF}, and the backscattering wavelength dependence. These uncertainties can lead to an error of 100% or more in the ozone concentration retrieval [4]. The correction is improved when fewer parameters have to be estimated. This is the case when more than two wavelength channels (elastic or Raman) are recorded. For example, the aerosol correction can be performed using additional Raman signals from atmospheric nitrogen [10]. Alternatively, in order to diminish or cancel the aerosol influence, a multiwavelength, and "dual-DIAL" schemes have been proposed [11,12]. Another way of solving this uncertainty is the application of the Raman DIAL approach: ON and OFF DIAL wavelengths are obtained by spontaneous Raman scattering from atmospheric oxygen and nitrogen. The advantage of the method is that the ratio of the backscatter coefficients is constant, that is, term B in Equation (4.2) is zero, and the error caused by differential aerosol extinction is twice smaller compared to the classical elastic DIAL. In this case, the aerosol influence may be neglected and species such as ozone may still be detected in regions with strong aerosol gradients in polluted air masses.

Some important trace gas species determined by DIAL are presented in the next section, considering the wavelength range (UV or IR) and the most commonly measured gas. Ozone will be discussed in a separate section at the end of this chapter.

4.4. DIAL MEASUREMENTS IN THE UV (200 TO 450 NM)

4.4.1. Nitrogen Oxides

Nitrogen oxides form when fuel is burned at high temperatures, as in a combustion process. The primary sources of NO_x, the sum of NO and NO_2, are motor vehicles, electric utilities, and other industrial, commercial, and residential sources that burn fuels. NO_x causes a wide variety of health and environmental impacts such as: ground-level ozone formation in the presence of volatile organic compounds and solar radiation, or acid rain in the presence of ammonia, moisture, and other compounds.

DIAL measurements of nitric oxide, NO, were first performed in the mid-IR with a frequency-doubled CO_2 laser system [13]. At present, it is preferably measured in the UV taking advantage of the strong absorption cross section of the γ bands around 226 nm and the availability of convenient tunable lasers. Due to the very narrow bands in the NO absorption spectrum, NO DIAL measurements are usually performed using tunable dye lasers [14–16].

Several other experimental setups have been employed for DIAL measurements of NO [16], among which one particular experimental configuration is based on a solid-state laser source [17]. This technique takes advantage of the second harmonic of an Nd:YAG laser pumped Ti:Sa laser, whose output near 790 nm is sum-frequency-mixed with the fundamental of another Nd:YAG laser. The resulting beam at 453.6 nm (ca. 30 mJ/pulse) is frequency-doubled in a barium borate (BBO) crystal. In order to match the very narrow absorption lines of NO, the linewidths are reduced to ca. 2 pm by using a grazing incidence grating in the resonator of the Ti:Sa laser, and an intracavity etalon is employed in the second Nd:YAG laser. The wavelength is switched pulse-to-pulse by tilting the tuning mirror in the Ti:Sa laser cavity. Two BBO doubling crystals are used for the 226.812 nm ON and the 226.824 nm OFF wavelength generation. A 2D mapping of a diesel engine plume containing about 100–300 ppb NO was reported, with a range resolution of 20 m, a measurement range of 200 m, and sampling time of 35 min.

Nitric dioxide, NO_2, is preferably measured in the UV near 445 nm because of the availability of tunable lasers, the relatively high absorption cross section, and the absence of overlapping absorption spectra from other pollutants. NO_2 was at first measured with flashlamp-pumped dye lasers [18] and then with excimer [14] or Nd:YAG [19,20] pumped dye lasers.

All-solid-state systems for NO_2 monitoring have been reported in Refs. [21,22]. The setup presented in Ref. [21] is identical to the one described in Ref. [17] for NO, except that the frequency doubling is not necessary. The reported detection limit is 0.2 ppm, with a spatial resolution of 12 m. The DIAL described in Ref. [22] also employs a transmitter based on mixing Ti:Sa and the Nd:YAG fundamental wavelengths. The spectral range is extended from UV to visible (250–480 nm) by using the second harmonic of the mixed wavelength and the second and the third harmonics of the Ti:Sa laser that are produced with two nonlinear crystals. The system was designed for NO_2, O_3, SO_2, and benzene and toluene measurements but only NO_2 measurements with lower detection limit of 28 ppb and a spatial resolution of 240 m have been reported so far.

4.4.2. Sulfur Dioxide, SO_2

The DIAL technique is easily adapted for measurements of sulfur dioxide in the 300 nm spectral region for several reasons including the availability of high-power lasers and the lack of narrow band interfering absorbing species. The presence of ozone must be taken into account, but it has a low differential absorption cross section when integrated over the linewidth of the SO_2 absorption. When using the suitable absorption lines in the solar-blind spectral region below 300 nm, the detection sensitivity is found to be about $5 \, mg/m^3$ [23].

Sulfur dioxide has been measured by DIAL since 1975, and many measurements have been reported since that time (see Ref. [24] for a review of the early measurements). Following the reduction of sulfur-containing fuels in the Western countries, the interest has shifted to ozone in urban case studies in these regions. There is still interest in the monitoring of SO_2 near industrial plants, or to measure the natural SO_2 emissions from volcanoes.

Early SO_2 measurements (1980) were reported using a scanning ground-based UV-DIAL as a means to map the surface impact of an industrial plume from a nearby power plant [25] (SRI, Menlo Park, CA). In another configuration, the lidar system was composed of two 7-Hz Nd:YAG pumped dye lasers, operated at 296.17 nm (ON-line) and 297.35 nm (OFF-line) [26] for the SO_2 determination, while a simultaneous measurement of the wind was taken with a 14×14 element phased-array Doppler sodar. These measurements were taken in Cubatao, an industrial center located 40 km southeast of Sao Paulo, an area known for its photochemical smog, which accumulates because of the coastal hills, the sea breeze, and the frequent temperature inversion layers that prevent the polluted air masses from being transported upwards. The measurement confirmed the most important accumulation of SO_2 at a height of ca. 600 m above the ground, which coincides with a temperature inversion layer. High aerosol concentrations that were associated with the accumulation of SO_2 were measured as well. The simultaneous measurement of wind together with the SO_2 DIAL data gave additional information on the SO_2 pollutant flux.

The mapping of a SO_2 plume from the volcano Etna, Italy, in September 1992 is also reported [27,28]. It was obtained with a system composed of two 20-Hz Nd:YAG pumped dye lasers, operating at 300.02 nm (ON-line) and 299.30 nm (OFF-line). The DIAL instrument is mounted on the board of a ship, and performs vertical measurements while the ship passes through the plume transversely. Ground-based wind measurements on the volcano site at 2500 and 3000 m above sea level (ASL) allowed the estimation of a total SO_2 flux of 1300 tons per day. Such geophysical gas emissions can be monitored using optical techniques including DIAL, but also by differential optical absorption spectroscopy (DOAS), diode laser spectroscopy, and gas correlation imaging [29,30].

A dual-DIAL method for the detection of SO_2 with accuracy of the order of ppb has been reported recently in Ref. [31]. The system is based on Nd:YAG pumped dye lasers operated at three different wavelengths: one wavelength is centered on a SO_2 absorption transition, and the other two are tuned apart from the absorption line. The multiwavelength method employed in this research reduces significantly the effect of ozone and aerosols interference on the SO_2 retrieval [32,33].

4.4.3. Chlorine, Cl_2

Some of the gases encountered in polluted atmospheres include Cl_2, HCl, and a number of chlorinated organic solvents, pesticides, and herbicides. Cl_2 is widely used in the chemical and plastics industries, water and sewage treatment plants, and swimming pools. Except in the vicinity of major pollution sources, atmospheric concentrations of chlorine compounds are low. Normal industrial emissions are not excessively high, but accidental release of the gas, which is shipped in large railroad tank cars, can be a threat to human health. These reasons motivated the first and only UV-DIAL measurement of Cl_2 from the Lund Institute of Technology [34]. Since high chlorine concentrations are only found in accidental situations, the measurement system is difficult to test.

Nevertheless, there have been conducted two types of tests which yielded realistic results: during the first, the Cl_2 gas was injected in an open-ended box, and during the second the gas was emitted as a Cl_2 plume and the laser beam was passed through the plume. In both cases, the DIAL instrument was located 350 m from the target. The measurements had to be performed quickly since Cl_2 is twice as heavy as air and therefore settles rapidly on the ground. The Cl_2 measurement sensitivity with the 298 to 308 nm wavelength pair is estimated to be 170 $\mu g/m^3$ for a 250 m integrating path. This can be compared with the NO measurement sensitivity of 20 $\mu g/m^3$ for the same integration path. With standard models, the differential extinction molecular offset is estimated as 0.32 mg/m^3, and the aerosol differential extinction as 0.36 mg/m^3 for 10 km visibility and 1.8 mg/m^3 for 4 km visibility.

4.4.4. Aromatic Hydrocarbons

Benzene (C_6H_6) and toluene (C_7H_8) can be measured by a UV-DIAL in the 240 to 260 and the 250 to 270 nm wavelength regions, respectively, where they show narrow absorption features. The wavelength pairs which minimize the interference of ozone, SO_2, NO_2, and other hydrocarbons are the 266.9 nm ON-line and the 266.1 nm OFF-line wavelengths for toluene, and the 252.9 nm ON-line and the 252.0 nm OFF-line wavelengths for benzene. The existing experimental configurations include Nd:YAG pumped dye lasers with frequency doubling [35]. Measurements of toluene plume during road tanker loading have been reported. The lidar measurement accuracy compared with ground-based point monitors is estimated to be of the order of 10 ppb. This compares with typical concentrations of 20 to 200 ppb in urban conditions, and 100 ppb to 1 ppm in industrial conditions.

4.4.5. Mercury, Hg

Lidar systems for Hg have been developed because volatile mercury is an important industrial pollutant that is known to present a hazard to human health. Mercury has strong narrow absorption bands near 250 nm (ca. 0.01 nm FWHM) that are well suited for UV-DIAL measurement. The detection limit is of the order of 2×10^{-3} to $3 \times 10^{-3} \mu g/m^3$ [36]. A recent review on field experiments regarding atomic mercury emissions from geothermal field in Iceland and Italy can be found in Ref. [30].

Mercury was also monitored in the mining sector of Almaden, Spain [37], with a system composed of two 20-Hz Nd:YAG pumped dye lasers, operating at 253.65 nm (ON-line) and 253.66 nm (OFF-line). Almaden has the world's largest natural deposit of mercury (estimated to contain about 250,000 tons before mining, which represents one third of the known resources on Earth). Atmospheric mercury concentrations from 0.1 to $5 \mu g/m^3$ have been detected close to the smelting plant, near the ventilation outlet from mines, and above old deposits of roasted cinnabar. Using the DIAL data combined with wind data, a total mercury flux into the atmosphere of 800 g/h from all sources was evaluated with 25% uncertainty.

4.5. DIAL MEASUREMENTS IN THE NEAR-IR (1 TO 5 μm)

4.5.1. Volatile Organic Compounds (VOCs)

It is now recognized that methane contributes strongly to the greenhouse effect, and therefore there is a need to identify the sources and the transport processes for this gas. Methane has suitable absorption bands for DIAL in the near-IR (1.5 to 4 μm) spectral region. With a $Co:MgF_2$ laser, the absorption of methane at the on-line wavelength of 1.6713 μm was demonstrated, but no attempt was made towards retrieval of CH_4 concentrations [38]. Airborne DIAL measurements of methane have been reported [39] where the DIAL wavelength pair used is 3.313 μm (ON-line) and 3.309 μm (OFF-line). Only integrated vertical profiles using the surface return were evaluated, but the system was able to detect methane plume downwind of a cattle feed lot. There have also been obtained range resolved profiles from a ground-based station [40,41].

DIAL measurements of methane have been reported in the case of a methane leakage. In this experiment, a Raman converter combined with a Ti:Sa laser pumped by the second harmonic of an Nd:YAG laser was used as a tunable source at 1.67 μm, and a CH_4 leakage of 6000 ppm at a distance of 130 m was successfully detected [42]. A variety of alkanes and alkenes were also reported to have been measured in the mid-IR [41].

4.5.2. Hydrogen Chloride, HCl

Hydrogen chloride can be measured with a deuterium fluoride (DF) laser that is line-tunable. A coincidence with the absorption spectrum of HCl is found at 3.636 μm (ON-line) and 3.698 μm (OFF-line). DIAL measurements of hydrogen chloride released from waste incineration ships have been performed [43,44]. Attempts toward DIAL measurement of hydrogen chloride were made with a $Co:MgF_2$ laser [38]. The absorption of HCl was demonstrated on an integrated path at the on-line wavelength of 1.7525 μm, but no absolute concentration was retrieved.

4.6. DIAL MEASUREMENTS IN THE MID-IR (5 TO 11 μm)

Most of the polyatomic molecules have well-defined and sharp features in the mid-IR (5 to 11 μm) spectral region. Therefore, this region is particularly attractive for species identification. Because of the much lower backscatter of the atmosphere in the mid-IR, the measurements are performed by using topographical targets. The lack of suitable laser sources and the low sensitivity of the detectors further complicate the use of the DIAL technique. In addition, the great number of absorption lines and their spectral broadening makes species' identification difficult because of the line overlap. At present, DIAL measurements in the mid-IR are performed mostly using line-tunable transverse excited atmospheric (TEA) CO_2 lasers, and rely on the coincidence of one of the laser lines with the absorption spectrum of the target species.

4.6.1. Sulfur Hexafluoride, SF_6

The artificial release of sulfur hexafluoride is used as an atmospheric tracer to map the diffusion of pollutants. It can be measured at 10.6 μm with the P_{18} (SF_6 ON-line) and the P_{30} (SF_6 OFF-line) lines of a CO_2 laser. An airborne coherent CO_2 DIAL has been developed [45,46]. The two CO_2 lasers are line-tunable between 9 and 11 μm, with pulse energy between 80 and 500 mJ, pulse duration of 100 nsec, and a repetition rate of 2 Hz. The detection sensitivity is estimated to be 20 ppb of SF_6. This system was used to map sulfur hexafluoride tracer gas released at locations from which the use of visible aerosol tracers is limited (e.g., nuclear power plants). Integrated SF_6 measurements have been reported with a hard target retroreflector used to enhance the measurement range to 5 km along a horizontal path (averaged over a few minutes). The system consists of two 30-Hz CO_2 lasers that deliver 60 to 200 mJ/pulse [47].

4.6.2. VOCs

Hydrazine (N_2H_4), monomethylhydrazine (MMH, $CH_3N_2H_3$), and unsymmetrical dimethylhydrazine (UDMH, $(CH_3)_2N_2H_2$) are known to be highly toxic and volatile hydrocarbons. Among other applications, they are used as rocket fuels. Thus, remote detection is needed in the case of accidental release. The DIAL measurement of these species has been tested in a laboratory configuration [48]. As for Cl_2, the conditions for remote measurement of toxic compounds are difficult to reproduce in test experiments. A dual CO_2 DIAL system is used in conjunction with a large optical tank, which contains the toxic hydrazine compounds. The CO_2 lasers used in the experiment have a linewidth of the order of $0.1 \, cm^{-1}$, output energy of 10 mJ/pulse, a pulse length of 100 nsec, and operate at 15 Hz repetition rate. Lidar returns have been obtained with a topographic target at a range of 2.7 km. The study concludes that the DIAL monitoring of hydrazine compounds is possible in real time (which is needed because of the high reactivity of this species in air) with a path-averaged sensitivity of 100 ppb over 5 km.

4.6.3. Ammonia, NH_3

Atmospheric ammonia is produced by a combination of natural and anthropogenic sources. Its concentration influences the rates of formation, transport, transformation, and removal of sulfur and nitrogen aerosols, and therefore plays an important role in the acid rain formation. Only closed-path integrated CO_2 DIAL measurements of ammonia have been reported [49]. Path-averaged ambient NH_3 concentrations between 5 ppb (the sensitivity limit) and 20 ppb could be detected over a range of 2.7 km. The concentrations of ambient NH_3 were found to be inversely correlated with the relative humidity.

4.7. TROPOSPHERIC OZONE AS A SPECIAL CASE STUDY

Tropospheric ozone is an important trace gas in atmospheric chemistry: it constraints the oxidation capacity of the atmosphere by playing an important role in the OH radical formation. Since it is the main driver of the photochemical processes, ozone contributes to the recycling of most of the natural and anthropogenic gasses emitted into the atmosphere. Tropospheric ozone is also an important greenhouse gas due to its strong IR absorption band. In the PBL, ozone has direct effect on human health and causes significant losses in agriculture. Finally, tropospheric ozone has increased considerably in the last decades because of the photochemical transformation of anthropogenically emitted ozone precursors such as nitrogen oxides, methane, carbon monoxide, and nonmethane hydrocarbons. Therefore, ozone is one of the trace gas species of important concern in the troposphere.

There are two bands in the ozone spectra suitable for DIAL measurements; the UV band centered at 256 nm and the mid-IR band lying between 960 and 1070 cm^{-1}. In the initial ozone DIAL developments in the early 1970s, the IR band was preferred mostly because of the availability of powerful CO_2 lasers operated around 9.5 µm. However, because of the weaker backscatter of the atmosphere, those lidars were operated only with topographical targets. Employing heterodyne detection makes real lidar-mode operation possible but the complexity of the system, the poor spatial resolution, and the appearance of powerful UV laser sources lead to the preferred use of UV ozone DIAL systems.

The ozone absorption spectrum [50–52] in the UV region (Figure 4.1) has a broad continuum from 200 to 350 nm, known as the Hartley band, with a maximum near 255 nm. On this continuum, weak diffuse Hartley bands spaced approximately $200\,cm^{-1}$ are superimposed. At wavelengths above 310 nm, the spectrum is dominated by vibrational Huggins bands. Wavelengths shorter than 250 nm are not suitable for lidar measurements mostly because of difficulties in producing and handling the laser radiation in this region of the spectra and because of the increased absorption by other

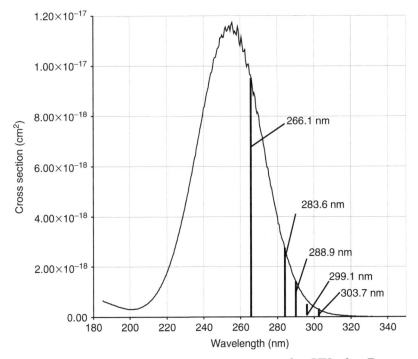

Figure 4.1 Ozone absorption spectrum in the UV: the Raman shifted wavelengths given in Table 4.2 are indicated in this figure.

species. Wavelengths near the maximum of the absorption spectra can give the largest differential cross sections yielding high sensitivity and high accuracy. However, these wavelengths will be strongly attenuated in the PBL and are therefore unsuitable for probing the high troposphere. The light sources in the wavelength band 280 to 300 nm are better suited for ozone measurements in the upper troposphere but have lower ozone differential absorption cross section. In any case, one wavelength pair is insufficient for an accurate retrieval of ozone concentrations up to the tropopause. Since most of the ozone DIAL instruments are operated in the UV spectral region, the so-called "solar-blind region," they may easily be operated under daylight conditions.

For the DIAL measurement of ozone in the stratosphere, wavelengths above 300 nm are used. In this wavelength region, ozone has a smaller absorption cross section, thus allowing lidar measurements at altitude range where higher ozone concentrations are present.

Since the UV ozone absorption spectrum lacks narrow and sharp structures, a large separation of the wavelength pair is required. However, such wavelength separation increases the systematic errors caused by the different extinction and backscatter properties of the atmosphere due to aerosols, interfering gases, and molecules at the DIAL wavelengths. The most significant systematic error in the regions with inhomogeneous aerosol distribution comes from the differences in the aerosol backscatter at the DIAL wavelengths [53]. Assuming that the molecular and aerosol volume backscattering coefficients vary inversely with the wavelength to the power of 4 and γ (where γ is the Angstrom coefficient), respectively, the differential backscatter error (term B in Equation 4.2) of the ozone measurement can be presented approximately as [9,55]

$$B \approx \frac{\Delta\lambda}{\lambda_{OFF}\Delta\sigma} \frac{(4-\gamma)}{2} \frac{d}{dR}\left(\frac{S}{1+S}\right) \tag{4.4}$$

since $\Delta\lambda = \lambda_{OFF} - \lambda_{ON} = \lambda_{OFF}$. S is the so-called aerosol backscatter ratio $S = \beta_{aer}/\beta_{mol}$ and $\Delta\sigma = \sigma_{ON} - \sigma_{OFF}$ is the differential absorption cross section of ozone.

The systematic error due to the aerosol extinction T_{aer} (term C in Equation 4.2) can be estimated assuming a power law dependence of the corresponding extinction coefficients $a_{aerON} = a_{aerOFF}(\lambda_{ON}/\lambda_{OFF})^K$ [9,54]:

$$T_{aer} = \frac{\alpha_{ON} - \alpha_{OFF}}{\sigma_{ON} - \sigma_{OFF}} \approx -\alpha_{OFF}K\frac{\Delta\lambda}{\lambda_{OFF}\Delta\sigma} \tag{4.5}$$

where K is the Angstrom coefficient. In Equations (4.4) and (4.5) both errors B and T_{aer} depend on the same parameter $\Delta\lambda/\lambda_{OFF}\Delta\sigma$, which in turn depends entirely on the ON and OFF wavelengths selection. Reducing the value of this parameter by choosing the proper wavelengths will lead to reducing the systematic errors due to aerosol influence. The systematic error T_X due to gases other than O_3 that absorb at the DIAL wavelengths is given by

$$T_X(R) = -\frac{\Delta\sigma_X}{\Delta\sigma_{O_3}}n_X(R) \tag{4.6}$$

where $n_X(R)$ is the concentration of the interfering gas X, with a differential absorption cross section $\Delta\sigma_X$. Interfering species such as SO_2, NO_2, and HCHO have absorption bands in the UV spectrum of the ozone DIAL wavelength region. They can be found at concentration values that are high enough in the mixed layer to cause significant systematic errors on the ozone concentration retrieve. Thus, the selection of wavelengths is a compromise between all those contradictory requirements [54–57] and depends on the available laser sources.

The existing DIAL systems for ozone measurements based on elastic backscatter from the atmosphere differ mostly on the employed laser source. In general, they can be classified into two major groups, as shown in Figure 4.2: systems based on tunable or fixed-frequency laser sources.

Tunable laser sources are attractive since they offer the possibility of selecting the DIAL pair of wavelengths so that optimum sensitivity is achieved with minimum aerosol and interfering gas influence. An additional advantage is the possibility to use the "null profiling" method for system

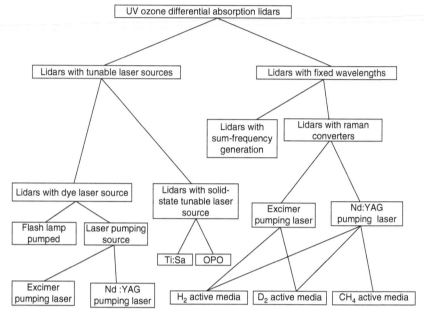

Figure 4.2 Classification of the UV ozone DIALs.

diagnostics and error evaluation [21]. This is obtained by tuning the two laser sources at the same wavelength. The tunable sources in all the early systems and in some of the recently developed are frequency-doubled dye lasers pumped either by the second harmonic of an Nd:YAG laser [54,55,57–59] or by an excimer (XeCl) laser [60]. The wavelength range in which the existing systems are operated is from 272 to 310 nm typically with a wavelength separation ranging from 5 to 20 nm. Most of the systems employ two separate dye lasers. The main drawbacks of the lidars based on dye lasers are the complex maintenance and the need for sophisticated feedback systems for maintaining the wavelength stability.

Solid-state tunable laser sources such as Ti:Sa (third harmonic) and optical parametric oscillator (OPO) still do not find wide application mostly because there are not yet any commercially available devices. The only lidar [61] based on a tripled Ti:Sa laser uses a flash lamp pumped laser in an oscillator–amplifier configuration. The ON and

OFF wavelengths are generated sequentially in a double oscillator. A rotating cube whose speed is synchronized with the oscillator Q switch makes the switching between the DIAL wavelengths. The OPO-based lidar transmitters are still in an experimental phase and use specially developed laser sources like the one reported in Ref. [62], where the ON and OFF wavelengths are produced by mixing the outputs of two IR OPOs with the third harmonic of an Nd:YAG laser. The OPOs are pumped by a part of the energy of the same Nd:YAG laser. The DIAL wavelengths are switched by sending the pump energy to one of the OPOs by a Pockels cell and a polarizing beam splitter on a shot-by-shot basis. The wavelengths generated by this system are 289 and 299 nm.

Most of the recently built systems are based on Raman-shifted outputs of a KrF excimer or a quadrupled Nd:YAG laser. Hydrogen and deuterium are the most frequently used Raman media [63,64]. Nonetheless, in some systems, deuterium hydride [65], methane [66], and nitrogen have been tested. The early systems with Raman converters used KrF as a pump source [4,67–70] because of its high pulse energy and repetition rates. The wavelengths produced by the Raman conversion of KrF radiation, suitable for ozone measurements in the troposphere, are the first 277.1 nm and the second 313.2 nm Stokes of hydrogen and the first 268.5 nm and the second 291.9 nm Stokes of deuterium. A combined conversion wavelength of 302.1 nm obtainable by the conversion in hydrogen of the first Stokes of deuterium was also proposed [68,71]. In some cases, the KrF pump wavelength at 248.5 nm was also used as a DIAL wavelength [4]. Commonly used wavelengths are 277.1, 291.9, and 313.2 nm. Despite the high energy and repetition rates of KrF lasers, their application as pump sources is limited because they use the highly toxic fluorine, with a lifetime of the gas mixture limited to several million shots. In addition, the beam cross section is rectangular and the lasers suffer from poor beam quality even with unstable resonators.

Quadrupled Nd:YAG lasers are widely used as pump sources for Raman converters in the newer DIAL systems since they are much smaller in size and easier to maintain.

They also have better long-term stability, beam quality, and energies sufficient for measurements at repetition rates reaching 100 Hz. The first Stokes of hydrogen (299.1 nm) and deuterium (288.9 nm) and the residual pump (266 nm) are most commonly used as DIAL wavelengths [72–75]. In the lidars reported in Refs. [76,77], the residual pump and the first and second Stokes of deuterium are used. Recently, efficient conversion in nitrogen was achieved and a DIAL system employing the residual pump and the first and second nitrogen Stokes was developed [78,79] and successfully operated in field measurements. The wavelengths that can be produced by Raman conversion of the fourth harmonic of Nd:YAG lasers are listed in Table 4.2.

Since a number of wavelengths can be produced in the solar-blind region, it is possible to select optimal wavelength pairs with maximum measurement sensitivity ($\gg \Delta\sigma_{O3}$) and minimum potential errors due to aerosols ($\ll T_{aer}$) and interfering gases ($\ll T_x$). The lower detection limit (LDL) is inversely proportional to the differential cross section $\Delta\sigma$ of ozone. The values of $1/\Delta\sigma$ in cm^2/mol are shown in Figure 4.3a for the different wavelength pairs that can be obtained by combining the wavelengths from Table 4.2. The values of the parameter $\Delta\lambda/\lambda_{OFF}\Delta\sigma$ are shown in Figure 4.3b and are proportional to the systematic errors caused by the aerosol differential extinction and backscatter for the same wavelength pairs. The second Stokes of deuterium is not considered because of the thermal sensitivity of the ozone cross section in the Huggins band and also because of the high daylight background that makes this wavelength suitable mostly for airborne or nighttime measurements. In Figure 4.3 the wavelength pairs with 266 nm as ON wavelength have lower LDL (low $1/\Delta\sigma$) and are less affected by aerosols (low $\Delta\lambda/\lambda_{OFF}\Delta\sigma$). The pair (288.9/299.1 nm), frequently used for tropospheric measurements, has the poorest LDL and is strongly affected by aerosol influence. The first and second Stokes wavelengths produced in nitrogen (283.6/303.7 nm) have more than twice better measurement sensitivity compared with 288.9/299.1 nm pair and are less affected by aerosol.

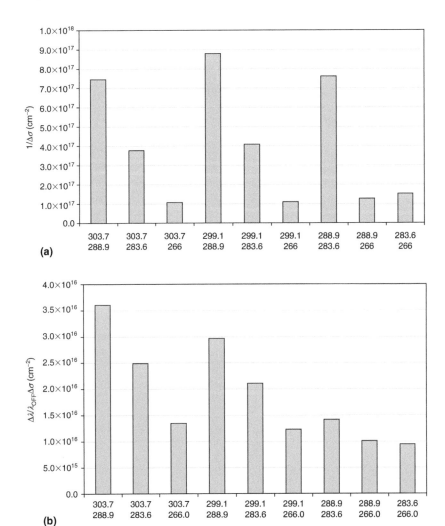

Figure 4.3 (a) The reciprocal of the differential ozone absorption cross-section $1/\Delta\sigma$ proportional to the LDL. The wavelength pairs on the horizontal axis are in accordance with Table 4.2. (b) The values of the parameter $\Delta\lambda/\lambda_{OFF}\Delta\sigma$, proportional to the systematic errors caused by the aerosol differential extinction and backscatter for the same wavelength pairs as in (a).

Table 4.2 Raman Shifted Wavelengths Suitable for Ozone DIAL
Observations in the Troposphere

Active gas	All	N_2	D_2	H_2	N_2	D_2
Raman component	Residual pump	I Stokes	I Stokes	I Stokes	II Stokes	II Stokes
Wavelength (nm)	266	283.6	288.9	299.1	303.7	316.3
Cross section (cm^2)	9.37×10^{-18}	2.92×10^{-18}	1.60×10^{-18}	4.48×10^{-19}	2.39×10^{-19}	4.38×10^{-20}

The values of the ratio $\Delta\sigma_X/\Delta\sigma_{O3}$ which is proportional to
the systematic error T_X (Equation 4.6) caused by interfering
gases for the case of SO_2, NO_2, and HCHO and the same
wavelength pairs are shown in Figure 4.4. These constituents
have absorption bands in the ozone DIAL wavelength region
and can be found in the mixed layer in concentration levels
high enough to cause significant systematic errors. The
biggest systematic errors are caused by SO_2 followed by

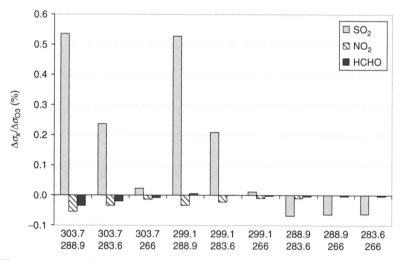

Figure 4.4 Influence of interfering gases on the ozone retrieval for
the wavelength pairs shown in Figure 4.3 the ratio $\Delta\sigma_X/\Delta\sigma_{O3}$ for
SO_2, NO_2, and HCHO is proportional to the systematic error T_X
(Equation 4.6).

NO_2. The SO_2 influence is smaller for pairs using 266 nm as ON wavelength. Therefore, those wavelength pairs are more suitable for measurements in the mixing layer where the SO_2 concentrations can reach high values. The presence of SO_2 causes a substantial error of -0.53 ppb O_3 per 1 ppb SO_2 for the frequently used wavelength pair 299.1/288.9 nm. This error is more than twice smaller for the pair 283.6/303.7 nm corresponding to the first and second Stokes of N_2.

For these reasons, we have decided to upgrade the EPFL mobile lidar with a transmitter based on Raman shifted in N_2 fourth harmonic of an Nd:YAG laser. The lidar is presented as an example of a UV ozone DIAL system. The optical layout of the lidar is shown in Figure 4.5. The residual pump (266 nm), and the first (283.6 nm) and second (303.7 nm) Stokes of nitrogen are used as three different DIAL wavelengths. In the

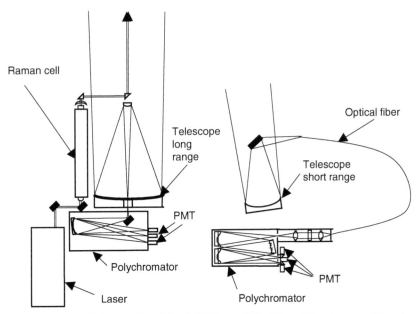

Figure 4.5 Schematic of the EPFL mobile lidar. The transmitter is based on Raman conversion of the fourth harmonic of an Nd:YAG laser in N_2 and simultaneously delivers three coaxial DIAL wavelengths: 266, 283.6, and 303.7 mm.

receiving part, a dual-telescope configuration is employed in order to reduce the dynamic range of the signals. The typical working distances for the "short range" and "long range" receivers are correspondingly 100 to 1200 m and 600 to above 7000 m.

The laser used in the transmitting part of the system is "Continuum-Powerlight-8000" with an output energy at the fourth harmonic (266 nm) of 100 mJ, repetition rate of 10 Hz, and pulse duration of 6 nsec (FWHM). The laser beam is linearly polarized with a beam diameter and divergence of 8 mm and 0.5 mrad, respectively. The cell is single pass, 1.2 m long, with a 15 mm thick, 600 mm focal length lens as an input window and a 15 mm thick, and flat output window. The cell is filled with pure N_2 at a pressure of 30 atm. The three wavelengths suitable for ozone DIAL measurements are generated in the N_2 converter. They are produced simultaneously and coaxially with a similar spatial distribution; therefore they probe the same volume of air at the same time, which reduces the systematic and random errors due to atmospheric turbulence and inhomogeneity. The output of the cell is recollimated by a 60 cm focal length plano-convex lens and transmitted into the atmosphere coaxially to the "long range" receiving telescope. The latter is 60 cm in diameter, f/8.5 Cassegrain type. The spectral separation of the signals in the "long range" receiving channel is carried out by a specially designed grating polychromator based on a UV-enhanced flat field imaging grating. The aperture ratios of the receiving telescope and the polychromator match without additional optics, which increases the total efficiency of the receiver. The field of view of the receiver is defined by an entrance diaphragm located in the focal plane of the telescope and is equal to 0.6 mrad with a 3 mm diaphragm, which ensures good initial daylight suppression. To further reject the daylight that may significantly affect the ozone retrieval, a solar-blind filter is placed before the diaphragm. The optical signals at the three wavelengths (266, 283.6, and 303.7 nm) are detected by three "Hamamatsu–5783-06" photosensor modules. To compensate for the spatial nonuniformity of the photosensor, an additional system consisting of a short focal length lens and a diffuser is used [80]. The "short range" receiver is equipped with a 20 cm, f/3 Newtonian telescope. The light collected by the telescope is delivered by a 2 mm diameter fused silica fiber to a polychromator for spectral separation. The latter is a commercial Czerny-Turner 500 mm grating monochromator converted to a polychromator. The resolution of the polyhromator is 1 nm/mm with 1200 grating/mm. The optical signals are detected by photosensor modules of the same type as in the "long range" receiver. A six-channel 20-MHz, 12-bit transient digitizer is used for the data acquisition. LabVIEW and Matlab programs on a PC perform the control over the transient recorder and the preliminary and final treatment of the data. The lidar is mounted on a van with autonomous cooling and air-condition systems and can be transported easily.

The system was operated in St Chamas, France, close to Marseille, during the international ESCOMPTE field campaign in the summer 2001. During one of the intensive observation periods (IOPs) of the campaign, 144 h of continuous measurements of ozone concentrations were taken with a time resolution of 4 min. The time sequence vertical ozone profile from this period is presented in Figure 4.6. The "short range" measurements were performed with a 266/283.6 nm wavelength pair while the "long range" data were taken with the 283.6/303.7 nm pair. Ozone concentrations simultaneously measured at ground level are shown at the bottom of the same picture separated from the lidar data by a white line.

The observation period can be divided into two distinctive subperiods separated on the figure with a red vertical line. During the first one, starting from the late evening of July 20th to the late afternoon of July 23rd, the predominant transport was from the northwest with average wind speed at ground level of 3.5 m/sec. Relatively clean air masses were transported to the observation site and moderate ozone concentration with mean value of approximately 60 ppb and slight diurnal variations were observed by the lidar in the first 1000 m. Except for the elevated values in the late afternoon of July 22nd low ozone concentrations of 20 to 40 ppb

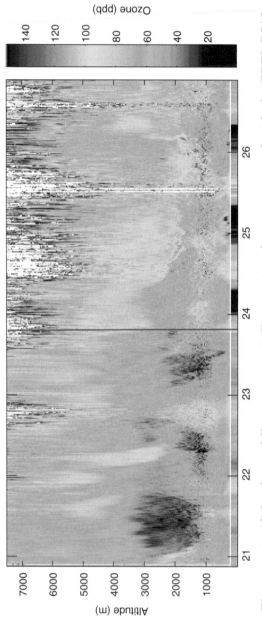

Figure 4.6 (Color figure follows page 398). Time series of ozone measured with the EPFL DIAL mobile system in St Chamas, France 2001. The numbers on the horizontal scale correspond to days of July.

were measured for the altitude range of 1000 to 3000 m for the almost entire period. The concentrations above 3000 m were higher than 60 ppb reaching values of up to 90 ppb.

During the second subperiod, the predominant transport was from the southwest, that is, from the industrial area with primary pollutants reacting to form higher ozone concentrations. The maximum ozone concentrations measured by the lidar during this period exceeded 110 ppb. A formation of a stable layer with constantly increasing in time ozone concentrations was registered. A layer initially formed between 3000 and 5000 m gradually descended and finally blended with the mixed layer at noon on 25th July when the maximum ground ozone concentrations exceeded 110 ppb.

Reduction or even canceling of the systematic errors due to aerosols can be achieved by using more than one wavelength pair as proposed for stratospheric measurements [81]. If we assume the use of three wavelengths $\lambda_1 < \lambda_2 < \lambda_3$, two wavelength pairs $\lambda_1-\lambda_2$ and $\lambda_2-\lambda_3$ can be formed. Then the DIAL equation can be written in the form

$$N = \frac{1}{2\Delta\sigma}[(A_{12} - kA_{23}) - (B_{12} - kB_{23}) - (C_{12} - kC_{23})]$$

(4.7)

where $\Delta\sigma = \sigma_{12} - k\sigma_{23}$ and A, B, and C have the same meaning as in Equation (4.2). The coefficient k is chosen in a way that it cancels the terms B and C (i.e., the systematic errors due to the aerosol). It is shown [81] that this will happen for values of $k = \Delta\lambda_{21}/\Delta\lambda_{32}$, if we assume a power law dependence of the extinction and backscatter coefficients. When $\Delta\lambda_{21} = \Delta\lambda_{32}$ or k is equal to 1, the best compensation is achieved.

It is important to note that the use of more wavelengths will increase the noise and the level of uncertainty in the lidar equation, reducing the accuracy and the operational range. So far, no operational tropospheric lidar systems applying three wavelengths have been reported. A potential candidate is the lidar based on converted in nitrogen fourth harmonic of Nd:YAG. The three wavelengths of this system (266, 283.6, and 303.7 nm) are practically equidistant and hence very suitable for this operational mode. The drawback of such wavelength combination is the strong attenuation of the 266 nm radiation, which limits the operational range to a maximum of 2 km.

By using spontaneous Raman scattering from atmospheric nitrogen and oxygen instead of Raleigh–Mie scattering in a DIAL system, the systematic errors caused by the differential aerosol backscattering (term B in Equation 4.2) can be eliminated since the ratio of the Raman backscattering coefficients is range independent. This method is known as the Raman-DIAL method. It has two approaches. The first is to use a single exciting wavelength and Raman signals from oxygen and nitrogen as ON and OFF DIAL wavelengths. With the second approach, the ON and OFF signals are produced by Raman scattering from nitrogen that is excited by two different wavelengths. The obvious advantage of the first approach is the use of only one laser source, which greatly simplifies the lidar design and operation. Furthermore, the aerosol extinction correction (term C in Equation 4.2) is twice smaller compared to an elastic DIAL and a Raman DIAL with two exciting sources. The reason for the latter is the fact that only backward differential extinction has to be accounted for. The drawbacks are the relatively small differential cross section defined by the small difference in the Raman shifts of oxygen and nitrogen leading to low sensitivity, and the fact that the oxygen signal is approximately three times lower because of the lower oxygen concentration. Lidars based on the single exciting wavelength approach are presented in Refs. [82,83]. In Ref. [82], the sensitivity of the method to both atmospheric and device perturbations by numerical simulation for two different laser sources (a quadrupled Nd:YAG and KrF laser) is investigated theoretically and experimentally. The aerosol optical effect in the planetary boundary layer (PBL), the statistical error, the photomultiplier tube afterpulse effect, the optical cross talk between the two Raman-shifted channels and among elastic and Raman signals are studied in detail. Simultaneous profiles of ozone and water vapor from this work are presented in Figure 4.7. They are measured using, respectively, the Raman nitrogen and oxygen shifted wavelengths as ozone DIAL pairs and the Raman water-vapor shifted wavelength for the water-vapor relative humidity.

The next section will give an example of the interest of combining monitoring techniques with interpretation tools in air pollution science, in particular with the advantage of having defined both measurements and model outputs with similar time and space resolution so that direct comparisons can be performed.

4.8. COMPARISON BETWEEN LIDAR MEASUREMENTS AND MODEL PREDICTIONS

Until very recently, local ground-based measurements were the essential, if not the only, results to be used for comparison with the model results. This is often an ambiguous way to check the accuracy of the models since ground-based measurements lack spatial resolution and can be influenced by local sources while the model results are averaged values over an entire model-grid box. Mesoscale air quality models (either the box or column models — Lagrangian formalism or the 3D-Eulerian models) are based on predefined volumes of typically 1 km range resolution on the horizontal plane and tens up to hundreds of meters resolution for the vertical extension of the model. The total height studied generally includes the top of the PBL and the lower part of the free troposphere. Lidar measurements being per definition volume averaged data, are particularly well suited for tropospheric air quality field studies. The lidar data have similar spatial and temporal resolution with the models and therefore can constitute an essential part of the database for the boundary conditions of the models. Furthermore, lidar data give an optimum way to test the accuracy of the model predictive results. We will see that the model results may also be used to underline the

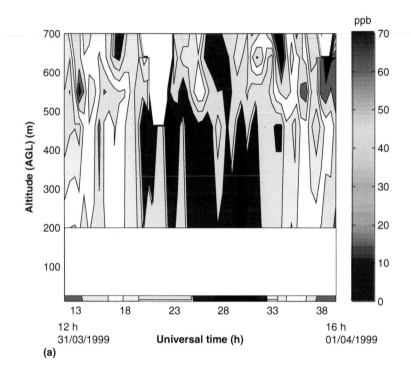

(a)

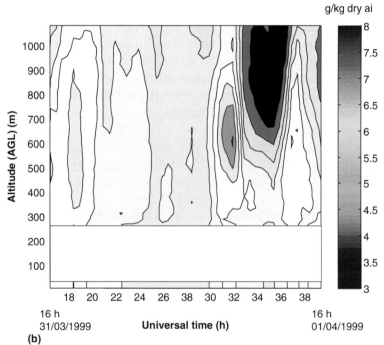

(b)

importance of the pump and probe OH technique, and to define a new indicator for the control of ozone formation.

4.8.1 Grenoble 1999

Grenoble city and its surroundings, in the French Alps, with more than 400,000 inhabitants, are often affected by periods of high ozone concentrations in the summer. In order to understand, predict, and elaborate air pollution abatement strategies against summer photochemical smog episodes, the city authorities and the local health authorities (ASCOPARG) conducted a major field experiment in the summer of 1999. Grenoble is located in the Y shape convergence of three deep valleys surrounded by mountains as high as 3000 m ASL. This topography is the reason for the very complex wind pattern in the region. The study was designed to control and validate the use of a 3D Eulerian mesoscale model specifically adapted for the region [84,85]. The air quality model MEteorological PHOtochemistry MODel (METPHOMOD) [84] was used to investigate the air dynamics and the air pollution in the Grenoble area. Two of the most important components of this model are the chemical mechanism and the transport algorithms. The chemical mechanism available is the socalled reactive atmospheric chemical mechanism (RACM) [86]. The use of this model requires a number of well-defined initial as well as boundary conditions.

The major field campaign involved standard groundbased stations, lidars, an aircraft, DOAS systems, and a microwave wind profiler and was carried out over the model domain in order to obtain the central database required for

Figure 4.7 (Color figure follows page 398). (a) Ozone time series taken with an UV Raman DIAL [82]. The ground valves are measured by a UV photometer. The Raman DIAL has shorter operational range mainly because of the weaker signals, but improved aerosal immunity and simpler transmitter design, allowing simultaneous daytime water vapour (b) measurements with minimal additional equipment.

the model validation. Ozone (O_3) and other pollutants such as NO_x, CH_2O, SO_2 and different VOCs as well as wind, temperature, solar radiation, and relative humidity were measured at ground level and also at different altitudes in the model domain using an instrumented aircraft (Metair Co.), two ozone lidars, two wind profilers (Degreane), and the dense ASCOPARG ground-based network.

Figure 4.8 is a 3D image of the model domain giving a better view of the topography of the region. In the figure, the wind field in the first layer (ground layer) of the model is given as green arrows. They represent the predicted wind field at 9 am on July 27, 1999. The formation of a plume of NO_2 is visualized in blue color, showing only NO_2 values that exceeded 5 ppb in the entire model domain.

Figure 4.9 shows the 2D map of predicted ozone in the ground layer of the model domain at 5 pm. A clear ozone formation downwind from the city center is seen over the southern suburb where the EPFL mobile lidar was located. Two other important ozone plumes are clearly seen in the

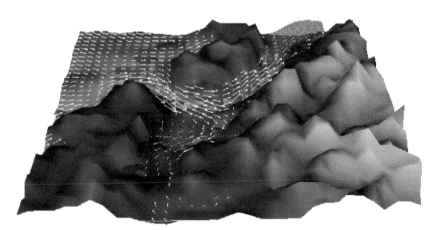

Figure 4.8 Topography of the Grenoble area as defined in the 3D Eulerian model. The wind fields in the first (ground) layer of the model for the case study on July 27, 1999 at 9 am are given as green arrows. The blue "plume" indicates NO_2 concentrations that exceed 5 ppb.

TIME : 27-JUL-1999 17:00

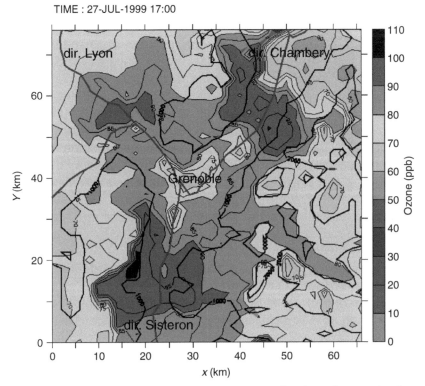

Figure 4.9 (Color figure follows page 398). Predicted ozone in the ground layer of the model domain at 5 pm on July 27, 1999.

northeast of Grenoble, coming from the city of Chambéry and in the northwest coming from the city of Lyon. Due to the ozone titration with intense primary emissions of NO_x in the city center, much lower ozone values were predicted over Grenoble.

Our DIAL instrument was located in the town of Vif (310 m ASL, semiurban area) about 20 km south of Grenoble. At that time the system was based on the standard combination of H_2/D_2 Raman shifted (first Stokes) wavelengths [74,80,87–89] pumped by two separate Nd:YAG fourth harmonic laser sources, that is, with the inherent drawbacks in the wavelengths selection as was previously explained. The lidar results were regularly compared with *in situ* aircraft ozone measurements above Vif. The mixing layer height was estimated from the aerosol backscatter signal at 299 nm and compared with a collocated wind profiler instrument. This allowed the estimation of the turbulence patterns and magnitudes in the complex topography of the valley.

In Figure 10 are given the 3-day time series of ozone over the entire air pollution episode, from July 25 to July 27, 1999. Both the lidar neasurements and the predicted results obtained by Eulerian model are shown. Note the regular ozone increase over the entire period, with the effect of NO_x titration at ground level in the morning hours, as well as ozone dry deposition at night. The formation of a residual ozone layer at higher altitudes is well seen in the measurements as an increase in the ozone concentration but was not predicted by the model.

The direct comparison between experimental and predictive results gives us "a first level of confidence" in the ability of the model to reproduce adequately the air quality over this region. In the future, the local public health authority will be able to propose air pollution abatement strategies that are directly "tested" with the model. The public health

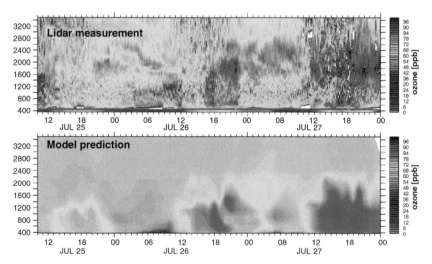

Figure 4.10 (Color figure follows page 398). Comparison between three-day lidar and model time-series, demonstrating ozone formation over Vif, south of Grenoble, from July 25 to July 27, 1999.

authority will know "a priori" the advantages or drawbacks of different possible air pollution abatement scenarios.

The direct comparison between lidar observations and model results is essential for model validation. Nevertheless, in such air quality models hundreds of chemical reactions are included to describe the chemistry, generally the homogeneous chemistry. Since this chemistry is highly nonlinear, such direct comparison between model and measurement results does not prove the perfect adequacy of the model to describe the reality. This is true even if both model and measurement results show good agreement. Therefore, there is a need for other experimental methods to find new ways to test the models, and the pump and probe OH lidar [90–92] is an attempt in this direction. This method starts from the simple idea that when ozone is measured in the UV by DIAL, it is indeed photodissociated, thus giving rise to another interesting atmospheric chemistry at conditions that are out of equilibrium. In this case, we do not measure anymore the *in situ* trace gas concentrations but we rather look at a perturbation technique where new chemical kinetics is gained.

4.8.2. Pump and Probe OH

The principle of the pump and probe lidar can be explained using the example of the OH radical, one of the most important radicals in relation to the greenhouse effect, stratospheric ozone depletion, tropospheric O_3, and acid rain formation. The OH concentration is only about 10^6 per cm^3 at best at say midday, summertime, 45°N, due to its high reactivity. Rather than looking at the more or less steady-state value of [OH] and comparing it with model results, it may be worthwhile looking at a perturbation method in which high concentrations of OH are created in a very short time, and the relaxation kinetics of OH are then followed back to equilibrium. This is like doing flash photolysis in the atmosphere. As compared to steady-state photolysis by sunlight, this approach gives additional kinetic information and simplification. Besides the kinetic simplifications, as one can rule out a number of slow reactions, atmospheric transport can also be less complex, and even eliminated if the temporal scale is small enough. Such a relaxation technique enables a much more direct check on at least an important fraction of the OH atmospheric chemistry.

As a first experiment, a powerful fourth harmonic Nd:YAG laser source is sent into the troposphere. At the 266 nm wavelength, an efficient photodissociation of ozone takes place resulting in the formation of highly excited oxygen atoms, the latter reacting extremely fast with water vapor to form the hydroxyl radical OH ("pump" step). The disappearance of the high transient concentrations of OH radicals due to chemical reaction can then be followed using laser-induced fluorescence (LIF). In this case, the "probe" step consists of the LIF measurement of the OH concentration time decay. This simple system can then be compared in a direct way with model calculations. In such a manner, the reactivity of hydroxyl radical (OH) with total VOCs, $\Sigma k_{VOC}[VOC]$, is directly obtained in the PBL.

As an important result, Figure 4.11 compares the direct experimental (Figure 4.11a) and box model (Figure 4.11b) OH lifetime observed under three different air quality conditions. A gas mixture is continuously flown through the pump and probe reaction cell, with a mixture of NO_x, VOCs, and O_3 added to the air stream at atmospheric pressure and temperature. VOCs are composed of propene, n-butane, and ethene in a ratio of 1:8:3 for total VOC concentrations varying between 125 and 210 ppb in accordance with real atmospheric conditions. NO_x and O_3 concentrations are controlled so that three stable gas mixtures are tested for typical moderately polluted, $rural_{NOx}$, and $rural_{VOC}$. In Figure 4.11a each data point is shown with its experimental uncertainty taking into account the uncertainties associated with the conversion from the LIF signal to the OH concentration, the statistical variability of the fluorescence signal and the determination of the trace gas concentrations estimated at 10% and assumed to be random. For each of the three air stream conditions, the time dependence of the [OH] natural logarithm is approximated by a linear fit. From this linear fit, an experimental lifetime for

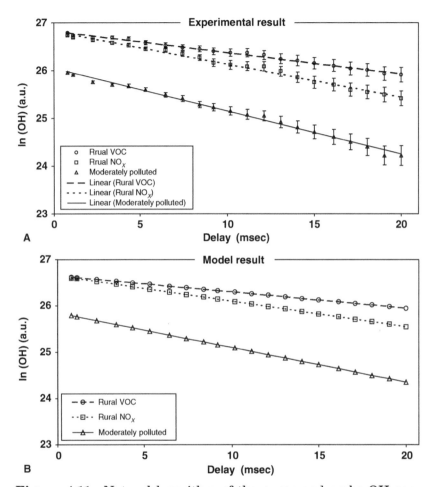

Figure 4.11 Natural logarithm of the pump and probe OH concentration versus time as measured (a) in the reaction cell for three NO_x/VOC/O_3 gas mixtures (respectively, the moderately polluted, rural$_{NO_x}$, and rural$_{VOC}$ conditions) and as predicted (b) with the RACM box model.

OH is obtained for each gas mixture. For the moderately polluted case where O_3 is set at lower initial concentration than in the two other cases, the initial $[OH]_0$ is measured with lower initial concentration than in the two other cases. Figure 4.11b presents the predicted results obtained for the same experimental conditions using RACM, the chemical solver used in the 3D Eulerian model. Both experimental and model OH lifetimes agree fairly well. A detailed discussion about this finding is given in Refs. [91,92].

For specific experimental conditions, the transport by the wind may be neglected, and the chemical processes governing the OH decay expressed by a detailed box model. The OH lifetime τ_{OH} can be expressed in terms of a simple pseudo-first-order chemical mechanism including reactions with CO, O_3, NO_x, and VOCs. The interesting aspect of this approach is that for the first time the direct total reactivity of VOCs with OH in real atmospheric conditions can be expressed.

Let us consider again the Grenoble 1999 case study for which the ozone formation was simulated. A new indicator is proposed for determining if tropospheric ozone production in a specific area of the domain is limited by VOCs or NO_x. The indicator $\Theta = \tau_{OH}^{VOC}/\tau_{OH}^{NO_x}$ describes the ratio of the lifetimes of OH against the losses by reacting with VOC and NO_x. Whereas $\tau_{OH}^{NO_x}$ can be obtained by conventional measurements, the new pump and probe OH approach makes it now possible to obtain τ_{OH}^{VOC} as well. Indicator values Θ above a threshold value of $\Theta = 0.2 \pm 50\%$ are representative of NO_x saturated conditions where an increase of NO_x emissions causes lower ozone production. For values below 0.01, the ozone production is very insensitive to changes of VOC emissions. The robustness of this indicator against several parameters such as temperature, humidity, photolysis, and initial ozone concentrations was initially tested in a box model and compared to the robustness of other earlier proposed indicators. In contrast to earlier proposed indicators, this new one is not based on photochemically produced long lived species but describes the instantaneous regime of an air parcel.

Figure 4.12 shows the ozone iso-concentration in a 2D plot given versus the primary emissions of VOCs and the primary emissions of NO_x, respectively. This is an Eulerian model result that clearly distinguishes two types of atmospheric chemical regimes: a regime where the ozone production is controlled by the presence of NO_x and another, where it is controlled by the presence of VOCs. For example for the "NO_x

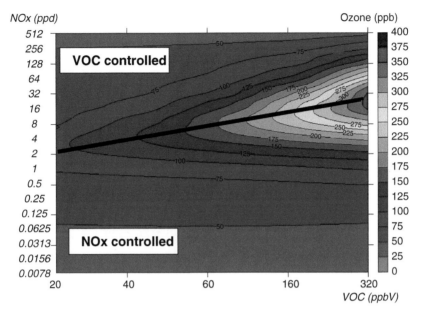

Figure 4.12 (Color figure follows page 398). Predicted ozone concentrations versus the primary emissions of VOCs and NO_x. Two regimes of ozone production are highlighted: one below the straight line where lower concentration of ozone will be produced only if reduced NO_x is emitted ("NO_x controlled regime"), and one above the line where only VOCs reduction strategies will lower the ozone production ("VOCs controlled regime").

controlled regime," lower concentration of ozone will be produced only if reduced NO_x are emitted.

In Figure 4.13 the indicator $\Theta = \tau_{OH}^{VOC}/\tau_{OH}^{NO_x}$ is directly used in the case of the Grenoble study. Now we are able to predict the regions in the ground layer of the model domain where the production of ozone is controlled by the presence of NO_x or VOCs: below $\Theta = 0.2$, ozone concentrations may decrease only if primary emissions of nitrogen oxides decrease. This result is a good example of the powerful combination of new experimental schemes, for example, the pump and probe experiment, together with interpretation tools such as air quality Eulerian model.

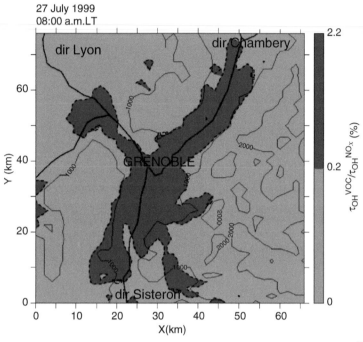

Figure 4.13 Two-dimensional map of the indicator $\Theta = \tau_{OH}^{VOC}/$ $\tau_{OH}^{NO_x}$ values in the case of the Grenoble air quality study. Values that are below $\Theta = 0.2$ are directly indicative of an ozone formation controlled by the presence of nitrogen oxides.

4.9. PERSPECTIVES

Lidar technology has matured over the last decade to an extent that many applications are becoming routine. Meteorological parameters like the wind and temperature, as well as trace gas concentrations, can be measured in short time periods with high spatial resolution in three dimensions, often at large distances. Plumes containing substances like chlorine, mercury, sulfur dioxide, hydrogen chloride, or ammonia can be tracked during their dispersion, and important air pollutants like nitric oxide, nitrogen dioxide, ozone, hydrocarbons, and others can be measured by lidar at strategic sites. This emphasizes one of the more significant uses of lidar data, namely the comparison with sophisticated 3D

Eulerian chemical transport models. Lidar measurements have played, and are continuing to play, an important role next to a multitude of other experimental techniques, in the validation and use of these mathematical models. Conversely, the models can help to point out in a measuring campaign which molecules or meteorological parameters are the best to be measured and where and when they are to be measured.

Future discoveries in lidar field are obviously linked with the development of new laser sources. One of the emerging domains of activity for DIAL application is in the IR wavelength region where new laser sources are developed. The idea of trace gases detection in the IR is not new, and in the past two decades high sensitivity semiconductor-laser absorption spectrometers have been developed for detection of various trace gases in the near-IR range [93–95]. Nevertheless, higher sensitivity and selectivity needed for the background concentrations monitoring can be achieved by working in the mid-IR (the so-called fingerprint region), where the molecular line strengths are the largest. The only available laser sources in the mid-IR so far were difficult to use or required liquid nitrogen cooling. More recently [96,97], quantum cascade (QC) laser sources [98,99] have been produced in a quasi-continuous wave (quasi-CW) mode with a repetition rate of up to several megahertz and a duty cycle of several percent at room temperature. The QC output wavelength is determined by the thickness of the active region and is independent of the band-gap. These lasers can be produced at any wavelength from 4.6 to 13 mm with a tuning range of the order of tens of cm^{-1} and an average power of 1 W or more. The first applications of QC lasers for trace gas sensing at room temperature with pulsed QC laser sources has been reported by applying photoelectric [100,101] or photoacoustic detection [102,103], as well as for routine field measurement [104].

ACKNOWLEDGMENTS

In this chapter, the authors have relied on the results of many groups as indicated by the references cited. We would, however, like to thank our colleagues from the EPFL lidar team who participated in obtaining and processing the data and especially Pablo Ristori, Ioan Balin, and Philippe Quaglia. We wish to acknowledge the valuable inputs of Hubert van den Bergh, Frank Kirchner, Olivier Couach, and Adelina Simeonova. This work was partly funded by the Swiss Federal Office for Education and Science and the Swiss National Research Foundation.

REFERENCES

1. R. M. Measures, *Laser Remote Sensing, Fundamentals and Applications* (Wiley, New York, 1984).
2. L. S. Rothman et al., The HITRAN molecular spectroscopic database: edition of 2000 including updates through 2001, *J. Quant. Spectrosc. Radiat. Transfer*, **82**, 5–44 (2003).
3. R. S. McDowell, Vibrational spectroscopy using tunable lasers, in *Vibrational Spectra and Structure*, J. R. Durig, Ed. (Elsevier, New York, 1981).
4. J. Boesenberg, Ed., Tropospheric Environmental Studies by Laser Sounding, TESLAS-EUROTRAC I Final Report, Max Planck Institut fuer Meteorologie, Hamburg, Germany, pp. 74–79 (1996)
5. R. E. Warren, Adaptive Kalman–Bucy filter for differential absorption lidar time series data, *Appl. Opt.*, **26**, 4755–4760 (1987).
6. R. E. Warren, Concentration estimation from differential absorption lidar using nonstationary Wiener filtering, *Appl. Opt.*, **28**, 5047–5051 (1989).
7. V. A. Kovalev and J. L. McElroy, Differential absorption lidar measurement of vertical ozone profiles in the troposphere that contains aerosol layers with strong backscatter gradients: a simplified version, *Appl. Opt.*, **33**, 8393–8401 (1994).
8. V. Kovalev, M. Bristow, and J. McElroy, Nonlinear-approximation technique for determining vertical ozone-concentration profiles with a differential-absorption lidar, *Appl. Opt.* **35**, 4803–4811 (1996).
9. E. V. Browell, S. Ismail, and S. T. Shipley, Ultraviolet DIAL measurement of O_3 in regions of spatially inhomogeneous aerosols, *Appl. Opt.*, **24**, 2827–2836 (1985).

10. V. Matthias, J. Bösenberg, and V. Wulfmeyer, Comparison of retrieval methods in tropospheric ozone DIAL: the impact of Raman measurements, in *Advances in Remote Sensing with Lidar*, A. Ansmann, R. Neuber, P. Rairoux, and U. Wandinger, Eds. (Springer, Berlin, 1996).

11. V. Kovalev and M. Bristow, Compensational three-wavelength differential-absorption lidar technique for reducing the influence of differential scattering on ozone-concentration measurements, *Appl. Opt.*, **35**, 4790–4797 (1996).

12. Z. Wang, J. Zhou, H. Hu, and Z. Gong, Evaluation of dual differential absorption lidar based on Raman-shifted Nd:YAG or KrF laser for tropospheric ozone measurements, *Appl. Phys. B*, **62**, 143–147 (1996).

13. N. Menyuk, D. Killinger, and W. E. DeFeo, Remote sensing of NO using a differential absorption lidar, *Appl. Opt.*, **19**, 3282–3286 (1980).

14. H. J. Kölsch, P. Rairoux, J. P. Wolf, and L. Wöste, Simultaneous NO and NO_2 DIAL measurement using BBO crystals, *Appl. Opt.*, **28**, 2052–2056 (1989).

15. M. Alden, H. Edner, and S. Svanberg, Laser monitoring of atmospheric NO using ultraviolet differential absorption techniques, *Opt. Lett.*, **7**, 543–545 (1982).

16. H. Edner, A. Sunesson, and S. Svanberg, NO plume mapping by laser-radar techniques, *Opt. Lett.*, **13**, 704–706 (1988).

17. R. Toriumi, H. Tai, H. Kuze, and N. Takeuchi, Tunable UV solid-state lidar for measurements of nitric oxide distribution, *Jpn. J. Appl. Phys.*, **38**, 6372–6378 (1999).

18. K. W. Rothe, U. Brinkmann, and H. Walther, Applications of tunable dye lasers to air pollution detection: measurement of atmospheric NO_2 concentrations by differential absorption, *Appl. Phys.*, **3**, 115–119 (1974).

19. K. Fredriksson and H. M. Hertz, Evaluation of the DIAL technique for studies on NO_2 using a mobile system, *Appl. Opt.*, **23**, 1403–1411 (1984).

20. N. Cao, T. Fujii, T. Fukuchi, N. Goto, and K. Nemoto, Estimation of differential absorption lidar measurement error for NO_2 profiling in the lower troposphere, *Opt. Eng.*, **41**, 218–224 (2002).

21. R. Toriumi, H. Tai, and N. Takeuchi, Tunable solid-state blue laser differential absorption lidar system for NO_2 monitoring, *Opt. Eng.*, **35**, 2371–2375 (1996).

22. M. Dell'Aglio, A. Kholodnykh, R. Lasandro, and O. De Pascale, Development of a Ti:sapphire DIAL system for pollutant monitoring and meteorological applications, *Opt. Lasers Eng.*, **37**, 233–244, (2002).

23. A. -L. Egeböck, K. A. Fredriksson, and H. M. Hertz, DIAL techniques for the control of sulfur dioxide emission, *Appl. Opt.*, **23**, 722–729 (1984).

24. K. A. Fredriksson, Differential absorption lidar for pollution mapping, in *Laser Remote Chemical Analysis*, R. M. Measures, Ed. (Wiley, New York, 1988).

25. J. G. Hawley, L. D. Fletcher, and G. F. Wallace, Ground-based ultraviolet differential absorption lidar (DIAL) system and measurements, in *Optical and Laser Remote Sensing*, D. K. Killinger and A. Mooradian, Eds. (Springer, Berlin, 1983).

26. U.-B. Boers, J. Glauer, S. Köhler, T. Bell, and C. Weitkamp, Measurment of Sulfur Dioxide, Aerosols, and Wind Fields Over Cubatao (Brazil) with the Remote Sensing System ARGOS, GKSS Report 96/E/22, Geesthacht, Germany (1996).

27. H. Edner, P. Ragnarson, S. Svanberg, E. Wallinder, R. Ferrara, R. Cioni, B. Raco, and G. Tadeucci, *J. Geophys. Res.*, **99**, 18–27 (1994).

28. H. Edner, DIAL measurements of natural and anthropogenic gas emissions, *Rev. Laser Eng.*, **23**, 93–96 (1995) (Special Issue on Laser Radar).

29. P. Weibring, H. Edner, S. Svanberg, L. Pantani, R. Ferrara, and T. Caltabiano, Monitoring of volcanic sulphur dioxide emissions using differential absorption lidar (DIAL), differential optical absorption spectroscopy (DOAS), and correlation spectroscopy (COSPEC), *Appl. Phys. B*, **67**, 419–426 (1998).

30. S. Svanberg, Geophysical gas monitoring using optical techniques: volcanoes, geothermal fields and mines, *Opt. Lasers Eng.*, **37**, 245–266 (2002).

31. T. Fujii, T. Fukuchi, N. Gao, K. Nemoto, and N. Takeuchi, Trace atmospheric SO_2 measurement by multiwavelength curve-fitting and wavelength-optimised dual differential absorption lidar, *Appl. Opt.*, **43**, 524–531 (2002).

32. T. Fukuchi, N. Goto, T. Fujii, and K. Nemoto, Error analysis of SO_2 measurement by multiwavelength differential absorption lidar, *Opt. Eng.*, **38**, 141–145 (1999).

33. T. Fukuchi, T. Fujii, N. Goto, K. Nemoto, and N. Takeuchi, Evaluation of DIAL measurement error by simultaneous DIAL and null profiling, *Opt. Eng.*, **40**, 392–397 (2001).

34. H. Edner, K. Fredriksson, A. Sunesson, and W. Wendt, Monitoring Cl_2 using a differential absorption lidar system, *Appl. Opt.*, **26**, 3183–3185 (1987).

35. M. J. T. Milton, P. T. Woods, B. W. Jolliffe, N. R. W. Swann, and T. J. McIlveen, Measurements of toluene and other aromatic hydrocarbons by differential-absorption lidar in the near-ultraviolet, *Appl. Phys. B*, **55**, 41–45 (1992).

36. H. Edner, G. W Faris, A. Sunesson, and S. Svanberg, Atmospheric atomic mercury monitoring using differential absorption lidar techniques, *Appl. Opt.*, **28**, 921–930 (1989).

37. R. Ferrara, B. E. Maserti, M. Andersson, H. Edner, P. Ragnarson, and S. Svanberg, Mercury degassing rate from mineralized areas in the Mediterranean basin, *Water Air Soil Pollut.*, **93**, 59–66, (1997).

38. N. Menyuk and D. K. Killinger, Atmospheric remote sensing of water vapor, HCl and CH_4 using a continuously tunable $Co:MgF_2$ laser, *Appl. Opt.*, **26**, 3061–3065 (1987).

39. E. E. Uthe, and N. B. Nielsen, Compact airborne system measures methane, *Laser Focus World*, December, 15–16, (1995).

40. M. J. T. Milton, T. J. Mc Iveen, D. C. Hanna, and P. T. Woods, A high-gain optical parametric amplifier tunable between 3.27 and 3.65 μm, *Opt. Commun.*, **93**, 186–190 (1992).

41. R. A. Robinson, P. T. Woods, and M. J. T. Milton, DIAL measurements for air pollution and fugitive loss monitoring, in *Air and Pollution Monitoring*, SPIE no. 2506, June 20–23, pp. 140–149 (SPIE, Munich, 1995).

42. K. Ikuta, N. Yoshikane, N. Vasa, Y. Oki, M. Maeda, M. Uchiumi, Y. Tsumura, J. Nakagawa, and N. Kawada, Differential absorption Lidar at 1.67 mm for remote sensing of methane leakage, *J. Appl. Phys.*, **38**, 110–114 (1999).

43. C. Weitkamp, The distribution of hydrogen chloride in the plume of incineration ships: development of new measurement systems, in *Wastes in the Ocean*, Vol. 3 (Wiley, New York, 1981).

44. C. Weitkamp, Infrared lidar measurement of the diffusion of hydrogen chloride from seaborne waste incineration, *Proceedings of the Fourth international Conference on Infrared Physics*, ETH Zürich, Switzerland, August 22–26, 1988, 218–226. Also: Report no. 88/E/45, GKSS Forschungszentrum, Geesthacht, Germany.

45. E. E. Uthe, Airborne CO_2 DIAL measurement of atmospheric tracer gas concentration distributions, *Appl. Opt.*, **25**, 2492–2498 (1986).

46. E. E. Uthe, Elastic scattering, fluorescent scattering, and differential absorption airborne lidar observations of atmospheric tracers, *Opt. Eng.*, **30**, 66–71 (1991).

47. T. Stuffler, *Entwicklung eines abstimmbaren Transmissionsmesssystems zum Nachweis atmosphöriches Verunreinigungen und dessen Einsatz bei Fernerkundunsmessungen* (in German), ISBN 3-89675-119-0 (Herbert Utz Verlag Wissenschaft, Munich, Germany, 1996).

48. N. Menyuk, D. K. Killinger, and W. E. DeFeo, Laser remote sensing of hydrazine, MMH, and UDMH using a differential-absorption CO_2 lidar, *Appl. Opt.*, **21**, 2275–2286 (1982).

49. A. P. Force, D. K. Killinger, W. E. DeFeo, and N. Menyuk, Laser remote sensing of atmospheric ammonia using a CO_2 lidar system, *Appl. Opt.*, **24**, 2837–2841 (1985).

50. A. Bass and R. Paur, The ultraviolet cross-sections of ozone: I. Measurements, in C. Zerefos and A. Ghazi, Eds., *Proceedings of the. Quadrennial Ozone Symposium*, Halkidiki, Greece, pp. 606–617 (D. Riedel, Dordrecht, 1984).

51. L. T. Molina and M. J. Molina, Absolute absorption cross section of ozone in the 185- to 350-nm wavelength range, *J. Geophys. Res.*, **91** (D13), 14501–14508 (1986).

52. D. Daumont, J. Brion, J. Charbonnier, and J. Malicet, Ozone UV spectroscopy I: absorption cross-section at room temperature, *J. Atmos. Chem.*, **15**, 145–155 (1992).

53. P. Völger, J. Bösenberg, and I. Schult, Scattering properties of selected model aerosols calculated at UV-wavelengths: implications for DIAL measurements of tropospheric ozone, *Beitr. Phys. Atmos.* 69, 177–187 (1996).

54. G. Megie and R. Menzies, Complementary of UV and IR differential absorption lidar for global measurements of atmospheric species, *Appl. Opt.*, **19**, 1173–1183 (1980).

55. J. Pelon and G. Mégie, Ozone monitoring in the troposphere and lower stratosphere evaluation and operation of a ground-based lidar station, *J. Geophys. Res.*, **87**, 4947–4955 (1982).

56. A. Papayannis, G. Ancellet, J. Pelon, and G. Mégie, Multi-wavelength lidar for ozone measurements in the troposphere and the lower stratosphere, *Appl. Opt.*, **29**, 467–476 (1990).

57. M. H. Proffitt and A. O. Langford, Ground-based differential absorption lidar system for day or night measurements of ozone throughout the free troposphere, *Appl. Opt.*, **36**, 2568–2585 (1997).

58. G. Mégie, G. Ancellet, and J. Pelon, Lidar measurements of ozone vertical profiles, *Appl. Opt.*, **24**, 3454–3463 (1985).
59. E. V. Browell, A. F. Carter, S. T. Shipley, R. J. Allen, C. F. Butler, M. N. Mayo, J. H. Siviter, and W. M. Hall, NASA multipurpose airborne DIAL system and measurements of ozone and aerosol profiles, *Appl. Opt.*, **22**, 522–534 (1983).
60. E. Durieux, L. Fiorani, B. Calpini, M. Flamm, L. Jaquet, and H. Van Den Bergh, Tropospheric ozone measurements over the great Athens area during the Medcaphot-trace campaign with a new shot-per-shot DIAL instrument: experimental system and results, *Atmos. Environ.*, **32**, 2141–2150 (1998).
61. J. P. Wolf, M. Douard, K. Fritzsche, P. Rairoux, G. Schubert, M. Ulbricht, D. Weidauer, and L. Wöste, 3D-monitoring of air pollution using "all solid state" lidar systems, SPIE, **2112**, 147–158 (1993).
62. T. H. Chyba, T. Zenke, R. Payne, C. Toppin, B. Thomas, D. Harper, N. Scott Higden, D. A. Ritchter, and J. Fishman, A Compact Ozone DIAL System, Selected Papers 20 ILC, Vichy, France, 10–14 July, pp. 137–140 (2000).
63. L. de Schoulepnikoff, V. Mitev, V. Simeonov, B. Calpini, and H. van den Bergh, Experimental investigation of high-power single-pass Raman shifters in the ultraviolet with Nd:YAG and KrF lasers, *Appl. Opt.*, **36**, 5026–5043 (1997).
64. M. Milton, G. Ancellet, A. Apituley, J. Bosenberg, W Garnuth, F. Castagoli, T. Trickl, H. Edner, L. Stefanutti, T. Schalberl, A. Sunesson, and C. Weitkamp, Raman-shifted laser sources suitable for differential-absorption lidar measurements of ozone in the troposphere, *Appl. Phys. B*, **66**, 105–113 (1998).
65. D. A. Haner and I. S. McDermid, Stimulated Raman shifting of the Nd:YAG fourth harmonic (266 nm) in H_2, HD and D_2, *IEEE J. Quantum Electron.*, **26**, 1292–1298 (1990).
66. V. Simeonov, V. Mitev, H. van den Bergh, and B. Calpini, Raman frequency shifting in $CH_4:H_2:Ar$ mixture pumped by the fourth harmonic of Nd:YAG laser, *Appl. Opt.*, **30**, 7112–7115 (1998).
67. O. Uchino, M. Tokunaga, M. Maeda, and Y. Miyazou, Differential-absorption-lidar measurement of tropospheric ozone with excimer-Raman hybrid laser, *Opt. Lett.*, **8**, 347–349 (1983).
68. V. S. Bukreev, S. K. Vartapetov, I. A. Vaselovskii, A. S. Galustov, Yu. M. Kovalev, A. M. Prokhorov, E. S. Svetogorov, S. S.

Khmelevtsov, and Ch. H. Lee, Excimer-laser-based lidar system for stratospheric and tropospheric ozone measurements, *Quantum Electron.*, **24** (6), 546–551 (1994).

69. U. Kempfer, W. Carnuth, R. Lotz, and T. Trickl, A wide range UV lidar system for tropospheric ozone measurements: development and application, *Rev. Sci. Instrum.*, **65**, 3145–3164 (1994).

70. V. S. Bukreev, S. K. Vartapetov, I. A. Vaselovskii, A. S. Galustov, Yu. M. Kovalev, E. S. Svetogorov, and S. S. Khmeltsov, Combined lidar system for stratospheric and tropospheric ozone measurements, *Appl. Phys. B*, **62**, 97–101 (1996).

71. W. Grant, E. Browel, N. Hygdon, and S. Ismail, Raman shifting of KrF laser-radiation for tropospheric ozone measurements, *Appl. Opt.*, **30**, 2628–2633 (1991).

72. L. Stefanutti, F. Castagnoli, M. Del Guasta, M. Morandi, V. M. Sacco, L. Zuccagnoli, S. Godin, G. Megie, and J. Porteneuve, The Antarctic ozone LIDAR system, *Appl. Phys. B*, **55**, 3–12 (1992).

73. J. A. Sunesson, A. Apituley, and D. P. D. Swart, Differential absorption lidar system for routine monitoring of tropospheric ozone, *Appl. Opt.*, **33**, 7045–7058 (1994).

74. B. Calpini, V. Simeonov, F. Jeanneret, J. Kuebler, V. Sathya, and H. van den Bergh, Ozone LIDAR as an analytical tool in effective air pollution management: The Geneva 96 Campaign, *Chimia*, **51**, 700–704 (1997).

75. A. D. Papayannis, J. Porteneuve, D. Balis, C. Zerefos, and E. Galani, Design of a new DIAL system for tropospheric and low stratospheric ozone monitoring in Northern Greece, *Phys. Chem. Earth*, **24**, 439–442 (1999).

76. G. Ancellet and F. Ravetta, Compact airborne lidar for tropospheric ozone description and field measurements, *Appl. Opt.*, **37**, 5509–5521 (1998).

77. J.-L. Baray, J. Leveau, J. Porteneuve, G. Ancellet, P. Kekhurt, F. Posny, and S. Baldy, Description and evaluation of a tropospheric ozone lidar implemented on an existing lidar in the southern subtropics, *Appl. Opt.*, **38**, 6808–6817 (1999).

78. V. Simeonov, B. Calpini, and H. van den Bergh, New Raman shifed sources for ozone DIAL applications, ILRC, **21** (S056), (2002).

79. V. Simeonov, B. Calpini, J. Balin, P. Ristori, R. Jimenez, and H. van den Bergh, UV ozone DIAL based on a N_2 Raman

converter design and results during ESCOMPTE field campaign, ILRC, **21** (S057), (2002).

80. V. Simeonov, G. Larchevêque, P. Quaglia, H. van den Bergh, and B. Calpini, The influence of the PMT spatial uniformity on LIDAR signal, *Appl. Opt.*, **38**, 5186–5190 (1999).

81. Z. Wang, J. Zhou, H. Hu, and Z. Gong, Evaluation of dual differential absorption lidar based on Raman-shifted Nd:YAG or KrF laser for tropospheric ozone measurements, *Appl. Phys. B*, **62**, 143–147 (1996).

82. B. Lazzarotto, M. Froud, G. Larchevêque, V. Mitev, P. Quaglia, V. Simeonov, A. Thompson, H. van den Bergh, and B. Calpini, Ozone and water-vapor measurements by Raman lidar in the planetary boundary layer: error sources and field measurements, *Appl. Opt.*, **40**, 2985–2997 (2001).

83. F. Tomasi, M. Perrone, and M. Protopapa, Monitoring O_3 with solar-blind Raman lidars, *Appl. Opt.*, **40**, 1314–1320 (2001).

84. S. Perego, A numerical mesoscale model for simulation of regional photosmog in complex terrain: model description and application during POLLUMET 1993 (Switzerland), *Meteor. Atmos. Phys.*, **70**, 43–69 (1999).

85. O. Couach, et al., Campagne de mesures intensives 1999 sur la région grenobloise, Ensemble des résultats et Analyse des Périodes d'Observations Intensives (POI), see http://lpas.epfl.ch/MOD/study/asco/rap99_final.pdf

86. W. R. Stockwell, F. Kirchner, M. Kuhn, and S. Seefeld, A new mechanism for regional atmospheric chemistry modelling, *J. Geophys. Res.*, **102** (D22), 25847–25879 (1997).

87. B. Lazzarotto, V. Simeonov, P. Quaglia, G. Larchevêque, H. van den Bergh, and B. Calpini, A Raman differential absorption lidar for ozone and water vapor measurement in the lower troposphere, *Int. J. Environ. Anal. Chem.*, **74**, 255–261 (1999).

88. P. Quaglia, G. Larchevêque, R. Jimenez, B. Lazzarotto, V. Simeonov, G. Ancellet., H. van den Bergh, and B. Calpini, Planetary boundary layer ozone fluxes from combined airborne, ground based lidars and wind profilers measurements, *Eur. J. Anal. Chem. (Analusis)*, **27** (4), 305–310 (1999).

89. K. Kourtidis, C. Zerefos, S. Rapsomanikis, V. Simeonov, D. Balis, E. Kosmidis, P. E. Perros, A. M. Thompson, J. Witte, B. Calpini, W. M. Sharobiem, A. Papayannis, N. Mihalopoulos, and R. Drakou, Regional levels of ozone in the troposphere over Eastern Mediterranean, *J. Geophys. Res.* 107, No. D18, 8140, doi: 10.1029/2000JD000140 (2002).

90. B. Calpini, F. Jeanneret, M. Bourqui, A. Clappier, R. Vajtai, and H. van den Bergh, Direct measurement of the total reaction rate of OH in the atmosphere, *Eur. J. Anal. Chem. (Analusis)*, **27**, 328–336 (1999).

91. F. Jeanneret, F. Kirchner, A. Clappier, H. van den Bergh, and B. Calpini, Total VOC reactivity in the planetary boundary layer, part one: estimation by a pump and probe OH experiment, *J. Geophys. Res.*, **106**, 3083–3094 (2001).

92. F. Kirchner, F. Jeanneret, A. Clappier, B. Kruger, H. van den Bergh, and B. Calpini, Total VOC reactivity in the planetary boundary layer, part two: a new indicator for determining the sensitivity of the ozone production to VOC and NO_x, *J. Geophys. Res.*, **106**, 3095–3110 (2001).

93. C. R. Webster, et al., Aircraft laser infrared absorption spectrometer (ALIAS) for in situ stratospheric measurement of HCl, N_2O, CH_4, and HNO_3, *Appl. Opt.*, **33**, 454–472 (1994).

94. H. I. Schiff, et al., A tunable diode laser system for aircraft measurement of trace gases, *J. Geophys. Res.*, **95**, 10147–10153, (1990).

95. G. W. Sachse, et al., Fast response, high precision carbon monoxide sensor using tunable diode laser technique, *J. Geophys. Res.*, **92**, 2071–2081 (1987).

96. J. Faist, F. Capasso, D. L. Sivco, C. Sirtori, A. L. Hutchinson, and A. Y. Cho, Quantum cascade laser, *Science*, **264**, 553 (1994).

97. J. Faist, F. Capasso, Quantum cascade laser, in *McGraw-Hill Yearbook of Science and Technology*, pp. 265–267 (McGraw Hill, New York, 1997).

98. D. Hofstetter, M. Beck, T. Aellen, and J. Faist, High-temperature operation of distributed feedback quantum-cascade lasers at 5.3 μm, *Appl. Phys. Lett.*, **78**, 396 (2001).

99. A. Müller, M. Beck, J. Faist, and U. Oesterle, Electrically tunable, room-temperature quantum-cascade lasers, *Appl. Phys. Lett.*, **75**, 1509 (1999).

100. K. Namjou, et al., Sensitive absorption spectroscopy with a room-temperature distributed-feedback quantum-cascade laser, *Opt. Lett.*, **23**, 219–221 (1998).

101. R. Jimenez, M. Taslakov, V. Simeonov, B. Calpini, F. Jeanneret, D. Hofstetter, M. Beck, J. Faist, and H. van den Bergh, Ozone detection by differential absorption spectroscopy at ambient pressure with a 9.6 μm pulsed quantum cascade laser, *Appl. Phys. B: Lasers Opt.*, **78**, 249–256 (2003).

102. S. W. Sharpe, et al., High resolution (Doppler limited) spectroscopy using quantum cascade distributed-feedback lasers, *Opt. Lett.*, **23**, 1396–1398 (1998).
103. B. A. Paldus, et al., Photoacoustic spectroscopy using quantum cascade lasers, *Opt. Lett.*, **24**, 178–180 (1999).
104. D. M. Sonnenfroh, et al., Application of balance detection to absorption measurements of trace gases with room-temperature, quasi-cw quantum-cascade lasers, *Appl. Opt.*, **40** (6), 812–820 (2001).

5

Resonance Fluorescence Lidar for Measurements of the Middle and Upper Atmosphere

XINZHAO CHU and GEORGE C. PAPEN

Department of Electrical & Computer Engineering,
University of Illinois at Urbana-Champaign, Urbana, IL, USA

5.1. INTRODUCTION

5.1.1. Lidar Study of the Middle and Upper Atmosphere

The middle and upper atmosphere is a complex and important region. It contains a wealth of important geophysical phenomena, for example, the Earth's coldest environment — the mesopause; polar mesospheric clouds (PMCs) and noctilucent clouds (NLCs); the meteoric metal layers of Na, K, Ca, Li, and Fe; the airglow layers of OH, O, and O_2; and planetary, tidal, and gravity wave activities that play vital roles in overall global atmosphere circulation. The growing interest in the middle atmosphere is also tied to this region's importance for detecting and understanding global climate change. Global climate change is not confined to the lower atmosphere, but extends into the middle and upper atmosphere as well. While greenhouse gases such as CO_2 and CH_4 help warm the lower atmosphere by absorbing infrared radiation, they are also efficient radiators of heat and result in cooling in the middle and upper atmosphere (Roble and Dickinson, 1989). Doubling the CO_2 concentration is predicted to cool the stratopause ($\sim$50 km) by 10 to 12 K and the mesopause region ($\sim$80 to 100 km) by 6 to 12 K (Portman et al., 1995). The observed increases in the geographic extent and brightness of PMCs and NLCs in the Polar Regions during the past decade are also believed to be evidence of cooling temperatures and increasing water-vapor concentration in the mesopause region caused by increases of greenhouse gases (Gadsden, 1990; Thomas, 1996). Therefore, the middle and upper atmosphere thermal structure and polar mesospheric clouds are recognized to be important and sensitive indicators of global climate change. Modeling the middle atmosphere is hampered by the lack of observational data and the uncertainties in the parameterization of gravity waves. Observational studies of the middle and upper atmosphere provide crucial tests of atmospheric general circulation models, which underlie our understanding of the global climate change, thereby leading to improvements in these models and our knowledge of the global atmosphere.

It is clear that data from the middle and upper atmosphere are vital for testing the atmospheric circulation models and studying complex atmospheric dynamics. However, it is challenging to collect data on temperature, wind, PMCs, and gravity waves from this region because the middle and upper atmosphere is usually inaccessible to balloon, aircraft, and most satellites. Optical and radio remote sensing instruments and *in situ* rocket measurements are the most common sources of observational data. Among these remote sensing technologies, lidar techniques are unique in their capabilities of providing high spatial and temporal resolution data on the temperature, wind, and constituent structure of the middle atmosphere region. Because of their daytime measurement capabilities and relatively low cost compared to rockets, lidars can have full seasonal and diurnal coverage of atmosphere measurements. These features make lidars important tools for the middle and upper atmosphere study.

5.1.2 Lidar Concepts and Classifications

Lidar stands for *light detection and ranging*. It is commonly referred to as *laser radar* because it is the optical equivalent of radar. Lidar studies began with searchlights for studying stratospheric aerosols and molecular density (Elterman, 1954). However, the use of a continuous wave (cw) light source prevented the measurement of the height variation of the density. Although it was possible to measure the height variation by separating the cw searchlight and receiving system in a bistatic configuration, modern lidars utilize pulsed laser sources in a monostatic configuration, which provide much better vertical resolution and much better beam collimation. Usually, a lidar consists of a transmitter, a receiver, and a data acquisition and control system. The lidar transmitter consists of a pulsed laser system equipped with spectral analyzing instruments and beam collimating optics to provide laser pulses with precise frequency, narrow spectral width, high energy, and small divergence. The lidar receiver includes an optical telescope to collect backscattered photons, filters to eliminate background light, and a photon detector to count the return photons. The data acquisition and control system is used to record the photon counts and to provide the timing and system control for the whole system.

For an atmospheric object to be detected by lidar methods, the object must interact with the light in such a manner that a sufficient number of photons are scattered back to the telescope, which are above the system noise level. When laser photons propagate in the atmosphere, they are scattered by different constituents and different mechanisms. The backscattered photons are collected by a telescope, filtered by optical filters, and detected by photomultiplier tubes or photodiodes. The range from the lidar system to the scatterer can be determined precisely by measuring the flight time of photons. This range is given by $R = c\Delta t/2$, where c is the speed of the light and Δt is the flight time. The factor of 1/2 accounts for the two-way propagation. Therefore, the photon count profiles recorded by the lidar receiver are range resolved. By combining knowledge of atmospheric dynamics with principles of light scattering of particles, the range-resolved atmosphere parameters (such as temperature, wind, density, backscatter coefficient, etc.) can be derived.

According to the physical mechanism of the particle-radiation interaction in the atmosphere, lidars can be classified into Mie scattering lidar, Raman scattering lidar, Rayleigh scattering lidar, resonance fluorescence lidar, and differential absorption lidar (DIAL). Lidars using Mie scattering (elastic scattering from small particles and aerosols), Raman scattering (inelastic scattering from atoms and molecules), and differential absorption of atoms and molecules are typically used in the lower atmosphere below 30 km where the backscatter signals from aerosols or molecules and the absorption from molecules are strong enough to derive useful information of atmosphere. Lidars using Rayleigh scattering (elastic scattering from atoms and molecules) are generally used in the altitude range of 30 to 90 km, where the atmosphere is aerosol-free but still contains enough gas molecules to produce a useful Rayleigh signal. Above approximately 85 km, because of the low atmospheric density, Rayleigh signal is typically too weak to derive any reliable information. Lidars using resonance fluorescence can be used in the range from 75 to 115 km where suitable constituents with large scattering cross-section exist to produce strong resonance fluorescence signals. The backscatter cross-section of resonance fluorescence is approximately 10^{14} times higher than Rayleigh backscatter cross-section. Although the density of tracer atoms or molecules is about 10^{10} times less than the atmosphere molecular density at the same altitude, the overall resonance fluorescence signal, which is proportional to the product of backscatter cross-section and tracer atom density, can be several orders of magnitude higher than Rayleigh backscatter signal for the same laser power and receiving telescope aperture. Thus, resonance fluorescence lidars can generate useful photon signals to derive atmospheric parameters in the altitude range of 75 to 115 km. This region is usually referred to as the MLT region (i.e., the mesosphere and lower thermosphere), and is also called the *mesopause region*.

5.1.3 Initial Developments of Resonance Fluorescence Lidar

The naturally existing metallic atoms in the mesopause region, such as Na, K, Li, Ca, and Fe, are believed to originate from the meteor ablation. Some electronic transitions of these atoms fall in the wavelength range of available laser sources. When a laser is tuned to a resonant line, an atom will absorb a laser photon and then re-emit through spontaneous emission. Since quenching (excited atoms returning to ground state without photon emission, usually caused by collisions with surrounding molecules) is not a problem in the middle and upper atmosphere because of the low density, fluorescence from these atoms and molecules can be detected. This resonance fluorescence signal can be used to determine the density of the metal species as well as measure the wind and temperature.

Since the first lidar observation of mesospheric Na atoms by Bowman et al. (1969), the Na density and column abundance have been measured by several groups using broadband resonance fluorescence lidars where the spectral width of the laser pulse was much larger than the absorption width of the resonance lines (Sandford and Gibson, 1970; Gibson and Sandford 1971, 1972; Blamont et al., 1972; Hake et al., 1972; Kirchhoff and Clemesha, 1973; Megie and Blamont, 1977). The variations of Na layers at various temporal and spatial scales were observed, including seasonal and diurnal variations, sporadic Na layers, and the variations induced by dynamical processes such as tides and gravity waves. Other metal species were also measured by the broadband resonance fluorescence lidars, for example, K (Felix et al., 1973), Li (Jegou et al., 1980), Ca and Ca^+ (Granier et al., 1985, 1989a; Qian and Gardner, 1995), and Fe (Granier et al., 1989b; Alpers et al., 1990; Bills and Gardner, 1990; Kane and Gardner, 1993a). Simultaneous observations of sporadic E, Na, Fe, and Ca^+ layers were also made by Gardner et al. (1993). By reducing the receiver bandwidth, daytime observations of the atmospheric Na layers were first reported by Gibson and Sandford (1972), and later by three different groups (Clemesha et al., 1982; Granier and Megie, 1982; Kwon et al., 1987).

Studies using broadband resonance fluorescence lidars provided key insights into mesospheric constituents. They also enabled the study of gravity and tidal waves by studying the tracer density variations (Batista et al., 1985; Gardner and Voelz, 1985; Senft and Gardner, 1991). However, broadband lidars cannot directly measure the temperature and wind, which are the key parameters for characterizing the middle and upper atmospheric dynamics. Scientific curiosity called for further developments to make the resonance fluorescence lidars capable of temperature and wind measurements in the mesopause region.

5.1.4 New Developments of Resonance Fluorescence Lidar

The first observation of Na ground-state hyperfine structure in the middle and upper atmosphere was reported by Gibson et al. (1979), establishing the possibility of mesopause temperature measurements with a resonance fluorescence lidar using a narrowband tunable laser. The atmosphere temperature can be determined by measuring the Doppler broadening and the wind can be determined by measuring the Doppler shift of laser-induced fluorescence from atmospheric metal atoms. In this study, a pulsed laser with linewidth of 0.1 GHz was scanned through sodium D_2 line with Doppler broadening. Although the absolute calibration of the laser frequency was not available, a least-square fit using theoretical model of Na spectroscopy gave the best estimate of the atmospheric temperature with an error of ± 15 K (Gibson et al., 1979; Thomas and Bhattacharyya, 1980; Thomas, 1995). The first Na temperature lidar that made routine observations was developed at the University of Bonn with an excimer-pumped pulsed dye-laser system (Fricke and von Zahn, 1985). Fricke and von Zahn (1985) were able to measure the temperature profiles through the mesopause region by tuning and monitoring the frequency of each laser pulse with the help of a wavelength meter and a Na vapor cell. An uncertainty of ± 5 K at the Na layer peak was obtained with 1-km vertical resolution and 10-min integration period. Their measurements at Andoya, Norway with this Na lidar led to an understanding of the temperature structure of a polar mesopause (Lübken and von Zahn, 1991).

A more advanced narrowband Na lidar system using a two-frequency technique was developed through a collaboration between the Colorado State University and the University of Illinois (She et al., 1990; Bills et al., 1991a). The lidar transmitter consisted of a cw frequency-stabilized ring dye laser pumped by an argon ion laser. The cw 589-nm laser was then amplified by a pulsed dye amplifier, which itself was pumped by an injection-seeded frequency-doubled Nd:YAG laser. Such laser system provided excellent frequency tuning accuracy, pulse-to-pulse stability, and high peak power. A temperature-sensitive ratio of the Na fluorescence signals at two frequencies was used to derive temperature from the Doppler-broadening linewidth. An accuracy of better than $\pm 3\,K$ at the Na layer peak was achieved with 1-km vertical resolution and 5-min integration period. Since a cw ring dye laser was used, the Doppler-free saturation–absorption features of a Na vapor cell could be used to fine-tune and lock laser frequency onto sodium D_2 line, which provided an absolute calibration of laser frequency. This technique not only improved the accuracy of temperature measurement, but also enabled the wind measurement by determining the Doppler shift of sodium lines (Bills et al., 1991a,b; She et al., 1992; She and Yu, 1994; Gardner and Papen, 1995). Initial horizontal wind measurements were made by tuning the lidar to four frequencies with a stabilized Fabry–Perot etalon (Bills et al., 1991b) and to three frequencies with an acousto-optic modulator (She and Yu, 1994). In currently implemented systems, two- and three-frequency techniques are most widely employed since they can provide higher temporal resolution and the fast switching between the two or three frequencies eliminates errors caused by Na density fluctuations during data collection. After the initial development, temperature and wind measurements were collected on a routine basis as well as during campaigns. The lidar beam was also coupled with large telescope through coude optics and was able to provide flexible beam pointing. This, in combination with the large receiver, enabled the first lidar studies of heat flux and momentum flux (Tao and Gardner, 1995; Gardner and Yang, 1998).

New lidars are continuously being proposed and developed. After a decade of development, the first report on mesopause temperature measurements using a narrowband alexandrite laser probing atmospheric potassium atoms was published by von Zahn and Höffner (1996). A laser-diode-injection-seeded pulsed ring alexandrite laser was used in the transmitter to achieve a narrowband system. The principle of deriving temperatures from a potassium lidar signal is very similar to the Na temperature lidar except the laser frequency was scanned through potassium D_1 line (instead of D_2) since the potassium D_2 line overlaps with a strong O_2 absorption line. Although potassium abundance is two orders of magnitude lower than sodium, this reduction can be offset by using an alexandrite laser with higher average power and much longer pulse length.

Both the Na and K temperature lidars utilize the Doppler broadening of metal atoms, so narrowband tunable lasers are required in the lidar transmitters to resolve the temperature-dependent Doppler-broadened atomic lineshape. A new concept for temperature measurement was proposed and developed over the past few years. This is the Boltzmann temperature technique. As proposed by Gelbwachs (1994), the temperature dependence of the population ratio on different energy levels of atomic iron (Fe) can be used to derive atmosphere temperatures. Motivated by this initial proposal and the need for polar atmosphere measurement, the University of Illinois at Urbana-Champaign (UIUC) developed a robust Fe Boltzmann temperature lidar using solid-state alexandrite laser technology (Chu et al., 2002a). The temperatures are determined by taking the ratio of the fluorescence photon signals from two independent Fe lines at 372 and 374 nm. This ratio is a very sensitive function of atmosphere temperature, and it obeys the Maxwell–Boltzmann distribution law under thermodynamic equilibrium. Since the ratio is from two transition lines instead of within one line, the lasers can be broadband with a comparable linewidth to Fe Doppler broadening ($\sim$1 GHz at 200 K). The lidar was then deployed on campaigns ranging from the geographic North Pole to the geographic South Pole to observe the polar atmosphere during both summer and winter. Significant scientific findings have been achieved through these measurements (Chu et al., 2001a,b, 2003a; Gardner et al., 2001; Pan et al., 2002).

Other developments in the resonance fluorescence lidars include creating an all-solid-state version of the Na lidar. Until now, the Na wind/temperature lidar is still the most accurate lidar in measuring temperatures. However, the ring dye laser used in the system is so sensitive that it requires a very stable operating environment. This precludes measurements in remote areas such as the Antarctica and the Arctic or deployment on aircraft. Two types of solid-state Na lidars have been developed. One is to mix the frequencies of two pulsed Nd:YAG lasers operating at 1.06 and 1.3 μm to produce 589-nm laser pulses (Kawahara et al., 2002). Another is to mix the frequencies of two cw single-mode Nd:YAG lasers to produce a cw 589-nm laser (Vance et al., 1998). This solid-state cw 589-nm laser, which replaces the ring dye laser, is then followed by a pulsed dye amplifier to produce 589-nm laser pulses (She et al., 2002b). Both types of lidar have been used to measure the mesopause temperatures in the Polar Regions.

5.1.5   Arrangement of this Chapter

In Section 5.2, we will concentrate on the technologies of the resonance fluorescence lidar developed in the past 20 years. These technologies include the developments of the Na wind/temperature lidar, the Fe Boltzmann temperature lidar, the K Doppler temperature lidar, and the solid-state Na Doppler lidar. A comparison of these lidar technologies is given at the end of Section 5.2. Section 5.3 provides a summary of the key scientific results obtained using resonance fluorescence lidars. These results include the characterization of the MLT region thermal structure, MLT region dynamics, mesospheric metallic layers and meteor trails, and the polar mesospheric clouds (noctilucent clouds). The chapter concludes in Section 5.4 with an outlook on the future of resonance lidar techniques.

5.2.   ADVANCED TECHNOLOGY OF RESONANCE FLUORESCENCE LIDAR

Since the development of broadband resonance fluorescence lidar in the late 1960s for atmosphere constituent density measurements, the most important developments in the lidar field are the new resonance fluorescence lidar techniques for the temperature and wind measurements in the middle and upper atmosphere. Wind measurements rely on the effect of Doppler frequency shift while there are mainly two concepts on how to measure temperatures in the mesopause region. One is to utilize the temperature dependence of Doppler broadening of electronic-dipole transition lines of the tracer gases, such as Na and K. We refer to this as the *Doppler technique*. Another is to utilize the temperature dependence of the ratio of populations on different energy levels of the tracer gases, such as Fe. We refer to this as the *Boltzmann technique*. These concepts and measurement principles will be described in detail in the Na wind/temperature lidar and Fe Boltzmann temperature lidar sections. First, we will introduce the lidar equations for the resonance fluorescence lidars.

5.2.1. Lidar Equations for Resonance Fluorescence Lidar

5.2.1.1. General Form of the Lidar Equation

The lidar equation is the basic equation in the field of laser remote sensing, which relates the received photon count obtained from a scattering region to the emitted laser photon numbers, the concentration of the scatterer, the interaction between the laser radiation and the scatterer, and the lidar system efficiency. The lidar equation is developed under two assumptions: the scattering processes are independent, and only single scattering occurs. For atmospheric scattering, due to the low atmosphere density, air particles are separated adequately and undergo random motion so that the contributions to the total scattered energy by many particles have no phase relations. This allows the total intensity to be calculated by simply adding the intensity from each scattering particle. This process is known as independent scattering. Single scattering implies that a photon is scattered only once. If a photon is scattered two or more times, then multiple scattering occurs. This will be excluded in our considerations since the probability for such multiple scattering to occur is negligibly small in the middle and upper atmosphere.

In general, the lidar interaction between the light photons and the particles is a scattering process. The expected received photon count is equal to the product of the number of transmitted photons, the probability that a transmitted photon is scattered, the probability that a scattered photon is received, and the system efficiency. The general lidar equation can be written as

$$N_S(\lambda, z) = \left(\frac{P_L(\lambda_L)\Delta t}{hc/\lambda_L}\right)\left(\beta(\lambda, \lambda_L, z)\Delta z\right)\left(\frac{A}{z^2}\right)$$
$$\times \left(\eta(\lambda, \lambda_L)T(\lambda_L, z)T(\lambda, z)G(z)\right) + N_B\Delta t \quad (5.1)$$

where

$N_S(\lambda, z)$ = expected photon counts detected at wavelength λ in the range interval $(z - \Delta z/2, z + \Delta z/2)$;

λ = detected photon wavelength (m);

$P_L(\lambda_L)$ = laser output power at laser wavelength λ_L (W);

λ_L = laser radiation wavelength (m);

Δt = integration time (s);

h = Planck's constant (6.626×10^{-34} J/s);

c = speed of light (2.99792458×10^8 m/s);

$\beta(\lambda, \lambda_L, z)$ = volume backscatter coefficient at range z under laser radiation at wavelength λ_L for the scattered photons falling into the wavelength λ (m^{-1} sr^{-1});

Δz = thickness of the range bin or range interval (m), which is related to the sampling time τ by $\Delta z = c\tau/2$;

A = receiving telescope aperture area (m^2);

z = range from the scatter to the lidar receiver;

$T(\lambda_L, z)$ = one-way transmittance of the atmosphere for wavelength λ_L from the lidar transmitter to the range z;

$T(\lambda, z)$ = one-way transmittance of the atmosphere for wavelength λ from the range z to the lidar receiver;

$\eta(\lambda, \lambda_L)$ = lidar optical efficiency for transmitted wavelength at λ_L and received wavelength at λ;

$G(z)$ = geometrical probability of radiation at range z reaching the detector, based on the geometrical considerations; and

N_B = expected photon counts per range bin per unit time due to background noise and detector dark counts.

The first parenthesis term in Equation (5.1) is the number of transmitted laser photons, in which the factor $P_{\mathrm{L}}\Delta t$ is the total laser energy and hc/λ_{L} is the energy for a single photon.

The second parenthesis term is the probability that a transmitted photon is backscattered by the scatters into a unit solid angle. This angular scattering probability is equal to the product of the volume backscatter coefficient $\beta(\theta = \pi)$ and the scattering layer thickness (Δz). The volume backscatter coefficient is the volume angular scattering coefficient $\beta(\theta)$ at the scattering angle $\theta = \pi$. The volume angular scattering coefficient $\beta(\theta)$ $(\mathrm{m}^{-1}\,\mathrm{sr}^{-1})$ is defined as the probability per unit distance traveled that a photon is scattered into wavelength λ in unit solid angle at the scattering angle θ. The volume total scattering coefficient β_{T} (m^{-1}) is the probability per unit distance traveled that a photon is scattered into wavelength λ in all directions. This can be obtained by integrating $\beta(\theta)$ over the 4π steradians. The volume backscatter coefficient can be expressed as

$$\beta(\lambda, \lambda_{\mathrm{L}}, z) = \sum_{i}\left[\frac{\mathrm{d}\sigma_i(\lambda_{\mathrm{L}})}{\mathrm{d}\Omega}n_i(z)p_i(\lambda)\right] \tag{5.2}$$

where

$\frac{\mathrm{d}\sigma_i(\lambda_{\mathrm{L}})}{\mathrm{d}\Omega} =$ the differential backscatter cross-section of an individual particle, that is, the angular scattering cross-section at the scattering angle $\theta = \pi$ under laser radiation at wavelength λ_{L};

$n_i(z) =$ the number density of scatter species i; and

$p_i(\lambda) =$ the probability of the scattered photons falling into the wavelength λ.

The differential scattering cross-section $\frac{\mathrm{d}\sigma_i(\lambda_{\mathrm{L}})}{\mathrm{d}\Omega}$ is defined as the ratio of the rate at which energy is scattered into a unit solid angle at an observation angle θ by an individual particle to the rate at which energy in the incident photon beam crosses a unit area perpendicular to its propagation direction. It has the dimension of the area and can be understood as the area of impinging flux that is scattered into a unit solid angle at an observation angle θ. It is also called angular scattering cross-section $\sigma(\theta)$. The total scattering cross-section is another quantity used in scattering theory and is defined as the total area of flux scattered in all directions by a scatterer. This can be found by integrating $\sigma(\theta)$ over 4π steradians.

The third parenthesis term in Equation (5.1) is the probability that a scattered photon is collected by the receiving telescope. This receiving probability is equal to the solid angle subtended by the receiver aperture to the scatterer divided by 4π steradians. Since the factor 4π has been incorporated into the angular scattering probability, the receiving probability is just equal to the solid angle A/z^2.

The fourth parenthesis term is the overall system efficiency. The lidar optical efficiency η includes the optical efficiencies of all the optics in the lidar transmitter and receiver (e.g., the beam splitters, mirrors, lenses, optical filters, etc.) as well as the quantum efficiency of the photon detector (such as a photomultiplier tube [PMT] or a photodiode). It can be expressed as the product of the transmitter efficiency $\eta_T(\lambda_L)$ and the receiver efficiency $\eta_R(\lambda)$:

$$\eta(\lambda, \lambda_L) = \eta_T(\lambda_L)\eta_R(\lambda) \tag{5.3}$$

This optical efficiency is usually dominated by the optical filter transmittance and the detector quantum efficiency. The atmospheric transmittance $T(\lambda_L, z)$ and $T(\lambda, z)$ depend on the wavelength and the traveling range of the photons. The geometrical form factor $G(z)$ mainly depends upon the overlap of the area of laser irradiation with the field of view of the receiver optics, but also depends on the details of the receiver optics.

5.2.1.2. Scattering Form of the Lidar Equation

There are a few different forms for the lidar equation, which are determined by the nature of the interaction between the laser radiation and the scatterer. Scattering form (for Rayleigh, Mie, and Raman scatterings) and fluorescence form (for resonance fluorescence) are two representative forms of the lidar equation. Both forms will be used in the data analysis for the resonance fluorescence lidars. The main differences between these forms lie in the volume backscatter coefficient and the overall system efficiency in the lidar equation.

The Rayleigh, Mie, and Raman scattering processes are instantaneous scattering processes; that is, there are no finite relaxation effects involved. For Rayleigh and Mie scattering, there is no frequency shift when the atmosphere particles are at rest, so $\lambda = \lambda_L$, $p(\lambda) = 1$, and the volume backscatter coefficient can be written as

$$\beta(\lambda, z) = \sum_i \left[\frac{d\sigma_i(\lambda)}{d\Omega} n_i(z) \right] \tag{5.4}$$

Therefore, the scattering form of the lidar equation for Rayleigh and Mie scatterings is

$$N_s(\lambda, z) = \left(\frac{P_L(\lambda)\Delta t}{hc/\lambda} \right) (\beta(\lambda, z)\Delta z) \left(\frac{A}{z^2} \right)$$
$$\times \left(\eta(\lambda) T^2(\lambda, z) G(z) \right) + N_B \Delta t \tag{5.5}$$

For Mie scattering, the summation in Equation (5.4) is mainly for different particle sizes and species, which vary in a large range. For Rayleigh scattering, the differential backscatter cross-section can be written as a common Rayleigh backscatter cross-section $\frac{d\sigma_i(\lambda)}{d\Omega} = \sigma_R(\lambda)$ and the air total number density can be written as $n_R(z) = \sum_i n_i(z)$ for the well-mixed atmosphere below 100 km. Thus, the Rayleigh volume backscatter coefficient can be simplified as

$$\beta(\lambda, z) = \sigma_R(\pi, \lambda) n_R(z) \tag{5.6}$$

The Rayleigh scattering is anisotropic. The relation between the volume angular scattering coefficient and the volume total scattering coefficient is determined by a phase function, which is defined as the ratio of energy scattered into a direction per unit solid angle to the average energy scattered in all directions per unit solid angle. The phase function for Rayleigh scattering is given by (Goody and Yung, 1989)

$$P(\theta) = 0.7629 \times (1 + 0.9324 \cos^2 \theta) \tag{5.7}$$

where θ is the scattering angle, and backscattering corresponds to $\theta = \pi$. The volume angular scattering coefficient can be derived from the volume total scattering coefficient β_T as

$$\beta(\theta) = \frac{\beta_T}{4\pi}P(\theta) = \frac{\beta_T}{4\pi} \times 0.7629 \times (1 + 0.9324\cos^2\theta) \quad (5.8)$$

The Rayleigh backscatter cross-section can be computed from quantum mechanics (Loudon, 1983; Measures, 1984; Goody and Yung, 1989). The volume total scattering coefficient can also be expressed through the atmosphere temperature and pressure (Cerny and Sechrist, 1980)

$$\beta_T(z) = 9.807 \times 10^{-23}\left(\frac{273}{T(z)}\right)\left(\frac{P(z)}{1013}\right)\frac{1}{\lambda^{4.0117}} \quad (5.9)$$

where T is the temperature in Kelvin, P is the pressure in millibars, and λ is the wavelength in centimeters. Substituting Equation (5.9) into (5.8) and letting $\theta = \pi$, we get the Rayleigh volume backscatter coefficient as (the wavelength has been changed to the unit of meters)

$$\beta(\pi, \lambda, z) = \sigma_R(\pi, \lambda)n_R(z)$$

$$= 2.938 \times 10^{-32}\left(\frac{P(z)}{T(z)}\right)\frac{1}{\lambda^{4.0117}} \quad (5.10)$$

Therefore, the Rayleigh scattering lidar equation is given by

$$N_S(\lambda, z) = \left(\frac{P_L(\lambda)\Delta t}{hc/\lambda}\right)(\sigma_R(\pi, \lambda)n_R(z)\Delta z)\left(\frac{A}{z^2}\right)$$
$$\times \left(\eta(\lambda)T^2(\lambda, z)G(z)\right) + N_B\Delta t \quad (5.11)$$

where the product of the backscatter cross-section and atmosphere number density is given by Equation (5.10).

5.2.1.3. Fluorescence Form of the Lidar Equation

The resonance fluorescence process involves absorption of a photon and spontaneous emission of a photon by each scattering atom. If the atom is collisionally deactivated (quenched), it will not contribute to the fluorescence intensity. Also, if the decay rate of the atom is too slow (i.e., the lifetime of the atomic excited state is too long) relative to the laser pulse duration, the saturation effect will be involved. (Details about the saturation effect will be discussed later.) Compared with the instantaneous scattering processes, such as Rayleigh and Mie scatterings that have an infinitely short duration, a few extra factors must be taken into account for the resonance fluorescence in the middle and upper atmosphere: (1) the laser spectral shape and linewidth; (2) the laser and signal polarization; (3) the laser and signal extinction due to the absorption of the resonance atom layers; (4) the finite lifetime of the excited state of the atoms; and (5) the laser pulse duration time and temporal shape. In the following treatment, the first two factors will be considered in the effective backscatter cross-section; the third factor will be considered as the extinction coefficient, which is part of the atmospheric transmission; and the last two factors will be considered as the saturation effect of the atomic layers in the middle and upper atmosphere. Since the collision rate is very low in the mesopause region due to the low atmosphere density, quenching is not a problem and will not be considered here for the measurements in the mesopause region.

5.2.1.3.1. Considerations for the effective backscatter cross-section

In the field of resonance fluorescence lidar, it is common to use the total effective scattering cross-section σ_{eff} instead of the differential backscatter cross-section. The total effective scattering cross-section σ_{eff} is defined as the ratio of the average photon number scattered by an individual atom (in all directions) to the total incident photon number per unit area. Since the resonance scattering is isotropic, the differential backscatter cross-section in Equation (5.2) can be replaced by the total effective scattering cross-section σ_{eff} divided by 4π. The σ_{eff} is determined by the convolution of the atomic absorption cross-section σ_{abs} and the laser spectral lineshape $g_L(\nu)$. The absorption cross-section σ_{abs} is the ratio of the average absorbed single-frequency photons per atom to the total incident photons per unit area, and is proportional to the probability of a single-frequency photon being absorbed by an atom:

$$\sigma_{abs}(\nu, \nu_0) = A_{ki} \frac{\lambda^2}{8\pi n^2} \frac{g_k}{g_i} g_A(\nu, \nu_0) \tag{5.12}$$

where i is the lower energy level; k is the higher energy level; A_{ki} is the transition probability for the transition from level k to level i, that is, the Einstein A coefficient (s^{-1}); λ is the laser wavelength; n is the refraction index; g_i and g_k are the degeneracy factors for the lower and upper energy levels, respectively; and g_A is the atomic transition lineshape. The absorption cross-section σ_{abs} is strongly dependent on the laser frequency ν and the atomic resonance frequency ν_0; that is, σ_{abs} depends on the lineshape g_A. There is a certain bandwidth around the resonance frequency. When the laser frequency falls in this bandwidth, resonance occurs, and the absorption cross-section becomes much larger. This bandwidth is mainly determined by the natural linewidth (due to the radiative lifetime of the atomic excited-state), the collisional broadening, and the Doppler broadening. For the atoms found in the mesopause region that we are interested in (such as Na, K, and Fe), their excited-state lifetime τ is longer than 16 ns; therefore, their natural linewidth is less than 10 MHz $(\sim\frac{1}{2\pi\tau})$. The natural linewidth and collisional broadening are both homogenous broadening and can be described by a Lorentzian lineshape

$$g_H(\nu, \nu_0) = \frac{\Delta\nu_H}{2\pi\left[(\nu - \nu_0)^2 + (\Delta\nu_H/2)^2\right]} \tag{5.13}$$

where $\Delta\nu_H$ is the homogenous broadened linewidth. Due to the low atmosphere density in the mesopause region, the broadening caused by the collisions between the atoms and the air molecules is negligible, so $\Delta\nu_H$ is mainly the natural linewidth. However, the atom thermal velocity v will cause Doppler frequency shift, and the velocity distribution of the atoms obeys the Maxwell's distribution law: $dN(v) = Ng_D(v)dv$, where

$$g_D(v) = \sqrt{\frac{M}{2\pi k_B T}} \exp\left(-\frac{Mv^2}{2k_B T}\right) \tag{5.14}$$

Where M is the mass of the atom, K_B is the Boltzmann Constant, and T is the temperature.

Equation (5.14) is for the ideal gas in the case when the average velocity is zero. Since the observed photon absorption or scattering is the overall statistical results over all the atoms with different velocities, the overall absorption cross-section would be the single atom natural lineshape integrating through the velocity distribution, that is, a statistical average absorption cross-section. This statistical average will cause the Doppler broadening of the transition line. The velocity distribution, essentially the Doppler broadening, is temperature dependent. For the typical temperature of 200 K in the mesopause region, the Doppler broadening linewidth is around 1 GHz. It is much wider than the natural linewidth so that the atomic absorption line is essentially Doppler broadened. The statistically averaged absorption cross-section for each atomic transition line is then given by

$$
\sigma_{\text{abs}}(\nu, \nu_0) = \sigma_0 \exp\left(-\frac{(\nu - \nu_0)^2}{2\sigma_{\text{D}}^2}\right) \tag{5.15}
$$

where

$\nu_0 =$ the resonance frequency of each atomic transition line;

$\sigma_{\text{D}} = \nu_0 \sqrt{\frac{k_{\text{B}}T}{Mc^2}}$ is the rms linewidth of Doppler broadening;

$\sigma_0 = \frac{1}{\sqrt{2\pi}\sigma_{\text{D}}} \frac{e^2}{4\epsilon_0 m_e c} f_{\text{ik}}$ is the peak absorption cross-section at resonance;

$f_{\text{ik}} = \frac{\epsilon_0 m_e c \lambda^2}{2\pi e^2} \frac{g_{\text{k}}}{g_{\text{i}}} A_{\text{ki}}$ is the absorption oscillator strength; and

$A_{\text{ki}} = \frac{\text{BF}_{\text{ki}}}{\tau_{\text{k}}}$ is the transition probability from level k to level i, where τ_{k} is the radiative lifetime of the upper level k, and BF_{ki} is the emission branching fraction for this transition.

When the laser has finite spectral linewidth, the total effective scattering cross-section is equal to the convolution of the atomic absorption cross-section and the laser lineshape. The laser lineshape, that is, its frequency distribution function g_{L} (ν) is normalized so that

$$
\int_0^\infty g_{\text{L}}(\nu, \nu_{\text{L}}) d\nu = 1 \tag{5.16}
$$

where ν_L is the central frequency of the laser radiation. It is usually referred as the laser frequency. The effective scattering cross-section is then given by

$$\sigma_{\text{eff}}(\nu_L, \nu_0) = \int_{-\infty}^{+\infty} \sigma_{\text{abs}}(\nu, \nu_0) g_L(\nu, \nu_L) d\nu \tag{5.17}$$

We approximate the laser spectral lineshape by a Gaussian lineshape with a rms width of σ_L:

$$g_L(\nu, \nu_L) = \frac{1}{\sqrt{2\pi}\sigma_L} \exp\left(-\frac{(\nu - \nu_L)^2}{2\sigma_L^2}\right) \tag{5.18}$$

Then the effective cross-section can be written as

$$\sigma_{\text{eff}}(\nu_L, \nu_0) = \frac{\sigma_D \sigma_0}{\sqrt{\sigma_D^2 + \sigma_L^2}} \exp\left(-\frac{(\nu_0 - \nu_L)^2}{2(\sigma_D^2 + \sigma_L^2)}\right) \tag{5.19}$$

The above derivation is intended for the resonance line, which displays no hyperfine structure. In general, if we have a resonance line with m hyperfine structures, we can express a similar relationship by summing up the weighted effect of each hyperfine line as below:

$$\sigma_{\text{eff}}(\nu_L) = \sum_{i=1}^{m} w_i \frac{\sigma_{Di} \sigma_0}{\sqrt{\sigma_{Di}^2 + \sigma_L^2}} \exp\left(-\frac{(\nu_{0i} - \nu_L)^2}{2(\sigma_{Di}^2 + \sigma_L^2)}\right) \tag{5.20}$$

where $w_i = I_i / \sum_{j=1}^{m} I_j$ is the weighting factor for the ith line, and I_i is the relative intensity of the ith line.

The polarization of resonance radiation is a direct consequence of the hyperfine structure of the transitions and of the Zeeman effect when a magnetic field is present. The Earth's magnetic field interacts with occupied atomic states to change their states of polarization. This results in a change in the emitted-radiation pattern from its zero-magnetic-field value. Thus, the relative intensities of the hyperfine components can be modified due to the Hanle effect, and they are dependent on the laser polarization. In the case of hyperfine structures for resonance fluorescence in the geomagnetic field, we count the polarization effect into the relative intensity I_i of the hyperfine components, which would affect the weighting factors w_i in Equation (5.20). Therefore, the total effective scattering cross-section can still be represented correctly by Equation (5.20) as long as proper weighting factors are used.

For the fluorescence form of the lidar equation, the volume backscatter coefficient given by Equation (5.2) will be modified as follows. The differential backscatter cross section in Equation (5.2) will be replaced by the total effective scattering cross-section given by Equation (5.19) or (5.20) over 4π for the resonance line without and with hyperfine structures, respectively. The $n_i(z)$ in Equation (5.2) will be replaced by the number density of the resonance constituent atoms $n_c(z)$. The $p_i(\lambda)$ in Equation (5.2) will be replaced by the branching ratio R_B, which is defined as the percentage of the fluorescence photons that fall in the absorption resonance wavelength λ, occcuping the entire fluorescence photons emitted by excited-atoms, when the excited atoms may return to several sub-ground-states or go to other excited states to emit photons at different wavelengths rather than the absorption resonance wavelength. Therefore, when the saturation effect of the atomic layers is not considered, the volume backscatter coefficient for the resonance fluorescence can be written as

$$\beta(\lambda, \lambda_L, z) = \frac{\sigma_{eff}(\lambda_L)}{4\pi} n_c(z) R_B(\lambda) \tag{5.21}$$

5.2.1.3.2 Extinction coefficient

The extinction due to the absorption of the atomic layers can be regarded as part of the atmospheric transmission T in the lidar equation (5.1). In the scattering form of the lidar equation, this transmission T only includes the loss due to the absorption and scattering by the lower atmosphere below the atomic layer in the mesopause region. In the resonance fluorescence case, the total transmission T can be written as the product of the lower atmosphere transmission T_a and the extinction coefficient E due to the absorption of the atomic layers:

$$T(\lambda, z) = T_a(\lambda) E(\lambda, z) \tag{5.22}$$

where the extinction coefficient is defined as the ratio of the transmitted laser power to the incident laser power at the atomic layers. It can be calculated from the effective cross-section and the atomic density:

$$E(z) = \exp\left(-\int_{z_{\text{bottom}}}^{z} \sigma_{\text{eff}}(\lambda, z) n_{\text{c}}(z) \mathrm{d}z\right) \qquad (5.23)$$

where z_{bottom} is the bottom altitude of the atomic layers.

Therefore, when the saturation effect of the atomic layers is not considered, the resonance fluorescence lidar equation is given by

$$\begin{aligned} N_{\text{S}}(\lambda, z) = {} & \left(\frac{P_{\text{L}}(\lambda)\Delta t}{hc/\lambda}\right)(\sigma_{\text{eff}}(\lambda)n_{\text{c}}(z)R_{\text{B}}(\lambda)\Delta z)\left(\frac{A}{4\pi z^2}\right) \\ & \times \left(\eta(\lambda)T_{\text{a}}^2(\lambda)E^2(\lambda, z)G(z)\right) + N_{\text{B}}\Delta t \end{aligned}$$

$$(5.24)$$

Here, the first parenthesis factor is the transmitted laser photon numbers during integration time; the second factor is the probability that a photon is scattered into the wavelength λ in all directions; the third factor is the probability that a scattered photon can be received by the receiving telescope, that is, the telescope area over the spherical surface area; the fourth factor is the overall system efficiency.

5.2.1.3.3   Saturation and optical pumping effects

For the case that laser pulse duration is in the nanosecond range, Equation (5.24) does not always hold the truth as the radiative lifetime of the atomic excited state is generally on the order of tens of nanoseconds. The finite lifetime of the atomic excited state will cause a saturation effect of the atomic layers. According to Equation (5.24), the number of backscattered photons should be proportional to the flux of the transmitted laser photons. However, for large laser intensities, the returned fluorescence photons are no longer proportional to the laser power, but less than what is expected from Equation (5.24). Saturation is defined as the deviation from this proportionality. The deviation can be caused by the following reasons. (1) Due to the finite lifetime, atoms at the excited states must relax before they can absorb another photon. In other words, during the lifetime of their excited state, excited atoms cannot absorb another photon. (2) Stimulated emission occurs with the photon emitted in the forward direction; that is, this photon is unobservable. Reasons (1) and (2) contribute about the same order of magnitude to these deviations (von der Gathen, 1991). (3) Optical pumping effect: when a narrowband laser is used to excite the populations from a certain ground state to excited states, the excited atoms can relax back to other ground states. This will cause a partial depopulation of the particular ground state in favor of the other substates. The optical pumping effect contributes only slightly to the deviations for short laser pulses, but becomes important only if the laser pulses are long compared to the radiative lifetime of the excited states.

Here we follow the analysis made by Megie et al. (1978) on the saturation effect. During the laser pulse duration Δt_L, the excited-state population (i.e., the total number density over all velocity distribution) n_e is not in the steady state, but is determined by the rate equation as below:

$$\frac{dn_e}{dt} = -\frac{n_e}{\tau_R} + (n - n_e)\frac{\sigma_{eff}N_L(t)T}{z^2\Omega} - n_e\frac{\sigma_{eff}N_L(t)T}{z^2\Omega} \qquad (5.25)$$

where τ_R is the radiative lifetime of the excited-state, n is the total number density of the atoms in the layer, i.e., the sum of the ground-state population $(n - n_e)$ and the excited-state population n_e, $N_L(t)$ is the total number of laser emitted photons per unit time, T is the one-way atmosphere transmission, and Ω is the solid angle illuminated by the laser beam. The Ω is related to the full divergence θ_L by

$$\Omega = \frac{\pi}{4}\theta_L^2 \qquad (5.26)$$

In order to derive an analytical solution of Equation (5.25), we assume that the temporal shape of the laser pulse can be represented by a rectangular shape with a width of Δt_L:

$$N_L(t) = \begin{cases} \frac{N_L}{\Delta t_L}, & 0 < t < \Delta t_L \\ 0, & t > \Delta t_L \end{cases} \qquad (5.27)$$

where N_L is the total number of emitted photons related to each pulse energy. By solving Equation (5.25), the excited-state population is given by

$$n_e(t) = \begin{cases} \dfrac{n\tau'}{2t_S}(1 - e^{-t/\tau'}) & t < \Delta t_L \\[3mm] \dfrac{n\tau'}{2t_S}(1 - e^{-\Delta t_L/\tau'})e^{-(t-\Delta t_L)\tau_R} & t > \Delta t_L \end{cases} \tag{5.28}$$

where t_S and τ' are defined as

$$t_S = \frac{z^2 \Omega \Delta t_L}{2\sigma_{eff}N_L T} \tag{5.29}$$

$$\frac{1}{\tau'} = \frac{1}{\tau_R} + \frac{1}{t_S} \tag{5.30}$$

Since the decay rate of the excited-state population through spontaneous emission is $n_e(t)/\tau_R$, the number of reemitted photons per unit time in the volume considered will be $(z^2 \Omega \Delta z)[n_e(t)/\tau_R]$. Its integration over the time is the number of reemitted photons, and can be used to replace the product of the first and the second factors in Equation (5.24). Therefore, the number of phonons received by the lidar during the integration time will be

$$N_S^{Sat}(\lambda, z) = \left(\frac{A}{4\pi z^2}\right)(\eta(\lambda)T(\lambda, z)G(z))(z^2 \Omega \Delta z)$$
$$\times \int_0^{\Delta z/c} \frac{n_e(t)}{\tau_R}\, dt + N_B \Delta t \tag{5.31}$$

If we consider the case that the integration time is much longer than the radiative lifetime of the atomic excited state and the laser pulse duration time, then the above integration can be extended to infinity so that

$$N_S^{Sat}(\lambda, z) = N_S(\lambda, z)\frac{1}{1 + (\tau_R/t_S)}\left\{1 - \frac{\tau_R}{\Delta t_L}\frac{(\tau_R/t_S)}{1 + (\tau_R/t_S)}\right.$$
$$\times \left[\exp\left(-\frac{\Delta t_L}{\tau}\left(1 + \frac{\tau_R}{t_S}\right)\right) - 1\right]\right\} + N_B \Delta t \tag{5.32}$$

where $N_S(\lambda, z)$ is the number of photons received when no saturation effect exists, and given by

$$N_S(\lambda,z) = N_L \sigma_{\text{eff}}(\lambda) n_c(z) \Delta z \left(\frac{A}{4\pi z^2} \right)$$
$$\times \left(\eta(\lambda) T^2(\lambda, z) G(z) \right) + N_B \Delta t \qquad (5.33)$$

Equation (5.32) shows that the saturation effect is independent of the Na number density and essentially determined by the ratio of the excited-state radiative lifetime to the saturation time (τ_R/t_S). In general, the smaller the ratio, the less the saturation. If $\tau_R = 0$ (such as the Rayleigh or Mie scattering), then $N_S^{\text{Sat}} = N_S$; that is, there is no saturation effect. From the definition of saturation time by Equation (5.29), we may find that the longer laser pulse duration Δt_L, lower laser pulse energy N_L, and smaller effective cross-section σ_{eff} will make the saturation time t_S longer linearly, while the beam divergence θ_L will affect the saturation time quadratically. If the beam divergence decreases by factor of 2, then the saturation time will decrease by factor of 4, which can cause much severe saturation effect.

Considering the Na lidar as an example, we have the Na radiative lifetime $\tau_R = 16.23$ nsec. Assuming the laser pulse energy of 50 mJ with a duration time of $\Delta t_L = 6$ nsec, the number of photons emitted per pulse is then given by $N_L = 1.48 \times 10^{17}$. For the beam divergence $\theta_L = 1$ mrad, atmospheric transmission $T = 0.7$, altitude around $z = 90$ km, and the effective cross-section $\sigma_{\text{eff}} = 10 \times 10^{-16}$ m^2, we get $t_S = 184$ nsec. Substituting these numbers into Equation (5.32), we obtain the ratio $N_S^{\text{Sat}}/N_S = 98.5\%$. If the beam divergence is compressed to $\theta_L = 0.4$ mrad, then $t_S = 29.4$ nsec and $N_S^{\text{Sat}}/N_S = 91.5\%$. When using the small beam divergence angle for daytime measurements, if we do not count in the saturation effect, the photon count will be approximately 9% less than expected. This saturation effect should not, therefore, be neglected.

In order to derive an analytical solution of the saturation effect, we made several approximations such as the rectangular pulse shape. When using different pulse shapes, for example, the Gaussian shape, the saturation results will be slightly different. However, Equation (5.32) still represents a good estimation of the saturation effect for actual situation if the Δt_{L} is carefully chosen. We would like to point out that the absolute saturation level would directly affect the constituent density measurement. However, the influence on the temperature and wind measurements is related to the relative saturation effects at different frequencies, not the absolute saturation. When a narrowband Na lidar is used, the considerations for the Na saturation computation will be quite different from above treatments that are more suitable for broadband lidars. More detailed analysis and numerical simulation results for the Na saturation can be found in Welsh and Gardner (1989), von der Gathen (1991), and Milonni et al. (1998, 1999).

5.2.1.4. Solutions for the Resonance Fluorescence Lidar Equation

If we include the saturation correction factor into the total effective scattering cross-section σ_{eff}, then the resonance fluorescence lidar equation can be generally written as Equation (5.24). By solving this equation, we can derive the number density of the constituent in the mesopause region as

$$n_{\mathrm{c}}(z) = \frac{N_{\mathrm{S}}(\lambda, z) - N_{\mathrm{B}}\Delta t}{\left(\frac{P_{\mathrm{L}}(\lambda)\Delta t}{hc/\lambda}\right)(\sigma_{\mathrm{eff}}(\lambda)R_{\mathrm{B}}(\lambda)\Delta z)\left(\frac{A}{4\pi z^2}\right)(\eta(\lambda)T_{\mathrm{a}}^2(\lambda)E^2(\lambda, z)G(z))}$$

$$(5.34)$$

If the atmosphere transmission and lidar system parameters are known, the absolute constituent number density can be obtained from Equation (5.34). However, the uncertainty in laser power, atmosphere transmission, and optical efficiency prevents the calculation of absolute number density of the constituent. The most common method for solving the lidar equation is to normalize the resonance fluorescence signals to the Rayleigh scattering signals at a common lower altitude (e.g., 35 km for the Na lidar signal) to obtain a relative number density profile. The relative number density is then modified by the absolute atmosphere number density at this lower altitude, which is either measured by other instruments or adopted from atmospheric models, to obtain the absolute number density of the constituent in the mesopause region. We rewrite the resonance fluorescence lidar equation and Rayleigh scattering lidar equation as below:

$$N_S(\lambda, z) = \left(\frac{P_L(\lambda)\Delta t}{hc/\lambda} \right) (\sigma_{\mathrm{eff}}(\lambda) n_c(z) R_B(\lambda) \Delta z) \left(\frac{A}{4\pi z^2} \right)$$
$$\times \left(\eta(\lambda) T_a^2(\lambda) E^2(\lambda, z) G(z) \right) + N_B \Delta t \qquad (5.35)$$

$$N_R(\lambda, z_R) = \left(\frac{P_L(\lambda)\Delta t}{hc/\lambda} \right) (\sigma_R(\pi, \lambda) n_R(z_R) \Delta z) \left(\frac{A}{z_R^2} \right)$$
$$\times \left(\eta(\lambda) T_a^2(\lambda, z_R) G(z_R) \right) + N_B \Delta t \qquad (5.36)$$

The Rayleigh scattering lidar equation is for the Rayleigh scattering signal at the normalization altitude z_R, and the one-way atmosphere transmission is $T_a(z_R)$. By normalizing Equation (5.35) with Equation (5.36), we obtain the relative number density as

$$\frac{n_c(z)}{n_R(z_R)} = \frac{N_S(\lambda, z) - N_B \Delta t}{N_R(\lambda, z_R) - N_B \Delta t} \frac{z^2}{z_R^2} \frac{4\pi \sigma_R(\pi, \lambda)}{\sigma_{\mathrm{eff}}(\lambda) R_B(\lambda)} \frac{T_a^2(\lambda, z_R) G(z_R)}{T_a^2(\lambda, z) E^2(\lambda, z) G(z)}$$
$$(5.37)$$

The normalization altitude z_R is usually chosen in the region free of aerosol particles to avoid any contributions from Mie scattering, typically above 30 km. The Rayleigh scattering signal at z_R should be of the same order of magnitude of the resonance fluorescence signals in the mesopause region in order to avoid any effects due to the signal dynamic range. We further assume $T_a(\lambda, z) = T_a(\lambda, z_R)$ (i.e., the extinction between z_R and z is negligible) and $G(z) = G(z_R) = 1$ (i.e., the receiving telescope can see the full laser beam at both z and z_R ranges). Thus, the number density of constituent in the mesopause region can be derived as

$$n_c(z) = n_R(z_R) \frac{N_S(\lambda, z) - N_B \Delta t}{N_R(\lambda, z_R) - N_B \Delta t} \frac{z^2}{z_R^2} \frac{4\pi \sigma_R(\pi, \lambda)}{\sigma_{\mathrm{eff}}(\lambda) R_B(\lambda)} \frac{1}{E^2(\lambda, z)} \qquad (5.38)$$

The resonance fluorescence and Rayleigh scattering photon counts, the ranges, the total effective scattering cross-section, and the extinction coefficient are measured from the lidar observations while the Rayleigh backscatter cross-section can be calculated from quantum mechanics. The atmosphere molecular number density $n_R(z_R)$ at the normalization altitude z_R is usually taken from atmosphere models. Therefore, the absolute number density of the constituent in the mesopause region can be computed from Equation (5.38).

An alternative way is to relate the product of the Rayleigh backscatter cross-section and the atmosphere molecular number density to the atmosphere temperature and pressure at the same altitude using the relationship of Equation (5.10). Then, by taking the temperature and pressure values from atmosphere models or from other instrument measurements, we can also compute the absolute number density of the constituent by the following solution:

$$n_c(z) = \frac{N_S(\lambda, z) - N_B \Delta t}{N_R(\lambda, z_R) - N_B \Delta t} \frac{z^2}{z_R^2} \frac{4\pi}{\sigma_{\mathrm{eff}}(\lambda) R_B(\lambda) E^2(\lambda, z)}$$
$$\times 2.938 \times 10^{-32} \left(\frac{P(z_R)}{T(z_R)} \right) \frac{1}{\lambda^{4.0117}} \tag{5.39}$$

where λ is the laser wavelength in meters, T is the atmosphere temperature in Kelvin, and P is the atmosphere pressure in millibars at the normalization range z_R.

5.2.2. Na Wind/Temperature Lidar

5.2.2.1. Introduction

Range-resolved high-resolution temperature profiles of the mesopause region can be obtained by active probing of the Doppler broadening of the hyperfine structure of the sodium D_2 line by a narrowband lidar. This idea was first proposed and demonstrated by Gibson et al. (1979) and Thomas et al. (1980), who were able to derive the temperature near the peak of Na layer with error of $\pm 15\,\mathrm{K}$. The first practical narrowband Na temperature lidar was developed at the University of Bonn with an excimer-laser pumped dye-laser system (Fricke and von Zahn, 1985). The lidar was deployed at Andoya, Norway, and routinely obtained Na temperature profiles above Andoya. The system performance was limited by the excimer-dye laser technology. Because the frequency of their pulsed laser was neither reproducible nor predictable, the relative frequency of each individual laser pulse was measured with a wavelength meter, which was calibrated to a sodium vapor cell, and the laser frequency was scanned over a relatively wide frequency range to adequately cover the Doppler-broadened Na spectrum. A least-square fit was then applied to the spectra using a theoretical model of sodium spectroscopy to derive temperature at each altitude. Using this technique, Fricke and von Zahn (1985) reported an uncertainty of ± 5 K at the layer peak with an integration period of 10 min and a vertical resolution of 1 km.

A more sophisticated narrowband Na lidar system using a two-frequency technique was developed through a collaboration between the Colorado State University (CSU) (She's group) and the University of Illinois at Urbana-Champaign (UIUC) (Gardner's group) (She et al., 1990; Bills et al., 1991a). The system employed a pulsed-dye-amplified, frequency-stabilized, single-mode, ring dye laser and the Doppler-free saturation–absorption spectroscopy of a Na vapor cell (She et al., 1990). The new Na lidar system and measurement techniques dramatically improved the resolution and precision of Na temperature measurements. Absolute temperature accuracies at the Na layer peak of better than ± 3 K with a vertical resolution of 1 km and an integration period of approximately 5 min were achieved (She et al., 1990). This measurement accuracy was actually limited by the theoretical and experimental uncertainties in the determination of the Na saturation spectrum. She and Yu (1995) did a detailed theoretical study and experimental measurement to determine the Doppler-free feature within ± 0.1 MHz. These studies improved the measurement accuracy to ± 1 K.

Since the cw ring laser frequency is locked to the Na D_2 line using the Doppler-free spectroscopy, the laser frequency can be calibrated to an absolute frequency reference. This also enabled the wind measurements in the mesopause region using four-frequency technique as demonstrated in Bills et al. (1991b) and using a three-frequency technique as demonstrated in She and Yu (1994). Error budgets for Na lidar measurements were presented in these early publications. Papen et al. (1995a) gave a detailed systematic error analysis for the two- and three-frequency techniques. The two- and three-frequency lidar techniques dominate the Na temperature and wind measurements today because they provide higher temporal resolutions, and the fast switching between two or three frequencies can help eliminate the bias caused by Na density fluctuations over the measurement time.

Chen et al. (1996) demonstrated the daytime temperature measurements of Na lidar using a Na Faraday anomalous dispersion optical filter (FADOF) to reject solar scattering noise in the CSU lidar receiver. A single etalon was employed in the UIUC lidar receiver to reject sky background noise enabling the full diurnal temperature measurements (Yu et al., 1997). The CSU group further improved the Na lidar receiver part and now is able to measure both temperature and wind through a complete 24-h diurnal cycle with a moderate receiving telescope (She et al., 2003). Concurrently, the UIUC group coupled the Na Doppler lidar transmitter with a large telescope of 3.5 m diameter through coude optics to make a steerable lidar with large power aperture product. This lidar enables more sophisticated atmosphere measurements, such as tracking the persistent meteor trails (Chu et al., 2000b), heat and momentum flux of gravity waves, and their dynamical influences on the atmosphere mean circulation (Tao and Gardner, 1995; Gardner and Yang, 1998; Zhao et al., 2003).

5.2.2.2. Measurement Principle (Doppler Technique)

The energy levels and spectrum of the alkali Na atoms have been studied in detail. The accuracy of the structure of the Na D_2 electric-dipole transition has been determined to a fraction of a megahertz (Arimondo et al., 1977). The energy levels of the ground state (3s $^2S_{1/2}$) and the first excited state (3p $^2P_{1/2}$

and 3p $^2P_{3/2}$) of sodium atoms are shown in Figure 5.1. The D_2 and D_1 lines are the transitions from 3p $^2P_{3/2}$ to 3s $^2S_{1/2}$ and from 3p $^2P_{1/2}$ to 3s $^2S_{1/2}$, respectively. The transition probabilities (Einstein A_{ki} coefficient) for the Na D_2 and D_1 lines are 0.616×10^8 and 0.614×10^8 s^{-1}, respectively. They correspond to the radiative lifetimes of $^2P_{3/2}$ and $^2P_{1/2}$ of 16.23 ± 0.03 and 16.29 ± 0.03 nsec (Jones et al., 1996; Oates et al., 1996; Volz et al., 1996). The Na D_2 line has a natural linewidth of 9.802 ± 0.022 MHz (Oates et al., 1996) and an oscillator strength f_0 of 0.6411. The Na nuclear angular momentum $I = 3/2$. Thus, the ground state $^2S_{1/2}$ splits into two hyperfine levels with total angular momentum $F = 1$ and $F = 2$. The excited states $^2P_{1/2}$ and $^2P_{3/2}$ split to two and four hyperfine levels with F values ranging from 1 to 2 and from 0 to 3, respectively. The D_2 line consists of two groups of lines, D_{2a} and D_{2b}, which are formed by the hyperfine splits of the Na ground state. The frequency separation between these two groups is 1771.6261 MHz (Kasevich et al., 1989). Each group contains three lines resulting from the hyperfine splits of the Na excite state of 3p $^2P_{3/2}$. The related numerical information of Na D_1 and D_2 lines is listed in Table 5.1. Here the zero-frequency reference is the weighted center of the D_2 six

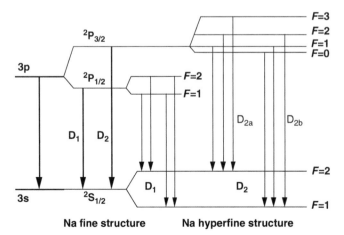

Figure 5.1 Energy level diagram of atomic Na. (Adapted from She, C.Y., and Yu, J.R., *Appl. Opt.*, 34, 1063–1075, 1995. With permission.)

Table 5.1 Parameters of the Na D_1 and D_2 Transition Lines

Transition Line	Central Wavelength (nm)	Transition Probability ($10^8 s^{-1}$)	Radiative Lifetime (nsec)	Oscillator Strength f_{ik}
D_1 ($^2P_{1/2} \rightarrow {}^2S_{1/2}$)	589.7558	0.614	16.29	0.320
D_2 ($^2P_{3/2} \rightarrow {}^2S_{1/2}$)	589.1583	0.616	16.23	0.641

Group	$^2S_{1/2}$	$^2P_{3/2}$	Offset (GHz)	Relative Line Strength[a]
D_{2b}	$F=1$	$F=2$	1.0911	5/32
		$F=1$	1.0566	5/32
		$F=0$	1.0408	2/32
D_{2a}	$F=2$	$F=3$	−0.6216	14/32
		$F=2$	−0.6806	5/32
		$F=1$	−0.7150	1/32

Doppler-Free Saturation–Absorption Features of the Na D_2 Line

f_a (MHz)	f_c (MHz)	f_b (MHz)	f_+ (MHz)	f_- (MHz)
−651.4	187.8	1067.8	−21.4	−1281.4

[a]Relative line strengths are in the absence of a magnetic field or the spatial average. When Hanle effect is considered in the atmosphere, the relative line strengths will be modified depending on the geomagnetic field and the laser polarization.

transitions, whose wavelength $\lambda_0 = 589.15826$ nm *in vacuo* (Martin and Zalubas, 1981). The relative oscillator strengths listed in the table are the values in the absence of a magnetic field.

For single Na atom, its absorption line has a Lorentzian shape given by Equation (5.13) with natural linewidth of about 10 MHz. Because of the Doppler effect, the absorption line will be broadened. To an atom moving away from the laser with velocity v_R, the laser frequency perceived by the atom appears to be shifted:

$$\nu' = \nu \left(1 - \frac{v_R}{c}\right) \tag{5.40}$$

where ν is the laser central frequency seen by an atom at rest, ν' is the Doppler-shifted laser frequency, v_R is the radial velocity of the atom ($v_R > 0$ if the atom moving away from the laser radiation source), and c is the speed of light. Since the velocity distribution obeys the Maxwell–Boltzmann distribution (given by Equation 5.14) under thermodynamic equilibrium, the detected average absorption cross-section of the Na (D_2) line will be the single atom absorption cross-section integrating over the velocity distribution. Thus, the absorption line is Doppler broadened. Since the collisional broadening at the mesopause region is negligible due to the low atmosphere density, the D_2 line has almost pure Doppler-broadening lineshape. Plotted in Figure 5.2(a) is the computed Na absorption cross-section for three different atmospheric temperatures at $v_R = 0$ m/s, and Figure 5.2(b) for three different radial wind velocities at $T = 200$ K. The Doppler-broadened line has a Gaussian shape given by

$$\sigma_{abs}(\nu) = \frac{1}{\sqrt{2\pi}\sigma_D} \frac{e^2 f}{4\epsilon_0 m_e c} \sum_{n=1}^{6} A_n \exp\left(-\frac{\left[\nu_n - \nu\left(1 - \frac{v_R}{c}\right)\right]^2}{2\sigma_D^2}\right)$$

(5.41)

where f is the oscillator strength of Na D_2 line, ν_n and A_n are the center frequencies and the line strengths of the six electric-dipole-allowed D_2 transitions, respectively, ν is the laser frequency, v_R is the radial velocity of the Na atom moving away from the laser source, and c is the light speed. The rms width of the Doppler-broadened line is σ_D given by

$$\sigma_D = \sqrt{\frac{k_B T}{M\lambda_0^2}}$$

(5.42)

where M is the mass of a single Na atom, k_B is the Boltzmann constant, λ_0 is the mean Na D_2 transition wavelength, and T is the temperature.

The Doppler-broadened linewidth is a sensitive function of temperature as shown in Equation (5.42) and Figure 5.2(a). If we measure this Doppler-broadened linewidth, then we can derive the temperature of the Na atoms. Since the Na atoms are in equilibrium with the surrounding atmosphere, the Na temperature would be equal to the atmosphere temperature in the mesopause region. Examining Equation (5.41) and Figure 5.2(b), the radial velocity v_R of Na atoms, and thus of

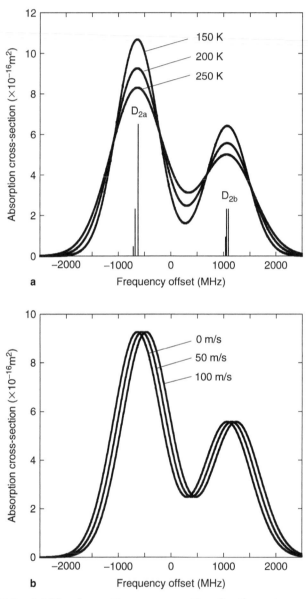

Figure 5.2 (a) Na absorption cross section for three temperatures at $v_R = 0$ m/s. (b) Na absorption cross section for three radial wind velocities at $T = 200$ K.

the atmosphere, can be derived by measuring the Doppler frequency shift of the central frequency of the Na D_2 line. Due to the finite linewidth of the laser beam, the total scattering cross-section σ_{eff} is the convolution of the absorption cross-section and the laser lineshape. The actual σ_{eff} depends on the laser lineshape. Under the assumption of Gaussian lineshape of the laser, σ_{eff} can be written as

$$\sigma_{eff}(\nu) = \frac{1}{\sqrt{2\pi}\sigma_e} \frac{e^2 f}{4\epsilon_0 m_e c} \sum_{n=1}^{6} A_n \exp\left(-\frac{\left[\nu_n - \nu\left(1 - \frac{v_R}{c}\right)\right]^2}{2\sigma_e^2} \right)$$

(5.43)

where $\sigma_e = \sqrt{\sigma_D^2 + \sigma_L^2}$ is the rms width of the total lineshape.

A sodium vapor cell with sufficient saturation vapor pressure can be maintained at reasonable temperature ($\sim60°C$). Under the simultaneous illumination of two counter-propagating beams at a saturated laser intensity, the Na fluorescence from the vapor cell shows the Doppler-free features at D_{2a} and D_{2b} peaks (f_a and f_b) as well as at the crossover resonance frequency f_c. These features are similar to those observed with the well-known Doppler-free saturated-absorption spectroscopy (Hansch et al., 1971). A measured Doppler-free spectroscopy in a Na vapor cell is shown in Figure 5.3 (Bills et al., 1991a). She and Yu (1995) studied the Doppler-free saturation fluorescence spectroscopy of Na atoms in great detail both theoretically and experimentally. They determined the frequencies of the dominant Doppler-free features to within $\pm0.1\,MHz$ from a simulated spectrum. The frequencies for D_{2a}, crossover, and D_{2b} resonances are -651.4, 187.8, and $1067.8\,MHz$, respectively, relative to the weighted center of the six D_2 hyperfine transition lines. These features are essential for the operation of the narrowband Na resonance fluorescence lidar because these Doppler-free dips or peaks can be used to lock the laser frequency precisely and provide an absolute frequency calibration for the lidar.

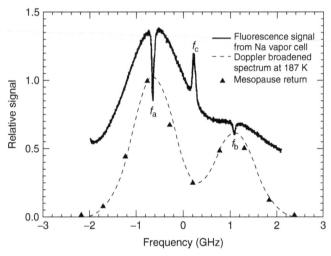

Figure 5.3 Doppler-free saturation–absorption spectroscopy measured from a Na vapor cell. (From Bills, R.E., Gardner, C.S., and She, C.Y., *Opt. Eng.*, 30, 13–21, 1991. With permission.)

5.2.2.2.1. Frequency-scanning method

Starting from the lidar equations (5.35) and (5.36), the measured Na photon counts and Rayleigh photon counts are given as

$$N_{\mathrm{Na}}(\lambda,z) = \left(\frac{P_{\mathrm{L}}(\lambda)\Delta t}{hc/\lambda}\right)(\sigma_{\mathrm{eff}}(\lambda)n_{\mathrm{Na}}(z)\Delta z)\left(\frac{A}{4\pi z^2}\right)$$
$$\times(\eta(\lambda)T_{\mathrm{a}}^2(\lambda)E^2(\lambda,z)G(z)) \qquad (5.44)$$

$$N_{\mathrm{R}}(\lambda,z_{\mathrm{R}}) = \left(\frac{P_{\mathrm{L}}(\lambda)\Delta t}{hc/\lambda}\right)(\sigma_{\mathrm{R}}(\pi,\lambda)n_{\mathrm{R}}(z_{\mathrm{R}})\Delta z)\left(\frac{A}{z_{\mathrm{R}}^2}\right)$$
$$\times(\eta(\lambda)T_{\mathrm{a}}^2(\lambda,z_{\mathrm{R}})G(z_{\mathrm{R}})) \qquad (5.45)$$

Equations (5.44) and (5.45) represent, respectively, the pure Na and Rayleigh photon counts as the background has been subtracted from the raw photon counts. From these two lidar equations, we obtain the measured total scattering cross-section as

$$\sigma_{\text{eff}}(\lambda, z) = \frac{C(z)}{E^2(\lambda, z)} \frac{N_{\text{Na}}(\lambda, z)}{N_{\text{R}}(\lambda, z_{\text{R}})} \tag{5.46}$$

where

$$C(z) = \frac{\sigma_{\text{R}}(\pi, \lambda) n_{\text{R}}(z_{\text{R}})}{n_{\text{Na}}(z)} \frac{4\pi z^2}{z_{\text{R}}^2} \tag{5.47}$$

Here we have assumed $T_a(\lambda, z) = T_a(\lambda, z_{\text{R}})$ and $G(z) = G(z_{\text{R}}) = 1$. It is apparent from Equation (5.46) that the wavelength-dependent lineshape of the total scattering cross-section σ_{eff} at each altitude z can be obtained by scanning laser frequency through the Na(D_2) resonance line. If we assume that the Na atom number density $n_{\text{Na}}(z)$ remains unchanged during the period of the laser frequency scan, then $C(z)$ is a constant for each altitude z. By fitting the theoretical function of the total effective scattering cross-section (Equation 5.43) to the measured fluorescence signal scanning over a selected range of frequencies (Equation 5.46), the Na temperature $T(z)$ at each altitude z can be derived from the fitting parameters. Thus, the range-resolved temperature profiles in the mesopause region can be achieved. This scanning method was first demonstrated by Fricke and von Zahn (1985). Shown in Figure 5.4 is an example taken from Fricke and von Zahn (1985). The laser frequency was scanned through a relative large range around the D_{2a} peak. The processed signals for the altitude interval 90.5 to 91.5 km were plotted as the relative intensities versus the relative wavelength in Figure 5.4. A best fit of the Doppler broadened line shape was applied to the data and plotted as a solid line in the figure. The convolution with the instrument profile was included. The fitting parameters give an atmospheric temperature of 195 K at the altitude of 90.5 to 91.5 km. The range-resolved temperature profile was then obtained by applying this method to the altitude range where the Na signals are large enough, as shown in Figure 5.5 (Fricke and von Zahn, 1985).

If the laser frequencies are calibrated against an absolute frequency reference, then the fitting method to the scanned

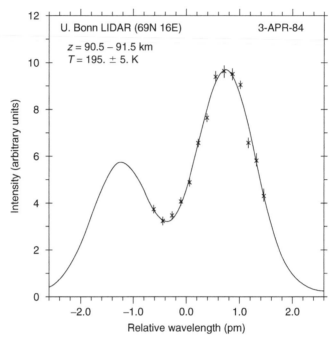

Figure 5.4   Relative intensities from the D_{2a} component of the Na hyperfine structure in the terrestrial atmosphere in the 91 ± 0.5 km range observed on April 3, 1984 at Bleik (69°N, 16°E) on Andoya Island. The solid line is the best fit of the Doppler widened line shape to the data. The convolution with the instrument profile function is included. It corresponds to an atmospheric temperature of 195 K. (From Fricke, K.H., and von Zahn, U., *J. Atmos. Terr. Phys.*, 47, 499–512, 1985. With permission.)

fluorescence signals can also derive the radial wind v_R as indicated by Equation (5.43).

5.2.2.2.2.   *Ratio techniques: two-frequency*

Scanning through a large frequency range reduces the temporal resolution and causes uncertainties because of inherent Na density fluctuations during the measurement time. Examining Figure 5.2(a), the signal levels at the D_{2a} peak and at the intermediate minimum between the D_{2a} and D_{2b} peaks are particularly sensitive to temperature. Since the responses

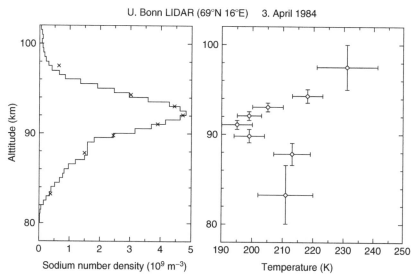

Figure 5.5 Na density and temperature profiles obtained on April 3, 1984 at Andoya (69°N, 16°E) by University of Bonn narrow-band Na Doppler lidar. The Na layer was subdivided into eight slices of about equal signal strength to obtain comparable signal/noise ratios throughout the layer. Note that the absolute value of the temperature scale is uncertain by about ±10 K because of the remaining uncertainty in the lineshape of the lidar transmitter. Error bars in the figure show statistical errors from the fitting procedure only. (From Fricke, K.H., and von Zahn, U., *J. Atmos. Terr. Phys.*, 47, 499–512, 1985. With permission.)

of the signal levels at these two frequencies to temperature changes are opposite to each other, the ratio of the fluorescence signals at the intermediate minimum and at the D_{2a} peak is a very sensitive function of temperature. In practice, the laser frequency is tuned to the Doppler-free feature at the D_{2a} peak f_a and at the crossover frequency f_c using a Na vapor cell. We define the normalized Na photon count as

$$N_{\mathrm{norm}}(f, z, t) = \frac{N_{\mathrm{Na}}(f, z, t)}{N_{\mathrm{R}}(f, z, t) E^2(f, z)} \tag{5.48}$$

From the lidar equations (5.44) and (5.45), we have

$$N_{\text{norm}}(f,z,t) = \frac{\sigma_{\text{eff}}(f)n_{\text{Na}}(z)}{\sigma_{\text{R}}(\pi,f)n_{\text{R}}(z_{\text{R}})} \frac{z_{\text{R}}^2}{4\pi z^2} \tag{5.49}$$

The ratio of the normalized Na photon counts at the crossover frequency f_c to the D_{2a} peak frequency f_a is then defined as

$$R_{\text{T}}(z) = \frac{N_{\text{norm}}(f_c,z,t_1)}{N_{\text{norm}}(f_a,z,t_2)} = \frac{\sigma_{\text{eff}}(f_c,z)n_{\text{Na}}(z,t_1)}{\sigma_{\text{eff}}(f_a,z)n_{\text{Na}}(z,t_2)} \approx \frac{\sigma_{\text{eff}}(f_c,z)}{\sigma_{\text{eff}}(f_a,z)} \tag{5.50}$$

To reach the last equality in Equation (5.50), two assumptions have been made: (1) the frequency difference between f_c and f_a is so small that their Rayleigh scattering cross sections at these two frequencies are the same; (2) the Na densities are the same at time t_1 and t_2. The σ_{eff} in Equation (5.50) is given by Equation (5.43). Plotted in Figure 5.6 is the ratio R_{T} versus the temperature T from 100 to 300 K for $v_{\text{R}} = 0$ m/s. Figure 5.6 illustrates that R_{T} is a sensitive function of temperature and can be used to accurately derive the Na temperature.

There are several reasons for choosing f_a and f_c as the two operational frequencies. (1) The ratio of the fluorescence signals at these two frequencies is a very sensitive function of

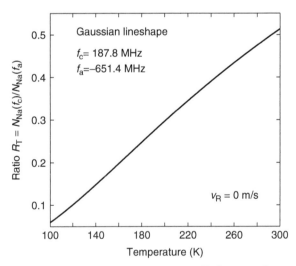

Figure 5.6 Calibration curve (R_{T} versus T) for two-frequency ratio technique of Na wind/temperature lidar.

temperature. (2) Since f_a is at the D_{2a} peak and f_c is near the minimum between the D_{2a} and D_{2b} peaks, where the slope of scattering cross-section versus frequency is close to zero, the fluorescence signal levels from the Na layer at these two frequencies are relatively insensitive to small frequency tuning errors of the laser. (3) Unlike the frequency (f_{min}) at the minimum between the D_{2a} and D_{2b} peaks, which depends on temperature, both f_a and f_c are independent of temperature and can be determined from the Doppler-free features to a very high precision (within $\pm 0.1\,\text{MHz}$) (She and Yu, 1995). These characteristics simplify the calculation of temperature from the measured ratio of the fluorescence signals and provide a good frequency calibration. Data are collected by tuning the laser to the D_{2a} peak and accumulating the photon returns from certain laser shots, and then tuning to the crossover resonance and accumulating the returns. This process was repeated continuously throughout the observation period. By taking the ratio of the photon counts for two frequencies collected at each altitude, one can derive a range-resolved temperature structure throughout the Na layer region. This is called the *two-frequency ratio technique*, which was first demonstrated by She et al. (1990) and Bills et al. (1991a). The temperature accuracy that is obtained with either the scanning method or the two-frequency ratio technique should be comparable as long as other conditions remain equal. The ratio techniques require precise and repeatable laser tuning as well as pulse-to-pulse stability. Its instrumentation will be described later. Since the laser switches only between two wavelengths, the temporal resolution of the two-frequency ratio technique can be much higher than the scanning method.

5.2.2.2.3. Ratio techniques: three-frequency

The Na atom velocity, that is, the wind velocity, has been involved in the effective cross-section given by Equation (5.43). Since the two-frequency measurements cannot derive the Na temperature, number density and the radial wind at the same time, the radial wind has to be assumed as zero when deriving the temperature from the ratio R_T. Typically, the vertical wind is less than $3\,\text{m/s}$, so the temperature error caused by this assumption would be less than 0.5 K (She et al., 1990). On the other hand, if we measure the fluorescence signals at three or more different frequencies, then we can derive the Na temperature, radial wind, and number density simultaneously. This is called the *multiple-frequency ratio technique* or *three-frequency ratio technique*, first proposed by Bills et al. (1991a) and She et al. (1992), and then demonstrated by Bills et al. (1991b) and She and Yu (1994). Bills et al. (1991b) tuned the laser frequency to each of four frequencies at f_a, $f_a + \Delta f$, f_c, and $f_a - \Delta f$. They derived the temperature using photon count profiles collected at f_a and f_c, and derived the radial wind using photon count profiles collected at $f_a + \Delta f$ and $f_a - \Delta f$. She and Yu (1994) tuned the laser frequency to each of three frequencies at f_a, f_c, and $f_a - \Delta f$, and then derived the temperature and radial wind at the same time.

The modern three-frequency ratio technique is to tune the laser frequency to each of the three frequencies at $f_a, f_a + \Delta f$, and $f_a - \Delta f$, where $\Delta f = 630\,\text{MHz}$. The temperature is derived from the following ratio:

$$
\begin{aligned}
R_T(z) &= \frac{N_{\text{norm}}(f_+, z, t_1) + N_{\text{norm}}(f_-, z, t_2)}{N_{\text{norm}}(f_a, z, t_3)} \\
&\approx \frac{\sigma_{\text{eff}}(f_+, z) + \sigma_{\text{eff}}(f_-, z)}{\sigma_{\text{eff}}(f_a, z)}
\end{aligned}
\tag{5.51}
$$

and the radial wind will be derived from the following ratio:

$$
R_W(z) = \frac{N_{\text{norm}}(f_-, z, t_2)}{N_{\text{norm}}(f_+, z, t_1)} \approx \frac{\sigma_{\text{eff}}(f_-, z)}{\sigma_{\text{eff}}(f_+, z)}
\tag{5.52}
$$

Plotted in Figure 5.7(a) and (b) are the ratio R_T versus temperature for the radial wind $v_R = 0\,\text{m/s}$, and the ratio R_W versus radial wind for the temperature $T = 200\,\text{K}$. Here, the radial wind $v_R > 0$ if the Na atoms move along the same direction as the laser photon propagation, that is, move away from the laser source. Figure 5.7(b) illustrates that the ratio R_W is a sensitive function of the radial wind. The ratio R_T in the three-frequency technique is not as sensitive as the ratio R_T in the two-frequency technique, but it is still sensitive enough for a good temperature measurement. Indeed, the

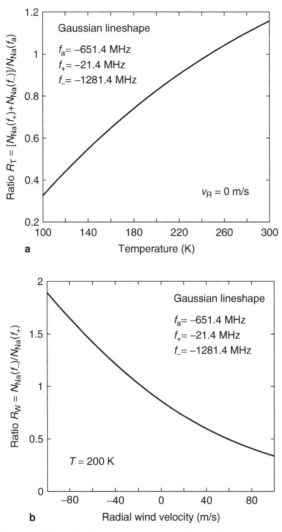

Figure 5.7 (a) Na wind/temperature lidar three-frequency ratio technique: R_T versus T. (b) Na wind/temperature lidar three-frequency ratio technique: R_W versus wind.

temperature and the radial wind are involved with each other in the ratios R_T and R_W, given by Equations (5.51) and (5.52), respectively. Therefore, it is necessary to solve Equations (5.51) and (5.52) simultaneously using iteration.

Both the two- and three-frequency ratio techniques require that the laser pulses have repeatable frequency and lineshape. The radial wind measurement also requires an accurate absolute frequency reference in order to derive the Doppler frequency shift. At a given temperature, the spectrum of the laser-induced fluorescence is the convolution of the laser lineshape with the thermal Doppler-broadened Na D_2 fluorescence spectrum. Therefore, one must use a narrowband laser system with a known lineshape function to experimentally retrieve the spectral information, and thereby the wind and temperature information, from the induced fluorescence of atmospheric Na atoms. The strength of the detected fluorescence signal strongly depends on the excitation frequency. The success of the ratio techniques relies on having a stable, tunable, and narrowband laser transmitter whose center frequency as well as lineshape can be characterized accurately against a known absolute frequency reference. Such a frequency marker should be conveniently available during the data acquisition phase of the lidar operation. Fortunately, it is convenient to utilize a Na vapor cell and the Doppler-free saturation–absorption spectroscopy to achieve this absolute frequency reference, as described earlier. The laser frequency can be locked to these Doppler-free features to have the absolute frequency calibration and to ensure the precise derivation of atmospheric temperature and radial wind from the convoluted laser-induced Na fluorescence.

5.2.2.2.4. Lidar wind measurement versus radar wind measurement

Although both the resonance fluorescence lidar and the radar utilize the effect of Doppler frequency shift to measure wind, the concepts of how to measure the Doppler frequency shift are quite different between the lidar and radar. Briefly, the resonance fluorescence lidar employs incoherent detection method; that is, it converts the frequency shift to the optical intensity information and then measures the frequency shift from the intensity ratios. The radar measures the frequency shift between the returned and transmitted beams directly using coherent detection method.

The radar scattering from charged particles is similar to the Rayleigh scattering of photons from the atmosphere molecules in the sense that they are both elastic scattering, and do not include any real resonance absorption. Suppose a particle has velocity $\vec{v}_1$ before the scattering and has new velocity $\vec{v}_2$ after the scattering. An incident photon with frequency ω_1 has a momentum of $\hbar \vec{k}_1$, where $k_1 = \omega_1/c$. The scattered photon with frequency ω_2 has a new momentum of $\hbar \vec{k}_2$, where $k_2 = \omega_2/c$. Considering the particle and the photon as one system, there is no external force interacting with this system. Therefore, we have the momentum conservation and energy conservation for this particle and photon system before and after the scattering, as shown in Equations (5.53) and (5.54), respectively:

$$m\vec{v}_1 + \hbar \vec{k}_1 = m\vec{v}_2 + \hbar \vec{k}_2 \tag{5.53}$$

$$\frac{1}{2}mv_1^2 + \hbar\omega_1 = \frac{1}{2}mv_2^2 + \hbar\omega_2 \tag{5.54}$$

Elimination of v_2 from Equation (5.54) with the help of Equation (5.53) gives

$$\omega_1 = \omega_2 + \vec{k}_1 \cdot \vec{v}_1 - \vec{k}_2 \cdot \vec{v}_2 + \frac{\hbar k_1^2}{2m} - \frac{\hbar k_2^2}{2m} \tag{5.55}$$

The last two terms on the right-hand side of Equation (5.55) are second-order terms, whose magnitudes are much smaller than the second and the third terms. Thus, the last two terms can be neglected altogether. For backscattered radar or backscattered Rayleigh signals, the returned photons propagate in the opposite direction of the transmitted signals, and the frequency shift is very small compared with the carrier frequency, so $\vec{k}_2 \approx -\vec{k}_1$. A single photon momentum $\hbar \vec{k}$ is much smaller than a particle's momentum, so the particle's velocity will only slightly change after scattering; thus, $\vec{v}_2 \approx \vec{v}_1$. Under these considerations, Equation (5.55) can be simplified, and we obtain the frequency shift of the returned signal relative to the transmitted beam as

$$\Delta\omega_{\text{radar}} = \omega_2 - \omega_1 = -2\vec{k}_1 \cdot \vec{v}_1 \tag{5.56}$$

Thus, in the radar scattering and the Rayleigh scattering cases, the frequency shift between the returned and transmitted signals is two times the Doppler frequency shift. The radar wind measurement relies on coherent detection; that is, taking the beat signal between the transmitted and the returned beams to obtain the frequency shift corresponding to the wind velocity. This is a direct measurement of the Doppler frequency shift.

The resonance fluorescence actually consists of two steps: first, the resonance absorption of an incident photon by an atom; and second, the spontaneous emission of a photon from the excited atom. Here, let us consider a simplified two-energy-level atom. Suppose the atom is in the ground state with energy E_1 and has velocity $\vec{v}_1$ before the absorption takes place. Absorption of a photon of energy $\hbar\omega_1$ promotes the atom into its excited energy level E_2. The photon has a momentum of $\hbar\vec{k}_1$, where $k_1 = \omega_1/c$. The absorption of the photon causes a recoil of the atom to a new velocity $\vec{v}_2$. The total momentum conservation and the energy conservation before and after the absorption give

$$m\vec{v}_1 + \hbar\vec{k}_1 = m\vec{v}_2 \tag{5.57}$$

$$E_1 + 1/2mv_1^2 + \hbar\omega_1 = E_2 + 1/2mv_2^2 \tag{5.58}$$

The resonance absorption frequency of an atom with zero velocity is given by

$$\omega_0 = (E_2 - E_1)/\hbar \tag{5.59}$$

Solving Equations (5.57) and (5.58) gives

$$\omega_1 = \omega_0 + \vec{k}_1 \cdot \vec{v}_1 + \frac{\hbar k_1^2}{2m} \tag{5.60}$$

Neglecting the last term on the right-hand side of Equation (5.60), we obtain the frequency shift of the perceived laser frequency relative to the transmitted laser frequency during the absorption as

$$\Delta\omega_{\text{abs}} = \omega_0 - \omega_1 = -\vec{k}_1 \cdot \vec{v}_1 \tag{5.61}$$

The excited atom then spontaneously emits a photon and decays back to its ground state. For backscattered photons received by a detector of the lidar, the emitted photon travels towards the lidar. Suppose the atom has a new velocity $\vec{v}_3$ and the returned photon has a momentum of $\hbar\vec{k}_2$, where $k_2 = \omega_2/c$. The momentum and energy conservations yield

$$m\vec{v}_2 = m\vec{v}_3 + \hbar\vec{k}_2 \tag{5.62}$$

$$E_2 + \frac{1}{2}mv_2^2 = E_1 + \frac{1}{2}mv_3^2 + \hbar\omega_2 \tag{5.63}$$

Solving Equations (5.62) and (5.63), we obtain

$$\omega_2 = \omega_0 + \vec{k}_2 \cdot \vec{v}_3 + \frac{\hbar k_2^2}{2m} \tag{5.64}$$

Recalling the similar discussion above, we have $\vec{k}_2 \approx -\vec{k}_1$ and $\vec{v}_3 \approx \vec{v}_1$. Thus, the frequency shift of the returned photon relative to the resonance frequency during spontaneous emission is given by

$$\Delta\omega_{\text{sp}} = \omega_2 - \omega_0 = -\vec{k}_1 \cdot \vec{v}_1 \tag{5.65}$$

We can obtain the frequency shift of the returned fluorescence photon frequency relative to the transmitted photon frequency from Equations (5.61) and (5.65) as follows:

$$\Delta\omega_{\text{overall}} = \omega_2 - \omega_1 = -2\vec{k}_1 \cdot \vec{v}_1 \tag{5.66}$$

For the two-energy-level atom, the returned fluorescence photon has two times the Doppler frequency shift relative to the transmitted photon frequency, which is similar to the radar scattering and Rayleigh scattering. However, the fluorescence photons have random phases, so the frequency shift cannot be detected by the coherent method used by the radar. As described in the three-frequency ratio technique above, the resonance fluorescence lidar actually converts the frequency information to the intensity signals of returned fluorescence photons and then derives the frequency shift from the intensity ratios. In other words, for each operating frequency of the laser, the lidar receiver only counts the total number of returned fluorescence photons no matter what frequencies they have. This is especially true when using the broadband optical filters in the lidar receiver. The filter function is flat in the frequency range we are interested in, so all returned fluorescence photons will be counted in the same way. Then the ratio of the photon counts collected at each operating frequency of the laser (e.g., $f_a \pm \Delta f$) is taken to give the frequency-shift information. The returned fluorescence photon number is proportional to the absorption cross-section of the atom given by Equation (5.41). Since only the number (intensity), not the frequency, of the fluorescence photons is considered in the wind measurements, the effective scattering cross-section will be given by Equation (5.43), which only includes the Doppler frequency shift associated with the resonance absorption process, but excludes the shift associated with the spontaneous emission. Therefore, for the resonance fluorescence lidar, the Doppler frequency shift derived from the photon intensity signals will be determined by Equation (5.61), rather than Equation (5.66).

The frequency shift of the fluorescence photons associated with the spontaneous emission needs to be considered when using narrowband optical filter (such as a Na FADOF, see Section 5.2.2.4) in the lidar receiver. In this case, the filter function is no longer flat at the frequencies of interest. So the filter function needs to be taken into account when computing the photon intensity ratio, and the frequency shift of the fluorescence photons will influence the transmission through the filter, and thus, the receiving efficiency. Nevertheless, the Doppler frequency shift derived from the photon intensity signals will still be determined by Equation (5.61), that is, only one-time Doppler frequency shift.

5.2.2.3. Na Doppler Lidar Instrumentation

As discussed earlier, a narrowband Na lidar is required in order to resolve the temperature and wind information from the laser-induced fluorescence signals. To illustrate the instrumentation of the modern Na wind/temperature lidar, a schematic diagram of the UIUC Na wind/temperature lidar

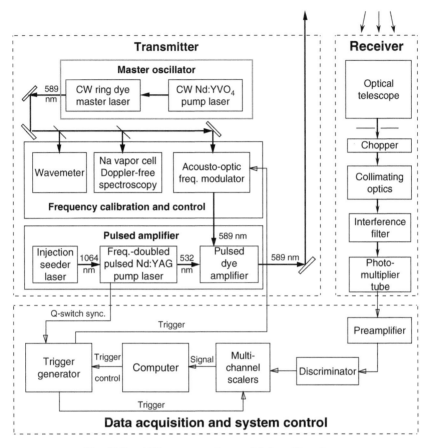

Na Wind/Temperature Lidar System

Figure 5.8 Schematic diagram of the University of Illinois Na wind/ temperature lidar system. The lidar consists of three subsystems: a transmitter, a receiver, and data acquisition and system control. The thick solid lines with solid arrows denote the optical path, while the thin solid lines with open arrows indicate the electronic signals. The major components of the lidar transmitter are the master oscillator, frequency calibration and control, and pulsed amplifier.

is plotted in Figure 5.8. The lidar system consists of three subsystems: a transmitter, a receiver, and a data acquisition and control system. The lidar transmitter is to provide laser pulses with high precision frequency, narrowband linewidth, high power, and small divergence. The lidar receiver is to collect backscatter photons, reject the solar background noise, and count the returned photon numbers. The data acquisition and control system interacts with both the transmitter and the receiver to record the photon counts and to provide the timing and system control for the whole lidar system. A list of the system parameters is given in Table 5.2.

Table 5.2 System Parameters for the University of Illinois Na Wind/Temperature Lidar

Lidar Transmitter		
Characteristics	Ring Dye Laser Master Oscillator	Pulsed Dye Amplifier
Model	Coherent 899-21	Spectra-Physics Quanta-Ray PDA-1
Optical pump laser	Coherent Nd:YVO$_4$ laser Verdi V-5 CW at 532 nm	Spectra-Physics Nd:YAG Laser Quanta-Ray GCR-3 pulsed at 532 nm
Pump power/energy per pulse	4 W	300 mJ
Gain medium	Dye Rhodamine 6G	RH640 + KR620
Pulse repetition rate	CW	30 Hz
Pulse length	–	6 nsec
Linewidth	500 kHz (rms)	60 MHz (rms)
Average output power	500–600 mW (350–400 mW at PDA input)	1.2–1.5 W (40–50 mJ per pulse)
Beam divergence	–	1 mrad FW at e^{-2}
Wavelength	589.1583 nm	589.1583 nm
Lidar Receiver		
Characteristics		
Telescope	SOR reflecting telescope	
Aperture diameter	3.5 m	
Aperture area	9.6 m^2	
Telescope focal length	315 m	
Field of view	1 mrad	
Interference filter bandwidth	3 nm (FWHM)	
PMT quantum efficiency	11%	
Range resolution	24 m	

The Na lidar transmitter can be further divided into three sections for ease of explanation: (1) master oscillator, (2) frequency calibration and control, and (3) pulsed amplifier. The master oscillator is a high-performance single-frequency cw ring dye laser, optically pumped by a frequency-doubled cw Nd:YVO$_4$ laser. (Originally, an argon ion laser was used to pump the ring dye laser.) The frequency calibration and control monitors the ring dye laser wavelength with a wavemeter, locks the laser frequency to the D$_{2a}$ Lamb dip or crossover peak of the Na D$_2$ line using the Doppler-free saturation–absorption spectroscopy of a Na vapor cell, and creates the two wing-operating frequencies for the three-frequency ratio technique by using an acousto-optic modulator. The pulsed amplifier provides pulsed amplification to the cw master laser to generate laser pulses needed for the lidar application. The pulsed amplification is achieved by the interaction of the cw master laser beam with the pump beam from a powerful frequency-doubled pulsed Nd:YAG laser. The laser pulses are then transmitted to the atmosphere. Return photons are collected by a receiving optical telescope, counted by a PMT, binned by multichannel scalers, and saved by a computer. Interference filters are used to reject most of the background skylight. Such a lidar system achieves the excellent laser pulse-to-pulse stability and provides a high degree of tuning accuracy, which are difficult to achieve with pulsed laser oscillators, but are essential for the temperature and wind measurements. In the following text, we discuss several key components for the system.

5.2.2.3.1. Master oscillator

The ring dye master laser (Model: Coherent 899-21) is a well-designed unidirectional-lasing ring cavity. The operation of unidirectional lasing prevents the linewidth broadening caused by the spatial hole burning. The ring cavity contains several frequency selection components. There are three passive frequency filters inside the ring dye cavity: a birefringent filer, a thin etalon, and a thick etalon. These filters altogether select a single longitudinal mode of the ring cavity. In theory, the single longitudinal mode should have bandwidth on the order of 1 Hz or less. However, due to environmental factors (such as the cavity length fluctuation due to vibrations, the change of air refraction index, or the change of dye jet thickness) that modulate the effective cavity optical length, the actual oscillation bandwidth is on the order of tens of megahertz if only the passive optical filters are used in the cavity. To achieve a better performance, an internal active frequency-stabilization servo loop is used to stabilize the effective cavity length. The servo loop locks the laser cavity frequency to an external temperature-stabilized confocal Fabry–Perot reference cavity. An error signal is produced if the laser frequency drifts from the lock point of the reference cavity. The fast varying error signal is separated from the slower signal and coupled to a piezoelectric-transducer-driven mirror (tweeter), while the slower varying signal is directed to a scanning Brewster plate. By rotating the Brewster plate and adjusting the tweeter, the effective cavity length can be stabilized and the cavity frequency can be locked to the reference-cavity. The resulting cavity is a single-frequency laser with TEM_{00} mode and rms bandwidth of 500 kHz.

The frequency of the ring cavity can be scanned in a range of several gigahertz by continuously varying the reference-cavity optical length with a rotating galvanometer-driven Brewster plate. The ring cavity frequency will then track the reference-cavity frequency due to the feedback of the internal frequency-stabilization servo loop. For the two-frequency ratio technique, the ring dye master laser must be locked respectively to the Na D_{2a} resonance frequency f_a and the crossover frequency f_c of the Doppler-free saturation–absorption spectroscopy of a Na vapor cell. For the three-frequency technique, the ring dye laser is locked to frequency f_a at all times, and the other two wing frequencies are generated by an acousto-optic modulator. Locking to a frequency is accomplished by first tuning the reference cavity to either f_a or f_c. Then the internal frequency servo-loop drives the ring cavity frequency to track the reference cavity. Thus, the ring dye master-laser frequency can be tuned to the Na D_{2a} line frequencies f_a and f_c with very high accuracies. This is an essential feature for the accurate temperature and wind measurements.

The dye jet used in the ring laser is highly sensitive to vibrations and the ring cavity is sensitive to the change of room temperatures, so the ring dye laser requires a well-controlled laboratory environment for operation. This is the main drawback for the applications of the Na wind/temperature lidar.

5.2.2.3.2. *Doppler-free spectroscopy for precise frequency calibration and control*

The experimental setup for the Doppler-free spectroscopy is illustrated in Figure 5.9. A small amount of ring dye laser beam is split into two beams by a cubic beam splitter. With the help of three mirrors, the two beams become two counter-propagating beams when going through a Na vapor cell. The Na cell is heated to and stabilized at 60°C in order to provide

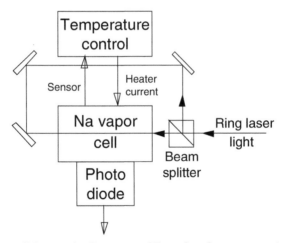

Figure 5.9 Schematic diagram of Doppler-free saturation–absorption spectroscopy experiment with a Na vapor cell. The ring laser beam enters from the right, and is split into two beams by a cubic beam splitter. With the help of three mirrors, the two beams pass through the Na vapor cell along the same line but in opposite directions. A photodiode is placed to the side of the cell to detect laser-induced fluorescence.

sufficient atomic sodium vapor density. A photodiode is placed adjacent to the Na cell to detect the laser-induced fluorescence. Under the simultaneous illumination of the two counter-propagating beams with intensities exceeding a saturation intensity ($\sim$0.4 mW/mm^2), the Na fluorescence spectrum exhibits two Doppler-free dips at the peak frequencies f_a and f_b of the Na D_{2a} and D_{2b} lines, and a Doppler-free peak at the crossover frequency f_c between D_{2a} and D_{2b}, as shown in Figure 5.3.

These Doppler-free features can be explained through a simplified three-level atomic energy system. Suppose the Na atom ground state is split into two levels ($|\,a\,\rangle$ and $|\,b\,\rangle$), and the Na atoms in these two states can be excited to state $|\,e\,\rangle$ by the photons with frequencies ν_a and ν_b, respectively. If a laser with frequency ν is incident on the Na vapor cell, the laser-induced fluorescence will only occur for those Na atoms with the correct velocity for ν to be Doppler-shifted to ν_0, where $\nu_0 = \nu_a$ or ν_b. When the laser intensity increases, the fluorescence intensity also increases but not in proportion to the intensity. The deviation from the linearity is called saturation. The saturation arises because (1) the population of the corresponding ground state decreases since the excited Na atoms can decay to another ground state (i.e., the optical pumping effect) and the population on the excited state is no longer zero, and (2) when the excited state population is not zero, the stimulated emission occurs, and it does not contribute to the fluorescence signal in the direction perpendicular to the laser beam because the stimulated emission photons are along the laser beam direction. Despite saturation, the laser-induced fluorescence spectrum follows the velocity distribution as ν is scanned. If the laser frequency $\nu < \nu_0$, the beam incident on the cell from the left interacts with atoms traveling towards the left with a radial speed v_R giving the shifted frequency $\nu' = \nu_0$, and the beam incident on the cell from the right interacts with atoms traveling towards the right with a same radial speed. As long as ν is not equal to ν_0, the two counter-propagating beams interact with two separate velocity groups of Na atoms, and the laser-induced fluorescence at a given laser intensity is simply double what it was with only one beam. This is also true for $\nu > \nu_0$. However, for $\nu = \nu_0$, the two counter-propagating beams interact with the same group atoms with zero radial velocity ($v_R = 0$). The population on the proper ground state will be reduced dramatically since the effective laser intensity is doubled. Therefore, the laser-induced fluorescence will be significantly reduced, resulting in a Lamb dip exactly at $\nu = \nu_0$, where $\nu_0 = \nu_a$ or ν_b. The dip exhibits a nearly natural linewidth and therefore is Doppler-free. As described above, the optical pumping effect reduces the population on the corresponding ground state so the reduction in the fluorescence intensity occurs at all frequencies across both D_{2a} and D_{2b} Gaussian spectra (with Lamb dips centered at ν_a and ν_b) except at the crossover frequency ν_c. At $\nu = \nu_c$, the atoms on the $|\,a\,\rangle$ state with a correct velocity to interact with the laser beam incident from the right will be excited to the $|\,e\,\rangle$ state. They can either decay back to $|\,a\,\rangle$ to continue interacting with the same laser beam, or decay to the $|\,b\,\rangle$ state to automatically be the correct velocity group to interact with the laser beam incident from the left. Thus, the optical pumping effect does not decrease the fluorescence intensity, and a sharp (relative) enhancement peak occurs in the laser-induced fluorescence spectrum at exactly $\nu = \nu_c$. This peak is also Doppler-free with nearly a natural linewidth.

The Lamb dip and crossover peak features in sodium saturation–absorption spectroscopy were first described by Hansch et al. (1971), and that paper still remains an illuminating reference. The actual Na atoms consist of four sublevels in the excited state, separated by 15 to 60 MHz, and six allowed transitions to the two ground sublevels. For each pair of allowed transitions, there exist two Lamb dips and a crossover peak. The natural linewidth of each transition is about 10 MHz, that is, the same order as the separation of the excited sublevels. Therefore, the transitions within the D_{2a}, D_{2b}, and crossover are not resolved. She and Yu (1995) did detailed experimental and theoretical studies of the Na Doppler-free saturation–absorption spectroscopy and determined the central frequencies of these Doppler-free features to ± 0.1 MHz as illustrated in Table 5.1. These Doppler-free features provide feedback for locking the ring dye master laser. The central frequencies of these Doppler-free features are independent of temperature, so they are an absolute frequency reference required for the wind and temperature measurements.

5.2.2.3.3. Acousto-optic modulator for precise frequency shift

The acousto-optic modulator (AOM) is used to generate the two wing frequencies $f_a \pm \Delta f$ required by the three-frequency ratio technique. This is a dual acousto-optic frequency shifter system. One is to shift the frequency output from the ring dye master laser (which is locked to the Lamb dip f_a) up by 630 MHz, and another is to shift it down by 630 MHz. The AOM is controlled by a computer through a trigger generator so that the operating laser frequency cycles between $f_a, f_+ = f_a + \Delta f$, and $f_- = f_a - \Delta f$. The idea of the AOM was first demonstrated by She and Yu (1994) and then employed in both the CSU and the UIUC Na lidar systems. A comprehensive review paper for the CSU Na wind/temperature lidar can be found in Arnold and She (2003). In the following, we provide an analysis of the operating principle of the acousto-optic modulator following Yariv (1997).

A piezoelectric transducer attached to the side of an acousto-optic crystal is driven by an electromagnetic field to convert the radio-frequency (RF) energy into acoustic energy. The vibration of this ultrasonic transducer produces a traveling acoustic wave with wave vector $\vec{k}_s$ to propagate across the crystal (Figure 5.10). The acoustic wave consists of a sinusoidal perturbation of the density of the AO crystal that travels at the sound velocity v_s. The variation in the density causes a change in its index of refraction because the index of refraction is higher in the compressed portions of the sound wave and lower in the rarefied region. A change in index of refraction results in partial reflection, so the acoustic wave can be represented by a series of partially reflecting plane mirrors, separated by the sound wavelength λ_s, that are moving with a wave vector $\vec{k}_s$ ($k_s = \omega_s / v_s$, where ω_s and v_s are the angular frequency and velocity of the acoustic wave, respectively). A laser beam with wave vector $\vec{k}_i$ ($k_i = \omega_i / c$, where ω_i and c are

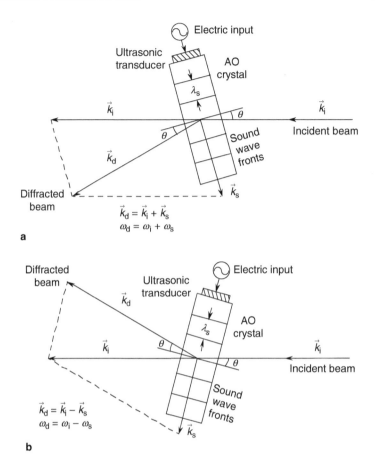

Figure 5.10 (a) The Bragg diffraction of the acousto-optic modulator (frequency up-shifting): the Bragg vector diagram and corresponding physical configuration for the diffraction of light from oncoming sound wave. (b) The Bragg diffraction of the acousto-optic modulator (frequency down-shifting): the Bragg vector diagram and corresponding physical configuration for the diffraction of light from retreating sound wave.

the laser angular frequency and speed of light, respectively) is incident on the AO crystal with a small angle θ relative to the sound wave front. According to the first-order Bragg diffraction condition (Yariv, 1988), if the incident laser beam, the acoustic wave and the angle θ satisfy the following equations:

$$k_s = 2k_i \sin \theta \tag{5.67}$$

or

$$2\lambda_s \sin \theta = \frac{\lambda_i}{n} \tag{5.68}$$

where n is the index of refraction for the laser beam, then part of the incident laser beam will be diffracted by the acoustic wave and exit the AO crystal with the same angle θ on the other side of the crystal as illustrated in Figure 5.10. The diffracted laser beam with a wave vector $\vec{k}_d$ ($k_d = \omega_d/c$, where ω_d is the laser angular frequency) will experience a Doppler frequency shift due to the moving of the acoustic wave at velocity v_s. The Doppler frequency shift between the diffracted beam and the incident beam is equal to

$$\Delta\omega = \omega_d - \omega_i = -(\vec{k}_i \cdot \vec{v}_s - \vec{k}_d \cdot \vec{v}_s) \tag{5.69}$$

Since $\vec{k}_d \cdot \vec{v}_s \approx -\vec{k}_i \cdot \vec{v}_s$, the above Doppler frequency shift can be written as

$$\Delta\omega = -2\vec{k}_i \cdot \vec{v}_s = \pm 2k_i v_s \sin \theta \tag{5.70}$$

where the "+" is for the oncoming sound wave shown in Figure 5.10(a), and the "−" is for the retreating sound wave shown in Figure 5.10(b). From Equation (5.67), it is easy to obtain that

$$\omega_s = 2k_i v_s \sin \theta \tag{5.71}$$

Therefore, the frequency shift is given by

$$\Delta\omega = \pm\omega_s \tag{5.72}$$

and the frequency of the diffracted laser beam is given by

$$\omega_d = \omega_i \pm \omega_s \tag{5.73}$$

that is, the laser frequency is shifted up or down by ω_s for the oncoming and retreating sound waves, respectively.

The AOM can also be interpreted through a particle picture of Bragg diffraction of light by sound. The acoustic wave can be represented by a stream of phonons with momentum $\hbar\vec{k}_s$ and energy of $\hbar\omega_s$, and the incident laser can be represented by a stream of photons with momentum $\hbar \vec{k}_i$ and energy $\hbar \omega_i$. In the case of Figure 5.10(a), an incident photon and a phonon are annihilated and simultaneously create a diffracted photon with a momentum $\hbar \vec{k}_d = \hbar \vec{k}_i + \hbar \vec{k}_s$ and energy $\hbar \omega_d = \hbar \omega_i + \hbar \omega_s$. These are required by the conservation of momentum and energy, respectively. Thus, the laser frequency is up-shifted to $\omega_d = \omega_i + \omega_s$. In the case of Figure 5.10(b), an incident photon is annihilated and simultaneously creates a new phonon and a diffracted photon with the momentum and energy of $\hbar \vec{k}_d = \hbar \vec{k}_i - \hbar \vec{k}_s$ and $\hbar \omega_d = \hbar \omega_i - \hbar \omega_s$, again required by the conservation of momentum and energy. Thus, the laser frequency is down-shifted to $\omega_d = \omega_i - \omega_s$. The up- or down-shifting is determined by the relative direction between the acoustic wave and the incident laser beam. When the acoustic wave travels towards or away from the incident laser beam, the laser frequency is up- or down-shifted, respectively.

The experimental setup for the AOM used in the UIUC Na wind/temperature lidar is illustrated in Figure 5.11. Two AO shifters are driven by an RF field with frequency f_s of 315 MHz. The laser beams at three frequencies f_a, f_+, and f_- need to be injected into the pulsed dye amplifier (PDA) with the same propagation direction and polarization direction independent of the final frequency. To achieve this goal, a

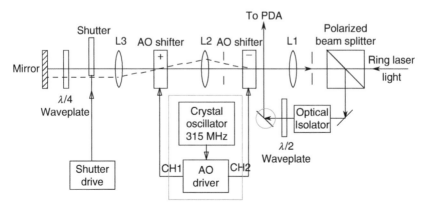

Figure 5.11 Schematic diagram of the acousto-optic modulator unit used in the UIUC Na wind/temperature lidar.

double-pass arrangement with two AO crystals separated by lenses is used. Lenses L_1, L_2, and L_3 are placed in such a way that the focal points of L_1 and L_3 are coincident with the two-time focal length points of lens L_2 on each side. Two AO crystals are placed at these two overlapping points. The incident ring laser beam from the right-hand side is horizontally polarized and passes the cubic polarizing beam splitter. It is focused by lens L_1 onto the AO shifter "−", then imaged by lens L_2 onto the AO shifter "+". Lens L_3 is used to collimate the divergent light. The mirror is used to reflect the beam back exactly along its original path. When the piezoelectric transducer is turned off, each AO crystal behaves as a dielectric medium, and the laser beam follows the solid line from the beam splitter to the reflecting mirror and reflects back through the same path from the mirror to the beam splitter. Since the beam passes through the $\lambda/4$ wave plate twice, the polarization of the reflected beam is rotated by $90°$. Thus, the return beam has perpendicular polarization and will be reflected by the beam splitter for going to the PDA.

When only AO shifter "−" is activated, the acoustic wave will diffract part of the incident beam from the right by the phonon creation process as shown in Figure 5.11(b). The resulting diffracted beam will exit the AO crystal to the left following the dashed line. Along the dashed line, it is imaged by lens L_2, passes the unactivated AO crystal "+", and collimated by lens L_3. It is then reflected by the mirror back to AO shifter "−". It is diffracted by the acoustic wave again and exits from the right following the solid line to the beam splitter. Since the laser beam is diffracted by the acoustic wave twice, the return beam has a net frequency shift of $\Delta f = -2\,f_s = -630\,\text{MHz}$, that is, down-shifted. When only AO shifter "+" is activated, the incident laser beam is diffracted twice by the phonon annihilation process as shown in Figure 5.11(a), leading to a net frequency shift of $\Delta f = 2\,f_s = 630\,\text{MHz}$, that is, up-shifted. In this case, the incident beam follows the solid line except between the AO crystal "+" and the mirror where it takes the dashed line path. With sufficient RF power to drive the piezoelectric transducer, the single-pass first-order diffraction efficiency is usually 85%. Thus, the double-pass diffraction efficiency is $(85\%)^2$, which is more than 70%. When either AO shifter "+" or AO shifter "−" is activated, the undiffracted beam ($\sim$15%) from the first pass has unshifted frequency. We do not want it to be reflected back to the beam splitter, so a shutter is placed in the undiffracted beam path to block the unshifted beam. The shutter and the AO drives are controlled by the computer through the trigger generator box to synchronize with the pulse timing: the shutter is open to let the master laser beam pass when both AO shifters are not activated, and then the shutter is closed to block the unshifted beam when one of the AO shifters is activated. This double-pass geometry ensures the optical beam exits from the beam splitter in the same direction (perpendicular to the incident beam) independent of final frequency, and allows the use of a lower acoustic modulation frequency by a factor of 2.

5.2.2.3.4. Pulsed amplification and frequency chirp measurement

After the laser beam exits the polarized beam splitter in Figure 5.11, it passes through an optical isolator, a $\lambda/2$ wave plate, and a periscope before entering the PDA. The isolator is to prevent optical feedback from PDA in order not to disturb the ring dye master laser. The half wave plate is to rotate the laser polarization from the perpendicular to the horizontal direction. The periscope then changes it again to a perpendicular polarization, which is required by the PDA dye cells to match the Brewster angle in order to have minimum loss. The periscope is used to match the PDA beam height and to provide the adjustment for the beam to enter the PDA with precise angle and position.

Pulsed amplification of the single-mode cw ring dye laser produces laser radiation combining high peak power with a bandwidth approaching the Fourier-transform limit. This is ideal for lidar application. As shown in Figure 5.12, the PDA is a three-stage amplifier, consisting of two transversely pumped dye cells as the first two stages and one longitudinally pumped dye cell as the last stage. The cw ring laser beam enters three dye cells at the Brewster angle. Without the

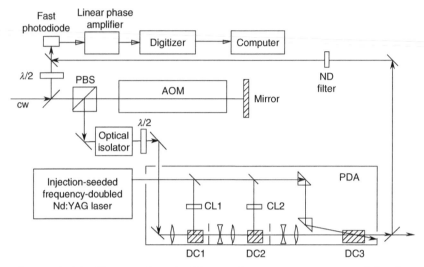

Figure 5.12 Proposed optical heterodyne measurement of the frequency shift and chirp of the pulsed dye amplifier (PDA). AOM is the acousto-optic modulator described in Figure 5.11, which is used to shift the cw beam up or down by 630 MHz. PBS is a polarized beam splitter. Two half-wave plates are used to rotate the polarizations of the cw beam and the amplified pulse to keep them match. DC1, DC2, and DC3 are three-stage dye cells of the PDA. Optical isolator is used to prevent the optical feedback from the PDA to the ring dye master laser. ND filter is a set of neutral density filters used to decrease the pulse peak amplitude in order to match the cw beam amplitude.

pump beam, the dye cells absorb most of the cw ring laser. However, a weak residual of the cw beam can still pass through the dye cells and exit the PDA. When a pump pulse from a frequency-doubled Nd:YAG laser (532 nm) is injected into the dye cells, the PDA will amplify the cw ring beam and produce a 589-nm laser pulse. Thus, the output of PDA is a quasi-pulsed laser beam with a high pulse peak and a very weak cw beam during idle period between pump laser pulses. The duration time (around 6 nsec) of PDA output pulse is determined by the pump laser. The spectrum of the PDA output is strongly influenced by the pulse shape of the pump laser and also is limited by the short pulse duration time (Fourier transform). When the pump laser is unseeded, the pulses acquire numerous large side bands stemming from mode beating in the Nd:YAG pump laser that is partially preserved in the amplification process. Thus, the PDA output pulses have significant shot-to-shot variations. In order to generate highly reproducible pulses, the pump laser is injection-seeded by a cw seed laser working at the fundamental wavelength (1064 nm). The injection-seeding reduces the possible number of modes that can lase and makes the Nd:YAG pump laser pulses to have nearly pure Gaussian shape with stable width and height. The resulting PDA output spectrum is nearly Fourier transform limited and highly reproducible. The rms bandwidth of the PDA output pulse is approximately 60 MHz, which corresponds to a full width at half maximum (FWHM) of 140 MHz.

For an ideal PDA, the output pulse would be a Fourier-transform-limited Gaussian lineshape centered precisely on the frequency of the cw input beam. However, the actual spectrum of the PDA output not only has a broadened linewidth (larger than the transform-limited) but also has a shift of the central frequency. These effects are mainly caused by three factors: the amplified spontaneous emission (ASE), unseeded Nd:YAG laser pulses, and nonlinear effects associated with the pulsed amplification process. The ASE is present even in the complete absence of the cw input beam. It is the result of spontaneous emission when the dye is excited by the pump laser. If it is amplified by successive pumped dye cells, it will add a broad spectral pedestal under the normal pulse lineshape, increasing the overall baseline. Fortunately, with careful alignment of the PDA to eliminate reflections along the beam path, ASE is typically reduced to a few percent of the output power. Thus, its effect is negligible. Unseeded or improperly seeded Nd:YAG pulses would produce a strong modulation of the PDA output with a much broader linewidth. Modern Nd:YAG laser seeding technique is well developed, and the seeding status can be monitored through Q-Switch build-up time or by Fabry–Perot etalon fringes.

Thus, well-seeded performance of the Nd:YAG laser can be assured.

The nonlinear effects associated with the pulsed amplification process are not completely identified, but the main factors are the optical phase perturbations during pulsed amplification, which can both broaden and shift the PDA output frequency spectrum (Eyler et al., 1997). The possible mechanisms that can cause the optical phase distortion include heating of the dye solvent, the intensity dependence of the refractive index, and the time dependence of the gain. The experimental and theoretical studies made by Gangopadhyay et al. (1994) and Melikechi et al. (1994) suggested that the principal source of the phase shifts is the time-varying susceptibility of the dye solution, which is a consequence of the changing excited-state population, that is, the time-varying gain. The phase distortion causes the instantaneous frequency of PDA output to vary with time (which is called the *frequency chirp*), resulting in broadened linewidth and shifted central frequency of the laser pulses. If the frequency shift cannot be measured precisely, it will cause a systematic error in wind and temperature measurements, especially in the wind measurement that is more sensitive to an absolute frequency shift.

There are two approaches to measure the frequency shift and linewidth broadening. One is to utilize the optical heterodyne technique to measure the shift and broadening directly. Another is to use an iodine vapor cell to convert the frequency information to intensity ratio signals. The optical heterodyne technique was developed by Fee et al. (1992), and applied by Gangopadhyay et al. (1994), Melikechi et al. (1994), and Eyler et al. (1997) to measure the frequency chirp and shift in the pulsed amplification of a cw single-mode laser. The basic idea is to use an AOM to shift the frequency of part of the cw beam over a range larger than a Fourier-transform-limited linewidth. This shifted beam provides a reference for heterodyne mixing. The rest of the cw beam is pulse-amplified, and then mixed with the frequency-shifted sample of the cw laser on a fast photodiode. The beat signal is then processed in the Fourier domain to reveal the phase and the instantaneous frequency of the pulse laser. Since an AOM has been used in the Na wind/temperature lidar, the following schematic (Figure 5.12) can make the optical heterodyne measurement of the PDA frequency chirp and shift. Before entering the AOM, a small amount of the cw beam is split to the fast photodiode as a reference. The rest of the cw beam is frequency-shifted by the AOM and pulse-amplified by the PDA. A small amount of PDA output pulses is attenuated and sent collinear with the reference beam onto a broadband fast photodiode detector. The output of the detector amplifier is digitized by a fast digital oscilloscope (with high sampling rate and wide bandwidth) and the resulting digitized signal is transferred to a computer for processing. The main frequency of the beat signal is equal to the sum of the AOM frequency shift (630 MHz) and the PDA frequency shift (on the order of 10 MHz). Both the PDA frequency shift and the pulse bandwidth can be derived from the beat signals since the bandwidth of the cw reference beam is very narrow compared with the pulse bandwidth.

Another method to measure the PDA frequency shift and chirp was developed by CSU group using an iodine vapor cell with edge technique (White, 1999). The molecular iodine has strong absorption bands in the vicinity of the Na D_{2a} line: one is about 1 GHz higher and another is about 1.5 GHz lower than the Na D_{2a} Doppler-free peak in frequency. Thus, these two absorption lines form an iodine transmission peak in between them. Plotted in Figure 5.13 is an example of this iodine transmission peak with respect to the three operating frequencies (f_a, f_+, and f_-) of the Na Doppler lidar when the iodine cell is heated to 80°C. At this temperature, the FWHM of this iodine transmission peak is about 1.1 GHz. It is fortunate that the Na f_a and f_- operating frequencies locate on the two edges of the iodine transmission peak, so the ratio of the transmission signals at these two frequencies is sensitive to the frequency shift in the transmitter. For example, if the transmitter has a positive frequency shift, the transmission at f_- should increase while the transmission at f_a should decrease. Since the two wing frequencies are generated by the AOM, the separation between the operating frequency is

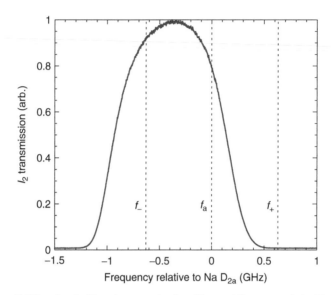

Figure 5.13 An iodine transmission line with respect to the three operating frequencies of the Na Doppler lidar. The spectrum was acquired using a 4-in. iodine cell at 80°C with cw laser light. The FWHM is about 1.1 GHz. (Courtesy of T. Yuan and C.-Y. She, Colorado State University, Fort Collins, Colorado. With permission.)

assumed to be constant regardless of any frequency shift imparted by the pulsed dye amplifier. Thus, the frequency shift can be derived from the intensity ratio, whose principle is similar to the wind measurement using the Na Doppler lidar described earlier. Since the location of the iodine peak is known, given a stable iodine temperature and other parameters, the frequency shift is measured in absolute frequency. Such technique has been employed successfully by the CSU group in the Na wind/temperature lidar to monitor the PDA frequency shift and to calibrate the wind measurements.

5.2.2.3.5. Receiving and photon counting

As illustrated in Figure 5.8, the Na lidar receiver consists of a receiving optical telescope, a field-stop, a chopper, collimating lenses, narrowband filters (interference filter), and a PMT working in photon-counting mode. The field-stop is used to limit the field of view in order to reject most of the sky background. The mechanical chopper is used to block the extremely strong scattering from the low altitudes ($\sim$0–15 km) to prevent the PMT being saturated. An alternative to the mechanical chopper is an electronic gain switch to vary the voltage between the PMT cathode and the first dynode to control the PMT gain in order to avoid the PMT saturation. The interference filter is used to reject the sky background while transmitting the signal photons, which is enough for the nighttime measurement. During daytime operation, an ultra-narrowband filter (e.g., a Faraday optical filter or a Fabry–Perot etalon) must be used to reject more sky background. The collected and filtered photons are then transmitted to detectors. The CSU system employs a fiber for transmitting the photons to the detector. This makes the receiver more flexible. Since the return signals are extremely weak, the highly sensitive PMT is used to count the photon number. The data acquisition and control system is a computer-controlled electronic system to time-gate the signals, record the photon counts, and save frequency calibration data.

5.2.2.4. Daytime Measurements

It is highly desirable to make the temperature and wind measurements in the mesopause region through the complete 24-h diurnal cycle since a large number of geophysical phenomena and long period changes of the Earth's atmosphere can be explored only if the instrument allows daylight operation with high quality data. Narrowband optical filters are necessary in daytime lidar measurements to reject the broadband background photons while passing the narrowband signals. Conventional interference filters cannot provide extremely narrow bandwidth (0.01 to 1 Å) with a highly stable transmission peak and a reasonably large field of view. Daytime operation of a broadband Na lidar was realized by using a Fabry–Perot etalon as a narrowband filter in the receiver (Granier and Megie, 1982; Kwon et al., 1987). However, only Na density was measured with such a system. In 1995, the CSU group introduced a FADOF into the narrowband Na lidar

receiver to reject the sky background, and realized the daytime measurements of Na temperature in the mesopause region (Chen et al., 1996). Later, the daytime measurements of
wind were also achieved with such lidar system (She et al.,
2003). The UIUC group employed an ultra-narrowband
Fabry–Perot etalon to help reject the skylight and also
achieved the daytime observations of the Na temperature
(Yu et al., 1997). Both the CSU and the UIUC groups reported
a large number of new scientific results from these 24-h full
diurnal cycle measurements, which will be summarized in
Section 5.3. Here we describe the instrumentation aspects.

5.2.2.4.1.   *Faraday anomalous dispersion optical filter (FADOF)*

FADOF is an atomic resonance filter based on the resonant Faraday effect. A Na vapor cell is placed within a permanent magnetic field between two crossed polarizers (Figure
5.14). The magnetic field is oriented parallel to the optical
axis. As explained in the following text, for light on Na resonance that was transmitted through the first polarizer, the Na
vapor rotates the plane of light polarization by 90° so that it
can pass through the second polarizer without loss. The refraction index n of a dilute Na atomic vapor at frequencies
close to its resonance absorption lines can be expressed as a
complex refraction index:

$$n = \sqrt{1 + \chi} \cong 1 + \frac{1}{2}\chi = 1 + \frac{1}{2}\chi' - i\frac{1}{2}\chi'' \qquad (5.74)$$

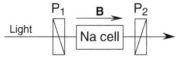

Figure 5.14   Schematic of the Faraday anomalous dispersion optical filter (FADOF). P_1 and P_2 are two crossed polarizers. A Na
vapor cell between P_1 and P_2 is housed in permanent magnets
such that a strong axial magnetic field $\vec{B}$ (along the direction of
the light propagation) permeates the Na cell.

where χ is the electric susceptibility of the Na vapor, $\chi = N\alpha$, N is the number density of the Na vapor, and α is the atomic polarizability. In the vicinity of an atomic resonance line, the electric susceptibility is given by $\chi = \chi' - i\chi''$, where

$$\chi' = \frac{Ne^2 f}{2m\omega\epsilon_0} \frac{\omega_0 - \omega}{(\omega_0 - \omega)^2 + (\gamma/2)^2} \tag{5.75}$$

$$\chi'' = \frac{Ne^2 f}{2m\omega\epsilon_0} \frac{\gamma/2}{(\omega_0 - \omega)^2 + (\gamma/2)^2} \tag{5.76}$$

Here, e is the electron charge, m is the electron mass, f is a dimensionless oscillator strength, ε_0 is the dielectric permittivity of vacuum, ω_0 is the Na atomic resonance frequency, ω is the light frequency, and γ is the transition probability of Na excited states, that is, $\gamma = A_{21} = 1/\tau_R$ (τ_R is the spontaneous lifetime of the Na excited states). Here χ' represents the dispersion near the resonance absorption lines, whereas χ'' represents the resonance absorption. When a linear polarized light with frequency ω close to the Na resonance frequency ω_0 propagates along the magnetic field $\vec{B}$ through the Na vapor, the linear polarization may be decomposed into a superposition of right- and left-circular polarized light. The static magnetic field $\vec{B}$ across the atomic vapor cell causes the Zeeman splitting of Na energy levels, which results in different Na resonance frequencies for right- and left-circular polarized light: $\omega_0^{\pm} = \omega_0 \pm \Delta\omega$, where $\Delta\omega$ is the Zeeman splitting. Therefore, according to Equation (5.74), the right- and left-circular polarized light will experience different indices of refraction at frequencies near ω_0 (the Na resonance frequency in absence of the magnetic field) as illustrated in Figure 5.15. The difference in the index of refraction will result in different propagating velocities and introduce phase shift between the two circular polarizations. The phase shift is given by

$$\Delta\varphi = 2\pi \frac{l\Delta n}{\lambda} \tag{5.77}$$

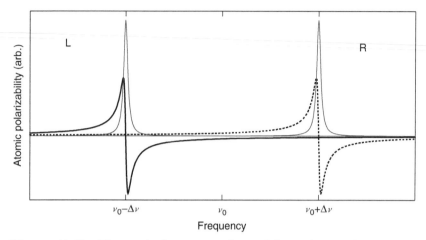

Figure 5.15 Theoretical atomic polarizability (thicker lines) and Lorentzian absorption (thin lines) curves of Na vapor in the presence of an axial magnetic field. Zeeman effect causes the right circular light has a higher resonance frequency while the left circular light has a lower resonance frequency. At ν_0 (the resonance frequency in absence of the magnetic field), there is no absorption for either component of the light, but there is a difference in polarizability (thus index of refraction). In the actual Faraday Na cell, all these features are thermally broadened. The magnetic field is set so that the Zeeman splitting is larger than the thermally broadened linewidth.

where l is the length of the Na cell, λ is the light wavelength, and Δn is the difference of the refraction index between the two circular polarizations. The resulting light will have a linear polarization rotated by certain angle from its initial linear polarization. This dispersive polarization rotation phenomenon is called the *resonant Faraday effect*. The degree of polarization rotation depends on the light frequency, the magnetic field strength, the vapor cell temperature (i.e., the Na density), and the vapor cell length. For a given frequency and cell length, the magnetic field and the temperature of the Na cell are adjusted so that the phase shift $\Delta\varphi = \pi$, corresponding to a rotation of polarization of $\pi/2$. The resulting light will pass the second polarizer without attenuation and result in nearly 100% transmission at the Na resonance frequency. Meanwhile, the polarization rotation for frequencies a few linewidth from the FADOF transmission peaks is vanishingly small. This results in the out-of-band rejection of the filter on the order of 10^{-5} determined by the extinction of polarizers. More detailed theory behind the FADOF can be found in Yeh (1982), Dick and Shay (1991), and Yin and Shay (1991).

The original Na FADOF consisted of a Na vapor cell of 0.76 cm long at 189°C in a 1750 G magnetic field (Chen et al., 1993). However, the Na vapor cell was easily degraded by the high temperature. Chen et al. (1996) improved the Na FADOF by using a 2.54-cm long Na vapor cell in a magnetic field of 1800 G in an oven operated at a temperature of 168°C. Plotted in Figure 5.16 is a measured transmission function of a Na FADOF operated at 168°C with 1850 G of magnetic field (Arnold and She, 2003). The peak transmission at the Na line center is 86%, and the FWHM of the peak is 1.9 GHz. One of the advantages for the Na FADOF is that the FADOF

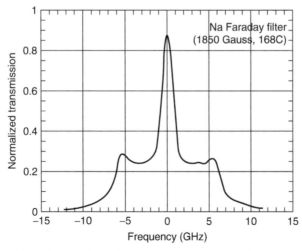

Figure 5.16 Normalized transmission curve of a CSU Na lidar Faraday filter. (From Arnold, K.S., and She, C.Y., *Contemp. Phys.*, 44, 35–49, 2003. With permission.)

does not need highly collimated light, and therefore, has large field of view. The combination of the ultra-narrow Na FADOF and the reduced beam divergence and receiver field of view makes it possible to reduce the detected daytime sky background to a level comparable with the typical values encountered at night, leading to a peak Na signal-to-background ratio of 15 at high noon (Chen et al., 1996). This signal-to-noise ratio enables the daytime temperature and wind measurements in the mesopause region.

Because of the hyperfine structure and the magnetic Zeeman splitting of the Na energy levels, an excited Na atom can decay via different energy paths and thus reradiate at a different frequency than that of the absorbed laser photon. With two ground states, four upper states and their magnetic substates, the selection rules from quantum mechanics allow many absorption/reemission possibilities, each having characteristic emission frequencies. All these transitions will be excited with different probabilities depending on the laser frequency, so the returned fluorescence is spectrally dispersed. This dispersion in returned fluorescence frequency is not important for wide bandwidth receivers. For nighttime configuration, when a broadband filter is used in the receiver, the filter transmission function is constant, and the receiving efficiency is the same over the spectral range of the Na D_2 Doppler lines. The nighttime filters with a bandwidth of hundreds of gigahertz pass all the dispersed light with the same transmission value. Therefore, the filter transmission function is canceled when taking the ratios (R_T and R_W) between fluorescence intensity signals at different frequencies. According to Equations (5.51) and (5.52), the ratios R_T and R_W are thus determined by the ratio of effective cross sections. However, for daytime configuration, the bandwidth of the Na Faraday filter is so narrow that the spectrally dispersed fluorescence will have different transmissions through the Faraday filter. For different laser exciting frequencies, the returned fluorescence will contain different portions of the 10 emission components, resulting in different receiving efficiencies. Therefore, the filter transmission function must be taken into account when calculating the ratios R_T and R_W. As discussed in Section

5.2.2.2, the Doppler frequency shift of the fluorescence should also be considered when computing the filter transmission function. By using the frequency-dependent scattering cross sections and taking into account the Hanle effect as well as the transmission function of the Faraday filter, one can deduce the calibration curve between the intensity ratio and temperature for the FADOF daytime measurements. Shown in Figure 5.17 are the calibration curves with and without FADOF for CSU Na wind/temperature lidar system taken from Chen et al. (1996). Notice that the intensity ratio for the FADOF channel at the same temperature is considerably smaller than that for the nighttime configuration. This is expected since FADOF transmission at the frequency f_c or $f_\pm$ is much lower than at the peak frequency f_a. The temperature of the mesopause region can be derived from the measured intensity ratio by using these calibration curves. The difference between the temperature determined from the nighttime configuration and daytime

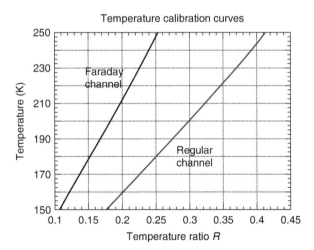

Figure 5.17 Calibration curves for the conversion of measured intensity ratio to temperature in the mesopause region for the regular channel (without Faraday filter) and the Faraday channel (with Faraday filter). (From Chen, H., White, M.A., Krueger, D.A., and She, C.Y., *Opt. Lett.*, 21, 1093–1095, 1996. With permission.)

configuration is within the experimental error (Chen et al., 1996).

5.2.2.4.2. Fabry–Perot etalon for daytime operation

The UIUC Na wind/temperature lidar can also make daytime observations (Yu et al., 1997). To reduce background noise from the bright daytime sky, the laser beam divergence and telescope field of view were reduced to 0.5 mrad. Further smaller divergence angles would cause serious saturation effects, which result in intolerable measurement distortion. A temperature stabilized, narrowband interference filter and a temperature-stabilized pressure-tuned Fabry–Perot etalon were employed in the daytime receiving telescope to reduce the optical bandwidth to 10 GHz (FWHM). These narrowband filters help reject the sky background. To optimize performance, different operating configurations and scenarios were used for day and night measurements. During the daytime, a 0.35-m diameter telescope was employed and only Na density and temperature profiles were measured. The major error source was signal photon noise and background noise. When the daytime data are averaged over a vertical range of 1 km and temporally for 30 min, the rms temperature errors between 85 and 100 km vary between 1.7 and 4.6 K. The smallest errors are at the peak of the Na layer near 92.5 km where the signal is strongest. Like the FADOF, the transmission function of the etalon is not constant over the spectral range. In order to derive accurate temperature, it is also necessary to derive the calibration curve after considering the transmission function. This requires precise knowledge of the etalon transmission function, including the central frequency, bandwidth, and the lineshape. However, the center of the etalon transmission band is not known accurately. There is no way to analytically correct for this transmission function and optical pumping effects. This can be solved only by an examination of the calculated temperatures near the night/ day transition. Any temperature jump between the two configurations is associated with the combined effects of optical pumping and the etalon transmission band curve. From the observed bias, the etalon central frequency is calculated and the data are reprocessed considering this central frequency.

5.2.2.5. Lidar Data and Error Analysis

5.2.2.5.1. Lidar data analysis

Here we briefly describe the lidar data analysis on the aspects of how to derive the atmosphere temperature, wind, and density information from the lidar photon count profiles. A standard preprocess of the raw photon counts includes the correction of the PMT and discriminator saturation effects, the correction of the chopper function or PMT gain effect, and the subtraction of the background photon counts. After the preprocess, the lidar equations for the Na resonance fluorescence and Rayleigh signals can be expressed as Equations (5.44) and (5.45), respectively. Notice that the background term no longer presents in these two equations as compared with the lidar equation (5.1). The Na photon counts are then normalized to the Rayleigh photon count at 35 km, which is obtained by averaging the Rayleigh signals between 30 and 40 km. This is defined as the normalized Na photon count given by Equation (5.48). With the Rayleigh term in the denominator, the time-dependent laser power and atmospheric transmittance can be canceled because the Rayleigh and Na photons resulted from the same laser pulses at nearly the same time. Further assumptions are made that the receiver has the same geometric factors for the Na signals and Rayleigh signal at Z_R. The receiver also has broadband filters so that the Na fluorescence and Rayleigh scattering photons experience the same filter transmission even if their frequencies are slightly different. Thus, these two factors can be canceled out and the normalized Na photon counts are then given by Equation (5.49).

To derive the atmosphere temperature and wind, the ratio of the normalized Na photon counts is calculated. For the two-frequency technique, the ratio for temperature R_T is calculated using the crossover frequency f_c and the D_{2a} peak frequency f_a, as shown in Equation (5.50). The frequency difference between f_c and f_a is very small, so the Rayleigh scattering cross sections at these two frequencies can be regarded as equal. Thus, the Rayleigh cross-section and atmosphere number density at Z_R can also be eliminated and the second equality in Equation (5.50) can be achieved. However, the Na densities at different times t_1 and t_2 are still involved in this equality. For the two-frequency technique, the laser is tuned between f_a and f_c. Approximately 1 min is required for tuning and 1 min for data acquisition at each frequency. At this interval length (1 min), the Na density fluctuations can bias the temperature measurement. To minimize this bias, the photon count profiles are weighted and averaged before taking the ratio R_T using the method described in She et al. (1990): four (three) consecutive photon count profiles obtained with the laser tuned to f_c (f_a) are added using weighting factors of 1/15, 1, 1, 1/15 (3/10, 1, 3/10) to form a single f_c (f_a) profile which corresponds to an equivalent data integration period of ~5 min. The weighting factors were chosen to minimize the effects of Na and atmospheric density variations. By using these weighting factors, only sixth- and higher-order temporal variations of the density influence the final temperature profiles. Detailed discussion about this average can be found in Appendix C of Papen et al. (1995a). Therefore, canceling the Na number density in Equation (5.50) is a good approximation. Now the ratio R_T is simply proportional to the ratio of the effective scattering cross sections at these two frequencies, shown as the third equality in Equation (5.50).

Since the lineshape of effective cross-section is a convolution of the atomic Doppler broadened lineshape and the laser lineshape, the ratio R_T will be influenced by the lineshape of the laser. Shown in Figure 5.18(a) is a comparison of three laser lineshape functions: a measured lineshape compared to the Lorentzian and Gaussian lineshapes. Using different laser lineshapes, the calculated theoretical ratios of R_T at the same temperatures are different, as plotted in Figure 5.18(b). The actual laser lineshape is very close to the Gaussian shape, and it is highly reproducible. The curve for measured laser lineshape in Figure 5.18(b) is served as a calibration curve, that is, when R_T is calculated from the

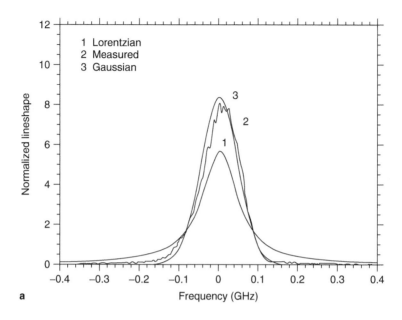

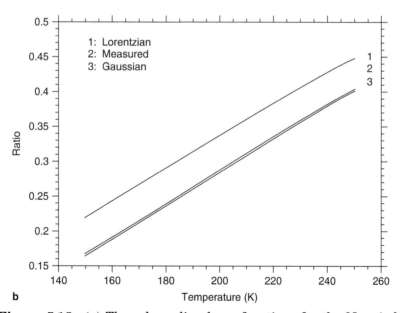

Figure 5.18 (a) Three laser lineshape functions for the Na wind/ temperature lidar: a measured lineshape along with two hypothetical line shapes with equal FWHM = 112 MHz. (b) The temperature calibration lines of these lineshapes. (From She, C.Y., Yu, J.R., Latifi, H., and Bills, R.E., *Appl. Opt.*, 31, 2095–2106, 1992. With permission.)

observational data of photon counts, the temperature can be then inferred from the curve accordingly. For the two-frequency technique, there is no wind information available, so the radial wind v_R is set to 0 m/s during the determination of temperatures. When the lidar beam points to zenith, the radial wind is usually less than 3 m/s so that the resulting temperature errors are negligible (<0.5 K).

For the three-frequency technique, the ring dye laser is locked to the Na Doppler-free feature at the D_{2a} peak f_a throughout the observations, and an AOM is used to shift the laser frequency by ± 630 MHz. The experiment scenario is that the outgoing lidar beam switches its frequency from the peak frequency f_a to the plus wing frequency f_+ then to the minus wing frequency f_-, and then repeats this sequence again. The data are accumulated at three frequencies accordingly. The frequency switching is very fast, either every single pulse (as in the CSU system) or every 50 pulses (as in the UIUC system), so the data at three frequencies are collected with very short intervals, such as 1/50 or 1 s. This interleaving of observations eliminates any biases due to Na density fluctuations. Therefore, when taking the ratio for temperature R_T and the ratio for wind R_W, the Na density terms can be safely canceled so these two ratios are given by Equations (5.51) and (5.52), respectively.

The radial wind velocities are determined by the measurement of the Doppler shift of the Na atoms along the laser beam path. A positive wind component moving along the laser beam, v_R, shifts the absorption spectrum of the atmospheric Na to higher frequency by v_R/λ, where λ is the laser wavelength. The Doppler frequency shift can be reflected from the change of the return signal ratio $R_W = N(f_-)/N(f_+)$. As the velocity increases ($v_R > 0$), the ratio of returns R_W decreases as shown in Figure 5.2(b). The maximum variation in R_W occurs for measurements taken at f_+ and f_- on each side of the D_{2a} resonance. Ignoring the contribution of the D_{2b} feature and using the analytic expression for the absorption cross-section, we find that the optimal frequencies for highest sensitivity of R_W to the wind variations are $f_{\pm} = f_a \pm \sqrt{2}\sigma_D$ (She et al., 1992; Papen et al., 1995a). For the atmospheric Na atoms, we choose $f_\pm = f_a \pm 630\,\text{MHz}$. The actual ratio R_W used for the UIUC Na wind/temperature lidar is more complicated than the simple ratio given by Equation (5.52). The ratio used is

$$R_{W2} = \frac{\ln\left(\frac{N_{\text{norm}}(f_-,t_2)}{N_{\text{norm}}(f_+,t_1)}\right)}{\ln\left(\frac{N_{\text{norm}}(f_-,t_2)N_{\text{norm}}(f_+,t_1)}{N_{\text{norm}}^2(f_a,t_3)}\right)} = \frac{\ln\left(\frac{\sigma_{\text{effNa}}(f_-,T,v_R)}{\sigma_{\text{effNa}}(f_+,T,v_R)}\right)}{\ln\left(\frac{\sigma_{\text{effNa}}(f_-,T,v_R)\sigma_{\text{effNa}}(f_+,T,v_R)}{\sigma_{\text{effNa}}^2(f_a,T,v_R)}\right)}$$

$$(5.78)$$

The fast switching between frequencies by the AOM allows the cancelation of the densities. This ratio is chosen to minimize the uncertainty in the measurement.

Sodium densities can also be calculated from the observations. Only one frequency at the D_{2a} peak frequency f_a is necessary, but a weighted average of all three frequencies is used to obtain a better estimate. The weighting is chosen to minimize the sensitivity of the density measurement to changes in the temperature and wind. The weighted cross-section is defined as

$$\sigma_{\text{eff\_wgt}} = \sigma_a + \alpha\sigma_+ + \beta\sigma_- \qquad (5.79)$$

where σ_f is the effective cross-section at frequency f. The weighting factors, α and β, are chosen so that

$$\frac{\partial\sigma_{\text{eff\_wgt}}}{\partial T} = 0; \quad \frac{\partial\sigma_{\text{eff\_wgt}}}{\partial v_R} = 0 \qquad (5.80)$$

After the weights are determined, the Na density is calculated by

$$n_{\text{Na}}(z) = 4\pi n_R(z_R)\sigma_R \frac{z^2}{z_R^2}$$
$$\times \frac{N_{\text{norm}}(f_a,z) + \alpha N_{\text{norm}}(f_+,z) + \beta N_{\text{norm}}(f_-,z)}{\sigma_a + \alpha\sigma_+ + \beta\sigma_-}$$

$$(5.81)$$

Once the ratios R_T and R_W are calculated from the observed photon counts, the temperature and radial wind can be inferred from the calibration curves similar to the ones shown in Figure 5.7(a) and (b), respectively. Since R_T and R_W depend on both temperature and wind, the solution for temperature and wind is iterative: by assuming 0 m/s wind, we calculate the temperature using R_T, and then calculate the wind using the calculated temperature and the ratio R_W. We then recalculate the temperature again using the computed wind. This is repeated until the values converge. After the temperature and wind are determined, the Na density can then be calculated using Equations (5.79)–(5.81).

5.2.2.5.2.   Lidar error analysis

There are several sources of measurement errors. They can be classified into two categories: systematic errors and random errors. Systematic errors are mainly caused by our imprecise information of the Na atom absorption cross-section $\sigma_{abs}(\nu)$, the laser absolute frequency calibration, the laser lineshape $g_L(\nu)$, the receiver filter function $\eta_R(\nu)$, and the geometric factor $G(z)$. Random errors are mainly caused by the photon noise, the detector shot noise, Na density fluctuations, and the random laser and electronic jitter. These system characterization errors and photon-counting errors will propagate and contribute to the errors of temperature, wind, and Na density measurements. Systematic errors determine the measurement accuracy (bias), and random errors determine the measurement precision (uncertainty).

A.   Systematic errors

a.   Determination of $\sigma_{abs}(\nu)$:   The accurate determination of $\sigma_{abs}(\nu)$ is mainly obstructed by the Hanle effect, the Na layer saturation, and the optical pumping effect. The $\sigma_{abs}(\nu)$ given in Equation (5.41) and the values of relative line strength A_n given in Table 5.1 are true for the case when Zeeman splitting of hyperfine structures (hfs) of Na atoms, saturation, and optical pumping effects are not present. However, in the geomagnetic field in the mesopause region, the Na hyperfine energy levels exhibit Zeeman splitting. The resulted frequency shift of the Na transition lines does not exceed 1.4 MHz, which is less than 0.3% of the hfs splitting and may be ignored in the temperature conversion procedure. Nevertheless, the existence of the geomagnetic field will cause the Hanle effect through the Zeeman splitting and the Hanle effect will modify the relative line strength A_n. Here, A_n is not the strength for a single transition line, but defined as a sum over the corresponding Zeeman transitions. This new relative line strength of Na atoms caused by the Hanle effect can be calculated using the equations derived by Fricke and von Zahn (1985) and Papen et al. (1995a) once the laser polarization, the strength, declination and inclination angles of the geomagnetic field are known. A_n shown in Table 5.1 are the spatially integrated values (5, 5, 2, 14, 5, 1). For the Urbana Atmospheric Observatory (UAO), the Hanle-effect-modified A_n values are (5, 5.48, 2, 15.64, 5, 0.98). The relative line strengths must be recomputed for each lidar site, beam direction, and beam polarization because of the large variations in the magnetic-field direction around the globe. If the Hanle effect is ignored, the temperature derived by using spatially averaged line strength will result in a bias of 1.4 K (Papen et al., 1995a).

For large laser pulse intensities, the intensity dependence of $\sigma_{abs}(\nu)$ becomes important because single atoms no longer respond in the linear fashion that is assumed for the calculation of the transition strengths. Saturation and optical pumping effects must be included to determine A_n for use in Equation (5.41). As described in Section 5.2.1.3, the saturation of Na density measurements results from the finite lifetime of the excited state. Once in the excited state, the Na atoms cannot absorb another photon before relaxing back to the ground state, and may be involved in stimulated emission further reducing the backscattered signal. If the laser linewidth is narrow enough that the laser excites Na atoms only from one individual ground state, for example, the $F = 2$ state, optical pumping occurs if the excited atoms decay to the $F = 1$ ground state. Optical pumping reduces the available Na density on $F = 2$ state for absorbing the incoming laser photons, and effectively reduces the absorption cross-section. Laser pulse lengths that are shorter than the lifetimes of the excited states do not suffer from optical pumping effects because no significant shift in the ground-state population can occur. As the laser pulse length increases, the error induced by optical pumping becomes more significant. For modern Na wind/temperature lidar, due to the short Nd:YAG laser pulse duration, the lidar outgoing pulse has a length around 6 to 7 nsec. It is much shorter than the Na 16.23 nsec lifetime. So, the optical pumping effect is not significant and will be excluded in the analysis.

The absolute saturation level directly affects the accuracy of the Na density measurement. However, the temperature and wind measurements are influenced by the relative saturation effect. In other words, if the Na layer saturation levels at different frequencies are the same, the saturation will not affect the temperature and wind measurements since the ratio of effective cross sections remains the same as nonsaturated. Unfortunately, the saturation at frequency near the absorption peak is higher than at wing frequencies or the crossover frequency. When the laser intensity increases, the ratio R_T for both two- and three-frequency techniques will be different from the nonsaturated ratio. The saturated R_T will always be higher than nonsaturated R_T, and thus the saturation will result in higher measured temperature. The effect of saturation on temperature measurement is not as sensitive as the effect on Na density measurement. Saturation effects have been considered by several groups. Megie et al. (1978) made the first quantitative analysis using a rate-equation model. Their method and results were very comprehensive and founded the basis for the Na saturation study, although they are more suitable for the broadband laser excitation instead of the narrowband laser excitation. Welsh and Gardner (1989), von der Gathen (1991), and Papen et al. (1995a) adapted Megie et al.'s (1978) rate-equation methods and applied them to the narrowband laser cases. Milonni et al. (1998) made further detailed analysis using density matrix simulations for laser pulse excitations with long, short, and intermediate duration time. Here, we give the results from von der Gathen (1991) in Figure 5.19: the saturation at maximum, minimum, and temperature correction in relation to the energy area density. The curves were computed under the conditions of pulse length 10 nsec, laser linewidth 130 MHz, atmospheric transmission 0.7, and Na temperatures 150, 200, and 250 K. Normal operation conditions of Na wind/temperature lidar have the laser pulse energy of 40 mJ, the laser

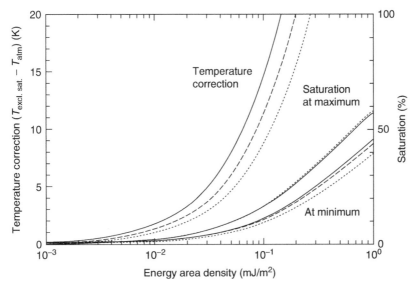

Figure 5.19 Na saturation at maximum, minimum, and temperature correction in relation to the energy area density. Basic assumptions are atmospheric temperature 200 K (dashed lines), laser linewidth 130 MHz, and pulse length 10 nsec. Further atmospheric temperatures are 150 K (dotted lines) and 250 K (solid lines). (From von der Gathen, P., *J. Geophys. Res.*, 96 (A3), 3679–3690, 1991. With permission.)

divergence of 1 mrad, and the laser linewidth of about 130 MHz. This gives the energy area density of 4.4×10^{-3} mJ/m² at 90 km. According to Figure 5.19, the estimated saturation for Na density is about 1% and the temperature correction is less than 0.5 K. This temperature bias caused by the saturation effect is less than the random errors due to photon noise.

b. Absolute laser frequency calibration and laser lineshape: The absolute laser frequency calibration and the laser lineshape influence the derived temperature and radial wind velocity because the effective backscatter cross-section $\sigma_{\mathrm{eff}}(\nu)$ is the convolution of the Na absorption cross section $\sigma_{\mathrm{abs}}(\nu)$ and the laser lineshape $g_{\mathrm{L}}(\nu, \nu_{\mathrm{L}})$. The factor $g_{\mathrm{L}}(\nu, \nu_{\mathrm{L}})$ contains the information of laser central frequency, linewidth, and lineshape. In the Na wind/ temperature lidar, the Doppler-free saturation–absorption features of a Na vapor cell serve as the absolute frequency reference to lock the CW ring dye laser. The PDA is seeded by the dye laser, so the frequency accuracy of the Na vapor cell response is supposed to transfer to PDA pulses. However, as described in Section 5.2.2.3, the central frequency of the PDA output pulse is shifted by the nonlinear effects associated with the pulse amplification. This frequency shift is on the order of 10 MHz. While this shift does not have a strong effect on temperature measurements, it can bias wind measurements. This frequency shift can be measured through carefully designed experiments as described in Section 5.2.2.3. Discrepancies between the lineshape (e.g., Gaussian shape) used in data reduction and the actual lineshape also introduce temperature and wind errors. The laser lineshape needs to be measured by using a narrowband etalon in order to derive accurate temperature and wind. As shown in Figure 5.18, since the actual laser lineshape is very close to be a Gaussian shape, the temperature error caused by using a Gaussian assumption is less than 0.5 K.

c. Receiver filter function and geometric factor: Accurate knowledge of the effective backscatter cross-section $\sigma_{\text{eff}}(\nu)$ is essential to both temperature and wind measurements. All the factors we considered above influence $\sigma_{\text{eff}}(\nu)$ by affecting either the determination of $\sigma_{\text{abs}}(\nu)$ or the laser frequency and lineshape. Our imprecise information of the effective backscatter cross-section $\sigma_{\text{eff}}(\nu)$ will result in systematic errors. In addition, our imprecise knowledge of the receiver filter function and geometric factor will also result in systematic errors. Through careful design of the laser and optical system, good arrangement of experimental setup (e.g., beam divergence), and precise calibration of laser frequency, lineshape, and filter transmission, these systematic errors can be eliminated or reduced to be within the limit of random errors caused by photon noise.

B. Random errors

Random errors for the lidar can be caused by the random uncertainty associated with laser and electronic jitter, random Na density fluctuations, and the shot noise associated with the photon-counting system. The laser and electronic jitter varies from system to system and also depends on the system alignment and environment stability. With proper laser and optics alignment in controlled lab environment, the jitter is usually quite small. The influence of Na density fluctuations can be minimized by proper weighting in two-frequency technique or eliminated by fast switching between three frequencies when taking data.

Shot noise ultimately limits the measurement precision of any photon-counting system because of the statistical nature of the photon-detection processes. The physics of the photon-counting process are well known. For the two-frequency ratio technique, the relative error of ratio R_T is given by

$$\frac{\Delta R_T}{R_T} = \frac{\left(1 + \frac{1}{R_T}\right)^{1/2}}{\left(N_{f_a}\right)^{1/2}} \left[1 + \frac{B}{N_{f_a}} \frac{\left(1 + \frac{1}{R_T^2}\right)}{\left(1 + \frac{1}{R_T}\right)}\right]^{1/2} \tag{5.82}$$

where B is the background photon count, N_{f_a} is the Na photon count at peak frequency, and $R_T = N_{f_c}/N_{f_a}$. For the three-frequency ratio technique, the relative error of ratio R_T is given by

$$\frac{\Delta R_T}{R_T} = \frac{\left(1 + \frac{1}{R_T}\right)^{1/2}}{\left(N_{f_a}\right)^{1/2}} \left[1 + \frac{B}{N_{f_a}} \frac{\left(1 + \frac{2}{R_T^2}\right)}{\left(1 + \frac{1}{R_T}\right)}\right]^{1/2} \tag{5.83}$$

while the relative error of ratio R_W is given by

$$\frac{\Delta R_W}{R_W} = \frac{\left(1 + \frac{1}{R_W}\right)^{1/2}}{\left(N_{f_+}\right)^{1/2}} \left[1 + \frac{B}{N_{f_+}} \frac{\left(1 + \frac{1}{R_W^2}\right)}{\left(1 + \frac{1}{R_W}\right)}\right]^{1/2} \tag{5.84}$$

where N_{f_+} is the Na photon count at the plus wing frequency and $R_W = N_{f_-}/N_{f_+}$.

C. Calculation of errors

Systematic errors and random errors can be converted into the errors of derived temperature and radial wind velocity by the use of differentials of the corresponding ratio. Here we take the two-frequency temperature ratio as an example to explain the calculation method. A similar method can be used to analyze the error propagation from the uncertainties of lidar system parameters and photon noise to the uncertainties of the temperature and wind measurements by the three-frequency ratio technique.

For the two-frequency technique, the ratio R_T used to derive temperature error is

$$R_T(f_a, f_c, T, v_R, \sigma_L) = \frac{\sigma_{eff}(f_c, T, v_R, \sigma_L)}{\sigma_{eff}(f_a, T, v_R, \sigma_L)} \qquad (5.85)$$

The temperature determined from Equation (5.85) is a function of R_T, f_a, f_c, σ_L, and v_R in implicit form. Therefore, the temperature errors are given by

$$\Delta T = \frac{\partial T}{\partial R_T} \Delta R_T + \frac{\partial T}{\partial f_a} \Delta f_a + \frac{\partial T}{\partial f_c} \Delta f_c + \frac{\partial T}{\partial \sigma_L} \Delta \sigma_L + \frac{\partial T}{\partial v_R} \Delta v_R$$
$$(5.86)$$

Implicitly differentiating Equation (5.86), we have

$$\Delta T = \Delta R_T \left(\frac{\partial R_T / \partial R_T}{\partial R_T / \partial T} \right) + \Delta f_a \left(\frac{\partial R_T / \partial f_a}{\partial R_T / \partial T} \right) + \Delta f_c \left(\frac{\partial R_T / \partial f_c}{\partial R_T / \partial T} \right)$$
$$+ \Delta \sigma_L \left(\frac{\partial R_T / \partial \sigma_L}{\partial R_T / \partial T} \right) + \Delta v_R \left(\frac{\partial R_T / \partial v_R}{\partial R_T / \partial T} \right) \qquad (5.87)$$

The derivatives of the ratio R_T with respect to each of the system parameters are

$$\frac{\partial R_{\rm T}}{\partial x} = R_{\rm T} \left[\frac{\partial \sigma_{\rm eff}(f_{\rm c})/\partial x}{\sigma_{\rm eff}(f_{\rm c})} - \frac{\partial \sigma_{\rm eff}(f_{\rm a})/\partial x}{\sigma_{\rm eff}(f_{\rm a})} \right] \tag{5.88}$$

where x is one of the parameters $(f_{\rm a}, f_{\rm c}, \sigma_{\rm L}, v_{\rm R}, T)$. If the laser lineshape is assumed to be Gaussian, the Na effective scattering cross-section $\sigma_{\rm eff}$ can be represented by Equation (5.43). By computing the partial differentials in Equation (5.87) with the help of Equation (5.43), the temperature errors introduced by photon noise and uncertainties in system parameters can be calculated. For example, the temperature error associated with the uncertainty in ratio $R_{\rm T}$ caused by photon noise is given by

$$\Delta T = \frac{\partial T}{\partial R_{\rm T}} \Delta R_{\rm T} = \frac{\Delta R_{\rm T}}{R_{\rm T}} \left[\frac{\partial \sigma_{\rm eff}(f_{\rm c})/\partial T}{\sigma_{\rm eff}(f_{\rm c})} - \frac{\partial \sigma_{\rm eff}(f_{\rm a})/\partial T}{\sigma_{\rm eff}(f_{\rm a})} \right]^{-1}$$
$$\tag{5.89}$$

Similar expression can be derived for other system parameters and measurement techniques.

The derivatives and their magnitudes were computed for an operating point $Q_{\rm T}$ ($T = 200\,{\rm K}$, $v_{\rm R} = 0\,{\rm m/s}$, $f_{\rm a} = -651.5\,{\rm MHz}$, $f_{\rm c} = 187.8\,{\rm MHz}$, $\sigma_{\rm L} = 60\,{\rm MHz}$) by Papen et al. (1995a). According to the computation results, the major source of temperature error is the uncertainty of laser linewidth, while the major source of wind error is the uncertainty of absolute laser frequency. An uncertainty of 4 MHz in laser linewidth will result in 0.46 K error in temperature. The temperature error is also sensitive to the laser frequency at $f_{\rm c}$ ($\sim$0.26 K per 2.45 MHz), but insensitive to the peak frequency $f_{\rm a}$. The wind error is sensitive to the uncertainties of the wing frequencies f_{+} and f_{-} ($\sim$0.7 m/s per 2.4 MHz) while insensitive to the uncertainty of laser linewidth. Because of the way the frequencies are chosen, the measurement errors are not sensitive to an error in the other measurement. For example, 1 m/s wind error results in temperature error by 0.017 K, while 1 K temperature error results in wind error by 0.087 m/s (Papen et al., 1995a).

5.2.3. Fe Boltzmann Temperature Lidar

5.2.3.1. Introduction

Knowledge of the temperature and constituent density in the mesopause region, especially in the polar mesopause region, is important for understanding a wide range of geophysical phenomena, such as the global climate change, the gravity wave effects on atmosphere dynamics, and polar mesospheric clouds. Currently, Na wind/temperature lidars are one of the most accurate remote sensing instruments used to probe the mesopause region. Unfortunately, the demanding environmental requirements of narrowband Na systems generally preclude their use at many important remote sites such as the North and South Poles. For global scale observations, the development of a robust temperature lidar is clearly attractive. To help address this measurement need, the University of Illinois in collaboration with the Aerospace Corporation and the National Center for Atmospheric Research (NCAR) developed a new lidar system — the Fe Boltzmann temperature lidar — for measuring temperature profiles from the middle stratosphere to the lower thermosphere, which can be deployed on research aircraft or operated at remote sites during both day and night (Chu et al., 2002a). This Fe Boltzmann temperature lidar was deployed to the North Pole in 1999 aboard the National Science Foundation/National Center for Atmospheric Research (NSF/NCAR) Electra aircraft and to the South Pole from 1999 to 2001 at a ground-based station. During these campaigns the instrument made the first lidar measurements of middle atmosphere temperature, Fe density, and PMC over both the North Pole and the South Pole during midsummer.

It is a completely different concept to use the Boltzmann technique to measure temperature. Unlike the Doppler technique that relies on the temperature dependence of the Doppler broadening of atomic resonance absorption lines, the Boltzmann technique relies on the temperature dependence of the Maxwell–Boltzmann distribution of atomic populations on different atomic energy levels under the assumption of thermodynamic equilibrium. The higher the temperature, the greater is the population in upper energy levels. The population ratio between the upper and lower energy levels is determined by the temperature T and the energy difference ΔE between these two levels; that is, the population ratio is proportional to $\exp(-\Delta E/k_B T)$, where k_B is the Boltzmann constant. This well-known concept has been applied to imagers for deriving atmosphere temperatures from airglow. Applying the Boltzmann concept to the lidar field is a revolution for resonance fluorescence lidar, compared to conventional Doppler lidars. It lowers the requirement on laser linewidth and allows the lidar to be constructed in a robust way to enable difficult measurements at remote sites and in aircrafts. One of the key points for a Boltzmann lidar is to find a constituent (atom or molecule) that satisfies the following conditions: (1) the energy difference between ground-state sublevels is large enough to be detected, and also small enough to ensure adequate population on the upper ground state; (2) the transition wavelengths from its ground states to excited states fall in the wavelength range of available laser sources. Gelbwachs (1994) first pointed out that the mesospheric Fe atoms have the energy level distribution suitable for this type of system. Based upon his suggestion, a solid-state Fe Boltzmann temperature lidar using twin solid-state alexandrite lasers was designed, developed, and deployed from the North Pole to the South Pole by the UIUC for both daytime and nighttime measurements (Chu et al., 2002a). Following the success of the first Fe Boltzmann temperature lidar, the Arecibo group constructed an Fe Boltzmann lidar using a single dye laser and obtained temperature profiles during nighttime operation (Raizada and Tepley, 2002). In the following, we will describe the principle, technology, and error analysis of Fe Boltzmann temperature lidar based upon the UIUC system.

5.2.3.2. Measurement Principle (Boltzmann Technique)

There are four natural isotopes of Fe atoms: ^{54}Fe, ^{56}Fe, ^{57}Fe, and ^{58}Fe. Among them, ^{56}Fe is the most abundant isotope

Table 5.3 Isotopic Data of Fe Atoms

	^{54}Fe	^{56}Fe	^{57}Fe	^{58}Fe
Z	26	26	26	26
A	54	56	57	58
Nuclear spin	0	0	1/2	0
Natural abundance	5.845%	91.754%	2.119%	0.282%
Standard Atomic weight		55.845 g/mol		

with natural abundance of 91.754%. The isotopic data of Fe atoms are summarized in Table 5.3. The isotope shifts of ^{54}Fe, ^{57}Fe, and ^{58}Fe with respect to ^{56}Fe are around 1 GHz for the two resonance lines 372 and 374 nm that we are interested. Since the energy difference between the two lowest ground states is about $416\,\text{cm}^{-1} = 1.248 \times 10^4\,\text{GHz}$, the contribution of these isotope shifts to the Boltzmann factor $\exp(-\Delta E/k_\text{B}T)$ is negligible (less than 1×10^{-4}). In addition, as shown below, the temperature determination is directly related to the effective cross-section ratio between 372 and 374 nm. Since the two lines have very similar isotope shifts, the effective cross-section ratio when considering all Fe isotopes is nearly identical to the ratio when considering ^{56}Fe only, which results in nearly identical Boltzmann temperature. Therefore, the influence of Fe isotopes to Boltzmann temperature is negligible. Here we only consider the ^{56}Fe atoms for the simplicity of description.

Since the nuclear spin of ^{56}Fe is zero, the Fe atoms have no hyperfine structures. The Fe Boltzmann temperature lidar relies on the unique energy level distribution diagram of the ground-state manifold of Fe atoms (see Figure 5.20). In thermodynamic equilibrium, the ratio of the populations in the $J = 3$ and $J = 4$ sublevels in the ground-state manifold is given by Maxwell–Boltzmann distribution law:

$$\frac{P_{374}(J=3)}{P_{372}(J=4)} = \frac{\rho_{\text{Fe}}(374)}{\rho_{\text{Fe}}(372)} = \frac{g_2}{g_1} \exp[-\Delta E/k_\text{B}T] \qquad (5.90)$$

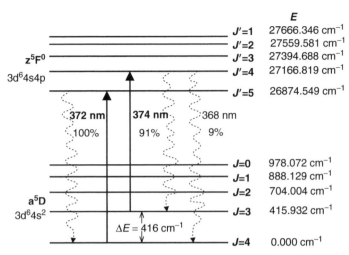

Figure 5.20 Energy level diagram of atomic Fe used for the Boltzmann technique. (From Chu, X., Pan, W., Papen, G.C., Gardner, C.S., and Gelbwachs, J.A., *Appl. Opt.*, 41, 4400–4410, 2002. With permission.)

where P_{372} and P_{374} are the populations in the two ground states with degeneracy factors $g_1 = 9$ and $g_2 = 7$, ρ_{Fe} (372) and ρ_{Fe} (374) are the associated Fe number densities, ΔE is the energy difference between the two levels ($\sim 416\,\mathrm{cm}^{-1}$), k_B is the Boltzmann constant, and T is the atmospheric temperature. The subscripts 372 and 374 refer to the excitation wavelengths of two transitions in nanometers (actually, $\lambda_{372} = 372.0993\,\mathrm{nm}$ and $\lambda_{374} = 373.8194\,\mathrm{nm}$ in vacuum). The line parameters for these two transitions are summarized in Table 5.4. At 200 K, the ratio of the populations is about 26. The temperature is then given by

$$T = \frac{\Delta E / k_B}{\ln\left(\dfrac{g_2}{g_1}\dfrac{\rho_{Fe}(372)}{\rho_{Fe}(374)}\right)} \tag{5.91}$$

The Fe densities in these two states can be measured using resonance fluorescence lidar techniques. The densities are proportional to the number of backscattered photon counts from Fe atoms (N_{Fe}) detected for each wavelength probed by the lidar. The detected Fe photon count is given by the lidar equation

Table 5.4   Fe Resonance Line Parameters

Transition wavelength λ	372.0993 nm	373.8194 nm
Degeneracy for ground state	$g_1 = 9$	$g_2 = 7$
Degeneracy for excited state	$g_1' = 11$	$g_2' = 9$
Radiative lifetime of excited state (ns)	61.0	63.6
Einstein coefficient A_{ki} (10^8 s^{-1})	0.163	0.142
Oscillator strength f_{ik}	0.0413	0.0382
Branching ratio R_B	0.9959	0.9079
σ_0 (10^{-17} m^2)	9.4	8.7

$$N_{Fe}(\lambda, z) = \left(\frac{P_L \Delta t T_a}{hc/\lambda} E(\lambda, z) \right) \sigma_{eff}(\lambda, T, \sigma_L) R_{B\lambda} \rho_{Fe}(\lambda, z)$$

$$\times \Delta z \left(E(\lambda, z) T_a \frac{A_R}{4\pi z^2} \eta \right) \tag{5.92}$$

where z is the altitude, P_L is the laser power, Δt is the temporal resolution, Δz is the vertical resolution, T_a is the one-way transmittance of the lower atmosphere, λ is the wavelength, c is the speed of light in vacuum, h is Planck's constant, $E(\lambda, z)$ is the extinction coefficient of the signal associated with Fe absorption, $\sigma_{eff}(\lambda, T, \sigma_L)$ is the effective scattering cross-section of the Fe transition (which is a function of temperature T, laser wavelength λ, and laser linewidth σ_L), $R_{B\lambda}$ is the branching ratio ($R_{B374} = 0.9079$, $R_{B372} = 0.9959$), A_R is the area of the receiving telescope, and η is the overall optical efficiency of the system. The effect of the radial velocity on the absorption cross-section has not been included, but it is negligible for zenith observations because the Doppler shift caused by the vertical wind velocity is typically less than 30 MHz, which is very small compared to the Doppler broadening and the laser linewidth ($\sim$1 GHz). The total detected signal also includes the background noise count $N_B(\lambda, z)$, which must be added to the Fe signal count given by Equation (5.92).

The current system uses two separate lasers and receivers because the two Fe lines are too far apart in wavelength to probe easily by tuning a single laser, yet they are too close to be separated in the receiving telescope using a dichroic beam splitter. To compensate for signal variations in each of these systems, the photon counts from the Fe layer at each wave-length are normalized by the Rayleigh counts (N_R) over a common lower altitude range, typically near 50 km:

$$N_R(\lambda, z_R) = \left(\frac{P_L \Delta t T_a}{hc/\lambda}\right) \sigma_R(\lambda) \rho_{atmos}(z_R) \Delta z \left(T_a \frac{A_R}{4\pi z_R^2} \eta\right)$$

$$(5.93)$$

where z_R is the normalization altitude, $\rho_{atmos}(z_R)$ is the atmospheric number density at normalization altitude, and $\sigma_R(\lambda)$ is the effective atmospheric Rayleigh backscatter cross-section that is 4π times the Rayleigh angular backscatter cross-section $\sigma_R(\theta = \pi)$; that is, $\sigma_R(\lambda) = 4\pi\sigma_R(\theta = \pi)$, where θ is the scattering angle. The product of atmospheric number density and effective Rayleigh backscatter cross-section can be expressed as

$$\sigma_R(\lambda)\rho_{atmos}(z_R) = 1.370 \times 10^{-30} \times \frac{273}{T(z_R)} \times \frac{P(z_R)}{1013} \times \frac{1}{\lambda^{4.0117}}$$

$$(5.94)$$

where $T(z_R)$ is the atmospheric temperature in Kelvin, $P(z_R)$ is the atmospheric pressure in millibars, and λ is the laser wavelength in meters. The Rayleigh-normalized background-corrected Fe signal count is defined as

$$N_{norm}(\lambda, z) = \frac{N_{Fe}(\lambda, z) + N_B(\lambda, z) - \hat{N}_B(\lambda)}{N_R(\lambda, z_R) + N_B(\lambda, z_R) - \hat{N}_B(\lambda)}$$

$$= \frac{z_R^2 E^2(\lambda, z) R_{B\lambda} \sigma_{eff}(\lambda, T, \sigma_L) \rho_{Fe}(\lambda, z)}{z^2 \sigma_R(\lambda)\rho_{atmos}(z_R)} \quad (5.95)$$

where $\hat{N}_B(\lambda)$ is the estimated background noise count. The Fe densities at each wavelength can be derived from Equation (5.95):

$$\rho_{Fe}(\lambda, z) = \frac{z^2}{z_R^2} \frac{N_{norm}(\lambda, z)\sigma_R(\lambda)\rho_{atmos}(z_R)}{E^2(\lambda, z) R_{B\lambda} \sigma_{eff}(\lambda, T, \sigma_L)} \quad (5.96)$$

By using Equation (5.90), the ratio of the normalized Fe signal counts at the two wavelengths (R_T) can be related to temperature

$$R_T(z) = \frac{N_{norm}(\lambda_{374}, z)}{N_{norm}(\lambda_{372}, z)}$$

$$= \frac{g_2}{g_1} \frac{R_{B374}}{R_{B372}} \left(\frac{\lambda_{374}}{\lambda_{372}}\right)^{4.0117} \frac{E^2(\lambda_{374}, z)}{E^2(\lambda_{372}, z)} \tag{5.97}$$

$$\times \frac{\sigma_{eff}(\lambda_{374}, T, \sigma_{L374})}{\sigma_{eff}(\lambda_{372}, T, \sigma_{L372})} \exp(-\Delta E/k_B T)$$

$$= 0.7221 R_E^2(z) R_\sigma \exp(-598.44/T)$$

where

$$R_E(z) = \frac{E(\lambda_{374}, z)}{E(\lambda_{372}, z)}, \quad R_\sigma = \frac{\sigma_{eff}(\lambda_{374}, T, \sigma_{L374})}{\sigma_{eff}(\lambda_{372}, T, \sigma_{L372})} \tag{5.98}$$

At 200 K, the 372 nm Fe signal is approximately 30 times stronger than the 374 nm signal so that $R_T \sim 1/30$. By solving Equation (5.97) for temperature, we obtain the final result

$$T(z) = \frac{\Delta E/k_B}{\ln\left[\frac{g_2}{g_1} \frac{R_{B374}}{R_{B372}} \left(\frac{\lambda_{374}}{\lambda_{372}}\right)^{4.0117} \frac{R_E^2(z) R_\sigma}{R_T(z)}\right]}$$

$$= \frac{598.44 K}{\ln\left[\frac{0.7221 R_E^2(z) R_\sigma}{R_T(z)}\right]} \tag{5.99}$$

Thus, by measuring the Fe signal level at the two wavelengths and computing $R_T(z)$ using Equations (5.95) and (5.97), the temperature can be derived from Equation (5.99) provided the ratios of the extinction and the effective backscatter cross sections are known.

As described earlier in the text, the two main concepts on how to measure temperature in the mesopause region by resonance fluorescence lidars are the Doppler and the Boltzmann techniques. No matter which method is used, the laser-induced fluorescence is the convolution of the laser lineshape with the thermal Doppler-broadened atomic absorption spectrum. Since the Doppler technique retrieves temperature from the shape of the Doppler-broadened atomic absorption spectrum, a narrow bandwidth laser ($\sim$0.1 GHz) is required in order to precisely resolve the atomic absorption spectrum from the convoluted laser-induced fluorescence lineshape. On the other hand, the Boltzmann technique does not depend on an individual Doppler-broadened atomic absorption spectrum, but relies on the population ratio between two energy levels, that is, photon count ratio between two Doppler-broadened atomic absorption lines. Therefore, the laser bandwidth can be much broader and comparable to the atomic Doppler broadening linewidth in the mesopause region ($\sim$1 GHz) as long as sufficient number of fluorescence photons can be excited from the mesopause region. This feature allows the relative broadband lasers (such as solid-state lasers) to be deployed for the temperature measurements, and also simplifies the calibration for daytime measurements using a Fabry–Perot etalon as a narrowband filter. These features enable the development of a robust temperature lidar and the measurements of temperature in the Polar Regions.

To compare the performance of the Fe Boltzmann temperature lidar with the Na wind/temperature lidar, we define the sensitivity as the normalized change in the ratio R_T per degree of temperature change:

$$S_T = \frac{\partial R_T/\partial T}{R_T} \tag{5.100}$$

The sensitivity is a measure of the relative change of the ratio R_T with temperature variation. For accurate temperature measurements, high sensitivity is required. Illustrated in Figure 5.21 is a comparison of the sensitivity of the ratio R_T for the Fe Boltzmann technique and the Na Doppler two-frequency technique, adapted from Papen and Treyer (1998). Over typical mesospheric temperature range (100 to 250 K) the sensitivity of the Na Doppler technique is approximately half of that of the Fe Boltzmann technique. For both techniques, as the temperature increases, the sensitivity decreases. A significant difference between the two techniques is their sensitivity to radial winds and to frequency tuning errors. As described in Section 5.2.2, the Na Doppler technique is sensitive to the radial winds and frequency tuning

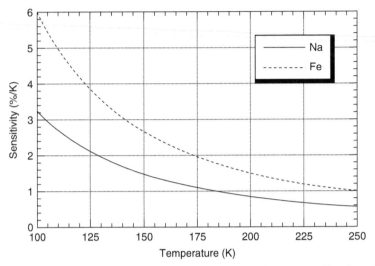

Figure 5.21 Comparison of sensitivity of the ratio R_T for the Fe Boltzmann technique and the Na Doppler technique (two-frequency). (From Papen, G.C., and Treyer, D., *Appl. Opt.*, 37, 8477–8481, 1998. With permission.)

errors. The nominal operating points for the broadband Fe system are at the optimal frequencies (peak frequencies), so the Fe Boltzmann lidar is insensitive to radial winds and frequency tuning errors. However, this robust insensitivity to radial winds also implies that the broadband Fe Boltzmann lidar system cannot be used for wind measurements.

5.2.3.3. Fe Boltzmann Lidar Instrumentation

A schematic of the system is shown in Figure 5.22. The system contains two laser transmitters and two optical receivers. Each laser transmitter consists of an injection-seeded, frequency-doubled, flash-lamp-pumped, pulsed alexandrite laser (Light Age, Inc., Model: PAL 101). The injection seed laser is a commercial tunable external cavity diode laser (Formerly EOSI, Inc., Model: 2010). The wavelength control of the seed laser is accomplished via a Burleigh wavemeter (Model: WA-1500) that has a frequency accuracy of $\pm 40\,\mathrm{MHz}$ at the fundamental wavelengths of 744 and 748 nm. This corresponds to a fre-

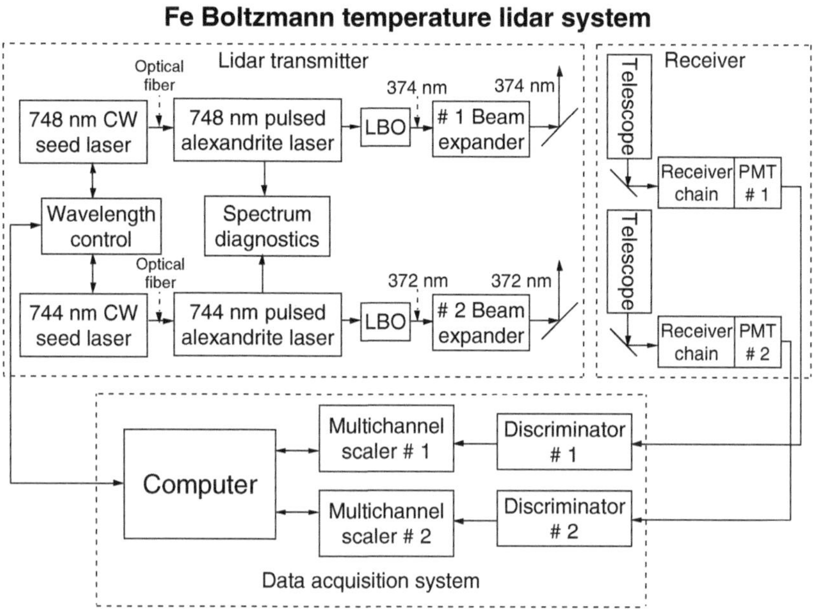

Fe Boltzmann temperature lidar system

Figure 5.22 Diagram of the University of Illinois Fe Boltzmann temperature lidar system. (From Chu, X., Pan, W., Papen, G.C., Gardner, C.S., and Gelbwachs, J.A., *Appl. Opt.*, 41, 4400–4410, 2002. With permission.)

quency accuracy of ± 80 MHz when the lasers are frequency doubled to probe the Fe fluorescence lines. Each seed laser wavelength is sequentially monitored using a computer-controlled flip mirror. The wavelength is adjusted via a software control loop. The elliptical output of the external cavity laser is corrected using an anamorphic prism pair and then the circularized output is fiber coupled for injection into the main laser cavity. The fiber coupling allows flexibility in locating the seed lasers with respect to the alexandrite lasers.

The frequency-doubled output of the alexandrite laser is expanded to decrease the divergence to about 350 μrad to enable daytime operation. The pulsed output of the alexandrite laser is monitored using a commercial pulsed laser spectrum analyzer (Burleigh Instruments, Inc., Model: RFP-3600). The pulsed output is not single frequency, but consists of three to four modes within a frequency range of approximately

800 MHz. The effective cross-section of each Fe transition is determined by scanning the wavelength of each laser through the Fe transition using the atmospheric returns. This will be discussed in detail in the next section.

Each optical receiver includes a commercial Meade 0.4-m LX200 Schmidt–Cassegrain telescope with custom coatings to enhance ultraviolet (UV) transmission. The field of view of the telescope is controlled via a pinhole at the focal plane and is set to match the outgoing divergence of the laser. A mechanical chopper is placed close to the focal plane to enable blanking of the low altitude laser returns. The return signal is collimated, passed through an interference filter and a pressure-tuned, temperature-stabilized Fabry–Perot etalon before being detected by a photon-counting PMT. The output signal from the PMT is then thresholded using a discriminator, and the photon counts are integrated using a multichannel scaler. A list of the system parameters is given in Table 5.5. Compared to narrowband Na systems, the Fe density at 372 nm is, on average, about a factor of 2 higher than Na and the effective cross-section is about an order of magnitude lower than Na. Thus, the average power required to produce an Fe signal count at 372 nm that is comparable to the Na signal count is roughly a factor of 5 higher for Fe assuming equal efficiencies for the rest of the system.

In the Section 5.2.2.4, we mentioned that a Fabry–Perot etalon used for the narrowband Na temperature lidar to reject the sky background had a calibration issue. Due to the hyperfine structures of the Na atoms and the optical pumping effect when narrowband laser excitation is used, the returned fluorescence photons have different frequencies than the transmitted laser photons. The spread of the returned photon frequency can be over 2 GHz. This requires a precise knowledge of the etalon transmission function in order to derive the temperature accurately. However, such issue does not exist in the Fe Boltzmann temperature lidar. This is because the Fe atoms have no hyperfine structure due to the zero nuclear spin. Thus, the returned fluorescence photons have the same frequency as the transmitted laser photons when using resonant frequency of laser excitation. The Rayleigh scattering

Table 5.5 System Parameters for the University of Illinois Fe
Boltzmann Temperature Lidar

Lidar Transmitter		
Characteristics	Channel 1 (374 nm)	Channel 2 (372 nm)
Alexandrite laser	Light Age, Inc., Model: PAL 101	
	Frequency doubled pulsed alexandrite laser	
Linewidth (FWHM) (MHz)	864	876
Pulse energy (mJ)	100	100
Pulse length (nsec)	55–65	55–65
Repetition rate (Hz)	33.2	34.1
Average power (W)	3	3
Beam divergence (mrad)	0.35	0.35
Injection-seeder laser	Formerly EOSI, Inc., Model: 2010	
	External cavity diode laser	
Wavelength (nm)	747.6390	744.1990
Output power (mW)	3	3
Linewidth (MHz) (50 msec)	<100 kHz	<100 kHz

Lidar Receiver		
Characteristics	Nighttime	Daytime
Telescope	Meade, Model: LX200 (f/10)	
	0.4064 m diameter	
	Schmidt–Cassegrain design	
Aperture area (m^2)	0.130	
Telescope focal length (mm)	4064	
Field-of-view (mrad)	1	0.5
Interference filter bandwidth (nm)	4.2	0.3
Interference filter peak transmission (%)	77	46
Fabry–Perot etalon bandwidth (GHz)	–	30
Etalon free-spectra-range (GHz)	–	650
Etalon finesse	–	22
Etalon spacing (μm)	–	230
Etalon peak transmission (%)	–	50
PMT quantum efficiency (%)	28	
Receiver optical efficiency (%)	58.7	18.0
Total receiver efficiency (%)	16.4	5.0

photons also have the same frequency as the transmitted laser
photons since the Doppler shift due to wind velocity in zenith
direction is negligible. Thus, the returned Fe fluorescence and

Rayleigh scattering photons have the same frequency and undergo through the same etalon transmission function. Therefore, the etalon filter function can be canceled out when normalizing the Fe signals to the Rayleigh signals as given by Equation (5.95).

5.2.3.4. Temperature and Error Analysis

The approach to derive the atmosphere temperature and Fe density information from the lidar photon count profile is similar to what has been used for the Na wind/temperature lidar. The calibration curve (R_T versus T) used for the Fe Boltzmann temperature lidar is plotted in Figure 5.23. Here, R_T is given by Equation (5.97). Once the ratio R_T is computed from the observed photon counts of the 374 and 372 nm channels, the temperature can be inferred from this calibration curve. Then the Fe density can be calculated from Equation (5.96) using derived temperature.

There are two major sources of error, viz. fluctuations in the signal levels associated with photon noise and fluctuations in the effective backscatter cross sections associated with laser tuning errors and laser linewidth fluctuations. The

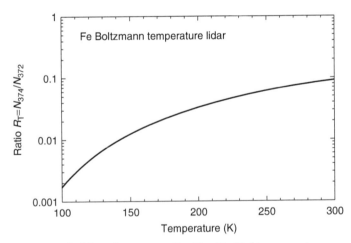

Figure 5.23 Calibration curve for the Fe Boltzmann temperature lidar: ratio R_T versus temperature T.

extinction is calculated from the measured Fe densities and temperatures and also includes errors. However, the two-way extinction correction is typically no more than about 1 to 2% at the top of the 372 nm Fe layer and much less than a percent for the 374 nm Fe layer. For this reason, extinction errors are small and can be neglected. Because the laser tuning errors and linewidth fluctuations are uncorrelated with signal photon noise, by using $\Delta T = (\partial T/\partial R_T)\Delta R_T + (\partial T/\partial R_\sigma)\Delta R_\sigma$, the total RMS temperature error can be derived from Equation (5.99) as

$$\Delta T_{\text{rms}} = \frac{T^2}{\Delta E/k_B} \sqrt{\left\langle \left(\frac{\Delta R_T}{R_T}\right)^2 \right\rangle + \left\langle \left(\frac{\Delta R_\sigma}{R_\sigma}\right)^2 \right\rangle} \qquad (5.101)$$

Since $\Delta E/k_B \sim 600\,\text{K}$, and T is in the range of 150 to 250 K, the total relative error for both R_T and R_σ in Equation (5.101) must be no larger than 1.5% for a $\pm 1\,\text{K}$ accurate temperature measurement at $T = 200\,\text{K}$. We now consider each of these two error sources separately.

The total effective backscatter cross-section is proportional to the effective absorption cross-section

$$\sigma_{\text{eff}}(\lambda, T, \sigma_L) = \sigma_{\text{effabs}}(\lambda, T, \sigma_L) \qquad (5.102)$$

We have accounted for the decay of the excited state to multiple ground states by including the appropriate branching ratios in the lidar equation. The effective absorption cross-section is the convolution of the atomic absorption cross section and laser lineshape

$$\sigma_{\text{eff}}(\lambda, T, \Delta\nu, \sigma_L) = \int_{-\infty}^{+\infty} \sigma_{\text{abs}}(\nu', \lambda_0, T) g_L(\nu', \lambda, \sigma_L) d\nu' \qquad (5.103)$$

where $\sigma_{\text{abs}}(\nu', \lambda_0, T)$ is the intrinsic atomic absorption cross-section centered at wavelength λ_0, and $g_L(\nu', \lambda, \sigma_L)$ is the normalized laser lineshape centered at wavelength λ, ν' is the frequency, and $\Delta\nu$ is the frequency difference between the Fe line center λ_0 and the laser line center λ. Assuming a Gaussian functional form for both the laser lineshape and the absorption cross-section, the effective cross-section is

$$\sigma_{\text{eff}}(\lambda, T, \Delta\nu, \sigma_{\text{L}}) = \frac{1}{\sqrt{2\pi}\sigma_\lambda} \frac{e^2}{4\epsilon_0 m_e c} f_\lambda \exp\left(-\frac{\Delta\nu^2}{2\sigma_\lambda^2}\right) \quad (5.104)$$

where f_λ is the oscillator strength of the transition ($f_{372} = 0.0413$ and $f_{374} = 0.0382$), ε_0 is the permittivity of free space, e is the charge of an electron, m_e is the mass of the electron, c is the velocity of light, and

$$\sigma_\lambda = \sqrt{\sigma_{\text{D}}^2 + \sigma_{\text{L}}^2} \quad (5.105)$$

where σ_{D} is the rms Doppler width of the intrinsic atomic absorption cross-section,

$$\sigma_{\text{D}} = \frac{1}{\lambda}\sqrt{\frac{k_{\text{B}}T}{m_{\text{Fe}}}} \quad (5.106)$$

and m_{Fe} is the mass of an Fe atom.

In practice, we tune the lasers to the peaks of the Fe lines so that uncertainties in the assumed values for the cross-section arise because of tuning errors ($\Delta\nu$) and uncertainties in the assumed values for the laser linewidths ($\Delta\sigma_{\text{L}}$). The ratio of the cross sections is given by

$$R_\sigma = 0.9249 \frac{\sigma_{372}}{\sigma_{374}} \exp\left(\frac{\Delta\nu_{372}^2}{2\sigma_{372}^2} - \frac{\Delta\nu_{374}^2}{2\sigma_{374}^2}\right) \quad (5.107)$$

While the temperature is required to evaluate the Doppler linewidths, the cross-section ratio is insensitive to temperature so that errors in R_σ associated with errors in assumed-temperature are negligible. Because the lasers are identical and the wavelength difference is small, we assume the laser linewidths are similar so that σ_{372} is approximately equal to σ_{374}. In this case, the cross-section ratio can be simplified

$$R_\sigma \cong \begin{cases} 0.9292\sqrt{1 + \frac{(\sigma_{\text{L}372}^2 - \sigma_{\text{L}374}^2)}{\sigma_\lambda^2}} \exp\left[\frac{(\Delta\nu_{372}^2 - \Delta\nu_{374}^2)}{2\sigma_\lambda^2}\right] & \sigma_{\text{L}} \ll \sigma_{\text{D}} \\ \\ 0.9249\sqrt{1 + \frac{(\sigma_{\text{L}372}^2 - \sigma_{\text{L}374}^2)}{\sigma_\lambda^2}} \exp\left[\frac{(\Delta\nu_{372}^2 - \Delta\nu_{374}^2)}{2\sigma_\lambda^2}\right] & \sigma_{\text{D}} \ll \sigma_{\text{L}} \end{cases}$$

$$(5.108)$$

If both the laser linewidths and the squares of the tuning errors are identical, then $0.9249 < R_\sigma < 0.9292$. For identical lasers, the cross-section ratio varies less than 0.5% as the laser line-width varies from zero to infinity. Because the cross-section ratio is insensitive to the absolute value of the laser linewidths when they are approximately equal, it is more important to know the linewidth difference rather than the absolute values of the individual linewidths. We can express the relative error in the cross-section ratio in terms of the laser tuning errors and uncertainties in the assumed values for the laser linewidths, or in terms of the linewidth difference.

$$\frac{\Delta R_\sigma}{R_\sigma} \cong \begin{cases} \frac{|\Delta\nu_{372}^2 - \Delta\nu_{374}^2|}{2\sigma_\lambda^2} \\ \sqrt{2}\left(\frac{\sigma_L}{\sigma_\lambda}\right)^2 \frac{\Delta\sigma_L}{\sigma_L} \\ \left(\frac{\sigma_L}{\sigma_\lambda}\right)^2 \frac{|\sigma_{L372} - \sigma_{L374}|}{\sigma_L} \end{cases} \tag{5.109}$$

By combining Equation (5.109) with Equation (5.101), we obtain

$$\Delta T_{rms}(K) \cong \begin{cases} \pm\frac{T^2}{2(\Delta E/k_B)} \frac{|\Delta\nu_{372}^2 - \Delta\nu_{374}^2|}{\sigma_\lambda^2} \\ \pm\frac{\sqrt{2}T^2}{(\Delta E/k_B)} \left(\frac{\sigma_L}{\sigma_\lambda}\right)^2 \frac{\Delta\sigma_L}{\sigma_L} \\ \pm\frac{T^2}{(\Delta E/k_B)} \left(\frac{\sigma_L}{\sigma_\lambda}\right)^2 \frac{|\sigma_{L372} - \sigma_{L374}|}{\sigma_L} \end{cases} \tag{5.110}$$

To compute the effective cross sections from Equation (5.104) or their ratio from Equation (5.107), we need to determine σ_λ for each line. This is done by scanning the lasers through the Fe fluorescence lines and recording the atmospheric returns as a function of frequency. A Gaussian function of frequency is fitted to the normalized signal counts to determine the total rms linewidths. The laser linewidths are computed from Equations (5.105) and (5.106). Plotted in Figure 5.24 is an example fit to 372 nm data. The results for the cross-section and laser parameters and the statistical errors of the fits are listed in Table 5.6.

There are two ways to determine the cross-section ratio and the errors. The cross sections can be computed from the

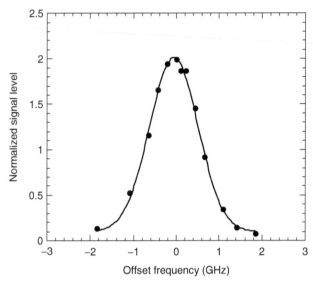

Figure 5.24 Plot of the effective backscatter cross section spectrum for the Fe 372-nm line obtained by the scanning of the lidar frequency and observation of the returns from the mesospheric Fe layer. (From Chu, X., Pan, W., Papen, G.C., Gardner, C.S., and Gelbwachs, J.A., *Appl. Opt.*, 41, 4400–4410, 2002. With permission.)

laser scans using Equation (5.104). In this case the relative error in the cross-section ratio is given by

$$\frac{\Delta R_\sigma}{R_\sigma} = \sqrt{\left(\frac{\Delta \sigma_{\text{eff}}(372)}{\sigma_{\text{eff}}(372)}\right)^2 + \left(\frac{\Delta \sigma_{\text{eff}}(374)}{\sigma_{\text{eff}}(374)}\right)^2} \qquad (5.111)$$

Table 5.6 Measured Cross Sections and Laser Parameters

Parameter	372 nm system	374 nm system
$\sigma_\lambda \pm \Delta\sigma_\lambda$ (MHz)	566.8 ± 16.7	562.1 ± 20.0
σ_D (MHz) at $T = 170\,\text{K}$	427.7	425.7
$\Delta\nu \pm \delta(\Delta\nu)$ (MHz)	59.7 ± 10.8	82.2 ± 13.8
$\sigma_L \pm \Delta\sigma_L$ (MHz)	371.9 ± 25.4	367.0 ± 30.7
$\sigma_{\text{eff}} \pm \Delta\sigma_{\text{eff}}$ ($\times 10^{-12}\,\text{cm}^2$)	0.770 ± 0.023	0.718 ± 0.026
$\Delta\sigma_\lambda/\sigma_\lambda$ (%)	2.95	3.56
$\Delta\sigma_L/\sigma_L$ (%)	6.83	8.37
$\Delta\sigma_{\text{eff}}/\sigma_{\text{eff}}$ (%)	2.99	3.62

By using the data in Table 5.6, the cross-section ratio is 0.9348 and the relative error is 4.7%, which according to Equation (5.101) yields an error of about ± 3.1 K at $T = 200$ K. Alternatively, we can assume the laser linewidths are identical and that the lasers are both tuned to the peaks of the fluorescence lines. In this case, the cross-section ratio is assumed to be 0.9292 when $\sigma_L \ll \sigma_D$ or 0.9249 when $\sigma_D \ll \sigma_L$ (Equation 5.108). In our case, where the laser linewidths are both approximately 370 MHz and do not satisfy either extreme, the assumed value of the cross-section ratio would be 0.9274 (Equation 5.107). According to the linewidth scan data in Table 5.6, tuning errors are comparable to the ± 80 MHz accuracy of the Burleigh wavemeter at the frequency-doubled UV lines. Tuning errors would introduce a relative error of less than 1% in the cross-section ratio, which corresponds to a temperature error of ± 0.7 K. Uncertainties in the laser linewidths (25 and 31 MHz) introduce a temperature error of ± 3.1 K. Alternatively, if the laser linewidths are similar, we can use the linewidth difference to compute the temperature error, rather than the uncertainties of the measured values. The laser linewidths differ by 4.9 ± 40 MHz. If we assume the laser linewidth difference is no larger than the uncertainty in the measured difference (40 MHz), then the relative cross-section error is less than 4.6%, which corresponds to a temperature error of ± 3.1 K. If the actual linewidth difference were assumed to be the measured value of 4.9 MHz, then the temperature error would be approximately ± 0.4 K.

The impact of photon noise is calculated using the same analysis for the Na lidar technique where winds and temperatures are determined by computing the ratio of normalized photon counts obtained at several frequencies within the Na D_2 fluorescence line. Because the Fe signals are normalized by the strong Rayleigh signal at lower altitudes near 50 km, the photon noise contributed by the Rayleigh signal is negligible. The dominant error source is photon noise contributed by the weaker 374 nm Fe channel. If we assume the Rayleigh normalizing signals for the 372 and 374 nm channels are comparable, the temperature ratio R_T is given approximately by

$$R_T \cong \frac{N_{Fe}(374,z) + N_B(374,z) - \hat{N}_B(374)}{N_{Fe}(372,z) + N_B(372,z) - \hat{N}_B(372)} \qquad (5.112)$$

The rms relative error in the temperature ratio associated with photon noise is easily calculated from Equation (5.112) by recognizing that the Fe signal and background counts are uncorrelated Poisson distributed random numbers. If we assume that the mean background counts on the 372 and 374 nm channels are approximately equal and $T = 200\,\text{K}$ so that $1/R_T \sim 30$, the rms relative error is given by

$$\sqrt{\left\langle \left(\frac{\Delta R_T}{R_T}\right)^2 \right\rangle} = \frac{\sqrt{1 + (1/R_T)}}{\sqrt{N_{Fe}(372,z)}} \sqrt{1 + \frac{[1 + (1/R_T^2)]}{[1 + (1/R_T)]} \frac{1}{\text{SBR}_{372}}}$$

$$\approx \frac{5.6}{\sqrt{N_{Fe}(372,z)}} \sqrt{1 + \frac{29}{\text{SBR}_{372}}}$$

$$(5.113)$$

where the signal-to-background ratio (SBR) is

$$\text{SBR}_{372} = \frac{N_{Fe}(372,z)}{\hat{N}_B(372)} \qquad (5.114)$$

By substituting Equation (5.113) into Equation (5.101), the rms temperature error associated with photon noise at $T = 200\,\text{K}$ is

$$\Delta T_{rms}(\text{K}) = \frac{\pm 372}{\sqrt{N_{Fe}(372,z)}} \sqrt{1 + \frac{29}{\text{SBR}_{372}}} \qquad (5.115)$$

At night when SBR_{372} is very large, approximately 140×10^3 Fe signal counts are required in each resolution cell of the stronger 372 nm channel to achieve a temperature accuracy of $\pm\,1\,\text{K}$. This requirement drops to about 15×10^3 counts for an accuracy of $\pm 3\,\text{K}$. During daytime when SBR_{372} can be much less than 1, the signal requirements increase substantially. For example if $\text{SBR}_{372} = 1$, the required signal level is 4×10^6 counts for $\pm\,1\,\text{K}$ accuracy. Long averaging times, typically several hours, are required to obtain accurate temperature measurements during the daytime. The error budget for the Fe Boltzmann lidar is summarized in Table 5.7.

5.2.3.5. Rayleigh Temperature Retrieval

Above the stratospheric aerosol layers ($\sim$30 km) and below the Fe layer, the received signal results from pure molecular scattering. Molecular scattering is proportional to the atmospheric density and the temperature profile can be derived from the relative atmospheric density profile using the Rayleigh lidar technique. The Fe lidar Rayleigh signal between 30 and 75 km can be used to study the atmospheric temperature and density profile. The use of high power lasers with the Rayleigh technique was pioneered by Hauchecorne and Chanin (1980). The data analysis approach is very similar to that employed by Elterman in the early 1950s to measure stratospheric temperatures with search light technique (Elterman, 1951, 1953, 1954). This involves integrating the relative density profile downward using a starting temperature at the highest altitude in combination with the hydrostatic equation and the ideal gas law. The starting temperature may be chosen from a model because when the equation has been integrated downward by about one and half scale heights (atmospheric scale height is the altitude range in which density decreases by the factor $1/e$), the calculated

Table 5.7 Fe Boltzmann Temperature Lidar Error Budget for $T = 200$ K and $\sigma_{\mathrm{L}} = 370$ MHz

Parameter	Temperature Error (K)	Value for $\Delta T_{\mathrm{rms}} = \pm 1$ K
Laser tuning errors $\Delta \nu$	$\pm 33 \frac{\lvert \Delta \nu_{372}^2 - \Delta \nu_{374}^2 \rvert}{\sigma_\lambda^2}$	100 MHz
Laser linewidth errors $\Delta \sigma_{\mathrm{L}}$	$\pm 95 \left(\frac{\sigma_{\mathrm{L}}}{\sigma_\lambda}\right)^2 \frac{\Delta \sigma_{\mathrm{L}}}{\sigma_{\mathrm{L}}}$	9 MHz
Laser linewidth difference $\lvert \sigma_{\mathrm{L}372} - \sigma_{\mathrm{L}374} \rvert$	$\pm 67 \left(\frac{\sigma_{\mathrm{L}}}{\sigma_\lambda}\right)^2 \frac{\lvert \sigma_{\mathrm{L}372} - \sigma_{\mathrm{L}374} \rvert}{\sigma_{\mathrm{L}}}$	13 MHz
Signal level N_{Fe} (372, z)	$\pm \frac{372}{\sqrt{N_{\mathrm{Fe}}(372,z)}} \sqrt{1 + \frac{29}{\mathrm{SBR}_{372}}}$	1.8×10^5 at $\mathrm{SBR}_{372} = 100$ 4.2×10^6 at $\mathrm{SBR}_{372} = 1$

temperature is relatively insensitive to the starting estimation. The hydrostatic equation

$$dP(z) = -\rho(z)g(z)dz \tag{5.116}$$

can be combined with the ideal gas law

$$P(z) = \frac{\rho(z)RT(z)}{M(z)} \tag{5.117}$$

and integrated to yield

$$T(z) = T(z_0)\frac{\rho(z_0)}{\rho(z)}\frac{M(z)}{M(z_0)} + \frac{M(z)}{R}\int_z^{z_0}\frac{\rho(z')g(z')}{\rho(z)}dz' \tag{5.118}$$

where

$T(z)$ = atmospheric temperature profile (K)
$P(z)$ = atmospheric pressure profile (mbar)
$\rho(z)$ = atmospheric mass density profile (kg/m^3)
$g(z)$ = gravitational acceleration (m/s^2)
$M(z)$ = mean molecular weight of the atmosphere
R = universal gas constant (8.31432 J/mol/K)
z_0 = altitude of the upper level starting temperature (m)

The atmospheric mass density $\rho(z)$ and number density $n(z)$ have the following relationship:

$$\rho(z) = n(z)M(z)/N_A \tag{5.119}$$

where N_A is the Avogadro constant. Substituting Equation (5.119) into Equation (5.118), we obtain

$$T(z) = T(z_0)\frac{n(z_0)}{n(z)} + \frac{M(z)}{R}\int_z^{z_0}\frac{n(z')M(z')g(z')}{n(z)M(z)}dz' \tag{5.120}$$

Below 100 km for the well-mixed atmosphere, the mean molecular weight of the atmosphere $M(z)$ is a constant. Thus, the $M(z')$ and $M(z)$ are canceled out in Equation (5.120), and we obtain the temperature derived from Rayleigh signals by

$$T(z) = T(z_0)\frac{n(z_0)}{n(z)} + \frac{M(z)}{R}\int_z^{z_0}\frac{n(z')g(z')}{n(z)}dz' \tag{5.121}$$

Notice that the atmosphere number density $n(z)$ appears as a ratio of densities, that is, the relative number density. Thus, to determine temperature, it is only necessary to measure the relative density, which can be derived from the Rayleigh photon count $N_R(z)$ at different altitudes given by Equation (5.93). The accuracy of this technique depends upon the Rayleigh photon count $N_R(z)$ and the accuracy of the upper level temperature estimate $T(z_0)$. The variance of derived temperature is given as below (Gardner, 1989):

$$\text{var}[T(z)] \approx \frac{T^2(z)}{N_R(z)} + \left\{ \text{var}[T(z_0)] + \frac{T^2(z_0)}{N_R(z_0)} \right\} \exp[-2(z_0-z)/H]$$

(5.122)

where H is the atmospheric scale height ($\sim 7\,$km for the upper stratosphere and the lower mesosphere).

Most middle atmosphere Rayleigh lidars employ frequency-doubled pulsed Nd:YAG lasers operating at 532 nm in the green region of the visible spectrum. Typical systems employ telescopes with diameters near 1 m and lasers with average powers levels in the 10 to 15 W range. These systems have power aperture products of about $10\,\text{W}\,\text{m}^2$. The Fe Boltzmann lidar operates in the near-UV where Rayleigh scattering is more than four times stronger than at 532 nm (Equation 5.94). By combining the molecular (Rayleigh) scattered signal from the 372 and 374 nm channels, temperatures can be derived from about 30 to about 75 km using the Rayleigh technique. The Fe Boltzmann temperature at the bottom of the Fe layer near 80 km, rather than a model estimate, can be used as the starting temperature for the retrieval. For our system, which employs two 3-W lasers and two 0.4-m diameter telescopes, the equivalent power aperture product at 532 nm is about $3\,\text{W}\,\text{m}^2$. Thus, the current Fe Boltzmann lidar system can derive temperature profiles with an accuracy and vertical resolution comparable to many existing Rayleigh lidars if the signals are integrated about three times longer. If the same 1-m diameter telescopes are used for the Fe system, the equivalent power aperture product at 532 nm will be about $18\,\text{W}\,\text{m}^2$, which is actually better than the 532-nm Rayleigh system.

5.2.3.6.   Temperature Measurement over the
              North and South Poles

As described previously, the Fe Boltzmann temperature lidar
can measure the temperature in the upper mesosphere and
lower thermosphere by the Boltzmann technique, and in the
upper stratosphere and lower mesosphere by the Rayleigh
technique. The combination of these two capabilities makes
the Fe Boltzmann temperature lidar able to measure the
temperature from 30 to 110 km. If further combining with
the balloon-sonde data from the ground to 30 km, the
temperature measurements will be able to cover the whole
altitude range from the ground to 110 km. These measurement capabilities were clearly demonstrated in the lidar observations from the North Pole to the South Pole made by the
University of Illinois (Chu et al., 2002a; Gardner et al., 2001;
Pan et al., 2002).

The Fe Boltzmann temperature lidar was first deployed
on the NSF/NCAR Electra aircraft to make observations of
Leonid meteor shower over Okinawa in November 1998 (Chu
et al., 2000a). In June and July 1999, the lidar system was
again deployed on the NSF/NCAR Electra to make temperature, Fe density, and PMC observations over the north polar
cap during the Arctic Mesopause Temperature Study (AMTS).
The Electra aircraft flew at varying altitudes from 6 to 8.5 km
with speed about 150 m/s. There were totally 10 flights during
AMTS campaign. AMTS began on June 16, 1999 with a ferry
flight from Broomfield, Colorado (40°N, 105°W) to Winnipeg,
Canada, and then to Resolute Bay, Canada (75°N, 95°W)
where the campaign was based. One roundtrip flight was
made to Sondrestromfjord, Greenland (67°N, 50°W) on June
19, 1999. The first flight to the North Pole was made on June
21, 1999, and then another three flights were made to the
geographic North Pole on July 1, July 2, and July 4, 1999.
On the last three flights to the geographic Pole, the return
flight path passed directly over the magnetic North Pole
(79°N, 105°W). On July 5, 1999, the system was flown to
Anchorage, Alaska (61°N, 150°W) and one additional flight
was made on July 8, 1999 to probe NLC over the Gulf of

Alaska. Finally, the lidar system was flown back to Broom-field, Colorado passing through Fort Collins on July 9, 1999. A total of 52 h of airborne lidar observations were made during AMTS. In November 1999, the Fe lidar system was installed in the Atmospheric Research Observatory on the ground, which is 488 m north of the geographic South Pole at the Amundsen–Scott South Pole Station. Observations began on December 2, 1999 and continued through the austral summer and winter from 1999 to 2001, until November 2001. More than 1700 h of temperature, Fe density, and polar mesospheric clouds observations were made at the South Pole during this 2-year deployment.

The mesospheric Fe chemistry is temperature dependent. The primary sink reaction $FeO + O_2 \rightarrow FeO_3$ on the layer bottom side proceeds most rapidly at low temperatures (Rollason and Plane, 2000). This reaction drives the seasonal variations in the Fe column abundance. Because mesopause temperatures are extremely low over the polar caps at midsummer, the Fe densities are also expected to be quite low. Thus, summertime observations over the polar caps, which are sunlit 24 h per day, represent the greatest challenge to making accurate temperature measurements with the Fe Boltzmann lidar. A composite profile of temperatures measured during the summer solstice periods over the North and South Poles is plotted in Figure 5.25(a). The corresponding Fe density profiles are plotted in Figure 5.25(b). During midsummer over the polar caps, the Fe layer is thin and the peak densities are typically quite low. The June 21, 1999 data over the North Pole are a notable exception. On this flight a sporadic Fe layer (Fe_s) formed near 106 km in lower thermosphere with peak densities exceeding $2 \times 10^5 \, cm^{-3}$. This prominent feature was observed on the three subsequent flights to the North Pole. However, the peak densities and abundances of the Fe_s on these later flights were much smaller. This dense sporadic layer provided an exceptionally strong backscatter signal so that excellent temperature data were obtained between 102 and 109 km. The thin layers observed at the South Pole near 91 km on December 24, 2000 and 98 km on December 29, 2000 are also sporadic layers, but the lower

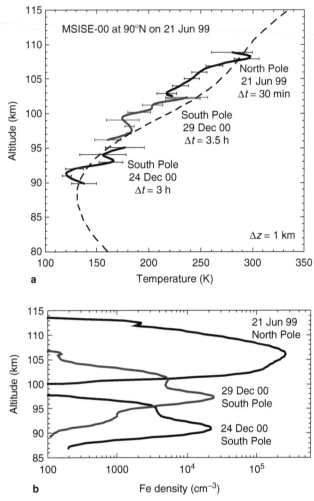

Figure 5.25 (a) Temperature profiles measured by the University of Illinois Fe Boltzmann temperature lidar over the North Pole on June 21, 1999 and over the South Pole on December 24, 2000 and December 29, 2000 along with MSISE-00 model data for June 21, 1999 at the North Pole. (b) The 372-nm ground-state Fe density profiles corresponding to the temperature profiles plotted in (a). (From Chu, X., Pan, W., Papen, G.C., Gardner, C.S., and Gelbwachs, J.A., *Appl. Opt.*, 41, 4400–4410, 2002. With permission.)

densities just above them are more typical. The combination of low densities and high background noise makes temperature observations difficult at midsummer over either pole. Even so, these airborne and ground-based profiles demonstrate that the lidar is capable of making useful temperature measurements, even in the daytime, whenever Fe densities (at 372 nm) are at least a few thousand atoms per cm^3.

The Polar Regions are more sensitive than elsewhere to global change effects associated with greenhouse gas warming. Profiles of atmospheric parameters and constituents at the geographic poles can provide a convenient means of validating and calibrating global circulation models. However, measurements of key parameters, such as temperature profiles, have only been conducted in the troposphere and lower stratosphere at the geographic poles with balloon borne sensors to altitudes less than 30 km. A major goal of the South Pole observations is to characterize the atmospheric temperature profile from the surface to the edge of space throughout the year. Plotted in Figure 5.26 is a composite temperature profile determined using balloon and Fe/Rayleigh lidar data collected on May 8, 2000, which illustrates this capability. The lower mesosphere is about 20 K warmer, and the upper mesosphere is about 20 K colder than predicted by the MSISE-00 model. By combining the Fe and Rayleigh lidar temperatures with the radiosonde balloon data, the atmospheric temperatures from the surface to 110 km were characterized for the first time at the South Pole throughout the year. The comparisons of the measured monthly mean temperature with the TIME-GCM model and MSIS00 model have revealed several significant differences. More detailed information about these results can be found in Section 5.3.

5.2.3.7. Polar Mesospheric Cloud Detection over Both Poles

Besides the capability of temperature measurements from 30 to 110 km, the Fe Boltzmann temperature lidar is highly capable of detecting PMCs in the polar region. PMCs and their visual counterparts NLCs are the thin water–ice layers

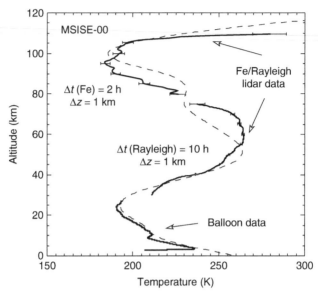

Figure 5.26 Composite temperature profile measured over the South Pole on May 8, 2000 by use of balloon and Fe-Rayleigh lidar data along with the MSISE00 model for May 8, 2000 at the South Pole. (From Chu, X., Pan, W., Papen, G.C., Gardner, C.S., and Gelbwachs, J.A., *Appl. Opt.*, 41, 4400–4410, 2002. With permission.)

occurring in the mesopause region ($\sim$85 km) at high latitudes ($\sim$poleward of 50°) in both hemispheres mainly during the 3 months surrounding summer solstice when the temperatures fall below the frost point ($\sim$150 K). PMCs/NLCs are sensitive tracers of middle and upper atmospheric water vapor and temperature. The observed increases in the geographic extent and brightness of PMCs and NLCs in the Polar Regions during the past few decades are believed to be the evidence of cooling temperature and increasing water-vapor concentration in the mesopause region caused by the increases of greenhouse gases. The possible link of PMCs to the global climate change has intensified the research on PMCs. However, the observations of PMCs are not easy, especially in the Polar Regions due to the continuous 24-h sunlight in the summer. The ground-based visual and photographic observations are limited to twilight conditions, so usually confined within 55°

to 65° latitudes. The space satellites can observe PMCs in the polar summer. Unfortunately, they have limited vertical resolution of about 2 to 3 km. Although *in situ* rocket experiments can measure PMC vertical distribution precisely, they are so sporadic that they cannot be used to characterize the seasonal and diurnal variations in PMC properties.

Lidar technology is unique in detecting PMCs because of its high vertical resolution, daytime measurement capability, and full seasonal and diurnal coverage. As demonstrated at the South Pole, the Fe Boltzmann temperature lidar is a powerful tool for observing PMCs. This is attributed to its daytime measurement capability, its short operating wavelength, and its high transmitted power. PMCs consist of water ice particles. Without PMC particles in the mesopause region, the Rayleigh scattering from pure atmosphere molecules decreases with height exponentially, and becomes negligibly small when reaching 80 to 85 km. When PMC particles are present in the mesopause region around 85 km, the Mie scattering from the PMC water–ice particles, which are much larger than atmosphere molecules, is significantly stronger than the Rayleigh scattering at the same altitude. Thus, a sharp, narrow peak will stand out in the lidar photon count profile. Plotted in Figure 5.27 is an example of a PMC profile obtained by the Fe Boltzmann temperature lidar at the South Pole on January 18, 2000. The photon counts have been converted to the volume backscatter coefficient. The challenge of detecting PMCs in the polar region comes from the high solar scattering background. Only when the PMC peak is higher than the background, it is possible to distinguish PMCs from background noise. The narrowband filters used in the Fe lidar receivers effectively reject the sky background and make it capable of PMC measurements in the daylight. The volume backscatter coefficient of PMCs is proportional to the sixth power of the particle radius while it is inversely proportional to the fourth power of light wavelength. Thus, the shorter wavelength of the Fe lidar at 374 nm will produce much stronger PMC backscattering, compared to normal Rayleigh lidar at 532 nm and the Na lidar at 589 nm. The background level and the PMC signal level together determine the lidar

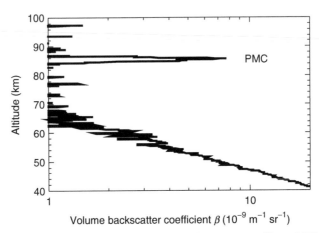

Figure 5.27 Volume backscatter coefficient profile of PMC and Rayleigh scattering of atmosphere molecules versus altitude obtained by the UIUC Fe Boltzmann temperature lidar between 22:38 and 23:00 UT on January 18, 2000 at the South Pole. (From Chu, X., Gardner, C.S., and Papen, G., *Geophys. Res. Lett.*, 28, 1203–1206, 2001a. With permission.)

detection sensitivity. The combination of its daylight measurement capability, short operation wavelength, and high output power ensures that the Fe Boltzmann temperature lidar has very high sensitivity for PMC detection in the polar summer. Extensive data on PMCs have been collected at the South and North Poles with the Fe Boltzmann temperature lidar. Several important scientific results and findings have merged from these data, such as the inter-hemispheric difference in PMC altitude (Chu et al., 2003a). More detailed description can be found in Section 5.3.

5.2.4. Potassium Doppler Lidar

5.2.4.1. Introduction

The geographic and seasonal dependence of the complex thermal structure of the mesopause region is a strong motivating factor for the development of transportable Doppler temperature lidars. The application of ring dye laser that is extremely sensitive to environmental conditions in the Na Doppler temperature lidar precludes its use for transportable systems. Fortunately, there are other neutral metal species besides Na, such as K, Ca, and Fe in the mesopause region. Among these species, the energy level diagram of the K atoms is very similar to that of the Na since both are alkali-metal atoms, and has been well studied (Arimondo et al., 1977). The potassium D_1 and D_2 line wavelengths (769.9 and 766.5 nm) fall in the fundamental gain range of several solid-state lasers including alexandrite, Ti:Sapphire, and Cr:LiSrAlF$_6$. Therefore, it is possible to develop robust Doppler temperature lidars based upon solid-state laser technology using potassium as a tracer. The idea was explored by Papen et al. (1995b) and Höffner and von Zahn (1995) simultaneously. Papen et al. (1995b) did a parametric study on a proposed K lidar system and pointed out that a potassium temperature and wind lidar had the potential for high-resolution atmosphere applications. Their work was mainly based on the K(D_2) line but the analysis for the K (D_1) line was added in their revised manuscript. At the same time, the University of Bonn/Leibniz Institute for Atmospheric Physics (IAP) shifted their alexandrite ring laser wavelength from the Na to the K, and started to construct a narrow-band K Doppler lidar (Höffner and von Zahn, 1995). The first useful K temperature profiles were obtained in 1995 (von Zahn and Höffner, 1996). Subsequently, the IAP K lidar was deployed to Arctic and other locations to study the global atmosphere temperatures. A similar narrowband K lidar based on a ring alexandrite laser was then built at Arecibo Observatory (18.3°N, 66.75°W) and began to make routine observations in the mesopause region (Friedman et al., 2002). In this section, we will discuss the measurement principles, instrumentation, and error analysis based upon the IAP K lidar. Other K lidars will only be mentioned briefly to compare their instrumentation with the IAP K lidar.

5.2.4.2. Measurement Principle

The partial energy levels of the potassium are shown in Figure 5.28. The transition from 4 $^2P_{1/2}$ to 4 $^2S_{1/2}$ is the D_1 line, while the transition from 4 $^2P_{3/2}$ to 4 $^2S_{1/2}$ is the D_2 line.

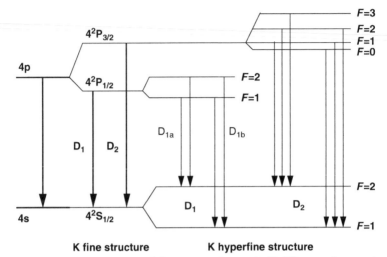

Figure 5.28   Energy level diagram of atomic K. The nuclear spin of K is 3/2.

The K (D_2) line at 766.491 nm cannot be used for ground-based lidar soundings as it is imbedded in a strong O_2 absorption line. Only the K (D_1) line at 769.898 nm can be used for ground-based lidar observations. Although it seems unfortunate that the D_1 resonance cross-section is only half that of the D_2 line, the Hanle effect is zero for D_1 resonance and need not be taken into account in either the experimental arrangement or the analysis.

The potassium D_1 line consists of four hyperfine structure lines in two groups D_{1a} and D_{1b}. However, the D_{1a} and D_{1b} peaks are significantly closer than those of the Na (D_2) line, and they are not resolved at typical mesospheric temperatures. This results in a nearly Gaussian shape for the K (D_1) backscatter cross-section for temperatures above 175 K. At lower temperatures, the K (D_1) lineshape becomes noticeably non-Gaussian. There are three natural isotopes of potassium ^{39}K, ^{40}K and ^{41}K. The abundance of ^{40}K is only 0.0117%, so it is ignored in the K lidar data analysis. The most abundant potassium isotope is ^{39}K with an isotopic abundance of 93.2581%. The presence of the minor isotope ^{41}K with an abundance of 6.7302% slightly modifies the shape of the K

(D_1) fine structure line and must be taken into account during data analysis. With respect to the ^{39}K line, the center of the ^{41}K line is shifted towards shorter wavelengths and the hyperfine structure transitions are somewhat closer together. The calculated backscatter cross sections are shown in Figure 5.29. The quantum number, frequency offsets, and relative line strength are listed in Table 5.8 for both isotopes.

The Doppler broadening of the K (D_1) resonance is temperature dependent. For a natural mixture of ^{39}K and ^{41}K, the variation of the FWHM of the K (D_1) line versus temperature is shown in Figure 5.30. At 200 K, the linewidth is 936 MHz. The mesopause temperature can be derived from the measured linewidth of the Doppler broadened D_1 resonance. To determine the FWHM of the received signal, the resonance

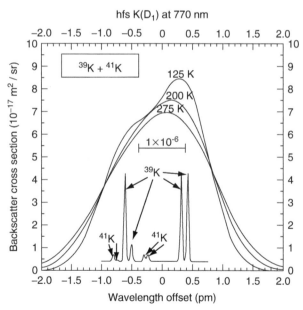

Figure 5.29 Variation of the backscatter cross section of the K (D_1) transition with temperature for ^{39}K + ^{41}K. The positions and relative strengths of the hyperfine transitions are indicated. The FWHM of these transitions is plotted with the estimated maximum bandwidth of the used laser (20 MHz). (From von Zahn, U., and Höffner, J., *Geophys. Res. Lett.*, 23, 141–144, 1996. With permission.)

Table 5.8 Quantum Numbers, Frequency Offsets, and Relative
Line Strength for K (D₁) Hyperfine Structure Lines

$^2S_{1/2}$	$^2P_{1/2}$	^{39}K (MHz)	^{41}K (MHz)	Relative Line Strength
$F=1$	$F=2$	310	405	5/16
	$F=1$	254	375	1/16
$F=2$	$F=2$	−152	151	5/16
	$F=1$	−208	121	5/16

Source: From von Zahn, U., and Höffner, J., *Geophys. Res. Lett.*, 23, 141–144, 1996.

backscatter cross-section, convoluted with the laser bandwidth and assumed Gaussian lineshape, is least-square-fitted
to the lidar-induced fluorescence lineshape. The temperature
is then derived from the parameters of the fit. By scanning the
laser frequency through the K (D₁) line and fitting the theoretical model to the fluorescence signals at each altitude, the
range-resolved atmospheric temperature profiles can be derived from the temperature-dependent line shape of the resonance fluorescence signals (von Zahn and Höffner, 1996).

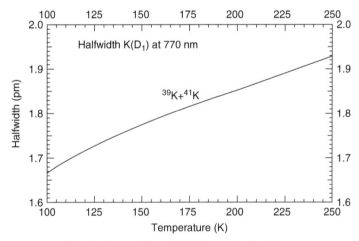

Figure 5.30 Variation of the FWHM of the backscatter cross section of K (D₁) with temperature for ^{39}K + ^{41}K. The gradient is about
0.0017 pm/K. (From von Zahn, U., and Höffner, J., *Geophys. Res.
Lett.*, 23, 141–144, 1996. With permission.)

This is similar to the original Na temperature system operated by the same group.

5.2.4.3. K Doppler Lidar Instrumentation

To achieve high spectral resolution for the scan, a narrowband laser source is needed. The IAP K lidar consists basically of (1) an alexandrite ring laser, which is injection-seeded by an external cavity diode laser; (2) a high-resolution Fabry–Perot etalon, which is used as both a wavelength meter and a spectrum analyzer to precisely measure the wavelength and the bandwidth of the each transmitted laser pulse; (3) a receiving telescope; and (4) a single photon counting device with computer. The whole system fits in a 20-ft standard container.

The pulsed alexandrite laser is configured in a unidirectional ring cavity that avoids the spatial-hole-burning problem suffered by linear cavities. With the help of injection-seeding from a single-frequency external cavity diode laser, the alexandrite laser can run in a nearly perfect single-longitudinal and single transverse mode (TEM_{00}) with a linewidth around 20 MHz (FWHM). Because of the low gain of the alexandrite crystal, the alexandrite laser pulse has much longer duration time (>200 nsec), compared to the high gain Nd:YAG laser ($\sim$ 7 nsec). This long pulse duration time allows for an ultra-narrow spectral width of the alexandrite laser and also helps avoid the saturation effect of potassium layer in the mesopause region. As a typical operation condition, the IAP K lidar has average output energy of 150 mJ per pulse with duration time of 250 nsec at a repetition rate of 33 Hz. The laser is tunable from pulse to pulse over a range greater than 50 pm. The alexandrite laser frequency is tuned by tuning the wavelength of the injection seed laser.

The seeder (diode laser) wavelength is controlled by a piezo-driven grating. The long-time stabilization of the seed laser is achieved by a lock-in technique with reference to a potassium vapor cell. However, in the IAP K lidar system, the diode laser is not locked to a Doppler-free spectroscopy. In fact, the seed laser is constantly scanned from one edge to the other of the K D_1 line. The alexandrite laser pulses then follow the seed laser and scan through the potassium D_1 line. The wavelength and bandwidth of each alexandrite laser pulse are measured in real time with a high-resolution etalon. The high-resolution etalon is a Fabry–Perot interferometer with 5.2 cm plate spacing. Its interference fringes are imaged onto a linear array of 1024 photodiodes. The data from these diodes are used to calculate the wavelength of each pulse in real time. The wavelength can be determined with 1 MHz resolution, and the bandwidth can be determined with a resolution of 20 MHz. About 99% of alexandrite laser pulses are in a TEM_{00} single mode and have a spectral width lower than the resolution of the spectrum analyzer (20 MHz).

The receiving telescope for the IAP K lidar uses a primary mirror of 80-cm diameter with the focus being fiber coupled to an optical bench. The single-photon counting system has an altitude resolution of 200 m. Depending on the measured wavelength of the output laser pulse, individual returned photon counts are stored to one of 18 different wavelength channels for later data processing, each being 0.16-pm wide (von Zahn and Höffner, 1996). Typically, the seed and the alexandrite lasers can scan through the entire potassium D_1 line in a few seconds. During a typical integration time of a few minutes, the returned data fill in all 18 channels with more shots towards the edge and less at the center of the D_1 line to compensate the low photon counts at the off-resonance wavelengths.

A narrowband K lidar developed by the Arecibo Observatory also uses an alexandrite ring laser, but the laser frequency is not scanned. Instead, the injection seed laser in the Arecibo system is locked to a Doppler-free feature near the center of the K (D_1) spectrum (Friedman et al., 2002). An AOM is then used to shift the frequency to achieve two wing-frequencies. During observations, the K lidar frequency is shifted between the peak frequency f_0 and the two wing-frequencies $f_\pm$ quickly. By taking the ratio between wing-frequency signals to the peak frequency signals, the atmospheric temperature is then derived from this ratio using the similar three-frequency ratio-technique initially developed for the Na Doppler lidar (She and Yu, 1994; Papen et al., 1995a).

Compared to the IAP K lidar system, the Arecibo K lidar using ratio-technique has the advantage of switching quickly only among three frequencies, so potentially has higher temporal resolution if other conditions are the same. However, the Arecibo system relies on the stability of the seed laser frequency, and the pulsed laser system tends to have a frequency chirp caused by the changing length of laser cavity due to variations of the refraction index of the laser rod during pulse formation. This results in a blue shift of the laser output frequency that can be tens of megahertz. This frequency shift results in a systematic offset in the temperature measurement of approximately 0.3 K/MHz (Friedman et al., 2003b). For typically 40 MHz frequency shift in the Arecibo K lidar, the laser chirp could introduce as large as 10 K uncertainties to the temperature measurements. Currently, the average laser chirp can be measured and removed from the temperature measurements at Arecibo (Friedman et al., 2003b). The IAP K lidar measures the wavelength and spectral width of each laser pulse in real time, so their system does not rely on the seed laser frequency and is immune to chirp influences. By scanning laser frequency through the K (D$_1$) line, they can derive the temperatures from the fitting parameters. The shortcoming is that the scan takes a longer time than the ratio-technique.

5.2.4.4. Daytime Measurements

To achieve daytime operation, the IAP K lidar employs a twin Faraday anomalous dispersion optical filter (FADOF) and a high-performance photon-counting avalanche photodiode (APD) (instead of a photomultiplier tube) along with reduced field of view (Fricke-Begemann et al., 2002). This improves the signal-to-noise ratio during daytime operations and allows ground-based lidar measurements of upper-atmospheric temperatures in full daylight.

In order to measure temperature accurately, the spectral transmission shape of the daylight filter must be uniform over the ~3 pm range required to scan the K (D$_1$) line, or its transmission must be measured with high accuracy for correction. With a suitable combination of the magnetic field strength

and vapor cell temperature of the FADOF, the spectral transmission curve of the IAP FADOF is uniform over the K lidar
scan range (D_1 line) as shown in Figure 5.31. The full theor-

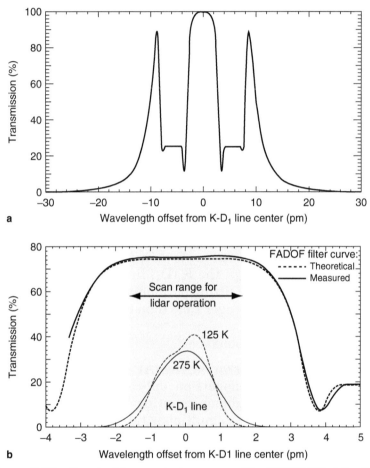

a

b

Figure 5.31   Spectral transmission curve of the K FADOF at the K D_1
line (770 nm). (a) Full theoretical shape. (b) Comparison of the theoretical shape of the center peak with the measured transmission shape
for the experimental configuration of the IAP K lidar (thick solid
curve) and spectral Doppler shape of the K D_1 fine structure line for
two different atmospheric temperatures (in arbitrary units). (From
Fricke-Begemann, C., Höffner, J., and von Zahn, U., *Geophys. Res.
Lett.*, 29 (22), doi: 10.1029/2002GL015578, 2002. With permission.)

etical transmission curve (Figure 5.31a) has a central peak of 6.7 pm (FWHM) and two side peaks of about 4 pm width at 10 pm distance. The central peak has a nearly uniform transmission of more than 99% within ±1.8 pm around its center. This uniform transmission function avoids the influence of the spectral filter shape on the temperature measurements. The FADOF filter shape is equivalent to that of a Fabry–Perot etalon with 10 pm FWHM, finesse of 50, and peak transmission of 85%. The measured transmission shape for the IAP potassium FADOF is compared with theoretical curve in Figure 5.31(b). The lower absolute value of the measured transmission (~75%) is due to losses in the optical components. The filter transmission modulation of 1% over the whole scan range requires a small correction of 0.5 K to the temperature. Therefore, any uncertainties in filter-shape determination have only negligible influence on the temperature results. In comparison to Na systems, the two peaks of the K (D_1) line are much closer than corresponding peaks D_{2a} and D_{2b} of the Na (D_2) line. Therefore, much smaller magnetic fields and lower vapor cell temperatures are required to achieve uniform transmission curve for K-FADOF. In fact, uniform transmission is not practical for Na systems. Thus, the nonuniform transmission curve of the Na-FADOF must be measured to high precision for calibration.

The operation of a FADOF requires linearly polarized light. However, because of the nature of resonance fluorescence and the depolarization in the fiber cable between the receiving telescope and the detection bench, the return light of the resonance lidar is unpolarized. Therefore, with a single FADOF, only half of the total backscatter signal can be detected. This can be avoided by using two parallel FADOFs with orthogonal polarizations as shown in Figure 5.32. The unpolarized signal from the receiving telescope is split by a polarizer into two beams with perpendicular polarizations. Each beam passes through one FADAF. After the analyzing polarizer of each FADOF, the two beams are recombined before detection.

With respect to the IAP K lidar configuration used for nighttime measurements, the background must be reduced at least three orders of magnitude to achieve a signal-to-noise

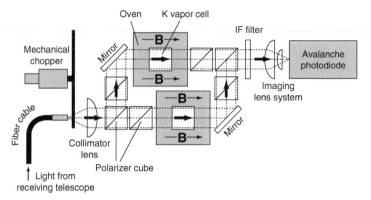

Figure 5.32 IAP K resonance fluorescence lidar detection bench with a twin FADOF system. B represents magnetic field. (From Fricke-Begemann, C., Höffner, J., and von Zahn, U., *Geophys. Res. Lett.*, 29 (22), doi: 10.1029/2002GL015578, 2002. With permission.)

ratio (SNR) greater than 1 for daytime operation. The FADOF reduces the solar background by two orders of magnitude, compared with an interference filter with 1 nm spectral width. A further reduction is achieved by reducing the field of view. The field of view of the IAP K lidar can be changed from ~790 to 192 µrad. This reduces the solar background by a factor of 17. The use of a photon-counting APD instead of a PMT also significantly enhances the quantum efficiency (QE) of the receiver. At 770 nm, the APD has QE of 68% while the PMT reaches only 15%. The signal enhancement with the APD by a factor of 4 reduces the statistical error of the lidar retrieval and improves the signal-to-noise ratio by a factor of 2. However, this type of low-noise APD (Perkin-Elmer SPCM-AQ) has a very small active area of only about 170 µm in diameter, requiring careful alignment of the imaging optics in front of the APD for proper operation.

The mobile IAP K Doppler lidar with a twin FADOF, in combination with a reduced field of view and a high-QE APD, allows ground-based lidar measurements of mesopause region temperatures in full daylight. Initial 24-h temperature data have been obtained on the islands of Tenerife (28°N) and Spitsbergen (78°N).

5.2.4.5. Temperature and Error Analysis

Temperatures are derived from the accurately measured shape of the K (D_1) fine-structure line. An example of measured lineshapes is shown in Figure 5.33 for the night of May 4 to 5, 1995 for different altitudes and an integration period of about 3 h. The least-square fits are calculated over altitude range of 2 km, but with 1-km separation from one another. Important parameters of the fits are listed on both sides of

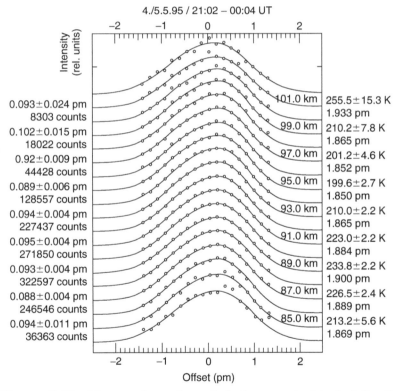

Figure 5.33 Measured number of photon counts (circles) for the 16 wavelength channels and computer fits (solid lines) for the K (D_1) fine-structure line at different altitudes. Information on the numbers on either side of the curves is given in the text. (From von Zahn, U., and Höffner, J., *Geophys. Res. Lett.*, 23, 141–144, 1996. With permission.)

each profile. On the left side, the upper values are the offsets of the line maxima from an arbitrary wavelength "zero." The lower ones are the total photon counts over the entire line. On the right side, the upper value gives the calculated temperature and its error. Errors are one-sigma errors of the least-square-fit excluding systematic errors. At the peak of the K layer, the error is about 2.2 K. The largest potential source for systematic errors is the accuracy of the wavemeter calibration used in the IAP K lidar (estimated contribution about ± 5 K). Due to the ultra-narrow laser bandwidth, the systematic error stemming from the convolution of laser bandwidth with the potassium backscatter cross-section is small, being less than 0.5 K. As von Zahn and Höffner (1996) pointed out, the Hanle effect of the K (D_1) resonance does not occur for linearly polarized laser light. Thus, the backscattered intensity is independent of the geomagnetic field.

Plotted in Figure 5.34 is an example of a measured temperature profile in the mesopause region. Both K isotopes have been taken into account for this temperature calculation. The data were acquired late in the night of May 4 to 5, 1995 and integrated over 3 h. The mesopause has a temperature of 200 K at an altitude of 96 km. In these observations, corrections of the calculated temperature due to saturation effects in the K layer are negligible. This is demonstrated in Figure 5.35 that shows the degree of saturation depending on the divergence of a 100 mJ laser beam. For the parameters of the IAP K lidar (beam divergence of 0.4 mrad and pulse duration time of 275 ns), the saturation is less than 0.2% for one of the hyperfine lines. Computer simulations show that even a saturation of 10% would change the temperature only about 0.2 K. The reason for such a small sensitivity of the calculated temperature to the saturation effect is because that the all three strongest hyperfine structure lines have the same line strength and therefore always exhibit the same degree of saturation. Only one hyperfine structure transition has a much smaller line strength, but since its spectral position is between the strong lines and because of its small line strength, it has only a marginal influence on the overall line shape and hence temperature determination (von Zahn and Höffner, 1996). For

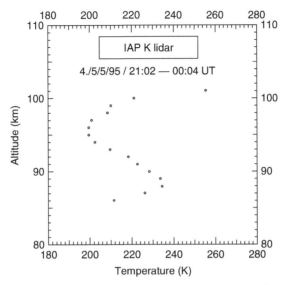

Figure 5.34 Typical temperature profile, integrated over 3 hours on May 4, 1995 for ^{39}K $+$ ^{41}K. The statistical 1σ-error at the maximum of the potassium layer is ± 2 K. (From von Zahn, U., and Höffner, J., *Geophys. Res. Lett.*, 23, 141–144, 1996. With permission.)

daytime observations, the lidar beam with a reduced divergence could cause saturation in K layer. Because saturation is more pronounced at the center of the K spectrum, it distorts the K lineshape. By scanning over a broad spectral range of the K (D$_1$) line, the IAP K lidar can actually detect this distortion effect on the lineshape and measure the saturation effect. In practice, the IAP K lidar runs at a relative lower power during daylight in order to avoid the saturation (Höffner, personal communication, 2003). For the daytime observations by the IAP K lidar, the systematic errors are slightly larger than in the nighttime. The statistical error per hour of integration can range from 1 to 20 K depending on the highly variable potassium densities at different locations and times (Höffner, personal communication, 2003).

When Papen et al. (1995b) analyzed the narrowband K lidar, the K abundance used in the calculation was taken from the measurement results of Megie et al. (1978), in which the Na to K abundance ratio was about 15. Assuming an 8-W

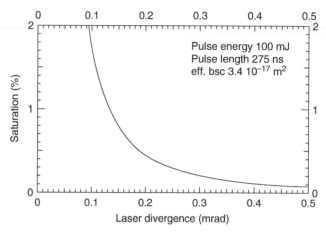

Figure 5.35 Dependence of the degree of saturation (in %) on the laser beam divergence for one of the strong hyperfine structure lines of ^{39}K, calculated for a laser pulse of 100 mJ energy and 275 nsec duration, and K layer height of 90 km. (From von Zahn, U., and Höffner, J., *Geophys. Res. Lett.*, 23, 141–144, 1996. With permission.)

766-nm K lidar, a 1-W 589-nm Na lidar and a ratio of Na to K density being 15, Papen et al. (1995b) concluded that the temperature performance of a K (D$_1$) lidar would be 37% of the Na lidar, while the K return photon counts would be about 55% of the Na counts. However, the IAP K lidar measurements show that the K abundance is about 80 times less than the Na abundance, at least five times lower than the number reported by Megie et al. (1978). This low K abundance has been confirmed by other K lidar measurements (Friedman et al., 2002) and twilight airglow measurements (Sullivan and Hunten, 1964). This unexpected low K density reduces the temporal and spatial resolution of the K lidar, and dramatically increases the integration time required for obtaining accurate temperature measurements. The higher pulse energy of the K lidar compared to the Na lidar can partially compensate the negative effects of the small K density. Although it is possible to measure the wind in the mesopause region using the K lidar as analyzed by Papen et al. (1995b), this has not been achieved to date because of the extremely low K density in the mesosphere.

5.2.5. Solid-State Na Doppler Lidar

5.2.5.1. Introduction

Three kinds of resonance fluorescence lidars have been developed so far for measuring the mesopause region temperatures: the Na Doppler technique, the K Doppler technique, and the Fe Boltzmann technique. Among them, the narrowband Na lidar currently provides the highest resolution and most accurate temperature measurements. This is because the combination of relatively high Na densities and large backscatter cross-section yields strong signals even with a relatively modest telescope of a few tens of centimeters in diameter. Unfortunately, the conventional sophisticated Na Doppler lidars require a well-controlled laboratory environment to maintain proper operation conditions, which precludes their deployment in aircrafts or at remote sites. It would be attractive to develop a narrowband Na lidar with all-solid-state lasers to take the advantages of both the Na lidar and the solid-state laser technology. Such a lidar would be robust and easy to operate for deploying to different geographical locations to study the global atmosphere. Three models of solid-state Na lidars have been tested so far: the sum frequency generation of two pulsed Nd:YAG lasers (Jeys et al., 1989; Chiu et al., 1994; Kawahara et al., 2002), the Raman-shifted and frequency-doubled alexandrite ring laser (Schmitz et al., 1995), and the sum-frequency generation of two cw Nd:YAG lasers followed by pulse amplification (Vance et al., 1998; She et al., 2002b). The two models with Nd:YAG lasers have been developed into practical lidar systems and performed temperature measurements in polar regions, while the alexandrite laser has been converted to a K lidar. Here we will briefly introduce the ideas of these solid-state Na lidar systems.

5.2.5.2. All-Solid-State Na Temperature Lidar

An interesting coincidence of nature is that the sum frequency of two appropriately tuned Nd:YAG lasers operating near 1064.591 and 1319.250 nm may be made resonant with the sodium D_2 transition wavelength at 589.1583 nm (Jeys et al., 1989). Shown in Figure 5.36 is a diagram of energy levels used in Nd:YAG lasers. The strong emission of 1064 nm is the lasing transition from the upper level $^4F_{3/2}$ to the lower level $^4I_{11/2}$ level. The weaker transition of 1319 nm is from the upper level $^4F_{3/2}$ to the lower level $^4I_{13/2}$ level. Jeys et al. (1989) were the first to obtain the sodium resonance radiation by sum-frequency mixing of these two Nd:YAG laser lines in LiNbO$_3$ crystal, although the 589 nm radiation was broadband with spectral width of about 2 GHz and contained multimodes. They made the lidar observations of the Na layer in the mesopause region using this 589 nm radiation and obtained sodium density profiles.

However, to make temperature measurements, a narrowband radiation at the sodium D_2 resonance is needed. The first all-solid-state single-frequency sodium resonance radiation was obtained by Chiu et al. (1994) using the sum-frequency mixing of the output from two injection-seeded pulsed Nd:YAG lasers. The overall layout of the Chiu et al. laser system is shown in Figure 5.37. The 1.319 μm Nd:YAG laser was configured as an injection-seeded, Q-switched mas-

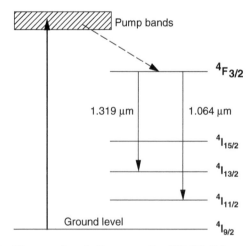

Figure 5.36 Energy level diagram for Nd:YAG laser.

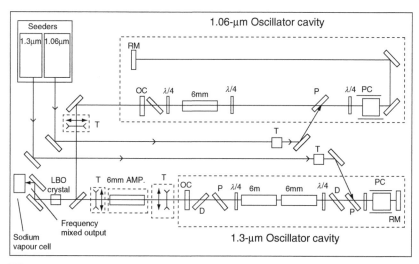

Figure 5.37 Layout of the laser system for all-solid-state single-mode sum-frequency generation of Na resonance radiation. RMs, rear mirrors; OCs, output couplers; PCs, Pockels cells; Ts, telescopes; Ps, polarizers; Ds, dichroic mirrors. (From Chiu, P.H., Magana, A., and Davis, J., *Opt. Lett.*, 19, 2116–2118, 1994. With permission.)

ter oscillator followed by a power amplifier. The $1.064\,\mu m$ Q-switched master oscillator has a typical folded, long-pulse cavity design. The 1.064 and the $1.319\,\mu m$ beams were combined on a common beam path by a dichroic mirror and then passed through a $5\,mm \times 5\,mm \times 15\,mm$ noncritically phase-matched, temperature-tuned LBO type I crystal to generate the 589 nm radiation. Both the 1.064 and the $1.319\,\mu m$ laser oscillators are well synchronized with temporal overlap adjustable to better than half a nanosecond. With input energies of 85 mJ at 1319 nm and 180 mJ at 1064 nm, the 589-nm radiation was generated as 65 mJ per pulse at a repetition rate of 20 Hz. The 589-nm radiation has a pulse duration time of 18 to 20 nsec. The measured linewidth of the 589-nm radiation was less than 100 MHz (FWHM) and the frequency stability was better than 100 MHz/h (Chiu et al., 1994). The injection seeders used in this laser system were the diode-pumped single-frequency cw Nd:YAG lasers with a continuous tuning range of more than 10 GHz. The seeder frequency can be tuned by adjusting the laser-crystal temperature with a Peltier thermoelectric cooler or by stressing the laser-crystal with a piezoelectric transducer bonded to the crystal. The 589-nm radiation frequency can be tuned over a 4 to 5 GHz range with a near-Gaussian beam profile in the near and far fields (Chiu et al., 1994).

No lidar observations of the mesopause region were ever reported with the Chiu et al. laser system. The Shishu University of Japan adopted the Chiu et al. laser techniques and developed a Na temperature/density lidar transmitter using the sum-frequency generation of two injection-seeded single-frequency pulsed Nd:YAG lasers (Kawahara et al., 2002). A different nonlinear crystal (BBO) was used in the Shishu University system, compared to the LBO crystal used in the Chiu et al. system. With input energies of 100 mJ at 1319 nm and 150 mJ at 1064 nm, Kawahara et al. generated the 589-nm laser pulses with energy of $\sim$40 mJ per pulse at a repetition rate of 10 Hz. The pulse duration time (28 nsec) is longer compared with the conventional Na lidar (about 7 nsec). This helps avoid the sodium layer saturation problem. The 589-nm laser bandwidth is estimated to be about 40 MHz. By tuning the injection-seeding laser wavelengths through controlling the temperature of seed laser crystal, the sum laser frequency can be tuned through sodium D_2 resonance line. The tuning error is within 100 MHz. During the operation, the 1319-nm seed laser wavelength was held at 1319.2012 nm while the 1064-nm seeder wavelength was tuned between 1064.6221 and 1064.6187 nm by adjusting crystal temperatures. The corresponding wavelength shift of mixing laser pulses was between 589.1589 nm (near sodium D_{2a} peak) and 589.1578 nm (near the minimum between D_{2a} and D_{2b}). The photon count ratio at these two wavelengths was then converted to temperatures using the two-frequency technique as described by She et al. (1990). The measured-temperature error due to the locking wavelength uncertainty is estimated to be ± 1 K.

Besides the 589-nm lidar beam, half of the output energy of the 1064-nm laser passed through a KDP crystal to produce laser pulses at 532 nm for Rayleigh temperature observations. This all-solid-state narrowband Na/Rayleigh temperature lidar was deployed to the Japanese Syowa Station (69°S, 39°E), Antarctica in the austral winters of 2000 to 2002. The lidar beams were pointed to zenith direction and the backscatter photons were collected by a 0.5-m diameter Dall–Kirkham Cassegrain telescope. The collected 589 and 532-nm signals were optically separated and then detected by two different PMTs. The lidar observations were limited to the nighttime periods from March to October. Even so, the Na/Rayleigh lidar has collected important temperature data in Antarctica and made significant contributions in characterizing the thermal structure of polar mesopause region and understanding the atmosphere dynamics (Kawahara et al., 2004).

A few nonlinear crystals have been used for this frequency mixing: $LiNbO_3$ (Jeys et al., 1989), LBO (Chiu et al., 1994), and BBO (Kawahara et al., 2002). Compared with $LiNbO_3$ crystal, the LBO and BBO crystals have much higher damage thresholds and larger phase-matching temperature range: about 4 to 5°C near 50°C for LBO and near 40°C for BBO compared to the 0.1°C range in the vicinity of 224°C for $LiNbO_3$. However, the mixing technique has a critical requirement on the spatial overlap of two Nd:YAG laser beams since the spatial overlap has a dramatic influence on the mixing conversion efficiency. The changes of seed laser alignment usually change the spatial overlap. In addition, the pulsed 589-nm laser is difficult to be locked to sodium D_2 Doppler-free spectrum, which results in the lack of absolute frequency reference for wind measurements. Therefore, it is unlikely that this all-solid-state narrowband Na lidar can be applied to mesospheric wind measurements.

Another idea to make all-solid-state Na lidar was tested on ring alexandrite lasers (Schmitz et al., 1995). The output of the alexandrite laser working at 791 nm can be Raman-shifted to 1178 nm (first Stokes line) using a H_2 Raman cell, and then frequency doubled to 589 nm. An alternate method is to use the first anti-Stokes line emerging from the H_2 Raman cell to convert the alexandrite laser to the 589 nm directly. However, a few serious shortcomings, such as the low energy conversion rate, low spectra purity, sensitive to environmental pressure and temperature, and a broadening of the laser spectral width, prevent this technique from being practical for the Na lidar applications. This idea has been abandoned, and the ring alexandrite laser has been converted to 770 nm for the K lidar applications as described in the Section 5.2.4.

5.2.5.3. Solid-State Na Wind/Temperature Lidar

Another idea to make solid-state Na lidar is to utilize the sum-frequency mixing of two cw single-mode Nd:YAG lasers operating at 1064 and 1319 nm to generate the 589-nm radiation in order to replace the ring dye master oscillator in the conventional Na wind/temperature lidar (Vance et al., 1998). The sum-frequency mixing of two cw single-mode Nd:YAG lasers in a doubly resonant congruent lithium niobate resonator generated TEM_{00} beam of single-frequency 589-nm radiation with power up to 400 mW as demonstrated in Vance et al. (1998). The two Nd:YAG lasers are based on a monolithic diode-pumped Nd:YAG ring cavity and incorporate both a relatively fast piezo-tuning mechanism (a 30-kHz response bandwidth) with a 30-MHz tuning range and a slow (1 GHz/s) thermal tuning mechanism with a tuning range of 60 GHz. Their narrow linewidth (5 kHz over 1 msec), TEM_{00} spatial mode, and low amplitude noise (0.05% rms between 10 Hz and 100 MHz) combined with their tenability makes them well suited for the generation of narrowband tunable radiation near 589 nm. The cw 589-nm beam still needs to be amplified by a PDA, which is pumped by an injection-seeded frequency-doubled pulsed Nd:YAG laser, to produce high peak power laser pulses for the lidar applications. Since the sum-frequency generated cw 589-nm radiation can be easily locked to the Doppler-free features of the sodium D_2 line using a Na vapor cell, the absolute frequency reference is available for the wind measurements. Therefore, it is still a Doppler wind and temperature lidar, and can make both temperature and wind measurements in the mesopause region.

Such a solid-state Na wind/temperature lidar has been developed by the Colorado State University in collaboration with the Colorado Research Associates (She et al., 2002b). This lidar is called the Weber lidar. Its transmitter consists of a sum-frequency generator (SFG) producing 100 to 200 mW cw tunable single-frequency light at 589 nm, an AOM that up- or down-shifts the SFG output by 630 MHz on command, and a PDA that is pumped by a 20 W frequency-doubled Nd:YAG laser working at a repetition rate of 50 Hz. The output 589-nm pulses have a bandwidth of 120 MHz and a power about 1 to 1.5 W. The solid-state master laser is much more robust than the ring dye laser, and is easy to operate once the alignment of frequency mixing is done. The Weber Na wind/temperature lidar has been successfully deployed to ALOMAR at Andoya, Norway (69°N, 16°E) (She et al., 2002b). With the large steerable twin telescopes at ALOMAR and demonstrated line-of-sight wind and dual-beam momentum flux measurement capability, our understanding of gravity wave dynamics in the polar region will be much enhanced by the Weber lidar.

To obtain an all-solid-state Na wind/temperature lidar, an option is to replace the PDA by an optical parametric amplifier (OPA) (She, private communication, 2003). This all-solid-state Na lidar can make both wind and temperature measurements with comparable accuracy and resolution to the conventional Na wind/temperature lidar. If the OPA could slow down the pulse (i.e., increase the pulse duration time), it would also help avoid the Na layer saturation problem.

5.2.6. Comparison of Na, Fe, K, and Rayleigh Lidar Techniques

As of today, four lidar techniques have been developed to measure temperature profiles in the middle and upper atmosphere: the narrowband Na Doppler lidar technique, the broadband Fe Boltzmann lidar technique, the narrowband K Doppler lidar technique, and the Rayleigh lidar technique. It will be worthwhile to compare their measurement capabilities and technical challenges for the applications in the atmosphere observations, as shown in Table 5.9.

Table 5.9 Comparison of Na, Fe, and K Lidar Performances

	Na	Fe	K
Wavelength in vacuum(nm)	589.158	372.099; 373.819	770.109
Peak backscatter cross	15.0×10^{-16}	0.94×10^{-16} (372 nm)	13.4×10^{-16}
section (m²) at 200 K	Na (D_2)	0.87×10^{-16} (374 nm)	K (D_1)
Abundance (cm^{-2})	5×10^9	10×10^9	6×10^7
Laser power (W)	1	3 ($\times$2)	3
Laser bandwidth FWHM (MHz)	140	800–1000	20
Pulse duration time (nsec)	6–7	60–100	250–275
Rayleigh range (km)	Up to 50	Up to 80	Up to 50
Daytime filter	Na-FADOF Fabry–Perot etalon	Fabry–Perot Etalon	K-FADOF

5.2.6.1. Narrowband Na Doppler Lidar Technique

The Na Doppler technique currently provides the highest resolution and most accurate temperature measurements of any system in the mesopause region. The combination of relatively high Na densities and large backscatter cross-section yields strong signals even with a relatively modest telescope of a few tens of centimeters in diameter. Besides temperature, the Na Doppler lidar is also capable of wind measurements in the mesopause region, as demonstrated by the UIUC and the CSU group. Because of the application of Na FADOF, which effectively rejects the sky scattering background and has absolute frequency calibration, the Na Doppler lidar is capable of both temperature and wind measurements in the mesopause region in both day and night, as achieved by the CSU group. However, to avoid the Na layer saturation effect, the transmitted laser power is limited to low power (1 to 2 W). Combined with its relative longer wavelength at 589 nm, the molecular (Rayleigh) scattering signal from Na lidars can only be used to derive Rayleigh temperatures to altitudes of 50 to 60 km with poor temporal resolution.

For the conventional Na wind/temperature lidar, the sophisticated laser technology requires a well-controlled laboratory environment to maintain proper operating conditions. Na systems cannot be easily deployed at remote sites and cannot be used for airborne observations. The full solid-state narrowband Na Doppler lidar has proved to be rugged enough for deployment in the Antarctica. However, it is unlikely to make wind measurements due to the lack of absolute frequency reference. The solidified Na wind/temperature lidar developed by CSU has made reliable temperature and wind measurements in Arctic. Once the frequency mixing is made more robust and reliable, this would be a good system for exploring the middle and upper atmosphere. However, this is still a complicated and sensitive system. It would be difficult to be deployed in the aircraft.

5.2.6.2. Fe Boltzmann Temperature Lidar Technique

The Fe Boltzmann temperature lidar does not require the narrowband lasers, but is comfortable with the broadband lasers with linewidth of about 1 GHz. The Fe Boltzmann lidar based upon the solid-state alexandrite lasers is rugged enough to be deployed at remote sites and in the aircrafts. The Fe Boltzmann temperature technique has higher sensitivity compared to the Na Doppler technique, as shown in Figure 5.21. Although limited by the lower Fe density on the 374 nm channel, the Fe Boltzmann lidar can still provide relatively accurate temperature measurements in the mesopause region. Because of its shorter wavelength and higher output power, the Fe lidar has much stronger molecular scattering and so the Rayleigh temperature measurement can be extended to the range 75 to 80 km. The combination of the Boltzmann technique and Rayleigh technique for the Fe Boltzmann lidar provides exceptionally wide altitude coverage for temperature measurements from 30 to 110 km. It is also capable of daytime temperature measurements. In addition, the Fe Boltzmann lidar has been proved to be a great tool for detecting the polar mesospheric clouds in the polar summer under 24-h sunlight. The combination of its daylight measurement capability, short operation wavelength, and high output power ensures that the Fe Boltzmann temperature lidar has very high sensitivity for PMC detection. Extensive data on PMC have been collected at the South and North Poles with the Fe Boltzmann temperature lidar. Several important scientific results and findings have merged from these data.

Due to the use of the broadband laser, the Fe Boltzmann lidar is not capable of wind measurements in the mesopause region. The biggest drawback to the Fe Boltzmann technique is the fact that the system is actually two complete lidar systems operating at 372 and 374 nm. The signal levels on the weak 374 nm channel limit the performance of the system. Typical Fe densities in this ground state at the peak of the layer vary from about 50 to 300 cm^{-3} so that daytime observations are doable but challenging and require long integration times. The observations conducted near summer solstice over the North and South Poles when the Fe densities are minimum and the background noise from the daytime sky is maximum, provided the toughest environmental tests of the instrument. In these first field campaigns, the impressive measurement capabilities of the Fe lidar have been demonstrated under the most extreme operating conditions.

5.2.6.3. Narrowband K Doppler Lidar Technique

The narrowband K Doppler lidars based upon solid-state alexandrite lasers operating in a pulsed ring configuration have been shown to be rugged enough to be deployed on ships and at remote sites to make temperature measurements. The low natural K abundance is partially compensated by the high detection efficiency of the avalanche photodiode and the high output power of the alexandrite laser. The development of twin K FADOFs enables the daytime temperature measurements. Therefore, as demonstrated by the IAP group, the K Doppler lidar has provided high quality temperature measurements in the mesopause region. In principle, the narrowband K Doppler lidar could make wind measurements. However, no wind measurement have been reported so far. Because the K resonance line is in the near IR region of the spectrum, the molecular scattering is weak and so the Rayleigh temperature capabilities of K lidars are limited to about 50 km.

5.2.6.4. The Rayleigh Lidar Technique

The Rayleigh technique, which depends on the relatively weak molecular scattering, requires a power aperture product of several hundred W m^2 to make useful temperature measurements at altitudes near 100 km. This requires 4-m class telescopes and 10 to 20 W lasers. While achievable, it is unlikely these systems will be deployed to remote locations like the South Pole anytime soon, and they certainly cannot be deployed on aircraft. However, its simplicity makes the Rayleigh technique easy to be built and used for atmosphere temperature and gravity wave study below 80 km, even with moderate laser power and telescope size.

5.3 KEY RESULTS OF LIDAR MEASUREMENTS IN THE MIDDLE AND UPPER ATMOSPHERE

The advanced resonance fluorescence lidar technologies have dramatically enhanced human's capabilities in the measurements of the middle and upper atmosphere, such as the temperature, wind, meteor, and polar mesospheric clouds in the mesopause region. These lidar measurements enable the systematic study of the atmospheric thermal structure, dynamics, and composition, as well as their relations to the global climate change. Significant progress has been made in these areas since the 1980s. It is impossible to cover all these results in this brief summary section. We will highlight several key results obtained by the resonance fluorescence lidars that are unique and important to the study of the middle and upper atmosphere.

5.3.1 Thermal Structure of the Middle and Upper Atmosphere

Temperature is one of the fundamentally important parameters for understanding the global atmosphere and long-term climate change. The range-resolved temperature profiles in the middle and upper atmosphere are one of the key results obtained by the advanced resonance fluorescence lidars. These lidar measurements provide, for the first time, the insight of the global thermal structure from the North Pole to the South Pole.

Many radiative, chemical and dynamic processes play important roles in establishing the thermal structure of the middle and upper atmosphere (Garcia, 1989; Roble, 1995). The mean background temperature is largely determined by the radiative forcing, which mainly includes the solar heating associated with absorption of solar UV radiation by O_3 and O_2 (Mlynczak and Solomon, 1993), and radiative cooling associated with infrared (IR) emissions of CO_2 (Andrews et al., 1987; Rodgers et al., 1992). The chemical heating from exothermic reactions contributes energy to the middle atmosphere (Mlynczak and Solomon, 1991; Reise et al., 1994) while energy can be lost from the atmosphere by airglow from excited photolysis products or by chemiluminescent emission from product species of exothermic chemical reactions. The middle atmosphere thermal structure is also strongly influenced by dynamic forcing. Lindzen (1981) pointed out that the momentum deposited by breaking gravity waves in the mesopause region slows down or reverses the zonal winds, resulting in a strong mean meridional circulation. Associated with this single-cell meridional circulation is strong upwelling at the summer pole, strong downwelling at the winter pole, and a meridional flow from summer pole to winter pole elsewhere. These vertical motions produce significant departures from radiative equilibrium through adiabatic heating and cooling (Walterscheid, 1981; Garcia and Solomon, 1985; Fritts and van Zandt, 1993; Fritts and Luo, 1995). The measurements of background thermal structure and dynamic behavior of the middle and upper atmosphere are essential for developing accurate atmosphere general circulation models, and help understand more about the energy budget in this region.

Since the first range-resolved temperature profile through the mesopause region was made by the first practical narrow-band Na Doppler lidar (Fricke and von Zahn, 1985), the thermal structures of the middle and upper atmosphere have been characterized by the lidars in the polar regions (Neuber et al., 1988; Lübken and von Zahn, 1991; von Zahn et al., 1996; Gardner et al., 2001; Chu et al., 2002a; Pan et al., 2002; Kawahara et al., 2002; Pan and Gardner, 2003), at the mid-latitudes (She et al., 1993, 2000, 2002a; Senft et al., 1994; Leblanc et al., 1998; Chen et al., 2000; States and Gardner, 2000a,b), and in the low-latitudes (Fricke-Begemann et al., 2002; Friedman, 2003a). The comparisons of the measurement results with the theoretical models improved our understanding of the middle and upper atmosphere, their relations to the global climate change, and their relations to the solar cycle effect (She et al., 2002c; She and Krueger, 2004).

5.3.1.1. Thermal Structure at Polar Latitudes

The middle and upper atmosphere at polar latitudes is a very interesting region of the global atmosphere. At the polar latitudes, the mesopause is coldest in summer and warmest in winter. The lowest temperatures of the entire terrestrial atmosphere are found at the polar summer mesopause. It also contains different layered phenomena, such as the PMCs, sporadic metal layers, and airglow layers. It is also a region of enhanced NO production through auroral processes. In particular, during winter, the NO can reach the stratosphere and act on the upper ozone layer. All these layers may or may not be related to the thermal structure of the atmosphere. In order to reach a deeper understanding of the processes acting in this interesting region, it is important to improve our knowledge of the thermal structure of the polar middle and upper atmosphere.

The University of Bonn Na Doppler lidar (Fricke and von Zahn, 1985) made nighttime temperature measurements at

Andoya, Norway (69°N, 16°E) in the winter conditions (Neuber et al., 1988) and late summer conditions (Kurzawa and von Zahn, 1990). Combining Na lidar technique with three other techniques: the passive falling sphere technique, the rocketborne mass spectrometer, and the rocketborne ionization gauge, Lübken and von Zahn (1991) presented the first standard atmosphere thermal structure of the mesopause region at 69°N based upon the 10-year data measured on 180 days. Most of the profiles were measured during winter and summer, few in early spring and during autumn. No lidar temperature measurements were available from May through July due to the permanent daylight during this period at Andoya, Norway. The altitude range of the temperature profiles was from 50 to 120 km. In total, nine monthly mean temperature profiles were obtained, as shown in Figure 5.38. The seasonal variations of the mesopause temperatures and mesopause altitudes are shown in Figure 5.39 and Figure 5.40. The polar mesopause is much colder and is at a lower altitude in summer than in winter, a fact known at least qualitatively since the pioneering work by Stroud et al. (1959). Lübken and von Zahn (1991) found the mesopause at approximately 98 km from October to March and at 88 km in the summer months. Both the mesopause temperature and mesopause altitude essentially had two values only: a "winter" value and a "summer" value. For the temperature, there was an extended period from September until April where the mesopause temperature was around 190 K and a shorter one during summer (end of May until middle of August) where the mesopause was much colder (typically 130 K). Lübken and von Zahn suggested that in the mesosphere the transition between winter and summer conditions (and vice versa) is faster than commonly anticipated. Lübken and von Zahn also compared the measured thermal structure with an empirical reference atmosphere, the COSPAR International Reference Atmosphere (CIRA) (Fleming et al., 1990), and found significant deviations. The measured temperature in summer is significantly lower than the CIRA temperature during the same period, but in winter the measured temperature in lower mesosphere is up to 20 K warmer than that of the CIRA. The

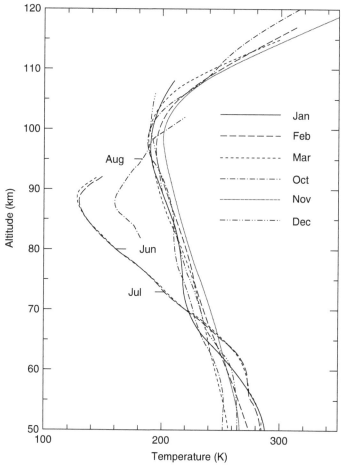

Figure 5.38 Smoothed monthly mean temperature profiles obtained by ground-based lidar and *in situ* density measurements at Andoya, Norway (69°N). (From Lübken, F.J., and von Zahn, U., *J. Geophys. Res.*, 96 (D11), 20841–20857, 1991. With permission.)

mean static stability was found to be 3.5 and 7.9 K/km in summer and winter, respectively. Lübken and von Zahn (1991) also found statistically significant (95% confidence level) positive correlations between solar flux and Na temperatures between 82 and 100 km.

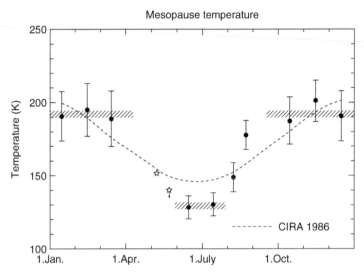

Figure 5.39 Mesopause temperature as a function of season at Andoya, Norway (69°N). The vertical bars give the variability (not error bars) of the temperature in that particular month. (From Lübken, F.J., and von Zahn, U., *J. Geophys. Res.*, 96 (D11), 20841–20857, 1991. With permission.)

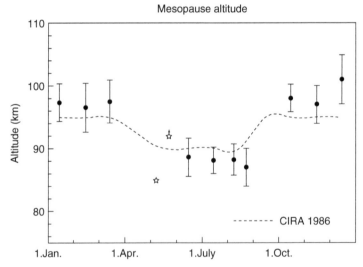

Figure 5.40 Altitude of the mesopause as a function of season at Andoya, Norway (69°N). Also shown are the error bars obtained from a polynomial fit. (From Lübken, F.J., and von Zahn, U., *J. Geophys. Res.*, 96 (D11), 20841–20857, 1991. With permission.)

The IAP narrowband K Doppler lidar (von Zahn and Höffner, 1996) was on board a ship (Polarstern) and made the mesopause temperature measurements from 80 to 105 km between 71°S and 54°N from late April to early July 1996 (von Zahn et al., 1996). Throughout the observations, the mesopause altitude was located at altitudes of either 100 ± 3 or 86 ± 3 km. The higher level was in general connected with winter conditions, while the lower with summer conditions. The high "winter" level extended from 71°S to 23°N, the low "summer" level extended from 24°N until the end of the field observations at 54°N. Shown in Figure 5.41 are the measured mesopause altitudes and temperatures versus latitudes during this shipborne campaign. From the latitudinally distributed observations, von Zahn et al. (1996) proposed that the mesopause altitude has a worldwide bimodal character. However, the mesopause temperature does not exhibit a bimodal character, but changes from the winter to summer state without noticeable jump. The mesopause temperatures measured in the southern hemisphere winter (von Zahn et al., 1996) are generally 20 K lower than previously measured in the northern hemisphere (Lübken and von Zahn, 1991). With the additional datasets from Fort Collins, USA (41°N, 105°W) by the CSU Na Doppler lidar and from Kühlungsborn, Germany (54°N, 12°E) by the IAP K Doppler lidar, She and von Zahn (1998) further investigated the concept of a two-level mesopause altitude. They concluded that the mesopause exists globally at only two altitudes: a winter state at an altitude near 100 km and a summer state near 88 km. But in contrast with the finding by von Zahn et al. (1996) that the southern mesopause in winter was colder than the northern sites, She and von Zahn (1998) concluded that in the winter state, the mesopause is at a nearly constant temperature of about 180 K, independent of latitude. Based upon these observational results, Berger and von Zahn (1999) used a three-dimensional nonlinear model of the middle-upper atmosphere (0 to 150 km) to explore the processes that lead to this two-level structure of the mesopause. In particularly, they studied the influences of photochemical heating processes, gravity wave momentum deposition, and vibrational excitation of CO_2 molecules by

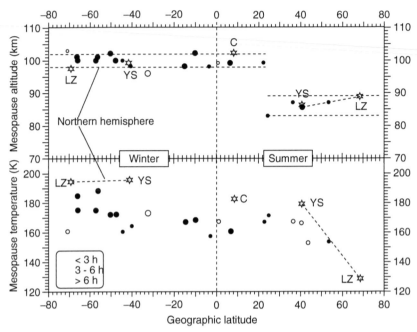

Figure 5.41 Mesopause altitude (upper panel) and temperature (lower panel) as function of latitude obtained by von Zahn et al. during their "Polarstern" cruise. The size of the dots increases with increasing length of the integration period. Dots are drawn hollow if during that night strong wave activity or strongly variable potassium densities raise some doubts as to the representativeness of the data as a climatological mean value. "LZ" refers to Lübken and von Zahn (1991), "YS" refers to Yu and She (1995), and "C" refers to Cole et al. (1979). (From von Zahn, U., Höffner, J., Eska, V., and Alpers, M., *Geophys. Res. Lett.*, 23, 3231–3234, 1996. With permission.)

collisions with O atoms on the global mesopause structure. They also calculated the effects of solar-induced tides locally on the diurnal variations of the mesopause altitude. The results of their numerical simulations suggest that two distinct mesopause altitude levels indeed exist worldwide even in terms of zonal mean states; the two-level feature is mostly due to the photochemical heating processes in combination with the gravity wave momentum deposition; and the two-level mesopause structure is not a local phenomenon of tidal wave activity.

The Shishu University all-solid-state narrowband Na Doppler lidar (Kawahara et al., 2002) made the measurements of the mesopause temperatures from 80 to 110 km between March and September during 2000 and 2001 at the Syowa Station (69°S, 39°E), Antarctica. Figure 5.42 illustrates the monthly mean temperature profiles along with MSIS90 model at the Syowa location and the lidar observation at the northern site Andoya (69°N, 16°E). The error bars shown in the figure are the day-to-day standard deviation since the temperature measurement uncertainty is considerably less than daily variability. No apparent monthly change of the

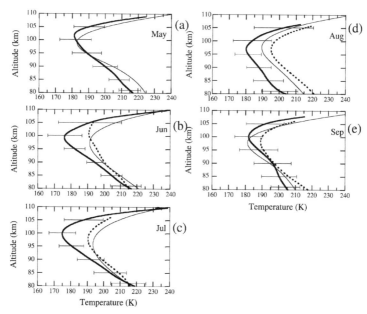

Figure 5.42 Monthly average Syowa (69°S) temperature (thick solid line) over 2 years (2000 and 2001) from May to September together with Andoya (69°N) data (dotted line) (Lübken and von Zahn, 1991) and MSIS model (thin solid line). (From Kawahara, T.D., Kitahara, T., Kobayashi, F., Saito, Y., Nomura, A., She, C.Y., Krueger, D.A., Tsutsumi, M., *Geophys. Res. Lett.*, 29, doi: 10.1029/2002GL015244, 2002. With permission.)

structures is seen in these 5 months. The mesopause altitudes are between 96 and 101 km, which is the same as the northern hemisphere winter level. The measured temperatures in the mesopause region (80 to 105 km) are, however, much lower than those at Andoya, and are also much lower than the MSIS model temperatures except in May and September. The spring (September) temperatures measured at Syowa shows closer agreement with the Andoya data and MSIS model. The measured winter mesopause altitudes at Syowa are 99 km within the day-to-day variation of ± 3 km, which are in good agreement with the observations at Andoya in wintertime. However, the averaged mesopause temperature in winter at Syowa is 175 K, about 20 K lower than those observed at Andoya in winter months (June, July, and August). The lower winter mesopause temperatures are consistent with southern hemisphere mesopause temperatures measured by a shipborne lidar (von Zahn et al., 1996). The Syowa results suggest the existence of a hemispheric difference. Though the cause of the observed asymmetry in mesopause temperature is not completely clear, Kawahara et al. (2002) suggested that it is partly due to the difference in gravity wave activities between the Arctic and the Antarctica.

The UIUC Fe Boltzmann temperature lidar (Chu et al., 2002a) was deployed from the geographic North Pole to the geographic South Pole from 1999 to 2001 to measure the temperature profiles between 30 and 110 km (Gardner et al., 2001; Chu et al., 2002a; Pan et al., 2002). With the daytime capability of this lidar system, it is possible to measure the mesopause region temperature around summer solstice at both poles under continuously sunlit (Chu et al., 2002a). An example has been shown in Figure 5.25 in the previous section. The measurement results from this Fe Boltzmann temperature lidar were combined with the balloon sonde observations at the geographic South Pole to produce the monthly mean winter temperature profiles from the surface to 110 km (Pan et al., 2002). Plotted in Figure 5.43 are the measured monthly mean temperatures along with the predictions of TIME-GCM (thermosphere–ionosphere–mesosphere-electrodynamics general circulation model) and

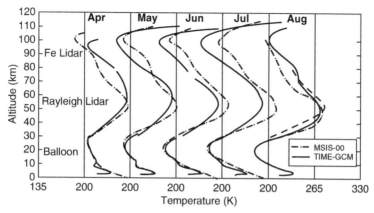

Figure 5.43 South Pole monthly mean temperature profiles for April through August (separated by 65 K for successive months). Lidar/balloon measurements are in solid lines, MSIS-00 model in dotted lines, and TIME-GCM model in dashed lines. (From Pan, W., Gardner, C.S., and Roble, R.G., *Geophys. Res. Lett.*, 29, doi: 10.1029/2002GL015288, 2002. With permission.)

the predictions of MSIS-00 (mass spectrometer incoherent scatter extended model). It is obvious from Figure 5.43 that the measured temperatures during midwinter (from May to July) in both stratopause and mesopause regions are 20 to 30 K colder than the current model predictions. Observations made at Syowa, Antarctica (69°S, 39°E) by the Shishu University Na temperature lidar also exhibit much colder mesopause temperatures in the winter (Kawahara et al., 2002). As plotted in Figure 5.44, the measured Syowa temperature in June by the Na lidar is similar to the measured South Pole temperature by the Fe lidar in the same month. Both observations are 20 to 30 K colder than TIME-GCM and MSIS-00 model predictions. Solar heating is absent during polar winter and the stable polar vortex in midwinter prevents the transport of warmer air from mid-latitudes into the south polar cap. After sunset in March, the stratosphere and mesosphere rapidly cool through thermal radiative emissions in the 15-μm band of CO_2. The wintertime radiative cooling is accompanied by downwelling over the south polar cap, which is associated

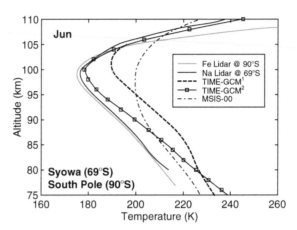

Figure 5.44 Comparison of the observed temperature at the South Pole and Syowa in June with the original TIME-GCM[1] predictions, and the TIME-GCM[2] predictions with weaker gravity wave forcing under June solstice conditions. (From Pan, W., Gardner, C.S., and Roble, R.G., *Geophys. Res. Lett.*, 29, doi: 10.1029/2002GL015288, 2002 and Kawahara, T.D., Kitahara, T., Kobayashi, F., Saito, Y., Nomura, A., She, C.Y., Krueger, D.A., Tsutsumi, M., *Geophys. Res. Lett.*, 29, doi: 10.1029/2002GL015244, 2002. With permission.)

with the mean meridional circulations driven primarily by the gravity wave forcing in the middle atmosphere. The down-welling heats the atmosphere through adiabatic compression and partially offsets the effects of radiative cooling, resulting in relatively warm winter temperatures in south polar cap (Garcia and Solomon, 1985). Observed colder winter temperatures in Antarctica suggest a weaker meridional circulation with subsequent weaker adiabatic warming than what is employed in models. When less gravity wave forcing was incorporated into TIME-GCM (TIME-GCM[2] in Figure 5.44), the weakened meridional circulation and down welling over the south polar cap produced colder mesopause temperatures as observed.

By combining the Fe and Rayleigh temperature measured by the UIUC Fe Boltzmann temperature lidar with the radiosonde balloon data, the atmospheric temperatures from the surface to 110 km were characterized for the first time at the South Pole throughout the whole year. Shown in Figure 5.45 is the University of Illinois South Pole temperature model (UISP-02) derived from the 2-year measurements at the South Pole (Pan and Gardner, 2003). Although high altitude lidar coverage above 70 km was limited during summer when the Fe densities were low and the background noise from solar scattering was high, the data were sufficient to characterize the dominant 12- and 6-month oscillations in the temperature profiles. The data show that the stratosphere and lower mesosphere between 10 and 60 km altitude are dominated by a 12-month oscillation, with the coldest temperatures in winter when solar heating is absent. The mesopause region between 70 and 100 km is dominated by 12- and 6-month temperature oscillations with maximum

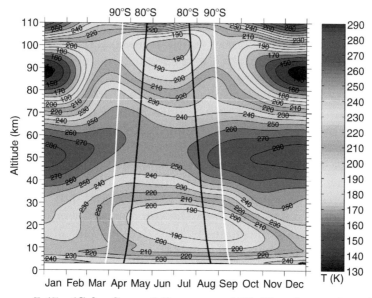

Figure 5.45 (Color figure follows page 398). The observed weekly mean temperature structure of the atmosphere above South Pole (UISP-02) plotted from 3 to 110 km. Polar nights (24 h darkness) occur between the white curve at 90°S and between the black curves at 80°S. The vertical resolution is 500 m. (From Pan, W., and Gardner, C.S., *J. Geophys. Res.*, 108 (D18), 4564, doi: 10.1029/2002JD003217, 2003. With permission.)

temperatures during the spring and fall equinoxes. The rapid response of this region to temperature changes as the sun rises and sets suggests that solar heating is stronger and upwelling is weaker than current model predictions.

5.3.1.2. Thermal Structure at Mid-Latitudes

Since the development of a new generation of the narrowband Na wind/temperature lidar through a collaboration between the Colorado State University and the University of Illinois (She et al., 1990; Bills et al., 1991a), observations of the thermal structure in the mesopause region at mid-latitudes were conducted routinely at Fort Collins, Colorado (41°N, 105°W) and at Urbana, Illinois (40°N, 88°W) for many years. The extensive observations and active researches by the CSU group and the UIUC group produced many scientific results. These results enhanced our knowledge of the background thermal structure and dynamic behavior in the mesosphere and lower thermosphere, which is essential for developing accurate global circulation and chemical models of the middle and upper atmosphere.

She et al. (1993) first reported two prevailing temperature minima observed at altitudes 86.3 ± 2.5 and 99.0 ± 2.9 km from their nighttime temperature profiles in the mesopause region, as shown in Figure 5.46. These two temperature minima appear to be a persistent phenomenon in the nighttime thermal structures at mid-latitudes, although the double minima are obvious during the spring and fall while one minimum is much weaker in the summer and winter. Similar to the bistable mesopause altitudes in the polar region reported by Lübken and von Zahn (1991), the mid-latitude mesopause also appears to be bistable with winter altitudes near 100 km and summer altitudes near 86 to 88 km (She et al., 1993; Bills and Gardner, 1993).

Senft et al. (1994) combined the nighttime observational results obtained at both Fort Collins and Urbana to provide the first comprehensive study of mesopause region temperatures at mid-latitude. Illustrated in Figure 5.47 are the contour plots of the seasonal variations of temperature structure in

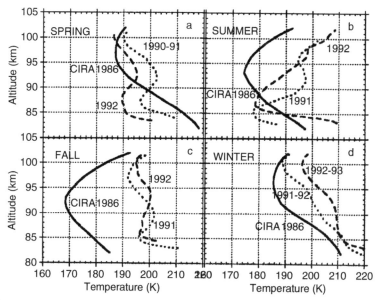

Figure 5.46 Seasonally averaged temperature profiles of the mesopause region obtained by the Colorado State University Na wind/temperature lidar over Fort Collins, Colorado, along with CIRA 1986 temperature profiles. (From She, C.Y., Yu, J.R., and Chen, H., *Geophys. Res. Lett.*, 20, 567–570, 1993. With permission.)

the mesopause region at Urbana, Illinois and Fort Collins, Colorado. The seasonal variations of the mesopause temperature and altitude are plotted in Figure 5.48. These data were obtained from the nightly mean temperature profiles at each site. The solid curves are annual plus semiannual harmonic fits to the data while the dashed curves are one standard deviation above and below the harmonic fits. The seasonal temperature variations and annual mean mesopause temperatures at these two sites are similar. The harmonic fits reach maximum values in winter of approximately 190 K and minimum values in summer of approximately 175 K. This annual variation ($\sim$15 K) is much smaller than the mesopause temperature variation ($\sim$60 K) at Andoya, Norway (69°N, 16°E) ranging from 190 K in winter to 130 K in summer (Lübken and von Zahn, 1991). The seasonal variations of the mesopause

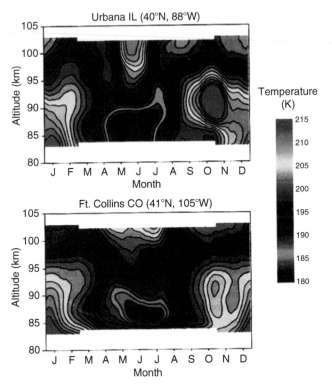

Figure 5.47 (Color figure follows page 398). Contour plots of the seasonal variations in the mesopause region temperature structure at Urbana, Illinois and Fort Collins, Colorado. The contour interval is 2 K. The plots were derived from the nightly mean temperature profiles by smoothing the data vertically and seasonally using Hamming windows with full widths of 2 km and 45 days, respectively. (From Senft, D.C., Papen, G.C., Gardner, C.S., Yu, J.R., Krueger, D.A., and She, C.Y., *Geophys. Res. Lett.*, 21, 821–824, 1994. With permission.)

heights at these two sites are significant. Both harmonic fits have strong annual components with minima near 86 km in June and maxima near 100 km from November to February. This mesopause altitude variations are quite similar to that observed by Lübken and von Zahn (1991) at Andoya, Norway. The aforementioned general features of the thermal structure in the mesopause region have been confirmed by more extensive nighttime datasets obtained at Fort Collins (She et al.,

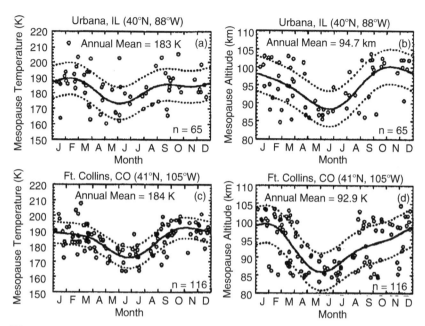

Figure 5.48 Seasonal variations of the mesopause temperatures and altitudes at Urbana (40°N, 88°W), Illinois and Fort Collins (41°N, 105°W), Colorado. The open circles are the values inferred from the nightly mean temperature profiles. (From Senft, D.C., Papen, G.C., Gardner, C.S., Yu, J.R., Krueger, D.A., and She, C.Y., *Geophys. Res. Lett.*, 21, 821–824, 1994. With permission.)

1995, 2000; Yu and She, 1995), although some details are slightly different. The mid-latitude climatology supports the concept of a two-level thermal structure with mesopause occurring at higher altitudes in winter and lower altitudes in summer, and the sharp transitions between them taking place in May and August (She et al., 2000).

The two prevailing temperature minima reported by She et al. (1993) appear to be associated with the mesosphere temperature inversion layers (MILs). The MILs are observed in low- to mid-latitude regions in two different altitude ranges at ~70 and ~95 km, which at midnight tend to be separated by the equivalent of one diurnal tidal wavelength (~25 km) (Meriwether and Gardner, 2000). Each warm air layer typically displays a thickness of $\sim$10 km. The lower MIL ($\sim$70 km) is a particularly characteristic and persistent feature of the winter mesosphere thermal structure with a typical temperature enhancement of about 10 to 25 K (Hauchecorne et al., 1987). The upper MIL is typically seen with amplitude ranging from 10 to 35 K and is also a persistent feature of the mesopause region (She et al., 1993; Yu and She, 1995). The lower MIL was first found by a rocket measurement (Schmidlin, 1976) then characterized mostly by Rayleigh lidars, while the upper MIL was characterized by the narrowband Na Doppler lidars.

The aforesaid measurements have been restricted to nighttime, including the thermal structure climatology and the temperature inversion layers. After the Na wind/temperature lidar was upgraded to daytime measurement capability by Chen et al. (1996) and Yu et al. (1997), it was highly desirable to investigate all these phenomena over diurnal coverage. When temperature profile was averaged over a complete diurnal cycle, the mesopause temperature profile showed little evidence of the prominent inversion layers. The temperature inversion layers frequently reported in nighttime measurements appear to be artifacts associated with incomplete sampling of the diurnal tide (States and Gardner, 1998).

The diurnal-mean climatology of the thermal structure in the mesopause region at the mid-latitude was then reported by States and Gardner (2000a) for Urbana, Illinois and by Chen et al. (2000) for Fort Collins, Colorado. Using more than 1000 h of lidar observational data obtained from February 1996 through January 1998 at Urbana, Illinois, States and Gardner (2000a) derived the seasonal variations of the thermal structure from 80 to 105 km, as shown in Figure 5.49, with (a) data covering the complete diurnal cycle and (b) data covering only the nighttime period. The nighttime mean climatology Figure 5.49(b) is very similar to the earlier datasets published by Senft et al. (1994), She et al. (1995), and Leblanc et al. (1998) using nighttime observational data. The main difference between the diurnal mean and nighttime mean thermal structures are the increased annual variability in

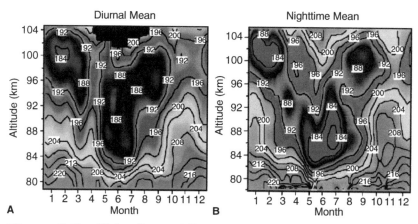

Figure 5.49 (Color figure follows page 398). Contour plot of the mesopause region annual temperature structure at Urbana, Illinois using (A) data covering the complete diurnal cycle and (B) data covering just the nighttime period. (From States, R.J., and Gardner, C.S., *J. Atmos. Sci.*, 57, 66–77, 2000. With permission.)

the nighttime data below 96 km and increased semiannual variability above 96 km, and inversion layers appearing only in the nighttime data. Plotted in Figure 5.50 are the comparison of mesopause altitude, width, and temperature between the diurnal mean and the nighttime mean. Here, the mesopause width is defined as the altitude range where the temperature is within 5 K of the temperature minimum. The quasi-two-level mesopause altitude is clearly presented here. According to the diurnal mean (solid line in Figure 5.50), the mesopause is near the high winter state (98 to 101 km) until about May 7 when it abruptly switches to the lower summer state (86 to 88 km). The mesopause remains in the summer state for about 70 days until about July 15 when it abruptly moves to about 96 km and then slowly increases over the next several months back to the winter state (~101 km). Both transitions last less than a week. For the nighttime data (dotted line), the mesopause transits to the summer state earlier in the spring and remains there longer. From the observational data shown in Figure 5.50(a),

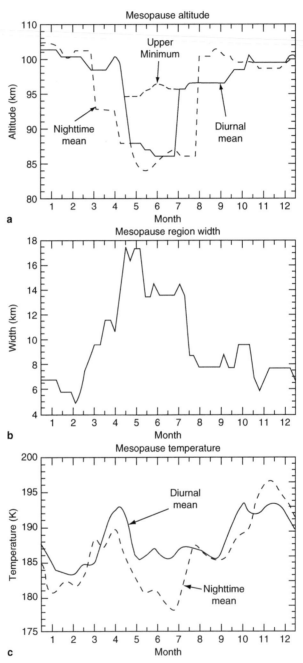

Figure 5.50 Annual variation of the (a) mesopause altitude,
(b) mesopause width, and (c) mesopause temperature at Urbana,
Illinois. (From States, R.J., and Gardner, C.S., *J. Atmos. Sci.*, 57,
66–77, 2000. With permission.)

States and Gardner (2000a) suggested that thermospheric heating forces the mesopause to a lower altitude during summer. This trend is interrupted when the adiabatic cooling in the lower mesopause region assumes control of the mesopause altitude in late spring. Without the strong summer adiabatic cooling below 90 km, the mesopause altitude would oscillate between a winter maximum of about 101 km and a summer minimum of about 96 km. The mesopause temperature derived from the diurnal mean is warmer than the nighttime data because of daytime solar heating as shown in Figure 5.50(c). The nighttime mesopause temperature averages 2.0 K colder than the diurnal mean, with biggest deviation occurring during summer time as expected. States and Gardner (2000a) also presented a brief summary of the energy budget for the thermal structure in the mesopause region. Plotted in Figure 5.51 is their annual mean temperature profile over diurnal-mean with their suggestions for the

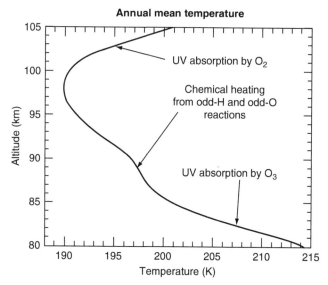

Figure 5.51 Mean background temperature averaged over both the diurnal and annual cycles at Urbana, Illinois (40°N, 88°W). (From States, R.J., and Gardner, C.S., *J. Atmos. Sci.*, 57, 66–77, 2000. With permission.)

main energy sources. This observed diurnal mean annual temperature structure is largely consistent with the assumption of radiative equilibrium between direct solar UV heating and radiative cooling by IR emission (States and Gardner, 2000a).

Chen et al. (2000) used 18 datasets of continuous temperature measurements covering a complete diurnal cycle over Fort Collins, Colorado from February 1997 to January 1999 to derive the seasonal variations of diurnal mean thermal structure in the mesopause region. Illustrated in Figure 5.52 are the temperature profiles for four seasons (spring, summer, fall, and winter) over the diurnal mean, nighttime mean, and daytime mean. The difference between annual diurnal, nighttime, and daytime means may be ascribed to a temperature wave propagating with a downward phase speed of ~0.8 km/h. The nighttime annual mean is colder than the

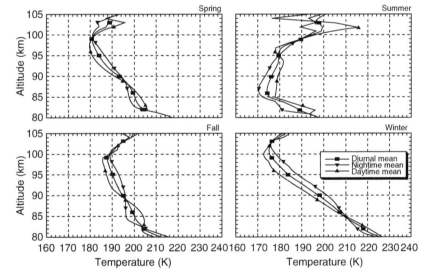

Figure 5.52 Profiles of seasonal averages of temperature means over 24 hours of data, means over nighttime (18:00 to 06:00 LST), and means over daytime (06:00 to 18:00 LST) from 18-day data obtained at Fort Collins, Colorado. (From Chen, S., Hu, Z., White, M.A., Chen, H., Krueger, D.A., and She, C.Y., *J. Geophys. Res.*, 105 (D10), 12371–12379, 2000. With permission.)

diurnal annual mean by no more than 2 K below 88 km and warmer by no more than 3 K above 88 k where the three means are nearly the same. Shown in Figure 5.53 are the mesopause altitude and temperature for the 18 datasets. In addition, the altitudes of the secondary minima above 91 km for the three summer data are marked by × in Figure 5.53(a). The winter-to-summer transitions, based on diurnal means, are abrupt and clear, further strengthening the concept of

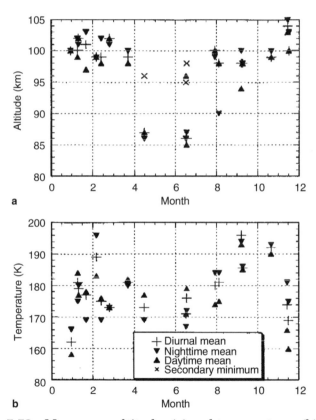

Figure 5.53 Mesopause altitudes (a) and temperatures (b) for the 18-day data taken at Ft. Collins, Colorado. The altitudes of the secondary minima for the three summer days are marked by × in (a). (From Chen, S., Hu, Z., White, M.A., Chen, H., Krueger, D.A., and She, C.Y., *J. Geophys. Res.*, 105 (D10), 12371–12379, 2000. With permission.)

two-level mesopause. Similar to the Urbana data, Chen et al. (2000) also observed the altitude of a secondary temperature minimum in the summer to be $\sim$96 km, which, without strong dynamic cooling in the summer, would be the summer mesopause altitude.

The full diurnal coverage of the Na Doppler lidar measurements of the thermal structure in the mesopause region enables the studies of thermal tides at the mid-latitudes. States and Gardner (2000b) used the same Na lidar dataset as in States and Gardner (2000a) to characterize the seasonal behavior of solar thermal tides in the mesopause region between 80 and 105 km. She et al. (2002) used the same 18 datasets of 24-h continuous lidar temperature observations as in Chen et al. (2000) to analyze the mean amplitude and phase of oscillations with 24, 12, 8, and 6 h periods for four seasons. Both observational results are compared to the recent global-scale wave model (GSWM) (Hagan et al., 1995). Good general agreements between the observed and model tidal phases were found for both sites, implicating the prevalence of migrating diurnal and semidiurnal tides. However, the GSWM consistently underestimates the diurnal and semidiurnal amplitude (States and Gardner, 2000b). These observations demonstrate the high potential of full-diurnal-coverage lidars for tidal studies.

5.3.1.3. Thermal Structure at Low Latitudes

The transportable IAP K Doppler lidar, operating at scanning K (D_1) line mode (von Zahn and Höffner, 1996), was located at the Observatorio del Teide (28°N, 17.5°W) on the island of Tenerife. Four campaigns were conducted in 1999. Using the data obtained in 55 nights, Fricke-Begemann et al. (2002) characterized the nocturnal mean temperatures and their seasonal variations. Shown in Figure 5.54 are the mean temperature profiles for the four campaigns. They provide the new evidence for the global two-level structure of the mesopause. Around midsummer, the mesopause is close to 86 km with about 183 K for a period of 4 weeks from the end of May to the end of June. The transition periods between the regular

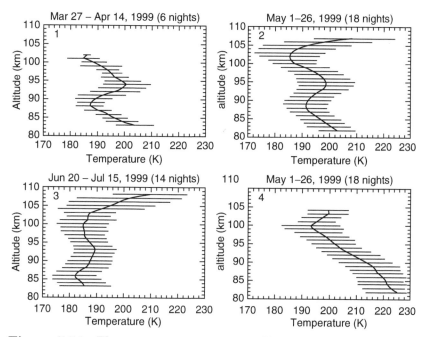

Figure 5.54 The mean temperature profiles obtained by the IAP K Doppler lidar at Tenerife (28°N) during 55 nights in 1999. The bars represent the rms variability of the nightly means. Values are given only for altitudes where more than 50% of the nights have reliable data. (From Fricke-Begemann, C., Höffner, J., and von Zahn, U., *Geophys. Res. Lett.*, 29 (22), doi: 10.1029/2002GL015578, 2002. With permission.)

(high) and the low altitude state of the mesopause are 2 weeks each. In November the observed regular mesopause is at 100 km with 192 K temperature. The thermal structure observed at this 28°N site is similar to what was observed at 41°N and 40°N.

The Arecibo K Doppler lidar, operating at three separate frequencies (Friedman et al., 2002), was operated routinely at Arecibo Observatory (18.35°N, 66.76°W). Based upon 74 nights temperature profiles of observing periods greater than 5 h, Friedman (2003a) derived the climatology of tropical thermal structure in the mesopause region. Plotted in Figure 5.55 is the annual mean thermal structure in

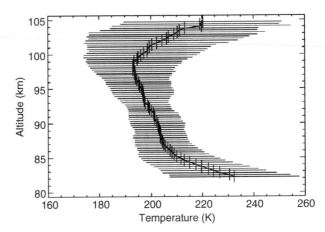

Figure 5.55 Annual mean mesopause temperature at Arecibo Observatory (18.35°, 66.75°W) based on the average of the K lidar data from 74 nights. The error bars represent the rms variability of the layer at each height. (From Friedman, J.S., *Geophys. Res. Lett.*, 30 (12), doi: 10.1029/2003GL016966, 2003. With permission.)

the mesopause region based on the average of the data from the 74 nights. The mean mesopause is at 98 km with a temperature of 193 K. This annual mean temperature profile is similar to what was seen at mid-latitudes. Shown in Figure 5.56 is the contour plot of the seasonal variation of thermal structure from 80 to 110 km at Arecibo. It presents a temperature minimum of 176 K at 100 km during mid-summer. The winter period minimum of about 190 K is found mostly in a broad flat distribution between 90 and 95 km, and less frequently near 100 km (Friedman, 2003a). These features are quite different from the 28°N observations made by Fricke-Begemann et al. (2002). Figure 5.57 shows the mesopause temperature and altitude observed at Arecibo over the 22 months. The high altitude of summer mesopause is clearly shown at the center of the plot, with only 3 of 29 nights from May through August have mesopause altitudes below 95 km. The harmonic fit to the mesopause temperature exhibits that the annual mean mesopause temperature is 179 K, and the mesopause temperature is 21 K warmer in winter than in summer. The temperature inversion layers

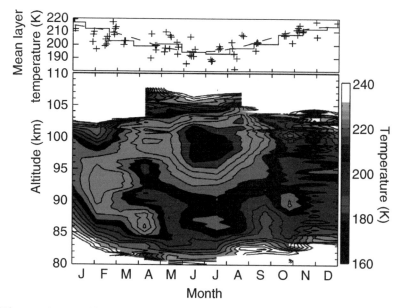

Figure 5.56 (Color figure follows page 398). Plotted in the upper panel are the nightly (+) and monthly (—) average temperatures along with a harmonic fit (- - -). Shown in the lower panel is the seasonal variation of mesopause region temperature above Arecibo. (From Friedman, J.S., *Geophys. Res. Lett.*, 30 (12), doi: 10.1029/2003GL016966, 2003. With permission.)

around 90 km are usually observed during the summer period with amplitudes often greater than 30 K. Inversions are not frequently observed in winter. Therefore, a dual minimum is observed in summer, with the upper level virtually always experiencing the lower temperature. The seasonal variation of the mesopause region temperature structure observed at Arecibo provides a challenge for both theoretical models and empirical studies. Both the summer mesopause altitude and the summer mean temperature near 176 K are consistent with the earlier observations by She and von Zahn (1998) and the model prediction by Berger and von Zahn (1999). But the broad flat temperature profile observed in February was not predicted by the model, and deserves further study.

Clemesha et al. (1997) proposed and demonstrated a new low-cost technique, using a comb laser, to determine the

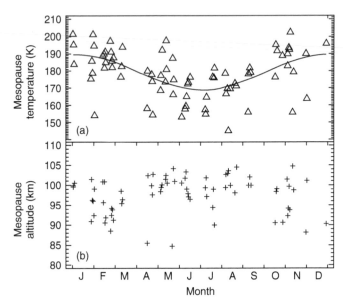

Figure 5.57 Shown, binned by date, are (a) the temperature and (b) the altitude of the mesopause for 67 nightly observations made by the Arecibo K Doppler lidar at Arecibo. (From Friedman, J.S., *Geophys. Res. Lett.*, 30 (12), doi: 10.1029/2003GL016966, 2003. With permission.)

Doppler temperature by a fairly simple modification to the dye laser transmitter of a broadband Na lidar. By replacing the front reflector with a low reflectivity Fabry–Perot interferometer, a broadband dye laser is modified to a comb laser with narrowband multi-line output where the spacing of the multi-line is tuned exactly equal to the separation of the D_{2a} and D_{2b} transition groups of Na (D_2) line. The lidar return from Na atom scattering is maximum when the laser lines coincide with the D_{2a} and D_{2b} hyperfine groups, and minimum when one of the lines lies approximately halfway between D_{2a} and D_{2b}. Therefore, the depth of modulation of the lidar response is a sensitive function of the Na atom temperature. By sweeping the Fabry–Perot interferometer through several free spectral ranges, the lidar response from atmosphere Na layers is recorded, and the temperature at each altitude is determined by least mean square fits to the lidar response versus

wavelength. With this simple, low-cost lidar system, Clemesha et al. (1997) were able to measure temperature profiles in
the mesopause region and claimed a potential precision of
$\pm 5\,$K with height resolution of 1 km and temporal resolution
of 6 min. The actual results give an estimate of $\pm 7\,$K error in
the absolute temperatures, and $\pm 2\,$K error in the relative
temperatures when the data are averaged over 10 min. Clemesha et al. (1999) reported the mesopause temperature profiles
measured at southern latitude Sao Jose dos Campos (23°S,
46°W), Brazil. As shown in Figure 5.58, the average temperature profile based on the measurements of 15 nights from July
to October 1998 reveals a mesopause temperature of 190 K at a
height of 102.5 km. Between 83 and 100 km, the measured
temperatures are typically 20 K warmer than the temperature
predicted by the CIRA86 standard atmosphere model for the
corresponding time period at 23°S location. This southern
mean temperature profile is very similar to the December
average at Fort Collins, supporting the suggestion of She and

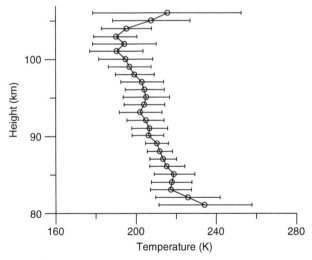

Figure 5.58 Average temperature profile for measurements made
by the INPE Na lidar at Sao Jose dos Campos (23°S, 46°W), Brazil
between July and October, 1998. (From Clemesha, B.R., Veselovskii,
I., Batista, P.P., Jorge, M.P.P.M., and Simonich, D.M., *Geophys. Res.
Lett.*, 26, 1681–1684, 1999. With permission.)

von Zahn (1998) that between 23°S and 23°N, the mesopause is probably always located at the high "winter" level.

5.3.2 Dynamics of the Middle and Upper Atmosphere

Studies of atmospheric dynamics depend on the availability of temperature and wind data. Na wind/temperature lidars can provide these data in the mesopause region and thus have a significant impact on the progress in dynamic studies. There are currently two types of experimental configurations for these studies: the coupled lidar beam with large aperture telescope through coude optics, which can be steered together, as the system developed by the UIUC group (Tao and Gardner, 1995; Zhao et al., 2003), and the dual-beam lidar system with two fixed beams and telescopes, as the system developed by the CSU group (She et al., 2003).

5.3.2.1. Heat, Momentum, and Na Flux

During the Airborne Lidar and Observations of Hawaiian Airglow (ALOHA-93) campaign (Gardner, 1995), the University of Illinois Na wind/temperature lidar was operated from the summit of Haleakala Mountain (20.7°N, 156°W) at Maui, Hawaii. As the system measured all three components of the wind and temperature, Tao and Gardner (1995) used the data obtained on October 21, 1993 to estimate the heat flux and heating rate profiles between 85 and 100 km. This is known as the first study of heat flux with a lidar. The momentum flux and mean flow acceleration in the troposphere, stratosphere, and mesosphere have been studied by numerous groups using the dual-beam radar technique pioneered by Vincent and Reid (1983). However, there was no report of heat flux measurements until Tao and Gardner (1995) because these radar observations did not include temperature measurements.

The Na lidar beam was coupled to a 0.8 m diameter Beam Director Telescope that provided full scanning capabilities for the lidar. The lidar was pointed at zenith (Z) and 15° off zenith to the north (N) and east (E) in the sequence of ZNZEZNZE. The laser was tuned to four frequencies: the Na D_{2a} peak frequency f_a, crossover frequency f_c, and 600 MHz above and below D_{2a} peak frequency ($f_\pm = f_a \pm 600$ MHz) in the sequence of f_-, f_a, f_+, and f_c. At each frequency the signal was integrated for 900 laser pulses over a period of 30 s at a range resolution of 48 m. Temperature, vertical wind, and Na density profiles were obtained at the zenith position with a temporal resolution of $\sim$7.5 min. Radial wind, temperature, and Na density profiles were obtained at the off-zenith positions with a temporal resolution of $\sim$15 min. With these data, Tao and Gardner (1995) calculated the Na density, temperature, vertical wind, zonal wind, meridional wind, and vertical shear of meridional wind shown as the contour plots in Figure 5.59. The temperature and wind perturbation profiles (T', u', v', and w') were computed by subtracting the corresponding mean profiles for the night. To improve the accuracy of the flux measurements, the perturbation profiles were smoothed using rectangular windows with full widths of 60 min and 4 km. Thus, the computed perturbations include the effects of waves with periods between about 2 and 10 h and vertical wavelengths between about 8 and 30 km. The heat fluxes were computed by calculating the covariances of temperature and wind perturbations. The results are plotted in Figure 5.60. Tao and Gardner (1995) did an analysis on the instrument errors and statistical variability in the measurement of heat fluxes. The error limits plotted in Figure 5.60 include the combined effects of instrument errors and the statistical uncertainties of the flux estimates. The zonal heat flux (Figure 5.60b) is positive everywhere, indicating strong eastward heat transport. Since the standard deviation of meridional flux measurement exceeds the mean (Figure 5.60c), we can only conclude that the meridional heat transport was weak. From the vertical heat flux, Tao and Gardner (1995) computed the heating rate associated with wave transport of thermal energy as illustrated in Figure 5.61 along with the rms error limit imposed by instrument errors and the statistical variability of the heat flux estimate. Because of the strong convergence of the vertical flux, the heating rate is positive between 87 and 97 km and negative at the extreme edges of the observation range.

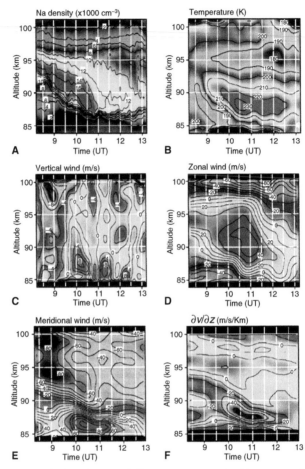

Figure 5.59 (Color figure follows page 398). Contour plots of (A) Na density, (B) temperature, (C) vertical wind, (D) zonal wind, (E) meridional wind, and (F) vertical shear of meridional wind as functions of time and altitude on October 21, 1993 at Haleakala, Maui by the UIUC Na wind/temperature lidar. To generate these plots the data were smoothed to resolution of 30 min and 1 km. (From Tao, X., and Gardner, C.S., *Geophys. Res. Lett.*, 22, 2829–2832, 1995. With permission.)

The vertical heat flux and heating rate reported by Tao and Gardner (1995) are substantially different from the values predicted by gravity wave theory (Schoeberl et al., 1983; Gardner, 1994). However, the measured values are statistic-

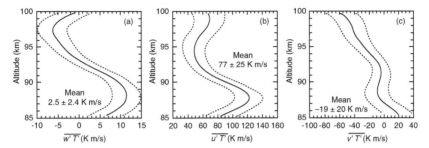

Figure 5.60 Vertical profiles of (a) vertical heat flux, (b) zonal heat flux, and (c) meridional heat flux obtained by the UIUC Na Doppler lidar at Haleakala. The dashed lines are 1 standard deviation error limits imposed by instrument errors and the statistical variability of the flux estimates. (From Tao, X., and Gardner, C.S., *Geophys. Res. Lett.*, 22, 2829–2832, 1995. With permission.)

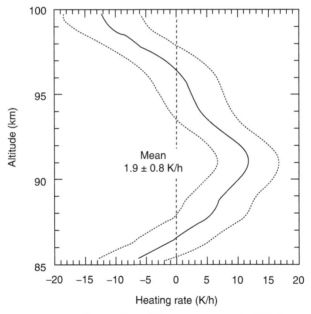

Figure 5.61 Vertical profile of heating rate (solid line) inferred from the UIUC Na wind/temperature lidar measurements at Haleakala in 1993. The dashed lines are the 1 standard deviation error limits. (From Tao, X., and Gardner, C.S., *Geophys. Res. Lett.*, 22, 2829–2832, 1995. With permission.)

ally significant and consistent with the observed short-term variations in the temperature profile (Figure 5.59b). The disagreement between the observation and theoretical predictions may be due to the short observational period ($\sim$5 h), which may bias the mean state of the mesopause region. In addition, the wind and temperature data were heavily averaged to reduce instrument errors to tolerable levels. Thus, the heat flux estimates include only the effects of waves with periods between 2 and 10 h and vertical wavelengths between 8 and 30 km. Fritts and Vincent (1987) have shown that in the upper mesosphere, waves with periods less than 2 h are responsible for the momentum deposition. If this is also true for the heat flux, then we need higher resolution data to compute the heat fluxes and heating rates. Because the dominant error source is photon noise, higher resolution heat flux measurements can be obtained only by substantially increasing the power aperture product of the lidar. Using larger telescopes is one of the best ways to improve the accuracy and resolution of the heat flux measurements.

Gardner and Yang (1998) used more extensive, higher-resolution observations made at the Starfire Optical Range (SOR) (35°N, 107°W) near Albuquerque, New Mexico to compute vertical heat flux and cooling rate profiles in the mesopause region. The University of Illinois Na wind/temperature lidar was interfaced with the SOR steerable telescope (3.5 m in diameter) to make nighttime measurements of temperature and all three wind-components between 82 and 102 km. The power-aperture product of the lidar was approximately 10 W m^2. A total of 65 h of measurement was conducted on eight different nights in November 1994, February 1995, and April 1995. The lidar and telescope were pointed at zenith (Z) and 15° off-zenith to the north (N), south (S), east (E), and west (W) in the sequence of NZEZSZWZ. The laser was tuned to the D_{2a} peak (f_a) and the wing frequencies $f_a \pm 600$ MHz. The laser was locked to f_a using the Doppler-free feature of a Na vapor cell, and to $f_a \pm 600$ MHz using a temperature-stabilized etalon (Bills et al., 1991b). Tuning was cycled rapidly through the three frequencies several times during the 2.5 min data accumulation period to minimize errors associated with Na density fluctuations. Due to the high signal levels, the vertical wind and temperature profiles computed from SOR data were obtained with temporal resolution of 2.5 min and vertical bin width of 480 m. Then the wind and temperature perturbations were computed by removing a linear trend in time at each altitude (Gardner and Yang, 1998). The mean observing period for each night was about 8 h. Thus, the computed heat fluxes include the effects of waves with periods between 5 min and 16 h, and vertical wavelengths between 1 and 24 km. Gardner and Yang (1998) also computed the dynamical cooling rate from the heat flux estimates. They did a detailed error analysis in the estimate of the heat flux and cooling rate. The seven-night mean heat flux profile, dynamical cooling rate, and their error limits are plotted in Figure 5.62. When averaged within 92.5 ± 6 km, the mean vertical heat flux was −2.29 ± 0.43 K m/s, the mean cross-correlation coefficient for the wind and temperature fluctuations was −0.12 ± 0.02, and the mean cooling rate was −30.9 ± 5.2 K/day. Despite the large geophysical variability of the nightly measurements,

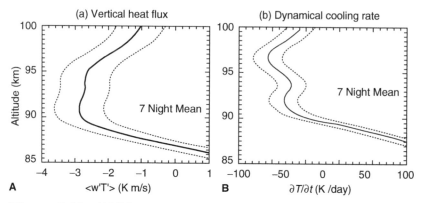

Figure 5.62 (A) Mean vertical heat flux and (B) mean dynamical cooling rate profiles obtained by the UIUC Na wind/temperature lidar at Starfire Optical Range, Albuquerque, New Mexico during 1994 to 1995 campaigns. The mean profiles were smoothed vertically using a 4 km FWHM Hamming window and exclude the February 3, 1995 data. (From Gardner, C.S., and Yang, W., *J. Geophys. Res.*, 103 (D14), 16909–16927, 1998. With permission.)

Gardner and Yang (1998) believed that these seven-night means are representative of the dynamical heat transport and cooling associated with dissipating gravity waves in the mesopause region. The measured cooling rate (-31 K/day) is consistent with theoretical predications (-36 K/day). The vertical wind and temperature data were acquired with exceptionally high spatial (480 m) and temporal (2.5 min) resolution, so the computed heat fluxes and cooling rates include the effects of the important short-period short-vertical-scale waves, which are expected to make the most significant contributions to momentum deposition and heat transport.

The vertical heat flux is defined as the expected value of the product of the vertical wind perturbation (w') and temperature perturbation (T'), that is, $\langle w'T' \rangle$. The measured perturbations from a lidar are the sum of the gravity wave perturbation (w' and T') and the instrumental errors (Δw and ΔT). Thus, the heat flux estimate is given by (Gardner and Yang, 1998):

$$\overline{\text{HF}} = \frac{1}{\tau L} \int_{z_0 - L/2}^{z_0 + L/2} dz \int_{t_0 - \tau/2}^{t_0 + \tau/2} dt (w' + \Delta w)(T' + \Delta T) \qquad (5.123)$$

where τ is the averaging period and L is the height range for sample average. Equation (5.123) is an unbiased estimator of $\langle w'T' \rangle$, and its variance is

$$\text{Var}(\overline{\text{HF}}) \approx \frac{\Delta z_{\text{HF}}}{L} \frac{\Delta t_{\text{HF}}}{\tau} [\langle (w')^2 \rangle + \langle (\Delta w)^2 \rangle][\langle (T')^2 \rangle + \langle (\Delta T)^2 \rangle] \qquad (5.124)$$

where $\Delta t_{\text{HF}} \approx 10$ min and $\Delta z_{\text{HF}} \approx 1$ km are the correlation time and vertical correlation length of the heat flux estimate. The vertical heat flux is a measure of the correlation between the vertical wind and the temperature perturbation. Because the temperature perturbations are approximately proportional to the vertical displacement, and w' is nearly orthogonal to T', the vertical heat flux is expected to be small. Typical values are a few Kelvin meter per second. To achieve statistically significant heat flux estimate, the standard deviation associated with Equation (5.124) must be significantly smaller than the mean heat flux value. The standard deviation is mainly caused by two types of errors: one is the instrumental errors, whose dominant source is photon noise; another is the statistical fluctuation due to the geophysical variability of gravity waves. The photon noise (instrumental) errors can be reduced by longer integration period and large spatial average. However, longer integration time will exclude many high frequency waves that make the most significant contributions to the momentum and heat fluxes. The most sufficient way to reduce photon noise but still preserve the high spatial and temporal resolutions is to use much large telescopes to improve the photon signal levels, as demonstrated in Gardner and Yang (1998). Nevertheless, even in the absence of the instrumental errors (mainly photon noise), that is, Δw and ΔT are zeros, the standard deviation due to the statistical fluctuations caused by the geophysical variability of gravity wave heat flux is still significant if the flux estimate is only averaged over a short period like $\tau = 8\,\mathrm{h}$. Several thousand independent measurements must be averaged in time and altitude to obtain statistically significant heat flux estimation. The same difficulties are also applied to the vertical flux of horizontal momentum of gravity waves, as recognized by Kudeki and Franke (1998) and Gardner and Yang (1998). These challenges require the heat flux and momentum flux measurements to be conducted with Na lidar coupled with large telescopes, such as 3.5 m or bigger diameter, and many independent measurements must be made to reduce the statistical fluctuations. An example of momentum flux lidar measurements can be found in Gardner et al. (1999).

Eddy transport is believed to play an important role in establishing the constituent structure of the middle atmosphere below the turbopause. Gravity waves contribute to this transport process by generating turbulence when they break. Dynamical transport occurs when dissipating (nonbreaking) gravity waves impart a net vertical displacement in the constituent as they propagate through a region. It can occur even in the absence of a concentration gradient. Unlike eddy transport, the effects of dynamical transport are usually neglected in chemical models of the middle and upper atmosphere. Over 400 h of high-resolution Na lidar data obtained at SOR were used to study the vertical dynamical transport of Na in the mesopause region between 85 and 100 km (Zhao, 2000; Liu and Gardner, 2004). The vertical Na flux was computed from Na and vertical wind data and it was related to the vertical heat flux in a simple way. The observed Na flux profile is consistent with theoretical predictions in the region between 85 and 93 km where the wave dissipation is strong. The direct measurements of the vertical dynamical flux of Na show that the dynamical transport is at least as important as eddy transport in the mesopause region and should be included in chemical models.

5.3.2.2. Atmospheric Instability and Gravity Wave Directions

Static (convective) and dynamic (shear) instabilities are believed to play major roles in gravity wave dissipation. The static (convective) stability of the atmosphere is characterized by the square of the buoyancy frequency N defined as

$$N^2 = \frac{g}{T}\left(\frac{\partial T}{\partial z} + \frac{g}{C_p}\right) \tag{5.125}$$

where g is the gravitational acceleration (9.5 m/s^2 in the mesopause region), T is the atmospheric temperature, and $C_p = 1004\,\mathrm{J\,K^{-1}\,kg^{-1}}$ is the specific heat at constant pressure. When N^2 is negative (i.e., the atmospheric lapse rate is larger than the adiabatic lapse rate), the atmosphere is unstable. Dynamic (shear) stability is characterized by Richardson number Ri:

$$Ri = \frac{N^2}{(\partial u/\partial z)^2 + (\partial v/\partial z)^2} = \frac{N^2}{S^2} \tag{5.126}$$

where u and v are the zonal and meridional wind profiles, respectively, and $S = [(\partial u/\partial z)^2 + (\partial v/\partial z)^2]^{1/2}$ is the total vertical shear of horizontal wind. Dynamic instability is induced by large vertical shears of the horizontal wind in combination with low static stability. The atmosphere is considered to be dynamically unstable when $0 < Ri < 1/4$.

Although the static and dynamic instabilities are important wave saturation mechanism in the mesopause region, there is still no detailed understanding of the instability due to the difficulties in obtaining temperature and wind data with sufficient spatial and temporal resolution and accuracy. Most studies of the instability properties in the mesopause region have been limited to theoretical work and lab experiments. The buoyancy period is about 5 min at mesopause heights and the viscous dissipation limit is about 1 km. Thus, to capture most gravity wave effects, at least 500-m vertical resolution and 2.5-min temporal resolution are needed. If Δz is the vertical resolution, the measurement rms errors in the temperature and horizontal wind must satisfy the following criteria according to Zhao et al. (2003):

$$\Delta T_{rms} \ll \Delta z \cdot \text{std}(\partial T'/\partial z)/\sqrt{2} \approx 3\text{K} \tag{5.127}$$

$$\Delta u_{rms} \ll \Delta z \sqrt{\langle S^2 \rangle}/2 \approx 6\,\text{m/s} \tag{5.128}$$

where "std" is the standard deviation, Δz is 500 m, the mean observed values for the rms temperature lapse rate are 7.7 K/km, and total wind shear is 23 m/s/km. These accuracy and resolution criteria present considerable observational challenges. Only by employing large telescope with high power Na wind/temperature lidar, the above demanding requirements can be satisfied, such as the observations made by the UIUC Na wind/temperature lidar at SOR (with a 3.5-m telescope in diameter), at Maui (with a 3.75-m telescope) and at UAO (with a 1-m telescope). The mean diurnal and annual temperature profiles at UAO demonstrate that in the absence of gravity wave and tidal perturbations, the background atmosphere is statically stable throughout the day and year (Gardner et al., 2002). Thin layers of instability can be generated only when the combined perturbations associated with tides and gravity waves induce large vertical shears in the horizontal wind and temperature profiles. Figure 5.63 shows the atmospheric instability during the night of February 18, 1999 at SOR. The structure and seasonal variations of static and dynamic instabilities in the mesopause region are

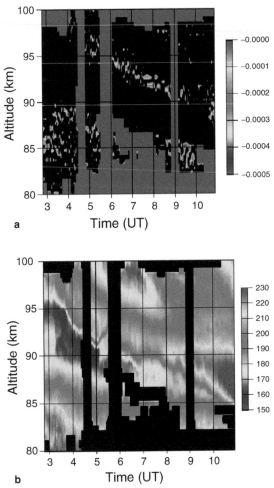

Figure 5.63 (Color figure follows page 398). (a) Convectively unstable regions for February 18, 1999, determined by N^2 calculated from 500-m 1.5-min resolution data taken by the UIUC Na wind/temperature lidar at Starfire Optical Range, New Mexico. The black region is where $N^2 > 0$; colored region is where $N^2 < 0$. No data are available in the purple region. (b) Temperature for the same night. (From Zhao, Y., Liu, A.Z., and Gardner, C.S., *J. Atmos. Sol. Terr. Phys.*, 65, 219–232, 2003. With permission.)

characterized at SOR (Zhao et al., 2003). The annual mean values of the buoyancy frequency N and the total vertical shear S in the horizontal winds are $0.021\,s^{-1}$ and $23\,m/s/km$, respectively. The probabilities of static and dynamic instabilities are maximum in midwinter averaging about 10 and 12%, and are minimum in summer with the average being around 7 and 5%. The instability probabilities vary considerably from night to night and the structure of the unstable regions is significantly influenced by atmospheric tides. Tides alone are usually not strong enough to induce instability but they can establish the environment for instabilities to develop. As the tidal temperature perturbations propagate downward, they reduce the stability on the topside of the positive temperature perturbation. Instabilities are then induced as gravity waves propagate through this layer with reduced static stability. In a region of reduced stability below the mesopause between 80 and 90 km, the temperature lapse rate is large, the buoyancy parameter N^2 is low, and the vertical heat flux is maximum. It suggests that this is a region of significant wave dissipation (Gardner et al., 2002).

With horizontal wind and temperature profiles simultaneously measured at SOR by the UIUC Na wind/temperature lidar, vertical wavelengths, intrinsic periods and propagation directions of the dominant monochromatic gravity waves in the mesopause region were extracted using the hodograph method (Hu et al., 2002). The results are illustrated in Figure 5.64. A total of 700 monochromatic gravity waves were determined from about 300 h of observations. Among them, 84.4% of the waves were propagating upwards. The mean vertical wavelength was 12.6 and 9.9 km for upward and downward propagating waves, respectively. The intrinsic period is about 10 h.

5.3.2.3. Tidal Study by Full-Diurnal-Cycle Lidar

The CSU Na wind/temperature lidar has been upgraded to a two-beam system capable of simultaneous measurements of temperature and wind in the mesopause region over full diurnal cycles, weather permitting (She et al., 2003; She, 2004). Although this lidar is a modest system with a power-aperture

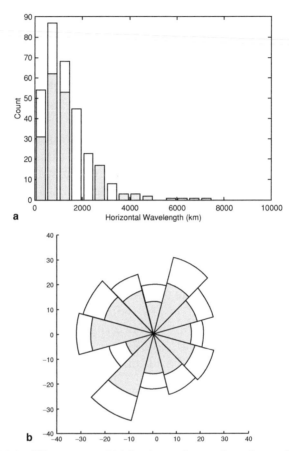

a Horizontal Wavelength (km)

b

Figure 5.64 Histogram of (a) horizontal wavelengths and (b) propagation direction derived from the UIUC Na wind/temperature lidar data taken at Starfire Optical Range, New Mexico from 1998 to 2000. In plot (b), north is up and east is right. Darker (lighter) shading indicates upward (downward) propagating waves. (From Hu, X., Liu, A.Z., Gardner, C.S., and Swenson, G.R., *Geophys. Res. Lett.*, 29 (24), 2169, doi: 10.1029/2002GL014975, 2002. With permission.)

product of only $0.6\,\mathrm{W\,m^2}$, good data quality is demonstrated by means of contour plots (Figure 5.65) depicting an 80-h continuous observation between August 9 and 12 showing the existence of atmospheric waves with different periods along with their coherence and interactions. The salient feature of

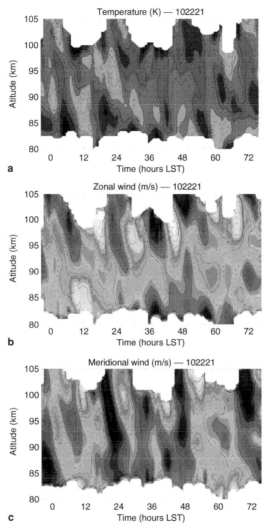

Figure 5.65 (Color figure follows page 398). (a) Mesopause region temperature, (b) zonal wind, and (c) meridional wind obtained by the CSU Na wind/temperature lidar between August 9 and 12, 2002 over Fort Collins, Colorado (41°N, 105°W). The origin of x-axis is 0:00 hours on August 9, 2002, LST. The ranges contours shown are approximately between (a) (160 K, 220 K), (b) (−90 m/s, 120 m/s), and (c) (−90 m/s, 90 m/s). (From She, C.Y., *J. Atmos. Sol. Terr. Phys.*, 66, 663–674, 2004. With permission.)

data with full-diurnal-cycle coverage lies in its ability to describe the vertical profiles of dynamical fields (temperature, zonal, and meridional winds) as a unique linear superposition of diurnal-mean and oscillations with different tidal periods, plus a residual term. In this manner, She (2004) investigated the variability of diurnal-means and of diurnal tidal amplitudes and phases. Using six datasets obtained between July 17 and August 12, each covering a full-diurnal-cycle, considerable day-to-day variability was found, as much as 20 K, 35 m/s, and 75 m/s for diurnal-mean temperature, zonal wind, and meridional wind, respectively, and 15 K and 40 m/s for the diurnal tidal amplitude of temperature and winds. The coherence of tidal excitation and of mean solar forcing is found to prevail over the variability, leading to the convergence to the climatological mean in a multiday observation. While a minimum of three full diurnal cycles appears to be required, the 6-day composite yields means and diurnal tides that resemble model predictions very well. Since the resulting amplitudes and phases of the observed diurnal oscillations agree well with the GSWM00 predictions, She (2003) concluded that the migrating diurnal tide contributes significantly to the observed oscillations with diurnal period over Fort Collins, Colorado (41°N, 105°W). These initial results demonstrate the high potential of the full-diurnal-cycle lidar systems for the atmospheric tides and dynamic studies.

5.3.3 Atmospheric Metallic Layers and Meteor Detection

5.3.3.1. Metal Layers and Mesospheric Chemistry

Meteoric ablation is believed to be the major source of the atomic metal layers that occur globally in the upper mesosphere (Plane, 1991). The resonance fluorescence lidar technique has been used to study the structure and chemistry of the mesospheric metal layers since Bowman et al. (1969) made the first lidar observations of Na layer in 1969. These studies are motivated by the increasing use of metal atoms (such as Na, K, and Fe) as tracers of atmospheric dynamics, chemistry, and temperature. Significant progress has also been made with laboratory investigations of the chemistry that controls these metal layers. The combined laboratory work and lidar observations have led to the development of models that are able to reproduce the characteristic features of the metal layers.

5.3.3.1.1. Na

The atomic Na is one of the most well-studied species in the middle and upper atmosphere. The broadband resonance Na lidars made remarkable measurements on the Na layers, as described in Section 5.1.3 (Bowman et al., 1969; Sandford and Gibson, 1970; Gibson and Sandford, 1971, 1972; Blamont et al., 1972; Hake et al., 1972; Kirchhoff and Clemesha, 1973; Megie and Blamont, 1977). Since the development of narrowband Na Doppler lidars, the Na layer has been studied more comprehensively combining with chemistry modeling (von Zahn et al., 1987; von Zahn and Hansen, 1988; Hansen and von Zahn, 1990; Plane et al., 1998, 1999; States and Gardner, 1999; She et al., 2000; Clemesha et al., 2002; Gong et al., 2002; Yi et al., 2002).

The IAP group measured the Na layer at Andoya (69°N, 16°E), Norway with their narrowband Na Doppler lidar and discovered the sporadic metal layers as a distinct phenomenon (von Zahn et al., 1987; von Zahn, 1988; Hansen and von Zahn, 1990). The rapid growth of narrow Na layers appears between 90 and 110 km altitude. The Na density increases (sometimes by large factors) within a few minutes in a narrow layer ($\sim$1 km of FWHM). Their appearance shows a strong, positive correlation with that of f-type E sporadic layers, but not showing a strong correlation with either riometer absorption or meteor showers (Hansen and von Zahn, 1990).

The University of Illinois operated a broadband Na density lidar (consisting of an excimer laser pumped dye laser) on board an aircraft and measured the summertime Na density profiles during both day and night flights to the Arctic Ocean during ANLC-93 campaign. The University of Illinois also operated another broadband Na lidar (consisting of an Nd:YAG laser pumped dye laser) at the South Pole station and measured the Na density in January 1995 and

1996 (Plane et al., 1998). Measurements at the South Pole reveal a Na layer that has less column abundance and are significantly higher and thinner than at mid-latitudes, as shown in Figure 5.66. Plane et al. (1998) modeled satisfactorily the characteristic features of the South Pole Na layer by choosing a proper rate coefficient for the reaction between sodium bicarbonate and atomic hydrogen. In particular, the model can reproduce the small-scale height of about 2 km observed on the underside of the Na layer.

Using the narrowband Na wind/temperature lidar, the CSU and the UIUC lidar groups made comprehensive measurements and studies of the Na layers at mid-latitudes in Fort Collins, Colorado (41°N) and Urbana, Illinois (40°N) (Kane and Gardner, 1993a; Plane et al., 1999; States and Gardner, 1999; She et al., 2000). Combining the Na data collected at both sites from 1991 through mid-1994, Plane et al. (1999) obtained the seasonal variations of the nighttime Na layers as shown in Figure 5.67. The Na density exhibits a strong annual variation at all heights between 81 and 107 km, with the

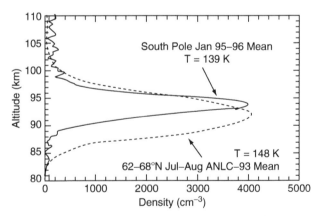

Figure 5.66 Mean Na density profiles computed by averaging all data collected at latitudes between 62° and 68°N during the ANLC-93 flights and mean Na density measured at the South Pole in January 1995 to 1996. (From Plane, J.M.C., Cox, R.M., Qian, J., Pfenninger, W.M., Papen, G.C., Gardner, C.S., Espy, P.J., *J. Geophys. Res.*, 103 (D6), 6381–6389, 1998. With permission.)

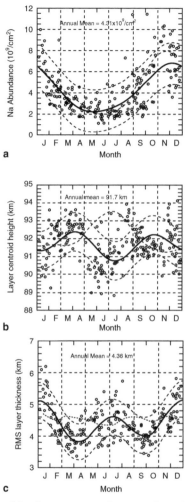

Figure 5.67 The Na layer (a) column abundance, (b) centroid height, and (c) rms width versus data obtained at Urbana, Illinois (40°N) and at Fort Collins, Colorado (41°N). The data were computed from the nightly mean Na density profiles. The thick solid curves are the least-square fits to the data including annual plus semiannual variations. The dashed curves are one standard deviation above and below the fits. The model predictions of these layer properties are shown as thick dotted curves. (From Plane, J.M.C., Gardner, C.S., Yu, J.R., She, C.Y., Garcia, R.R., and Pumphrey, H.C., *J. Geophys. Res.*, 104 (D3), 3773–3788, 1999. With permission.)

column abundance of the layer peaking in early winter and then decreasing by nearly a factor of 4 to a midsummer minimum (Figure 5.67a). The Na layer centroid altitude exhibits strong semiannual variations, with the altitude peaking at spring and fall equinox and minimizing at midsummer (Figure 5.67b). The annual mean profiles of the measured nighttime Na density and temperature are plotted in Figure 5.68. The annual mean Na density profile has a peak density around $3000\,cm^{-3}$ near 92 km. The fundamental features of the nighttime Na layers were confirmed by the 8-year (1991 to 1999) nocturnal Na density measurements at Fort Collins (She et al., 2000) and further measurements (1996 to 1998) at Urbana, Illinois (States and Gardner, 1999). Plane et al. (1999) constructed a seasonal model of the Na layer incorporating the laboratory studies of the pertinent neutral and ionic reactions of the metal Na. The modeled annual mean Na density profile is also plotted in Figure 5.68 for comparison. The model was able to reproduce many observed features of the Na layer remarkably well, including the monthly variation in column abundance and layer shape. Both the observations and the model show that the Na density and temperature are highly correlated below 96 km (with correlation coefficient equal to 0.8 to 0.95), mostly as a result of the influence of odd oxygen/hydrogen chemistry on the partitioning of Na between atomic Na and its principal reservoir species $NaHCO_3$.

With full diurnal coverage, States and Gardner (1999) further studied the seasonal and diurnal variations of the Na layer at 40°N. They found that the seasonal variations in Na density and abundance are influenced primarily by changes in mesopause region temperatures, which are coldest during midsummer. Below 95 km, reactions leading to the bicarbonate sink and the liberation of Na from this reservoir dominate Na chemistry. These reactions and their temperature dependencies are responsible for the large annual variations in Na abundance from 2.3×10^9 to $5.3 \times 10^9\,cm^{-2}$. The annual mean Na density profiles for both the diurnal mean and nighttime mean are plotted in Figure 5.69. The peak Na density for the annual mean is $3040\,cm^{-3}$ at 91.8 km. The

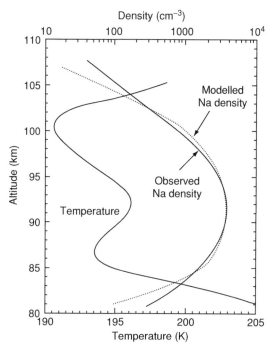

Figure 5.68 Annual mean profiles of the measured Na density (thin solid curve) and temperature (thick solid curve) at 40°N and 41°N. The modeled mean Na density profile (thin dotted curve) is shown for comparison. (From Plane, J.M.C., Gardner, C.S., Yu, J.R., She, C.Y., Garcia, R.R., and Pumphrey, H.C., *J. Geophys. Res.*, 104 (D3), 3773–3788, 1999. With permission.)

largest difference between the diurnal and nighttime means is on the underside of the layer, between 76 and 85 km, where the density is larger for the diurnal mean. This difference is associated with photochemistry, which affects the daytime Na densities. The measured diurnal variations of Na layers are plotted in Figure 5.70 (States and Gardner, 1999). Although there is little evidence of direct tidal perturbations in Na density, 24-h oscillations dominate the diurnal variations. The Na abundance is maximum at sunrise, and the peak-to-peak diurnal variation averages more than 30%. The combined effects of photoionization above 90 km and photochemistry below 85 km induce a strong 24-h oscillation in the layer

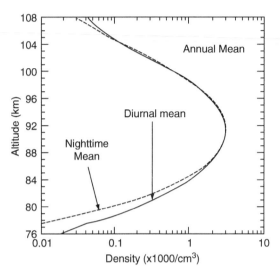

Figure 5.69 Annual mean Na density profiles for observations covering the complete diurnal cycle (solid curve) and observations covering just the nighttime period (dashed curve) at 40°N. (From States, R.J., and Gardner, C.S., *J. Geophys. Res.*, 104 (D9), 11783–11798, 1999. With permission.)

centroid altitude. The peak-to-peak variation exceeds 1 km. The lowest centroid height is at local noon when Na densities below 85 km are maximum and photoionization above 90 km is the strongest.

Clemesha et al. (2002) presented the simultaneous measurements of meteor winds and Na density in the 80 to 110 km region at Sao Jose dos Campos (23°S, 46°W) and studied the variations in Na density profiles. They found that the observed vertical oscillations in the Na isopleths are closely correlated with the meridional winds, which confirms their earlier conclusion (Batista et al., 1985) that the Na density oscillations are driven by the solar diurnal and semidiurnal tides. Clemesha et al. (2002) claimed that their conclusion is in direct opposition to those of States and Gardner (1999) that the observed diurnal variations of Na density were the result of solar zenith angle driven changes in photo- and ion-chemistry, but not related to tides. The difference of tidal influence on Na variations may be related to the different amplitude of

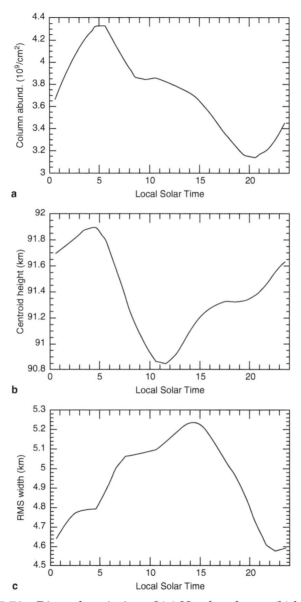

Figure 5.70 Diurnal variation of (a) Na abundance, (b) layer centroid height, and (c) rms layer width at 40°N. These parameters were derived from the annual mean day data. (From States, R.J., and Gardner, C.S., *J. Geophys. Res.*, 104 (D9), 11783–11798, 1999. With permission.)

tides at the locations. Further investigations in both observations and modeling efforts are needed.

Recently, systematic studies of sporadic Na layers (Na_S) were made by Gong et al. (2002) and Yi et al. (2002) with broadband Na lidar systems at Wuhan (31°N, 114°E), China. The sporadic Na layers at Wuhan show some different features than the low and high latitudes. The Na_S layers at Wuhan often exhibit broader layer widths and a longer formation time than those at low and high latitudes (Yi et al., 2002). Gong et al. (2002) found a fairly good correlation between the sporadic Na layer and the sporadic E layers. Through analyzing the observed perturbations during most of the Na_S developments, Gong et al. (2002) suggested that the role of dynamic processes of atmosphere in the Na_S formation should not be ignored. Further investigations are needed to understand the formation mechanism of sporadic Na layers.

5.3.3.1.2. Fe

The first lidar observations of the atomic Fe layers were performed by Granier et al. (1989b) at the Observatoire de Haute-Provence (44°N, 6°E), France. It was followed by the lidar observations of normal and sporadic Fe layers at 40°N (Bills and Gardner, 1990) and at 69°N (Alpers et al., 1990). The Fe layers exhibited considerably different properties in these three measurements, especially on the Fe column abundance values and sporadic Fe occurrence. Densities at the normal Fe layer peak vary from $3000\,cm^{-3}$ (Granier et al., 1989b) to $7000\,cm^{-3}$ (Alpers et al., 1990) to approximately $15,000\,cm^{-3}$ (Bills and Gardner, 1990), while the Fe column abundance is 3×10^9, 9×10^9, and $15 \times 10^9\,cm^{-2}$ for the three measurements, respectively. Sporadic Fe layers were observed by Bills and Gardner (1990) and Alpers et al. (1990), but not shown in Granier et al.'s (1989b) measurements.

Kane and Gardner (1993a) presented the first climatology of the seasonal variations of mesospheric Fe layer at mid-latitude 40°N. The nighttime averages of the Fe layer column abundance, centroid altitude, and rms width, summarized from 75 nights of Fe lidar data from 1989 to 1992, are shown in Figure 5.71 (Kane and Gardner, 1993a). The

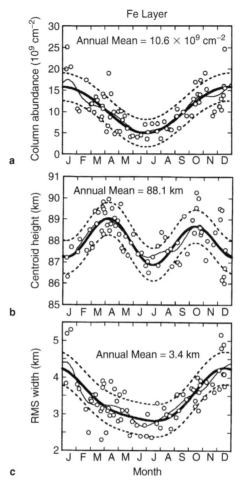

Figure 5.71 Nightly averages of the Fe layer (a) column abundance, (b) centroid height, and (c) RMS width plotted versus date at Urbana, Illinois (40°N). The annual mean Fe layer abundance, centroid height, and width are $10.6 \times 10^9\,\mathrm{cm}^{-2}$, 88.1 m, and 3.4 m, respectively. The thin curves are running means of the data computed using a Hamming window with a full width of 90 days. The dark curves are annual plus semiannual fits to the data, with the dashed curves lying 1 standard deviation above and below the harmonic fit. (From Kane, T.J., and Gardner, C.S., *J. Geophys. Res.*, 98, 16875–16886, 1993. With permission.)

annual mean Fe column abundance is $10.6 \times 10^9 \, \text{cm}^{-2}$ with the mean centroid altitude at 88.1 km and rms width of 3.4 km. Kane and Gardner (1993a) also provided a climatology of the Na layer seasonal variations: the annual mean Na column abundance is $5.35 \times 10^9 \, \text{cm}^{-2}$ with the mean centroid altitude at 92.1 km and rms width of 4.4 km. These data show that the Fe layer column abundance is about twice the Na layer abundance, while the Fe peak altitude is about 4 km lower than the Na peak, and the Fe layer RMS width is about 24% narrower than the Na layer width. The Fe layer abundance shows much stronger annual variations than the Na layer, but similar annual variations in the layer centroid altitudes. Both layers reach maximum centroid height around spring and fall equinox.

Sporadic Fe layers occurred in about 27% of the total observation time (Kane and Gardner, 1993a). Sporadic Fe and Na layers were further studied by Alpers et al. (1994) with Bonn University's narrowband Fe lidar and Na lidar at polar, middle, and low latitudes. The comparison between Fe_S and Na_S shows that the overall occurrence rate for Fe_S is much higher than for Na_S; both the normal and sporadic Fe layers are considerably more dynamic than the Na layers; Fe_S layer shape is typically broader, slower growing, and longer lasting than Na_S layer; the correlation with E_S appears to be weaker for Fe_S than for Na_S (Alpers et al., 1994).

The mesospheric Fe layers were measured at both the North Pole and the South Pole by the University of Illinois Fe Boltzmann temperature lidar (Gardner et al., 2001; Chu et al., 2002a). A huge sporadic Fe layer was observed at the North Pole on the summer solstice (June 21, 1999) with peak density of $2.3 \times 10^5 \, \text{cm}^{-3}$ at 106 km (Figure 5.25). The peak sporadic Fe density is comparable to the meteor trail density observed during Leonid meteor shower (Chu et al., 2000a), but its width ($\sim$10 km) is much wider than a meteor trail ($\sim$250 m). Although this huge sporadic Fe layer is likely linked to sporadic E layer, its abnormal high peak density is not yet completely understood. Further modeling and observation efforts are needed. The annual variation of Fe density at the South

Pole was characterized by the UIUC Fe lidar as shown in Figure 5.72. The South Pole Fe layer has a significant higher peak altitude ($\sim$92 km) in summer compared to its winter peak altitude around 88 km.

Raizada and Tepley (2003) present the annual variation of the mesospheric Fe layer for a low latitude at Arecibo Observatory (18.35°N, 66.75°W). The comparison between Arecibo results and the mid-latitude results (Kane and Gardner, 1993a) shows that the annual variation of the height and width of the Fe layers at Arecibo are essentially anticorrelated (opposite) with the mid-latitude observations, while the Fe column abundance lags its mid-latitude counterpart by roughly 3 months. These features are illustrated in Figure 5.73. Raizada and Tepley (2003) suggested that the latitudinal differences in Fe layer properties result from

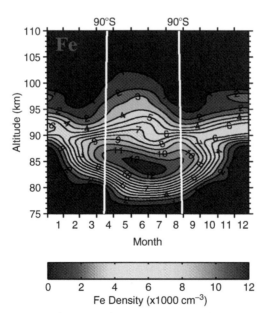

Figure 5.72 (Color figure follows page 398). Seasonal variations of Fe density observed by the UIUC Fe Boltzmann temperature lidar at the South Pole (90°S).

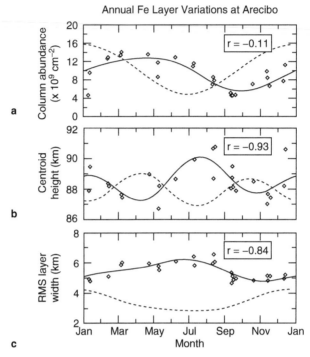

Figure 5.73 Annual variation of (a) column abundance, (b) centroid height, and (c) rms width for Fe measured at Arecibo (18.35°N). A fit (solid line) to the data points is shown together with a comparison to similar observations at 40°N latitude (dashed line). The correlation coefficients, r, are listed in each panel. (From Raizada, S., and Tepley, C.A., *Geophys. Res. Lett.*, 30 (2), doi: 10.1029/2002GL016537, 2003. With permission.)

variations in the distribution of mesospheric O_3 and O_2, which in turn affect the Fe chemistry at the two different latitudes. A one-dimensional steady-state model of the mesospheric Fe layer has been developed by Helmer et al. (1998) for the mid-latitude 40°N. It considered three processes thought to be most influential in affecting the atomic Fe layer: the Fe deposition via meteor ablation, the vertical distribution of the ablated material, and the partitioning of the metal along 10 constituent species as a result of neutral and ion gas phase reactions in dynamic steady state. Most observed features of the Fe layer can be successfully reproduced by the model (Helmer et al., 1998).

5.3.3.1.3. K

The K layer was first measured by Felix et al. (1973). Fifteen nights of lidar observations made by Megie et al. (1978) did not show significant seasonal variations of the K column and peak densities. After the development of narrow-band K Doppler lidar, comprehensive measurements of the K layer and its seasonal variations were made by the IAP group (Eska and Höffner, 1998; Eska et al., 1998, 1999; Fricke-Begemann et al., 2002) and the Arecibo group (Friedman et al., 2002). Eska et al. (1998) present the seasonal variations of the mesospheric K layers over Kühlungsborn, Germany (54°N, 12°E) based on the 110 nights data obtained between June 1996 and June 1997. The measured nightly mean K layer column abundance, centroid height, and rms width are plotted in Figure 5.74. The averaged K column abundance is 4.4×10^7 cm^{-2}, with peak altitude at 90.5 km and rms width of 4.0 km. The nightly mean column and peak densities exhibit dominantly semiannual variations with maxima in summer and winter. The centroid height varies semiannually too, attaining highest altitudes during the equinoxes. The rms layer width shows a strong annual variation with a maximum width in winter. The seasonal variations of the Na and Fe layers (Kane and Gardner, 1993a) are compared with the K layer in Figure 5.74. For all three metal layers, their centroid heights are dominated by semiannual variations with quite similar phases. The column abundances are dominated by a semiannual variation in the case of the K layer but are dominated by an annual variation for the Na and Fe layers. The layer rms widths are dominated by annual variations in the case of the K and Fe layers, while the Na layer exhibits a semiannual variation (Eska et al., 1998).

Observations of the nighttime K layer were performed on the German research vessel *Polarstern* from March to June

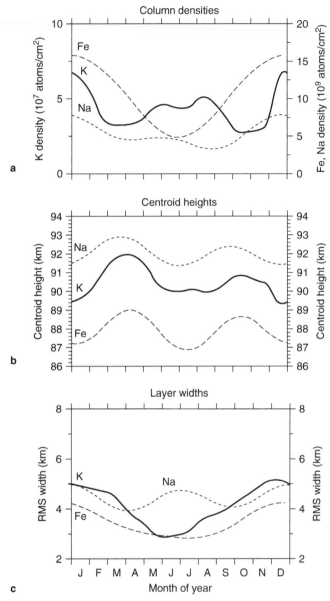

Figure 5.74 (a) Column density, (b) centroid height, and (c) layer
RMS width for K (solid curve), Na (dotted curve), and Fe (dashed
curve) plotted versus month of the year. (From Eska, V., Höffner, J.,
and von Zahn, U., *J. Geophys. Res.*, 103 (A12), 29207–29214, 1998.
With permission.)

1996. The K densities were obtained between 71°S and 45°N. The nightly mean peak densities varied from 140 cm^{-3} in the equatorial region to 10 cm^{-3} in the Antarctic, and the column abundances decreased from 1.2×10^8 to 1.3×10^7 cm^{-2} for low to high latitudes (Eska et al., 1999). The global mean peak height of the normal K layer was found to be 88.3 km, which is lower than what reported in Eska et al. (1998) for Kühlungsborn (54°N) observations. A one-dimensional model of the K layer was developed, which includes meteoric deposition, vertical transport through eddy diffusion, and a full chemical scheme (Eska et al., 1999). This model was able to reproduce very satisfactorily the seasonal behavior of the K layer at 54°N if the wintertime deposition flux of the metal was reduced by 30% compared to the summer. The mid-latitude ratio of K to Na was about 1%, much less than either the chondritic or cosmic ratios of the two metals (~8 or 6%, respectively). The most likely reason is that potassium vaporizes less efficiently from meteoroids than sodium. The model was generally very successful in reproducing the latitudinal variations in the K layer (Eska et al., 1999).

The observations of the K layer at Arecibo Observatory (18.3°N, 66.75°W) show quite different features of the K layer compared to the Kühlungsborn results (Friedman et al., 2002). At 18°N, there is less variability in the layer width and a generally lower peak height with no winter minimum. More sporadic K layers were observed at this low-latitude site than were seen at higher latitude. However, the semiannual variation of K column abundance was seen at both the low and mid-latitudes. Further studies are needed to understand the differences in the behavior of the K layer at different latitudes.

5.3.3.1.4. Ca and Ca$^+$

The Ca and Ca$^+$ layers have been studied by lidars at mid-latitudes 44°N (Granier et al., 1985, 1989a), 40°N (Gardner et al., 1993; Qian and Gardner, 1995), and 54°N (Alpers et al., 1996; Gerding et al., 2000) and at a low-latitude 18°N (Tepley et al., 2003). All the lidar systems used for these measurements consist of either Nd:YAG laser pumped or excimer laser pumped dye lasers (broadband) working at 423 nm for Ca and at 393 nm for Ca^+. The comprehensive observations of the Ca and Ca^+ layers were obtained by ground-based lidars at Kühlungsborn (54°N, 12°E), Germany between December 1996 and December 1998. Gerding et al. (2000) studied the data on 112 nights of Ca and 58 nights of Ca^+ during this period. The Ca layer has an average column abundance of $2.1 \times 10^7 \, cm^{-2}$ with a mean peak density of $22 \, cm^{-3}$ at 89.9 km altitude. The Ca^+ layer has a column abundance of $4.9 \times 10^7 \, cm^{-2}$ with a mean peak density of $29 \, cm^{-3}$ at 91.9 km. The high variability of the layer profiles was found as a basic feature of the Ca layer and Ca^+ layer. The column density of Ca is influenced by both annual and semiannual variations with a maximum in summer and autumn/winter. Ca^+ shows even more obvious short-time variations of sporadic layers. Gerding et al. (2000) also presented a one-dimensional steady-state chemistry model of the nighttime Ca and Ca^+ layers, based on the laboratory studies of CaO reaction kinetics. This model can reproduce satisfactorily the characteristic features of the annual mean layers and can provide a possible explanation for the unusual seasonal variation of the Ca layer, which exhibits a pronounced summertime enhancement around 87 km.

These metal layer observations and chemistry modeling provide us more and deeper understanding of the atmosphere constituents, their variations, and their relations to the temperature, solar activity, and atmosphere dynamics.

5.3.3.2. Meteor Trail Detection by Lidar

The daily influx of extraterrestrial matter into the Earth's upper atmosphere has been deduced to be of the order of 100 tons per day. The ionization and ablation of meteors give rise to a number of research topics including radar meteor echo, sporadic-E formation, and optical meteor investigation. In order to understand all these processes, it is necessary to know the origin, the influx rate, the chemical composition, size distribution, and incoming velocities of this extraterrestrial matter as well as the temporal and spatial variations of these properties. The lidar observations give us the capability to quantify the abundances of a number of elements in these meteor trails and to allow a characterization of the ablation processes. Meteoric ablation is also believed to be the major source of the neutral metal layers of Na, Fe, K, Ca, and Li, etc. Lidar studies of the meteor trails also seek further support for this belief. If this belief is indeed true, one would expect that the abundances of the metals in this region and their temporal and spatial variations should reflect, in some fashion, the properties of the incoming meteoroid population (von Zahn et al., 2002).

Lidar observations of five short-lived meteor trails were first reported by Beatty et al. (1988) with a broadband Na lidar at Urbana (40°N), Illinois. She et al. (1991) reported about a single meteor trail observed with a narrowband Na Doppler lidar. Kane and Gardner (1993b) reported a comprehensive study of 101 Na meteor trails and five Fe meteor trails observed during 4 years at Urbana, Illinois; Arecibo, Puerto Rico; and near Hawaii. They found that the vertical distribution of these 106 meteor trails is approximately Gaussian shape with a centroid height of 89.0 ± 0.3 km and an rms width of 3.3 ± 0.2 km. The 1996 Leonid meteor trails were studied by the IAP narrowband K lidar (Höffner et al., 1999) and a March 1997 meteor shower was studied by simultaneous and common-volume observations of a K and a Ca lidar (Gerding et al., 1999). Grime et al. (1999) reported a first two-laser-beam observation of a meteor trail on March 23, 1998 by a broadband Na lidar at Arecibo. The lidar was operating with two beams probing different volumes of the Na layer separated zonally by 15.7 ± 0.8 km. A single meteor trail was observed near 89 km altitude in both lidar field of views with a 310 ± 50 s temporal displacement. The observational separation suggests a westward zonal wind of 50 ± 10 m/s, while trail dispersion yielded an upper bound for the total diffusion coefficient of 2.6 ± 0.5 m^2/s, which is consistent with dispersion seen in other meteor trails.

The 1998 Leonid meteor shower was comprehensively studied by a few different lidars at different locations: Chu et al. (2000a) observed 18 Fe meteor trails over Okinawa (26.3°N, 127.7°E) with an airborne Fe Boltzmann temperature lidar; Chu et al. (2000b) observed seven persistent Na meteor trails at SOR (35°N, 106.5°W) with a steerable Na wind/temperature lidar coupled with a 3.5 m diameter optical telescope; von Zahn and Bremer (1999) observed the 1998 Leonid meteor trails by a cluster of K, Ca, and Fe lidars. Additional studies of the 1998 Leonid meteor showers observed at SOR using the Na lidar data and imagers were reported by Kelley et al. (2000), Grime et al. (2000), and Drummond et al. (2001). A comprehensive summary of the lidar observations of meteor trails was given by von Zahn et al. (2002).

The University of Illinois Fe Boltzmann temperature lidar was installed on the NSF/NCAR Electra aircraft and flew above Okinawa on the night of November 17/18, 1998 as part of the 1998 Leonid Multi-Instrument Aircraft Campaign (Chu et al., 2000a). During the 8-h Leonid flight, 18 meteor trails were observed with this Fe lidar. One of the most dense meteor trails probed by the airborne lidar over Okinawa is plotted in Figure 5.75 with the peak density reaching $2.33 \times 10^5 \, cm^{-3}$. The average altitude of the 18 trails from the high velocity (72 km/s) Leonid meteors, 95.67 ± 0.93 km, is approximately 6.7 km higher than previously observed for slower (~30 km/s) sporadic meteors (Kane and Gardner, 1993b). This height difference is consistent with the assumption that meteor ablates when atmospheric drag reaches a critical threshold. The average abundance of the trails is 10% of the abundance of the background Fe layer. Observations suggested that the 1998 Leonid shower did not have a significant impact on the abundance of the background Fe layer (Chu et al., 2000a).

SOR is located on the Kirtland Air Force Base near Albuquerque, New Mexico. The facility is operated by the Air Force Research Laboratory, Directed Energy Directorate and includes a 3.5-m astronomical telescope. The University of Illinois Na wind/temperature lidar is coupled to this telescope through the coude optics so that the beam can be pointed in any direction above 5° elevation. A 1-m diameter portion of

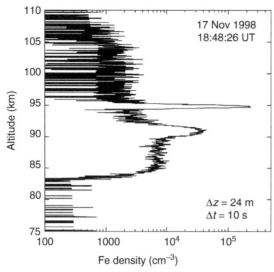

Figure 5.75 Fe meteor trail profile probed by the UIUC airborne lidar on November 17, 1998 over Okinawa. The vertical resolution is 24 m and integration period is 10 s. The dense narrow layer at 94.8 km is a meteor ablation trail, while the broader layer at 91 km is a sporadic Fe layer. (From Chu, X., Pan, W., Papen, G., Gardner, C.S., Swenson, G., and Jenniskens, P., Characteristics of Fe ablation trails observed during the 1998 Leonid meteor shower, *Geophys. Res. Lett.*, 27, 1807–1810, 2000a. With permission.)

the telescope primary mirror is used to project the laser beam while the remainder is used for collecting the backscattered light and focusing it onto the detector. The beam divergence is approximately 1 mrad. At 100 km range, the beam diameter (full width at e^{-2}) is 100 m. The lidar operates at 30 pps and the laser output power varies between 1 and 1.5 W. For this Leonid experiment, the range resolution was 24 m and the integration period was 15 s. The lidar was operated in the normal scanning mode to measure radial wind, temperature, and Na density profiles at zenith and 10° off-zenith to the north, east, south, and west. These data are used to derive profiles of the three wind components of the atmosphere as well as the temperature and Na density profiles. An observer was positioned outside the telescope dome to look for persistent trails associated with fireballs. When one was spotted, the telescope operator was apprised of the approximate azimuth and elevation angle and the telescope was moved to that position. By using a bore sighted video camera mounted on the telescope and the lidar profile data, the telescope operator positioned the lidar beam on the persistent trail. In this way, seven persistent trails were tracked and probed with the lidar for as long as 30 min. Visible CCD cameras, all-sky airglow imagers, and meteor radar also collected correlative data at SOR during the meteor shower. The persistent trails observed on 17/18 November at SOR were given descriptive names that characterized their visual appearances in the video and CCD images. Figure 5.76 is a false color CCD image of the one of

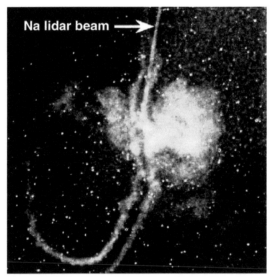

Figure 5.76 (Color figure follows page 398). CCD image of the persistent chemiluminescence trail (Diamond Ring) observed at 09:30 UT November 17, 1998 at the Starfire Optical Range, New Mexico. The straight line near the top center of the image is the UIUC Na lidar beam. (From Chu, X., Liu, A.Z., Papen, G., Gardner, C.S., Kelley, M., Drummond, J., and Fugate, R., *Geophys. Res. Lett.*, 27, 1815–1818, 2000b. With permission.)

these trails (Diamond Ring) observed at 09:30 UT on 17 November. The bright trail is chemiluminescence associated with a variety of emissions including Na and OH airglow (Kelley et al., 2000). The corresponding lidar profile for the diamond-ring trail is shown in Figure 5.77. The peak density of the Na trail reached 3.41×10^5 cm^{-3}. The duration time of the persistent Na meteor trails observed at SOR ranges from 3 min to more than 30 min, with average peak altitude of 94.0 $\pm$ 1.6 km (Chu et al., 2000b). The average Na abundance within the trails was 52% of the background Na layer abundance, which suggests the corresponding masses of the meteors were from 1 g up to 1 kg. Elevated temperatures were observed by the lidar in the bright chemiluminescent meteor

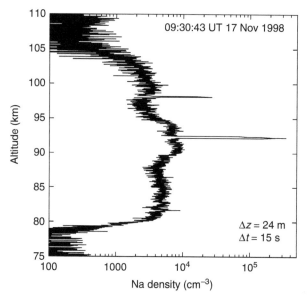

Figure 5.77 Na meteor trail profile of the Diamond Ring shown in Figure 5.76, probed by the UIUC Na wind/temperature lidar at SOR during 1998 Leonid meteor shower. The vertical resolution is 24 m and the integration period is 15 s. (From Chu, X., Liu, A.Z., Papen, G., Gardner, C.S., Kelley, M., Drummond, J., and Fugate, R., *Geophys. Res. Lett.*, 27, 1815–1818, 2000b. With permission.)

trails. Approximately 3 min after ablation, the temperature at the tube walls was about 20 to 50 K warmer than the tube core and background atmosphere (Chu et al., 2000b). Neither chemical nor frictional heating provides a satisfactory explanation for the observations.

5.3.4 Polar Mesospheric Clouds (Noctilucent Clouds)

5.3.4.1. Historical Perspective

Polar mesospheric clouds (PMCs) and their visual counterparts noctilucent clouds (NLCs) are the thin scattering layers (water–ice particles) occurring in the mesopause region (~85 km) at high latitudes (~poleward of 50°) in both hemispheres mainly during the three months surrounding summer solstice when the temperatures fall below the frost point (~150 K) (Chu et al., 2003). NLCs were first identified as a distinct atmospheric phenomenon in 1885 when the summernight unusually bright clouds came to people's attention throughout Europe, Great Britain, and Russia (Backhouse, 1885). The idea that PMC particles are composed of water ice was first suggested by Humphreys (1933), and then later taken up by Hesstvedt (1961, 1962), Chapman and Kendall (1965), Charlson (1965, 1966), Reid (1975), and Gadsden (1981). More comprehensive models including microphysics of ice formation were developed by Turco et al. (1982), Jensen and Thomas (1988, 1994), Jensen et al. (1989), Thomas (1996), Klostermeyer (1998, 2001), Rapp et al. (2002), Berger and von Zahn (2002), and von Zahn and Berger (2003a). Satellite observations provided the first evidence of the water–ice composition of PMC particles and the enhanced water vapor layer in the PMC region (Hervig et al., 2001; Summers et al., 2001; Stevens et al., 2001).

Due to the sudden appearance of NLCs in 1885 and their close relations to the mesopause conditions on temperature and water vapor (Thomas, 1991), PMCs/NLCs are generally regarded as a sensitive tracer of middle and upper atmospheric water vapor and temperature, and possibly as an indicator of long-term global climate change (Thomas, 1991; Avaste, 1993). Since the first observation of NLCs (Backhouse, 1885), PMCs/NLCs have been studied from ground by numerous visual, photographic, spectroscopic, photometric, and polarimetric observations for more than a century as reviewed by Gadsden and Schröder (1989). NLCs are generally visible only within the latitude range of 50 to 65° at night when the mesopause region is still sunlit. The silver-blue NLCs are typically cirrus-like clouds rippled by complex waves and billows, and can occasionally provide a spectacular visual display (Gadsden and Schröder, 1989). PMCs/NLCs have also been investigated by *in situ* rocket experiments (Witt, 1969; Gumbel and Witt, 1998) and by space observations from the satellites OGO-6, SME, NIMBUS 7, UARS, SPOT 3, MSX, and METEOSAT 5 (Donahue et al., 1972; Thomas and Olivero, 1989; Thomas et al., 1991; Evans et al., 1995; Debrestian et al., 1997; Carbary et al., 1999; Gadsden, 2000) in both hemispheres. Satellite observations in the 1970s led to the discovery of extensive scattering layers (PMCs) near 85 km over the entire north and south polar caps during local summers (Donahue et al., 1972). NLCs are believed to be the "ragged edges" of these much more persistent and extensive PMC layers that are visible at lower latitudes. The geographic coverage, occurrence frequency, and brightness of PMCs/NLCs have been increasing in the past decades. Recently, NLCs have been observed at mid-latitude such as Utah (41°N) (Taylor et al., 2002; Wickwar et al., 2002). The observed increases in the geographic extent and brightness of PMCs and NLCs in polar regions during the past decades have been related to the decreasing mesopause temperature and increasing mesospheric water vapor concentration possibly caused by the rising level of mesospheric CO_2 and CH_4, respectively (Thomas, 1996). The possible link of PMC to the global climate change has intensified the research on PMCs.

The passive optical observations from ground and satellite observations from space have made great contributions to the study of PMC/NLC. However, these observations either have limited vertical resolution of about 2 to 3 km or difficulties in covering the 24-h diurnal cycles. Although *in situ* rocket experiments can measure PMC vertical distribution precisely, they are so sporadic that they cannot be used to characterize the seasonal and diurnal variations in PMC properties. The active lidar sounding technology is unique in detecting and studying PMCs because of its high vertical resolution, daytime measurement capability, full seasonal and diurnal coverage, and atmospheric temperature and gravity wave measurement capability. The ability to detect PMC throughout the 24-h diurnal cycle is one of the key advantages that lidars have for PMC/NLC studies, since passive observations are limited to twilight conditions and satellite observations are either limited by their Sun-synchronous orbits or the need for illumination of the NLC region by the Sun.

The first lidar observation of NLCs in the northern hemisphere was made by Hansen et al. (1989) at Andoya (69°N, 16°E), Norway with the University of Bonn narrowband Na Doppler lidar, and the first lidar observations of PMCs in the southern hemisphere was made by Chu et al. (2001a,b) at the South Pole (90°S) with the University of Illinois Fe Boltzmann temperature lidar. Numerous lidar observations have been made in the northern hemisphere at the North Pole (90°N) and the Gulf of Alaska (58°N, ~150°W) (Gardner et al., 2001), Svalbard (78.2°N, 15.4°E) (Höffner et al., 2003), Andoya (69°N, 16°E) (Hansen and von Zahn, 1994; Langer et al., 1995; Nussbaumer et al., 1996; Lübken et al., 1996; von Cossart et al., 1997, 1999; von Zahn et al., 1998; von Zahn and Bremer, 1999; Baumgarten et al., 2002a,b; Fiedler et al., 2003), Sondrestrom (67.0°N, 50.9°W) (Thayer et al., 1995, 2003; Hecht et al., 1997; Gerrard et al., 1998), Poker Flat (65°N, 147°W) (Collins et al., 2003), Juliusruh (54.6°N, 13.4°E) (von Cossart et al., 1996), Kühlungsborn (54°N, 12°E) (Alpers et al., 2000, 2001), Aberystwyth (52.4°N, 4.1°W) (Thomas et al., 1994), and Logan (41.7°N, 111.8°W) (Wickwar et al., 2002). Lidar observations of PMCs have also been made in the southern hemisphere at the South Pole (90°S) (Chu et al., 2001a,b, 2003; Gardner et al., 2001), Rothera (67.5°S,

68°W) (Chu et al., 2004), and Davis (68.6°S, 78°E) (Klekociuk et al., 2003). These lidar observations provide precise information on PMC/NLC altitude, layer vertical structure, and backscatter coefficient as well as valuable information on particle size and shape, seasonal and diurnal variations, and wave signatures.

5.3.4.2. PMC Characteristics Measured by Lidar

The detection of PMCs/NLCs by lidar arises from the Mie scattering of transmitted laser photons by PMC ice particles. Without PMC particles in the mesopause region, the Rayleigh scattering from pure atmosphere molecules decreases exponentially with height due to the decreasing number density of atmosphere molecules. The Rayleigh scattering becomes negligibly small at altitudes above 80 to 85 km. When PMC particles present in the mesopause region around 85 km, the Mie scattering from the PMC water–ice particles that are much larger than atmosphere molecules is significantly stronger than the Rayleigh scattering at the same altitude. Thus, a sharp, narrow peak will stand out in the lidar photon count profile. Figure 5.27 provides an example of PMC profile obtained by the Fe Boltzmann temperature lidar at the South Pole on January 18, 2000. The photon counts have been converted to the volume backscatter coefficient. The narrow peak at 85.5 km is a PMC layer with peak volume backscatter coefficient approximate 8×10^{-9} m^{-1} sr^{-1}.

The lidar detection limit/threshold of PMCs is primarily determined by the volumetric backscatter properties of the PMC particles and the performance capabilities of the lidar system for background noise compression, laser wavelength, and power. The challenge of detecting PMCs in the polar region comes from the high solar background. Only when the PMC backscatter peak is higher than the background, is it possible to distinguish PMC from background noise. Narrowband filters are needed in the lidar receivers to effectively reject the sky background. The volume backscatter coefficient of PMC is approximately inversely proportional to the fourth power of light wavelength. Thus, the shorter wavelength,

such as the Fe lidar operating at 374 nm, will produce much
stronger PMC backscattering, compared to normal visible
light, such as the Rayleigh lidar operating at 532 nm and the
Na lidar at 589 nm. The background level and the PMC signal
level together determine the lidar detection sensitivity. The
combination of daytime measurement capability, short operation wavelength, and high output power is necessary to make
a lidar capable of PMC detection during the polar summer.
The volume backscatter coefficient of PMCs is approximately
proportional to the sixth power of the particle radius, which
means the larger PMC particles will produce stronger backscatter. For the lidar wavelength ranges from near UV to
visible light, PMC particles can be detected by lidar only
when their radius in excess of about 20 nm (von Cossart
et al., 1999). Therefore, lidars observe the larger radii of the
water–ice particles in the mesopause region.

Commonly, the parameters used in the lidar field to characterize PMCs are the volume backscatter coefficient, the
backscatter ratio, the total backscatter coefficient, the centroid altitude, and the layer rms width. The definition of
volume backscatter coefficient $\beta(z)$ is the same as Equation
(5.4). The backscatter ratio $R(z)$, the total backscatter coefficient β_{Total}, the centroid altitude Z_C, and the layer rms width
σ_{rms} are defined as

$$R(z) = \frac{\text{Total signal}(z)}{\text{Molecular signal}(z)} = \frac{\beta_{\text{PMC}}(z) + \beta_R(z)}{\beta_R(z)} \tag{5.129}$$

$$\beta_{\text{Total}} = \int \beta_{\text{PMC}}(z)\mathrm{d}z \tag{5.130}$$

$$Z_C = \frac{\sum_i z_i \beta_{\text{PMC}}(z_i)}{\sum_i \beta_{\text{PMC}}(z_i)} \tag{5.131}$$

$$\sigma_{\text{rms}} = \sqrt{\frac{\sum_i (z_i - Z_C)^2 \beta_{\text{PMC}}(z_i)}{\sum_i \beta_{\text{PMC}}(z_i)}} \tag{5.132}$$

The data processing methods and the equations used to compute these parameters from lidar photon count profiles could
be found in Chu et al. (2003).

5.3.4.2.1. Mean characteristics

Numerous lidar and satellite observations have shown that the PMC/NLC occurrence varies with latitude and season. Although PMCs/NLCs have been observed by lidar from the North Pole to 41°N and from the South Pole to 67.5°S, the PMC/NLC occurrence in polar latitudes is more frequent, persistent, and stronger while the mid-latitude PMCs/NLCs are less frequent, more sporadic, and weaker. PMCs/NLCs usually occur in the summer, from late May to the end of August in the northern hemisphere and from late November to the end of February in the southern hemisphere. This is because the PMCs contain water–ice particles that only form in the saturation region. The very small amount of water vapor in the mesopause region (typically 3 to 5 ppmv) requires very cold temperatures to reach saturation. Lidar, rocket, and satellite measurements have shown that the polar mesopause temperature is coldest in summer but warmer in winter. This is caused by the atmosphere dynamic interaction — upwelling in summer and downwelling in winter resulting from the global mean meridional circulation. These upwelling and downwelling motions result in adiabatic cooling and heating in summer and winter, respectively. The upwelling is strongest at the poles and becomes weaker towards lower latitudes, so the polar summer mesopause temperature is coldest in the atmosphere, often lower than 150 K in the polar region, which is usually the frost point at mesopause.

Systematic lidar measurements of PMCs can reveal the mean characteristics of PMC layers. Shown in Figure 5.78 are the histograms of the measured hourly mean PMC peak backscatter ratio $R_{\max}$, peak volume backscatter coefficient $\beta_{\max}$, total backscatter coefficient β_{Total}, layer centroid altitude Z_{C}, and layer rms width σ_{rms} obtained by the Fe Boltzmann temperature lidar during the 1999 to 2000 and 2000 to 2001 austral summer seasons at the South Pole (Chu et al., 2003a). More than 430 h worth of data were collected at the South Pole. The mean peak backscatter ratio, peak volume backscatter coefficient, total backscatter coefficient, centroid altitude, and layer RMS width are 58.4, $3.75 \times 10^{-9}\,\mathrm{m^{-1}\,sr^{-1}}$,

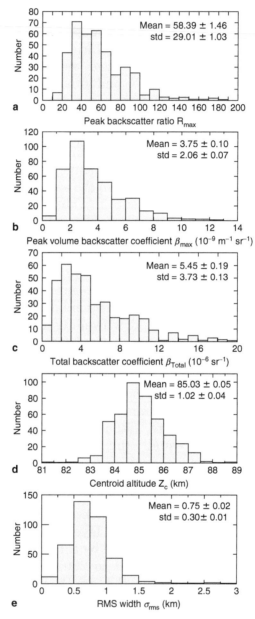

Figure 5.78 Histogram of PMC characteristics obtained by the Fe Boltzmann temperature lidar in the 1999 to 2001 summer seasons at the South Pole. (From Chu, X., Gardner, C.S., and Roble, R.G., *J. Geophys. Res.*, 108 (D8), 8447, doi: 10.1029/2002JD002524, 2003. With permission.)

$5.45 \times 10^{-6}\,\mathrm{sr}^{-1}$, 85.03 km, and 0.75 km, respectively. The overall PMC occurrence frequency at the South Pole from 1999 to 2001 is 67.4%. Besides the South Pole, extensive lidar measurements of PMCs/NLCs have been made at Rothera by the same Fe Boltzmann temperature lidar, at Svalbard by the IAP K Doppler lidar, and at Andoya and Sondrestrom by two Rayleigh lidars. The lidar detection of PMC/NLC made at other locations is much less frequent. A comparison of mean characteristics of PMCs observed by lidars at the South Pole, Rothera, Svalbard, Andoya, and Sondrestrom is given in Table 5.10.

When comparing the PMC brightness (expressed as the peak volume backscatter coefficient), the wavelength dependence of the backscatter coefficient must be taken into account. We assume a lognormal size distribution of spherical ice particles with radius of 40 nm and a width of 1.4 as used in Höffner et al. (2003). Then the color ratios of 374/770 and 374/532 nm are 7.3 and 2.3, respectively. These two ratios are used when converting the 532 nm (for Andoya and Sondrestrom) and 770 nm (for Svalbard) backscatter coefficient to 374 nm as of the South Pole and Rothera. The PMC layer's FWHM was converted to rms width by assuming a Gaussian distribution of the layer vertical structure so that $\mathrm{FWHM} = \sqrt{8\ln 2}\,\sigma_{\mathrm{rms}}$. Most of the locations exhibit PMCs around 83 to 85 km. Table 5.10 indicates that the PMCs at higher latitudes (such as the South Pole and Svalbard) usually have higher occurrence frequency, higher brightness, and higher centroid altitude compared to lower latitude PMCs.

Table 5.10 Comparison of Mean Characteristics of PMC Observed by Lidars at the South Pole, Rothera, Svalbard, Andoya, and Sondrestrom

	South Pole (90°S)	Rothera (67.5°S)	Svalbard (78°N)	Andoya (69°N)	Sondrestrom (67.5°N)
$\beta_{\max}$ ($\times 10^{-9}\,\mathrm{m}^{-1}\,\mathrm{sr}^{-1}$)	3.75	0.98	2.85	2.19	2.17
Z_{C} (km)	85.03 ± 0.05	83.7 ± 0.25	83.6	83.3	82.9
σ_{rms} (km)	0.75 ± 0.02	0.83 ± 0.06	0.72	0.51	0.38
Occurrence frequency (%)	67.4	17.3	77	43	20

5.3.4.2.2. Geographic difference in PMC altitudes

A significant hemispheric difference in PMC altitudes was found through our lidar observations over both poles (Chu et al., 2001a, 2003a; Gardner et al., 2001). The mean altitude, computed from 437 h of PMCs observed in two seasons at the South Pole, is 85.03 km, which is approximately 2 km higher than the commonly observed PMC altitude (∼83 km) in the northern hemisphere. To further explore the geographic differences in PMC altitudes, we plot in Figure 5.79 the mean PMC altitudes measured by lidars and ground-based triangulation versus latitude for both the southern and northern hemispheres. The lidar data were from the South Pole (90°S) (Chu et al., 2003); Davis (68.6°S) (Klekociuk et al., 2003); Rothera (67.5°S) (Chu et al., 2004); North Pole

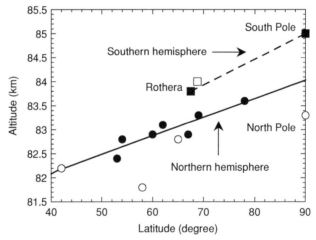

Figure 5.79 PMC altitude versus latitude in both hemispheres. Circles are for the northern hemisphere, and squares are for the southern hemisphere. Filled circles indicate large data sets and are used to obtain the linear fit (solid line). Open circles indicate limited data sets and are excluded in the fit. The open square is for the Davis data. See text for references. (From Chu, X., Nott, G.J., Espy, P.J., Gardner, C.S., Diettrich, J.C., Clilverd, M.A., and Jarvis, M.J., *Geophys. Res. Lett.*, 31, L02114, doi: 10.1029/2003GL018638, 2004. With permission.)

(90°N), and Gulf of Alaska (58°N) (Gardner et al., 2001); Svalbard (78°N) (Höffner et al., 2003); Andoya (69°N) (Fiedler et al., 2003); Sondrestrom (67°N) (Thayer et al., 2003); Poker Flat (65°N) (Collins et al., 2003); Juliusruh and Kühlungsborn (55/54°N) (von Cossart et al., 1996; Alpers et al., 2000, 2001); and Logan (42°N) (Wickwar et al., 2002). Three other northern data points were adapted from von Zahn and Berger (2003b), in which they summarized the triangulation measurements of NLC altitudes from several sources including Grahn and Witt (1971) for 62°N, Gadsden and Schröder (1989) for 60°N, and Jesse (1896) for 53°N. The lidar data at 90°S, 67.5°S, 78°N, 69°N, 67°N, 65°N, and 55/54°N were averaged over many hours for one or more seasons. There were only a few hours of observations at 90°N, 65°N, 58°N, and 42°N. These four points (open circles) were excluded in the linear fit (solid line) to the northern PMC altitudes, but are plotted for reference. The data at 90°N, 65°N, and 58°N fall below this trend line, while the data from 42°N lie on the line. The two southern points (filled squares) were obtained from many days of measurements by the same Fe lidar at the South Pole and Rothera. The single Davis (68.6°S) observation (open square), obtained on January 9, 2002, is slightly above the dashed line linking the Rothera and South Pole means. Both the southern and northern hemisphere trend lines indicate that PMC altitudes are generally higher at higher latitudes. What is more interesting in Figure 5.79 is that the southern line is significantly higher than the northern line. The altitude difference between the two hemispheres is about 1 km for similar latitudes. We conclude from these measurements that there is a significant inter-hemispheric difference in PMC altitudes. The southern hemisphere PMCs are about 1 km higher than their northern counterparts observed at similar latitudes.

The formation and evolution of PMC particles are a complex process that depends on the temperature, water vapor, and vertical wind structures of the mesopause region. PMC particles actually consist of water ice, and the water ice particles are only formed in the supersaturation region where water vapor pressure P_w is higher than saturation water

vapor pressure P_s, or in other words, the atmosphere temperature T is lower than the frost point T_s. Once they fall into the
subsaturation region where $P_w < P_s$, the ice particles will be
destroyed quickly by sublimation due to higher temperatures
at lower altitudes (Thomas, 1991). Therefore, PMC altitudes
will be determined primarily by the altitude of the supersaturation region. To study the possible causes for this hemispheric difference in PMC altitudes, TIME-GCM simulation
data are used to represent the temperature, water vapor, and
vertical wind structures of atmosphere at the poles. The
supersaturation regions at the South and the North Poles
predicted by TIME-GCM are plotted in Figure 5.80 along
with the two seasons of PMC data at the South Pole. The
lines denote the boundary where the atmosphere temperature
is equal to the frost point. The scattered points in Figure 5.80

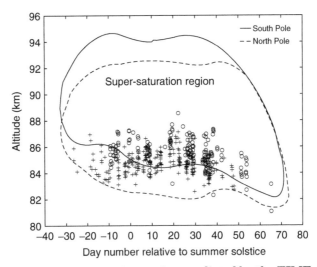

Figure 5.80   Supersaturation region predicted by the TIME-GCM at
the South and North Poles. The lines denote the boundary where the
atmosphere temperature is equal to the frost point. Scattered data
points are South Pole PMC data obtained by the Fe Boltzmann temperature lidar in (o) 1999 to 2000 season and (+) 2000 to 2001 season.
(From Chu, X., Gardner, C.S., and Roble, R.G., *J. Geophys. Res.*, 108
(D8), 8447, doi: 10.1029/2002JD002524, 2003. With permission.)

are the lidar data of PMC at the South Pole. They mainly stay within the saturation region clustered along its bottom. Comparing the saturation regions at the South and North Poles, we find that the South Pole saturation region is generally 2 km higher than the North Pole saturation region. This hemispheric difference in super-saturation region altitudes is approximately equal to the hemispheric difference in PMC altitudes found by the lidar measurements over both poles. Once PMC particles form in the saturation region around mesopause, under gravitational pull, the PMC particles will slowly settle down through the supersaturation region while growing in size and mass. The particles also experience a buoyancy force mainly due to the upwelling atmosphere over the summer pole. When the buoyancy force of the upwelling atmosphere balances the gravitational force on the PMC particle, PMC particles will stay around that altitude, which is several kilometers below mesopause, and PMC layers will be observed there. Because atmospheric density is smaller at higher altitude and this buoyancy provided by the upwelling air is proportional to the product of atmospheric density and vertical wind, a higher vertical wind is required to maintain the South Pole PMCs staying at 2 to 3 km higher altitudes than the North Pole if PMC particle mass and drag coefficient are similar at both poles. Checking with the TIME-GCM predictions, the vertical wind on the 13th day after summer solstice is about 3.5 cm/s around 85 km at the South Pole, which is larger than the vertical wind of 2.5 cm/s around 83 km at the North Pole. The South Pole mesopause ($\sim$91 km) is about 2 km higher than the North Pole mesopause ($\sim$89 km), just like the South Pole saturation region is about 2 km higher than that of the North Pole, as shown in Figure 5.80. The PMC layers are higher at the South Pole because they begin forming in the higher saturation region near the higher mesopause. As they grow in size and slowly fall to lower altitudes within the supersaturation region, the larger vertical wind of upwelling atmosphere at the South Pole would provide sufficient buoyancy to maintain the PMC particles at altitudes that are about 2 km higher than at the North Pole. Thus, the South Pole PMCs are expected to be higher than the North Pole PMC as we observed. Since the upwelling is strongest at the pole and becomes weaker towards lower latitudes, it is not surprising that the PMC appears to be lower at lower latitudes as shown in Figure 5.79.

Checking with the simulation of TIME-GCM, the hemispheric differences in mesopause altitudes and upwelling wind are mainly caused by the solar flux in January being 6% greater than the solar flux in July because of the Earth's orbital eccentricity where the Earth is closest to the Sun on January 3 and farthest from the Sun on July 5. The increased solar heating at the South Pole during austral summer results in a warmer stratopause that causes a dynamic adjustment. The dynamic adjustment results in higher mesopause altitude and stronger upwelling vertical winds in summer at the South Pole than at the North Pole. These differences cause the supersaturation region to be at higher altitudes over the South Pole than over the North Pole (Chu et al., 2003). Besides the hemispheric differences in PMC altitudes, the lidar observations we made in two different years also indicate major inter-annual differences that the PMC in 1999 to 2000 season is about 0.8 km higher than the 2000 to 2001 season. Undoubtedly, the model solar forcing variation alone cannot account for the observed inter-annual variability at the South Pole. It is most likely to be dynamically forced and related to gravity wave and planetary wave variability. Nevertheless, it is believed that the main process responsible for the hemispheric differences is related to the difference in solar heating caused by the Earth's orbital eccentricity and that there is considerable dynamic variability superimposed the measurements.

5.3.4.2.3. *Inter-annual and seasonal variations*

The backscatter coefficient and centroid altitude of PMC observed at the South Pole exhibit considerable seasonal variations. Figure 5.81 shows the smoothed total backscatter coefficient, centroid altitude, and rms width for the two-season data taken at the South Pole. The smoothing full width is 11 days. It clearly shows that the 2000 to 2001 season PMCs are

Figure 2.1 Terawatt white-light beam in the atmosphere. Photograph taken at the Jena institute. The upper edge of the planetary boundary layer (∼1.5 km) is visible via enhanced aerosol scattering (Terramobile©).

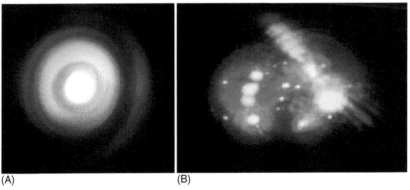

(A)　　　　　　　　　　(B)

Figure 2.5 (A) Filament image on a screen (laser energy 8 mJ, pulse duration 120 fsec) showing conical emission. (B) Multiple filamentation in a TW beam (350 mJ, 80 fsec).

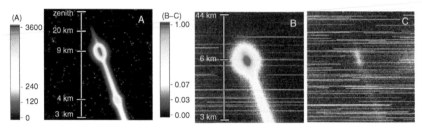

Figure 2.7 Long-distance filamentation and control of nonlinear optical processes in the atmosphere. Pictures of the fsec-laser beam propagating vertically, imaged by the CCD camera of the 2-m telescope at the Thüringer Landessternwarte Tautenburg. (A) Fundamental wavelength, visible up to 25 km through two haze layers appearing as multiple-scattering halos. (B and C) Blue light (390 to 490 nm) generated by two 600 fsec pulses of same initial peak power, with negative (GVD precompensating) and positive chirps. The stripes across the images are due to the star motion during the acquisition time of several minutes. Note that the white-light generation requires GVD precompensation. (Derived from Kasparian J. et al., *Science*, 301, 61, 2003. With permission.)

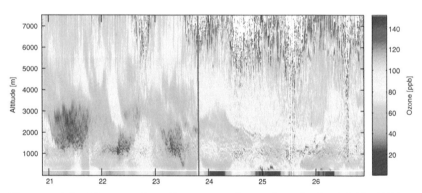

Figure 4.6 Time series of ozone performed with the EPFL DIAL mobile system located in St Chamas, France, in summer of 2001.

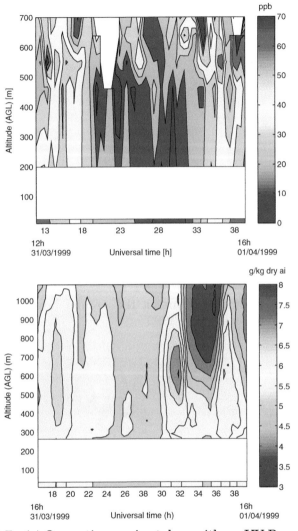

Figure 4.7 (a) Ozone time series taken with an UV Raman DIAL [82]. The ground valves are measured by a UV photometer. The Raman DIAL has shorter operational range mainly because of the weaker signals, but improved aerosal immunity and simpler transmitter design, allowing simultaneous daytime water vapour (b) measurements with minimal additional equipment.

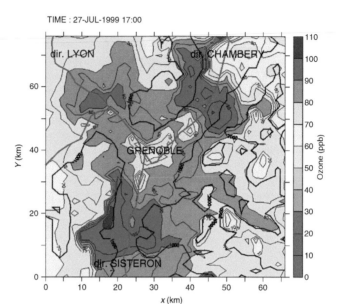

Figure 4.9 Predicted ozone in the ground layer of the model domain at 5 pm on July 27, 1999.

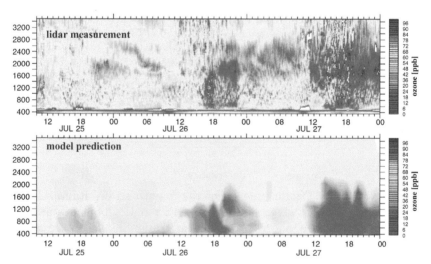

Figure 4.10 Direct comparison between lidar measurements and predicted results of a 3-day time series of ozone formation over Vif, south of Grenoble, from July 25 to July 27, 1999.

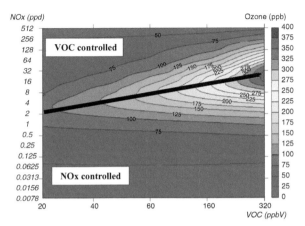

Figure 4.12 Predicted ozone concentrations versus the primary emissions of VOCs and NO_x. Two regimes of ozone production are highlighted: one below the straight line where lower concentration of ozone will be produced only if reduced NO_x is emitted ("NO_x controlled regime"), and one above it where it is only applying VOCs reduction strategies that will lower the ozone production ("VOCs controlled regime").

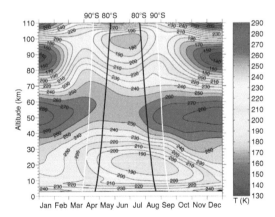

Figure 5.45 The observed weekly mean temperature structure of the atmosphere above South Pole (UISP-02) plotted from 3 to 110 km. Polar nights (24 h darkness) occur between the white curve at 90°S and between the black curves at 80°S. The vertical resolution is 500 m. (From Pan, W., and Gardner, C.S., *J. Geophys. Res.*, 108 (D18), 4564, doi: 10.1029/2002JD003217, 2003. With permission.)

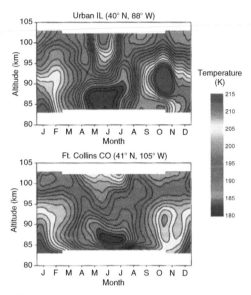

Figure 5.47 Contour plots of the seasonal variations in the mesopause region temperature structure at Urbana, Illinois and Fort Collins, Colorado. The contour interval is 2 K. The plots were derived from the nightly mean temperature profiles by smoothing the data vertically and seasonally using Hamming windows with full widths of 2 km and 45 days, respectively. (From Senft, D.C., Papen, G.C., Gardner, C.S., Yu, J.R., Krueger, D.A., and She, C.Y., *Geophys. Res. Lett.*, 21, 821–824, 1994. With permission.)

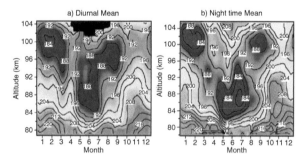

Figure 5.49 Contour plot of the mesopause region annual temperature structure at Urbana, Illinois using (a) data covering the complete diurnal cycle and (b) data covering just the nighttime period. (From States, R.J., and Gardner, C.S., *J. Atmos. Sci.*, 57, 66–77, 2000. With permission.)

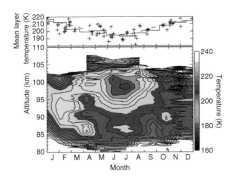

Figure 5.56 Plotted in the upper panel are the nightly (+) and monthly (—) average temperatures along with a harmonic fit (- - -). Shown in the lower panel is the seasonal variation of mesopause region temperature above Arecibo. (From Friedman, J.S., *Geophys. Res. Lett.*, 30 (12), doi: 10.1029/2003GL016966, 2003. With permission.)

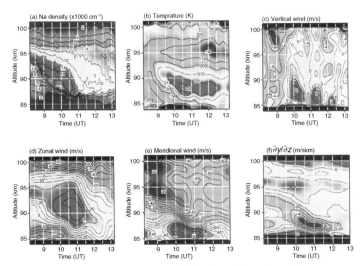

Figure 5.59 Contour plots of (a) Na density, (b) temperature, (c) vertical wind, (d) zonal wind, (e) meridional wind, and (f) vertical shear of meridional wind as functions of time and altitude on October 21, 1993 at Haleakala, Maui by the UIUC Na wind/temperature lidar. To generate these plots the data were smoothed to resolution of 30 min and 1 km. (From Tao, X., and Gardner, C.S., *Geophys. Res. Lett.*, 22, 2829–2832, 1995. With permission.)

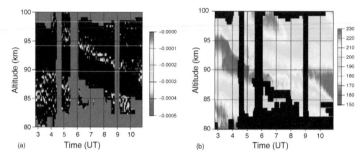

Figure 5.63 (a) Convectively unstable regions for February 18, 1999, determined by N^2 calculated from 500-m 1.5-min resolution data take by the UIUC Na wind/temperature lidar at Starfire Optical Range, New Mexico. The black region is where $N^2 > 0$; colored region is where $N^2 < 0$. No data are available in the purple region. (b) Temperature for the same night. (From Zhao, Y., Liu, A.Z., and Gardner, C.S., *J. Atmos. Sol. Terr. Phys.*, 65, 219–232, 2003. With permission.)

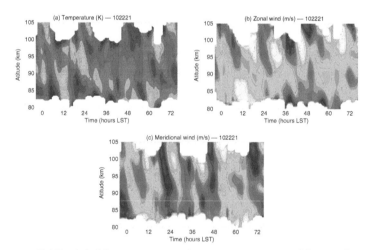

Figure 5.65 (a) Mesopause region temperature, (b) zonal wind, and (c) meridional wind obtained by the CSU Na wind/temperature lidar between August 9 and 12, 2002 over Fort Collins, Colorado (41°N, 105°W). The origin of x-axis is 0:00 hours on August 9, 2002, LST. The ranges contours shown are approximately between (a) (160 K, 220 K), (b) (−90 m/s, 120 m/s), and (c) (−90 m/s, 90 m/s). (From She, C.Y., *J. Atmos. Sol. Terr. Phys.*, in press, 2003. With permission.)

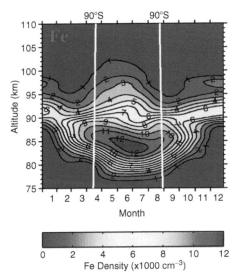

Figure 5.72 Seasonal variations of Fe density observed by the UIUC Fe Boltzmann temperature lidar at the South Pole (90°S).

Figure 5.76 CCD image of the persistent chemiluminescence trail (Diamond Ring) observed at 09:30 UT November 17, 1998 at the Starfire Optical Range, New Mexico. The straight line near the top center of the image is the UIUC Na lidar beam. (From Chu, X., Liu, A.Z., Papen, G., Gardner, C.S., Kelley, M., Drummond, J., and Fugate, R., *Geophys. Res. Lett.*, 27, 1815–1818, 2000. With permission.)

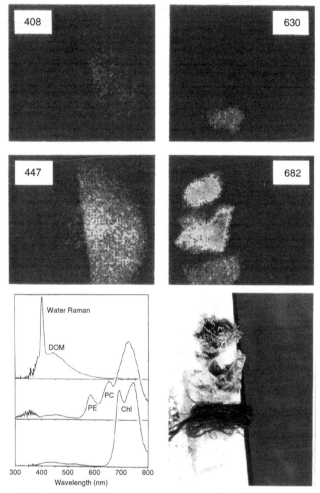

Figure 6.11 Multicolor imaging of three types of benthic vegetation, extracted from Mediterranean water and placed on a concrete quay next to the deep water. A photograph of the algae is shown in the lower right part of the figure, and typical spectra for water (top), the red alga (middle), and the two green algae (bottom) are given to the left of the photograph (DOM, distributed organic matter; PE, phycoerythrine; PC, phycocyanine; Chl, chlorophyll). Individual fluorescence images recorded at the indicated center wavelengths (in nm) are seen in the upper part of the figure. (From Alberotanze, L. et al., *EARSEL Adv. Remote Sens.*, 3, 102, 1995. With permission.)

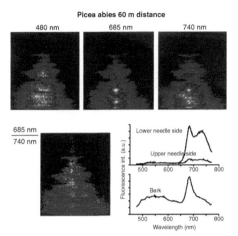

Figure 6.12 Multicolor pushbroom fluorescence imaging of a spruce tree in 60 m distance form the lidar system. The three upper-row images were simultaneously recorded by successively scanning line by line the linearly dispersed excitation light across the tree. A ratio image is also shown, as well as illustrations of point spectra from the same object. (From Johansson, J. et al., *J. Plant Physiol.*, 148, 632, 1996; Svanberg, S., *Optics and Photonics News*, September 1996, p. 19. With permission.)

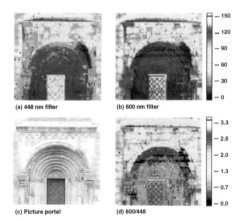

Figure 6.14 Fluorescence lidar imaging of the northern gate of the Lund Cathedral. A dual-filter approach has been taken, and individual images as well as a ratio image are shown. (From Weibring, P. et al., *Appl. Opt.*, 40, 6111, 2001. With permission.)

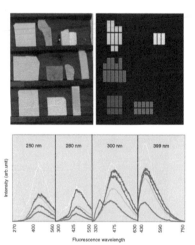

Figure 6.15 Arrangement of Italian marble samples studied by a fluorescence lidar system. Four different excitation wavelengths were employed to perform a multivariate discrimination between the different samples. Stone spectra are shown in the lower panel in different shades, also used to indicate corresponding evaluated pixels in the processed image (upper right). (From Weibring, P. et al., *Appl. Opt.*, 42, 3583, 2003. With permission.)

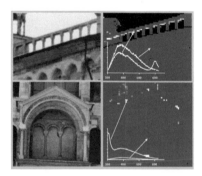

Figure 6.16 Data from fluorescence lidar imaging at the Parma Cathedral and Baptistery (Italy). The lower sector pertains to a section of the Cathedral, and the fluorescence image is processed for the detection of surface chemical treatment. The upper sector shows data from the Baptistery and here the fluorescence recordings are processed to detect the presence of biodeteriogens. (From Weibring, P. et al., *Appl. Opt.*, 42, 3583, 2003; Lognoli, D., *Appl. Phys. B*, 76, 457, 2003. With permission.)

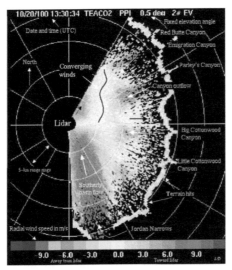

Figure 7.54 Plan-position indicator (PPI) scan at 0.5° elevation of the radial Doppler wind velocity measured by the NOAA ETL TEA CO₂ coherent lidar positioned near the center of the Salt Lake City basin on October 20, 2000. The circular grid lines represent 5 km range intervals. A convergence zone is indicated to the northeast of the lidar at ranges between about 5 and 12 km.

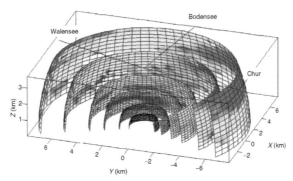

Figure 7.55 Radial wind velocity measured by the TWL system during the 1999 Mesoscale Alpine Program shown on the surface of a series of concentric sections of spheres at various ranges. The radial wind velocities are in the range of ± 15 m/sec. The negative velocities (toward the lidar) are in blue, the positive velocities (away from the lidar) are in red.

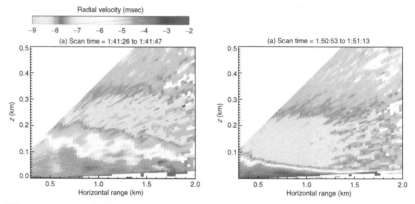

Figure 7.56 NOAA HRDL RHI scan display from the CASES-99 campaign showing passage of a strong boundary layer density current. The data on the right was taken 10 min after the data on the left. Velocities are given by the color bar at the top of the left panel. Negative radial velocities indicate flow towards the lidar.

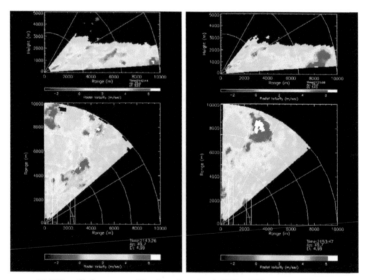

Figure 7.57 Coherent Technologies' WindTracer RHI and PPI radial velocity measured at the Dallas Fort Worth Airport in July, 2000. The horizontal range is 10 km and the velocity scale is −3 to +7 m/sec. The data in the right panel were taken 10 min after the data in the left panel.

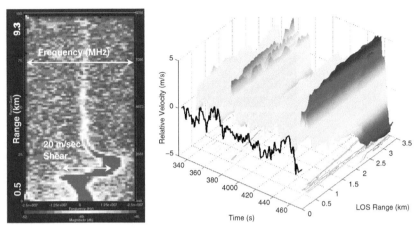

Figure 7.64 Forward-looking ACLAIM CAT detection: range-resolved Doppler spectrum from the 1998 Electra flight tests (left) and radial velocity time series onboard NASA's DC-8 from the 2001 CAMEX-IV flight tests compared to the *in situ* true air speed sensor (right).

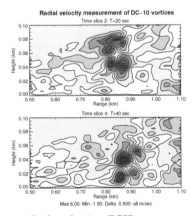

Figure 7.66 Two radial velocity RHI scans, separated by 20 sec, through a pair of DC-10 wake vortices from Coherent Technologies 2 μm pulsed coherent Doppler lidar. The scan was perpendicular to vortex axis and the core's dipole signature is clearly evident in the data. Red colors represent velocities away from the lidar and blue toward the lidar, The velocity contours are separated by 0.5 m/sec.

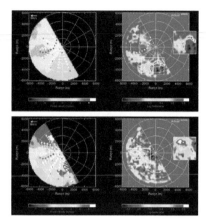

Figure 7.69 WindTracer radial velocity (left) and aerosol back-scatter coefficient (right) estimates from measurements conducted at Dugway Proving Ground, Utah in September, 2000 over a 12 km diameter circular area. The time difference between the data in the top and bottom panels is approximately 20 min. The ability to detect, track, and predict the aerosol plume is clearly evident in the data.

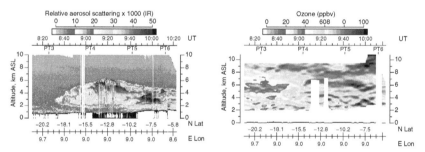

Figure 8.1 Latitudinal distribution of ozone over western Pacific Ocean obtained during Pacific Exploratory Mission (PEM West B) in February to March 1994.

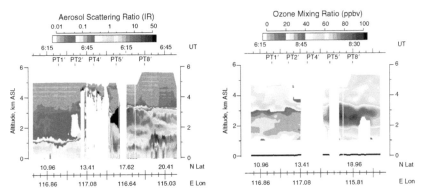

Figure 8.2 Biomass burning plume over Atlantic Ocean arising from biomass burning in central part of western Africa observed on October 14, 1992, during the TRACE-A mission.

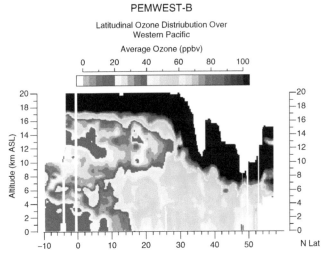

Figure 8.3 Pollution outflow from China over South China Sea (right side of figure) with clean tropical air on south side of front (left side of figure) observed on February 21, 1994, during PEM West B.

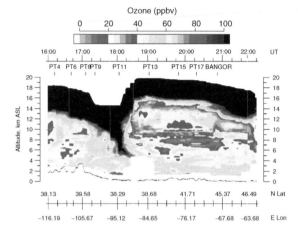

Figure 8.4 Ozone distribution observed on October 13, 1997, on a flight across the United States during the SASS (Subsonic Assessment) Ozone and Nitrogen Experiment (SONEX). Stratospheric intrusion clearly evident on the left side of the figure, and low ozone air from tropics transported to midlatitudes can be seen in the upper troposphere on the right.

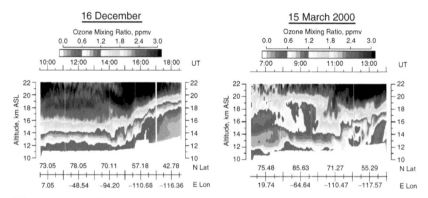

Figure 8.5 Ozone cross sections in stratosphere measured in winter of 1999/2000 during SOLVE mission. The change in ozone number density in the Arctic polar vortex due to chemical loss during winter is clearly evident at latitudes north of 72°N.

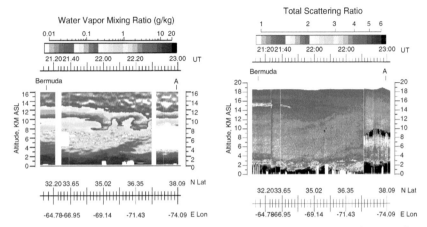

Figure 8.6 LASE measurements of water vapor (left) and aerosols and clouds (right) across troposphere on an ER-2 flight from Bermuda to Wallops on July 26, 1996, during Tropospheric Aerosol Radiative Forcing Experiment (TARFOX).

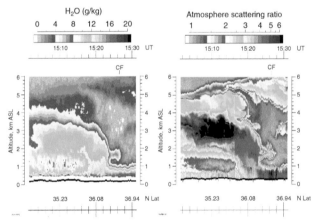

Figure 8.7 Water vapor and aerosol cross section obtained on July 14, 1997, on a flight across a cold front during Southern Great Plains (SGP) field experiment conducted over Oklahoma.

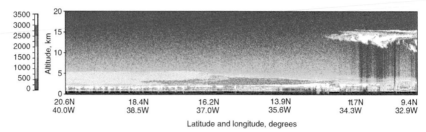

Figure 9.5 LITE 532-nm nighttime raw data showing Saharan dust over the Atlantic Ocean on Sepember 17, 1994. A layer of dust (below 5 km) is located above a cloud-capped marine boundary layer (≈1 km). At the right, a layer of cirrus is seen at an altitude of 15 km.

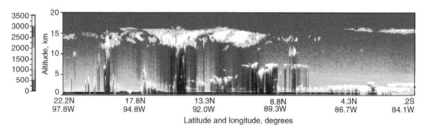

Figure 9.6 LITE 532-nm nighttime raw data showing multiple cloud layers observed over the tropical Pacific Ocean on September 14, 1994.

Figure 9.11 Simulated CALIOP 532-nm observations (parallel channel).

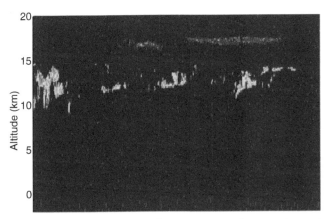

Figure 9.12 Simulated CALIOP 532-nm observations (perpendicular channel).

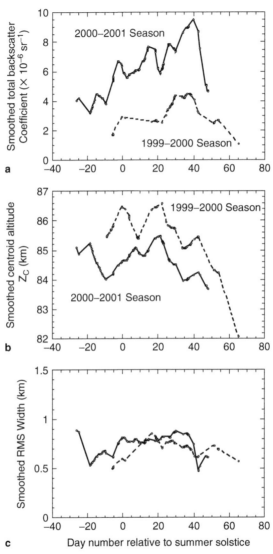

Figure 5.81 Eleven-day smoothed PMC parameters versus day number at the South Pole: (a) smoothed total backscatter coefficient; (b) smoothed centroid altitude; and (c) smoothed RMS width. Circles are the smoothed data points, dashed line is for the 1999 to 2000 season and solid line is for the 2000 to 2001 season. (From Chu, X., Gardner, C.S., and Roble, R.G., *J. Geophys. Res.*, 108 (D8), 8447, doi: 10.1029/2002JD002524, 2003. With permission.)

much brighter than the ones in the 1999 to 2000 season, and both seasons reach maximum brightness in the period of 25 to 40 days after summer solstice, which is also the period when the PMC occurrence probabilities are maximum during both seasons.

Significant seasonal trends emerge from the scattered PMC altitude data as shown in Figure 5.81(b). During both seasons, since 10 days before summer solstice, the altitudes generally increase as the season progresses, reach maximum values around 10 to 20 days after summer solstice, and then decrease throughout the remainder of the season. PMC altitudes exhibit similar descending rate from about 18 to 50 days after solstice in both seasons. PMC altitudes decrease by about 64 m/day from day 18 to day 53 after solstice in the 1999 to 2000 season (Chu et al., 2001a), while the decreasing rate is 65 m/day from day 18 to day 48 for the 2000 to 2001 season data. At the beginning of the 2000 to 2001 season, PMC altitudes start at a relatively high value of 85 km, and then decrease to slightly below 84 km around 10 days before solstice. Due to the lack of data in that period, this feature does not appear in the 1999 to 2000 season data. There is no obvious seasonal trend in the PMC layer rms width as shown in Figures 5.81(c). The rms width is approximately homogeneous through the PMC seasons.

The PMCs at Andoya (69°N) do not show the same seasonal trends as the South Pole (Fiedler et al., 2003). Similar analysis for PMC seasonal variations has not been done for other locations. It is an open question whether the seasonal variations of PMC at lower latitudes are different from the South Pole. Further investigations are needed. If they are indeed different, it would indicate differences in atmosphere dynamics between the pole and lower latitudes.

5.3.4.2.4. Diurnal variations

The full-diurnal-coverage of lidar detection of PMCs allows the study of PMC diurnal variations. Significant diurnal variations have been observed in PMC backscatter coefficients and altitudes at Andoya (69°N) (von Zahn et al., 1998) and at the South Pole (Chu et al., 2001b) as shown in Figure 5.82.

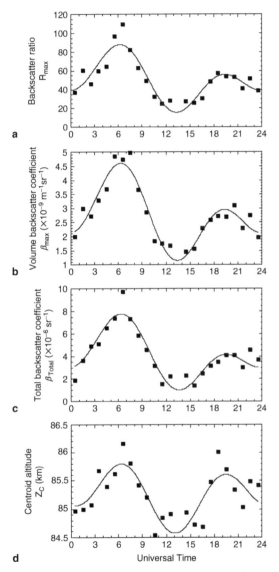

Figure 5.82 Diurnal variations of the South Pole PMC (a) peak backscatter ratio, (b) peak volume backscatter coefficient, (c) total backscatter coefficient, and (d) layer centroid altitude plotted versus UT. The solid curves are the 12 and 24 h harmonic fits. (From Chu, X., Gardner, C.S., and Papen, G., *Geophys. Res. Lett.*, 28, 1937–1940, 2001b. With permission.)

A clear in-phase relationship between backscatter coefficients and altitudes was found at the South Pole (Chu et al., 2001b). In contrast, a clear out-of-phase relationship was observed at Andoya (von Zahn et al., 1998). The observed diurnal variation at Andoya was attributed to a stable semidiurnal tide (von Zahn et al., 1998), while the diurnal variation at the South Pole was attributed to tidal variations in temperature and vertical wind caused by a zonally symmetric tide (Chu et al., 2001b).

More data obtained in the second season at the South Pole allow us to further investigate the diurnal variations. An overall average of all data in both seasons versus UT hour shows that there is an in-phase relation for the 12-h oscillation and an out-of-phase relation for the 24-h oscillation between PMC backscatter coefficient and centroid altitude. The backscatter coefficient variation has similar amplitudes in semidiurnal and diurnal variations, but the altitude variation has a larger amplitude in diurnal variation than in semidiurnal variation. The observed overall diurnal oscillations at the South Pole may be explained as the following: the positive-correlation between backscatter coefficient and the altitude is caused by the in-phase semidiurnal variation, while the anti-correlation between the maxima of backscatter coefficient and altitude is determined by the out-of-phase diurnal variation.

The PMC total backscatter coefficients are plotted versus PMC centroid altitude in Figure 5.83(a) for the South Pole. The circles are for 1999 to 2000 season, and the crosses are for 2000 to 2001 season. The plot is quite different from those observed at Andoya shown in Figure 5.83(b) (Fiedler et al., 2003). At Andoya, the largest backscatter coefficients are observed at the lowest altitudes, which is consistent with the current theory of PMC particles forming at higher altitudes near the mesopause and then falling to lower altitudes as they grow in size and mass. At the South Pole, the distribution is roughly symmetric, and PMCs exhibit the largest backscatter coefficients near the center of the altitude distribution around 85 km. The backscatter coefficients tend to be smaller both above and below the mean altitude. The aforementioned

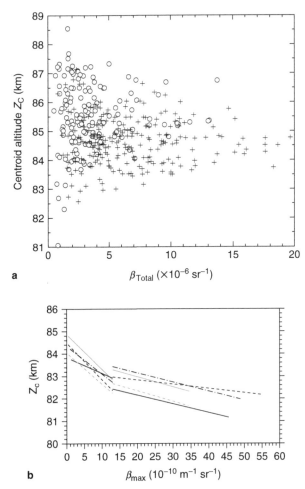

Figure 5.83 (a) PMC total backscatter coefficients versus centroid altitudes for the 1999 to 2000 (○) and 2000 to 2001 (+) summer seasons at the South Pole. (From Chu, X., Gardner, C.S., and Roble, R.G., *J. Geophys. Res.*, 108 (D8), 8447, doi: 10.1029/2002JD002524, 2003.) (b) Relation between altitude and brightness of PMC above ALOMAR at Andoya (69°N, 16°E), Norway. Linear fits for each year are shown: 1997 (black solid line), 1998 (black dashed line), 1999 (black dashed–dotted line), 2000 (gray dashed line), and 2001 (gray solid line). The brightness has been subdivided into two ranges, and the border between them is $\beta_{max} = 13 \times 10^{-10}\,\mathrm{m}^{-1}\,\mathrm{sr}^{-1}$. (From Fiedler, J., Baumgarten, G., and von Cossart, G., *J. Geophys. Res.*, 108 (D8), 8453, doi: 10.1029/2002JD002419, 2003. With permission.)

differences observed between the South Pole and lower latitudes need modeling investigations to understand.

5.3.4.2.5.   Wave signature in PMCs

Gerrard et al. (1998) observed wave structures in NLCs with a period of 2 to 3 h, and wave perturbations in stratosphere at the same time. Their data indicate that the increasing stratospheric wave perturbations may cause the decreasing of NLC backscatter values. They suggested that wave activity has a negative effect on NLCs (either by inducing sublimation on existing NLC particles, reducing their radius, or by limiting the creation of new NLC particles). Hecht et al. (1997) also observed an NLC sublimating due to the passage of an acoustic gravity wave through 85 km. Thayer et al. (2003) reported that gravity-wave signatures were routinely observed in the NLC detection. Upon estimating stratospheric wave activity in the lidar data, they observed stronger cloud backscatter during low gravity-wave activity and weak cloud backscatter during high gravity-wave activity.

5.3.4.2.6.   PMC particle size, number density, and shape

Multicolor lidar observations can determine PMC/NLC particle size based upon the assumptions of spherical water ice particles and monomodal lognormal size distribution. Most results show that the diameter of PMC particles is 20 to 51 nm, the width is 1.2 to 1.6, and the number density is 82 to $610 \, cm^{-3}$ (von Cossart et al., 1997, 1999; Alpers et al., 2000). Alpers et al. (2001) reported a strange NLC that showed a lack of backscattering at 770 nm while showing strong scattering at three other shorter wavelengths. This phenomenon cannot be explained by backscattering on any size distribution of homogeneous water–ice spheres. Baumgarten et al. (2002a) deduced the shape of PMC particles with a polarization lidar technique. They measured the polarization state of 532 nm laser light backscattered from NLC particles. The observed depolarization, averaged over the altitude range of 84.2 to 85.5 km, was $1.7 \pm 1.0\%$. The layer of enhanced depolarization was centered 1 km above the maximum of the PMC layer. A comparison of the observed depolarization with that calculated for cylinder-shaped PMC particles shows that the observed depolarization can be explained by the presence of elongated particles with a length-over-diameter ratio larger than 2.5. This experiment indicates that the depolarizing particles occur primarily in a region where PMC particles grow.

5.3.4.2.7. PMC and PMSE

Simultaneous and common volume lidar/radar/rocket observations show that PMCs and PMSEs (polar mesosphere summer echoes) can be either tightly coupled or loosely coupled (Nussbaumer et al., 1996; von Zahn and Bremer, 1999; Stebel et al., 2000). In the tightly coupled cases, PMCs and PMSEs occur at the same time, and PMCs exist along the bottom of PMSEs. Goldberg et al. (2001) showed that for loosely coupled cases, PMCs and PMSEs can be either temporally or spatially separated. These observations suggest that PMCs and PMSEs may have a common origin. However, PMSEs prefer large number density but small subvisible charged aerosols, whereas PMCs/NLCs prefer small number density but relatively large visible particles.

5.3.4.2.8. PMC and Fe depletion in summer at the South Pole

PMCs are a nearly persistent phenomenon at the South Pole in summer. The observed PMCs are commonly located around 85 km, which overlaps in altitude with the mesospheric Fe layers produced by the ablation of meteoroids entering the atmosphere. The Fe Boltzmann temperature lidar can observe both the Fe layer (by resonance fluorescence scattering) and the PMC (by Mie scattering) with high vertical resolution. When PMCs occur, the Fe data would be "contaminated" by the PMCs around 85 km. By taking the difference between 372- and 374-nm channel densities and then scaling back to the 372-nm density, we obtained the pure Fe density shown in Figure 5.84 as an example on January 19, 2000 (Chu et al., 2002b). An obvious Fe bite-out occurs at the PMC peak altitude. Plane et al. (2004) explained this phenomenon by the

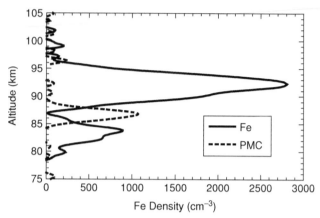

Figure 5.84 Simultaneous observations of the Fe density (black solid line) and PMC (gray dashed line) by the UIUC Fe Boltzmann temperature lidar at the South Pole on January 19, 2000. An Fe bite-out occurs at the altitude of PMC around 85 to 86 km.

efficient uptake of Fe atoms on the surface of the PMC ice particles. Combining the laboratory measurements of Fe uptake rate on the ice surface with the lidar observations of PMC at the South Pole, Plane et al. (2004) developed a one-dimensional mesospheric model to simulate the Fe layer under the influence of PMC ice particles. The model reproduces the Fe bite-out, higher peak height (~92 km), and overall depletion of the layer in the presence of the PMCs in January, and the higher Fe density and lower peak height (below 90 km) of the Fe layer in the absence of the PMCs during February. This is compelling evidence for heterogeneous removal of Fe on the ice particles causing these Fe depletion events.

5.4 CONCLUSIONS AND FUTURE OUTLOOK

Atmospheric lidar research and application is a field that integrates science, technology, and engineering. Lidar has proven to be a unique technology for studying the middle and upper atmosphere. Exciting advances in lidar design and development have been proposed and implemented in recent years. Advances in resonance fluorescence lidar technology, including the development of the narrowband Doppler technique lidars (Na and K) and the Boltzmann technique lidar (Fe), enabled temperature and wind measurements in the middle and upper atmosphere as well as the atmospheric constituent and cloud particle detections. Over the past decade, these resonance fluorescence lidars along with Rayleigh lidars have made great contributions to the understanding of the global atmosphere thermal structure, dynamics, composition, and their relations to the global climate change. Exciting lidar measurements of the middle and upper atmosphere have been made from the geographic North Pole to the geographic South Pole. Compared with 20 years ago, our knowledge about the MLT thermal structures, the MLT dynamics, the polar mesospheric clouds (noctilucent clouds), the influences of gravity waves on the MLT region, the atmospheric metal layers, and mesospheric metal chemistry has been greatly enhanced by lidar observations in the middle and upper atmosphere.

These sophisticated Doppler lidars (Na and K) and Boltzmann lidars (Fe) have both advantages and disadvantages in their technologies, measurement capabilities, and operational complexities. The narrowband Na wind/temperature lidars based on the ring dye laser and pulsed dye amplifier have now reached technical maturity. They currently provide the highest resolution and most accurate temperature and wind measurements in the mesopause region. The dominant error source is the signal photon noise, which can be largely reduced by increasing the product of laser power and telescope aperture. Thus, coupling Na lidar with a large aperture telescope will significantly improve the temporal and spatial resolution, as demonstrated by the system developed by the University of Illinois at SOR and Maui. The full-diurnal-cycle Na wind/temperature lidar has been achieved at the Colorado State University by adding a Na FADOF narrowband filter in the lidar receiver. This system is making great contributions to the study of atmosphere dynamics. However, both the full diurnal cycle and large aperture Na wind/temperature lidars require sophisticated instrumentation, a well-controlled laboratory working environment, and high-level operation/ maintenance. These shortcomings preclude their application in aircrafts, on ships, or at remote sites.

The all-solid-state narrowband Na Doppler lidar based on the frequency mixing of two pulsed Nd:YAG lasers has proven to be rugged enough for deployment in Antarctica. However, it is currently only capable of temperature measurements, and unlikely to make wind measurements due to the lack of absolute frequency calibration. The solid-state Na wind/temperature lidar is a good alternative to the conventional Na wind/temperature lidar. It reduces the system complexity by replacing the ring dye laser with the frequency mixing of two cw Nd:YAG lasers while maintaining the measurement capability and accuracy. Further improvements in the stability of frequency mixing are needed and the possible replacement of pulsed dye amplifier (PDA) by optical parametric amplifier (OPA) is worthy of further investigation. Once the OPA successfully replaces the PDA, the Na wind/ temperature lidar will be a full-solid-state system that would be more robust.

A possible solution to simplify the Na wind/temperature lidar is to develop interferometric receivers with broadband resonance fluorescence lidars. Traditional lidar receivers only collect and collimate the return signals onto a PMT, and then count the detected photons. The idea for interferometric receivers is to employ a high-resolution etalon in a lidar receiver. By measuring the Doppler broadening and Doppler frequency shift of the interference fringes, the atmospheric temperature and wind can be derived. This is basically the same as the interferometer principle for spectral imaging of the atmosphere (Shepherd, 2002). However, to obtain the range information of the returned photon signals, the CCD device needs to have the capability to rapidly switch between columns. These devices are being investigated and developed by Michigan Aerospace, but it may take years before they reach a usable state. Once this CCD device is practical for use, the lidar transmitters could be much simpler than their current narrowband system. Actually, a broadband lidar beam would be required to illuminate the Na Doppler broadened line. Broadband lidar can also help avoid the Na layer saturation problem, which is experienced by the narrowband Na Doppler lidars. The price would be the much weaker return signals compared to narrowband systems. The high quantum efficiency of CCD devices (~90%) could partially compensate for the weak signals. Such broadband lidar systems may experience significant difficulties in daytime measurements, however.

The study of the middle and upper atmosphere requires more high-resolution lidar measurements at different locations all over the world. Thus, transportable lidar systems are desirable. The narrowband K Doppler lidar based on a solid-state ring alexandrite laser has been containerized, as demonstrated by the IAP, and is transportable for temperature measurements at different locations. The application of K FADOF makes the K lidar capable of temperature measurements during both day and night. Although it is possible to make wind measurements, the low K densities in the mesosphere result in relatively poor resolution and accuracy for wind measurements. Rayleigh Doppler lidar is being proposed for making wind measurements. It is either based on an iodine filter edge technique or on a Fabry–Perot etalon for measuring the Doppler frequency shift. Since the Rayleigh signals decrease exponentially as the altitude increases, a pulsed laser with high energy and short wavelength would be required to perform good wind measurements above 50 km.

The Fe Boltzmann temperature lidar based on the solid-state alexandrite lasers has proven to be robust enough to be deployed at remote sites and in aircraft. The lidar measurements of the MLT thermal structure, polar mesospheric clouds, and Fe layers from the North Pole to the South Pole have been accomplished by the University of Illinois with this type of lidar system. Because of its shorter wavelength, higher output power, and daytime capability, the Fe Boltzmann temperature lidar is very suitable for PMC/NLC detection in the polar region and can also provide Rayleigh temperature measurements over a wide range of altitudes. However, the biggest drawback to the Fe Boltzmann technique is the signal level on the weak 374-nm channel, which is about 30 times weaker than the 372-nm channel at 200 K. This weak signal level limits the resolution of Fe Boltzmann lidar temperature measurements. Although the daytime measurement of MLT temperature is feasible, it requires long integration times.

To overcome this problem and maintain the large altitude coverage for temperature measurements as well as achieving wind measurements, a narrowband Fe Doppler lidar operating only at 372 nm would be attractive. This lidar combines the solid-state alexandrite ring laser technologies with existing acousto-optic modulation techniques to enable temperature and wind measurements in the mesopause region using Fe resonance fluorescence. Although the Fe atomic backscatter cross-section is about an order of magnitude smaller than that of Na, the Fe density is at least two times higher than the Na density. Since alexandrite ring laser pulses have long duration time (>200 ns) and saturation is not a problem for the Fe layer, the Fe lidar power could be many times higher than the Na lidar. Thus, the overall product of the backscatter cross-section, constituent density and laser power for the narrowband Fe Doppler lidar can be comparable to that of a Na wind/temperature lidar. Since the Fe lidar operates at UV wavelength and has high power, it will still have good altitude coverage for Rayleigh temperature measurements. Therefore, the narrowband Fe Doppler lidar is expected to have virtually continuous temperature coverage from 30 to 110 km. It will also be able to measure wind in the mesopause region. Such measurements will be available for both daytime and nighttime with Fabry–Perot etalons employed in the lidar receiver. Such an Fe Doppler lidar will be robust enough to be deployed in aircraft and at remote sites, and to be transportable for measurements of temperature and wind all over the globe.

The current and proposed resonance fluorescence lidars are all based upon the mesospheric metal layers Na, Fe, and K. These layers exist from approximately 75 to 120 km. Out of this range, the neutral Na, Fe, and K number densities are too low to make reliable temperature or wind measurements. Nevertheless, it is desirable to measure the temperature and species above 120 km for upper atmosphere studies.

Therefore, new atoms or molecules with appropriate number density distribution, and suitable energy level diagram for available laser wavelengths, need to be identified for developing new lidars. Several ideas have been proposed. One is the N_2^+ molecule when aurora occurs in the upper atmosphere (Garner and Dao, 1995; Collins et al., 1997). Another is atomic He in the thermosphere (Kerr et al., 1996; Gerrard et al., 1997). These molecules or atoms have a significant number density above 150 km, and it would be possible to detect them using resonance fluorescence lidar techniques. Atomic oxygen O and hydrogen H are also very important species in the upper atmosphere, especially for atmospheric chemistry. However, neither atomic O nor H has been measured with high range resolution. The importance of these constituents is the motivation for the investigation of far-ultraviolet (FUV) lidars. The lines at 121.6 nm for H and 130.5 nm for O are two potential candidates for measuring the densities with resonance fluorescence lidar techniques. The challenge is to build the FUV laser. Since the lower atmosphere strongly absorbs FUV light, such lidars would need to be space-borne. Other neutral and ionic molecules in the middle and upper atmosphere, such as N, NO, NO_2, CO, CO_2, O_2, N_2, N_2^+, O^+, He^+, are also new candidates for lidar detection.

Although lidars can provide high-resolution data, a lidar only probes a single point in the atmosphere. It would be ideal to have a wider detection range as well as high resolution. Correlated observations from ground-based clusters of collocated lidars, imagers, radars, and other remote sensing equipment provide unique vertical and horizontal measurement capabilities that can aid in resolving the outstanding fundamental issues in atmospheric science.

Resonance fluorescence lidar techniques are now an established measurement technology for high-resolution and high-accuracy studies of the middle and upper atmosphere. It is clear that given the important contributions these instruments have made to atmosphere science in their brief lifetime, many more new and significant results will be achieved by existing and new proposed lidars.

ACKNOWLEDGMENTS

The chapter of Resonance Fluorescence Lidar for Measurements of the Middle and Upper Atmosphere was finished in October 2003. Authors Chu and Papen have drawn on the experience of several years work at the Department of Electrical & Computer Engineering in the University of Illinois at Urbana-Champaign, and are grateful to all colleagues in the research team during this period, especially to Dr Chester S. Gardner for his leadership and encouragement. The authors are also indebted to Dr Chiao-Yao She, Dr Ulf von Zahn, Dr Franz-Josef Lübken, Dr Josef Höffner, Dr Jonathan Friedman, and Dr Takuya Kawahara, who have contributed material to this chapter. The authors gratefully acknowledge Dr Steve J. Franke, Dr Patricia M. Franke, Dr Erhan Kudeki, Dr Xing Gao, and Jing Tang for their fruitful discussion during the preparation of this manuscript and their proofreading of the manuscript.

REFERENCES

Alpers, M., J. Höffner, and U. von Zahn, Iron atom densities in the polar mesosphere from lidar observations, *Geophys. Res. Lett., 17,* 2345–2348, 1990.

Alpers, M., J. Höffner, and U. von Zahn, Sporadic Fe and E layers at polar, middle and low latitudes, *J. Geophys. Res. 99 (A8),* 14971–14985, 1994.

Alpers, M., J. Höffner, and U. von Zahn, Upper atmosphere Ca and Ca^+ at mid-latitudes: first simultaneous and common-volume lidar observations, *Geophys. Res. Lett., 23,* 567–570, 1996.

Alpers, M., M. Gerding, J. Höffner, and U. von Zahn, NLC particle properties from a five-color lidar observation at 54°N, *J. Geophys. Res., 105,* 12235–12240, 2000.

Alpers, M., M. Gerding, J. Höffner, and J. Schneider, Multiwavelength lidar observation of a strange noctilucent cloud at Kühlungsborn, Germany (54°N), *J. Geophys. Res., 106,* 7945–7953, 2001.

Andrews, D. G., J. R. Holton, and C. B. Leovy, *Middle Atmosphere Dynamics*, Academic, San Diego, CA, 1987.

Arimondo, E., M. Inguscio, and P. Violino, Experimental determinations of the hyperfine structure in the alkali atoms, *Rev. Mod. Phys., 49,* 31–75, 1977.

Arnold, K. S., and C. Y. She, Metal fluorescence lidar (light detection and ranging) and the middle atmosphere, *Contemp. Phys., 44,* 35–49, 2003.

Avaste, O., Noctilucent clouds, *J. Atmos. Terr. Phys., 55,* 133–143, 1993.

Backhouse, T. W., The luminous cirrus clouds of June and July, *Meteorol. Mag., 20,* 133, 1885.

Batista, P. P., B. R. Clemesha, D. M. Simonich, and V. W. J. H. Kirchhoff, Tidal oscillations in the atmospheric sodium layer, *J. Geophys. Res., 90,* 3881–3888, 1985.

Baumgarten, G., K. H. Fricke, G. von Cossart, Investigation of the shape of noctilucent cloud particles by polarization lidar technique, *Geophys. Res. Lett., 29 (13),* doi: 10.1029/2001GL013877, 2002a.

Baumgarten, G., F. J. Lübken, and K. H. Fricke, First observation of one noctilucent cloud by a twin lidar in two different directions, *Ann. Geophys., 20,* 1863–1868, 2002b.

Beatty, T. J., R. E. Bills, K. H. Kwon, and C. S. Gardner, CEDAR lidar observations of sporadic Na layers at Urbana, Illinois, *Geophys. Res. Lett., 15,* 1137–1140, 1988.

Berger, U., and U. von Zahn, The two-level structure of the mesopause: a model study, *J. Geophys. Res., 104 (D18),* 22083–22093, 1999.

Berger, U., and U. von Zahn, Icy particles in the summer mesopause region: three-dimensional modeling of their environment and two-dimensional modeling of their transport, *J. Geophys. Res., 107* (A11), 1366, doi: 10.1029/2001JA000316, 2002.

Bills, R. E., and C. S. Gardner, Lidar observations of mesospheric Fe and sporadic Fe layers at Urbana, Illinois, *Geophys. Res. Lett., 17,* 143–146, 1990.

Bills, R. E., C. S. Gardner, and C. Y. She, Narrowband lidar technique for sodium temperature and Doppler wind observations of the upper atmosphere, *Opt. Eng., 30,* 13–21, 1991a.

Bills, R. E., C. S. Gardner, and S. J. Franke, Na Doppler/temperature lidar: initial mesopause region observations and comparison with the Urbana medium frequency radar, *J. Geophys. Res., 96 (D12),* 22701–22707, 1991b.

Bills, R. E., and C. S. Gardner, Lidar observations of the mesopause region temperature structure at Urbana, *J. Geophys. Res., 98 (D1),* 1011–1021, 1993.

Blamont, J. E., M. L. Chanin, and G. Megie, Vertical distribution and temperature of the nighttime atmospheric sodium layer obtained by laser backscatter, *Ann. Geophys. 28,* 833–838, 1972.

Bowman, M. R., A. J. Gibson, and M. C. W. Sandford, Atmospheric sodium measured by a tuned laser radar, *Nature, 221,* 456–457, 1969.

Carbary, J. F., G. J. Romick, D. Morrison, L. J. Paxton, C. I. Meng, Altitudes of polar mesospheric clouds observed by a middle ultraviolet imager, *J. Geophys. Res., 104,* 10089–10100, 1999.

Cerny, T., and C. F. Sechrist, Jr., Calibration of the Urbana Lidar System, Aeronomy Report No. 94, University of Illinois, August 1, 1980.

Chapman, S., and P. C. Kendall, Noctilucent clouds and thermospheric dust: their diffusion and height distribution, *Q. J. R. Metorol. Soc., 91,* 115–131, 1965.

Charlson, R. J., Noctilucent clouds: a steady-state model, *Q. J. R. Metorol. Soc., 91,* 517–523, 1965.

Charlson, R. J., A simple noctilucent cloud model, *Tellus, 18,* 451, 1966.

Chen, H., C. Y. She, P. Searcy, and E. Korevaar, Sodium-vapor dispersive Faraday filter, *Opt. Lett., 18,* 1019–1021, 1993.

Chen, H., M. A. White, D. A. Krueger, and C. Y. She, Daytime mesopause temperature measurements using a sodium-vapor dispersive Faraday filter in lidar receiver, *Opt. Lett., 21,* 1093–1095, 1996.

Chen, S., Z. Hu, M. A. White, H. Chen, D. A. Krueger, and C. Y. She, Lidar observations of seasonal variation of diurnal mean temperature in the mesopause region over Fort Collins, Colorado (41°N, 105°W), *J. Geophys. Res., 105 (D10),* 12371–12379, 2000.

Chiu, P. H., A. Magana, and J. Davis, All-solid-state single-mode sum-frequency generation of sodium resonance radiation, *Opt. Lett., 19,* 2116–2118, 1994.

Chu, X., W. Pan, G. Papen, C. S. Gardner, G. Swenson, and P. Jenniskens, Characteristics of Fe ablation trails observed during the 1998 Leonid meteor shower, *Geophys. Res. Lett., 27,* 1807–1810, 2000a.

Chu, X., A. Z. Liu, G. Papen, C. S. Gardner, M. Kelley, J. Drummond, and R. Fugate, Lidar observations of elevated temperatures in bright chemiluminescent meteor trails during the 1998 Leonid shower, *Geophys. Res. Lett., 27,* 1815–1818, 2000b.

Chu, X., C. S. Gardner, and G. Papen, Lidar observations of polar mesospheric clouds at South Pole: seasonal variations, *Geophys. Res. Lett., 28,* 1203–1206, 2001a.

Chu, X., C. S. Gardner, and G. Papen, Lidar observations of polar mesospheric clouds at South Pole: diurnal variations, *Geophys. Res. Lett., 28,* 1937–1940, 2001b.

Chu, X., W. Pan, G. C. Papen, C. S. Gardner, and J. A. Gelbwachs, Fe Boltzmann temperature lidar: design, error analysis, and initial results at the North and South Poles, *Appl. Opt., 41,* 4400–4410, 2002a.

Chu, X., and C. S. Gardner, Pole-to-Pole: lidar observations of middle and upper atmosphere temperature and polar mesospheric clouds over the North and South Poles, *Proc. SPIE, 4893,* 223–236, 2002b.

Chu, X., C. S. Gardner, and R. G. Roble, Lidar studies of interannual, seasonal, and diurnal variations of polar mesospheric clouds at the South Pole, *J. Geophys. Res., 108 (D8),* 8447, doi: 10.1029/2002JD002524, 2003.

Chu, X., G. J. Nott, P. J. Espy, C. S. Gardner, J. C. Diettrich, M. A. Clilverd, and M. J. Jarvis, Lidar observations of polar mesospheric clouds at Rothera, Antarctica (67.5°S, 68.0°W), *Geophys. Res. Lett., 31,* L02114, doi: 10.1029/2003GL018638, 2004.

Clemesha, B. R., D. M. Simonich, P. P. Batista, and V. W. J. H. Kirchhoff, The diurnal variation of atmospheric sodium, *J. Geophys. Res., 87,* 181–185, 1982.

Clemesha, B. R., M. P. P. Martins Jorge, D. M. Simonich, and P. P. Batista, A new method for measuring the Doppler temperature of the atmospheric sodium layer, *Adv. Space Res., 19,* 681–684, 1997.

Clemesha, B. R., I. Veselovskii, P. P. Batista, M. P. P. M. Jorge, and D. M. Simonich, First mesopause temperature profiles from a fixed southern hemisphere site, *Geophys. Res. Lett., 26,* 1681–1684, 1999.

Clemesha, B. R., P. P. Batista, and D. M. Simonich, Tide-induced oscillations in the atmospheric sodium layer, *J. Atmos. Sol. Terr. Phys., 64,* 1321–1325, 2002.

Collins, R. L., D. Lummerzhein, and R. W. Smith, Analysis of lidar systems for profiling aurorally excited molecular species, *Appl. Opt., 36,* 6024–6034, 1997.

Collins, R. L., M. C. Kelley, M. J. Nicolls, C. Ramos, T. Hou, T. E. Stern, K. Mizutani, and T. Itabe, Simultaneous lidar observations of a noctilucent cloud and an internal wave in the polar mesosphere, *J. Geophys. Res., 108 (D8),* 8435, doi: 10.1029/2002JD002427, 2003.

Debrestian, D. J., et al., An analysis of POAMII solar occultation observations of polar mesospheric clouds in the southern hemisphere, *J. Geophys. Res., 102*, 1971–1981, 1997.

Dick, D. J., and T. M. Shay, Ultrahigh-noise rejection optical filter, *Opt. Lett., 16*, 867–869, 1991.

Donahue, T. M., B. Guenther, and J. E. Blamont, Noctilucent clouds in daytime: circumpolar particulate layers near the summer mesopause, *J. Atmos. Sci., 29*, 1205–1209, 1972.

Drummond, J. D., B. W. Grime, C. S. Gardner, A. Z. Liu, X. Chu, and T. J. Kane, Observations of persistent Leonid meteor trails 1. Advection of the "Diamond Ring", *J. Geophys. Res., 106* (A10), 21517–21524, 2001.

Elterman, L. B., The measurement of stratospheric density distribution with the search light technique, *J. Geophys. Res., 56*, 509–520, 1951.

Elterman, L., A series of stratospheric temperature profiles obtained with the searchlight technique, *J. Geophys. Res., 58*, 519–530, 1953.

Elterman, L., Seasonal trends of temperature, density, and pressure to 67.6 km obtained with the searchlight probing technique, *J. Geophys. Res., 59*, 351–358, 1954.

Eska, V., and J. Höffner, Observed linear and nonlinear K layer response, *Geophys. Res. Lett., 25*, 2933–2936, 1998.

Eska, V., J. Höffner, and U. von Zahn, Upper atmosphere potassium layer and its seasonal variations at 54°N, *J. Geophys. Res., 103* (A12), 29207–29214, 1998.

Eska, V., U. von Zahn, and J. M. C. Plane, The terrestrial potassium layer (75–110 km) between 71°S and 54°N: observations and modeling, *J. Geophys. Res., 104* (A8), 17173–17186, 1999.

Evans, W. F., L. R. Laframboise, K. R. Sine, R. H. Wiens, and G. G. Shepherd, Observation of polar mesospheric clouds in summer, 1993 by the WINDII instrument on UARS, *Geophys. Res. Lett., 22*, 2793–2796, 1995.

Eyler, E. E., A. Yiannopoulou, S. Gangopadhyay, and N. Melikechi, Chirp-free nanosecond laser amplifier for precision spectroscopy, *Opt. Lett., 22*, 49–51, 1997.

Fee, M. S., K. Danzmann, and S. Chu, Optical heterodyne measurement of pulsed lasers: toward high-precision pulsed spectroscopy, *Phys. Rev. A, 45*, 4911–4924, 1992.

Felix, F., W. Keenliside, G. Kent, and M. C. W. Sandford, Laser radar observations of atmospheric potassium, *Nature, 246*, 345–346, 1973.

Fiedler, J., G. Baumgarten, and G. von Cossart, Noctilucent clouds above ALOMAR between 1997 and 2001: occurrence and properties, *J. Geophys. Res., 108 (D8)*, 8453, doi: 10.1029/2002JD002419, 2003.

Fleming, E. L., S. Chandra, J. J. Barnett, and M. Corney, Zonal mean temperature, pressure, zonal wind and geopotential height as functions of latitude, *Adv. Space Res., 10(12)*, 11–59, 1990.

Fricke, K. H., and U. von Zahn, Mesopause temperatures derived from probing the hyperfine structure of the D_2 resonance line of sodium by lidar, *J. Atmos. Terr. Phys., 47*, 499–512, 1985.

Fricke-Begemann, C., J. Höffner, and U. von Zahn, The potassium density and temperature structure in the mesopause region (80–105 km) at a low latitude (28°N), *Geophys. Res. Lett., 29 (22)*, doi: 10.1029/2002GL015578, 2002.

Friedman, J. S., S. C. Collins, R. Delgado, and P. A. Castleberg, Mesospheric potassium layer over the Arecibo Observatory, 18.3°N 66.75°W, *Geophys. Res. Lett., 29 (5)*, doi: 10.1029/2001GL013542, 2002.

Friedman, J. S., Tropical mesopause climatology over the Arecibo Observatory, *Geophys. Res. Lett., 30 (12)*, doi: 10.1029/2003GL016966, 2003a.

Friedman, J. S., C. A. Tepley, S. Raizada, Q. H. Zhou, J. Hedin, and R. Delgado, Potassium Doppler-resonance lidar for the study of the mesosphere and lower thermosphere at the Arecibo Observatory, *J. Atoms. Sol.-Terr. Phys., 65*, 1411–1424, 2003b.

Fritts, D. C., and R. A. Vincent, Mesospheric momentum flux studies at Adelaide, Australia: observations and a gravity wave-tidal interaction model, *J. Atmos. Sci., 44*, 605–619, 1987.

Fritts, D. C., and T. van Zandt, Spectral estimates of gravity wave energy and momentum fluxes: energy dissipation, acceleration, and constraints, *J. Atmos. Sci., 50*, 3685–3694, 1993.

Fritts, D. C., and Z. Luo, Dynamical and radiative forcing of the summer mesopause circulation and thermal structure 1. Mean solstice conditions, *J. Geophys. Res., 100 (D2)*, 3119–3128, 1995.

Gadsden, M., The silver-blue cloudlets again: nucleation and growth of ice in the mesosphere, *Planet. Space Sci., 29*, 1079–1087, 1981.

Gadsden, M., and W. Schröder, *Noctilucent Clouds*, Springer-Verlag, Berlin, 1989.

Gadsden, M., A secular change in noctilucent cloud occurrence, *J. Atmos. Terr. Phys., 52*, 247–251, 1990.

Gadsden, M., Polar mesospheric clouds seen from geostationary orbit, *J. Atmos. Sol. Terr. Phys., 62*, 31–36, 2000.

Gangopadhyay, S., N. Melikechi, and E. E. Eyler, Optical phase perturbations in nanosecond pulsed amplification and second-harmonic generation, *J. Opt. Soc. Am. B, 11*, 231–241, 1994.

Garcia, R. R., and S. Solomon, The effect of breaking gravity waves on the dynamics and chemical composition of the mesosphere and lower thermosphere, *J. Geophys. Res., 90*, 3850–3868, 1985.

Garcia, R.R., Dynamics, radiation and photochemistry in the mesosphere — implications for the formation of noctilucent clouds, *J. Geophys. Res., 94*, 14605–14615, 1989.

Gardner, C. S., and J. D. Voelz, Lidar measurements of gravity wave saturation effects in the sodium layer, *Geophys. Res. Lett., 12*, 765–768, 1985.

Gardner, C. S., Sodium resonance fluorescence lidar applications in atmospheric science and astronomy, *Proc. IEEE, 77*, 408–418, 1989.

Gardner, C. S., T. J. Kane, D. C. Senft, J. Qian, and G. C. Papen, Simultaneous observations of sporadic E, Na, Fe, and Ca^+ layers at Urbana, Illinois: three case studies, *J. Geophys. Res., 98*, 16865–16873, 1993.

Gardner, C. S., Diffusive filtering theory of gravity wave spectra in the atmosphere, *J. Geophys. Res., 99*, 20601–20622, 1994.

Gardner, C. S., and G. C. Papen, Mesospheric Na wind/temperature lidar, *Rev. Laser Eng., 23*, 131–134, 1995.

Gardner, C. S., Introduction to ALOHA/ANLC-93: the 1993 airborne lidar and observations of the Hawaiian airglow/airborne noctilucent cloud campaigns, *Geophys. Res. Lett., 22*, 2789–2792, 1995.

Gardner, C. S., and W. Yang, Measurements of the dynamical cooling rate associated with the vertical transport of heat by dissipating gravity waves in the mesopause region at the Starfire Optical Range, *J. Geophys. Res., 103 (D14)*, 16909–16927, 1998.

Gardner, C. S., K. Gulati, Y. Zhao, and G. Swenson, Measuring gravity wave momentum fluxes with airglow imagers, *J. Geophys. Res., 104 (D10)*, 11903–11915, 1999.

Gardner, C. S., G. C. Papen, X. Chu, and W. Pan, First lidar observations of middle atmosphere temperatures, Fe densities, and polar mesospheric clouds over the North and South Poles, *Geophys. Res. Lett., 28*, 1199–1202, 2001.

Gardner, C. S., Y. Zhao, and A. Z. Liu, Atmospheric stability and gravity wave dissipation in the mesopause region, *J. Atmos. Sol. Terr. Phys., 64*, 923–929, 2002.

Garner, R. C., and P. Dao, Molecular nitrogen fluorescence lidar for remote sensing of the auroral ionosphere, *J. Geophys. Res., 100,* 14131–14140, 1995.

Gelbwachs, J. A., Iron Boltzmann factor lidar: proposed new remote-sensing technique for mesospheric temperature, *Appl. Opt., 33,* 7151–7156, 1994.

Gerding, M., M. Alpers, J. Höffner, and U. von Zahn, Simultaneous K and Ca lidar observations during a meteor shower on March 6–7, 1997 at Kühlungsborn, Germany, *J. Geophys. Res., 104 (A11),* 24689–24698, 1999.

Gerding, M., M. Alpers, U. von Zahn, R. J. Rollason, and J. M. C. Plane, Atmosphere Ca and Ca$^+$ layers: midlatitude observations and modeling, *J. Geophys. Res., 105 (A12),* 27131–27146, 2000.

Gerrard, A. J., T. J. Kane, D. D. Meisel, J. P. Thayer, and R. B. Kerr, Investigation of a resonance lidar for measurement of thermospheric metastable helium, *J. Atmos. Sol. Terr. Phys., 59,* 2023–2035, 1997.

Gerrard, A. J., T. J. Kane, and J. P. Thayer, Noctilucent clouds and wave dynamics: observations at Sondrestrom, Greenland, *Geophys. Res. Lett., 25,* 2817–2820, 1998.

Gibson, A. J., and M. C. W. Sandford, The seasonal variation of the night-time sodium layer, *J. Atmos. Terr. Phys., 33,* 1675–1684, 1971.

Gibson, A. J., and M. C. W. Sandford, Daytime laser radar measurements of the atmospheric sodium layer, *Nature, 239,* 509–511, 1972.

Gibson, A. J., L. Thomas, and S. K. Bhattachacharyya, Laser observations of the ground-state hyperfine structure of sodium and of temperatures in the upper atmosphere, *Nature, 281,* 131–132, 1979.

Goldberg, R. A., et al., DROPPS: a study of the polar summer mesosphere with rocket, radar and lidar, *Geophys. Res. Lett., 28,* 1407–1410, 2001.

Gong, S. S., G. T. Yang, J. M. Wang, B. M. Liu, X. W. Cheng, J. Y. Xu, and W. X. Wan, Occurrence and characteristics of sporadic sodium layer observed by lidar at a mid-latitude location, *J. Atmos. Sol. Terr. Phys., 64,* 1957–1966, 2002.

Goody, R. M., and Y. L. Yung, *Atmospheric Radiation: Theoretical Basis,* 2nd edition, Oxford University Press, Oxford, 1989.

Grahn, S., and G. Witt, Photogrammetric Triangulation of Diffuse Objects in Space: Part II, Report AP-4, Institute of Meteorology, University of Stockholm, April 1971.

Granier, C., and G. Megie, Daytime lidar measurements of the mesospheric sodium layer, *Planet. Space Sci., 30*, 169–177, 1982.

Granier, C., J. P. Jegou, and G. Megie, Resonant lidar detection of Ca and Ca^+ in the upper atmosphere, *Geophys. Res. Lett., 12*, 655–658, 1985.

Granier, C., J. P. Jegou, and G. Megie, Atomic and Ionic Calcium in the Earth's upper atmosphere, *J. Geophys. Res., 94 (D7)*, 9917–9924, 1989a.

Granier, C., J. P. Jegou, and G. Megie, Iron atoms and metallic species in the Earth's upper atmosphere, *Geophys. Res. Lett., 16*, 243–246, 1989b.

Grime, B. W., T. J. Kane, S. C. Collins, M. C. Kelley, C. A. Kruschwitz, J. S. Friedman, and C. A. Tepley, Meteor trail advection and dispersion: preliminary lidar observations, *Geophys. Res. Lett., 26*, 675–678, 1999.

Grime, B. W., T. J. Kane, A. Liu, G. Papen, C. S. Gardner, M. C. Kelley, C. Kruschwitz, and J. Drummond, Meteor trail advection observed during the 1998 Leonid shower, *Geophys. Res. Lett., 27*, 1819–1822, 2000.

Gumbel, J., and G. Witt, *In situ* measurements of the vertical structure of a noctilucent cloud, *Geophys. Res. Lett., 25*, 493–496, 1998.

Hagan, M. E., J. M. Forbes, and F. Vial, On modeling migrating solar tides, *Geophys. Res. Lett., 22*, 893–896, 1995.

Hake, R. D. Jr., D. E. Arnold, D. W. Jackson, W. E. Evans, B. P. Ficklin, and R. A. Long, Dye-laser observations of the nighttime atomic sodium layer, *J. Geophys. Res., 77*, 6839–6848, 1972.

Hansch, T. W., I. S. Shahin, and A. L. Schawlow, High-resolution saturation spectroscopy of the sodium D lines with a pulsed tunable dye laser, *Phys. Rev. Lett., 27*, 707–710, 1971.

Hansen, G., M. Serwazi, and U. von Zahn, First detection of a noctilucent cloud by lidar, *Geophys. Res. Lett., 16*, 1445–1448, 1989.

Hansen, G., and U. von Zahn, Sudden sodium layers in polar latitudes, *J. Atmos. Terr. Phys., 52*, 585–608, 1990.

Hansen, G., and U. von Zahn, Simultaneous observations of noctilucent clouds and mesopause temperatures by lidar, *J. Geophys. Res., 99*, 18989–18999, 1994.

Hauchecorne, A., and M. L. Chanin, Density and temperature profiles obtained by lidar between 35 and 70 km, *Geophys. Res. Lett., 7*, 565–568, 1980.

Hauchecorne, A., M. L. Chanin, and R. Wilson, Mesospheric temperature inversion and gravity wave dynamics, *Geophys. Res. Lett., 14,* 935–939, 1987.

Hecht, J. H., J. P. Thayer, D. J. Gutierrez, D. L. McKenzie, Multi-instrument zenith observations of noctilucent clouds over Greenland on July 30/31, 1995, *J. Geophys. Res., 102,* 1959–1970, 1997.

Helmer, M., J. M. C. Plane, J. Qian and C. S. Gardner, A model of meteoric iron in the upper atmosphere, *J. Geophys. Res., 103 (D9),* 10913–10925, 1998.

Hervig, M., R. E. Thompson, M. NcHugh, L. L. Gordley, J. M. Russell III, and M. E. Summers, First confirmation that water ice is the primary component of polar mesospheric clouds, *Geophys. Res. Lett., 28,* 971–974, 2001.

Hesstvedt, E., Note on the nature of noctilucent clouds, *J. Geophys. Res., 99,* 1985–1987, 1961.

Hesstvedt, E., On the possibility of ice cloud formation at the mesopause, *Tellus, 14,* 290–296, 1962.

Höffner, J., and U. von Zahn, Mesopause temperature profiling by potassium lidar: recent progress and outlook for ALOMAR, in *Proc. 12th ESA Symp. on European Rocket and Balloon Programmes and Related Research,* edited by B. Kaldeich-Schürmann, ESA SP-370, 403–407, 1995.

Höffner, J., U. von Zahn, W. J. McNeil, and E. Murad, The 1996 Leonid shower as studied with a potassium lidar: observations and inferred meteoroid sizes, *J. Geophys. Res., 104 (A2),* 2633–2643, 1999.

Höffner, J., C. Fricke-Begemann, and F. J. Lübken, First observations of noctilucent clouds by lidar at Svalbard, *Atmos. Chem. Phys. Discuss., 3,* 521–549, 2003.

Hu, X., A. Z. Liu, C. S. Gardner, and G. R. Swenson, Characteristics of quasi-monochromatic gravity waves observed with Na lidar in the mesopause region at Starfire Optical Range, NM, *Geophys. Res. Lett., 29 (24),* 2169, doi: 10.1029/2002GL014975, 2002.

Humphreys, W. J., Nacreous and noctilucent clouds, *Mon. Weather Rev., 61,* 228–229, 1933.

Jegou, J.-P., M.-L. Chanin, G. Megie, and J. E. Blamont, Lidar measurements of atmospheric lithium, *Geophys. Res. Lett., 7,* 995–998, 1980.

Jensen, E. J., and G. E. Thomas, A growth-sedimentation model of polar mesospheric clouds: comparisons with SME measurements, *J. Geophys. Res., 93,* 2461–2473, 1988.

Jensen, E. J., G. E. Thomas, and O. B. Toon, On the diurnal variation of noctilucent clouds, *J. Geophys. Res., 94*, 14693–14702, 1989.

Jensen, E. J., and G. E. Thomas, Numerical simulations of the effects of gravity waves on noctilucent clouds, *J. Geophys. Res., 99*, 3421–3430, 1994.

Jesse, O., Die Höhe der leuchtenden Nachtwolken, *Astron. Nachrichten, 140*, 161–168, 1896.

Jeys, T. H., A. A. Brailove, and A. Mooradian, Sum frequency generation of sodium resonance radiation, *Appl. Opt., 28*, 2588–2591, 1989.

Jones, K. M., P. S. Julienne, P. D. Lett, W. D. Phillips, E. Tiesinga, and C. J. Williams, Measurement of the atomic Na (3P) lifetime and of retardation in the interaction between two atoms bound in a molecule, *Europhys. Lett., 35*, 85–90, 1996.

Kane, T. J., and C. S. Gardner, Structure and seasonal variability of the nighttime mesospheric Fe-layer at midlatitudes, *J. Geophys. Res., 98*, 16875–16886, 1993a.

Kane, T. J., and C. S. Gardner, Lidar observations of the meteoric deposition of mesospheric metals, *Science, 259*, 1297–1300, 1993b.

Kasevich, M. A., E. Riis, S. Chu, and R. G. Devoe, RF spectroscopy in an atomic fountain, *Phys. Rev. Lett., 63*, 612–615, 1989.

Kawahara, T. D., T. Kitahara, F. Kobayashi, Y. Saito, A. Nomura, C. Y. She, D. A. Krueger, M. Tsutsumi, Wintertime mesopause temperatures observed by lidar measurements over Syawa station (69°S, 39°E), Antarctica, *Geophys. Res. Lett., 29*, doi: 10.1029/2002GL015244, 2002.

Kawahara, T. D., C. S. Gardner, and A. Nomura, Observed temperature structure of the atmosphere above Syowa Station, Antarctica (69°S, 39°E), *J. Geophys. Res., 109*, D12103, doi: 10.1029/2003JD003918, 2004.

Kelley, M. C., C. S. Gardner, J. Drummond, T. Armstrong, A. Liu, X. Chu, G. Papen, C. Kruschwitz, P. Loughmiller, B. Grime, and J. Engelman, First observations of long-lived meteor trains with resonance lidar and other optical instruments, *Geophys. Res. Lett., 27*, 1811–1814, 2000.

Kerr, R., J. Noto, R. S. Lancaster, M. Franco, R. J. Rudy, R. Williams, and J. H. Hecht, Fabry-Perot observations of helium 10830 Å emission at Millstone Hill, *Geophys. Res. Lett., 23*, 3239–3242, 1996.

Kirchhoff, V. W. J. H., and B. R. Clemesha, Atmospheric sodium measurements at 23°S, *J. Atmos. Terr. Phys., 35*, 1493–1498, 1973.

Klekociuk, A., M. M. Lambert, and R. A. Vincent, Characteristics of a polar mesospheric clouds observed by Rayleigh lidar at Davis, Antarctica, paper presented at IUGG Conference, Sapporo, Japan, 30 June–11 July 2003.

Klostermeyer, J., A simple model of the ice particle size distribution in noctilucent clouds, *J. Geophys. Res., 103*, 28743–28752, 1998.

Klostermeyer, J., Effect of tidal variability on the mean diurnal variation of noctilucent clouds, *J. Geophys. Res., 106*, 9749–9755, 2001.

Kudeki, E., and S. J. Franke, Statistics of momentum flux estimation, *J. Atmos. Sol. Terr. Phys., 60*, 1549–1553, 1998.

Kurzawa, H., and U. von Zahn, Sodium density and atmospheric temperature in the mesopause region in polar summer, *J Atmos. Terr. Phys., 52*, 981–993, 1990.

Kwon, K. H., C. S. Gardner, D. C. Senft, F. L. Roesler, and J. Harlander, Daytime lidar measurements of tidal winds in the mesospheric sodium layer at Urbana, Illinois, *J. Geophys. Res., 92*, 8781–8786, 1987.

Langer, M., K. P. Müller, and K. H. Fricke, Rayleigh lidar detection of aerosol echoes from noctilucent cloud altitudes at the Arctic Circle, *Geophys. Res. Lett., 22*, 381–384, 1995.

Leblanc, T., I. S. McDermid, P. Keckhut, A. Hauchecorne, C. Y. She, and D. A. Krueger, Temperature climatology of the middle atmosphere from long-term lidar measurements at middle and low latitudes, *J. Geophys. Res., 103*, 17191–17204, 1998.

Lindzen, R. S., Turbulence and stress owing to gravity-wave and tidal breakdown, *J. Geophys. Res., 86*, 9707–9714, 1981.

Liu, A. Z., and C. S. Gardner, Vertical dynamical transport of mesospheric constituents by dissipating gravity waves, *J. Atmos. Sol. Terr. Phys., 66*, 267–275, 2004.

Loudon, R., *The Quantum Theory of Light*, 2nd edition, Oxford University Press, New York, 1983.

Lübken, F. J., and U. von Zahn, Thermal structure of the mesopause region at polar latitudes, *J. Geophys. Res., 96 (D11)*, 20841–20857, 1991.

Lübken, F. J., K. H. Fricke, and M. Langer, Noctilucent clouds and the thermal structure near the Arctic mesopause in summer, *J. Geophys. Res., 101*, 9489–9508, 1996.

Martin, W. C., and R. Zalubas, Energy levels of sodium Na i through Na xi, *J. Phys. Chem. Ref. Data*, 10, 153–196, 1981.

Measures, R. M., *Laser Remote Sensing — Fundamentals and Applications*, Wiley-Interscience, New Year, 1984.

Measures, R. M. (ed.), *Laser Remote Chemical Analysis*, Wiley-Interscience, New York, 1988.

Megie, G., and J. E. Blamont, Laser sounding of atmospheric sodium interpretation in terms of global atmospheric parameters, *Planet. Space Sci., 25*, 1093–1109, 1977.

Megie, G., F. Bos, J. E. Blamont, and M. L. Chanin, Simultaneous nighttime lidar measurements of atmospheric sodium and potassium, *Planet. Space Sci., 26*, 27–35, 1978.

Melikechi, N., S. Gangopadhyay, and E. E. Eyler, Phase dynamics in nanosecond pulsed dye laser amplification, *J. Opt. Soc. Am. B, 11*, 2402–2411, 1994.

Meriwether, J. W., and C. S. Gardner, A review of the mesosphere inversion layer phenomenon, *J. Geophys. Res., 105 (D10)*, 12405–12416, 2000.

Milonni, P. W., R. Q. Fugate, and J. M. Telle, Analysis of measured photon returns from sodium beacons, *J. Opt. Soc. Am. A, 15*, 217–233, 1998.

Milonni, P. W., H. Fearn, J. M. Telle, and R. Q. Fugate, Theory of continuous-wave excitation of the sodium beacon, *J. Opt. Soc. Am. A, 16*, 2555–2566, 1999.

Mlynczak, M. G., and S. Solomon, Middle atmosphere heating by exothermic chemical reactions involving odd-hydrogen species, *Geophys. Res. Lett., 18*, 37–40, 1991.

Mlynczak, M. G., and S. Solomon, A detailed evaluation of the heating efficiency in the middle atmosphere, *J. Geophys. Res., 98 (D6)*, 10517–10541, 1993.

Neuber, R., P. von der Gathen, and U. von Zahn, Altitude and temperature of the mesopause at 69°N latitude in winter, *J. Geophys. Res., 93*, 11093–11101, 1988.

Nussbaumer, V., K. H. Fricke, M. Langer, W. Singer, and U. von Zahn, First simultaneous and common volume observations of noctilucent clouds and polar mesosphere summer echoes by lidar and radar, *J. Geophys. Res., 101*, 19161–19167, 1996.

Oates, C. W., K. R. Vogel, and J. L. Hall, High precision linewidth measurement of laser-cooled atoms: resolution of the Na 3p $^2P_{3/2}$ lifetime discrepancy, *Phys. Rev. Lett., 76*, 2866–2869, 1996.

Pan, W., C. S. Gardner, and R. G. Roble, The temperature structure of the winter atmosphere at South Pole, *Geophys. Res. Lett., 29*, doi: 10.1029/2002GL015288, 2002.

Pan, W., and C. S. Gardner, Seasonal variations of the atmospheric temperature structure from 3 to 110 km at South Pole, *J. Geophys. Res., 108 (D18)*, 4564, doi: 10.1029/2002JD003217, 2003.

Papen, G. C., W. M. Pfenninger, and D. M. Simonich, Sensitivity analysis of Na narrowband wind–temperature lidar systems, *Appl. Opt., 34*, 480–498, 1995a.

Papen, G. C., C. S. Gardner, and W. M. Pfenninger, Analysis of a potassium lidar system for upper-atmospheric wind–temperature measurements, *Appl. Opt., 34*, 6950–6958, 1995b.

Papen, G. C., and D. Treyer, Comparison of an Fe Boltzmann temperature lidar with a Na narrow-band lidar, *Appl. Opt., 37*, 8477–8481, 1998.

Plane, J. M. C., The chemistry of meteoric metals in the Earth's upper atmosphere, *Int. Rev. Phys. Chem., 10*, 55–106, 1991.

Plane, J. M. C., R. M. Cox, J. Qian, W. M. Pfenninger, G. C. Papen, C. S. Gardner, P. J. Espy, Mesospheric Na layer at extreme high latitudes in summer, *J. Geophys. Res., 103 (D6)*, 6381–6389, 1998.

Plane, J. M. C., C. S. Gardner, J. R. Yu, C. Y. She, R. R. Garcia, and H. C. Pumphrey, Mesospheric Na layer at 40°N: modeling and observations, *J. Geophys. Res., 104 (D3)*, 3773–3788, 1999.

Plane, J. M. C., B. J. Murray, X. Chu, and C. S. Gardner, Removal of meteoric iron on polar mesospheric clouds, *Science, 304*, 426–428, 2004.

Portman, R. W., G. E. Thomas, S. Solomon, and R. R. Garcia, The importance of dynamical feedbacks on doubled CO_2-induced changes in the thermal structure of the mesosphere, *Geophys. Res. Lett., 22*, 1733–1736, 1995.

Qian, J., and C. S. Gardner, Simultaneous lidar measurements of mesospheric Ca, Na, and temperature profiles at Urbana, Illinois, *J. Geophys. Res., 100 (D4)*, 7453–7461, 1995.

Raizada, S., and C. Tepley, Iron Boltzmann lidar temperature and density observations from Arecibo — an initial comparison with other techniques, *Geophys. Res. Lett., 29 (12)*, doi: 10.1029/2001GL014535, 2002.

Raizada, S., and C. A. Tepley, Seasonal variation of mesospheric iron layers at Arecibo: first results from low-latitudes, *Geophys. Res. Lett., 30 (2)*, doi: 10.1029/2002GL016537, 2003.

Rapp, M., F.-J. Lübken, A., Müllemann, G. E. Thomas, and E. J. Jensen, Small-scale temperature variations in the vicinity of NLC: experimental and model results, *J. Geophys. Res., 107 (D19)*, 4392, doi: 10.1029/2001JD001241, 2002.

Reid, G. C., Ice clouds at the summer polar mesopause, *J. Atmos. Sci., 32*, 523–535, 1975.

Reise, M., D. Offermann, and G. Brasseur, Energy released by recombination of atomic oxygen and related species at mesopause heights, *J. Geophys. Res., 99,* 14585–14594, 1994.

Roble, R. G., and R. E. Dickinson, How will changes in carbon dioxide and methane modify the mean structure of the mesosphere and thermosphere? *Geophys. Res. Lett.,* 16, 1441–1444, 1989.

Roble, R. G., Energetics of the mesosphere and thermosphere. The upper mesosphere and lower thermosphere: a review of experiment and theory, *Geophys. Monogr., 87,* 1–22, 1995.

Rodgers, C. D., F. W. Taylor, A. H. Muggeridge, M. Lopez-Puertas, M. A. Lopez-Valverde, Local thermodynamic equilibrium of CO_2 in the upper atmosphere, *Geophys. Res. Lett., 19,* 589–592, 1992.

Rollason, R. J., and J. M. C. Plane, The reactions of FeO with O_3, H_2, H_2O, O_2, and CO_2, *Phys. Chem., 2,* 2335–2343, 2000.

Sandford, M. C. W., and A. J. Gibson, Laser radar measurements of the atmospheric sodium layer, *J. Atmos. Terr. Phys., 32,* 1423–1430, 1970.

Schmidlin, F. J., Temperature inversion near 75 km, *Geophys. Res. Lett., 3,* 173–176, 1976.

Schmitz, S., U. von Zahn, J. C. Walling, and D. Heller, Alexandrite lasers for temperature sounding of the sodium layer, in *Proc. 12th ESA Symp. on European Rocket and Balloon Programmes and Related Research*, edited by B. Kaldeich-Schürmann, ESA, SP-370, 395–402, 1995.

Schoeberl, M. R., D. F. Strobel, and J. P. Apruzese, A numerical model of gravity wave breaking and stress in the mesosphere, *J. Geophys. Res., 88,* 5249–5259, 1983.

Senft, D. C., and C. S. Gardner, Seasonal variability of gravity wave activity and spectra in the mesopause region at Urbana, *J. Geophys. Res., 96,* 17229–17264, 1991.

Senft, D. C., G. C. Papen, C. S. Gardner, J. R. Yu, D. A. Krueger, and C. Y. She, Seasonal variations of the thermal structure of the mesopause region at Urbana, IL (40°N, 88°W) and Ft. Collins, CO (41°N, 105°W), *Geophys. Res. Lett., 21,* 821–824, 1994.

She, C. Y., H. Latifi, J. R. Yu, R. J. Alvarez II, R. E. Bills, and C. S. Gardner, Two-frequency lidar technique for mesospheric Na temperature measurements, *Geophys. Res. Lett., 17,* 929–932, 1990.

She, C. Y., J. R. Yu, J. W. Huang, C. Nagasawa, and C. S. Gardner, Na lidar measurements of gravity wave perturbations of wind, density, and temperature in the mesopause region, *Geophys. Res. Lett., 18,* 1329–1331, 1991.

She, C. Y., J. R. Yu, H. Latifi, and R. E. Bills, High-spectral-resolution fluorescence light detection and ranging for mesospheric sodium temperature measurements, *Appl. Opt., 31*, 2095–2106, 1992.

She, C. Y., J. R Yu, and H. Chen, Observed thermal structure of a midlatitude mesopause, *Geophys. Res. Lett., 20*, 567–570, 1993.

She, C. Y., and J. R. Yu, Simultaneous three-frequency Na lidar measurements of radial wind and temperature in the mesopause region, *Geophys. Res. Lett., 21*, 1771–1774, 1994.

She, C. Y., and J. R. Yu, Doppler-free saturation fluorescence spectroscopy of Na atoms for atmospheric application, *Appl. Opt., 34*, 1063–1075, 1995.

She, C. Y., J. R. Yu, D. A. Krueger, R. Roble, P. Keckhut, A. Hauchecorne, and M.-L. Chanin, Vertical structure of the midlatitude temperature from stratosphere to mesopause (30–105 km), *Geophys. Res. Lett., 22*, 377–380, 1995.

She, C. Y., and U. von Zahn, Concept of a two-level mesopause: support through new lidar observations, *J. Geophys. Res., 103 (D5)*, 5855–5863, 1998.

She, C. Y., S. Chen, Z. Hu, J. Sherman, J. D. Vance, V. Vasoli, M. A. White, J. R. Yu, and D. A. Krueger, Eight-year climatology of nocturnal temperature and sodium density in the mesopause region (80 to 105 km) over Fort Collins, CO (41°N, 105°W), *Geophys. Res. Lett., 27*, 3289–3292, 2000.

She, C. Y., S. Chen, B. P. Williams, Z. Hu, and D. A. Krueger, Tides in the mesopause region over Fort Collins, Colorado (41°N, 105°W) based on lidar temperature observations covering full diurnal cycles, *J. Geophys. Res., 107 (D18)*, 4350, doi: 10.1029/2001JD001189, 2002a.

She, C. Y., J. D Vance, B. P. Williams, D. A. Krueger, H. Moosmüller, D. Gibson-Wilde, and D. Fritts, Lidar studies of atmospheric dynamics near polar mesopause, *EOS, 83 (27)*, 289–293, 2002b.

She, C. Y., J. Sherman, J. D. Vance, T. Yuan, Z. Hu, B. P. Williams, K. Arnold, P. Acott, D. A. Krueger, Evidence of solar cycle effect in the mesopause region: observed temperatures in 1999 and 2000 at 98.5 km over Fort Collins, CO (41°N, 105°W), *J. Atmos. Sol. Terr. Phys., 64*, 1651–1657, 2002c.

She, C. Y., J. Sherman, T. Yuan, B. P. Williams, K. Arnold, T. D. Kawahara, T. Li, L. F. Xu, J. D. Vance, P. Acott, and D. A. Krueger, The first 80-hour continuous lidar campaign for simultaneous observation of mesopause region temperature and wind, *Geophys. Res. Lett., 30 (6)*, 1319, doi: 10.1029/2002GL016412, 2003.

She, C. Y., and D. A. Krueger, Impact of natural variability in the 11-year mesopause region temperature observation over Fort Collins, CO (41°N, 105°W), *Adv. Space Res.*, *34*, 330–336, 2004.

She, C. Y., Initial full-diurnal-cycle lidar observations: variability in mesopause region diurnal-means and diurnal tides over Fort Collins, CO (41°N, 105°W), *J. Atmos. Sol. Terr. Phys.*, *66*, 663–674, 2004.

Shepherd, G. G., *Spectral Imaging of the Atmosphere*, Academic Press, New York, 2002.

States, R. J., and C. S. Gardner, Influence of the diurnal tide and thermospheric heat sources on the formation of mesospheric temperature inversion layers, *Geophys. Res. Lett.*, *25*, 1483–1486, 1998.

States, R. J., and C. S. Gardner, Structure of the mesospheric Na layer at 40°N latitude: seasonal and diurnal variations, *J. Geophys. Res.*, *104 (D9)*, 11783–11798, 1999.

States, R. J., and C. S. Gardner, Thermal structure of the mesopause region (80–105 km) at 40°N latitude. Part I: seasonal variations, *J. Atmos. Sci.*, *57*, 66–77, 2000a.

States, R. J., and C. S. Gardner, Thermal structure of the mesopause region (80–105 km) at 40°N latitude. Part II: diurnal variations, *J. Atmos. Sci.*, *57*, 78–92, 2000b.

Stebel, K., V. Barabash, S. Kirkwood, J. Siebert, and K. H. Fricke, Polar mesosphere summer echoes and noctilucent clouds: simultaneous and common-volume observations by radar, lidar and CCD camera, *Geophys. Res. Lett.*, *27*, 661–664, 2000.

Stevens, M. H., R. R. Conway, C. R. Englert, M. E. Summers, K. U. Grossmann, and O. A. Gusev, PMCs and the water frost point in the Arctic summer mesosphere, *Geophys. Res. Lett.*, *28*, 4449–4452, 2001.

Stroud, W. G., W. Nordberg, W. R. Bandeen, F. L. Bartman, and R. Titus, Rocket-grenade observations of atmospheric heating in the Arctic, *J. Geophys. Res.*, *64*, 1342–1343, 1959.

Sullivan, H. M., and D. M. Hunten, Lithium, sodium, potassium in the twilight airglow, *Can. J. Phys.*, *42*, 937–956, 1964.

Summers, M. E., R. R. Conway, C. R. Englert, D. E. Siskind, M. H. Stevens, J. M. Russell III, L. L. Gordley, and M. J. McHugh, Discovery of a water vapor layer in the Arctic summer mesosphere: implications for polar mesospheric clouds, *Geophys. Res. Lett.*, *28*, 3601–3604, 2001.

Tao, X., and C. S. Gardner, Heat flux observations in the mesopause region above Haleakala, *Geophys. Res. Lett.*, *22*, 2829–2832, 1995.

Taylor, M. J., M.Gadsden, R. P. Lowe, M. S. Zalcik, and J. Brausch, Mesospheric cloud observations at unusually low latitudes, *J. Atmos. Sol. Terr. Phys., 64*, 991–999, 2002.

Tepley, C. A., S. Raizada, Q. Zhou, and J. S. Friedman, First simultaneous observations of Ca^+, K, and electron density using lidar and incoherent scatter radar at Arecibo, *Geophys. Res. Lett., 30 (1)*, doi: 10.1029/2002GL015927, 2003.

Thayer, J. P., N. Nielsen, and J. Jacobsen, Noctilucent cloud observations over Greenland by a Rayleigh lidar, *Geophys. Res. Lett., 22*, 2961–2964, 1995.

Thayer, J. P., M. Rapp, A. J. Gerrard, E. Gudmundsson, T. J. Kane, Gravity-wave influences on Arctic mesospheric clouds as determined by a Rayleigh lidar at Sondrestrom, Greenland, *J. Geophys. Res., 108 (D8)*, 8449, doi: 10.1029/2002JD002363, 2003.

Thomas, G. E., and J. J. Olivero, Climatology of polar mesospheric clouds 2. Further analysis of Solar Mesosphere Explorer data, *J. Geophys. Res., 94*, 14673–14681, 1989.

Thomas, G. E., R. D. McPeters, and E. J. Jensen, Satellite observations of polar mesospheric clouds by the solar backscattered ultraviolet spectral radiometer: evidence of a solar cycle dependence, *J. Geophys. Res., 96*, 927–939, 1991.

Thomas, G. E., Mesospheric clouds and the physics of the mesopause region, *Rev. Geophys., 29*, 553–575, 1991.

Thomas, G. E., Is the polar mesosphere the miner's canary of global change?, *Adv. Space Res., 18*, 149–158, 1996.

Thomas, L., and S. K. Bhattacharyya, Mesospheric temperatures deduced from laser observations of the Na D_2 line profile, *Proc. 5th ESA-PAC Symp. Europ. Rocket & Balloon Programmes & Related Res.* (ESA SP-152), 49–50, 1980.

Thomas, L., A. K. P. Marsh, D. P. Wareing, and M. A. Hassan, Lidar observations of ice crystals associated with noctilucent clouds at middle latitudes, *Geophys. Res. Lett., 21*, 385–388, 1994.

Thomas, L., Lidar methods and applications, in *Spectroscopy in Environmental Science*, Chapter 1, edited by R. J. H. Clark and R. E. Hester, Wiley & Sons Ltd, New York, 1995.

Turco, R. P., O. B. Toon, R. C. Whitten, R. G. Keesee, and D. Hollenbach, Noctilucent clouds: simulation studies of their genesis, properties and global influences, *Planet. Space Sci., 30*, 1147–1181, 1982.

Vance, J. D., C. Y. She, and H. Moosmüller, Continuous-wave, all-solid-state, single-frequency 400-mW source at 589 nm based on doubly resonant sum-frequency mixing in a monolithic lithium niobate resonator, *Appl. Opt., 37,* 4891–4896, 1998.

Vincent, R.A., and I. M. Reid, HF Doppler measurements of mesopheric gravity wave momentum fluxes, *J. Atmos. Sci., 40,* 1321–1333, 1983.

Volz, U., M. Majerus, H. Liebel, A. Schmitt, and H. Schmoranzer, Precision lifetime measurements on NaI 3p $^2P_{1/2}$ and 3p $^2P_{3/2}$ by beam-gas-laser spectroscopy, *Phys. Rev. Lett., 76,* 2862–2865, 1996.

von Cossart, G., P. Hoffmann, U. von Zahn, P. Keckhut, and A. Hauchecorne, Mid-latitude noctilucent cloud observations by lidar, *Geophys. Res. Lett., 23,* 2919–2922, 1996.

von Cossart, G., J. Fiedler, U. von Zahn, G. Hansen, U. P. Hoppe, Noctilucent clouds: one- and two-color lidar observations, *Geophys. Res. Lett., 24,* 1635–1638, 1997.

von Cossart, G., J. Fiedler, and U. von Zahn, Size distributions of NLC particles as determined from 3-color observations of NLC by ground-based lidar, *Geophys. Res. Lett., 26,* 1513–1516, 1999.

von der Gathen, P., Saturation effects in Na lidar temperature measurements, *J. Geophys. Res., 96 (A3),* 3679–3690, 1991.

von Zahn, U., P. von der Gathen, and G. Hansen, Forced release of sodium from upper atmospheric dust particles, *Geophys. Res. Lett., 14,* 76–79, 1987.

von Zahn, U., and T. L. Hansen, Sudden neutral sodium layers: a strong link to sporadic E layers, *J. Atmos. Terr. Phys., 50,* 93–104, 1988.

von Zahn, U., and J. Höffner, Mesopause temperature profiling by potassium lidar, *Geophys. Res. Lett., 23,* 141–144, 1996.

von Zahn, U., J. Höffner, V. Eska, and M. Alpers, The mesopause altitude: only two distinctive levels worldwide? *Geophys. Res. Lett., 23,* 3231–3234, 1996.

von Zahn, U., G. von Cossart, J. Fiedler, and D. Rees, Tidal variations of noctilucent clouds measured at 69°N latitude by ground-based lidar, *Geophys. Res. Lett., 25,* 1289–1292, 1998.

von Zahn, U., and J. Bremer, Simultaneous and common-volume observations of noctilucent clouds and polar mesospheric summer echoes, *Geophys. Res. Lett., 26,* 1521–1524, 1999.

von Zahn, U., J. Höffner, and W, J. McNeil, Meteor trails as observed by lidar, in *Meteors in the Earth's Atmosphere,* edited by E. Murad and I. P. Williams, Cambridge University Press, Cambridge, UK, pp. 149–187, 2002.

von Zahn, U., and U. Berger, Persistent ice cloud in the midsummer upper mesosphere at high latitudes: three-dimensional modeling and cloud interactions with ambient water vapor, *J. Geophys. Res., 108 (D8),* 8451, doi: 10.1029/2002JD002409, 2003a.

von Zahn, U., and U. Berger, The altitude of noctilucent clouds: groundbased observations and their interpretation through numerical modeling, *Proc. 16th ESA Symp. European Rocket & Balloon Programmes and Relat. Res.,* St. Gallen, Switzerland, June 2003b.

Walterscheid, R. L., Dynamical cooling induced by dissipating internal gravity waves, *Geophys. Res. Lett., 8,* 1235–1238, 1981.

Welsh, B. M., and C. S. Gardner, Nonlinear resonant absorption effects on the design of resonance fluorescence lidars and laser guide stars, *Appl. Opt., 28,* 4141–4152, 1989.

White, M. A., A Frequency-Agile Na Lidar for the Measurement of Temperature and Velocity in the Mesopause Region, Ph.D. thesis, Colorado State University, Fort Collins, Colorado, 1999.

Wickwar, V. B., M. J. Taylor, J. P. Herron, B. A. Martineau, Visual and lidar observations of noctilucent clouds above Logan, Utah, at 41.7°N, *J. Geophys. Res., 107 (D7),* doi: 10.1029/2001JD001180, 2002.

Witt, G., The nature of noctilucent clouds, *Space Res. IX,* 157–169, 1969.

Yariv, A., *Quantum Electronics*, 3rd edition, John Wiley & Sons, New York, 1988.

Yariv, A., *Optical Electronics in Modern Communications*, 5th edition, Oxford University Press, New York, 1997.

Yeh, P., Dispersive magnetooptic filters, *Appl. Opt., 21,* 2069–2075, 1982.

Yi, F, S. Zhang, H. Zeng, Y. He, X. Yue, J. Liu, H. Lv, and D. Xiong, Lidar observations of sporadic Na layers over Wuhan (30.5°N, 114.4°E), *Geophys. Res. Lett., 20 (9),* doi: 10.1029/2001GL014353, 2002.

Yin, B., and T. M. Shay, Theoretical model for a Faraday anomalous dispersion optical filter, *Opt. Lett., 16,* 1617–1619, 1991.

Yu, J. R., and C. Y. She, Climatology of a midlatitude mesopause region observed by a lidar at Fort Collins, Colorado (40.6°N, 105°W), *J. Geophys. Res., 100 (D4),* 7441–7452, 1995.

Yu, J. R., R. J. States, S. J. Franke, C. S. Gardner, and M. E. Hagan, Observations of tidal temperature and wind perturbations in the mesopause region above Urbana, IL (40°N, 88°W), *Geophys. Res. Lett., 24,* 1207–1210, 1997.

Zhao, Y., Stability of the Mesopause Region: Influence of Dissipating Gravity Waves on the Transport of Heat, Momentum and Constituents, Ph.D. thesis, University of Illinois at Urbana-Champaign, Urbana, Illinois, 2000.

Zhao, Y., A. Z. Liu, and C. S. Gardner, Measurements of atmospheric stability in the mesopause region at Starfire Optical Range, NM, *J. Atmos. Sol. Terr. Phys., 65*, 219–232, 2003.

6

Fluorescence Spectroscopy and Imaging of Lidar Targets

SUNE SVANBERG

Department of Physics, Lund Institute of Technology,
Lund, Sweden

6.1. INTRODUCTION

Our perception of the world is largely based on our vision, where colors are perceived as a result of the different reflectance properties of materials. Sunlight, or the sky, has a well-defined spectral distribution and when impinging on a surface, those parts of the spectrum that match some transition of the molecules are absorbed, leaving the others the freedom to scatter and eventually reach our eye. Through the interaction with the different types of cones in the retina and the subsequent processing in the brain, the perception of color occurs. The eye is sensitive to a very limited range of the electromagnetic spectrum: from 400 to 700 nm. A grass lawn being identified as green is due to chlorophyll absorption in the blue and red spectral regions, while the red color of blood is the result of strong hemoglobin absorption all through the visible spectrum up to 600 nm. Outside the visible range, selective absorption occurs, in the ultraviolet (UV) and infrared (IR) ranges, but here the features must be recorded by different types of electronic detectors. Satellite remote sensing using whiskbroom or pushbroom multispectral sensors is widely used for earth resource monitoring and classification.[1,2]

    It is well known that the fluorescence phenomenon can provide information in addition to that obtained by absorptive or reflective spectroscopy. Things that are hidden to the naked eye or normal photography are revealed using fluorescence techniques. The use of fluorescence in microscopy,[3] forensic science,[4] art inspection,[5] and tissue diagnostics[6] is well known. All these applications are carried out in indoor controlled environments, although sometimes competing background light, for example, operation lamps, make pulsed excitation and time-gated detection necessary.[7]

From the discussion just given it seems evident that fluorescence techniques should also be able to provide valuable diagnostic information in the outdoor environment. Clearly, distances are frequently large and the background light conditions are hard to control in this case, calling for special techniques. This chapter deals with such techniques, fluorescence lidar techniques.[8] Lidar (*light detection and ranging*) was first developed for atmospheric probing of aerosols and trace gases. The special feature of an atmospheric lidar is to provide range resolution in a radar-like fashion. A pulsed laser transmitter is used, and backscattered radiation is collected with an optical telescope and recorded by a photodetector connected to a transient digitizer. Atmospheric lidar techniques are discussed in Refs. 9, and chapter 3 of this book. A particular powerful variety is the differential absorption lidar (dial) technique for trace species detection, which is reviewed in, for example, Refs. 10, and chapter 4 of this book.

A straightforward variety of the lidar technique is when the beam is directed toward a solid target and the distance to that target is determined by measuring the time delay for the detection of the distinct echo (laser range finder). The echo is detected in the elastic backscattering of the laser light; that is, the detector is equipped with a filter letting the backscattered laser light pass, but blocking the ambient light for a better signal-to-background ratio. If instead the elastic backscattering is blocked, and the fluorescence induced by the laser pulse hitting the target is wavelength dispersed and detected, a fluorescence lidar system has been established. To suppress background light, which now enters the wide wavelength region needed for capturing the fluorescence light distribution, the detection is restricted to a narrow temporal window corresponding to the time of the arrival of the pulse. In this way, fluorescence can be detected remotely, even in the presence of full daylight.

This chapter on fluorescence lidar on solid or liquid targets is organized as follows. The phenomenon of fluorescence is described and illustrations of spectra are given in the next section. Then, in Section 6.3, scenarios and equipment for fluorescence lidar measurements are described. Point monitoring as well as imaging techniques are described. Different fluorescence lidar applications are then illustrated in Section 6.4. We will start with marine monitoring, including oil and phytoplankton studies. Terrestrial vegetation is considered and, finally, fluorescence studies of historical monuments are presented. For the convenience of the author, most examples are taken from research activities of the Lund Institute of Technology group. In the final section, possible future developments are discussed.

6.2. FLUORESCENCE

In contrast to free atoms and molecules, solids and liquids exhibit broad absorption and emission spectra because of the strong intermolecular interactions. The principal arrangement of the energy levels and the basic absorption and emission processes are shown in Figure 6.1.[11] By electronic excitation, visible or UV photons can be absorbed by the molecules of the target, with promotion of an outer electron to a higher orbit. Each energy level corresponding to a particular electron arrangement has a vibrational structure, while the rotational structure, typical for free molecules, is suppressed by the intermolecular interaction, which also smears the vibration structure. In complex molecules, there is a large number of vibrational levels. The net result is that not much can be inferred from the absorption and emission spectra with respect to structure. Most of the molecules in the ground state are in the lowest vibrational levels, and from here they can be excited provided that the photon energy absorbed falls within the distribution of excited-state sublevels. In contrast to the situation for free molecules, a fixed frequency laser can be used for the excitation, provided that the photon energy falls within the broad excitation band. Following the excitation there is a very fast (picosecond time scale) radiationless relaxation down to the lowest substate, where the molecules remain for a typical excited-state fluorescence lifetime, which is typically of the order of some nanoseconds. The decay can then occur to different sublevels of the ground state giving

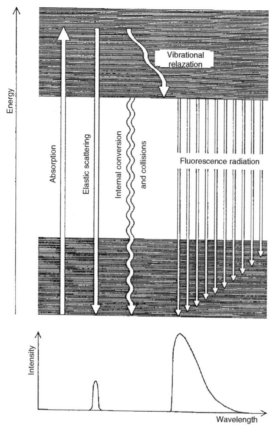

Figure 6.1 Schematic molecular energy level diagram and absorption and emission/fluorescence processes. In the lower part of the figure, the resulting fluorescence spectrum is shown, with a strongly filter-attenuated elastic scattering peak included. (From Andersson, P.S. et al., *Lasers Med. Sci.*, 2, 41, 1987. With permission.)

rise to a distribution of fluorescence light, which reflect the lower-state level distribution. Because of the radiationless transitions, the fluorescence distribution is independent of the exact excitation wavelength. However, the overall intensity depends on the excitation, which can be more or less efficient. Actually, by fixing the detection channel and varying the excitation wavelength, an excitation spectrum can be recorded. Thus, fluorescence spectroscopy provides information on the system studied from the excitation and fluorescence spectra.

Further, the lifetime is also characteristic. Frequently, natural materials consist of a mixture of different molecules, which results in the lifetime being dependent on the fluorescence wavelength. Fluorescence spectroscopy is discussed in a number of monographs and reviews.[12–14]

Like in all remote sensing applications, a good basis is needed in "ground-truth" measurements, where standard spectra are recorded under well-controlled conditions. Spectra can be recorded using a point-monitoring setup of the kind described in Figure 6.2.[15] The fluorosensor depicted operates with a pulsed nitrogen laser ($\lambda = 337$ nm) as the primary excitation source. Frequency conversion to longer excitation wavelengths can also be accomplished in a compact dye laser unit. The radiation is focused into an optical fiber, the distal end of which is brought to the sample under study. Laser-induced fluorescence is collected by the same fiber and is brought back to the entrance slit of a spectrometer after passing a dichroic mirror, reflecting the laser light by transmitting the red-shifted fluorescence light. In the focal plane of the spectrometer, there is a CCD detector, preceded by a time-gated image intensifier. The gate is opened electronically only to accept light during a narrow time window, normally set to 100 nsec. Background light can be efficiently eliminated

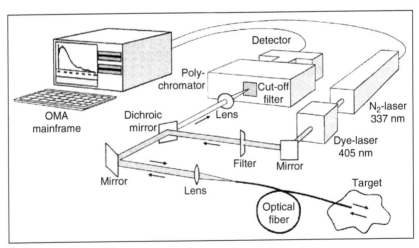

Figure 6.2 Fluorosensor for fiber-optic point monitoring. (From af Klinteberg, C., et al., *Proc. SPIE*, 2926, 1996. With permission.)

in this way. Following each laser shot, a full spectrum covering 350 to 800 nm is recorded and multiple shots can be averaged to increase the signal-to-noise ratio. The system can not only be used in the laboratory but can also be brought into the field with electric power supplied from a small motor generator and with a sufficiently long optical fiber to be brought in contact with the natural surface to be studied. Examples of spectra of living vegetation are shown in Figure 6.3,[16] as recorded with a setup of the type shown in Figure 6.2. The chlorophyll fluorescence is clearly visible for 405 nm excitation with peaks at 690 and 735 nm, while signals due to leaf structures and wax layer appear as weak bands in the blue–green spectral region. On the contrary, this latter fluorescence is very strong for 337 nm excitation when the chlorophyll signals are absent, since the UV light does not penetrate the protective layer. The fluorescence signals from vegetation depend in a complex way on many parameters as discussed, e.g., in Refs. 17, 18. For example, the intensity ratio between the two dark red peaks is related to the concentration of leaf chlorophyll. Stress due to draught, ozone, or excess UV exposure is reflected in the spectra.

With the development of violet and blue diode lasers, very compact fluorosensor constructions have become possible. The system shown in Figure 6.4a utilizes a continuous wave (CW) detection scheme with fluorescence being dispersed and detected in an integrated spectrometer.[19] Spectra recorded by bringing the fiber tip in contact with different parts of a beech leaf are shown in Figure 6.4b.[19]

Most frequently, the time-integrated fluorescence following pulsed or CW excitation is recorded by the fluorosensor. However, since all molecular species and materials have characteristic temporal decay patterns, time-resolved detection of fluorescence can provide additional diagnostic information of the sample under study. Time-resolved studies of the excited-state lifetimes have been performed in laboratory measurements for oil samples and leaves.[20] Remote fluorescence measurements must provide means to isolate temporal delays due to excited-state lifetime from temporal delays due to larger physical distance in a three-dimensional target, such as a vegetation canopy. We will return to these aspects later.

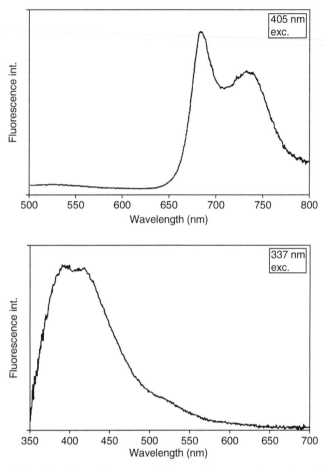

Figure 6.3 Spectra of spruce (*Picea abies* growing at an Italian test site at Camporgiano) recorded with a fiber-optic fluorosensor using 337 and 405 nm excitation. (From Edner, H. et al., *EARSEL Adv. Remote Sens.*, 1, 119, 1992. With permission.)

6.3. REMOTE FLUORESCENCE RECORDING

6.3.1. General Considerations

In order to facilitate remote fluorescence recordings, the fluorescence detection techniques must be integrated with the ranging capability of a lidar system. Two scenarios for remote measurements from a land mobile system are included in

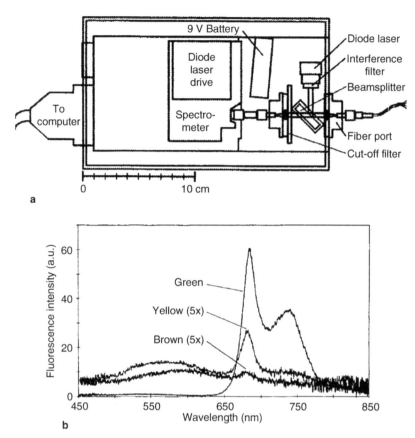

Figure 6.4 (a) Compact blue fluorosensor based on a violet diode laser. (b) Fluorescence spectra from different parts of a beech leaf, showing green, yellow, or brown color in reflectance. (From Gustafsson, U. et al., *Rev. Sci. Instrum.*, 71, 3004, 2000. With permission.)

Figure 6.5; one indicating aquatic monitoring via a folding mirror and one showing vegetation monitoring. Typical recorded spectra are also included for 355 nm excitation. Water fluorescence is dominated by the blue–green distribution due to distributed organic matter and with a faint structure peaking at 690 nm due to microscopic algae (plankton) and the O–H stretch Raman signal of water peaking at 404 nm. The vegetation spectral illustration shows signals due to fully green and slightly yellowing beech leaves, with

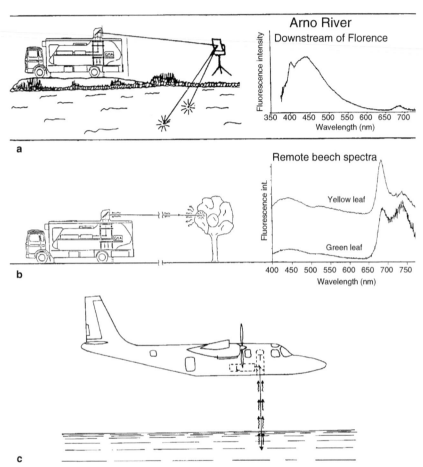

Figure 6.5 Scenarios for fluorescence lidar remote monitoring: (a) land-mobile system in aquatic monitoring; (b) corresponding vegetation monitoring; and (c) airborne fluorosensing. (From Svanberg, S., Differential absorption lidar, in M. Sigrist (ed.), *Air Monitoring by Spectroscopic Techniques*, Wiley, New York, 1994; Celander, L. et al., Göteborg Institute of Physics Reports, GIPR-149, 1978. This work is used with permission from John Wiley & Sons, Inc.)

the typical change in the relative intensities of the 690 and 735 nm peaks as discussed above. The measurement distance was in both cases of the order of 50 m. Included in Figure 6.5 is also a scenario with an airborne fluorescence lidar system.

The airborne implementation allows the coverage of large areas and is a preferred arrangement especially for marine monitoring.

Fluorescence recordings can be performed in a chosen point, as illustrated above, or in an imaging mode with very high information content. Both aspects will be illustrated in this chapter.

6.3.2. Instrumentation

A fluorescence lidar system consists of, like all other lidar systems, a laser transmitter, receiving telescope, and detection system. A diagram of a possible setup is shown in Figure 6.6.[22] Based on this figure, we will discuss the different components in a system in general terms.

6.3.2.1. Transmitters

Generally, a pulsed laser operating in the UV region is the source of fluorescence excitation. In principle, a UV flash lamp can also be utilized. As a matter of fact, the pulse energy in such a flash can be larger than that from a laser, but it is difficult to combine high energy with a short pulse length. Thus, flash-lamp-based systems have problems with the ambient background due to sunlight or daylight, since a very long recording time is needed to collect the induced fluorescence and during this time also a substantial background accumulates. While a flash-lamp system yields good signals at night, it cannot compete with a short-pulse UV laser during daytime. These types of fluorosensor considerations are discussed in Ref. 23, and we will return to this topic in Section 6.3.2.4.

The most common fluorescence lidar transmitters are Nd:YAG lasers and excimer lasers. An Nd:YAG laser primarily transmits at 1.064 μm, and the frequency tripled or quadrupled output at 355 and 266 nm, respectively, is utilized for inducing fluorescence. Pulse energies range from a few millijoules to hundreds of millijoules or more in a pulse length of 5 to 10 nsec and at a repetition rate rarely exceeding a few tens of Hertz. These solid-state lasers are pumped by flashlamps. A frequency-tripled Nd:YAG laser is included in the set

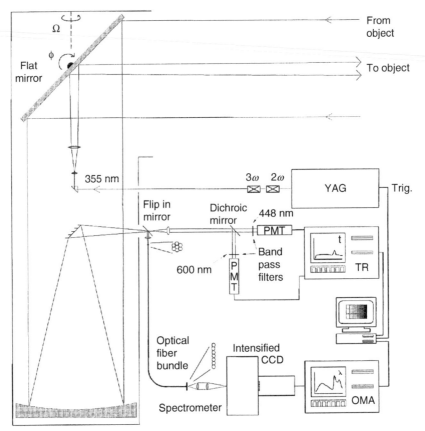

Figure 6.6 Schematic diagram of a fluorescence lidar system with filter-based and optical multichannel analyzing detection. (From Weibring, P. et al., *Appl. Opt.*, 40, 6111, 2001. With permission.)

up shown in Figure 6.6. By using a tuneable laser source, recordings taken for a number of excitation wavelengths can help in the discrimination of target materials. An optical parametric oscillator (OPO), pumped by an Nd:YAG laser, provides great flexibility.[24] When high repetition rates are needed excimer lasers are useful. They operate on a halogen/ inert gas mixture as the active medium; in a fast electric discharge, excited dimers (excimers) of the type KrF, XeCl, and XeF are formed, giving rise to emission at 249, 308, and 351 nm, respectively. Pulse energies typically range from a few to tens of millijoules in pulses of 10 to 20 nsec duration. Repetition rates of a few hundreds of hertz are achievable and are useful in airborne scanning/imaging applications. Nitrogen lasers, using N_2 molecules, operate in a similar way as the excimer lasers and emit at 337 nm at low pulse output energies. Compact sealed-off devices with battery operation can be achieved and are useful in short-range lidar fluorosensors useful in industrial applications.[25]

6.3.2.2. Optics

The laser radiation transmitted toward the target is shaped using a beam expander, which can be a Galileian telescope with quartz lenses. Such a telescope does not have an internal focal point, and thus an air breakdown due to very high optical field strengths is avoided. By adjusting the telescope, the beam divergence of the radiation is set. The divergence is chosen quite large in the case of simultaneous imaging (to be discussed later). For point monitoring the spot on the target is kept small but still finite to avoid very high target irradiance. The type of detection also influences the desirable size of the target spot (as will be discussed further in the next section). By using cylinder optics, it is also possible to distribute the laser light along a line for line-imaging (pushbroom) applications.

Fluorescence is recorded with a telescope that should be large enough to allow a reasonable measurement range. Typical diameters could range from 10 to 50 cm. All-reflective optics is preferred to achieve an achromatic system allowing fluorescence detection over the full visible spectral range. A steerable folding mirror, as indicated in Figure 6.6, is convenient for selection of the measurement point. In imaging airborne applications, a fast scanning mirror performs a perpendicular (whiskbroom) sweep, while the movement of the platform provides the longitudinal sweep.

6.3.2.3. Detectors

Two types of detection systems for point monitoring are indicated in Figure 6.6. A beam splitter, preferably of a dichroic type for improved photon economy, is used for sending the light to two individual photomultiplier tubes (PMTs) preceded by properly selected band-pass filters. The PMTs are connected to the inputs of a dual-channel transient digitizer, and the range-localized fluorescence echoes are digitally integrated. While the individual channel signals are strongly influenced by laser pulse energy, range, and angle of incidence, the ratio of the two signals, constituting a dimensionless quantity, is immune to such influences.

Alternatively, and preferably, the fluorescence collected from the target can be focused down into a fiber bundle and be conducted to the entrance slit of a spectrometer. The individual fibers can be rearranged from a circular shape to a linear shape to match the slit. We note that the diameter of the illuminated spot must be kept small enough to allow all the light to be collected by the fiber bundle. The CCD detector of the spectrometer is preceded by a single- or double-stage image intensifier, which can be time gated to accept only light with the right delay, thus suppressing ambient light.

Imaging can be performed by sequential scanning of the beam over the target or by simultaneous monitoring of an image line across the target, or even by simultaneous monitoring of the full target. We will now discuss the conditions for these different types of imaging.

6.3.2.4. Strategies for Fluorescence Imaging

Fluorescence imaging is governed by primarily two considerations: stationary or moving targets, and signal-to-background conditions. Three different measurement strategies are schematically illustrated in Figure 6.7. Since sunlight or daylight has substantial intensity over the whole visible region, it constitutes a strong background competing with the induced fluorescence. Gated detection of a pulsed signal provides a first remedy by concentrating the signal to a gate time, during which few background photons impinge. We have

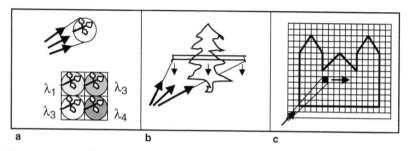

Figure 6.7 Scenarios for remote fluorescence spectroscopy and fluorescence imaging with special consideration of background: (a) simultaneous four-color imaging; (b) pushbroom imaging; and (c) scanning pointwise imaging.

already discussed this advantage of pulsed experiments over CW measurements in Section 6.3.2.1. However, the exciting beam target imprint cannot be too large, since then the specific surface irradiation becomes too low compared to the competing background. Thus, there is a limit to the diameter to which the beam can be expanded in simultaneous imaging (a), to be further discussed in connection with figure 6.11. For larger targets the photons have to be concentrated along a line in a pushbroom scanning scenario, where the recording is made line by line (b), or to a point, which is sequentially moved (whiskbroom scanning) to cover the full scene (c). Clearly, the target is required to be fixed for the scanning of types (b) and (c).

6.4. ILLUSTRATIONS OF FLUORESCENCE LIDAR APPLICATIONS

6.4.1. Marine Monitoring

Marine monitoring using ship- or airborne systems have been performed for a long time. Initial work was much focused on laser bathymetry[8,9,26,27] directed toward shallow water depth measurements using elastic backscattering. Depths down to tens of meters could be achieved, depending on the type of ocean waters. Some work related to fish school detection was also performed.[28,29] In order to get a more favorable relation between the strong surface reflection and the week sea floor

return, polarized laser light is employed, and crossed polarization detection is used, strongly suppressing the largely
polarized surface return while favoring the depolarized
sea-floor signal.

A range-resolved signal from the water column is also
obtained in bathymetric measurements, and carries information on the amount of scattering particles in the water. The
inelastic process of Raman scattering also has an effect in
these types of measurements. Due to the full concentration
of water the O–H stretch vibrational mode at a Stokes shift of
about $3400\,cm^{-1}$, that is, at $404\,nm$ if a frequency-tripled
Nd:YAG laser at $355\,nm$ is employed, is quite prominent.
Using a high spectral resolution detection system, it is possible to analyze the detailed shape of the Raman signal, which
is temperature dependent. This is because the aggregation of
water molecules depends on the temperature. This effect has
been utilized in ship-borne water temperature measurements.[30,31]

Much work on water fluorescence recording has been
performed. Land-based, ship-borne, and airborne campaigns
have been pursued. An important aspect of water fluorescence
monitoring is the assessment or the general level of dissolved
organic matter (DOM) in water (sometimes referred to as
Gelbstoff). Water fluorescence for different samples studied
in the laboratory is shown in Figure 6.8a for laboratory measurements using nitrogen-laser excitation at $337\,nm$. It can be
clearly seen that distilled water produces little DOM signal in
the blue–green spectral region, while river water yields a
strong signal. By normalizing the DOM signal to the back-
ground-free water Raman signal a convenient build in reference is obtained, and a knowledge of the effective sampling
volume is no longer required. Similar spectra, but now
recorded remotely at the Arno river in the scenario shown in
Figure 6.5a, are displayed in Figure 6.8b for $355\,nm$ excitation. The increased level of DOM, rising by a factor of 2 in
Raman-normalized units, when moving from a measurement
location upstream of Florence to the river mouth at the Tyr-
renian Sea is evident.

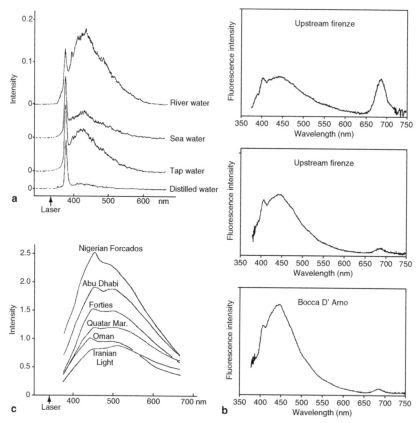

Figure 6.8 (a) Laboratory spectra obtained for different water samples using a nitrogen laser excitation source. (From Celander, L. et al., Göteborg Institute of Physics Reports, GIPR-149, 1978; Svanberg, S., Laser fluorescence spectroscopy in environmental monitoring, in S. Martellucci and A.N. Chester (eds.), *Optoelectronic for Environmental Science*. Plenum Press, New York, 1990.) (b) Remotely recorded water spectra from the Arno River, excited at 355 nm from a distance of about 40 m. (From Edner, H. et al., *EARSEL Adv. Remote Sens.*, 1, 42, 1992.) (c) Laboratory spectra for different crude oils, excited at 337 nm. (From Celander, L. et al., Göteborg Institute of Physics Reports, GIPR-149, 1978; Weibring, P. et al., *Appl. Opt.*, 40, 6111, 2001. With permission.)

Figure 6.8b also provides a good introduction to the remote monitoring of phytoplankton. The signal peaking at 690 nm is due to chlorophyll a, present in all algae. It can be seen that the Raman-normalized concentration is reduced by a factor of almost 10 when moving toward the sea. Some algae also contain the accessory pigments phycocyanine and phycoerytrine, helping to utilize the predominantly available blue–green light penetrating larger depths. The associated fluorescence peaks occur at about 660 and 590 nm, respectively.

Coastal zones can be monitored from land-based systems since the coupling into the sea for nominally almost 90° angle of incidence is enhanced, because the waves form a prismatic action on light hitting the steep wave slopes.[34,35]

Several airborne fluorescence lidar systems for water column measurements have been developed and tested.[36–39] In principle, range-resolved fluorescence data can be taken by gating the image intensifier at different delays. Such data from a ship-borne campaign in the Mediterranean Sea are included in Figure 6.9,[40] which also shows surface waters of different types.

The strongest signals received in aquatic monitoring are from oil spills covering the surface. At the same time as a strong oil fluorescence is replacing the DOM signal in the blue–green region the water Raman signal disappears, since even a layer only a few micrometers thin becomes optically thick for the exciting UV radiation. Thus, oil spills have a very clear spectral signature. Laboratory studies of different types of crude oils and lighter fractions show that oils of specific origins can be identified (see Figure 6.8c). A special challenge is to identify weathered oil that displays a modified fluorescence profile or oil that has sunk after extended time in the water. Results from airborne oil-slick monitoring campaigns are given in Refs. 41–44.

6.4.2. Vegetation Monitoring

Vegetation is the main object of air- and satellite-borne remote sensing. Fluorescence monitoring can supply additional information. Early works are discussed in Refs. 21, 32, 45, 46. An

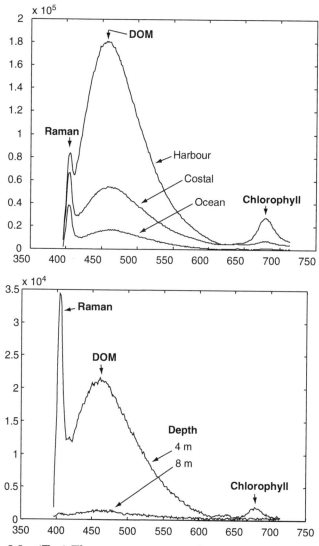

Figure 6.9 (Top) Fluorescence curves for surface waters recorded in ship-borne measurements in the Mediterranean Sea, when approaching the port of Civitavecchia. (Bottom) Depth-resolved fluorescence curves of Mediterranean water. (From Cecchi, G. et al., CNR Report RR/OST/01.03 (CNR-IFAC "Nello Carrara", Firenze, 2003). With permission.)

earlier review of this subfield is provided in Ref. 49. We will first consider the remote recording of point spectra of vegetation. Figure 6.10 shows four such spectra[48,49] as recorded with a system of the kind shown in Figure 6.6. Poplar, cypress, and plane-tree spectra are shown as integrated from 100 laser shots and at distances of 64, 125, and 210 m, respectively. In addition, a single-shot cypress spectrum captured at 125 m distance is shown for comparison with the corresponding 100-shot spectrum. It can be seen that the main fluorescence features are still well discernible.

Vegetation fluorescence monitoring can be extended to multicolor imaging as demonstrated in Refs. 50, 51. When the area imaged is limited to 1 m in diameter it is possible to use the method indicated in Figure 6.7a. By using split-mirror Cassegrainian optics adopted in the fluorescence

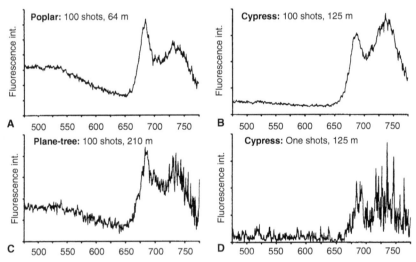

Figure 6.10 Fluorescence spectra of poplar, cypress, and plane-tree recorded at different ranges employing 100 laser shots. Spectra for cypress are shown using 100 as well as one laser shot. (From Andersson, M. et al., Proc. ISPRS Symposium on Physical Measurements and Signatures in Remote Sensing, Cal d'Isère, 1994; With permission.)

lidar telescope, it is possible to simultaneously generate four fluorescence images, each recorded through a band-pass filter. The reason why a good target characterization can be obtained with as few as four wavelengths is that a vegetation spectrum shows a rather limited structure. Thus, by choosing the wavelengths to match the important spectral features, the rough spectrum in all spatial points can be effectively obtained. The images are arranged as four quadrants, and the whole structure is fitted onto a gated image intensifier preceding the CCD detector. This type of system has been used to image agricultural plants as well as branches of trees. A spectacular example is shown in Figure 6.11,[34] where a special case of benthic algae extracted from the Mediterranean has been chosen. Three algae are placed on a concrete quay next to the deep water and are imaged from above via a folding mirror by a mobile lidar system placed about 20 m away from the samples. A photograph of the scene with the algal specimen is shown in the lower right part of the figure. The spectra from the two green algae and the red alga (with additional peaks due to phycoerythrine [PE] and phycocyanine [PC]) are given to the left of the photograph. The four fluorescence images are shown separately, recorded at filter center wavelengths of 408, 447, 630, and 682 nm, respectively. The red alga is clearly distinguished from the green ones, and more trivially, the water signature in the Raman scattering passing the 408 nm filter is evident.

As a last example for fluorescence lidar monitoring of vegetation we show in Figure 6.12[52,53] the result of a push-broom imaging of a spruce tree, 60 m away, using the approach indicated in Figure 6.7b. The output of the frequency-tripled Nd:YAG laser is shaped into a fan-like beam using cylinder optics, and the excitation line is moved from the tree top to its root using a vertical scan of the mobile lidar system roof-top mirror. Again, the fluorescence light is simultaneously monitored in several colors with sufficient signal in the image pixels to allow daytime monitoring.

As mentioned earlier it is also possible to utilize the excited-state temporal decay characteristics in remote vegetation monitoring.[54] By first monitoring the elastic range-resolved

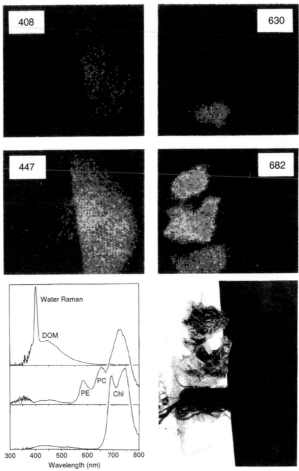

Figure 6.11 (Color figure follows page 398). Multicolor imaging of three types of benthic vegetation, extracted from Mediterranean water and placed on a concrete quay next to the deep water. A photograph of the algae is shown in the lower right part of the figure, and typical spectra for water (top), the red alga (middle), and the two green algae (bottom) are given to the left of the photograph (DOM, dissolved organic matter; PE, phycoerythrine; PC, phycocyanine; Chl, chlorophyll). Individual fluorescence images recorded at the indicated center wavelengths (in nm) are seen in the upper part of the figure. (From Alberotanza, L. et al., *EARSEL Adv. Remote Sens.*, 3, 102, 1995. With permission.)

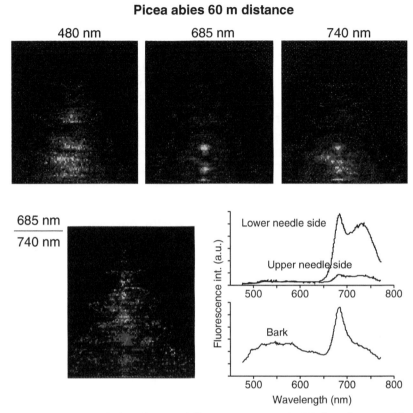

Figure 6.12 (Color figure follows page 398). Multicolor pushbroom fluorescence imaging of a spruce tree in 60 m distance form the lidar system. The three upper-row images were simultaneously recorded by successively scanning line by line the linearly dispersed excitation light across the tree. A ratio image is also shown, as well as illustrations of point spectra from the same object. (From Johansson, J. et al., *J. Plant Physiol.*, 148, 632, 1996; Svanberg, S., *Optics and Photonics News*, September 1996, p. 17. With permission.)

backscattering from a canopy, into which the laser beam successively penetrates, and then recording the fluorescence lidar return for some wavelengths, it is possible to separate lifetime effects from propagation effects.

6.4.3. Historical Monuments

Historical buildings form an important part of our cultural heritage, and their maintenance and restoration need substantial resources. Stone surfaces are subject to degradation not only due to the action of weather and air pollutants, but also because of the invasion of biodeteriogens. Photographic documentation of architectural monuments is important, but as in the case of vegetation, fluorescence monitoring can reveal additional features not visible in normal reflectance monitoring. Thus, the fluorescence lidar technique has recently been applied as a tool in the management of historical monuments.[55] Sometimes it is attempted to slow down the degradation of building facades by applying surface treatments, which are invisible to the eye, but easily recognizable through fluorescence.[56]

Figure 6.13 is shown as a first example of building fluorescence monitoring.[22]. Here, the northern gate of the Lund Cathedral (Sweden) is shown with six individual spectra, recorded at selected spots. It is possible to see different shapes of stone spectra and, in particular, the clear signatures of biodeteriogen invasion. Since the object is large it is not possible to use any other imaging strategy than that shown in Figure 6.7c. A 2-h recording time is no basic limitation for a fixed target. The result of an $8 \times 8\,\mathrm{m}^2$ scan of the gate, comprising in total 6400 image points, is shown in Figure 6.14.[22] Here, the two-filter-channel detection mode indicated in Figure 6.6 has been chosen by employing a laser spot size on target of 8 cm. Two separate recordings, at 448 and 600 nm, are shown in a false-color representation. Red corresponds to the highest intensity, yellow and green to lower intensities, and blue to the lowest intensities, as indicated in the color bars. Such images are clearly sensitive to variations in the laser intensity and to the surface topography, and thus artifacts can easily occur. However, by dividing the two images by each other, pixel by pixel, a ratio image is shown which is free of such influences. Basically, the ratio is only sensitive to the shape of the spectrum. Much more information can be extracted by recording the full spectral signature using the

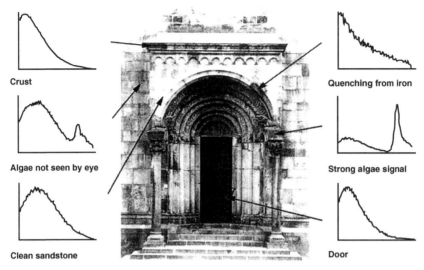

Figure 6.13 Photograph of the northern gate of the Lund Cathedral and six remotely recorded fluorescence spectra. (From Weibring, P. et al., *Appl. Opt.*, 40, 6111, 2001. With permission.)

optical multichannel analyzer mode of the fluorescence lidar system. Multivariate techniques,[57,58] such as principal component analysis, can then be employed to classify the different pixels, and different stone types and vegetation types can be discerned as illustrated in Ref. 22.

As discussed above, excitation at a number of wavelengths increases the discrimination power in fluorescence analysis. An example is given in Figure 6.15,[24] which shows an arrangement of Italian marble slabs placed at 60 m distance from the fluorescence lidar system. The full arrangement was scanned for the laser wavelengths 250, 280, 300, and 399 nm, obtained from an OPO arrangement. Spectra of different stone types are shown in the lower panels. In the right-hand part of the figure, it is indicated in different shades, of the "brick-like" processed fluorescence image pixels that different types of stones can be consistently identified by combining data for the four excitation wavelengths.

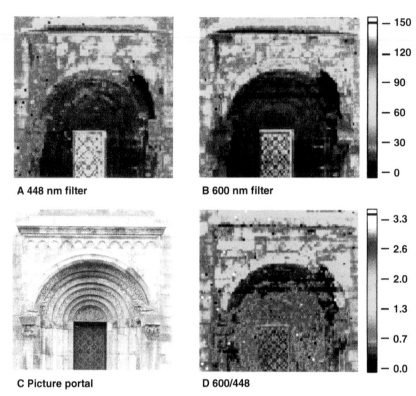

A 448 nm filter   **B 600 nm filter**

C Picture portal   **D 600/448**

Figure 6.14 (Color figure follows page 398). Fluorescence lidar imaging of the northern gate of the Lund Cathedral. A dual-filter approach has been taken, and individual images as well as a ratio image are shown. (From Weibring, P. et al., *Appl. Opt.*, 40, 6111, 2001. With permission.)

In a recent experiment, the Cathedral and the Baptistery in Parma (Italy) were studied with fluorescence lidar techniques.[59,60] Examples from the extensive data recorded are shown in Figure 6.16. In the lower part of the figure, a section of the front wall of the cathedral is shown. Here, the fluorescence image has been processed for the detection of surface treatment agents. From the spectral insert, it is clear that the treated areas strongly deviate in spectral shape from the stone areas. A ratio of the intensities at 400 and 445 nm was formed to identify the treated areas. A section in the

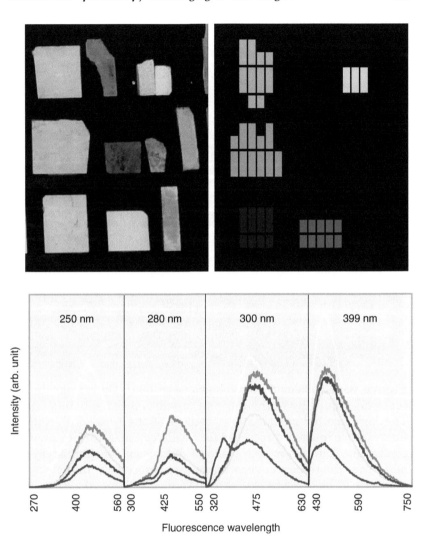

Figure 6.15 (Color figure follows page 398). Arrangement of Italian marble samples studied by a fluorescence lidar system. Four different excitation wavelengths were employed to perform a multivariate discrimination between the different samples. Stone spectra are shown in the lower panel in different shades, also used to indicate corresponding evaluated pixels in the processed image (upper right). (From Weibring, P. et al., *Appl. Opt.*, 42, 3583, 2003. With permission.)

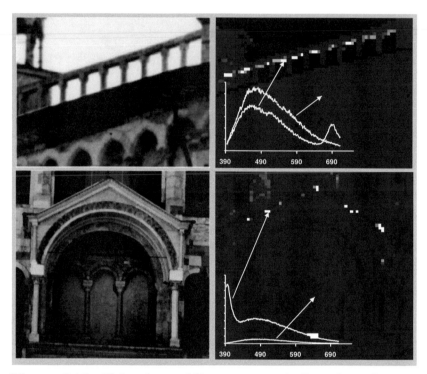

Figure 6.16 (Color figure follows page 398). Data from fluorescence lidar imaging at the Parma Cathedral and Baptistery (Italy). The lower sector pertains to a section of the Cathedral, and the fluorescence image is processed for the detection of surface chemical treatment. The upper sector shows data from the Baptistery and here the fluorescence recordings are processed to detect the presence of biodeteriogens. (From Weibring, P. et al., *Appl. Opt.*, 42, 3583, 2003; Lognoli, D., *Appl. Phys. B*, 76, 457, 2003. With permission.)

upper part of the Baptistery is shown in the upper part of Figure 6.16. Here, the presence of biodeteriogens is instead studied. As shown in the inserts, areas with algal growth (mostly on the lower side of the balustrade) have a strongly deviating spectrum as compared to the clean marble areas. A ratio image of the recorded intensities at 685 and 645 nm was formed to localize the areas invaded by biodeteriogens.

Very recently we have conducted similar measurements at the Coliseum and the Lateran Baptistry in Rome (Italy).

6.5. DISCUSSION

The rapid developments in laser and detector technology have increased the potential of fluorescence lidar systems and opened up new application fields such as the monitoring of historical monuments. Multispectral fluorescence data combined with powerful multivariate analysis techniques have enabled improved characterization of remote targets. The main limitation of fluorescence lidar is its limited range, which makes satellite implementation difficult. As a new possibility sunlight-induced fluorescence is being developed for vegetation monitoring (see, e.g., Ref. 61), where the reflectance data are modified due to fluorescence transfer for energy toward the red part of the spectrum. A particularly sensitive way of observing this natural fluorescence is to measure the contrast of the Fraunhofer lines, a technique earlier used for passive oil slick detection.

Presently there is a very fast development in fluorescence *microscopy* techniques[3] with the introduction of many powerful analysis techniques. Imaging fluorescence lidar can, in contrast, be considered as *macroscopy*. Also, on an intermediate size level, medical fluorescence imaging[62] is developing rapidly. The techniques are all manifestations of the general topic of *multispectral imaging*.[63] A strong cross-fertilization between the traditionally different fields can be expected, leading to an accelerated enhancement of the capabilities over a wide spectrum of applications.

ACKNOWLEDGMENTS

The author gratefully acknowledges a stimulating collaboration with a large number of colleagues and graduate students in the field of remote fluorescence monitoring. This work was supported by the Swedish Space Board, the Swedish Natural Sciences Research Council, the Knut and Alice Wallenberg

Foundation and the European Community within the EUREKA-LASFLEUR (EU380), and the Access to Large Scale Facility/Research Infrastructure programs.

REFERENCES

1. H.S. Chen, *Space Remote Sensing Systems* (Academic Press, Orlando, 1985).
2. R.M. Hoffer, Biological and physical considerations in applying computer-aided analysis techniques to remote sensor data, in P.H. Swain and S.M. Davis (eds.), *Remote Sensing — The Quantitative Approach* (McGraw-Hill, New York, 1978).
3. B. Herman and J.J. Lemasters, *Optical Microscopy — Emerging Methods and Applications* (Academic Press, Orlando, 1993).
4. E.R. Menzel, *Laser Detection of Fingerprints*, 2nd edn (Marcel Dekker, New York, 1999).
5. V. Zafiropulus and C. Fotakis, Lasers in the conservation of painted artwork, in M. Cooper (ed.), *Laser Cleaning in Conservation: an Introduction* (Butterworth Heinemann, Oxford, 1998), Chapter 6.
6. S. Andersson-Engels, C. af Klinteberg, K. Svanberg, and S. Svanberg, In vivo fluorescence imaging for tissue diagnosis, *Phys. Med. Biol.* **42**, 815 (1997).
7. K. Svanberg, I. Wang, S. Colleen, I. Idvall, C. Ingvar, R. Rydell, D. Jocham, H. Diddens, S. Bown, G. Gregory, S. Montán, S. Andersson-Engels, and S. Svanberg, Clinical multi-colour fluorescence imaging of malignant tumours — initial experience, *Acta Radiol.* **38**, 2 (1998).
8. R.M. Measures, *Laser Remote Sensing: Fundamentals and Applications* (Wiley, New York, 1984).
9. R.M. Measures (ed.), *Laser Remote Chemical Analysis* (Wiley-Interscience, New York, 1988).
10. S. Svanberg, Differential absorption lidar, in M. Sigrist (ed.), *Air Monitoring by Spectroscopic Techniques* (Wiley, New York, 1993), Chapter 3.
11. P.S. Andersson, E. Kjellén, S. Montán, K. Svanberg, and S. Svanberg, Autofluorescence of various rodent tissues and human skin tumour samples, *Lasers Med. Sci.* **2**, 41 (1987).

12. S. Udenfriend, *Fluorescence Assay in Biology and Medicine*, Vols. I and II (Academic Press, New York, 1962, 1969).
13. E.L. Wehry (ed.), *Modern Fluorescence Spectroscopy*, Vols. 1 and 2 (Plenum, New York, 1976).
14. J.R. Lakowicz, *Principles of Fluorescence Spectroscopy* (Plenum, New York, 1983).
15. C. af Klinteberg, A.M.K. Nilsson, I. Wang, S. Andersson-Engels, S. Svanberg, and K. Svanberg, Laser-induced fluorescence diagnostics of basal cell carcinomas of the skin following topical ALA application, *Proc. SPIE* 2926 (1996).
16. H. Edner, J. Johansson, S. Svanberg, E. Wallinder, M. Bazzani, B. Breschi, G. Cecchi, L. Pantani, B. Radicati, V. Raimondi, D. Tirelli, G. Valmori, and P. Mazzinghi, Laser-induced fluorescence monitoring of vegetation in tuscany, *EARSEL Adv. Remote Sens.* **1**, 119 (1992).
17. H.K. Lichtenthaler and U. Rinderle, The role of chlorophyll fluorescence in the detection of stress conditions in plants, *CRC Crit. Rev. Anal. Chem.* **19**, Suppl. 1, S29 (1988).
18. A.J. Hoff and J. Deisenhofer, Photophysics of photosynthesis, *Phys. Rep.* **287**, 1 (1997).
19. U. Gustafsson, S. Pólsson, and S. Svanberg, Compact fiber-optic fluorosensor using a continuous-wave violet diode laser, *Rev. Sci. Instrum.* **71**, 3004 (2000).
20. P. Camagni et al., Fluorescence response of mineral oils: spectral yield vs absorption and decay time. *Appl. Opt.* **30**, 26 (1991).
21. L. Celander, K. Fredriksson, B. Galle, and S. Svanberg, Investigation of Laser-induced Fluorescence with Applications to Remote Sensing of Environmental Parameters, Göteborg Institute of Physics Reports, GIPR-149 (1978).
22. P. Weibring, Th. Johansson, H. Edner, S. Svanberg, B. Sundnér, V. Raimondi, G. Cecchi, and L. Pantani, Fluorescence lidar imaging of historical monuments, *Appl. Opt.* **40**, 6111 (2001).
23. P.S. Andersson, S. Montán, and S. Svanberg, Remote sample characterization based on fluorescence monitoring, *Appl. Phys.* B **44**, 19 (1987).
24. P. Weibring, H. Edner, and S. Svanberg, Versatile mobile lidar system for environmental monitoring, *Appl. Opt.* **42**, 3583 (2003).
25. S. Montán and S. Svanberg, A system for industrial surface monitoring utilizing laser-induced fluorescence, *Appl. Phys.* B **38**, 241 (1985).

26. H.H. Kim, Airborne bathymetric charting using pulsed blue-green lasers, *Appl. Opt.* **16**, 46 (1977).
27. J. Banic, S. Sizgoric, and R. O'Neill, Airborne scanning lidar bathymetry measures water depth, *Laser Focus* **23** (2), 40 (1987).
28. K. Fredriksson, B. Galle, K. Nyström, S. Svanberg, and B. Öström, Underwater Laser-Radar Experiments for Bathymetry and Fish-School Detection, Göteborg Institute of Physics Reports GIPR-162 (1978).
29. K. Fredriksson, B. Galle, K. Nyström, S. Svanberg, and B. Öström, Marine Laser Probing: Results of a Field Test, Meddelanden från Havsfiskelaboratoriet, No. 245 (1979).
30. D.A. Leonard, B. Caputo, and F.E. Hoge, Remote sensing of subsurface water temperature by Raman scattering, *Appl. Opt.* **18**, 1732 (1979).
31. V. Raimondi and G. Cecchi, Lidar field experiments for monitoring seawater temperature, *EARSEL Adv. Remote Sens.* **3** (3), 84 (1995).
32. S. Svanberg, Laser fluorescence spectroscopy in environmental monitoring, in S. Martellucci and A.N. Chester (eds.), *Optoelectronic for Environmental Science* (Plenum Press, New York, 1990), p. 15.
33. H. Edner, J. Johansson, S. Svanberg, E. Wallinder, G. Cecchi, and L. Pantani, Fluorescence lidar monitoring of the Arno River, *EARSEL Adv. Remote Sens.* **1**, 42 (1992).
34. L. Alberotanza, P.L. Cova, C. Ramasco, S. Vianello, M. Bazzani, G. Cecchi, L. Pantani, V. Raimondi, P. Ragnarson, S. Svanberg, and E. Wallinder, Yellow substance and chlorophyll monitoring in the Venice Lagoon using laser-induced fluorescence, *EARSeL Adv. Remote Sens.* **3**, 102 (1995).
35. V.V. Fadeev and T.A. Dolenko, Concept, Methods and Tools for Laser Monitoring of Coastal Sea Waters, Proc. LAT2002, Moscow, June 22–27, 2002, Paper LThC4.
36. G.A. Chapelle, L.A. Franks, and D.A. Jessup, Aerial testing of a KrF laser-based fluorosensor, *Appl. Opt.* **22**, 3382 (1983).
37. F.E. Hoge, Ocean and terrestrial lidar measurements, in R.M. Measures (ed.), *Laser Remote Chemical Analysis* (Wiley-Interscience, New York, 1988), p. 409.
38. M. Bazzani, B. Breschi, G. Cecchi, L. Pantani, D. Tirelli, G. Valmori, P. Carlozzi, E. Pelosi, and G. Torzillo, Phytoplankton monitoring by laser induced fluorescence, *EARSeL Adv. Remote Sens.* **1**, 106 (1992).

39. R. Barbini et al., Differential lidar fluorosensor system used for phytoplankton bloom and seawater quality monitoring in Antarctica, *Int. J. Remote Sens.* 22, 369 (2001); A. Palucci, Synoptic Study of the Antarctic Ross Sea with the ENEA Lidar Fluorosensor, Proc. LAT2002, Moscow, June 22–27, 2002, Paper LThC1.

40. G. Cecchi, M. Bazzani, C. Cuzzi, D. Lognoli, I. Mochi, L. Pantani, V. Raimondi, R. Carlà, R. Cappadona, B. Breschi, D. Novelli, Th. Johansson, P. Weibring, H. Edner, and S. Svanberg, Probing the Marine Environment with Fluorescence Lidars — Comparison of Three Fluorosensors in a Field Campaign. CNR Report RR/OST/01.03 (CNR-IFAC "Nello Carrara", Firenze 2003).

41. R.A. O'Neill, L. Buja-Bijunas, and D.M. Rayner, Field performance of a laser fluorosensor for the detection of oil spills, *Appl. Opt.* **19**, 863 (1980).

42. H. Amann, Laser spectroscopy for monitoring and research in the ocean, *Phys. Scripta* **T78**, 68 (1998).

43. I.V. Boychuk, T.A. Dolenko, V.V. Fadeev, M. Kompitsas, and R. Reuter, Real abilities and problems in laser monitoring (in situ) of oil pollution in coastal marine waters, in R. Reuter (ed.), *Lidar Remote Sensing of Land and Sea*, EARSEL Proc., Vol. 1, p. 115 (2001).

44. R. Reuter, H. Wang, R. Willkomm, K.D. Loquay, T. Hengstermann, and A. Braun, A laser fluorosensor for maritime surveillance: measurement of oil spills, *EARSEL Adv. Remote Sens.* **3**, 152 (1995).

45. F.E. Hoge, R.N. Swift, and J.K. Yungel, Feasibility of airborne detection of laser-induced fluorescence of green terrestrial plants, *Appl. Opt.* **22**, 2991 (1983).

46. E.W. Chappelle, F.M. Wood, W.W. Newcomb, and J.E. McMurtrey III, Laser-induced fluorescence of green plants. 3. LIF spectral studies of five major plant types, *Appl. Opt.* **24**, 74 (1985).

47. S. Svanberg, Fluorescence lidar monitoring of vegetation status, *Phys. Scripta* **T58**, 79 (1995).

48. M. Andersson et al., Proc. ISPRS Symposium on Physical Measurements and Signatures in Remote Sensing, Val d'Isère, 1994, p. 835.

49. H. Edner, J. Johansson, P. Ragnarson, S. Svanberg, and E. Wallinder, Remote monitoring of vegetation using a fluorescence lidar system in spectrally resolving and multi-spectral imaging modes, *EARSEL Adv. Remote Sens.* **3**, 193 (1995).

50. H. Edner, J. Johansson, S. Svanberg, and E. Wallinder, Fluorescence lidar multicolor imaging of vegetation, *Appl. Opt.* **33**, 2471 (1994).

51. H. Edner, J. Johansson, S. Svanberg, H.K. Lichtenthaler, M. Lang, F. Stober, Ch. Schindler, and L.O. Björn, Remote multi-colour fluorescence imaging of selected broad-leaf plants, *EARSEL Adv. Remote Sens.* **3**, 2 (1995).

52. J. Johansson, M. Andersson, H. Edner, J. Mattsson, and S. Svanberg, Remote fluorescence measurements of vegetation spectrally resolved and by multi-colour fluorescence imaging, *J. Plant Physiol.* **148**, 632 (1996).

53. S. Svanberg, Real-world applications of laser spectroscopy, *Optics and Photonics News*, September 1996, p. 19.

54. Z.G. Cerovic, Y. Goulas, M. Gorbunov, J.M. Briantais, L. Camenen, and I. Moya, Fluorosensing of water-stress in plants — diurnal changes of the mean lifetime and yield of chlorophyll fluorescence, measured simultaneously and at distance with a tau-lidar and a modified pam-fluorometer, in maize, sugar-beet, and kalanchoe. *Remote Sens. Environ.* **58**, 311 (1996).

55. V. Raimondi, G. Cecchi, L. Pantani, and R. Chiari, Fluorescence lidar monitoring of historical buildings, *Appl. Opt.* **37**, 1089 (1998).

56. G. Ballerini, S. Bracci, L. Pantani, and P. Tiano, Lidar remote sensing of stone cultural heritage: detection of protective treatments, *Opt. Eng.* **40**, 1579 (1997).

57. K.R. Beebe and B. Kowalski, An introduction to multivariate calibration and analysis, *Anal. Chem.* **59**, 1607A (1987).

58. K. Esbensen, T. Midtgaard, S. Schonkopf, and D. Guyoyf, *Multivariate Analysis — A Training Package* (CAMO ASA, Oslo, 1994).

59. P. Weibring, D. Lognoli, R. Chiari, G. Cecchi, H. Edner, T. Johansson, L. Pantani, S. Svanberg, D. Tirelli, and M. Trambusti, Lidar remote sensing of the parma cathedral and baptistery, *Proc. SPIE* 4402–16 (2001).

60. D. Lognoli, G. Cecchi, I. Mochi, L. Pantani, V. Raimondi, R. Chiari, Th. Johansson, P. Weibring, H. Edner, and S. Svanberg, Fluorescence lidar imaging of the Parma Cathedral and Baptistery, *Appl. Phys. B* **76**, 457 (2003).

61. I. Moya, L. Camenen, G. Latouche, C. Mauxion, S. Evain, and Z.G. Cerovic, An instrument for the measurement of sunlight excited plant fluorescence, *Photosynthesis* **42**, 65 (1998).

62. S. Andersson-Engels, K. Svanberg, and S. Svanberg, Medical fluorescence imaging, in J. Fujimoto (ed.), *Medical Laser Applications*, to appear.
63. S. Svanberg, *Multi-Spectral Imaging — From Astronomy to Microscopy — From Radiowaves to Gammarays,* to appear.

7

Wind Lidar

**SAMMY W. HENDERSON, PHILIP GATT,
DAVID REES, and R. MILTON HUFFAKER**

Coherent Technologies Inc,
Louisville, CO

7.1. INTRODUCTION

Lidar systems for wind velocity measurements have demonstrated their efficacy in many different applications where spatially resolved measurements of atmospheric wind velocity from a remote location are needed. Examples include: measurement of aircraft true airspeed, detection and tracking of clear air turbulence, wind shear, gust fronts, aircraft wake vortices, and measurement of atmospheric wind profiles from the ground up to 50 km altitude.

Before the invention of the laser, the first LIght Detection And Ranging (lidar) optical systems utilized intense flashlamps. Today it is understood that laser energy is utilised and these systems are sometimes referred to as laser radar (or "ladar") to distinguish them from the earlier flashlamp-based lidar techniques. Most atmospheric scientists use lidar when referring to laser-based systems. To be consistent, the lidar acronym is used throughout this chapter.

The first remote atmospheric wind measurements using the lidar technique were demonstrated in the last half of the 1960s, shortly after the invention of suitable lasers. All lidar systems for wind velocity measurement operate by transmitting narrow bandwidth laser light into the atmosphere. Moving molecules and aerosol particles entrained in the atmosphere scatter a small portion of that light back to the lidar receiver and impart a frequency shift onto the scattered light according to the Doppler effect. This Doppler shift is estimated, by appropriate signal processing, to provide estimates of the wind velocity along the line of sight.

There are two fundamental methods used to determine the Doppler shift. In the first method, which is called coherent (or heterodyne) detection, the return signal is optically mixed with a local oscillator laser and the resulting beat-signal's frequency is, except for a fixed offset, the Doppler shift due to the moving particles. As a result of its near quantum-limited sensitivity, even with noisy detectors, most of the wind measurement lidar systems demonstrated prior to the late 1980s used the coherent detection method. Coherent detection lidar systems have matured significantly over the past 35 years, with compact operational systems using eye-safe solid-state lasers now being commercially available.

In the second method, no local oscillator laser is used; instead an optical frequency discriminator or spectrum analyzer converts an optical frequency change (i.e., Doppler shift) into a change in power or power spatial distribution, which is in turn directly detected. In these direct detection (or incoherent) lidar systems, the return optical signal is filtered or resolved into its spectral components prior to detection. Examples of optical discriminators include: the transmission edge of a molecular absorption, the edge of a transmission fringe of an optical interferometer, or fringe pattern imaging of the output of an optical interferometer. Early versions of these systems were demonstrated in the 1970s, and by 1989 a direct detection ground-based Doppler lidar system demonstrated wind measurements to 50 km altitude using molecular backscatter. A direct detection system is currently being developed in Europe for eventual application to measure winds globally from a space-based platform.

In this chapter we summarize the history, describe the theory and architecture, and provide example measurements for both the coherent and direct detection wind lidar architectures. Although it resulted in the chapter being relatively larger the authors decided to include a significant amount of background and theory in this work for completeness, since no previous consolidated work on wind lidar systems exists. Our hope is that this chapter will be a useful building block and reference tool for those interested in the history and fundamentals of coherent and direct detection wind lidar systems.

The chapter is organized as follows: In Section 7.2, a historical overview of lidar systems used for the wind measurement application is provided. Section 7.3 describes the fundamental principles of operation of wind lidar systems and the basic governing physics. In Section 7.4, detailed theory for the operation of both coherent and direct detection lidar system types is developed and presented. System architectures and example Doppler lidar systems are provided in Section 7.5. In Section 7.6, wind measurement applications are discussed and example measurements are provided. Finally, a summary and future outlook is given in Section 7.7. Sections 7.3 and 7.4 cover the fundamental principles and theory of wind lidar in detail. For readers more interested in system architectures and measurements, these two sections can be skipped without significantly impacting the readability of the following sections.

7.2. BACKGROUND

A brief history of the development and utilization of lidar systems for wind measurement applications is presented in this section. The presentation is not meant to be exhaustive, but rather to provide enough background information and references to allow the interested reader to investigate in greater detail. The background of the development of coherent detection lidar systems is presented in Section 7.2.1 followed by the background for direct detection lidar in Section 7.2.2.

7.2.1. Coherent Lidar Systems

The earliest efforts in optical heterodyne detection and coherent lidar began shortly after the invention of the laser and utilized the two reliable lasers then in existence, the ruby laser and the helium–neon (HeNe) laser.[1-5] Much of the earliest work in coherent lidar was funded by the military for nonwind measurement applications and will not be described here — examples can be found in Jelalian's text.[6] After its development in the mid-1960s,[7] the CO_2 laser became the laser of choice for most coherent lidar systems. The primary reasons for using these lasers include: energy efficiency, the ability to produce stable single frequency output, operation in both continuous wave (CW) and pulsed with energies of several Joules; acceptable atmospheric transmission; and eye safety.

The history of some of the early CO_2-based coherent wind lidar systems is summarized in previous publications.[8-12] The first reported coherent lidar wind measurements, using a focused CW CO_2-laser-based system, were of laboratory air flow at short range by Huffaker.[13,14] CW CO_2 systems have been used for applications ranging from the measurement of wind profiles reaching to heights of a few hundred meters to the measurement of the properties of aircraft-induced wake vortices.[15-18] A CW CO_2 system, developed by a group of researchers led by Vaughan in England, was utilized extensively for many years for airborne wind speed and aerosol backscatter measurements.[19-21]

By the early 1970s, pulsed CO_2 systems began to be the choice for many wind measurement applications due to the increased range capability. The first pulsed CO_2 coherent lidar systems demonstrated in the early 1970s used the master oscillator power amplifier (MOPA) architecture.[22,23] Several groups still use MOPA systems for boundary layer wind measurements.[24-26] In order to obtain higher pulse energy in a practical size instrument, the transverse excited atmospheric-pressure (TEA) CO_2 laser transmitter was utilized by several researchers beginning in the early 1980s.[27] These TEA CO_2-based systems are still in use by several groups today. A system producing up to 1 J pulse energy at 20 Hz has been operational since the mid-1980s by a U.S. group now led by Hardesty and has been used in many ground and airborne wind measurement programs.[28-32]

In addition to wind measurements, other early coherent systems using TEA CO_2 transmitters were utilized for applications ranging from measurements of the effects of atmospheric refractive turbulence on coherent lidar performance to measurements of the atmospheric backscatter at 9.25 and 10.6 μm.[33,34] A recent example of a TEA CO_2-based coherent lidar system is the airborne pulsed system developed and utilized for wind measurements in a joint effort between the French and German groups led by Flamant and Werner, respectively.[35] Examples of CO_2-based coherent lidar systems and example measurements using these systems are presented in Sections 7.5.1.2 and 7.6.

The advent of high-power diode lasers in the mid-1980s led to renewed interest in using solid-state lasers for coherent lidar systems. Diode-pumped solid-state transmitters offer significant improvement in efficiency, size, and lifetime compared to CO_2 lasers or flashlamp-pumped solid-state lasers. The shorter wavelength of solid-state lasers compared to CO_2 lasers also allows improved range resolution for equivalent wind velocity resolution. The continuous tuning range of many solid-state lasers permit operation at wavelengths that suffer low absorption in the atmosphere, allowing long-range measurements at lower power or pulse energy.

The disadvantages of shorter wavelengths include: stricter requirements on optical surface quality and optical alignment to maintain high heterodyne mixing efficiency and increased sensitivity to atmospheric refractive turbulence.

A group led by Kane et al.[36] at Stanford University was the first to publish results describing a 1.06 μm coherent lidar using Nd:YAG lasers in 1985. In 1988, a 1.06 μm lidar was developed by a group led by Kavaya and Henderson[37] at Coherent Technologies, demonstrating the first long-range wind measurements with a Nd:YAG transmitter. By 1991, that system had been upgraded to produce up to 200 mJ per pulse at 20 Hz[38] and was used for wind and hard target measurements as well as the first aircraft wake vortex measurements using a pulsed lidar.[39,40] The pulse energy was further increased to 1 J and the system was used to make wind measurements from the ground to an altitude of 26 km at Kennedy Space Center in conjunction with the launch and landing of the space shuttle.[41]

Because of the significant eye safety advantage, 2-μm wavelength solid-state lidar systems have been utilized much more frequently than the 1.06-μm systems. The first 2.09-μm coherent lidar using flashlamp-pumped Tm,Ho:YAG lasers was demonstrated in 1990 by a group led by Henderson et al.[42,43] The demonstrated ability to make wind measurements to approximately 20 km horizontal range and the detection of the return from a mountainside at a range of 145 km with only 20 mJ pulse energy and a 20-cm diameter aperture attested to the good beam quality, high overall system efficiency, and high atmospheric transmission of this early solid-state lidar. The first diode-pumped pulsed 2-μm lasers, with energy sufficient for use in wind lidar systems, were demonstrated by Suni and Henderson in 1991.[44] Using this technology, the first all-diode-pumped Tm:YAG lidar system operating at 2.01 μm was developed by the group at Coherent Technologies for airborne operation and flown in 1994 on the NASA 737 aircraft to demonstrate the ability to detect microburst windshear.[45,46] An improved design was used to demonstrate vector velocity measurements from an airborne platform to a precision of ~10 cm/sec in 1996.[47] A similar design has been used successfully for several years by a group at the U.S. National Oceanic and Atmospheric Administration (NOAA) for high-resolution boundary layer wind measurements.[48–53]

Higher-pulse energy (10-mJ), 100-Hz, 2-μm lidar systems were also developed under NASA and U.S. Air Force funding for airborne wind measurement applications.[54] In the NASA application, the 10-mJ diode-pumped transceiver was used to demonstrate the detection of clear air turbulence several kilometers ahead of aircraft and the measurement of more favorable winds above or below the flight path.[55] In the Air Force application, the transceiver was used for the measurement of wind profiles below the aircraft for improved parachute drop precision and ballistics accuracy.[56] Even higher-energy diode-pumped 2 μm transmitters have been developed for potential application in space-based wind measurement lidar systems,[57–59] but only a limited number of these systems have been integrated into working lidar systems.[60] During the past few years, significant progress has been achieved in increased measurement capability,[61,62] reduction in size, reduction in prime power requirements, and increased reliability of the 2 μm solid-state lidar systems. An all-diode-pumped 2 μm lidar transceiver is now utilized in a commercially available lidar system called the WindTracer®.[63,64] This system is presented as an example in Section 7.5.1.3.

More recently, coherent lidar systems operating near the 1.5 μm wavelength have been demonstrated. The primary motivation is further improvement in the compactness, efficiency, and cost of coherent lidar systems. The 1.5 μm wavelength has the advantage that it allows access to components that have had significant development investments for application in the telecommunications industry — for example, fibers and fiber couplers, master/local oscillators using diode lasers, optical isolators, and detectors. There are four primary types of lidar systems operating near the 1.5 μm wavelength that have been demonstrated to date and the primary differentiator is the transmitter that is used. The four transmitter types are pulsed Raman-shifted Nd:YAG systems operating at 1.56 μm,[65,66] bulk pulsed Er:glass lasers operating at 1.54 μm,[67] Er:YAG lasers at 1.6 μm,[68] and direct use of Er:glass fiber lasers operating either CW or pulsed.[69–73] The first three transmitter types are capable of generating high pulse energies while the fiber-based transmitter is only suitable for either CW or low-energy high-PRF pulsed operation. Examples of fiber-based coherent lidar systems are presented in Section 7.5.1.3.

Novel coherent lidar waveforms have also been demonstrated. A coherent lidar waveform that allows the coherent measurement of atmospheric winds with simultaneous high range and velocity resolution has been demonstrated.[74–76] The coherent lidar technique is similar to the pulse-pair radar waveform. The short pulses allow good range resolution and the time between the first and second pulse of the pair allows precise velocity (phase) measurements. Numerous new measurement capabilities are expected using this, and other multiple-pulse coherent radar waveforms, in coherent lidar systems.

7.2.2. Direct Detection Wind Lidar Systems

As with the development of coherent wind lidar systems described in Section 7.2.1, the development of the Direct Detection Doppler wind Lidar (DDL) technique has also been tied to the development and availability of suitable reliable high-power pulsed lasers.[77] Three distinct laser technologies have been considered seriously as the laser source — the excimer laser operating in the 300 nm regime,[78] lasers based on the Nd:YAG technology operating at 1064 nm and its second, third, and fourth harmonics[77,79–81] and those based on the alexandrite laser operating in the 700 to 800 nm regime and its harmonics.[82] While each of these laser techniques is capable of achieving on the order of 1 J pulse energies, it is other considerations that have led to the Nd:YAG and alexandrite lasers being used in preference to the excimer laser. Although the excimer laser is versatile and powerful, it requires considerable ancillary equipment, including consumable gas supplies such as chlorine or fluorine, for successful long-term operation.

The injection-seeded, high-pulse-energy single-longitudinal-mode Nd:YAG laser, developed by Byer and colleagues[83,84] in the early 1980s, was instrumental for application in long-range wind lidar sensors. The CW single-frequency diode-pumped Nd:YAG laser, now used to seed the high-energy laser oscillators of most DDL systems, can be reliably stabilized, providing the stable and narrow bandwidth laser source that is essential for practical Doppler wind lidar measurements.[77,79–81,85–87]

The fact that the DDL technique can exploit both the aerosol (Mie) and the molecular (Rayleigh) backscattered signal provides additional operational flexibility. However, it is the capability of the molecular DDL to measure in the very low aerosol backscatter regions occasionally found in the troposphere (but an endemic feature of the middle stratosphere) that gives it a unique capability for wind measurement under any middle-atmospheric meteorological condition. Thus, for scientific exploitation of the Doppler wind lidar technique for the study of wind circulation within the stratosphere, the direct detection technique using the molecular scattered signal has been widely used.[86,87]

Preliminary analyses of the DDL technique and measurements using aerosol backscatter in the atmospheric boundary layer were performed in the 1970's and the early 1980's.[88–90] Application of the technique for long-range measurements (space based) was analyzed in some detail by Abreu in 1979.[91] He showed that it might be possible to exploit the Nd:YAG laser, operating at the frequency-doubled wavelength of 532 nm, to create a wind lidar system with a useful performance. However, at that time, Nd:YAG lasers were neither reliably single-mode, nor were they adequately stable in wavelength for a practical wind lidar system.

Several of the critical technical steps toward development of a practical DDL system were demonstrated by Hays et al.,[92] Rees et al.,[93–95] and McLeese and Margolis.[96] In particular, the development of key technological elements required to implement direct detection was advanced with the development of the stable and robust Fabry–Perot etalons, required for the high-resolution spectral analysis of lidar signals. Meanwhile, the "passive" remote sensing of winds in the stratosphere,[92,94] mesosphere and in the thermosphere[92,97] was demonstrated by sensing the Doppler shifts of upper atmosphere emission and/or absorption spectra.

During the mid-1980s the DDL technique was investigated by a combination of theoretical studies and practical tests using early systems.[98–101] These proved to be valuable steps toward the development and subsequent demonstration of a practical DDL system. An investigation of the feasibility of the near-UV DDL technique for wind measurements was reported by Rees and McDermid.[102]

In 1989, the Service d'Aeronomie[79,103] of CNRS (France) was the first group to bring together the various technical elements to exploit the DDL technique for long-range wind measurements using the "double-edge" technique at the Nd:YAG frequency-doubled wavelength of 532 nm. Following the initial measurements by CNRS, the DDL technique was further developed and exploited using "full fringe imaging" by a group at the University of Michigan[81,104] and by another group who developed a wind lidar for the Artic Lidar Observatory for Middle Atmospheric Research (ALOMAR) Observatory in Norway.[80,86,105,106] During this period, groups at the NASA Goddard Space Flight Center (GSFC)[87,107–111] Aricebo,[112] and University of Geneva[113–115] were also involved in the further development and exploitation of the DDL double-edge technique.

An important derivative wind lidar technique that follows the principles of the DDL is that of the laser-induced fluorescence lidar, LIF (e.g., She and Yu[116]). LIF-based Doppler measurement is an important technique, since it can be used to monitor atmospheric winds at mesopause and lower thermosphere levels (90 to 100 km), where the air density and aerosol concentrations are far too low for the molecular or aerosol signals to be used for wind measurement. One LIF lidar system, which is in operation at ALOMAR,[117] observes shifts in the sodium spectrum, to measure atmospheric dynamics as well as climatology[118,119] in the upper mesosphere and lower thermosphere (80 to 110 km). The LIF technique is described elsewhere,[120] so it will not be covered here.

Other direct detection wind measurement techniques that do not utilize the Doppler shift have been demonstrated. These include: tracking aerosol cloud motion[121] and the measurement of partial transit time through a pair of light sheets.[122]

The DDL technique has recently been chosen by the European Space Agency (ESA) as the basis of the AEOLUS space-based wind lidar mission.[123] There are two primary reasons for this decision. The recent development of stable, single-mode, diode-pumped, injection-seeded, high-power Nd:YAG lasers has created the option for an efficient eye-safe laser, operating in the near-UV ($\lambda = 355\,\text{nm}$). At this wavelength, it is possible to combine efficient molecular and aerosol detection channels to measure winds from space at all altitudes from the surface to at least 20 km, by means of the development of a single space-qualified laser. For the AEO-LUS Mission, the molecular signal will be analyzed by a double-edge molecular spectrometer.[103,107,111] However, for analysis of the narrow aerosol signal, the Mach–Zehnder and Fizeau interferometers are currently being considered.[124] While neither technique has been widely used for lidar wind measurements, detailed theoretical analyses show that the performance is expected to be comparable to that of equivalent double-edge or fringe-imaging systems, while their optical output (spectrum) may readily be coupled to available detectors of high quantum efficiency.[124–126] Details of DDL systems and techniques will be described in Sections 7.4.2 and 7.5.2.

7.3. DOPPLER WIND LIDAR PRINCIPLE OF OPERATION

The detailed implementation, principle of operation, signal-to-noise ratio, and velocity precision of coherent and direct detection lidar, are somewhat different. However, both techniques are very similar at a fundamental level. Both types of systems: have a stable narrow-bandwidth high-power laser source, experience attenuation by the atmosphere, utilize atmospheric backscatter (i.e. particles entrained in the atmosphere) to reflect the signal, utilize a receiver (collection optics and detector) to collect the signal, analyze the collected signal to measure the Doppler frequency shift, resulting in an estimate of the line-of-sight (LOS) or radial wind velocity, and sometimes use multiple lines of sight to estimate vector wind velocity.

This section provides a conceptual overview of Doppler wind lidar at a level that does not require distinction between the coherent and direct detection approaches. The emphasis is to provide a heuristic presentation of the Doppler wind lidar concept. A top-level overview of the concept is presented in Section 7.3.1.

A simplified lidar equation, based upon physical insight, which describes the amount of total power collected by the lidar system, is described in Section 7.3.2. Receiver performance varies depending upon noise and efficiency. The performance of a perfect receiver is bounded by the quantum-limit, which is described in Section 7.3.3. Atmospheric backscatter and attenuation are discussed in Sections 7.3.4 and 7.3.5, respectively. Radial velocity estimation precision and approaches to measure vector winds by combining the measurements from multiple LOS are presented in Sections 7.3.6 and 7.3.7. Laser speckle and its effect on lidar receiver performance are described in Section 7.3.8. Finally, background light generated by thermal and solar sources is described in Section 7.3.9.

7.3.1. Concept Overview

Figure 7.1 depicts a generalized bistatic lidar configuration. The sensor can be monostatic (shared transmit/receive optics) or bistatic (separate transmit/receive optics). Monostatic lidar systems simplify transceiver alignment and provide good performance over extended ranges. These systems usually employ an off-axis telescope to minimize transmitter

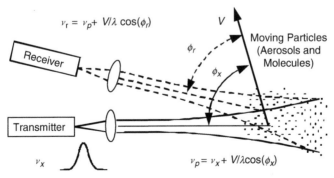

Figure 7.1 Generic wind lidar concept. Solid and dashed lines represent transmit beam and receiver field of view, respectively.

feedback or interference. Bistatic transceivers mitigate the feedback problem, but can limit the range over which the receiver field of view and the volume illuminated by the transmitter overlap. If the receiver field of view is sufficiently large, as can be achieved with a DDL, the range over which the overlap is poor can be minimized to the extreme near-field.

In either configuration, a transmitted laser beam, with frequency $\nu_x = c/\lambda$ (Hz) is scattered by a volume of particles (aerosol particles and molecules), whose mean velocity, V (m/s), is the mean wind speed, where $c = 2.998 \times 10^8$ m/s is the speed of light, and λ (m) is the optical wavelength. Barring relativistic effects, the laser transmitter frequency, as seen by a particle in its reference frame, is given by $\nu_p = \nu_x + (V/\lambda) \cos(\phi_x)$. Likewise, the frequency of the reflected laser light, observed in the receiver reference frame, is given by $\nu_r = \nu_p + (V/\lambda)\cos(\phi_r)$. Therefore, the frequency difference or Doppler shift is given by

$$\nu_d = \nu_r - \nu_x = \frac{2}{\lambda} \frac{V(\cos \phi_x + \cos \phi_r)}{2} \tag{7.1}$$

For the monostatic condition $\phi_r = \phi_x = \phi$, the Doppler frequency shift reduces to the familiar formula, $\nu_d = 2V \cos(\phi)/\lambda = 2V_r/\lambda$, where the radial velocity, V_r, is the component of the velocity projected onto the transceiver's line of sight.

A portion of the scattered radiation is collected by the lidar receiver, which has a means to measure the Doppler frequency shift. In coherent detection lidar, the signal is mixed with a local oscillator beam then detected with a photodetector. The resulting beat signal is a radio frequency (RF) photocurrent whose frequency, which can be estimated using digital or analog frequency discrimination techniques, is a direct measure of the Doppler shift. In DDL, the frequency of the Doppler-shifted laser radiation is typically measured using an optical frequency analyzer (optical interferometer, sharp edge molecular absorption filter, etc.) and the detected power or power spatial distribution transmitted through the optical analyzer provides a measure of the Doppler shift. Therefore, coherent detection lidar is less sensitive to intensity distortion and bias and more sensitive to phase distortion, whereas the opposite is true for DDL sensors.

The velocity estimates from a properly designed coherent Doppler lidar system are not biased and to first order the precision does not depend upon the wind velocity. In DDL systems, certain calibration errors may result in systematic velocity error and bias. By careful design, the errors and biases can be minimized to avoid significantly impacting the measurement accuracy. Also, in some designs the discriminator output signal strength, is a function of the Doppler shifted signal frequency. Thus care must be taken to properly design and calibrate the receiver. In well-designed DDL systems, the discriminator signal can usually be made linear over the nominal operating range of wind velocities.

Pulsed (amplitude modulated) or chirped (phase-modulated) laser sources provide a means of measuring the velocity as a function of range, $R = ct_r/2$, where t_r is the round-trip pulse time-of-flight. For these range-gated systems, the "effective" interaction volume is defined by the beam diameter and a characteristic length which is on the order of the smaller of the depth of focus and the waveform range resolution, ΔR, where $\Delta R = c\Delta T/2$ for a pulsed single-frequency waveform or $\Delta R = c/2B$ for a phase modulated waveform whose bandwidth is B.

If a scanner is added to a range-gated system, the radial velocity can be measured as a function of position in three-dimensional (3D) space. The measurement of radial wind velocity in multiple directions, combined with wind flow-field constraints, allows the vector wind to be estimated. These concepts are explained in greater detail in Section 7.3.7.

7.3.2. Lidar Equation

In both coherent and direct detection Doppler lidar systems, the precision with which the frequency (velocity) can be measured depends on backscattered power that is collected by the receiver. The lidar equation describes the collected optical power in terms of the transceiver, target, and atmospheric parameters. It is useful, for heuristic purposes, to first consider a monostatic lidar whose range resolution, ΔR, is shorter than range variations in atmospheric backscatter or attenuation coefficients. Figure 7.2 shows the key elements of the lidar. The result is an easily understood lidar equation that has the same parameter dependence as the more general lidar equations that are derived in Section 7.4.

The lidar equation predicts the amount of optical signal power, $P_s(W)$, collected by the receiver aperture from a target echo at range R, through a cascade of multiplicative operators on the laser pulse energy, $E_x(J)$. First the energy is attenuated by transmit optical losses, η_{ox} and transmit truncation loss η_{Tx}. The reason for separating these transmission losses into two terms will become apparent in Section 7.4. This energy is then attenuated by atmospheric extinction, T, along the transmit path. A portion of the remaining energy is scattered back towards the lidar by aerosol particles and molecules.

The amount of power reflected, P_r (W), by these particles, at any instant in time, is governed by the product of the backscatter coefficient β ($\text{m}^{-1} \text{ sr}^{-1}$), the incident pulse power ($P_i = E_i/\Delta T$) and the range extent (or resolution) of the pulse, $\Delta R = c\Delta T/2$, or more simply $P_r = E_i c\beta/2$. So to first order, the reflected power is a function of the pulse energy, not the peak pulse power. For example, for fixed energy pulses, longer pulses simultaneously interact with more particles but have lower irradiance compared to shorter pulses, which interact with fewer particles with higher irradiance, the result is equal

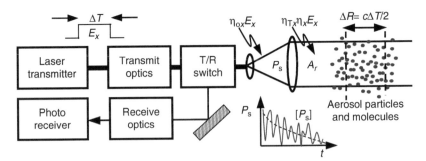

Figure 7.2 Simplified diagram of a monostatic pulsed lidar.

signal power. The backscattered intensity, $\psi = \eta_{ox}\eta_{Tx}TE_x\beta c/2$ (W/sr), is attenuated by atmospheric extinction, T, along the receive path and the power collected by the aperture, with area A_r, is $P_s = A_r I_s$, where $I_s = T\psi/R^2$ (W/m^2) is the receiver-plane signal irradiance. Combining these terms yields

$$P_s = \eta_{ox}\eta_{Tx}E_xT^2(c\beta/2)(A_r/R^2), \tag{7.2}$$

where P_s is the power collected by the primary aperture. This received power is proportional to the mean photon arrival rate. The number of photons integrated in time T_{int} is given by

$$N_s = P_sT_{int}/h\upsilon = E_s/h\upsilon, \tag{7.3}$$

where E_s is the received energy and $h = 6.626 \times 10^{-34}$ J s is Planck's constant.

The parameter dependence shown in Equation (7.2) holds for both coherent and direct detection systems. The specific response of the receiver to the total received power (its efficiency, η_R) depends on whether a direct or coherent receiver is used and on the specific design details of the receiver.

The different receiver designs and their response to the received power are covered in Section 7.4. In general, direct detection receiver performance is characterized by the receiver signal-to-noise ratio (SNR) defined as the ratio of the total power in the directly detected signal to the noise power signal due to fluctuation, background light or electrical noise. In coherent detection receivers, the carrier-to-noise ratio (CNR), defined as the ratio of the mean RF signal power to the mean noise power, defines performance. For either receiver type, SNR or CNR is a function of the received signal power, P_s, and the receiver noise. In a well-designed receiver, these quantities approach the quantum limit as governed by fluctuations in the signal itself.

7.3.3. Quantum-Limited Optical Detection

Detection of optical signals is stochastic in nature because of the random photon arrival and annihilation time (generation of photoelectrons), which produces fluctuations in the number of integrated photoelectrons in time T_{int}. This "shot-noise" is characterized by an SNR, which is the ratio of the signal power (mean squared) and the signal variance (i.e., $\text{SNR} = \langle N_s \rangle^2 / \text{var}[N_s]$). Barring speckle noise (see Section 7.3.8), the number of integrated photons follows a Poisson distribution,[127,128] where the mean and variance are equal. Therefore, the *quantum-limited* SNR for this case is given by

$$\text{SNR}_{\text{QL}} = \langle N_s \rangle^2 / \text{var}[N_s] = \langle N_s \rangle \qquad (7.4)$$

This limit can, in principle, be exceeded if the detected signal photons can be made to arrive and be detected more uniformly.[129]

The statistics of the photon arrival time depends upon the target type. Stationary point targets produce photon echoes, which have no speckle modulation and are therefore Poisson distributed. On the other hand, dynamic diffuse targets produce speckle-modulated signals — signals having a negative binomial distribution (i.e., a Poisson deviate whose mean is modulated by a Gamma distributed random process).[128] The SNR of a speckle-modulated signal is given by

$$\text{SNR}_{\text{QL}} = M_e \langle N_s \rangle / (M_e + \langle N_s \rangle) \qquad (7.5)$$

where M_e is the level of total (or effective) speckle diversity (Section 7.3.8). This expression is more general than the Poisson expression (i.e., Equation 7.4). For the case of no speckle modulation M_e is infinity and the Poisson expression results. In the high signal count limit ($N_s \gg M_e$), the SNR saturates at the speckle diversity limit, M_e, and in the low signal count limit ($N_s \ll M_e$) the SNR saturates to that of a Poisson distribution, N_s. For example, when $N_s = (0.1, 1.0, \text{or } 10)M_e$, the SNR is $\approx(0.1, 0.5, \text{or } 0.91)M_e$. Thus there is not much benefit in having $N_s \gg M_e$. Multiple photons in each diversity mode are degenerate and no new information is contained in these excess photons. On the other hand, if N_s is less than M_e, then the full benefit of speckle diversity is not achieved, and the SNR is limited to $N_s < M_e$.

Later, in Section 7.4, it will be shown that both coherent detection CNR and direct detection SNR can approach this quantum limit. However, upon detection, even with state-

of-the-art electronics, poor detection efficiency, optical interference, and electrical noise inevitably result in a performance
less than the quantum limit.

7.3.4. Backscatter Coefficient

As shown in Equation (7.2), the strength of the backscattered
signal is proportional to the volume backscatter coefficient,
β $(\text{m}^{-1}\,\text{sr}^{-1})$, which depends upon the type of scattering involved and the relative angle between the transmitter and
receiver, $\phi = \phi_x - \phi_r$. Because molecules are much smaller
than typical wavelengths, their scattering is characterized
by molecular scattering, which has a λ^{-4} dependence. Molecular scattering strength is a function of air density and temperature and consequently it has a generally predictable
altitude dependence. For altitudes below 100 km, an approximate model for the molecular backscatter coefficient is[21]

$$\beta_M(\lambda) = (\lambda_0/\lambda)^{4.09} \exp(-z/z_0) \times 10^{-7}\,\text{m}^{-1}\,\text{sr}^{-1} \qquad (7.6)$$

where $\lambda_0 = 1.064\,\mu\text{m}$ and $z_0 = 8\,\text{km}$ are a reference wavelength
and altitude, respectively. This produces $\beta_M \approx 8.9 \times 10^{-6}$ and
$6.3 \times 10^{-9}\,\text{m}^{-1}\,\text{sr}^{-1}$ at 0.355 and 2.09 μm, respectively. The
wavelength dependence, $\lambda^{-4.09}$, differs from pure molecular
scattering, $\lambda^{-4.0}$, to account for dispersion of the index of
refraction of air.

Aerosol scattering, due to larger particles, which are
comparable in size to the wavelength, is more complex than
molecular scattering with wavelength dependence between λ^{-2}
and λ^{-1}, from the UV to the far IR.[130] For comparative trade
studies, a group of scientists, studying the feasibility of a Global
Tropospheric Wind Sounder (GTWS) in conjunction with
NASA, has developed upon a standard set of models for the
aerosol backscatter coefficient at several key wavelengths.[131]
These models along with the molecular model (Equation 7.6)
are shown in Figure 7.3. Two aerosol models are shown: the
"background" and "enhanced" models. The background model
is meant to represent the geometric mean of a lognormal
distribution of "background" aerosol distribution. These
models were derived from measurements during the GLObal

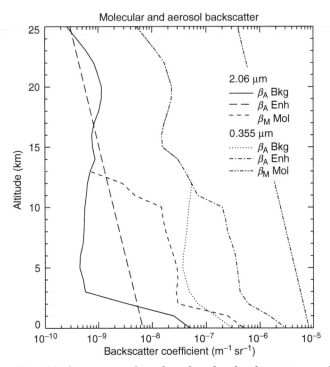

Figure 7.3 Median aerosol and molecular backscatter coefficient models at 355 nm and 2.06 μm.

Backscatter Experiment, GLOBE.[132] The enhanced model corresponds to the geometric mean of a lognormal distribution of the variable backscatter in excess of the "background" mode, sometimes referred to as the convective mode.

Aerosol backscatter variability is much stronger than molecular backscatter variability, since aerosol constituents and densities are weather, pollution, and environment dependent whereas molecular number densities are relatively stable and uniform. The data shown in Figure 7.4 demonstrate this variability. These figures represent data collected at various sites in the continental United States during 1995 and 1996 using a pulsed 2.09 μm coherent lidar.[133] The large number of individual points represents the collected data and the lines represent the 10, 50, and 90 percentiles of the data set. These data show that at $\lambda \approx 2.09$ μm, the median bound-

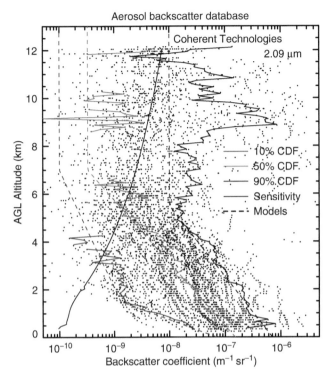

Figure 7.4 Summary plots of a clear-air 2 µm aerosol backscatter database collected over the period 1995 to 1996 in the continental United States (courtesy of Coherent Technologies).

ary layer aerosol backscatter coefficient is $\beta_A \approx 1 \times 10^{-7}\,\mathrm{m}^{-1}\,\mathrm{sr}^{-1}$. The altitude dependence is strong, reducing to a background level of approximately $3 \times 10^{-10}\,\mathrm{m}^{-1}\,\mathrm{sr}^{-1}$. The data in Figure 7.4 show about two orders of magnitude between the 10 and 90 percentiles aerosol backscatter levels at altitudes above 2 km.

There are additional scattering processes,[134] namely Brillouin and Raman scattering, that have to be accounted for in DDL receivers. The Brillouin scattering is highly symmetric about the laser spectral source and therefore, it does not significantly corrupt the DDL wind measurement. While the spectrum due to Raman scattering is complex, the total intensity of the Raman spectrum is of order 10^{-2} of the molecular scattered signal. The Raman spectrum is also divided into many individual lines that are well separated from the molecular/aerosol-scattered signal. As a result, the individual and faint Raman lines are normally very strongly suppressed by the combination of the interference and etalon filters used to reject background (sky) light.

7.3.5. Atmospheric Extinction

Extinction also plays a role in lidar performance. The one-way transmission is given by Beer's law,[134] $T = \exp\left[-\int_0^z \alpha(r)dr\right]$, where $\alpha(r)$ is the wavelength-dependent extinction coefficient along the path. The extinction coefficient is a function of both aerosol and molecular absorption and scattering and the number density of these atmospheric constituents. Accurate high spectral resolution altitude-dependent models are required to achieve adequate estimation fidelity, since both the laser frequencies and molecular absorption have relatively narrow spectral width.

Numerical algorithms, such as the Air Force Research Laboratory's FASCODE,[135] achieve adequate resolution through the use of an extensive molecular database, such as HITRAN.[136] FASCODE predictions of the extinction profile at key laser wavelengths for the Rural Mid-Latitude Summer Atmosphere model with 23 km visibility are shown in Figure 7.5. These data suggest that the strongest absorption occurs for the 355 nm, 532 nm, and 10.6 µm laser wavelengths. UV extinction is dominated by ozone absorption and molecular scattering. At 10.6 µm the extinction is dominated by carbon dioxide and water vapor continuum absorption. Extinction is much weaker at selected wavelengths in the near to far IR, which avoid molecular absorption features.

7.3.6. Radial Velocity Precision

In Section 7.4, detailed descriptions of the radial velocity precision achievable using both coherent and direct detection lidar systems are provided. Precision refers to the variation in estimates of the velocity from an ensemble of measurement realizations and is distinct from resolution which is related to

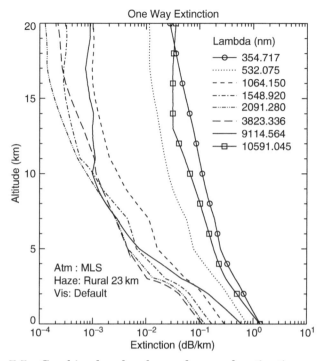

Figure 7.5 Combined molecular and aerosol extinction profiles at key lidar wavelengths, as predicted by FASCOD3 and HITRAN2K.

how well two closely spaced velocities can be resolved. In this section, the focus is on the fundamental limitations due to the spectral width of the signals, which is governed by the transmitted pulse width and molecular or aerosol velocity distributions within the measurement volume.

The variance of the estimate of the mean of a random variable x, from N independent samples, is given by the following statistical lower bound:

$$\mathrm{var}[\hat{\bar{x}}] = \delta x^2 / N \tag{7.7}$$

where δx is the standard deviation of the random variable, and the symbol, $\hat{\bar{x}}$, represents an estimate of the mean of x. A key point to be understood is that the N samples must be independent. Samples that are correlated (or degenerate) with other samples that have already been utilized do not contribute to the reduction in the variance of the estimate of the mean. As will be shown in Section 7.4, this concept is important in understanding the effect of speckle-modulated signals on the velocity measurement performance lidar systems.

Gagne et al.[137] and later Rye and Hardesty[138] have related this statistical bound to the lower bound of performance of mean velocity estimators for direct and coherent detection lidar systems. The statistical bound is given by

$$\text{var}[\hat{\overline{V}}_r] = \delta V_r^2/N_s = (\lambda \delta \nu/2)^2/N_s \qquad (7.8)$$

where $\hat{\overline{V}}_r$ is the mean radial velocity estimate, $\delta \nu$ is the signal frequency spectrum standard deviation width, $\delta V_r = (\lambda \delta \nu/2)$ is the resulting effective velocity standard-deviation width, and N_s is the total number of accumulated receiver-plane signal photons, which is the product of the number of signal photons per range gate, the number of range gates averaged, N_g, and the number of independent pulse echoes averaged, N_p. True velocity estimation variance will be degraded at least by the factor $1/\eta_R$, where η_R is the overall photoreceiver efficiency including detector quantum efficiency. Both coherent and direct detection systems are subject to this lower bound. Actual performance relative to this bound is described in Sections 7.4.1 and 7.4.2 for coherent and direct detection lidars, respectively.

For a narrow velocity spread target, the signal spectral width is ultimately limited by the intrinsic spectral width of the pulse. In practice, for atmospheric lidar, the signal spectral width is governed by a combination of Brownian motion (due to momentum transfer in particle or molecule collisions) and wind turbulence within the measurement volume (Figure 7.6). For Brownian motion, the radial velocity is Gaussian distributed, with standard deviation,

$$\delta V_r = \sqrt{k_B T/m} \qquad (7.9)$$

where $k_B = 1.3806 \times 10^{-23}$ J/K is Boltzmann's constant, and T (K) and m (kg) are the temperature and mass of the particle, respectively. The dominant atmospheric molecules are nitro-

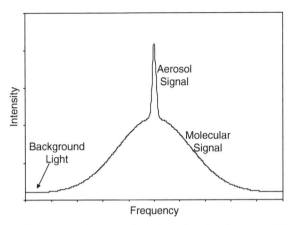

Figure 7.6 Conceptual aerosol and molecular backscatter spectral distributions.

gen (~78%) and oxygen (~21%). The mass of an N_2 molecule is ~4.65×10^{-26} kg, therefore at $T \approx 300K$, the velocity standard deviation is 298 m/sec. For the much heavier aerosol particles, Brownian velocities are small. For example, for 1 to 2 μm diameter aerosol particles, the Brownian velocity standard deviation is of order 1 mm/sec. This is typically insignificant compared to the spectral width due to the pulse Fourier transform limit or wind turbulence within the measurement volume. One exception would be high-spatial-resolution long-coherence time lidar measurements where only one or a few aerosol particles are in the measurement volume. This is usually achieved with a short-range, tightly focused beam using a CW or long-pulse waveform.

Small aerosol particles are sufficiently buoyant in the atmosphere that they move with the local wind (net average motion of the surrounding molecules). Wind turbulence in the measurement volume typically varies, between a few centimeters per second and a few meters per second depending on the turbulence level and the lidar measurement volume. In coherent lidar systems, operating at near-IR wavelengths, the aerosol signal is typically used due to its narrower spectrum. Although the total power in the molecular scatter may be similar to that of the aerosol particle scatter (e.g., see Figure

7.3 for the 2.06 μm wavelength), it is spread over a bandwidth that is two to three orders of magnitude larger than the aerosol signal, resulting in minute excess noise in the signal spectrum, which, due to the spectral filtering in the coherent receiver, can usually be ignored. In direct detection lidar systems, either aerosol or molecular signals, or both, are used.

For equal strength returns, the precision of the molecular signal is much worse due to its broader spectrum, since the velocity estimate precision is proportional to the signal spectral width (see Equation 7.8). Nevertheless, molecular lidar systems can have an advantage when aerosol number densities are very low.

7.3.7. Vector Wind Velocity Estimation

Vector (u, v, w) wind velocity estimates require radial velocity measurements from at least three independent Lines-Of-Sight LOS. The most accurate approach to obtaining a vector wind measurement at a given point in space is to view that same point from three or more directions. For short-range measurements, this can be accomplished by using multiple apertures from the same lidar system, but for longer-range measurements multiple lidar systems are typically required. With constraints on the wind field, for example, the vertical wind being zero, it is sufficient to use two lidar systems from different LOS. In cases where it is impractical to use multiple lidar systems, additional constraints on the wind flow field must be assumed in order to extract useful vector winds.

One widely used approach for measuring the vector winds is the velocity–azimuth display (VAD) scan pattern (Figure 7.7). In a VAD scan, three or more independent LOS are achieved by scanning the beam in azimuth at a fixed elevation angle. Range-dependent radial velocity, V_r, estimates for each azimuth position are estimated and applied to a least-squares algorithm to estimate the parameters of a flow-field model. With just three LOS, accurate estimation of the winds in a zeroth-order 3D mean wind field is possible — that is, a wind field that has no gradients. A first-order model, with linear velocity gradients, requires at least six LOS, etc.

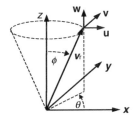

Figure 7.7 Velocity–azimuth display 3D scan geometry for a ground-based laser vector wind profiler.

For the zero[th]-order wind field assumption with $N_{los} \geq 3$ uniformly distributed azimuthal LOS, the [u, v, w] (see Figure 7.7) vector velocity precisions can be shown to be given by

$$[\text{var}[u], \text{var}[v], \text{var}[w]]$$
$$= [2/\sin^2(\phi), 2/\sin^2(\phi), 1/\cos^2(\phi)] \frac{\text{var}[V_r]}{N_{LOS}} \qquad (7.10)$$

where ϕ is the VAD cone half-angle and N_{LOS} is the number of LOS.

In severe velocity turbulence, these VAD processing algorithms can produce substantial error. With sufficient number of LOS and density of the measurements, wind vector velocities can accurately be extracted from the radial measurements even in strong turbulence. As an example, in the highly turbulent wind field case, it is possible, with sufficient scanline density, to map velocity turbulence patterns or eddies. This is done by spatially correlating successive scans to estimate velocity motion (see Figure 7.57). Several examples of VAD wind measurement profiles are provided in Section 7.6.

7.3.8. Laser Speckle Effects

Both coherent and direct detection systems typically employ narrowband laser sources, which results in a random speckle pattern at the receiver plane for diffuse aerosol particle or molecular targets.[139] Laser speckle is an interference pattern resulting from the superposition of numerous small spatially incoherent reflectors (or scatterers) diffusely distributed in a volume. Each scatterer produces a Huygens wavelet with random amplitude and phase. In the receiver plane, the net field is the superposition of these wavelets. This superposition produces a spatially correlated speckle irradiance and phase (Figure 7.8). The spatial size of the correlated areas (or coherence radius) can be calculated using the Van Cittert–Zernike theorem that is described later. Within a speckle coherence area (or "speckle lobe"), the phase is relatively constant and efficient mixing with a uniform phase local oscillator (LO) beam is possible. Exceptions occur when the irradiance goes to zero, where the phase varies significantly. These are known as "branch-points."

When the number of scattering particles is large, the probability density functions (PDFs) of the real and imaginary parts of the net complex-field approach a zero mean Gaussian density function with equal variances. Under this assumption, it can be shown that the distribution of the amplitude, phase, and irradiance density functions are Rayleigh, uniform $(0–2\pi)$, and exponential, respectively. A new realization of the speckle field will result when new scatterers are illuminated

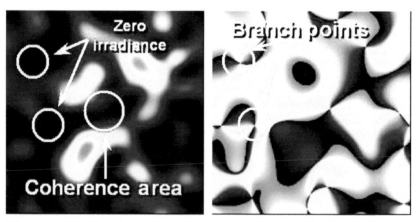

Figure 7.8 Receiver-plane laser speckle amplitude (left) and phase modulo 2π (right), for a Gaussian target-plane beam. The coherence radius is equal to the focused transmit beam radius. Branch points in the phase occur where the amplitude goes through zero.

or when the relative phases of the scatterers, as seen from the lidar location, change by order π rad. This phase change can result from a number of physical changes including: changes of order of $\lambda/4$ in the differential ranges of the illuminated scatterers, the frequency of the lidar changing, or the polarization of the lidar changing.

For a pulsed lidar, the speckle pattern at the receiver plane will evolve as the pulse propagates through the scatterers. This results in a new independent speckle field for each range resolution cell of length $c\Delta T/2$.

The strong fluctuation in the speckle amplitude or irradiance (squared amplitude) is seen in the left panel of Figure 7.8. Since the PDF of the irradiance is exponential, the standard deviation of the irradiance fluctuations is equal to the mean value. To mitigate these fluctuations, the irradiance can be incoherently averaged over multiple independent speckles (over different independent spatial points or independent speckle field realization at the same spatial point at different times). The PDF of the resulting averaged irradiance is known to follow a gamma density function (Figure 7.9), for which the irradiance SNR is given by

$$\text{SNR}_I = \langle I \rangle^2 / \text{var}[I] = M_e \qquad (7.11)$$

where $M_e = N_p N_g M$ is the total or effective diversity, N_p is the number of independent pulse echoes (or "shots") averaged, N_g is the number of independent range gates averaged, and M is Goodman's "M" parameter, which represents the product of the number of independent spatial, polarization ($M_p \leq 2$) and spectral diversities.

As described in Section 7.4.1, the single-frequency single-detector coherent lidar system only detects a single spatial mode of the signal field (i.e., $M \approx 1$) and the net detected signal power, therefore, fluctuates with an exponential PDF. A good design choice, for a fixed average power laser, in coherent systems is to trade the pulse-energy and PRF such that a few (three to seven) uncorrelated shots are averaged for a measurement. This mitigates the high single-shot probability of a low signal or fade (note the significantly

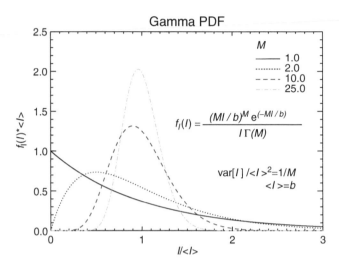

Figure 7.9 Mean-normalized irradiance probability density function, plotted parametrically in the diversity parameter M.

reduced probability of deep fades shown in Figure 7.9 as the diversity increases from 1 to 10), while maximizing the coherent energy per shot and thus maximizing the measurement range. Multiple channel receivers[140,141] and multiple frequency sources[142] have also been employed to increase coherent lidar signal diversity.

The polarization insensitivity, high etendue ($A\Omega \gg \lambda^2$), and the large number of range gates and shots averaged, characteristic of direct detection Doppler wind lidar systems results in very large effective diversity ($M_e = N_p N_g M_p A\Omega/\lambda^2$) where M_p represents the degree of polarization diversity and $A\Omega/\lambda^2$ is the spatial diversity. Typically for DDLs, M_e is much larger than the total number of integrated signal photons. Therefore, for DDL, the noise contribution of speckle can be very small compared to the Poisson noise due to the photon detection process.

The spatial coherence of a speckle field is characterized by the field spatial autocorrelation function, which describes the correlation of the field between two points: P_1 and P_2. A useful measure of the spatial correlation is the complex coherence factor (CCF), which is a normalized spatial auto-correlation function given by

$$\mu_{12} \equiv \langle U_1 U_2^* \rangle / \sqrt{I_1 I_2} \tag{7.12}$$

where U_1 and U_2 are the complex field amplitudes at receiver-plane coordinates $[x_1, y_1]$ and $[x_2, y_2]$, and I_1, and I_2 are the corresponding irradiances. The CCF for laser reflected light from a rough surface has been derived by Goodman[143] (Equations 5.6–5.10) and can be found in numerous other optics books. His expression is given by

$$\mu_{12}(\Delta x, \Delta y)$$
$$= \frac{\exp(-j\psi) \iint I_R(\xi, \eta) \exp(j2\pi(\Delta x \xi + \Delta y \eta)/\lambda R) d\xi d\eta}{\iint I_R(\xi, \eta) d\xi d\eta} \tag{7.13}$$

where $I_R(\xi, \eta)$ is target-plane reflected irradiance and ψ is usually an unimportant phase. The denominator integral is a normalization factor representing the total reflected power.

This expression is known as the Van Cittert–Zernike theorem,[143] which states that the correlation function of the field, resulting from a spatially incoherent source, is proportional to the 2D Fourier transform of the source irradiance distribution. It should be noted that the integral equation above is also the integral equation used to calculate the diffraction pattern at range R resulting from a coherent source with the given illumination pattern and focused at range R (i.e., a spherical wave with curvature centered at the observation plane R), with the exception that the irradiance is used to calculate the CCF and the field is used to calculate the diffraction pattern. This means that the CCF has the same sputial dependence as the diffraction-limited spot of a constant phase field distribution whose amplitude matches the target-plane irradiance distribution function of the incoherent source in the target plane.

Many definitions for this characteristic coherence area, A_c, can be found in the literature. For a circularly symmetric source, $A_c = \pi \rho_c^2$, where ρ_c is the coherence radius. Common examples for circularly symmetric sources are the $e^{-1/2}$, e^{-1}, and e^{-2} radii of the CCF. Goodman uses the following definition to quantify the coherence area:

$$A_{cg} \equiv \int\int |\mu_{12}(\Delta x, \Delta y)|^2 \mathrm{d}\Delta x \; \mathrm{d}\Delta y \tag{7.14}$$

By substituting Equation (7.13) into Equation (7.14) and employing Parseval's theorem, it can be shown that

$$A_{cg} = (\lambda R)^2 \frac{\int\int I_R^2(\xi, \eta)\mathrm{d}\xi \; \mathrm{d}\eta}{\left[\int\int I_R(\xi, \eta)\mathrm{d}\xi \; \mathrm{d}\eta\right]^2} \tag{7.15}$$

For a uniformly illuminated target the coherence area simplifies to

$$A_{cg}^u = (\lambda R)^2 / A_R = \lambda^2 / \Omega_R \tag{7.16}$$

where A_R is the area of the incoherent source and Ω_R is the solid angle subtended by the source at range R. The coherence area, A_{cg}, of the speckle field due to a uniformly illuminated area, A_R, at range R, is on the order of the diffraction-limited spot area of that source if it were coherent.

In terms of a Gaussian beam lidar, the target-plane irradiance distribution is Gaussian, and therefore in the receiver plane the average field correlation is also Gaussian, given by[144]

$$\mu_{12}(\rho) = \exp\left(-j\psi\right) \exp\left(-(\rho^2 / 2(\lambda R / \pi \omega_R)^2)\right) \tag{7.17}$$

where ρ is the transverse distance, ω_R is the e^{-2} target-plane irradiance radius at range R, and λ is the optical wavelength. Note this expression can be found by direct evaluation of Equation (7.13). For this Gaussian target illumination, the average coherence radius of the speckle field (i.e., the e^{-1} point of a circular Gaussian CCF) at range R is $\rho_c \sqrt{2}\lambda R / \pi \omega_R$. This radius is also $\sqrt{2}$ times the transmitter-plane beam radius, ω, for a focused Gaussian beam that produces a beam radius ω_R at range R. Note, however, that if Goodman's coherence area definition is used (i.e., Equation 7.14) the resultant coherence radius (i.e., the radius for which $\pi\rho_{cg}^2 = A_{cg}$)

is $\rho_{cg} = \omega$. Therefore, for a monostatic lidar, in the near field of a transmitted beam the coherence radius is much smaller than the transmit beam size, whereas in the far-field $\rho_{cg} = \omega$.

The spatial speckle diversity[145] obtained by integrating the speckle field over both circular and square aperture areas is shown in Figure 7.10, along with three solid lines corresponding to a simple approximate model ($M = [1+(A_a/A_{cg})^b]^{1/b}$ with $b = 0.866$), and the reference lines $M = 1$ and $M = A_a/A_{cg}$. A general result for any aperture shape is that whenever the aperture area is large compared to the coherence area the spatial diversity is given by A_a/A_{cg}.

7.3.9. Background Light

The amount of background light power collected by the aperture depends upon the radiation source (e.g., direct viewing of a thermal blackbody or reflected blackbody radiation, such as a sunlit-cloud) and the scattering and absorption between the source and the receiver. The background spectral radiance, L_{bg} (W m^{-2} m^{-1} sr^{-1}), of a blackbody source or reflector can be calculated using Planck's equation by assuming an equivalent

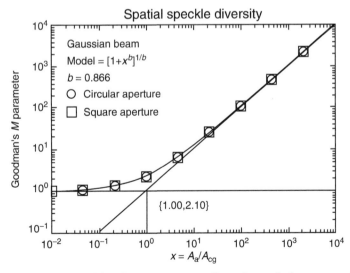

Figure 7.10 Speckle diversity as a function of the normalized aperture area.

Lambertian source at temperature T_{bb} with emmisivity, ε. Planck's equation is given by[146]

$$L_{bg} = \frac{\varepsilon 2hc^2}{\lambda^5[\exp(hv/k_B T_{bb}) - 1]} \tag{7.18}$$

In Figure 7.11, the spectral radiance computed with LOWTRAN, an AFRL Geophysics Laboratory model, for vertical viewing of a cloudless daytime with the sun at a 77 degree elevation angle, is co-plotted with black-body source models. LOWTRAN provides the most realistic estimate for the scattered solar spectral radiance by aerosols and molecules in the sky. For comparison, two blackbody radiator models are also shown in the figure. The first represents a moonless night sky, using a 293°K blackbody radiator with an emmisivity of 100%. For the day sky, the sum of the moonless sky and a 5900°K blackbody with an very weak emmisivity, $\varepsilon = 2 \times 10^{-6}$, provides a rough estimate to the LOWTRAN data over some spectral regions.

The blackbody curves are not adequate estimates for much of the spectrum, because they do not include the finer features of the sun's emission spectra or the wavelength-dependent absorption, transmission, and scattering by the atmospheric aerosols and molecules. Rayleigh and Mie scattering of the solar radiation from the atmosphere causes the sky radiance at the longer wavelengths to be significantly reduced relative to the simple blackbody model. At wavelengths shorter than approximately 320 nm ozone absorption, and then O_2 absorption, progressively reduce the intensity of sunlight penetrating the atmosphere, so that below about $\lambda = 280$ nm the intensity of sunlight penetrating below the stratopause is negligible.

The amount of detector-integrated background light is found by integrating the spectral radiance over the receiver aperture, the receiver field of view, and the filter bandwidth; including the receiver optics efficiency and transmission losses from the source to the receiver.

For narrowband systems, the solution to this integral is well approximated by

$$P_{bg} = TA_r \Omega \Delta\lambda L_{bg},$$ (7.19)

where $\Delta\lambda$ is the width of the optical narrowband filter and T is the one-way atmospheric transmission. From diffraction theory, $A_r\Omega \geq \lambda^2$ with equality achieved by a diffraction-limited receiver. The number of photons accumulated in an integration time, T_{int}, is given by

$$N_{bg} = P_{bg}T_{int}/h\upsilon$$ (7.20)

In the next two subsections these approximations are applied to yield estimates of the background photon arrival rate for both direct and coherent detection receivers.

7.3.9.1. Direct Detection

The molecular back-scattered signal provides an excellent source for the direct detection wind lidar at visible light wavelengths. However, during daytime, for exactly the same reason, sunlight scattered from the lower atmosphere provides a signal against which the back-scattered lidar signal must compete.

Receiver performance reduction results from two different effects: (1) a reduction in the receiver's dynamic range due to the integrated photons, and (2) an increased noise level from fluctuations of the detected background photons (shot-noise). The former is usually of insignificant consequence compared to the latter. In Figure 7.12 the number of background photons per microsecond is shown for the radiance spectra provided in Figure 7.11 assuming a diffraction-limited receiver ($A\Omega = \lambda^2$) and a narrowband filter whose width is $\lambda/1000$.

In order to calculate the photon arrival rate for a given lidar design, the actual receiver field of view (FOV) and filter bandwidth must be taken into account. For example, with a 50 cm diameter telescope, the diffraction limit at 355 nm (λ/D) is 0.7 μrad. Typical direct detection designs at 355 nm use FOVs on the order of 100 μrad, so the number of background counts will increase by four orders of magnitude from that shown in the figure to $\sim$100 photons per microsecond from the cloudless daytime sky. The rate for a sunlit cloud would be higher. Using a properly designed Fabry–Perot etalon (see

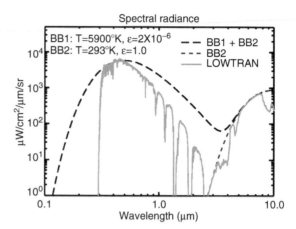

Figure 7.11 Radiance from the ground looking upward for a cloudness day time sky (solid). Also shown for comparison are models for the radiance of a 293°k blackbody source (thermal – BB2) and the combination of a 293°k blackbody source with a weakly-reflected 5900°k source (solar + thermal – BB1 + BB2).

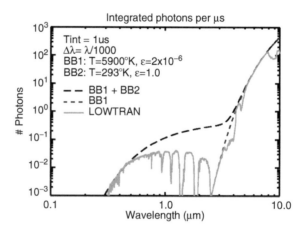

Figure 7.12 Number of background photons per microsecond for a diffraction-limited receiver with a narrowband filter width equal to $\lambda/1000$, for a daytime cloudless sky (solid). Also shown for comparison are models for the photons per microsecond that would result from the solar and blackbody radiance presented in Figure 7.11.

Section 7.4.2.3) as a second bandpass filter stage following the first narrowband filter can reduce the bandpass to roughly the bandwidth of the signal itself for molecular backscatter.

For example, for a molecular scattered signal at 355 nm the spectral width required to transmit the majority of the signal is 5–10 GHz (see Section 7.3.6) which is ~2–4 pm at 355 nm. With this narrowband filter, the background counts can be reduced by approximately two orders of magnitude, resulting in 1–10 photons per microsecond. If this is still too high for the given measurement, the rate could be reduced further by narrowing the receiver FOV. The photo-electron count rate of a given system resulting from the background photons will depend on the overall efficiency of the receiver. The most significant background light issue for direct detection lidars occurs when viewing sunlit clouds or direct solar glints. The photon arrival rates in both of these situations is significantly higher than the daytime clear sky at most visible and UV wavelengths.

It has been shown in the analysis by Rees and McDermid[102] and shown practically by Rees et al.[106] at $\lambda = 0.532\,\mu$m that the daytime "background" signal can be reduced to an essentially negligible level for many measurement applications.

7.3.9.2. Coherent Detection

Thermally generated background light does not typically pose a significant problem for a coherent receiver. For example, Siegman[147] showed that the noise equivalent temperature (NET) of an extended unpolarized thermal source is given by

$$\text{NET} = h\upsilon/k_\text{B}\ln(1 + \varepsilon\eta_\text{r}) \tag{7.21}$$

where NET is the source temperature, which produces one coherently detected interference photoelectron per reciprocal bandwidth or per integration time and η_r is the receiver's total combined electrical and optical efficiency. This expression is derived from Planck's equation (i.e., Equation 7.18) and Equations (7.19) and (7.20), using $A\Omega = \lambda^2$, $\Delta\lambda = \Delta\upsilon\lambda^2/c$, and $\Delta\upsilon = 1/T_\text{int}$. In the visible (0.55 μm), near-IR (2 μm), and far-IR (10 μm), the NET is ≈37,000, 10,000, and 2,000 K, respectively, assuming a perfect single polarization coherent receiver, $\eta_r = 0.5$ and an emmisivity of 1.

Siegman's result is independent of both the receiver diameter and the receiver bandwidth. The first is true, because of Siegman's antenna theorem, which states that the effective etendue of a coherent receiver is governed by the diffraction limit, $A\Omega \approx \lambda^2$ (see Section 7.4.1.3 for more detail). This results in a fixed amount of spatially integrated background light being detected by the coherent receiver independent of the aperture area and detector FOV. The second is true because the inherent noise of the coherent receiver is one detected photon per second per unit bandwidth. As the bandwidth is increased the number of detected background photons increases commensurate with the increase in the inherent coherent detection noise.

The sensitivity of a coherent receiver operating in the visible or near IR is not significantly degraded even when viewing the sun directly (T $\sim$ 5900 K). Even so, an appropriate optical bandpass filter will likely be required for direct solar viewing, to avoid damage due to the focused solar radiation collected by the lidar aperture.

7.4. DOPPLER WIND LIDAR THEORY OF OPERATION

In this section, we present the basic lidar theory that will allow system performance to be calculated and also provide physical insight into system performance. Coherent and direct detection Doppler wind lidar theory is presented in Sections 7.4.1 and 7.4.2, respectively. Key results and differences between the two sensor architectures are summarized in Section 7.4.3. Sections 7.4.1 and 7.4.2 contain a significant amount of detailed theory and mathematics. For readers who do not desire to initially go through this level of detail, it is recommended that the summaries of these two sections (7.4.1.9 and 7.4.2.7) and Section 7.4.3 be read prior to moving into Section 7.5.

7.4.1. Coherent Detection Doppler Wind Lidar Theory

In this section, the coherent detection process is described. The concept of mixing a local oscillator beam with the received signal to provide both near-quantum-limited detection and a direct frequency down conversion from the optical band to the RF band is described in Sections 7.4.1.1 and 7.4.1.2. It is shown that optimum performance is achieved through matched-filter detection. A generalized CNR and antenna efficiency expression is developed in Section 7.4.1.3 in terms of transceiver and target parameters (pulse energy or power, aperture diameter, receiver efficiency, target range, backscatter coefficient, etc.). This generalized CNR expression is then applied to several specific lidar configurations in Section 7.4.1.4.

A discussion of the effects of misalignment and refractive turbulence is presented in Sections 7.4.1.5 and 7.4.1.6. Analytic expressions describing velocity precision are presented in Section 7.4.1.7. The performance of an example 2 μm ground-based coherent lidar is presented in Section 7.4.1.8. Key concepts and results are summarized in Section 7.4.1.9.

7.4.1.1. Heterodyne Detection Concept Overview

In the previous section, the lidar range equation (Equation 7.2) was presented. This first-order model predicts the optical signal power collected by the receiver aperture. Although the total power received is equivalent for both types of systems, coherent and direct detection receivers employ characteristically different methods of utilizing this power to estimate the Doppler frequency shift, from which the wind velocity is derived.

The basic elements of a typical coherent receiver are illustrated in Figure 7.13. In this receiver, the signal field is collected by a receiver aperture with area A_r, transmitted through the receiver optics and then mixed (added) with an optical LO beam in a beam splitter. This combined field is focused onto a high-speed detector, which is usually, but not always, larger than the LO field extent, to produce a photocurrent which carries the signal information.

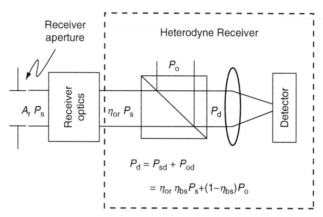

Figure 7.13   Heterodyne receiver functional diagram.

When the signal field is mixed (added) with an optical LO field, the combined field power, and hence the photocurrent, is sinusoidally modulated by the interference of these two fields. The lower frequency component of this modulated signal, which is detected by the finite bandwidth detector, is typically an RF signal, whose frequency and phase is the difference frequency and phase between the two beams. The frequency of the modulated signal is therefore the Doppler shift imparted on the received signal by the target's velocity plus an arbitrary offset or carrier frequency, which is typically established by on offset between the transmit and local oscillator beam frequencies. In addition to providing a means of directly detecting the signal field frequency, it will be shown, below, that the local oscillator beam provides optical gain for the signal beam through the mixing process. For a well-designed coherent detection receiver, this optical gain, combined with narrowband matched-filter detection of the signal, results in near quantum-limited performance.

7.4.1.2.   Heterodyne Detection CNR and Related Concepts

Frequency or velocity estimation accuracy is directly related to the strength of the detected signal power relative to the receiver noise power. In coherent detection receivers, the signal strength is characterized by the power of the modulation signal at the difference frequency. Following RF communication and radar nomenclature, the ratio of these two powers is called the carrier-to-noise ratio, CNR, which is given by

$$\text{CNR}(t) = \langle i_{\text{h}}^2(t) \rangle / \langle i_{\text{n}}^2(t) \rangle \tag{7.22}$$

where i_{h} is the heterodyne signal photocurrent at the difference frequency between the two fields and i_{n} is the receiver noise current.

Later in this section, it is shown that a principal component impacting the power in the heterodyne signal photocurrent, and hence the CNR, is how much signal power the receiver aperture collects. The lidar range equation (Equation 7.2) is a first-order model predicting this optical signal power level in terms of the transceiver, target, and atmospheric parameters. In this section, theoretical models of the CNR are derived based upon receiver-plane calculations, in terms of this optical signal power. It is shown that for well-designed receivers, the CNR can approach the quantum limit. A theory that predicts the CNR using target-plane fields rather than receiver-plane fields is also presented in the following subsection, and these two theories are shown to be equivalent.

7.4.1.2.1. Field and Photocurrent Mathematics

A real electric field, U (V/m), with carrier frequency ν, amplitude A, and phase ϕ, can be written as

$$U(x,y,z,t) = A(x,y,z,t)\cos\left(2\pi\nu t + \phi(x,y,z,t)\right) \tag{7.23}$$

where (x, y, z) is a spatial coordinate and the field is assumed to propagate along the z-axis. Furthermore, the detector is assumed to be in the $z = 0$ plane.

The field irradiance is the field power density averaged over a time scale that is long compared to the optical carrier cycle time but short compared to amplitude fluctuation time scales. The irradiance is given by

$$I(x,y,z,t) = A^2(x,y,z,t)/2z_{\text{m}} \tag{7.24}$$

where z_m is the impedance of the propagation medium and the factor of 1/2 results from time averaging over the many optical carrier oscillations. For free space $z_m = z_0 = 1/\varepsilon_0 c = 376.7$ ohm (here $\varepsilon_0 = 8.85 \times 10^{-12}$ F/m is the permittivity of free space).

Mathematical descriptions of the field properties (irradiance, auto- and cross-correlations, etc.) are often much simpler if complex, rather than real, field representations are used. The complex field representation, U, of the real electric field, U, is called the "analytic-signal" representation of U. That is

$$U(x,y,z,t) = A(x,y,z,t)\exp(i2\pi\nu t + i\phi(x,y,z,t)) \qquad (7.25)$$

where $i = \sqrt{-1}$.

Using this complex representation, the irradiance is simply

$$I(x,y,z,t) = |u(x,y,z,t)|^2 \qquad (7.26)$$

where

$$u(x,y,z,t) = A(x,y,z,t)\exp(i\phi(x,y,z,t))/\sqrt{2z_m} \qquad (7.27)$$

is the field's normalized *complex amplitude*, which carries the signal information (target reflectance, phase, and Doppler), but does not include the relatively unimportant transmit carrier frequency. The normalization factor of $1/\sqrt{(2z_m)}$ permits the simple square-modulus relationship in Equation (7.26).

Most high-performance lidars employ photovoltaic detectors to avoid the 3 dB excess loss associated with recombination noise associated with photoconductive detectors.[148] For these devices the photocurrent is proportional to the spatially integrated irradiance

$$i(t) = \iint \Re(x,y)I(x,y,0,t)dxdy \qquad (7.28)$$

where

$$\Re(x,y) = e\eta_q(x,y)/h\upsilon \qquad (7.29)$$

is the detector responsivity (A/W), $e = 1.602 \times 10^{-19}$ C is the electron charge, η_q (electrons/photon) is the detector quantum efficiency (QE), and the integration limits are infinite in the detector-plane. A detector with nonuniform responsivity, or that is finite in size, is captured by the spatially dependent responsivity. For uniform responsivity detectors over a finite area, such that $\Re(x, y) = \Re$ inside the detector area (A_d and is zero outside, the photocurrent is given by

$$i(t) = \Re P_{A_d}(0, t) \tag{7.30}$$

where P_{A_d} (W) is the field power, or spatially integrated irradiance, falling within the detector area. That is,

$$P_{A_d}(z, t) = \int\int_{A_d} I(x, y, z, t)\mathrm{d}x\mathrm{d}y \tag{7.31}$$

7.4.1.2.2. Heterodyne Efficiency and Wideband CNR

We begin this analysis by assuming an arbitrary nonuniform detector responsivity and show that the CNR is maximized when the spatial signal field is matched (i.e., is proportional in amplitude and has equal phase) to the "receiver spatial mode" defined by the responsivity or quantum efficiency weighted LO field distribution. Later, we relax this generalization and consider the special, but often typical, case of a uniform responsivity detector.

In coherent detection, the detector-plane field complex amplitude, $\boldsymbol{u}_d$ is the sum of a local oscillator, $\boldsymbol{u}_{od}$, and the received signal, $\boldsymbol{u}_{sd}$, fields at the detector location. The corresponding space–time dependent irradiance, I_d, has three additive components (see Figure 7.14): a baseband (signal spectrum is centered about 0 Hz) LO irradiance, I_{od}, a baseband signal irradiance I_{sd}, and an RF heterodyne or interference irradiance, I_h, corresponding to the interference between the LO and the signal fields.

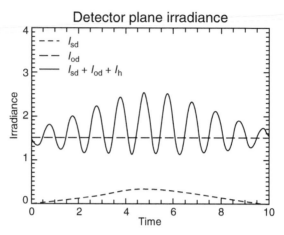

Figure 7.14   Coherent detection RF and baseband signals.

$$I_d(x,y,t) = |u_d(x,y,t)|^2 = |u_{od}(x,y,t)|^2 + |u_{sd}(x,y,z)|^2$$
$$+2\,\mathrm{Re}[u_{od}(x,y,t)^* u_{sd}(x,y,t)]$$
$$= I_{od}(x,y,t) + I_{sd}(x,y,t) + I_h(x,y,t) \qquad (7.32)$$

where Re[x] denotes the real part. In coherent detection the interference irradiance term, I_h, carries the signal information (amplitude and phase), while in direct detection, I_{sd} is the signal. The heterodyne signal irradiance is given by

$$I_h(x,y,t) = 2\sqrt{I_{od}(x,y,t)I_{sd}(x,y,t)}$$
$$\times \cos[2\pi\Delta\nu t + \Delta\phi(x,y,t)] \qquad (7.33)$$

where $\Delta\nu$ and $\Delta\phi$ are the frequency and phase difference between the LO and signal fields, respectively. If a finite frequency difference, $\Delta\nu$, exists between the signal and LO fields, the time-averaged heterodyne signal irradiance term is zero, therefore the mean detector-plane irradiance is $I_{sd} + I_{od}$.

It can be shown, through conservation of power arguments, that the two interference signals in the two output ports of the beam splitter are negatives of each other. This fact can be exploited in "dual-port" differential coherent receivers to mitigate the beam splitter loss, η_{bs}, and efficiently utilize all of the signal and LO photons, while also suppressing LO irradiance noise or other common-mode noise.[149,150]

7.4.1.2.2.1. Signal Current and Power: In single pixel coherent receivers, the detector is usually much larger than the LO beam extent. Furthermore, these detectors typically have very uniform detector responsivity. In coherent imaging applications employing dense detector arrays, the individual detector elements (pixels) may in fact be smaller than the LO beam. In this section, the infinite uniform responsivity detector approximation is not assumed, in order to expose the reader to the complete physics. However, in later sections, these two assumptions are employed to simplify the mathematics.

From Equations (7.28) and (7.32), the heterodyne signal photocurrent is given by

$$i_h(t) = 2\text{Re}\left[\iint \Re(x,y)\boldsymbol{u}_{od}^*(x,y,t)\boldsymbol{u}_{sd}(x,y,t)\text{d}x\text{d}y\right] \quad (7.34)$$

This current is typically an RF signal whose time-averaged signal power is

$$\langle i_h^2(t)\rangle = 2\left|\iint \Re(x,y)\boldsymbol{u}_{od}^*(x,y,t)\boldsymbol{u}_{sd}(x,y,t)\text{d}x\text{d}y\right|^2 \quad (7.35)$$

where a factor of 1/2 arising from averaging over many cycles of the RF current, was used.

7.4.1.2.2.2. Noise Power: The total noise power is the sum of the powers from many constituents. In a well-designed lidar the dominant noise power is the shot-noise due to the random generation of photoelectrons by the interaction of the local oscillator laser with a photovoltaic detector. For a typical, classical, local oscillator laser the number of photons arriving in a time interval, T_{int}, follows a Poisson distribution. Likewise, the number of generated electrons also follows a Poisson distribution, whose mean is degraded by the detector quantum efficiency. This random process is known as shot-noise. The shot-noise power, can be derived with the knowledge that the variance of a Poisson random variable is equal to the mean.[146] This produces

$$\langle i_{sn}^2(t) \rangle = ei_o(t)/T_{int}, \tag{7.36}$$

where

$$i_o(t) = \iint \Re(x,y)|\boldsymbol{u}_{od}(x,y,t)|^2 dxdy \tag{7.37}$$

is the local oscillator induced photocurrent. This random noise has a white noise spectral density, S_{no}, since no matter how short the time integration interval is, the number of electrons is still a Poisson random process. Using the relation that the noise-equivalent-bandwidth of a perfect baseband integrator is $1/2\ T_{int}$, the shot-noise expression can be written in its standard form

$$\langle i_{sn}^2(t) \rangle = S_{no}(t)B = 2ei_o(t)B, \tag{7.38}$$

where B is the receiver noise-equivalent bandwidth.

It should be noted that although shot-noise is attributed to the local oscillator in a semiclassical analysis such as presented here, it has been shown through a full quantum analysis that the shot-noise is actually due to zero-point field fluctuations of the vacuum.[151] It is interesting that the vacuum fluctuations, which are responsible for more subtle effects such as the Lamb shift or Casmir effect, are so easily observable as shot-noise in heterodyne or homodyne receivers.

Other noise terms present in the coherent receiver include shot-noise from the signal, background light and detector dark current, electronics amplifier noise, interference noise terms resulting from the beating of the LO with the background light and unwanted laser reflection; and even the shot-noise from the heterodyne signal photocurrent itself. In typical systems, the LO power is very large compared to that of other light sources so the shot-noise from other light sources is negligible. The extra beating terms and the dark current shot-noise are also usually very small compared to the electronics noise. The composite sum of these extra noise terms is an excess noise that can be collected into a single noise term represented by a noise spectral density, S_n (A^2/Hz).

In the remainder of this analysis, an amplitude and phase stable CW local oscillator is assumed and therefore the shot-noise is temporally stationary and the time dependence will be removed. In a typical coherent lidar, the other added noise is usually dominated by preamplifier noise. So for white-noise over the bandwidth of operation, we can write

$$\langle i_n^2 \rangle = 2ei_oB + S_nB = 2ei_oBF_h, \qquad (7.39)$$

where F_h is the receiver excess noise factor representing the ratio of the total noise power to the LO shot-noise power. For example, when the shot-noise power is 90% of the total noise power, the excess noise factor is $10/9 = 1.11$.

An important lidar design trade is the tradeoff between minimizing the excess noise factor, F_h, by increasing the LO power and detector saturation, which results in a reduction in "small signal" quantum efficiency with increasing local-oscillator power. For typical near-IR, InGaAs photodiodes and pre-amplifier circuits, this trade is usually optimized for LO powers on the order of 1.0 mW.

Solid-state CW lasers exhibit relaxation oscillations resulting in additional noise with spectral content on the order of several hundred kilohertz for typical solid-state lasers. This noise is usually disregarded in coherent receiver designs because it is a near-baseband noise that can be filtered out from the RF signal current.

7.4.1.2.2.3. Wideband CNR: Combining Equations (7.22), (7.29), (7.35), (7.37), and (7.39) produces the following wide-band CNR expression:

$$\text{CNR}(t) = \frac{\eta_{or}\eta_{bs}P_s(t)}{h\nu BF_h}$$

$$\times \frac{\left| \iint \eta_q(x,y)\boldsymbol{u}_{od}^*(x,y)\boldsymbol{u}_{sd}(x,y,t)dxdy \right|^2}{\iint \eta_q(x,y)|\boldsymbol{u}_{od}(x,y)|^2 dxdy \iint |\boldsymbol{u}_{sd}(x,y,t)|^2 dxdy}$$

$$(7.40)$$

where

$$\eta_{or}\eta_{bs}P_s(t) = \iint |u_{sd}(x,y,t)|^2 dxdy \tag{7.41}$$

was used, P_s is the total signal power collected by the receiver aperture (i.e., Equation 7.2) and $\eta_{or}\eta_{bs}$ account for the losses in the receiver optics train.

Given an LO field and quantum efficiency distribution, the maximum CNR is obtained when the signal mode matches the receiver mode defined by the quantum efficiency weighted LO field. To see this, we can rewrite Equation (7.40) as follows:

$$\text{CNR}(t) = \frac{\eta_r \eta_h(t)P_s(t)}{h\nu BF_h} = \frac{\eta_R\, P_s(t)}{\eta VB} \tag{7.42}$$

where $\eta_R = \eta_r\eta_h/F_h$ is the total receiver efficiency and

$$\eta_h(t) = \frac{\left|\iint \eta_q(x,y)\boldsymbol{u}_{od}^*(x,y)\boldsymbol{u}_{sd}(x,y,t)dxdy\right|^2}{\iint |\eta_q(x,y)\boldsymbol{u}_{od}(x,y)|^2 dxdy \iint |\boldsymbol{u}_{sd}(x,y,t)|^2 dxdy} \tag{7.43}$$

is the heterodyne efficiency,

$$\eta_r = \eta_{or}\eta_{bs}\hat{\eta}_q \tag{7.44}$$

is the receiver efficiency less heterodyne efficiency and

$$\hat{\eta}_q = \frac{\iint |\eta_q\boldsymbol{u}_{od}(x,y)|^2 dxdy}{\iint \eta_q|\boldsymbol{u}_{od}(x,y)|^2 dxdy} \tag{7.45}$$

is an effective heterodyne quantum efficiency, which is temporally stationary due to the CW LO assumption.

The CNR and heterodyne signal current are both maximized when the heterodyne efficiency is maximized. Schwartz's inequality states that $\left|\int f*g\right|^2 \leq \int |f|^2 \int |g|^2$ and equality holds if f is proportional to g. Therefore, the heterodyne efficiency is maximized, and equal to one, when the signal field is spatially matched to the "receiver spatial mode," defined by the quantum efficiency weighted LO field or vice versa. Therefore, the heterodyne efficiency represents the fraction of the total optical signal power, P_{sd}, which is matched to (or coupled into) the "receiver spatial mode," and a heterodyne receiver is a single mode receiver, only sensitive to incoming power that is matched to the receiver spatial mode. A simple example is the case of a uniform detector responsivity over the entire irradiance distribution of the LO. In this case, the heterodyne efficiency is the fraction of the signal power that is matched to the "spatial mode" of the local oscillator. As a second example, assume a receiver made up of a single mode fiber with an LO beam coupled into the fiber and with the output being incident on a uniform quantum efficiency detector. In this case, the heterodyne efficiency is simply that fraction of the total signal power that is coupled into the single mode fiber.

7.4.1.2.2.4. Uniform Responsivity Detector: For the special case of a uniform detector responsivity, $\Re = \eta_q e / h\nu$, over a finite detector area, A_d and zero responsivity outside this area, the above mathematical descriptions can be written in a more common form. In particular, the signal photocurrent (i.e., Equation 7.34) can be written as

$$i_h(t) = 2\Re \sqrt{\eta_h(t) P_{sd}(t)(\eta_{od} P_{od})} \cos\left(2\pi \Delta\nu t + \Delta\theta(t)\right) \quad (7.46)$$

where $\Delta\nu$ is the difference or beat frequency between the LO and signal field, (this frequency is also called the RF carrier frequency), $\Delta\theta$ is the net phase from the spatial integration of the two fields at time t, η_{od} is the detector LO truncation efficiency representing the fraction of the LO power that is distributed over the detector area, given by

$$\eta_{od} = \frac{\iint_{A_d} |\boldsymbol{u}_{od}(x,y)|^2 dx dy}{\iint |\boldsymbol{u}_{od}(x,y)|^2 dx dy} \quad (7.47)$$

and η_h is the heterodyne efficiency as described above, but which, in this special case of a uniform finite detector, can be rewritten as

$$\eta_{\mathrm{h}}(t) = \frac{\left|\iint_{A_{\mathrm{d}}} \boldsymbol{u}_{\mathrm{od}}^{*}(x,y)\boldsymbol{u}_{\mathrm{sd}}(x,y,t)\mathrm{d}x\mathrm{d}y\right|^{2}}{\iint_{A_{\mathrm{d}}} \left|\boldsymbol{u}_{\mathrm{od}}(x,y)\right|^{2}\mathrm{d}x\mathrm{d}y \iint \left|\boldsymbol{u}_{\mathrm{sd}}(x,y,t)\right|^{2}\mathrm{d}x\mathrm{d}y} \tag{7.48}$$

The expression for the CNR (i.e., Equation 7.42) remains unchanged and $\hat{\eta}_{\mathrm{q}} = \eta_{\mathrm{q}}$.

For the special, but often typical, case of an infinite (relative to the LO field distribution) uniform detector, the heterodyne efficiency can also be written as (see Equation 7.43 using $i_{\mathrm{o}} = \Re P_{\mathrm{od}}$ and $i_{\mathrm{s}} = \Re P_{\mathrm{sd}}$)

$$\eta_{\mathrm{h}}(t) = \frac{\langle i_{\mathrm{h}}^{2}(t)\rangle}{2i_{\mathrm{o}}(t)i_{\mathrm{s}}(t)} \tag{7.49}$$

where i_{s} is the baseband or direct detection signal current. Note that this expression assumes the receiver gain is the same at baseband as it is in the heterodyne signal band. Thus, in principle, the heterodyne efficiency can be directly measured if all three currents can be measured independently.

7.4.1.2.3. Narrowband or Matched-Filter CNR: The matched-filter, also called "narrowband" carrier-to-noise ratio, $\mathrm{CNR}_{\mathrm{n}}$, is the CNR produced when the receiver is a matched-filter receiver (i.e., the effective integration time is equal to the signal coherence time). This matched-filter receiver produces the maximum possible CNR.[152] For the matched-filter receiver, the noise-equivalent or narrowband bandwidth is $B_{\mathrm{n}} = 1/\tau_{\mathrm{c}}$, where τ_{c} is the coherence time of the signal (see Section 7.4.1.7).

From Equation (7.42), the narrowband CNR is given by

$$\mathrm{CNR}_{\mathrm{n}}(t) = \frac{\eta_{\mathrm{R}}P_{\mathrm{s}}(t)\tau_{c}}{h\nu} = \eta_{\mathrm{r}}N_{\mathrm{sc}}(t) \tag{7.50}$$

where N_{sc} represents the number of detector-plane photons accumulated within the coherent integration time, τ_{c}. Therefore, a perfect coherent receiver, $\eta_{\mathrm{R}} = 1$, is quantum-limited. Equation (7.50) should not be interpreted that the $\mathrm{CNR}_{\mathrm{n}}$ is quantized by the signal photon number. It is simply the result of algebraic manipulation. Because of the LO "amplification" described below, the actual number of signal photo-electrons is quite substantial even when the mean signal photon number is much less than 1.

7.4.1.2.4. Signal Amplification: A key feature of the heterodyne receiver, evident from this analysis (i.e., Equation 7.46), is that the signal matched to the receiver mode, given by $\eta_h P_{sd}$, gets "amplified" by the local oscillator power, that is incident on the detector area, $\eta_{od} P_{od}$. Therefore in the integration time, T_{int}, the total RMS number of heterodyne signal photoelectrons, n_s generated is given by

$$n_s = T_{int} i_{h_{rms}}/e = \eta_q \sqrt{2\eta_h N_{sd}(\eta_{od} N_{od})} \qquad (7.51)$$

where N_{sd} and N_{od} are, respectively, the total number of signal and LO photons incident at the detector plane in the integration time. As shown in Figure 7.15, *a significant number of signal photoelectrons are generated for each signal photon. Even when less than one signal photon arrives during the signal integration time, there are still many signal electrons generated from the mixing of the weak signal field with the strong LO field; however, the signal is weak compared to the local oscillator shot-noise, thus difficult to detect. Neverthless, there is still a signal.*

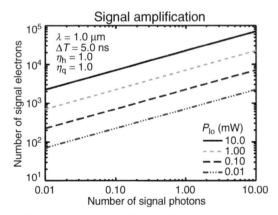

Figure 7.15 Heterodyne signal amplification by assuming perfect heterodyne efficiency and detector quantum efficiency.

For the number of signal-generated photoelectrons to overcome the variation in the integrated photoelectrons due to the LO shot-noise and other excess noise captured by F_h, $1\eta_R = F_h/\eta_r\eta_h$ signal photons must be presented to the receiver (telescope primary) during the coherent integration time of the matched-filter receiver. Put another way, in shot-noise-limited operation ($F_h = 1$) the number of detector-plane signal photons coupled into the receiver mode and detected by the detector must be 1 in order for the LO amplified signal to match the shot noise. In the full quantum treatment (not presented here) this is due to an average of one-noise-photon per coherent integration time in the receiver mode (from the zero-point fluctuations of the vacuum) being amplified by the LO.

7.4.1.2.5. Coherent Lidar CNR Parameter Dependence: The heterodyne detection receiver CNR equation (Equation 7.42) can be combined with the simplified lidar equation for P_s (Equation 7.2) to show the overall parameter dependence of the coherent lidar CNR. The result is

$$\text{CNR}(R) = \frac{\eta_a\eta_r\eta_{ox}}{F_h} \frac{T^2 E_x}{h\nu B} \frac{c\beta}{2} \frac{A_r}{R^2} \qquad (7.52)$$

where $\eta_a = \eta_h\eta_{Tx}$ is defined as the transmitter antenna efficiency and η_{ox} and η_{Tx} represent the transmitter optical efficiency and the truncation efficiency (fraction of power not truncated), respectively. This CNR expression is exact in the short pulse limit when the parameters in the equation do not vary significantly over the pulse range length $c\Delta T/2$. The general CNR expression, which does not make the short pulse assumption, is developed in the following section.

Maximum possible sensor range is achieved by designs that maximize the antenna efficiency. For example, if the aperture truncates the transmit beam too little, then the beam size at the target is large, resulting in a low heterodyne efficiency, η_h, due to speckle-induced phase variation of the signal across the aperture. This leads to a lower than possible CNR for the aperture. On the other hand, if the aperture truncates the transmit beam too heavily, in order to maximize the area of the speckle lobes, excessive transmitter loss ensues and performance falls short of the optimum.

As shown in Equation (7.40), the receiver-plane heterodyne efficiency is a function of the received signal field at the detector location, which in turn depends on the target illumination pattern, the target scattering properties, the optical path from the target, optical aberrations in the lidar receiver, etc. Therefore, although it can be calculated, the calculation of the received field at the detector plan can be rather difficult in many practical cases. In the following section, a method for calculation of the antenna and heterodyne efficiency in the target plane is derived. In most cases, this target-plane calculation is simpler than the detector-plane calculation.

7.4.1.3. Target-Plane Formalism — General CNR and Antenna Efficiency

A target-plane formalism for the antenna efficiency and the theory of designs that maximize this efficiency is the primary subject of this section. The term "antenna" is borrowed from radar terminology — in this context, it refers to the optical antenna or telescope aperture. As described in the previous section, the antenna efficiency is the product of aperture truncation efficiency, η_{TX}, and the heterodyne efficiency, η_h. Maximizing this product maximizes the CNR.

The general range-dependent CNR for a coherent lidar is first developed by summing signals from randomly distributed particles in the target-volume. This produces a simple target-volume overlap integral representation for the coherent CNR, utilizing the concept of a backpropagated local oscillator (BPLO) beam.[147] The target-volume overlap integral calculation based on the BPLO concept vastly simplifies the mathematics of coherent laser radar theory, since only one-way propagation integrals are required. It is a powerful conceptual and mathematical tool for analyzing truncated and untruncated lidar CNR performance,[153] optimal transceiver configuration,[154] and signal coherence properties.[155] The target-volume calculation using the BPLO concept also greatly simplifies space–time numerical experiments to simulate

lidar signals, which are needed for studies of refractive turbulence effects[156,157] or for estimating the coherent laser
radar cross section of complex targets.

7.4.1.3.1.   CNR and Antenna Efficiency using the Target-Plane Calculations

For heuristic purposes, the development of the target-
plane CNR formalism begins with the very simple case of a
single point scatterer. In order to focus on the physical interpretation, some constant scalar factors, such as receiver and
atmospheric transmission, are ignored during the development. At the end of the analysis these factors will be reinserted
into the final equations. Furthermore, the simplified receiver
shown in Figure 7.16 is assumed. In particular, to simplify the
mathematics, the LO field and detector are assumed to be in a
relayed aperture (or pupil-plane) of the lossless receiver optics,
rather than the image plane after focusing optics. It can be
shown that the general results developed here are directly
applicable to a more complex optical train.

7.4.1.3.1.1.   Simplified Single Particle Calculation:
Consider the coherent photocurrent resulting from a single
stationary small particle whose irradiance backscattering
cross section is σ_π (m^2/sr), where $\sigma_\pi = \sigma/4\pi$ and σ (m^2) is the
particle's laser radar cross section.[158–161] The complex amplitude
scattering coefficient, s_c, ($\sqrt{(\text{m}^2/\text{sr})}$) has an amplitude, $|s_c| = \sqrt{\sigma_\pi}$
and an arbitrary random phase between zero and 2π. For this
scattering particle, the reflected field is a Huygens spherical

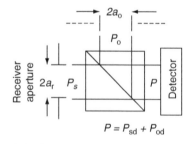

Figure 7.16   Simplified pupil-plane coherent receiver.

wavelet. In the receiver-plane, assuming free-space propagation, this field is given by

$$u_{s}(x,y,t) = \frac{s_{c}u_{x}(x_{t}, y_{t}, z_{t}, t - R/c)}{R} \exp[-ikR] \qquad (7.53)$$

where $u_{x}(x_{t}, y_{t}, z_{t}, t - R/c)$ represents the transmit field's complex amplitude at the particle location (x_{t}, y_{t}, z_{t}), retarded in time by the time-of-flight from the target to the receiver (R/c), R is distance from the particle location (x_{t}, y_{t}, z_{t}) to a point $(x, y, 0)$ in the receiver plane, and k is the wavenumber. Thus the signal field at the current time in the receiver plane, is proportional to the target-plane field at a past time $(t - R/c)$. For a nonstationary particle, k would be replaced by $k + \delta k$ where δk is the wavenumber Doppler shift.

Since, for now, we have assumed no losses and a pupil-plane receiver, then $u_{sd} = u_{s}$ and the heterodyne photocurrent (Equation 7.34) is given by

$$i_{h}(t) = (2e/h\nu)\text{Re}[s_{c}\ u_{x}(x_{t}, y_{t}, z_{t},\ t - R/c)$$
$$\times \iint n_{q}(x, y)u_{od}^{*}(x, y, t)\exp[-ikR]/Rdxdy] \qquad (7.54)$$

The target range is given by $R = [(x - x_{t})^{2} + (y - y_{t}^{2}) + z_{t}^{2}]^{1/2}$ and under the Paraxial approximation, it simplifies to $R \approx z_{t} + [(x - x_{t})^{2} + (y - y_{t})^{2}]^{1/2}/2z_{t}$. Using this approximation in the quickly varying exponent and $R = z_{t}$ in the slowly varying denominator, this expression simplifies to

$$i_{h}(t) = (2e/h\nu)\text{Re}[(s_{c}u_{x}(x_{t}, y_{t}, z_{t},\ t - R/c)/z_{t})\exp[-ikz_{t}]$$
$$\times \iint n_{q}(x, y)u_{od}^{*}(x, y, t)\times \exp[-ik((x - x_{t})^{2}$$
$$+ (y - y_{t})^{2})/2z_{t}]dxdy] \qquad (7.55)$$

This expression can be written as

$$i_{h}(t) = (2e/h\nu)$$
$$\times \text{Re}[i\lambda s_{c}u_{x}(x_{t}, y_{t}, z_{t}, t - R/c)u_{b}\ (x_{t}, y_{t}, z_{t}, t - R/c)]$$
$$\qquad (7.56)$$

where

$$u_b(x_t, y_t, z_t, t - R/c) = (1/i\lambda z_t) \exp[-ikz_t]$$

$$\times \iint \eta_q(x,y) u_{od}^*(x,y,t)$$

$$\times \exp[-ik((x-x_t)^2 + (y-y_t)^2)/2z_t] dx dy$$

$$(7.57)$$

is recognized to be the field defined by the Huygens–Fresnel propagation integral of a quantum efficiency weighted conjugate local oscillator, at the position of the scattering particle delayed in time by the time-of-flight from the receiver to the target.

This field is known as the "virtual" or backpropagated LO (BPLO), which can be interpreted as the target-plane (or particle-position) complex field of a source beam located in the detector-plane given by

$$u_b(x,y,0,t) = \eta_q(x,y) u_{od}^*(\boldsymbol{x}, \boldsymbol{y}, \boldsymbol{t}) \qquad (7.58)$$

The BPLO is a beam that propagates backwards through space and time as implied by the conjugate operator. Thus the heterodyne signal at the current time, t, is proportional to the target-plane scattered field at a past time, $t - R/c$, and the target-plane BPLO field at that same earlier time, $t - R/c$, since the BPLO beam at time $t - R/c$ is proportional to the LO field at the current time.

This time-keeping is important when considering disturbances along the path, such as time-varying refractive turbulence. For example, suppose a time-varying phase disturbance occurs at range $z = R$. The disturbance seen by the transmit and BPLO beams is virtually the same, since the time when the transmit and BPLO beams interact with this disturbance is the same. On the other hand, allow a time-varying disturbance to occur at the transceiver plane. This disturbance at time $t = 0$, impacts the transmit beam at $R = 0$, while it is the disturbance at time $t = 2R/c$ that impacts the BPLO beam at $R = 0$.

The heterodyne signal photocurrent power is given by

$$\langle i_h^2(t) \rangle = 2\lambda^2(e/h\nu)^2 \sigma_\pi I_x(x_t, y_t, z_t, t - R/c)$$
$$\times I_b(x_t, y_t, z_t, t - R/c) \qquad (7.59)$$

where I_x and I_b are the target-plane transmit and BPLO irradiances, respectively. By substituting this expression and Equation (7.39) into the CNR definition of Equation (7.22), and including the optical losses neglected in the above analysis, the CNR from a single small particle can be shown to be given by

$$CNR_\sigma(t) = \frac{\lambda^2(e/h\nu)^2 \sigma_\pi}{ei_oBF_h} I_x(x_t, y_t, z_t, t - R/c)$$
$$\times I_b(x_t, y_t, z_t, t - R/c) \qquad (7.60)$$

using

$$I_x(x_t, y_t, z_t, t - R/c)$$
$$= \eta_{ox}T(R)P_x(t - 2R/c)I_{nx}(x_t, y_t, z_t, \ t - R/c) \qquad (7.61)$$

for a collocated transmitter. If the transmitter were not collocated with the receiver, then the $2R$ in the above expression would be replaced by path-length from the transmitter to the target, and then to the receiver. Also using

$$I_b(x_t, y_t, z_t, t - R/c) = \eta_{or}T(R)I_{nb}(x_t, y_t, z_t, t - R/c)$$
$$\times P_{od}(t) \frac{\iint |\eta_q(x,y)\boldsymbol{u}_{od}(x,y,t)|^2 dxdy}{\iint |\boldsymbol{u}_{od}(x,y,t)|^2 dxdy} \qquad (7.62)$$

and

$$i_o(t) = (e/h\nu)P_{od}(t) \frac{\iint \eta_q(x,y)|\boldsymbol{u}_{od}(x,y,t)|^2 dxdy}{\iint |\boldsymbol{u}_{od}(x,y,t)|^2 dxdy} \qquad (7.63)$$

produces

$$CNR_\sigma(t) = \frac{\eta_r \eta_{ox} T^2(R) \sigma_\pi P_x(t - 2R/c)}{h\nu BF_h}$$
$$\times \lambda^2 I_{nx}(x_t, y_t, z_t, t - R/c)I_{nb}(x_t, y_t, z_t, t - R/c) \qquad (7.64)$$

where the receiver efficiency, η_r as defined in Equation (7.44), has the effective heterodyne quantum efficiency, $\hat{\eta}_q$, as defined in Equation (7.45), imbedded in it to account for a nonuniform detector response over the LO beam area. The factor $T(R)$ is the one-way transmission for the transmit and BPLO beams. P_x is the transmitter laser power before losses and P_{od} is the detector-plane LO laser power. I_{nx} and I_{nb} (m^{-2}) are the target-plane irradiances of the transmit and BPLO beams normalized by their total power *prior to aperture truncation*, such that their area integrals before the aperture are unity. The area integrals of these normalized beams anywhere after the aperture, are simply the respective aperture truncation factors, η_{Tx} and η_{Tb}, representing the fraction of laser power lost through aperture truncation. For example, for a Gaussian beam centered in a circular aperture with no central obscuration, the truncation efficiency is given by

$$\eta_T = 1 - \exp\left[-2/\rho_T^2\right] \tag{7.65}$$

where $\rho_T = \omega_0/a$ is the aperture truncation ratio, ω_0 (m) is the Gaussian beam e^{-2} irradiance radius, and a (m) is the circular aperture radius.

This concept of describing the heterodyne signal photocurrent and CNR using a target-plane overlap of the transmitted field and a BPLO field was first described by Siegman.[147] Even though the simplified analysis presented here assumed a pupil-plane receiver, this technique also applies to more complicated receiver systems, that is, those that have focusing optics, stops, etc., provided the BPLO is backpropagated from the detector through the receiver optics train and out through the receiver telescope. This is a very powerful technique that applies to non-diffraction-limited systems as well, including systems with optical aberrations and those that have beams that propagate through refractive turbulence.[153] It greatly simplifies the analysis of the effects of speckle dynamics on heterodyne receivers and numerical modeling of receiver performance.

7.4.1.3.1.2. Volume of Particles: At this point in the analysis, we assume the normalized target-plane irradiance distributions are temporally static. That is to say, that they do not have a temporal dependence. This could be the case if temporal fluctuations in the refractive turbulence can be ignored over the round-trip time to the target ($2R/c$). Therefore, we drop the temporal dependence of the two target-plane normalized irradiance distributions, I_{nx} and I_{nb}.

Scattering from a volume of random diffuse scatters produces a composite photocurrent, which is the coherent sum of the photocurrents from each scatter. Due to the random position, scattering amplitude, and phase from each particle in the target-volume, the CNR of the combined signals, averaged over an ensemble of random realizations of the particle scattering amplitudes and positions, is the sum (integral) of the CNRs from each particle.[162] For uniform extinction over the beam width, the ensemble target-average CNR as a function of time ($t = 2R/c$) can be written as

$$\text{CNR}(t) = \frac{\eta_r \eta_{ox}}{h\nu BF_h}\lambda^2 \int_0^\infty P_x(t - 2z/c)T^2(z)$$

$$\times \iint \beta_A(x,y,z)\, I_{nx}(x,y,z)I_{nb}(x,y,z)\mathrm{d}x\mathrm{d}y\mathrm{d}z \quad (7.66)$$

where, $\beta_A = \rho_a |s_c|^2 = \rho_a \sigma_\pi$, ρ_a is the volume density of the particles and σ_π is the average scattering cross section of the particles.

Consider the case of a hard target, $\beta(x,y,z) = \delta(z - R)$ $\rho_\pi(x,y)$. In this case, the range integral collapses, and we can write

$$\text{CNR}(t) = \frac{\eta_r \eta_{ox}}{h\nu BF_h}\lambda^2 P_x(t - 2R/c)T^2(R)$$

$$\times \iint \rho_\pi(x,y)I_{nx}(x,y,z)I_{nb}(x,y,z)\mathrm{d}x\mathrm{d}y \quad (7.67)$$

Note, the CNR, as described above, is represented as a function of time, where it is understood that time and range are interchangeable using $t = 2R/c$. It should be emphasized that these CNR expressions, written in terms of an overlap integral of the target-volume irradiances, represent the average CNR over an ensemble of targets (aerosols, hard targets, etc.). Due to speckle fluctuations, any single realization of the CNR can be significantly higher or lower than the average CNR (see Section 7.3.8). If there is a sufficiently large number of particles in the overlap volume, quasi-Gaussian complex signal statistics apply and the fluctuations in the CNR obey exponential statistics with a standard deviation equal to the mean.[143]

If the backscatter is uniform over the beam area, then the aerosol target CNR can also be written in the following form, which emphasizes the antenna efficiency. That is

$$\text{CNR}(t) = \frac{\eta_r \eta_{ox}}{h\nu BF_h} \frac{A_r}{R^2}$$

$$\times \int_0^\infty P_x(t - 2z/c)T^2(z)\beta_A(z)\eta_a(z)\mathrm{d}z \qquad (7.68)$$

where

$$\eta_a(z) = \frac{\lambda^2 R^2}{A_r} \iint I_{nx}(x,y,z)I_{nb}(x,y,z)\mathrm{d}x\mathrm{d}y \qquad (7.69)$$

is the receiver antenna efficiency[163,164] relative to the collection aperture area, $A_r (\text{m}^2)$. It can be shown that $\eta_a = \eta_h \eta_{Tx}$, where η_h is the heterodyne efficiency as defined in Section 7.4.1.2. The purpose of writing the CNR in this form is to aid system design, since for a given telescope aperture, the CNR is maximized for designs which maximize the antenna efficiency.

In this analysis, we could have defined the normalized transmit and BPLO irradiances to be normalized to unity after the truncating aperture, rather than before, as some authors do. In which case, the definition for the CNR would have an η_{Tx} in the numerator and the antenna efficiency equation would have η_{Tx} in the denominator.

7.4.1.3.2. Short-Pulse and Long-Pulse Limits: In the short-pulse limit (i.e., when the pulse length is shorter than changes in any of the other integrand functions), Equation 7.68 becomes

$$\text{CNR}(R) = \frac{\eta_r \eta_a \eta_{ox}}{F_h} \frac{E_x T^2}{h\nu B} \frac{c\beta_A}{2} \frac{A_r}{R^2} \qquad (7.70)$$

where $\int P(2z/c)\mathrm{d}z = cE_x/2$ was used. Note the agreement with Equation (7.52).

In the CW (or long-pulse limit) the CNR integral (Equation 7.68) must be evaluated explicitly.

7.4.1.3.3. Seigman's Antenna Theorem: Physical insight into the heterodyne efficiency can be obtained from Siegman's antenna theorem.[147] This theorem states that heterodyne receivers have an effective etendue, ξ_{eff}, that is the same as the diffraction limit regardless of what the FOV defined by the BPLO is. This can be shown by

$$\xi_{\mathrm{eff}} = \eta_{\mathrm{h}} A_{\mathrm{r}} \Omega \approx \lambda^2 \tag{7.71}$$

where A_{r} is the receiver aperture area, η_{h} is the heterodyne efficiency, Ω is the solid angle of the receiver's main lobe defined by the BPLO, and λ^2 is the etendue of a diffraction limited receiver. This theorem implies that the coherent signal power loss from a non-diffraction-limited receiver FOV, due to defocus, refractive turbulence, or imperfect receiver optics, is given by $\eta_{\mathrm{h}} \approx \lambda^2/A_{\mathrm{r}}\Omega$.

Since a diffraction-limited receiver's angular resolution is governed by $\Omega \approx \lambda^2/A_{\mathrm{r}}$, the diffraction-limited receiver etendue is given by $A_{\mathrm{r}}\Omega \approx \lambda^2$. Therefore, from Siegman's antenna theorem, coherent receivers cannot be designed with larger than diffraction-limited field of view without incurring an offsetting heterodyne efficiency loss. Direct detection receivers are not bound by the same constraint, and can be designed with larger than diffraction-limited etendues ($A_{\mathrm{r}}\Omega \gg \lambda^2$), without incurring an offsetting efficiency loss.

7.4.1.3.4. Summary: In this section and section 7.4.1.2, we have presented the theory which describes the heterodyne signal photocurrent and CNR using both detector- and target-plane formalisms (see the CNR expressions in Equations 7.40 and 7.66). These two formalisms are equivalent as was shown for the special case of a short-pulse lidar (i.e., Equations 7.52 and 7.70). Understanding both theories is key to being able to conceptualize performance of various lidar designs, different key insights are obtained from the two forms of the CNR expressions.

In the following subsections, the above theory is applied to calculate the antenna efficiency and CNR of truncated and untruncated lidar systems.

7.4.1.4.   Antenna Efficiency Calculations

In this section, the target-volume calculation derived in the previous section is applied to calculate and compare the antenna efficiency, η_a, of various lidar configurations, including: bistatic and monostatic, truncated and untruncated, and Gaussian and uniform beam lidars. The bistatic analysis is limited to near collinear and collocated transmitter and receiver apertures which are separate, and not necessarily the same size. The bistatic condition with a focused uniform LO beam is considered, since it yields perfect (100%) heterodyne efficiency when the target-volume transmit beam size is much smaller than the diffraction limit of the receiver, which could be produced if the transmitter aperture were much larger than the receiver aperture. Perfect efficiency results are achieved in this case since the focused LO beam perfectly matches the spherical receiver-plane signal field.

In a monostatic condition, it is shown that the antenna (or also heterodyne, in this case) efficiency for a uniform circular LO and transmit beam is 46.0%. This is improved to 47.1% by pretruncation of an optimally sized Gaussian LO. The reduction from 100% for the monostatic systems occurs because the target-plane spot size cannot be made arbitrarily small, resulting in spatial speckle fluctuations (random intensity and phase variations) of the scattered field over the receiver aperture. For a monostatic, truncated Gaussian beam lidar, the highest antenna efficiency reported is 43.8%.[154] This is achieved when the transmit beam waist is 81.5% of the aperture radius for which 95.1% of the beam power is transmitted, thus the heterodyne efficiency for this design is ~46.1% (43.8%/95.1%).

The study of untruncated Gaussian beam lidar performance has practical application in sensor development because, for most transceiver designs, the small beam (i.e., before telescope expansion) is reasonably modeled as an untruncated Gaussian beam. The theory developed for these systems can be compared to measurements to estimate relative sensor performance prior to integration of the beam-expanding telescope. This untruncated beam analysis also serves as an upper bound for the performance of practical truncated beam lidar systems by employing a Gaussian beam LO with the same radius. The heterodyne efficiency for bistatic and matched monostatic Gaussian beam lidar is 100% and 50%, respectively, which are the upper bounds on the performance of truncated designs. Furthermore, efficiency loss, due to receiver misalignment and refractive turbulence, have closed form solutions for an untruncated Gaussian beam lidar, which are reasonably accurate approximations for weakly truncated systems typical of optimal lidar designs.

7.4.1.4.1. Truncated Systems Antenna Efficiency

In the following analysis a uniform responsivity detector, which is much larger than the LO beam field extent, is assumed. This is reasonable, since most high performance detectors exhibit a relatively uniform responsivity over the physical detector area and the diameter of typical high-speed detectors are on the order of 25 to 100 μm, whereas focused LO beam diameters can be much smaller. This assumption greatly simplifies the analysis. The general lidar equation presented in Section 7.4.1.3 can be used to calculate the antenna efficiency when the detector responsitivity is not uniform.

Many practical lidar systems employ Gaussian laser beams and finite telescope apertures. A critical design consideration is how to fill the hard transmit aperture with a quasi-infinite transverse Gaussian beam and how to design the LO beam relative to the receiver aperture, such that the CNR is maximized for the given aperture selection. This turns out to be a tradeoff between the receiver heterodyne efficiency and transmitter truncation efficiency, which is captured by the antenna efficiency. Compared to a two-way propagation calculation of the transmitted and reflected beams, the target plane formalism presented in Section 7.4.1.3 greatly simplifies the calculation of this antenna efficiency.

Optimal designs are characterized in terms of the truncation ratios for the transmitted and BPLO beams, $\rho_T = \omega_0/a$, where ω_0 is the e^{-2} radius of the Gaussian beam (BPLO or transmit) irradiance and a is the aperture (receiver or transmitter) radius. If the truncation ratio is too small the full aperture is not utilized and the diffraction spread of the beams is large, resulting in a low heterodyne efficiency. On the other hand, if the truncation ratio is too large, excessive truncation occurs resulting in a lower antenna efficiency. Therefore, optimization of the antenna efficiency, is a trade between diffraction and aperture truncation. One of the earliest works on this truncated Gaussian beam topic was developed by Thomson and is covered in an FAA report by Huffaker et al.[165]

For a truncated Gaussian beam there is no closed form solution to the target-volume beam irradiance function needed in Equation (7.69), although, an efficient series expansion has been developed,[166] which can be utilized in numerical integration algorithms.[163] Consequently, far-field optimization is possible through numerical techniques. However, when the truncation ratio is small, $\rho_T < 0.5\%$, the transmit beam is virtually untruncated and very little error is produced using an untruncated Gaussian beam model to produce analytic expressions valid at all ranges. In addition, in the near-field, where the diffraction effects have not yet strongly modified the intensity distribution, the infinite overlap integral of Equation (7.69) can be approximated by a finite integral over the aperture of undiffracted beams to produce analytic expressions valid in the near-field. Otherwise numerical techniques are required.

In this section, the performance of truncated lidar systems is investigated. First, a bistatic lidar sensor with a focused diffraction-limited uniform circular LO field is considered, since it should produce unit mixing efficiency when mixed with a spherical wave signal field emanating from a small diffuse target at the receiver's focus. Next monostatic truncated Gaussian beam lidar is investigated, to assess the performance of practical systems. In this context, monostatic refers to systems for which both the transmitter and receiver share the same size aperture. The LO and transmit beams, however, are not necessarily matched in size. Bistatic implies the transmitter and receiver are not necessarily collocated or matched in size. However, in the bistatic case, the transmitter and receiver are assumed to be approximately coaxial, such that their overlap in the target volume is circular and concentric.

7.4.1.4.1.1. Bistatic Heterodyne Efficiency: Consider a near-collinear bistatic transceiver with a uniform circular receiver-plane BPLO field of radius a_r, matched and aligned to a circular aperture and focused at the target range. Allow that the target-plane transmit beam is an untruncated Gaussian beam with e^{-2} irradiance radius, ω_{Rx}. This hypothetical target-plane beam is generated using a large transmit aperture, or a transmit aperture that is much closer to the target than the receiver so that transmit aperture truncation effects are negligible. For this combination of fields, the antenna efficiency would also be the heterodyne efficiency, since the transmit truncation efficiency is unity, and it is expected to converge to unity in the limit as ω_{Rx} approaches zero, because the receiver-plane signal field should be a spherical wave with radius of curvature matched to the focused BPLO beam curvature.

Figure 7.17 shows the variation of the antenna efficiency (also heterodyne efficiency in the case) of this receiver as the target plane beam size is increased from zero to ~5x the diffraction-limited resolution of the receiver aperture. For larger target-plane beam sizes ($x > 0$ in Figure 7.17) the efficiency is reduced, because the spatial coherence of the receiver-plane field is reduced due to speckle and poor mixing results. From the Van Cittert–Zernike theorem (see Section 7.3.8), the complex coherence factor is Gaussian with coherence radius, $\rho_{cg} = \lambda R/\pi\omega_{Rx}$. This coherence radius decreases as the target-plane transmit beam radius increases, resulting in a reduction in mixing efficiency.

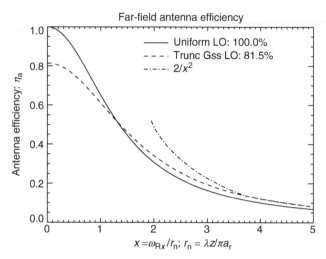

Figure 7.17 Far-field, bistatic, antenna efficiency for uniform circular and a truncated Gaussian BPLO fields mixing with the speckle field produced from a Gaussian target-plane irradiance of radius ω_{Rx}. Since there is no truncation by the aperture in this case, the antenna and the heterodyne efficiencies are the same.

The antenna efficiency, is most easily derived from the target-plane calculation of Equation (7.69). For this example, the normalized far-field BPLO and transmit beam irradiance functions are Airy and Gaussian beams, respectively. These functions are given by

$$I_{nb}(\rho) = J_1(2\pi a_r \rho)/\pi\rho^2 \quad \text{and} \quad I_{nx}(\rho) = (2/\pi\omega_{Rx}^2)\exp(-2(\rho/\omega_{Rx})^2)$$

$$(7.72)$$

where $J_1(x)$ is the Bessel function of the first kind of order 1. Substituting these two expressions into Equation (7.69) and performing the integration produces

$$\eta_a(z) = \frac{2}{x^2}(1 - \exp(-x^2)[I_0(x^2) + I_1(x^2)])$$ $$(7.73)$$

where $x = \omega_{Rx}/r_n = a_r/\rho_{cg}$, $r_n = \lambda R/\pi a_r$ is a normalized target-plane radius representing a scale size on the order of the diffraction limit of the receiver, and $I_n(x)$ is the modified Bessel function of the first kind of order n.

As shown in Figure 7.17, the mixing efficiency is unity for $x = 0$ or when the transmit beam is small compared to a diffraction-limited BPLO beam. This validates the physical intuition that unit antenna and mixing efficiency should occur under this condition.

The detector-plane interpretation of this result is that there is an optimal Gaussian beam size to mix with the Airy diffraction pattern of the detector-plane signal from a point target. For smaller local oscillator beam sizes, parts of the return signal do not mix with the LO. On the other hand, for larger local oscillator beams, the entire signal is well mixed; but there is excess shot-noise from the exterior regions of the LO where no signal present. The best situation is achieved if the LO is made to be an Airy pattern matching that of the signal. This condition corresponds to the uniform LO case described above, leading to unity mixing efficiency.

In the limit of $x = \infty$ (i.e., an infinite aperture or zero range with a finite aperture), the efficiency converges to $2/x^2$. This implies that when mixing a fine grain speckle field with a uniform LO field over a circular aperture, the mixing efficiency converges to $\eta_a = 2A_{cg}/A_r$, where $A_r = \pi a_r^2$ and $A_{cg} = \pi \rho_{cg}^2$ with ρ_{cg} being the Goodman coherence radius of the speckle lobes as defined in Section 7.3.8.

The factor of 2 is simply a consequence of the definition of the coherence radius, which was based upon independent irradiance areas rather than field areas. Note that if the e^{-1} radius of the complex coherence function is used (see Equation 7.17) $\rho_c = \sqrt{2}\rho_{cg}$ and $\eta_a = A_c/A_r$. The coherent lidar only detects one speckle lobe (coherence area), even when a large number of lobes are present over the receiver aperture.

Practical systems typically use a Gaussian detector-plane LO beam, which, when backpropagated through the transceiver, is truncated by the transmit aperture. Therefore, the receiver-plane BPLO distribution is a truncated Gaussian, rather than a uniform circular, beam. For an infinitesimally small target-plane Gaussian beam, maximum CNR is achieved for a target-plane BPLO with maximum on-axis intensity. This is achieved with an aperture truncation ratio $\rho_T = 89.3\%$,[167] to produce a maximum efficiency of 81.5%. This performance is also shown in the figure.

7.4.1.4.1.2. Monostatic Antenna Efficiency: In the previous section, it was shown that the antenna efficiency is unity for a uniform circular receiver-plane BPLO distribution with an infinitesimally small target-plane transmit beam. Monostatic lidar systems employ a common shared transmit–receive telescope. Consequently, the target-plane transmit beam size is limited by diffraction and 100% efficiency cannot be achieved. It seems reasonable to postulate that the maximum monostatic antenna efficiency is obtained when both the transmit and BPLO receiver-plane beams are uniform circular beams. For this condition, Equation (7.69) can be used to show that the far-field efficiency is 46.0% $(1-16/3\pi^2)$. However, even higher efficiency (47.1%) can be obtained with a uniform transmit beam and an optimally pretruncated Gaussian LO, because the truncated Gaussian LO is a slightly better match to the speckle field spatial-autocorrelation or mutual intensity function of the backscattered signal. The truncation ratio, which optimizes the result, is approximately 134% as shown below.

The antenna efficiency for four monostatic circular aperture lidar designs is shown in the lower panel of Figure 7.18; each differing in how the LO beam is generated. The heterodyne and truncation efficiencies (the two factors of the antenna efficiency) are shown in the top panel of the Figure. The simplest design, first described by Rye,[168] is commonly referred to as the Wang design.[169,170] In this design, the detector-plane LO is an untruncated Gaussian beam, which, when backpropagated to the aperture, matches the transmit beam. In this design both the transmit and BPLO beam are truncated by the aperture and the optimal truncation ratio for an extended far-field target is 80.2%. At this truncation ratio the antenna and truncation efficiencies are 40.1 and 95.6%, respectively. The majority of the reduction from the 46.0% obtained for the matched uniform beam case can be attributed to the squared truncation efficiency loss, 91.4%.

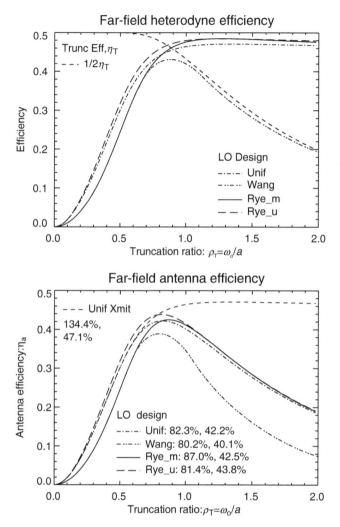

Figure 7.18 (Bottom) Antenna efficiency and (Top) far-field, extended-target, monostatic lidar, heterodyne, and truncation efficiency for various Gaussian beam lidar designs vs. truncation ratio.

Obviously, better performance could be achieved if the LO beam were pretruncated before the detector, such that the BPLO is not truncated by the aperture. In this case, the target-plane BPLO is unchanged, however, less shot-noise is produced, since less LO light is detected. In his paper, Rye recognized this fact and presented two designs corresponding to a matched and unmatched BPLO. These two designs are also shown in the figure. For the matched design, originally determined by Priestly,[171] the optimum truncation ratio is 87.0% yielding an antenna efficiency of 42.54%. The equation governing the performance curve is the same as the equation for the Wang design divided by the truncation efficiency to account for the fact that the BPLO is not truncated. This extends the peak to slightly larger ρ_T, 87.0%, instead of 80.2. For the unmatched design, Rye showed that the optimum LO truncation ratio is 118.6% and the optimum transmit beam truncation ratio is 81.5%, yielding a peak antenna efficiency of 43.8%. Also shown in the figure, for reference, is the case for a circular uniform BPLO beam. As shown, the optimum truncation ratio is 82.3% to yield an antenna efficiency of 42.2%, which is indeed slightly less than Rye's unmatched Gaussian LO design.

Frehlich[164] has shown that the optimal LO for a circular aperture system is very close to a Gaussian field truncated by a circular aperture such that the BPLO field at the primary aperture has no truncation. Also shown in Figure 18 is a design with an efficiency of 47.1%, achieved with a uniform transmit beam and a pre-truncated, $\rho_T = 134.4\%$, Gaussian LO beam. For this curve the horizontal axis of the figure is the LO beam truncation ratio rather than the transmit beam truncation ratio. For this case, the peak at $\rho_T = 134.4\%$ is very weak. If the pre-truncated Gaussian LO beam size becomes very large, the efficiency converges to 46.0%, matching the uniform transmit and BPLO monostatic, condition described above. This assumes there is sufficient LO power available so that shot-noise-limited operation is maintained.

7.4.1.4.2. Untruncated Gaussian Beam CNR and Antenna Efficiency

In this section, theory is developed for untruncated Gaussian beam lidar configurations that have both matched and unmatched (or general) transmit and BPLO beam sizes. This is useful, as it allows for analytic expressions that can be used to gain physical insight and for fast first-order performance prediction calculations of coherent lidar systems. Additionally, during sensor assembly, it is important to measure performance, prior to telescope installation. For these measurements, the transmit and BPLO beams are often well modeled as untruncated Gaussian beams.

7.4.1.4.2.1.   Short-Pulse Limit Bistatic Systems: For an untruncated Gaussian beam

$$I_n = 2 \exp(- 2\rho^2/\omega_R^2)/\pi\omega_R^2 \qquad (7.74)$$

where ρ is the transverse radial distance and

$$\omega_R^2 = \omega_0^2\left((1 - R/F)^2 + (R/R_r)^2\right) \qquad (7.75)$$

is the target-plane Gaussian beam radius, F (m) is the radius of curvature of the transmit beam and $R_r = \pi\omega_0^2/\lambda$ is the beam Rayleigh range. Substituting Equation (7.74) into Equation (7.66) using Equation (7.75) and invoking the short-pulse limit, produces the following target-plane parameter CNR expression:

$$\mathrm{CNR}_{sp}^{ub} = \frac{\eta_r\eta_{ox}}{F_h} \frac{2\lambda^2}{(\pi\omega_{Rx}^2 + \pi\omega_{Rb}^2)} \frac{T^2E_x}{h\nu B} \frac{c\beta_A}{2} \qquad (7.76)$$

where ω_{Rx} and ω_{Rb} are the target-plane transmit and BPLO beam radii, respectively. The subscript sp and superscript ub on CNR are used to indicate short pulse and untruncated bistatic Gaussian beams respectively.

This target-plane formalism portrays an inverse dependence on the combined target-plane BPLO and transmit beam areas. This interpretation aids the understanding of the effects of refractive turbulence, which will be discussed in Section 7.4.1.6.

This can be rearranged into the following form:

$$\mathrm{CNR}_{sp}^{ub} = \frac{\eta_r\eta_{ot}}{F_h} \frac{\omega_{dRb}^2}{(\omega_{Rx}^2 + \omega_{Rb}^2)} \frac{T^2E_x}{h\nu B} \frac{c\beta_A}{2} \frac{2\pi\omega_0^2}{R^2} \qquad (7.77)$$

where $\omega_{dRb} = \lambda R/\pi\omega_0$ is the diffraction-limited focused BPLO beam radius and ω_0 is the receiver-plane BPLO beam radius.

Thus, the CNR is maximized when the BPLO is diffraction limited: $\omega_{Rb} = \omega_{dRb}$ and with $\omega_{Rx} \ll \omega_{Rb}$. Under this condition the CNR is well approximated by

$$\text{CNR}_{sp}^{ub} \approx \frac{\eta_r \eta_{ox}}{F_h} \frac{T^2 E_x}{h \nu B} \frac{c\beta_A}{2} \frac{2\pi\omega_0^2}{R^2}. \tag{7.78}$$

Comparing this maximum CNR condition to the generalized beam short-pulse expression (i.e., Equation 7.70 with $\eta_a = 1$, corresponding to a uniform LO filling the aperture and a small target-plane transmit spot size, cf. Figure 7.17 with $x = 0$) results in the conclusion that the effective collection area of the untruncated Gaussian beam lidar is $2\pi\omega_0^2$. That is to say that at the maximum CNR, an untruncated Gaussian beam receiver produces the same CNR as a uniformly illuminated aperture of radius $\sqrt{2}\omega_0$. This permits the definition of an untruncated Gaussian beam antenna efficiency and corresponding CNR expression given by

$$\text{CNR}_{sp}^{ub} = \frac{\eta_r \eta_{ox} \eta_a^u}{F_h} \frac{T^2 E_x}{h \nu B} \frac{c\beta_A}{2} \frac{2\pi\omega_0^2}{R^2} \tag{7.79}$$

where

$$\eta_a^u = \frac{\omega_{dRb}^2}{(\omega_{Rx}^2 + \omega_{Rb}^2)} \tag{7.80}$$

This untruncated Gaussian beam antenna efficiency has a different interpretation than its truncated beam counterpart. In this case, it is the ratio of the CNR to the maximum possible CNR for a given receiver-plane LO distribution, whereas for the truncated system, the antenna efficiency is the product of the heterodyne efficiency and the transmit truncation efficiency. In both cases the maximum efficiency is unity.

It is instructive to visualize these results in the receiver plane. In the instance of random speckle fields the antenna efficiency is an ensemble average over many speckle field realizations. From the Van Cittert–Zernike theorem (Section 7.3.8), the receiver-plane correlation function is a Gaussian function with correlation radius, $\rho_{cg} = \lambda R/\pi \omega_{Rx}$. In terms of this coherence radius, the untruncated antenna efficiency becomes

$$\eta_a^u = \frac{1}{(\alpha_b^2 + \omega_0^2/\rho_{cg}^2)} \qquad (7.81)$$

where

$$\alpha_b = \omega_{Rb}/\omega_{dRb} \qquad (7.82)$$

is the ratio of the target-plane BPLO beam radius to the diffraction-limited focused BPLO beam radius. Many references define this diffraction ratio as the beam's M^2 value, where $M^2 = \alpha_b$. In the limit of infinite signal coherence radius, the efficiency saturates to $1/\alpha_b^2$ and unit efficiency is achieved if the sensor is diffraction limited.

7.4.1.4.2.2. Short-Pulse Limit Monostatic Systems: For a matched beam lidar $\omega_{Rx} = \omega_{Rb}$, the CNR becomes (cf. Equation 7.77)

$$\text{CNR}_{sp}^{um} = \frac{\eta_r \eta_{ox} \eta_a^{um}}{F_h} \frac{T^2 E_x}{h\nu B} \frac{c\beta_A}{2} \frac{2\pi\omega_0^2}{R^2}, \qquad (7.83)$$

where now the antenna efficiency is given by (Equation 7.80 with $\omega_{Rx} = \omega_{Rb}$)

$$\eta_a^{um} = 1/2\alpha_b^2 \qquad (7.84)$$

Thus, for an untruncated matched-beam diffraction-limited Gaussian beam lidar, the maximum efficiency is 50% compared to the 46.0% for a monostatic truncated beam lidar with transmit and BPLO beams that uniformly fill the aperture. Recall that the effective receiver radius of the untruncated Gaussian beam lidar is $\sqrt{2}\omega_0$, so the performance of a uniformly filled truncated-circular-aperture lidar, whose aperture radius is $\sqrt{2}\omega_0$, can have about 92% (46/50) of the efficiency of a matched-beam untruncated Gaussian beam lidar with Gaussian beam $1/e^2$ diameter of ω_0. Deviation from diffraction-limited operation, $\alpha > 1$, also reduces the efficiency.

Substituting Equations (7.75) and (7.82) into Equation (7.84) produces an alternate form of the antenna efficiency, which provides physical insight into near-field efficiency losses:

$$\eta_a^{um} = \left(2 + 2(1 - R/F)^2 (R_r/R)^2\right)^{-1} \tag{7.85}$$

This form, which was first published by Thomson and Boynton,[172] shows that perfect monostatic efficiency (50%) is achieved when the transmit beam is focused on the target-plane. Near-field losses are weighted by the ratio of the transmit beam Rayleigh range to the target range. For a collimated system, near perfect (50%) efficiency is achieved when $R \gg R_r$.

7.4.1.4.2.3. CW or Long-Pulse Limit Monostatic Systems: When the pulse length is long compared to the sensor depth of focus the *CW limit* is approached. In this limit, the pulse power is constant over the focusing volume in Equation (7.66). With the assumption that the aerosol backscatter coefficient and atmospheric transmission losses are constant over the depth of focus, Equation (7.66) reduces to

$$\text{CNR}_{CW}^{um} = \frac{\eta_r \eta_{ox}}{F_h} \frac{P_x \beta_A T^2 \lambda}{h\nu B} \left(\frac{\pi}{2} + \arctan\left(\frac{R_r}{F}\right)\right) \tag{7.86}$$

where $\int dz/\pi\omega_R^2 = (\pi/2 + \arctan(R_r/F))/\lambda$ was used. This result was first published by Sonnenschein and Horrigan[173] through a more complicated analysis without the aid of the BPLO concept. Their expression applies to a photoconductive detector and is, therefore, a factor of 2 smaller than Equation (7.86).

Focusing well inside the Rayleigh range restricts the measurement to a well-defined volume $\pi\omega_R^2\Delta R$, and provides 3 dB CNR enhancement over a collimated CW lidar. Equation (7.86) lacks a transmit beam diameter dependence. Physically, this is because the product of the interaction length and the irradiance is independent of the transmit beam diameter. That is, larger transmit beams produce a higher irradiance at the focus, but shorter interaction lengths.

The effective measurement volume length (i.e., the length of the volume responsible for the vast majority of the signal strength) is found by equating the short-pulse target-plane untruncated Gaussian beam CNR equation (Equation 7.76) with Equation (7.86), using the matched condition $\omega_{Rx} = \omega_{Rb} = \omega_R$, assuming a tight focus, $F \ll R_r$ and using $E_x c/2 = P_x \Delta R$. This analysis produces an effective interaction length of $\Delta R \approx \pi R_r'$, where $R_r' = \pi \omega_R^2 / \lambda$ is the Rayleigh range of the focused spot. Assuming uniform backscatter, the signal contribution from aerosol particles outside this range is reduced due to a poor mixing efficiency. A general analysis of the measurement volume for a focused untruncated Gaussian beam CW lidar was published by Kavaya and Suni.[174]

7.4.1.5. Alignment Errors

The above special case analyses assumed perfect alignment between the BPLO and the transmit beam. Misalignment can occur from improper design, assembly or from an excessive lag angle in scanning systems. For general misalignment, the efficiency is the ratio of the BPLO and transmit beam overlap integral with misalignment to itself with perfect alignment. For a monostatic matched BPLO untruncated Gaussian beam lidar, the alignment efficiency can be written as

$$\eta_{\delta\rho} = \exp\left[-\delta\rho^2/\omega_R^2\right] \tag{7.87}$$

where $\delta\rho$ is the transverse displacement between the transmit and BPLO beam at range R. In the far-field this can be written as

$$\eta_{\delta\theta}^{FF} = \exp\left[-\delta\theta^2/\theta_d^2\right] \tag{7.88}$$

where $\delta\theta$ is the angular displacement and $\theta_d = \lambda/\pi\omega_0$ is the diffraction angle. When the beam alignment error is a full diffraction angle, $\delta\theta = \theta_d$, the loss is 4.3 dB and only 1 dB for $\delta\theta = \theta_d/2$.

Scanning of the lidar output beam leads to an angular misalignment of the return beam with the receiver due to the lag time (i.e., the round-trip time to the target and back). This misalignment is called the lag angle misalignment because the angle of the return signal lags that of the receiver due to the scanner rotation. The lag angle can be shown to be given by

$$\delta\theta_L = (2R/c)\Omega_s \qquad (7.89)$$

where Ω_s is the angular rate of the scanner in radians per second and $2R/c$ is the pulse time-of-flight. A scan rate of $20°/\text{sec}$ ($0.35\,\text{rad/sec}$) results in a lag angle misalignment of $11.6\,\mu\text{rad}$ at $5\,\text{km}$ range. From Equation (7.88), a diffraction-limited coherent lidar, with $\omega_0 = 5\,\text{cm}$ scanning at $20°/\text{sec}$, would suffer a $3.6\,\text{dB}$ loss for returns from a $5\,\text{km}$ range. There are various lag angle compensation approaches to mitigate the lag angle loss, many of which are designed for a fixed target range as the lag angle is range dependent.[175] Note that only rotations cause lag angle misalignment, a pure transverse platform translation does not cause a lag angle misalignment.

7.4.1.6. Refractive Turbulence Effects

A formal treatment on the effects of refractive turbulence is beyond the scope of this chapter. For a more complete treatment, the reader is referred to the works of Fried,[177] Yura,[178] Clifford and Wandzura,[179] and Frehlich and Kavaya.[153] The general effect of refractive turbulence is to distort the phase front of the propagating beam. This in turn produces spatial and temporal irradiance scintillation in the target plane and an average target-plane beam size that is larger than diffraction-limited beam size. However individual realizations can have a focusing effect which creates small regions of intense irradiance greater than can be produced by the diffraction limit.

General refractive turbulence effects can be calculated analytically or numerically using Equation (7.66). Analytic solutions usually assume a bistatic condition, where the transmit and BPLO beams propagate through independent turbulence paths. Numerical solutions propagate the beams through a series of random phase screens to estimate the target-plane irradiance function and can be applied to either bistatic or monostatic conditions. In this section, we first describe an analytic solution for an untruncated Gaussian beam lidar, assuming the bistatic condition. This bistatic assumption produces a lower bound for system performance. Next we provide a heuristic description of the performance of monostatic lidars relative to the bistatic assumption.

7.4.1.6.1. Bistatic Condition

For the bistatic condition, the ensemble average target-plane beam radius due to focusing, diffraction and turbulence is given by (see, e.g., Frehlich and Kavaya[153] Equation 177 with $\sigma = \omega/\sqrt{2}$ and $F_{TE} = F$)

$$\omega_R^2 = \omega_0^2\left((1 - R/F)^2 + (R/R_r)^2\right) + (\lambda R/\pi\rho_0)^2, \qquad (7.90)$$

where ρ_0 is the transverse coherence radius of a coherent source propagating from the target to the receiver. The last term corresponds to an increased "average" beam size, over the diffraction limit due to refractive turbulance. The increased average beam size reduces the CNR (cf., Equation 7.80). For a Kolmogorov turbulence spectrum and a spherical wave, ρ_0 is given by (see, e.g., Frehlich and Kavaya[153] Equation 165)

$$\rho_0^{5/3} = 2.91k^2 \int_0^R C_n^2(z)(1 - z/R)^{5/3}\mathrm{d}z \qquad (7.91)$$

where $C_n^2(z)$ is the refractive index structure constant profile (Figure 7.19) and $k = 2\pi/\lambda$ is the wave number. Physically, this is the receiver-plane transverse radius over which the field from a target-plane point source is coherent. This coherence radius is proportional to Fried's[177] coherence diameter $r_0 \approx 3.181\rho_0$.

Since $k = 2\pi/\lambda$, ρ_0 has a $\lambda^{6/5}$ dependence assuming a weak dependence between C_n^2 and λ, which is true from the visible to the far-IR band. For example, for a given propagation path, a 10.6 μm lidar would have a coherence radius that is approximately seven times that of a 2.1-μm lidar. This can be significant in applications where ρ_0 is comparable to the receiver aperture radius.

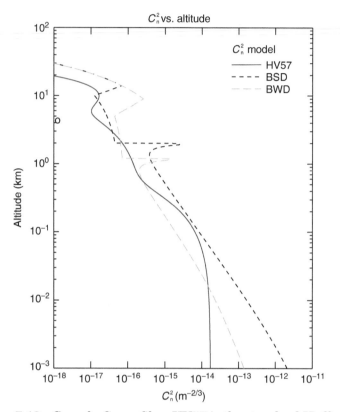

Figure 7.19   Sample C_n profiles. HV57 is the standard Huffnagel–Valley 5/7[176] model and BSD and BWD are the NOAA Boulder Summer Day and Winter Day models.

Substituting Equations (7.90) and (7.82) into Equation (7.84) produces (Frehlich and Kavaya's[153] Equation 190 with $\sigma_{RE} = \sigma_{TE} = \sigma_{LO} = \omega_0/\sqrt{2}$, $\sigma_R = \infty$, $F_{TE} = F_{RE} = F$ and $\eta_{Ho} = (\omega_0/\sigma_R)^2\,\eta_a^{um}$ or Targ et al.'s[180] Equation 4)

$$\eta_a^{um}(R) = \left(2 + 2(1 - R/F)^2(z_r/R)^2 + 2(\omega_0/\rho_0(R))^2\right)^{-1}.$$

$$(7.92)$$

In the far-field and for strong path-integrated turbulence (e.g., long low altitude horizontal propagation paths) the antenna efficiency is dominated by the last term $(\omega_0/\rho_0)^2$. Under this condition, the CNR (Equation 7.83) becomes

$$\text{CNR}_{\text{sp}}^{\text{um}} = \frac{\eta_r \eta_{\text{ox}}}{F_h} \frac{T^2 E_x}{h\nu B} \frac{c\beta_A}{2} \frac{\pi\rho_0^2}{R^2} \tag{7.93}$$

Thus, in the far-field and under strong turbulence, the lidar behaves as if the transmit beam diameter is ρ_0 rather than ω_0.

7.4.1.6.1.2. Refractive Turbulence Efficiency Approximation for the Bistatic Condition

The refractive turbulence efficiency, η_{rt}, is the ratio of the CNR with and without turbulence. Since the CNR is inversely proportional to the beam area (i.e., Equation 7.76), the loss due to refractive turbulence is the ratio of the undistorted beam area to the turbulence broadened beam area or, equivalently: the ratio of the two antenna efficiencies (Equations 7.85 and 7.92):

$$\eta_{\text{rt}} = \frac{1 + (1 - R/F)^2 (R_r/R)^2}{1 + (1 - R/F)^2 (R_r/R)^2 + (\omega_0/\rho_0)^2}, \tag{7.94}$$

which is similar to Yura's[178] Equation 41. In the far-field, $R = F$, the turbulence efficiency is

$$\eta_{\text{rt}}^{\text{ff}} = \frac{1}{1 + (\omega_0/\rho_0)^2}. \tag{7.95}$$

7.4.1.6.2. Monostatic Condition

The independent-path or bistatic assumption, used above, represents a CNR or efficiency lower bound. Under the monostatic condition (i.e., correlated transmit and BPLO paths), the CNR is higher. Two physical effects are responsible for this improvement.

The first is beam wander, which dominates at low or intermediate levels of refractive turbulence. In this case the instantaneous beam size is approximately equal to the diffraction-limited beam size, even though the average beam size due to wander is significantly larger. Slowly varying tilts affect the BPLO and transmit beams equally resulting in little reduction in CNR over the diffraction limit. This second physical effect occurs at higher turbulence levels where the beam breaks up and forms hot spots (i.e., scintillation). For the monostatic condition, both the BPLO and transmit beams form nearly identical hotspots. Since the antenna efficiency and CNR are proportional to the target-plane beam irradiance (c.f. Equations 7.80 and 7.76), the presence of these hot-spots results in a higher monostatic CNR compared to that predicted by the average beam profile and, in some cases, even that of the diffraction limit.

This monostatic phenomenon can be described through reciprocal path arguments, where the wavefront tilt and focusing effects are self-correcting.[153,179] In the low to moderate turbulence regime, this effect can result in an efficiency that is 5 to 10 dB higher than that predicted by the first-order approximation of Equation (7.94).

For very strong refractive turbulence, the Central-Limit theorem applies and the field distribution converges to a Complex-Gaussian distribution. In this regime, the CNR can be twice as large as predicted by the bistatic condition, through what is called the enhanced backscatter effect.[181, 182] For intermediate refractive turbulence, the enhancement can be even larger.[156, 183, 184]

7.4.1.7. Pulsed Coherent Lidar Velocity Estimation Accuracy

Lower bounds on the frequency estimate precision of the coherent lidar signal (therefore, target velocity) are described in numerous publications. Added noise and fluctuations of the signal amplitude and phase prevent coherent lidar systems from achieving the fundamental statistical bound presented in Section 7.3.6 (see Equation 7.7). For a signal in noise, it can be shown that the variance of any unbiased estimator cannot be less than a particular bound known as the Cramer–Rao lower bound (CRLB).[185] By making certain assumptions about the statistics of the signal and noise, specific analytic lower bounds, for the variance of mean frequency (velocity) estimators, have been derived and published.[186-189] Approximate analytic representations are very useful in that they allow the key parameter dependence to be understood, but they must be used carefully to avoid confusion with the exact CRLB. Empirical models for various mean frequency estimators have also been investigated through numerical simulation (e.g., see Frehlich and Yadlowsky[190]) and have been shown to agree very well with measured lidar data. Simulations are advantageous because the experiment designer can control the critical parameters: CNR, mean wind velocity, wind turbulence strength or spectral width, velocity search interval, and the number of pulses averaged.

In the following sections, a physical description of what causes the velocity variance is presented along with analytic approximations to the CRLB for mean frequency estimators under several different signal assumptions. This discussion is culminated by a presentation of an analytic expression, which agrees with the intuition developed by physical arguements and Monte Carlo simulations. Finally, a "rule of thumb" is given for the requirement on the CNR_n and the number of independent samples that must be averaged to allow the detection of a signal in noise so that a "good" estimate (i.e., not an outlier or anomaly, known as a bad estimate) can be made.

7.4.1.7.1. Description of the Velocity Measurement Error

The velocity measurement error, σ_V, that can be achieved with a coherent lidar depends on the target type (or signal statistics) as described in the following subsections.

7.4.1.7.1.1. Negligible Dynamic-Speckle — Zero-Depth Constant-Velocity Targets: In his book, Van-Trees[185] derives the following CRLB for velocity estimate variance, which is valid for fixed-velocity zero-depth hard targets with infinite (large compared to the pulse width) coherence times, such that the target induced spectral broadening is much less than the pulse RMS spectral width, $\delta\nu_p$:

$$\mathrm{var}[\hat{\bar{V}}_r]_{\mathrm{CNR}} \geq \left(\frac{\lambda}{2}\right)^2 \frac{1}{8\pi^2 \delta t^2} \frac{1}{M_e} \frac{1 + \mathrm{CNR_n}}{\mathrm{CNR_n^2}} \tag{7.96}$$

where $M_e = N_p N_g$ is the total diversity – the combined number of independent pulses, range gates, frequencies, detectors, and polarizations. The velocity precision is directly proportional $1/\delta t$, where δt is the standard deviation or RMS temporal width of the transmitted pulse, since the ability to measure frequency improves linearly with the measurement time. For a waveform consisting of a single transform-limited Gaussian pulse, whose full-width-half-maximum power width is ΔT, the RMS width is given by

$$\delta t = \Delta T / \sqrt{8 \ln (2)} = 1/4\pi \delta \nu. \tag{7.97}$$

Therefore we can write

$$\mathrm{var}[\hat{\bar{V}}_r]_{\mathrm{CNR}} \geq \left(\frac{\lambda}{2}\right)^2 \frac{2\delta \nu^2}{M_e} \left(\frac{1}{\mathrm{CNR_n}} + \frac{1}{\mathrm{CNR_n^2}}\right) \tag{7.98}$$

where, it is assumed that the CNRs in each diversity mode are equal. This CRLB exhibits two key points: 1) At high $\mathrm{CNR_n}$ it is twice the statistical bound, and (2) there is a break-point in performance near $\mathrm{CNR_n} = 1$, that is for $\mathrm{CNR_n} \ll 1$ performance scales as $1/\mathrm{CNR_n^2}$ whereas when $\mathrm{CNR_n} \gg 1$ performance scales with $1/\mathrm{CNR_n}$. The authors do not have a good physical understanding of the first point, but it is thought to be due to the need to measure both in-phase and quadrature components of a signal to unambiguously determine the signal phase or velocity. The two measurements result in twice the variance of each measurement alone. The factor of two has been validated to be correct through numerical simulation. The second point is fundamentally a result of the "one detected photon per coherent integration time" noise of the coherent receiver. When $\mathrm{CNR_n}$ is unity, the signal is equal to the noise and, in general, $\mathrm{CNR_n}$ indicates the total number of signal photons that are detected in the signal coherence time. For $\mathrm{CNR_n} \gg 1$ the shot noise of the signal itself dominates the inherent receiver noise resulting in the $1/\mathrm{CNR_n}$ signal shot noise term in the CRLB expression. For $CNR_n \ll 1$ the coherent receiver noise dominates the signal resulting in the $1/CNR_n^2$ dependence, with the 1 in the numerator indicating the one detected noise photon per coherent integration time.

The high CNR_n dependence predicted by the above equation is not typically achieved for coherent aerosol lidar because random speckle noise dominates performance (the subject of the next subsection), however, as shown below it does describe the low CNR_n dependence for coherent aerosol lidar, since the dominant noise source is LO shot-noise, not speckle noise.

7.4.1.7.1.2. Dynamic-Speckle-Limited Signals — Distributed Targets and High-CNR_n: In the high CNR_n regime, the hard-target CRLB (Equation 7.98) is not typically achieved. This is due to the fact that speckle noise in the signal eventually dominates performance, resulting in velocity estimate variance saturation at high CNR_n. For the case of a distributed aerosol target, this can be physically understood by examining the simulated signal and the range-gated signal spectra, as shown in Figure 7.20, and realizing that the velocity estimate is produced by estimating the mean frequency of the spectrum in a given range gate.

For each range resolution cell, a new set of randomly distributed scatterers contribute to the signal. This produces amplitude and, more significantly, phase variations. The phase variations are more important because the frequency (i.e., the number of cycles) in any given range gate is dominated by the phase derivative. If the signal shown in the figure is examined closely, the number of full cycles of the heterodyne signal in a range gate varies from one range gate to the next, which produces a mean frequency deviation from the true average frequency. This is also evidenced by the phase plot shown in the figure and the range-gated single-realization spectra shown in the right panel. Even in range gates where the speckle intensity is large, the random phase variation of the speckled signal, within the range gate, results in the signal frequency *actually varying* from range gate to range gate. The phase variations in this 1D speckle signal are strongest where the intensity is

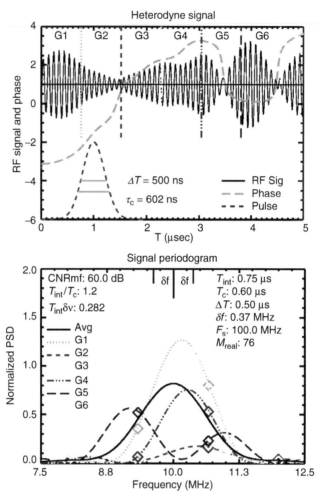

Figure 7.20 (Top) High CNR, 10 MHz coherent laser radar RF signal and phase, for a 500 nsec pulse, and 600 nsec coherence time, τ_c (see Equation 7.100) parsed into six 750 nsec range gates G1–G6, with $T_{int}\delta\nu = \sqrt{8}$. (Bottom) Normalized periodograms for each zero-padded range gate and the 10,000-shot average spectrum (solid). The diamond symbols on the spectral plots are the corresponding nonzero padded periodograms. The signal's mean frequency and ±1 standard deviation widths are indicated by the three vertical lines at the top of the figure. The RF signal in gates G1, G4, and G6 is larger than it is in G2, G3, and G5, which results in the larger spectra shown on the lower plot. The center or peak

small compared to the mean, similar to the branch points in the 2D speckle pattern shown in Figure 7.8. This phase or frequency variation is present even if all of the scattering particles are frozen with respect to each other in space, that is, no velocity spread between the scatters is required.

For measurement times short compared to the signal coherence time, the velocity variance is physically understood as the properly scaled variance of the phase derivative of the speckle modulated signal (see dashed curve in top panel of Figure 7.20) – i.e., $\text{var}[\hat{\overline{V}}_r] \geq (\lambda/2\sqrt{M_e})^2 \text{var}[d\theta/dt]$. The speckle-induced signal phase variation results in a velocity measurement precision that saturates at high CNR. This is because once the signal is sufficiently large, compared to the shot-noise of the receiver, additional signal photons do not improve the estimate of the phase derivative or frequency deviation with time, as all the photons within a pulse range resolution cell are degenerate (all having the same phase vs. time characteristic). Therefore, in the high CNR_n regime the signal frequency with time is dependent on the speckle-phase variations and additional signal photons will not change the frequency vs. time.

This is a consequence of the speckle noise introduced by the pulse propagating through range-distributed targets. It should be noted that mean frequency (or velocity) estimation algorithms that have soft temporal filter responses can accurately sense the phase at longer and longer measurement times with increasing CNR_n, resulting in a variance that slows significantly but that does not fully saturate at high CNR_n. Frehlich and Yadlowski[190] show that a maximum likelihood estimator for mean frequency agree with the exact CRLB and do not saturate at high CNR. Single point scatterers or other zero-depth fixed-velocity hard targets do not exhibit this high

---

frequency of each gate's spectra is dominated by speckle-induced phase noise within the range gate, not receiver shot-noise. Gates with constant phase (i.e., G6) have less frequency error than gates with high phase derivatives (e.g., G1, G2, and G5), since the frequency deviation is the phase derivative.

CNR saturation effect. For these targets, velocity variance continues to improve as CNR_n increases, according to Equation (7.98).

An approximate model that describes this high-CNR saturation velocity precision was given by Doviak and Zrnic.[191] Their high CNR model is given by

$$\text{var}[\hat{\overline{V}}_r]_{\text{sat}} \geq \left(\frac{\lambda}{2}\right)^2 \frac{\delta\nu^2}{2M_e} \tag{7.99}$$

where $\delta\nu$ is the spread of frequencies due to the pulse spectral width and the spread of velocities within the measurement volume.

This spectral width is often characterized by the signal coherence time, which for a Gaussian spectrum is measured by the $\exp[-1]$ time of the Gaussian autocorrelation function (i.e., the Fourier transform of the signal power spectrum). For a Gaussian spectrum, the two widths are related by

$$\tau_C = 1/\sqrt{2}\pi\delta\nu. \tag{7.100}$$

To test Doviak and Zrnic's high-CNR velocity saturation model a numerical simulation was devised. In this simulation an infinite CNR lidar, signal with a finite coherence time was generated synthetically. Then velocity estimates from this signal were generated using a pulse pair velocity estimator with a measurement time much smaller than the signal coherence time. The results are shown in Figure 7.21. These simulations demonstrate that Doviak and Zrnic's model is valid if the diversity is moderate to high ($M_e > 5$) and if the measurement time is short compared to the signal coherence time. An empirical model based upon the simulation data was developed. This model is

$$\text{var}[\hat{\overline{V}}]_{\text{sat}} \geq \left(\frac{\lambda}{2}\right)^2 \frac{\alpha^2(M_e)\delta\nu^2}{2M_e}$$

$$\text{with } \alpha(M_e) = \frac{M_e^2 + \pi - 1}{M_e^2}, \tag{7.101}$$

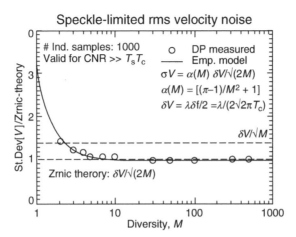

Figure 7.21 Dynamic speckle-limited rms velocity noise vs. diversity level, for the pulse-pair velocity estimator with infinite CNR. The data are normalized to the Doviak and Zrnic model, showing good agreement for diversity, M, greater than 5. For unit diversity, the measured data are $\sim\pi$ higher than the Zrnic model.

where $\alpha(M_e)$ is a factor which saturates to ~ 1 for large M_e, but is $\sim\pi$ for unit diversity. The factor of $\sim\pi$ increase in noise for unit diversity is attributed to the large phase excursions during a deep speckle fade (see Figure 7.20) which are eliminated when the signal diversity is sufficiently high.

For an aerosol target, with negligible target velocity spread, the spectrum width is governed by the transmitted pulse width, because the power spectrum width is the same as the energy-spectrum (i.e., the magnitude-squared of the Fourier transform of the transmitted pulse) width. This is because the heterodyne signal is the convolution of the transmitted pulse and white-noise (aerosols with random location and random scattering amplitude). Therefore, the power spectrum is the product of the pulse's energy spectrum and the spectrum of white noise (i.e., a constant), which is just the spectrum of the transmitted pulse. Thus the pulse's energy spectrum width and the aerosol-signal power spectrum width are equal, $\delta\nu = \delta\nu_p$. For a Gaussian transmitted pulse of δt duration, $\delta\nu_p = 1/4\pi\delta t$ and the signal coherence time is given by

$$\tau_C = 1/\sqrt{2}\pi\delta\nu_p = \sqrt{8}\delta t = \Delta T/\sqrt{\ln(2)} \qquad (7.102)$$

If there is appreciable target velocity spread (aerosol motion) within the measurement volume the spectral width will be larger than $\delta\nu_p$ and the coherence time will be smaller than $\sqrt{8}\delta t$.

7.4.1.7.1.3. Composite Models: In this section, we present two models for the combined effects of a finite CNR and signal coherence time. The first is a variance summation model based upon the concept that noise variances add to constitute the total variance. The second model is Levin's approximation for the CRLB.[186] These two models are compared analytically and graphically.

Variance Summation Model: A composite analytic model incorporating the effect of speckle saturation and the finite-CNR variance (CRLB for a perfect amplitude modulated sine wave in Gaussian noise) is given by

$$\text{var}[\hat{\bar{V}}_r] \approx \text{var}[\hat{\bar{V}}_r]_{\text{CNR}} + \text{var}[\hat{\bar{V}}_r]_{\text{sat}}, \qquad (7.103)$$

where the total variance is modeled as the sum of the two variances from independent noise sources, which is the correct model from a statistical perspective.

Levin Model: Levin's approximation of the CRLB[186] has been published by Frehlich[189] for generalized $T_{\text{int}}\delta\nu$. Rye and Hardesty[187] have also contributed to this knowledge base. Below we present a summary of key results derived from the Rye and Hardesty paper with respect to the CRLB for the Levin algorithm with $T_{\text{int}}\delta\nu \gg 1$. In their paper, Rye and Hardesty show that Levin's approximation of the CRLB is given by the statistical bound multiplied by an excess noise factor $F_e(\alpha_D)$. That is,

$$\text{var}[\hat{\bar{V}}_r] = \left(\frac{\lambda}{2}\right)^2 \frac{\delta\nu^2}{N_S} F_e(\alpha_D) \qquad (7.104)$$

where

$$N_S = M_e N_{SC} \tag{7.105}$$

is the total number of photons collected during the measurement and α_D is the "so-called" degeneracy factor given by

$$\alpha_D = \sqrt{2}CNR_n = \sqrt{2}N_{SC}/\eta_R \tag{7.106}$$

The physical significance of this degeneracy factor, instead of others (i.e., CNR_n) is that it represents the peak of the normalized, un-windowed signal spectrum, where the spectrum is normalized to unit-noise spectral density.

The excess noise factor for Levin's approximation of the CRLB is given by[187]

$$F_e(\alpha_D) = \left[\frac{\alpha_D}{\sqrt{2\pi}} \int_{-\infty}^{\infty} \frac{\exp(-x^2)\mathrm{d}x}{1 + \alpha_D \exp(-x^2/2)} \right]^{-1} \tag{7.107}$$

This is a minimum when α_D is near unity (3.23 to be precise, corresponding to a CNR_n of 2.31). Physically, it states that when the CNR is greater than ~ 2, the photoelectrons become degenerate and little additional new information can be obtained by a stronger coherent signal. Rather, more diversity, M_e, is needed (instead of more degenerate photons) if the measurement error is to be significantly improved. This is the same conclusion that can be drawn from the variance-sum model described above.

Model Comparison: The performance predicted by the variance sum model is shown in the top panel of Figure 7.22 for a $1\,\mu s$ pulse width, which is assumed to dominate the spectral width (note, in strong wind turbulence a coherence time shorter than $1\,\mu s$ may be dominant). Therefore, for the aerosol target case, $\delta\nu\delta t = 1/4\pi$, speckle noise dominates at a CNR_n of about $7\,dB$ and velocity precision saturates with increasing CNR_n. For the $\delta\nu\delta t = 1/40\pi$ hard-target case, performance continues to improve with increasing CNR_n until saturation ensues at about $26\,dB$.

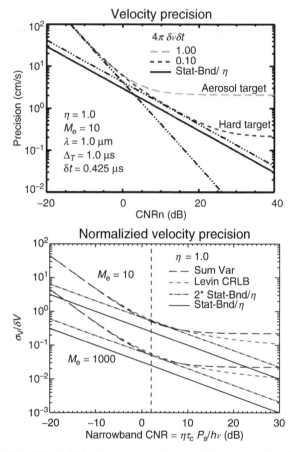

Figure 7.22 (Top) Velocity estimate precision as a function of the narrowband CNR for two saturation velocity cases with $M_e = 10$: (long dash) aerosol limit $\delta\nu\delta t = 1/4\pi$ and slowly rotating hard target with $\delta\nu\delta t = 1/40\pi$ (short dash). Also shown are the two asymptotic forms of the infinite coherence time signal performance (dot–dot–dot–dash) and the statistical bound (solid). (Bottom) Relative velocity precision for $M_e = 10$ and 1000, as a function of the narrowband CNR. Also shown is the statistical bound (solid) and this bound multiplied by the minimum square-root excess noise factor, 2.0 (long dash).

In the bottom panel of the same figure, the performance curves of an aerosol lidar are shown for two diversity levels ($M = 10$ and 1000) as compared to the statistical bound. The performance is closest to the statistical bound shown by the black curves when $CNR_n = 2$. At this CNR_n value, the velocity precision exceeds the statistical bound by a factor of 2 and it is also two times higher than the precision achieved at very high CNR_n. This precision excess noise factor of 2 above the statistical bound is indicated by the short-dash line on the figure. Deviation from this optimum CNR_n value results in worse performance, relative to the statistical bound, but the CNR_n dependence is relatively weak for significant CNR_n deviations. For example, when CNR_n is between -8.4 and $14.4\,dB$, the estimate precision excess noise factor above the statistical bound is only twice as large as it is at the minimal CNR_n point.

These plots show very good agreement between the variance-sum model and Levin's estimate of the CRLB for CNRn $< 10\,dB$. At higher CNR_n, Levin's model is better than the saturation limit of Doviak and Zrnic. The numerical simulations illustrated in Figure 21, show agreement with the variance sum model at high CNR for a pulse pair velocity estimation algorithm with the measurement time short compared to the signal coherence time. It is obvious from the signal phase vs. time plot of Figure 20, that processing algorithms that average the phase over longer measurement times will result in somewhat lower velocity variance. Therefore, the difference in the high CNR predictions of the Levin model does not represent a problem, but only indicates that lower bounds are not always achieved with any given signal processing approach. It should be emphasized that neither of these approximate models represents the exact CRLB, which can be numerically calculated from the covariance matrix of the lidar signals assuming the signal has a Gaussian joint probability density.[189]

For signal degeneracy significantly greater than $1/\eta_R$ (which is equivalent to one *detected* signal photon per coherent integration time) better measurement performance is obtained by reducing the degeneracy and increasing signal diversity. This can be achieved for example, by increasing the laser PRF and reducing the laser energy, assuming the average power remains constant. In practice, this is achieved by operating at $CNR_n \sim 1$ and using multiple pulses, frequencies, or detectors to increase the number of independent speckles that can be viewed within the allowable measurement time.

7.4.1.7.2 Signal Detection Figure of Merit and Fraction of Good Estimates: Sensor maximum measurement range is the range corresponding to a specified, typically 50%, fraction of good estimates. For an unbiased estimator, the fraction of good estimates, which is the complement of the anomaly probability, depends upon three factors: (1) the single-pulse CNR, (2) the number of averaged independent pulse echoes, N_p, and (3) the number of velocity search bins, M_b (i.e., the velocity search interval divided by velocity resolution, $\Delta V = [8 \ln(2)]^{1/2} (\lambda/2) \delta f)$. A mathematical expression for the anomaly statistics has been published by Van Trees[185] and Shapiro,[192] however, their results are left in the form of intractable integrals, which have slow numerical convergence. Nevertheless, integral approximations exhibit a $1/CNR_n\sqrt{N_p}$ dependence in the weak signal regime, when the $CNR_n \ll 1$.

Therefore, in the weak signal regime, a useful rule of thumb, or figure-of-merit (FOM), for describing the maximum measurement range is given by

$$FOM = \sqrt{N_p}CNR_n \qquad (7.108)$$

This FOM is representative of the fact that in the weak-signal regime the anomaly probability follows this $CNR_n\sqrt{N_p}$ dependence, as does the signal detectivity, provided a few (three to seven) shots are averaged to minimize the probability of signal loss due to deep speckle fades. Consider an averaged signal spectra whose noise spectral density is normalized to unity. The statistics of the spectrum follow the Gamma density function and have a normalized standard deviation equal to $1/\sqrt{N_p}$. Since the height of the signal above the unit noise floor is $\alpha = \sqrt{2}CNR_n$, the detectivity (i.e., the ratio of the signal height above the noise floor to the noise standard deviation) is $CNR_n\sqrt{(2N_p)} = FOM\sqrt{2}$.

Empirical results from lidar data (cf. Figure 7.23) suggest that approximately 50% good estimates are achieved when

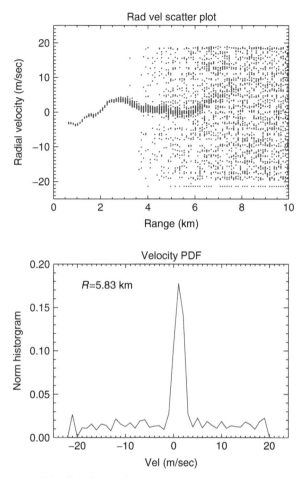

Figure 7.23 Wind lidar velocity scatter plot for 50 consecutive velocity estimates using 100 pulse averaging (Top) and velocity estimate PDF from a sample population of 1000 data products at $R = 5.83$ km (Bottom) corresponding to approximately 50% good estimates. These data demonstrate that the spread of the "good" estimates increases with decreasing CNR (increasing range), as predicted by the CRLB, and that the outliers are nearly uniformly distributed over the velocity search interval. The data were collected from a ground-based, $2\,\mu$m, $2\,$mJ, $10\,$cm aperture, $500\,$Hz lidar at an elevation angle of $25°$.

FOM ≈ 2, for 32 velocity search bins. These same data show that the probability of an anomaly increases dramatically for smaller FOMs. This value of the FOM, which produces 50% good estimates, is only an approximation for typical lidar parameters. The calculation of the lidar parameters required for good estimates for general conditions is difficult. Empirical results for a larger regime can be found in published literature, Dabas[193] and Frehlich and Yadlowsky.[190]

7.4.1.8. Example System Performance Estimation

In this section, a $2\,\mu$m ground-based lidar example is described. First, the efficiency is tabulated. Next, the theory of the previous subsections is applied to compare CNR performance of truncated and untruncated monostatic Gaussian beam lidar. These performance predictions also trade the transceiver focus and refractive turbulence strength.

7.4.1.8.1. Efficiency Calculation

Figure 7.24 shows the key elements of a typical monostatic coherent lidar, for which the receiver efficiency is estimated to be 51.6%, and, assuming a well-designed and optimally truncated lidar, the antenna efficiency relative to the aperture area is 40.1%. The 40.1% antenna efficiency is only possible if the lidar is well aligned and if the optics are of high quality (order of $\lambda/10$ RMS wavefront errors). The effect of wavefront aberrations on the lidar is described by Rye[168] and Spiers.[194] Depolarization losses from imperfect optics is conservatively estimated to be 10%. A 90% far-field misalignment efficiency has also been included to produce a *total efficiency* of 18.6%.

7.4.1.8.2. CNR Calculation

Turbulence and extinction free model predictions of the CNR for a matched beam truncated (Equations 7.68 and 7.69) and untruncated (Equations 7.83 and 7.84) monostatic Gaussian beam lidar are shown in Figure 7.25. The plot displays curves for both a collimated ($F = \infty$) and Rayleigh-range focused ($F = R_r \approx 2.5\,\text{km}$) beam lidar assuming the efficiency tabulated in Figure 7.24. For this CNR_n prediction, a median

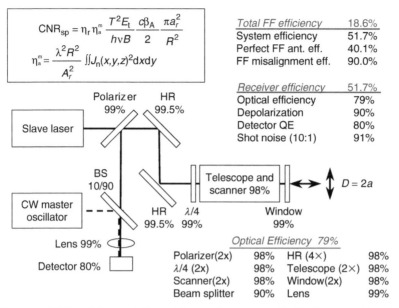

$$CNR_{sp} = \eta_r \eta_a^m \frac{T^2 E_t}{h\nu B} \frac{c\beta_A}{2} \frac{\pi a_r^2}{R^2}$$

$$\eta_a^m = \frac{\lambda^2 R^2}{A_r^2} \iint J_n(x,y,z)^2 dx dy$$

Total FF efficiency	18.6%
System efficiency	51.7%
Perfect FF ant. eff.	40.1%
FF misalignment eff.	90.0%

Receiver efficiency	51.7%
Optical efficiency	79%
Depolarization	90%
Detector QE	80%
Shot noise (10:1)	91%

Polarizer 99% HR 99.5%

Slave laser

BS 10/90

CW master oscillator

HR 99.5% $\lambda/4$ 99%

Telescope and scanner 98% $\leftrightarrow$ $D = 2a$

Window 99%

Lens 99%

Detector 80%

Optical Efficiency 79%

Polarizer(2x)	98%	HR (4×)	98%
$\lambda/4$ (2x)	98%	Telescope (2×)	98%
Scanner(2x)	98%	Window(2x)	98%
Beam splitter	90%	Lens	99%

Figure 7.24 Coherent lidar block diagram and efficiency estimate for a truncated Gaussian beam monostatic lidar with a matched BPLO (i.e., the Wang design).

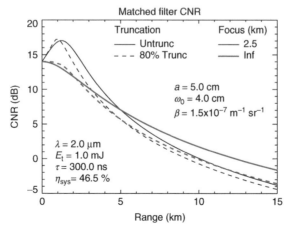

Matched filter CNR

Truncation	Focus (km)
—— Untrunc	—— 2.5
- - - 80% Trunc	—— Inf

$a = 5.0$ cm
$\omega_0 = 4.0$ cm
$\beta = 1.5 \times 10^{-7}$ m^{-1} sr^{-1}

$\lambda = 2.0$ μm
$E_t = 1.0$ mJ
$\tau = 300.0$ ns
$\eta_{sys} = 46.5$ %

Figure 7.25 Monostatic lidar, narrowband CNR without turbulence and atmospheric extinction loss. Thick (thin) lines correspond to a collimated (2.5 km focus) lidar.

boundary layer 2 μm aerosol backscatter coefficient ($\beta_A = 1.5 \times 10^{-7} \, \mathrm{m}^{-1} \, \mathrm{sr}^{-1}$) was assumed. Other key model parameters are listed in the figure. In the far-field, $R \gg R_r$, and without turbulence, the collimated untruncated beam CNR is 3 dB (1 dB for an optimally truncated beam) higher than the focused beam. Similarly, there is a 2 dB = 1/62.6% difference in the far-field CNR between the untruncated and truncated beam lidar.

The additional loss due to extinction is wavelength dependent (cf. Figure 7.5). In the planetary boundary layer and in the near-IR, the one-way extinction can be less than 0.2 dB/km. The additional loss due to refractive turbulence (Equation 7.94) can be quite severe, as shown in Figure 7.26. The figure shows the turbulence efficiency for three levels of C_n^2 and two focus conditions ($F = 2.5 \, \mathrm{km}$ and ∞). It is emphasized that this turbulence efficiency is a lower bound. The effects of beam wander and scintillation discussed in Section 7.4.1.6 can result in less loss than predicted in the figure.

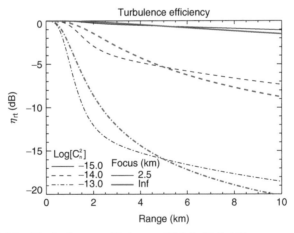

Figure 7.26 Turbulence efficiency. Thick (thin) lines correspond to a collimated (2.5 km focus) lidar.

7.4.1.9. Coherent Detection Doppler Wind Lidar
Theory Summary

In the previous sections, the theory and performance of a coherent lidar was described. In this section we summarize the key concepts and point to the key equations describing the coherent receiver.

The coherent receiver is a single spatial mode receiver only detecting that fraction of the signal power that is in the receiver field mode. The receiver field mode is the quantum efficiency weighted LO field. For a large detector with uniform quantum efficiency, the receiver field mode is simply the LO field itself. The *heterodyne efficiency*, which is defined in Equation (7.43), is the overlap integral between the signal and receiver field modes, representing that fraction of the signal power that is coupled into the receiver field mode. For high heterodyne efficiency, the amplitude, and more importantly, phase of the signal field must match well that of the LO field.

The coherent receiver amplifies the signal power. Each photon that is coupled into the receiver field mode results in a large oscillation of the number of photons incident on the detector at the difference frequency between the signal and LO fields. Exactly how large depends on the LO power. This photon number oscillation carries the phase and amplitude information embedded in the signal and is converted into an electron number oscillation (current) by the photodetector. This signal amplification is described in Equation (7.51) and illustrated in Figure 7.15.

The noise power that the LO-amplified signal competes with in a well-designed coherent receiver is the local oscillator shot-noise (fluctuation in the arrival rate of the photons from the LO). In the full quantum treatment (not included in this work), this fluctuation is attributed to the amplification (by mixing with the LO field) of the one photon per second, per unit, bandwidth fluctuation of the zero-point energy of the vacuum (vacuum field fluctuations) that are coupled into the receiver field mode. In a well-designed coherent receiver this noise dominates other noise sources, resulting in the signal sensitivity of one detected photon. The shot-noise power is described in Equation (7.38).

The coherent receiver narrow bandwidth carrier to noise ratio (CNR_n) is equivalent to the number of *detected* signal photons in the coherent integration time. The optimal CNR is obtained with a receiver bandwidth that matches the bandwidth of the signal itself (also known as matched-filter or narrow band detection). The ability to ensure that the bandwidth is sufficiently narrow (well matched to the signal spectral width) is aided by the conversion of the signals to RF currents by the heterodyne detector. Analog or digital signal processing of these RF signals allows optimal filtering to be achieved. The narrow bandwidth CNR is described in Equation (7.50).

Target plane calculations are an alternate, and in many cases simpler, way to calculate the CNR, heterodyne efficiency, and antenna efficiency of lidar systems. The antenna efficiency is defined as the heterodyne efficiency times the truncation efficiency (the fraction of transmitter power that is not truncated by the primary aperture of the lidar). The receiver-plane calculations, derived in Section 7.4.1.2, require knowledge of the signal field at the detector plane, so that the coupling (or overlap) of the signal field with the receiver field mode, or heterodyne efficiency, can be calculated. The target-plane formalism, which is derived in Section 7.4.1.3, is an alternate formalism which reduces the problem to that of calculating the transmitted and *virtual* backpropagated local oscillator (BPLO) irradiances at the target location, and, in turn, estimating the overlap function of the two irradiances. The target-plane formalism is used to derive the general lidar equation, Equation (7.66), and the antenna (and heterodyne) efficiency for targets with uniform backscatter across the beam irradiance, Equation (7.69).

The antenna efficiency of monostatic lidar systems is less than or equal to 50%, with practical designs being in the 40 to 47% efficiency range, assuming well-designed systems with low aberrations and little misalignment. Many examples of the heterodyne and antenna efficiency for various lidar configurations are calculated using the target-plane formalism in Section 7.4.1.4.

Target-plane calculations allow the effects of beam distortions and misalignments to be more easily calculated. By reducing the problem to that of calculating the irradiances of the transmit and BPLO beams at the target location, the effects of optical aberrations, atmospheric refractive turbulence, and misalignment are more easily calculated. Section 7.4.1.5 describes the loss due to common misalignments. An upper bound to the impact of refractive turbulence was presented in terms of an efficiency loss (Equation 7.94). In many cases, the loss is significantly less than this upper bound, as described in Section 7.4.1.6.

Total far-field efficiencies of 10 to 20% are practical in coherent lidar systems. The total efficiency includes that of the transmit optics, the receive optics, the heterodyne efficiency, misalignment, the detector quantum efficiency, and others. A breakdown of the individual efficiencies making up the total efficiency for a realistic coherent lidar design is provided in Section 7.4.1.8.

A good estimate of the velocity precision obtainable in a coherent lidar is given by an analytic model that includes the effects of signal shot-noise, LO shot-noise, and speckle saturation. This model is described in Section 7.4.1.7 and Equation (7.103). Moderate diversity levels ($M_e = 3$–7) are required to mitigate deep speckle fading which can result in missed detection even at moderately high CNR_n. Actual performance depends upon the particular velocity estimation algorithm (i.e., spectral peak detection, first spectral moment, pulse-pair, Capon, autoregressive, Levin, etc.), the fraction of good estimates and the distribution of the bad estimates. The performance for these and other algorithms has been investigated by Frehlich and Yadlowsky[190] through Monte Carlo simulation techniques, under a variety of M_e and CNR_n conditions.

The velocity precision for a coherent lidar can approach the statistical bound of velocity measurement performance as long as the total photon count is well matched to the signal diversity. In a coherent receiver, minimal velocity estimate variance is achieved when the measurement spans the coherence time and at least one photon is received per coherence time. Photon rates in excess of one per coherence time (i.e., degeneracy >1) are inefficient. This means that for a fixed average power laser, an agile laser PRF or agile frequency is optimal, such that the degeneracy level is approximately one. For cases where there is excess energy, it is better to increase diversity and reduce the degeneracy. For cases where the energy is not sufficient to produce $CNR_n \sim 1$, it is better to reduce the diversity and increase the energy per pulse. The most efficient level of degeneracy is somewhat forgiving, in that for $1 < CNR_n < 10$ near optimal performance is achieved, and performance within 3 dB of optimal is achieved with CNR_n between approximately -8 and 14 dB. In this regime, diversity and degeneracy can be traded evenly provided the diversity is at least 3 to 7, because here velocity variance scales linearly with $1/(NCNR_n)$. This range allows for efficient sensor design even with variable backscatter coefficient conditions. This is described in greater detail in Section 7.4.3.

7.4.2. Direct Detection Doppler Wind Lidar Theory

Many of the basic principles and underlying theory related to the incoherent or direct detection Doppler Lidar (DDL) have already been addressed in the previous subsection dealing with coherent detection systems.

However, important differences occur for the following reasons:

- For both the fringe imaging and double-edge techniques used to implement direct detection, a direct spectral analysis is performed on the backscattered laser light, as opposed to the signal mixing and heterodyne detection of the Doppler shift. A Fabry–Perot etalon has mainly been favored as the means to perform this spectral analysis.
- Since the molecular backscattered signal is strong at wavelengths below 1 μm, it is possible to use the molecular signal for wind measurement, either alone, or in combination with the aerosol backscattered signal.
- Optical systems used for direct detection are generally less challenging than those required for coherent detec-

tion. Much of this is due to the lack of a need to accurately match wavefronts from the local oscillator and backscattered signal at the heterodyne mixer. As a result, it is possible to use a large-aperture "light bucket" for the receiver, since it is not generally necessary to use diffraction-limited optics. However, careful and often complicated instrument design is needed to obtain high efficiency and more complicated calibration procedures are required. For example, in the edge-filter designs, differences in the electro-optic throughput (beam transmission, detector responsivity, electronic gain, etc.), between two or more signal paths, must all be known (calibrated) to very high precision.

• The pulse duration of lasers typically used for direct detection is generally of order 5 to 10 nsec. The advantage is that this provides inherently excellent range resolution (1 to 2 m). The disadvantage is that the Fourier transform of the pulse length creates a limiting spectral bandwidth of the laser spectrum of order 50 to 100 MHz. This implies, particularly for measurements of the aerosol scattered signal, a reduced inherent sensitivity for a given received signal level. A second advantage of the short pulse is that it helps to mitigate speckle noise when the range gate (or integration range) is much larger than the pulse width.

• Coherent detection sensitivity is high because the LO amplifies the signal. Direct detection receivers do not usually employ optical signal amplifiers, therefore high performance can only be achieved with high-efficiency low-noise photoreceivers. Fortunately, in the UV and visible photon-counting receivers (photomultiplier tubes or PMTs) and low-noise charge-coupled devices (CCDs) are commercially available, with quantum efficiencies between 10 and 80%.

The computation of the coherent-detection CNR, or direct detection SNR, and velocity error can be related to comparable factors relating to the power of the laser, the area of the receiver, the efficiency of the entire detection process, and the scattering cross-section within the sample volume. Thus the analysis of the direct detection technique is distinct super-ficially, rather than fundamentally.

7.4.2.1. Direct Detection Concept Overview

A DDL receiver conceptual diagram is illustrated in Figure 7.27. The backscattered signal irradiance is collected by a receiver aperture with radius a_r then passed through a nar-rowband filter to reduce background thermal radiation. The filtered signal is then passed through an optical frequency analyzer or frequency discriminator. The two most popular optical frequency analyzer techniques fall into one of two classes: edge-filters and fringe-imaging.

In fringe-imaging, the optical spectrum is separated into multiple channels, each of which is detected separately, whereas edge techniques employ one or more filter edges to create a discriminator function. In practice, the frequency analyzer for either technique is usually a Fabry–Perot etalon. Both techniques transform the Doppler shift into an irradi-ance variation, and an estimate of the frequency change is accomplished by processing the detected signals. The perform-ance of these systems can be related to the direct detection power signal-to-noise ratio (SNR), defined as

$$\text{SNR} = \langle i_s \rangle^2 / \langle i_n \rangle^2 \tag{7.109}$$

where i_s is the signal photocurrent and i_n is the receiver noise current. As will be derived later in this section, for a photon-counting receiver, this becomes

$$\text{SNR} = \frac{\eta_r \eta_d \eta_D N_s^2}{(N_s + N_n) F_d}, \tag{7.110}$$

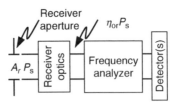

Figure 7.27 Direct detection receiver conceptual diagram.

where N_s the total photon number is the same as it was for the coherent receiver. The detector truncation efficiency, η_d, an overlap efficiency, representing the overlap between the receiver field of view and the target-plane transmit beam. The direct detection frequency discriminator efficiency, η_D, depends upon the discriminator type (i.e., fringe-imaging or edge detection), therefore, having various representations. The efficiency due to the discriminator optics is kept separate in this analysis because it is highly dependent upon the particular frequency discriminator design. F_d is a detector-gain excess noise factor and N_n represents the number of noise equivalent photons accumulated during the measurement or integration time. Note, the same symbol, η_r, for coherent detection is used in direct detection, however, the beam splitter efficiency used in coherent detection to mix the signal with the LO beam is not present in direct detection. Recall that this beam splitter efficiency can be made to near unity through a dual-port receiver design. Therefore,

$$\eta_r = \eta_q \eta_{or} \qquad (7.111)$$

In a direct detection receiver, η_{or} most likely includes a narrowband interference filter to reject background light, which is not needed in coherent detection (see Section 7.3.9).

It should be noted that many authors define the direct detection SNR as the square root of the above expression (e.g., see Equation 14 of Reference 107). This results in an SNR expression that is proportional to $\sqrt{N_s}$ in the signal shot-noise limited (SSNL) operating regime, where N_s dominates the other noise. We choose to use the expression of Equation (7.109) since the shot-noise limited SNR ($N_s F_d \gg N_n$) is proportional to the number of signal photons (i.e., the quantum limit). The SNR, which depends upon the signal power as predicted by the lidar equation (Equation 7.2) and receiver efficiencies, are the subject of the next section. The velocity estimation error for edge-filter and fringe-imaging techniques is developed in Section 7.4.2.4.

7.4.2.2. Direct Detection SNR

In Section 7.3.2, a specific example of a short-pulse monostatic lidar was presented to give physical insight into the parameter dependence of the lidar equation, which describes the collected signal power, P_s, in terms of the transceiver, target, and atmosphere parameters. Coherent and direct detection receivers utilize the return power in different ways to estimate the Doppler shift (wind speed). In this section, theory describing the wideband SNR and matched-filter SNR, which characterizes direct detection performance, is presented. It is shown that the matched-filter SNR can equal the quantum-limited SNR for a lossless and noiseless receiver.

7.4.2.2.1. Wideband SNR

For a detector with uniform responsivity, $\Re$, the pregain direct detection signal photocurrent is given by

$$i_s = \Re \int\int_{A_d} I_s \, da = \eta_r \eta_d \eta_D e P_s / h\nu \qquad (7.112)$$

where P_s is the available optical signal power at the input to the receiver (i.e. Equation 7.2).

The noise current is a combination of shot-noise from the signal, background radiation, and dark current, plus thermal noise from the amplifier and other electronics. That is,

$$\langle i_n^2 \rangle = 2eB[i_s + i_{bg} + i_d]G^2 F_d + S_n B \qquad (7.113)$$

where B is the receiver noise equivalent-bandwidth, NEB. The first three terms represent shot-noise from the signal, background light, and equivalent pregain dark current, all of which are multiplied by the preamplifier gain-squared and the excess noise factor F_d, which accounts for signal fluctuations due to nonuniform gain.[195] The expression for the background light current, i_{bg}, is similar to that for the signal current with potentially different efficiency. The excess noise factor, which is similar to the coherent detection excess shot-noise factor, is usually gain-dependent and depends on the detector characteristics. For example, PMTs have extremely high gain ($G \approx 10^6$ electrons/photon) with a very low excess noise factor $F_d \approx 1.0$. Silicon avalanche photodiodes (APDs), on the other hand, have linear gains on the order of 100 to 1000 with gain dependent excess noise factors between 4 and 10. The last term represents thermal noise from the receiver amplifier and other electronics, where S_n (A^2/Hz) is a noise spectral density.

From Equation (7.109), the SNR is given by

$$\text{SNR} = \frac{(Gi_s)^2}{2eB[i_s + i_{bg} + i_d]G^2F_d + S_nB} \tag{7.114}$$

The primary purpose of the gain, G, is to overcome the constant thermal noise spectral density, S_n. Sufficient gain or, equivalently, sufficiently low additive noise allows the receiver to be signal shot-noise limited (SSNL). This condition occurs when the noise is dominated by signal shot-noise $(2eBi_sG^2 F_d)$. Under this SSNL condition, the SNR becomes linearly proportional to the collected power. That is,

$$\text{SNR}_{ssnl} = \frac{\eta_r\eta_d\eta_D P_s}{2h\nu BF_d} = \frac{\eta_r P_s}{2h\nu B}. \tag{7.115}$$

Hence, like a coherent receiver (Equation 7.42), except for differences in efficiency and for the factor of 1/2, which is compensated by the fact that a DDL needs half the bandwidth of a CD lidar, an SSNL DD receiver has an SNR that is proportional to the quantum limit. If the receiver is not signal-shot-noise limited, the SNR is proportional to the received signal power squared, whereas with coherent detection the CNR is always proportional to the received power (see Equation 7.42). Note that when the receiver noise is dominated by the electronics noise, the excess noise factor is inconsequential.

7.4.2.2.2. Narrowband or Matched-Filter SNR

For a direct detection receiver, the signal is a baseband (signal spectrum is centered about 0 Hz) signal. For a perfect integrator, with integration time T_{int}, the noise equivalent bandwidth (NEB) is $1/2T_{int}$.[146] Under this narrowband or matched-filter, condition and using Equation (7.112), the SNR given in Equation (7.114) can be written as

$$\text{SNR}_n = \frac{(\eta_r \eta_d \eta_D N_s)^2}{(\eta_r \eta_d \eta_D N_s + \eta_{bg} N_{bg} + n_d)F_d + n_a/G^2},$$

$$= \frac{\eta_R}{(N_s + N_n)F_d} \tag{7.116}$$

where N_{bg} is the equivalent number of background light photons and η_{bg} represents an overall receiver efficiency for background light. Likewise, n_d and n_a are the equivalent number of dark current and amplifier noise electrons in the integration time, $T_{int} = 1/2B$. The total number of noise photons, $N_n = (\eta_{bg} N_{bg} + n_d + n_a/G^2 F_d)/\eta_r \eta_d \eta_D$ is, therefore, increased by the signal detection efficiency $(\eta_r \eta_d \eta_D)$ and when amplifier noise limited is reduced by G^2, demonstrating the importance of low noise detector gain.

For SSNL operation, or $N_s \gg N_n$, the matched-filter SNR is similar to that of a coherent receiver (i.e., Equation 7.50). That is,

$$\text{SNR}_{n\text{-ssnl}} = \frac{\eta_r \eta_d \eta_D}{F_d} N_s = \eta_R N_S \tag{7.117}$$

Thus, a direct detection receiver can be nearly quantum-limited, provided sufficiently near noiseless gain can be achieved. This requirement is difficult to achieve in the near to far infrared, due to a lack of commercially available, low-noise, high-gain, high quantum efficiency photodetectors in this wavelength band. At shorter wavelengths (UV to just beyond the visible spectrum) where the molecular signal is very strong, very low-noise high-speed photon-counting detectors (PMTs and silicon APDs) are available, and near quantum-limited direct detection is possible, if excessive background light can be avoided (cf. Figure 7.12). It is noted that in a photon-counting system, the excess noise factor is effectively one, since digital counts are recorded and the count error is not dependent on the excess noise factor. However, when an integrating receiver (i.e., a CCD camera) is used, the net SNR does depend on F_d as indicated in Equation (7.116).

In Figure 7.28, the direct detection SNR (Equation 7.116) for various typical APD gains and excess noise factors is compared with the coherent detection CNR (Equation 7.50). Under signal shot-noise limited conditions the direct detection SNR is proportional to N_s, the number of signal photons. For any direct detection receiver, when N_s is sufficiently small some other noise source dominates and the SNR becomes proportional to N_s^2. Failure to design the direct detection lidar system so that the signal counts dominate the noise counts results in rapidly degrading performance as the signal counts decrease.

7.4.2.3. The Fabry–Perot Etalon

Many direct detection Doppler wind lidar receivers employ the Fabry–Perot etalon as a frequency analyzer. To understand the performance of these receivers, a review of the fundamental principles of a Fabry–Perot etalon is needed. The theory for a Fabry–Perot etalon consisting of two parallel-plane surfaces separated by optical path length d, each

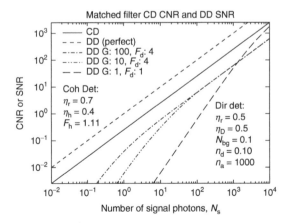

Figure 7.28 Direct detection SNR vs. received power. The perfect DD curve corresponds to an ideal noiseless photon-counting receiver, whereas the $G = 1$, $F = 1$ curve corresponds to a p–i–n photodetector. The other DD curves correspond to typical silicon and InGaAs APDs.

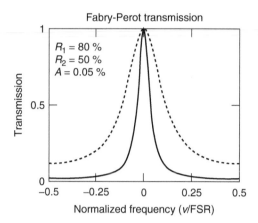

Figure 7.29 Fabry–Perot transmission function over one free spectral range (FSR) for $A = 0.05\%$, $T = 1 - R - A$, and $R = 80\%$ (solid), and $R = 50\%$ (dash), corresponding to $\Im = 14$ and 4.4, respectively.

with reflectance R, transmittance T, and absorptance A ($T + R + A = 1$), can be found in any number of texts.[196,197] In this text, we highlight a few key principles necessary for understanding the basic "frequency analyzer" concept. The reader is referred to other texts for a more detailed description of etalon theory.

The frequency-dependent transmission through the Fabry–Perot etalon is a periodic function over the free spectral range, $\text{FSR} = c/2d$ (Figure 7.29). This function is the well-known *Airy function* given by

$$T(\Delta) = \frac{T_{\mathrm{m}}}{1 + F_{\mathrm{FP}} \sin^2 (\Delta/2)} \tag{7.118}$$

where Δ is the normalized frequency ($\Delta = 2\pi\nu/\text{FSR}$) and $T_{\mathrm{m}} = T_{\mathrm{FP}}^2/(1 - R_{\mathrm{FP}})^2$ is the peak transmission through the etalon. The quantity $F_{\mathrm{FP}} = 4R_{\mathrm{FP}}/(1 - R_{\mathrm{FP}})^2$ is the *coefficient of finesse*. The reflectivity finesse is defined as $\Im = (\pi/2)\sqrt{F_{\mathrm{FP}}}$ and is much like the "Q" of an electronic oscillator, in that the FWHM of the transmission fringe is given by $\Delta\nu = \text{FSR}/\Im$.

The Fabry–Perot etalon can be used to create the transmission edge for edge-detection systems, as the transmission filters in a two-channel system, or to resolve the spectrum in fringe-imaging systems. In the fringe-imaging application, the light at the entrance to the etalon is assumed to be diverging, such that different frequency components of the signal map to unique spatial locations (rings) in the image plane. The irradiance distribution of the throughput, along the radial dimension in the image-plane, is the convolution of the receiver-plane signal-spectrum with the interferometer's transmission function (i.e., the Airy function). When illuminated by divergent light over one or more FSRs, the etalon throughput efficiency, defined as the mean transmission across one FSR, can be found by integrating $T(\Delta)/2\pi$ between 0 and 2π. Thus the etalon efficiency (η_D in a fringe-imaging DDL receiver) is given by

$$\eta_{FP} = T_m/\sqrt{1 + F_{FP}}$$
$$= (1 - R_{FP} - A_{FP})^2/(1 - R_{FP}^2). \qquad (7.119)$$

Thus, there exists a tradeoff between resolution and throughput or efficiency, for a fixed optical path difference of the etalon. Higher resolution requires higher reflectivity, yet higher throughput requires lower reflectivity. A reflectivity of 0.8 with negligible absorption results in a throughput efficiency of only 11%. For this reason, relatively low reflectivity etalons are typically employed in DDL systems. To improve this efficiency, some designs utilize a light recirculator, whereby the reflected light's angle of incidence is scrambled and recirculated back into the etalon. Example recirculator concepts will be described in Section 7.5.2.3. Given the practical limitations, these recirculators can improve efficiency by a factor of 3 to 5, depending upon the single-pass efficiency and the recirculator design.

7.4.2.4. Velocity Measurement Accuracy

In this section, we describe the velocity measurement performance of several optical analyzer types used in DDL systems. For heuristic purposes, the analysis is done for idealized analyzers so that the key parameter dependencies can be understood without adding too much complexity to the analysis. Relevant published literature on real systems is cited in the individual subsections below.

7.4.2.4.1. Generalized Frequency Estimate Variance

Consider the generalized frequency analyzer, with arbitrary discriminator response, shown in Figure 7.30. In this analyzer, a frequency (or velocity $V = \lambda \nu/2$) estimate is produced from the discriminator signal S, for which there is a one-to-one mapping between the signal offset frequency, $\nu_s = (\nu - \nu_0)$, and the discriminator signal S, where ν_0 is an offset frequency representing a zero Doppler reference frequency.

At a given signal frequency, the variance of the frequency estimate, $\hat{\nu}_s$, is given by

$$\text{var}[\hat{\nu}_s] = \text{var}[\hat{S}]/\beta_D^2, \tag{7.120}$$

where $\hat{S}$ is the estimate of S and $\beta_D = dS/d\nu$ is the discriminator slope at frequency ν_s. Therefore, steeper slopes (i.e., large β_D) produce less frequency estimate error for the same discriminator signal noise variance. So to estimate the velocity error variance, one needs to know β_D and the variance of the discriminator signal estimate. In general, the discriminator signal is a function of two or more photocurrents. For the case

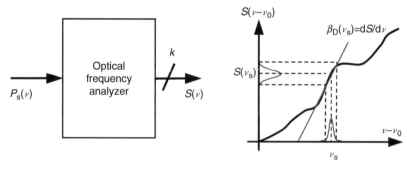

Figure 7.30 Frequency analyzer notional block diagram (left) and notional arbitrary single discriminator function (right).

of two photocurrents, the discriminator signal variance is given by[198]

$$\mathrm{var}[\hat{S}(I_1,\ I_2)] = \left(\frac{\partial S}{\partial I_1}\right)^2 \mathrm{var}[I_1] + \left(\frac{\partial S}{\partial I_2}\right)^2 \mathrm{var}[I_2]$$

$$+ 2C[I_1,\ I_2]\left(\frac{\partial S}{\partial I_1}\right)\left(\frac{\partial S}{\partial I_2}\right), \tag{7.121}$$

where $C[I_1,I_2]$ is the covariance of the two photocurrents, which is zero if the fluctuations in I_1 and I_2 are uncorrelated. This is usually the case as the fluctuations are dominated by separate independent photon and amplifier noise.

7.4.2.4.2. Single-Edge Frequency Discriminator

In this section, the performance of the simplest optical frequency analyzer, the single-edge fixed-slope discriminator is evaluated (see Figure 7.31). A Fabry–Perot etalon is usually employed as the edge filter because very sharp edges can be realized in a relatively compact device. The etalon is locked (e.g., temperature controlled) to the zero-Doppler laser frequency, ν_0, such that the frequency of the transmitted light is matched to the mid-point of the quasilinear transmission edge of the etalon. This approach has been applied to wind velocity measurements at the $1\,\mu\mathrm{m}$ wavelength[108] and performance predictions have been developed for ideal conditions.[199,200]

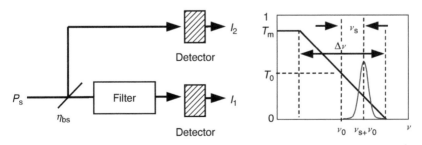

Figure 7.31 Single-edge frequency discriminator functional diagram and filter transmission.

In this section, a first-order analysis of the single-edge technique assuming an idealized edge filter as shown in Figure 7.31 is presented. This will allow the key performance parameter dependencies to be shown without complicating the analysis with specific nonlinear transmission profiles such as that obtained with the edge of a transmission fringe of a Fabry–Perot etalon.

In the single-edge direct detection lidar, the received signal power (Equation 7.2) is divided into two channels, one signal (I_1) and one reference (I_2), using a beam splitter with signal transmission, η_{bs}. After the beam splitter, the signal passes through an edge filter, which acts as the frequency discriminator. The discriminator slope drives system performance, since higher sensitivity to velocity changes is achieved with a higher slope. The limit on discriminator slopes is governed by the constraint that $\Delta\nu$ must be large enough to accommodate the signal spectral width plus the velocity search interval. Uncertainties with regard to this sensitivity (i.e., slope) require the system to be calibrated.

The filter transmission function at the offset frequency, $\nu_s = \nu - \nu_0$ (cf., Figure 7.31), is given by

$$T_s = T(\nu_s) = (T_0 - T_m \nu_s/\Delta\nu) \tag{7.122}$$

where T_m is the peak or maximum filter transmission and $T_0 = T(\nu_0)$ is the transmission at frequency $\nu = \nu_0$. Therefore, the signal offset frequency is proportional to the transmission T_s,

$$\nu_s = \Delta\nu(T_0 - T_s)/T_m \tag{7.123}$$

The transmission, T_s, can be measured by the ratio of two signal currents given by

$$I_2 = (1 - \eta_{bs})\Re_2 \eta_{or} \eta_d P_s \tag{7.124}$$

and

$$I_1 = \eta_{bs}\Re_1 \eta_{or} \eta_d P_s \int T(\nu_s)S_{ns}(\nu_s)d\nu_s, \tag{7.125}$$

where $S_{\text{ns}}(\nu_{\text{s}})$ is the unit-power normalized signal spectrum, $\nu_{\text{s}} = \nu - \nu_0$, and $\Re_1$ and $\Re_2$ are the two detector responsivities. If $T(\nu)$ is a linear function over the entire signal spectrum, the integral is just T_{s}. Therefore, assuming a linear edge,

$$I_1 = \eta_{\text{bs}}\Re_1\eta_{\text{or}}\eta_{\text{d}}T_{\text{s}}P_{\text{s}} \tag{7.126}$$

The ratio of these two currents is proportional to the filter transmission, T_{s}. This ratio is the discriminator signal, S, given by

$$S = I_1/I_2 = \frac{\eta_{\text{bs}}}{(1 - \eta_{\text{bs}})}\frac{\Re_1}{\Re_2}T_{\text{s}}$$

$$= \frac{\eta_{\text{bs}}}{(1 - \eta_{\text{bs}})}\frac{\Re_1}{\Re_2}(T_0 - T_{\text{m}}\nu_{\text{s}}/\Delta\nu) \tag{7.127}$$

This expression can be inverted to provide a unique solution for the unknown offset signal frequency estimate

$$\hat{\nu}_{\text{s}} = \Re_1\eta_{\text{bs}}T_0 - \hat{S}\frac{\Delta\nu}{T_{\text{m}}}\frac{\Re_2(1 - \eta_{\text{bs}})}{\Re_1\eta_{\text{bs}}}$$

$$= \Re_1\eta_{\text{bs}}T_0 - \hat{S}/\beta_{\text{D}} \tag{7.128}$$

where β_{D} is the discriminator sensitivity,

$$\beta_{\text{D}} = \frac{\text{d}S}{\text{d}\nu} = \frac{\text{d}S}{\text{d}\nu_{\text{s}}} = \frac{-T_{\text{m}}}{\Delta\nu}\frac{\eta_{\text{bs}}}{(1 - \eta_{\text{bs}})}\frac{\Re_1}{\Re_2} \tag{7.129}$$

Clearly, system parameter (i.e., η_{bs}, filter transmission, detector responsivity, and other losses) knowledge is needed for calibration.

The frequency estimate variance is related to the discriminator signal estimate variance through Equations (7.120) and (7.121). This produces

$$\text{var}[\hat{S}] = (I_1/I_2)^2(\text{var}[I_1]/I_1^2 + \text{var}[I_2]/I_2^2)$$

$$= (I_1/I_2)^2(1/\text{SNR}_{I_1} + 1/\text{SNR}_{I_2}) \tag{7.130}$$

where $\mathrm{SNR}_I = \langle I \rangle^2/\mathrm{var}[I]$ and

$$\mathrm{var}[\hat{\nu}_\mathrm{s}] = \mathrm{var}[\hat{S}]/\beta_\mathrm{D}^2$$
$$= (\Delta\nu T_\mathrm{s}/T_\mathrm{m})^2(1/\mathrm{SNR}_{I_1} + 1/\mathrm{SNR}_{I_2}) \tag{7.131}$$

The reader is reminded that the SNR is not only a function of the number of signal and noise photons but also the level of speckle diversity (Section 7.3.3). As long as the speckle diversity is greater than N_s then the SNR is dominated by the receiver noise (photon plus amplifier). Otherwise speckle noise dominates. Under signal shot-noise limited conditions

$$\mathrm{SNR}_1 = \eta_{r_1}\eta_{d_1}\eta_\mathrm{bs}T_\mathrm{s}N_\mathrm{s}/F_{d_1} \ \ \text{and} \ \ \mathrm{SNR}_2 = \eta_{r_2}\eta_{d_2}(1 - \eta_\mathrm{bs})T_\mathrm{s}N_\mathrm{s}/F_{d_2} \tag{7.132}$$

and the frequency estimate variance becomes

$$\mathrm{var}[\hat{\nu}_\mathrm{s}] = \frac{(\Delta\nu T_\mathrm{s}/T_\mathrm{m})^2}{N_\mathrm{s}}\left(\frac{F_{d_1}}{\eta_{r_1}\eta_{d_1}\eta_\mathrm{bs}T_\mathrm{s}} + \frac{F_{d_2}}{\eta_{r_2}\eta_{d_2}(1 - \eta_\mathrm{bs})}\right) \tag{7.133}$$

This variance is minimized when the beam splitter transmission is set to its optimum value given by

$$\eta_{\mathrm{bs}_\mathrm{opt}} = \frac{\sqrt{\eta_{r_2}\eta_{d_2}/F_{d_2}}}{\sqrt{T_\mathrm{s}\eta_{r_1}\eta_{d_1}/F_{d_1}} + \sqrt{\eta_{r_2}\eta_{d_2}/F_{d_2}}} \tag{7.134}$$

In a typical design, T_0 is equal to $T_\mathrm{m}/2$ to provide equal room for both positive and negative offset frequencies and the optimal beam splitter transmission is given by the above equation with $T_\mathrm{s} = T_0 = T_\mathrm{m}/2$. Using this beam splitter value results in

$$\mathrm{var}[\hat{\nu}_\mathrm{s}] = \frac{\Delta\nu^2}{N_\mathrm{s}}\frac{T_\mathrm{s}^2\left(\sqrt{T_\mathrm{m}\eta_{r_1}\eta_{d_1}/F_{d_1}} + \sqrt{2\eta_{r_2}\eta_{d_2}/F_{d_2}}\right)^2}{T_\mathrm{m}^3\eta_{r_1}\eta_{d_1}\eta_{r_2}\eta_{d_2}/F_{d_1}F_{d_2}} \tag{7.135}$$

Finally, if $\eta_{r_1} = \eta_{r_2} = \eta_r$, $\eta_{d_1} = \eta_{d_2} = \eta_d$, and $F_{d_1} = F_{d_2} = F_d$, this simplifies to

$$\text{var}[\hat{\nu}_\text{s}] = \frac{\Delta\nu^2}{N_\text{s}} \frac{F_\text{d}}{\eta_\text{r}\eta_\text{d}T_\text{m}} T_\text{s}^2 \left(\frac{\sqrt{T_\text{m}} + \sqrt{2}}{T_\text{m}}\right)^2 \qquad (7.136)$$

or in terms of the signal frequency (Equation 7.122)

$$\text{var}[\hat{\nu}_\text{s}] = \frac{\Delta\nu^2}{N_\text{s}} \frac{F_\text{d}}{\eta_\text{r}\eta_\text{d}} \frac{(1/2 - \nu_\text{s}/\Delta\nu)^2 (\sqrt{T_\text{m}} + \sqrt{2})^2}{T_\text{m}}, \qquad (7.137)$$

where the last product term represents the reciprocal discriminator efficiency, η_D. Therefore, for an SSNL receiver, the estimated variance is proportional to the signal frequency. As the frequency approaches the edge limit (i.e., $\nu_\text{s} \to \Delta\nu/2$ or $T_\text{s} \to 0$) the error approaches zero. This contradicts intuition, which might lead one to think that performance is improved with more signal transmitted through the filter edge rather than less signal. The performance going to zero is a relative term. The performance for any signal spectral width is always above the statistical bound for that spectral width. While it is true that the performance at the lower transmission portion of the edge is improved compared to operation at the center, that performance is still significantly worse than what could be achieved by using a narrower edge whose width is twice the distance that the signal is from the minimum transmission, thereby re-centering the signal in a narrower edge. Also, if the receiver is not noiseless, SSNL operation will not be achieved when T_s approaches 0, and the minimum velocity variance will degrade as the noise overcomes the signal, leading to a different value for the minimum variance ν_s.

Performance is optimized by maximizing the peak transmission, T_m, and the total number of incident photons, N_s, and by minimizing the width of the discriminator function, $\Delta\nu$, subject to the constraint that $\Delta\nu$ must be large enough to encompass the signal spectrum plus enough margin for the velocity search interval. If a Gaussian spectrum is to be passed through this edge, and the edge is made wide enough to avoid too much spill over beyond the linear portion of the response, one might choose $\Delta \nu = 6\,\delta \nu$, where $\delta \nu$ is the standard deviation of the Gaussian signal spectrum ($\sim 1\%$ of energy outside linear slope at zero velocity). In this case, if $\nu_s \approx \nu_0$, and $T_m \approx 1$, the velocity estimate variance is approximately $52F_d/(\eta_q \eta_d)$ times the statistical variance limit (Equation 7.8).

7.4.2.4.3. Double-Edge Frequency Discriminator

It should be noted that most of the DDL community refers to variants of the two-channel discriminator (to be described in the next section) as a double-edge discriminator since the two channels are operating on the "edges" of the spectrum. In fact, the double-edge lidar systems discussed in Section 7.5.2 are actually variants of the two-channel receiver. However, these authors distinguish between the two designs, since in the two-channel receiver the signal photons are separated into two distinct spectral bands or channels while in the double-edge receiver the signal is divided into two parts and each is passed through one of two oppositely sloped edge filters. In its most basic form, the double-edge receiver can be represented by the functional diagram shown in Figure 7.32.

Two oppositely sloped quasilinear discriminator edges are used for the two receiver channels in the double-edge design. In most cases, etalon transmission fringes are used to create the edges. The etalons are locked together such that the edges cross near the transmit wavelength and have a local quasi-linear transmittance profile throughout the desired

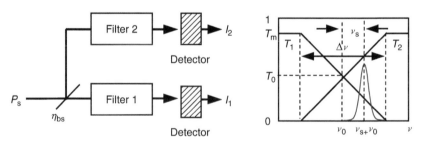

Figure 7.32 Double-edge frequency discriminator functional diagram and filter transmission.

Doppler bandwidth. The discriminator signal is derived from the difference of the two signal currents normalized by the sum. In this way 50% of the available photons are always utilized, independent of the Doppler shifts within the linear-range of the filter. This is very beneficial when the signals are weak and approaching the noise level. With a third signal detector, not passing through either filter, the three unknowns (Doppler, aerosol signal strength, and molecular signal strength) can be determined. This resolves the uncertainty experienced with the single-edge design, thus enabling more accurate measurements. This double-edge discriminator has been applied to both wind lidar[113] and hard target ladar,[201] and performance predictions have been previously developed for ideal conditions.[199,200]

Having seen the dependence of various system and signal parameters in the previous section, in this text several relaxing assumptions are applied early in the analysis for simplification. These assumptions include that the two detector responsivities are equal and the two filters have symmetric (equal but opposite slopes) transmission functions and are crossing each other at the half transmission point corresponding to the zero-Doppler frequency, ν_0. In addition, the beam splitter transmission, η_{bs}, is 50% and the signal spectral width, $\delta\nu$, is smaller than the filter width, $\Delta\nu$, so that the entire spectrum "sees" the linear edge.

It is possible to conceive of designs that eliminate the beam splitter by using a single filter, which transmits to one detector and reflects to the other. For this design, reduction by a factor of 2 in the frequency variance is possible compared to the design shown in Figure 7.32.

For these assumptions, the two filter transmission functions are

$$T_{s_1} = T_m(1/2 - \nu_s/\Delta\nu)$$
$$\text{and} \quad T_{s_2} = T_m - T_{s_1} = T_m(1/2 + \nu_s/\Delta\nu) \tag{7.138}$$

It can be shown that the optimal discriminator function is the ratio of the difference to the sum of the two photocurrents. That is,

$$S = \frac{I_\Delta}{I_\Sigma} = \frac{I_1 - I_2}{I_1 + I_2} = \frac{T_{s_1} - T_{s_2}}{T_{s_1} + T_{s_2}} = \frac{2\nu_s}{\Delta\nu} \tag{7.139}$$

and the discriminator sensitivity is

$$\beta_D = 2/\Delta\nu \tag{7.140}$$

which is roughly a factor of 2 more than for the single-edge. Furthermore, it is independent of the peak edge transmission, T_m. However, T_m impacts the SNR. The variance of the discriminator signal can be found from Equation (7.121). This produces

$$\text{var}[\hat{S}] = \gamma^2(1/\text{SNR}_{I_1} + 1/\text{SNR}_{I_2}) \tag{7.141}$$

where

$$\gamma = 2I_1I_2/(I_1 + I_2)^2 \tag{7.142}$$

is a standard deviation weighting factor. Therefore,

$$\text{var}[\hat{\nu}_s] = \text{var}[\hat{S}]/\beta^2$$
$$= (\gamma\Delta\nu/2)^2(1/\text{SNR}_{I_1} + 1/\text{SNR}_{I_2}) \tag{7.143}$$

Under SSNL conditions, and assuming the receiver has sufficient diversity such that the SNR is dominated by Poisson photon noise, the two current SNRs are given by

$$\text{SNR}_{I_1} = \eta_{r_1}\eta_{d_1}T_{s_1}N_s/2F_{d_1}$$
$$\text{and} \quad \text{SNR}_{I_2} = \eta_{r_2}\eta_{d_2}T_{s_2}N_s/2F_{d_2} \tag{7.144}$$

and $\gamma = 2T_{s_1}T_{s_2}/(T_{s_1} + T_{s_2})^2$. If we assume $\eta_{r_1} = \eta_{r_2} = \eta_r$, $\eta_{d_1} = \eta_{d_2} = \eta_d$ and $F_{d_1} = F_{d_2} = F_d$, this expression simplifies to

$$\text{var}[\hat{\nu}_s] = \frac{\Delta\nu^2}{N_s} \frac{F_d}{\eta_r\eta_d(T_{s_1} + T_{s_1})} \frac{T_{s_1}T_{s_2}}{(T_{s_1} + T_{s_2})^2} \tag{7.145}$$

But with the symmetric filter assumption, $T_m = T_{s_1} + T_{s_2}$ and $T_{s_2} = T_m - T_{s_1}$, the frequency estimate variance becomes

$$\text{var}[\hat{\nu}_s] = \frac{\Delta\nu^2}{N_s} \frac{F_d}{\eta_r\eta_d T_m} \frac{T_{s_1}(T_m - T_{s_1})}{T_m^2} \tag{7.146}$$

Using Equation (7.138) for the filter transmission function produces

$$\text{var}[\hat{\nu}_s] = \frac{\Delta\nu^2}{N_s} \frac{F_d}{\eta_r\eta_d} \frac{\left((1/2)^2 - (\nu_s/\Delta\nu)^2\right)}{T_m}, \tag{7.147}$$

where the last product term represents the reciprocal discriminator efficiency, η_D. This result reveals a few key points. First, relative to the single-edge receiver (Equation 7.137), the frequency estimate variance at $\nu_s = 0$ and for $T_m = 1$, is about a factor of $5.8 = (1 + \sqrt{2})^2$ smaller. In addition, the velocity variance is maximum at $\nu_s = 0$ and minimum at $\nu_s = \pm\Delta\nu/2$, whereas for the single edge this minimum variance only occurs at the zero transmission frequency. In fact, for a noiseless double-edge receiver, the velocity variance approaches zero at the two extreme edges of the sensing range. Again, at these extreme points, the signal levels approach zero and the SSNL assumption will be invalid due to some, perhaps small, but present receiver noise source. Also, as before, better performance can always be achieved with a narrower edge.

It is possible to obtain a double-edge response using a single discriminator device by using an etalon for the frequency discriminator. Assuming that the etalon is very low loss, the reflected signal plus the transmitted signal always equals unity and a perfect double-edge is formed. This has the advantage over the double-edge that uses two filters in that none of the photons are lost due to the transmission edge and this more efficient use of the available photons results in a factor of 2 improvement in the frequency estimate variance.

In order to achieve a velocity precision equal to the statistical lower bound (Equation 7.8) at $\nu_s = 0$, the efficiency factor $(F_d/\eta_r\eta_d T_m)$ would have to be unity and the spectrum of the signal would have to be such that its standard-deviation width $\delta\nu \approx \Delta\nu/2$ for the dual-filter. However, in that case, only 68% of the signal photons would be influenced by the filter, so the statistical bound cannot be achieved. On the other hand, for the single filter implementation of the double edge, the statistical bound is approached when $\delta\nu \approx \Delta\nu/2\sqrt{2}$. In this case, the bulk of the spectrum ($\pm 3\delta\nu$) just barely fits under the filter response function and near perfect performance can be achieved, however, there is little margin to accommodate Doppler shifts.

The above analysis could have also been developed using an analysis based upon application of Equation (7.121) with $S = I_\Delta/I_\Sigma$. This was the approach taken by Rye,[199] which should produce the same result. However, in Rye's analysis, the covariance between I_Δ and I_Σ was neglected and therefore his result (Rye Equation 5) is only valid at $\nu_s = 0$, where the correlation is zero, since $I_\Delta = 0$.

In addition, an analysis based upon the optimal least-squares combination of two frequency estimates, $\nu_1 = I_1/(I_1+I_2)$ and $\nu_2 = I_2/(I_1+I_2)$ also produces the same result; however, the analysis is more complicated due to the nonzero covariance between the estimates of ν_1 and ν_2.

7.4.2.4.4.   Two-Channel Frequency Analyzer

One problem with the double-edge discriminator, designed for molecular signals, is that if the transmission functions are not linear, the discriminator sensitivity, β_D, becomes dependent upon the shape and strength of the aerosol signal spectra. This produces an error that can be calibrated out, if the relative strength of the aerosol signal spectra compared to the molecular signal is known *a priori*. This, however, is difficult since the aerosol backscatter coefficient is variable. The other problem is that the velocity estimate variance depends upon the edge-width, not the signal spectral width. Variants of the two-channel discriminator depicted in Figure 7.33 have been employed to yield improved performance.[113] The small "dead" band in the middle prevents the narrowband aerosol signal from being detected and biasing the molecular signal velocity estimates and, as will be shown, the velocity variance depends upon the signal spectral yielding performance closer to the statistical bound.

While the mathematics describing the two-channel discriminator are similar to those double-edge discriminator, the two designs and their performance are distinctly different.

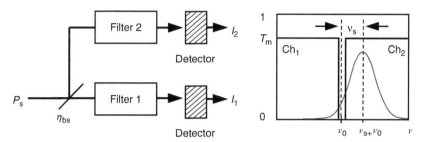

Figure 7.33 Two-channel frequency discriminator functional diagram and filter transmission.

As mentioned in the previous section and repeated here, most of the DDL community refers to variants of the two channel discriminator as a double-edge discriminator since the two channels are operating on the "edges" of the spectrum. The double-edge system is discussed in Section 7.5.2 as a variant discussed of the two-channel receiver.

Two-channel variants for the aerosol signal can also be conceived, whereby the two channels are high-resolution spectrally separated narrowband filters straddling the laser-broadened (50 to 100 MHz for 10 to 5 nsec pulses) aerosol signal spectra. Calibration errors from the molecular signal are minimized by the narrow bandwidth of the filters. In this section, we focus on the performance of a two-channel signal optimized for molecular scattering, with the understanding that this analysis is readily extended to include any two-channel design.

For the molecular signal, the signal spectrum is Gaussian with standard deviation $\delta\nu$:

$$S_s(\nu_s) = N_s \exp\left[(\nu_s/\sqrt{2}\delta\nu)^2\right]/\delta\nu\sqrt{2\pi}. \tag{7.148}$$

For a sufficiently narrow gap, $\Delta\nu_g/\delta\nu \ll 1$, the two photocurrents can be approximated as

$$I_1 \approx \int_{-\infty}^{0} S_s(\nu_s)d\nu_s \quad \text{and} \quad I_2 \approx \int_{0}^{\infty} S_s(\nu_s)d\nu_s. \tag{7.149}$$

Like the double-edge receiver, the two-channel discriminator signal is defined as the ratio of the sum and difference currents, $S = (I_1 - I_2)/(I_1 + I_2) = I_\Delta/I_\Sigma$. For the Gaussian signal spectrum, the discriminator signal is given by

$$S = -\mathrm{erf}[\nu_{\mathrm{sn}}] \qquad (7.150)$$

where $\nu_{\mathrm{sn}} = \nu_{\mathrm{s}}/\sqrt{2}\delta\nu$ is a normalized signal frequency. The sensitivity $\beta_{\mathrm{D}} = \mathrm{d}S/\mathrm{d}\nu$ is given by

$$\beta_{\mathrm{D}} = (-\sqrt{2/\pi}/\delta\nu)\exp[-\nu_{\mathrm{sn}}^2] \qquad (7.151)$$

Using an analysis similar to that in the previous subsection (i.e., Equation 7.143), the following signal shot-noise limited performance is produced:

$$\mathrm{var}[\hat{\nu}_{\mathrm{s}}] = \frac{\delta\nu^2}{N_{\mathrm{s}}} \frac{\pi F_{\mathrm{d}}}{\eta_{\mathrm{r}}\eta_{\mathrm{d}}T_{\mathrm{m}}}(1 - \mathrm{erf}[\nu_{\mathrm{sn}}]^2)\exp[2\nu_{\mathrm{sn}}^2] \qquad (7.152)$$

For small $\nu_{\mathrm{s}}/\delta\nu$ a Taylor series expansion about $\nu_{\mathrm{sn}}=0$ produces the following approximation:

$$\mathrm{var}[\hat{\nu}_{\mathrm{s}}] \approx \frac{(\delta\nu^2 + (1 - 2/\pi)\nu_{\mathrm{s}}^2)}{N_{\mathrm{s}}} \frac{F_{\mathrm{d}}}{\eta_{\mathrm{r}}\eta_{\mathrm{d}}} \frac{\pi}{T_{\mathrm{m}}}, \qquad (7.153)$$

where the last product term represents the reciprocal discriminator efficiency, η_{D}, when $\nu_{\mathrm{s}}=0$. It is worth noting that if the beam splitter were eliminated through an alternate design, the estimate variance would decrease by a factor of 2. This expression shows that the statistical bound is approached to within a factor of $\sim 1/\eta_{\mathrm{R}}$ when $\nu_{\mathrm{s}=0}$, where $\eta_{\mathrm{R}} = (1/\pi)\eta_{\mathrm{r}}\eta_{\mathrm{d}}T_{\mathrm{m}}/F_{\mathrm{d}} = \eta_{\mathrm{r}}\eta_{\mathrm{d}}\eta_{\mathrm{D}}/F_{\mathrm{d}}$ is the overall efficiency (or $\eta_{\mathrm{R}} = (2/\pi)\eta_{\mathrm{r}}\eta_{\mathrm{d}}T_{\mathrm{m}}/F_{\mathrm{d}}$ if the beam splitter were eliminated)

Unlike the double-edge receiver, the variance of this two-channel frequency analyzer increases with nonzero Doppler shifts rather than decreasing. For example, at $\nu_{\mathrm{s}}/\delta\nu=1$ and 2, the variance increases by a factor of ~ 1.4 and 4.9, respectively.

The most significant benefit of the two-channel frequency analyzer is that the spectral width, which appears in the velocity variance equation, is the width of the signal spectra itself, $\delta\nu$, not the discriminator width $\Delta\nu$. This is a significant benefit over the single- and double-edge discriminators discussed above, making it much more practical to approach the statistical lower bound of performance. The difficulty arises in trying to make a two-channel device that has good photon throughput and small spectral gap, $\Delta\nu_g$.

7.4.2.4.5. Multichannel Fringe-Imaging Spectral Analyzer

The edge discriminators discussed are both difficult to calibrate and sensitive to molecular/aerosol signal contamination, resulting in a calibration (i.e., β_D) error. The fringe imaging Doppler lidar (or spectral analyzer) reduces both of these problems. Optimizing the etalon reflectivity and use of recirculators (see Section 7.5.2) allows good etalon throughput. It will be shown later that the two-channel discriminator is a special case of a fringe-imaging receiver.

The basic concept of the fringe-imaging discriminator is to utilize a high-resolution interferometer to produce a spatial irradiance distribution, which is representative of the receiver-plane signal spectrum. This irradiance pattern is imaged by an imaging detector array. The mean frequency is then estimated by one of a variety of methods, which include: the location of the irradiance peak, the first-moment of the irradiance distribution, and matched-filter techniques. In practice, a Fabry–Perot etalon is often used though other interferometers have been successfully employed.[125] This section focuses on the Fabry–Perot etalon interferometer. The basic concepts developed herein also apply to any designs that allow the spectrum of the return signal to be measured.

For generic fringe-imaging lidars, Abreu et al.[104] and McGill and Spinherne[107] have shown that the following equation applies

$$\text{var}[\hat{\nu}_\text{s}] = \left[\sum_j \text{SNR}_j \left(\frac{1}{N_{\text{s}_j}} \frac{\partial N_{\text{s}_j}}{\partial \nu} \right)^2 \right]^{-1}, \tag{7.154}$$

where N_{s_j} and SNR_j are the number of signal photons and SNR in channel j, respectively, and the summation is over all channel elements. Equation (7.154) is equivalent to McGill's Equation (12) with $dN_j/d\lambda$ replaced by $(\lambda^2/c)dN_{\text{s}_j}/d\nu$. Also, McGill uses an SNR that is the square root of the SNR used in this chapter.

Under signal shot-noise limited conditions $(\text{SNR}_j = \eta_\text{R} N_{\text{s}_j})$, the performance of a high-resolution fringe-imaging sensor can approach the statistical limit de-rated by the receiver efficiency, $\eta_\text{R} = \eta_\text{r} \eta_\text{d} \eta_\text{D}/F_\text{d}$, where, in this case, the discriminator efficiency, η_D, is the Fabry–Perot throughput efficiency, η_FP (see Equation 7.119), including any recirculator gain (see Section 7.5.2.3). This can be understood by considering a highly resolving interferometer with FWHM much smaller than the receiver-plane signal spectra width, $\delta \nu$, and FSR much greater than $\delta \nu$ (i.e., $\mathfrak{J} \gg 1$). Assume the irradiance distribution is measured by a detector array with pixel size corresponding to a frequency channel bandwidth $\Delta \nu_\text{c} \ll \Delta \nu$. Further, assume a Gaussian receiver-plane signal photon spectrum given by Equation (7.148). The number of photons in each signal channel is the spatial integral of the irradiance distribution, which is given by

$$N_j = \eta_\text{r} \eta_\text{d} \eta_\text{FP} \int_{\nu_j - \Delta \nu_\text{c}/2}^{\nu_j + \Delta \nu_\text{c}/2} N_\text{s}(\nu)d\nu \approx \eta_\text{r} \eta_\text{d} \eta_\text{FP} \Delta \nu_\text{c} N_\text{s}(\nu_j). \tag{7.155}$$

Therefore, Equation (7.154) can be written as

$$\text{var}[\hat{\nu}_\text{s}] = \frac{F_\text{d}}{\eta_\text{r} \eta_\text{d} \eta_\text{FP}} \left[\sum_j \Delta \nu_\text{c} N_\text{s}(\nu_j) \left(\frac{1}{N_\text{s}(\nu_j)} \frac{\partial N_\text{s}(\nu_j)}{\partial \nu_j} \right)^2 \right]^{-1}.$$
$$\tag{7.156}$$

For a Gaussian spectrum, it can be shown that $[(1/N)\mathrm{d}N/\mathrm{d}\nu] = -\nu_s/\delta\nu^2$ and in the limit as $\Delta\nu_c$ approaches zero, the summation can be approximated as an integral,

$$\text{var}[\hat{\nu}_s] \approx \frac{\delta\nu^4}{\int_\infty \nu_s^2 N_s(\nu)\mathrm{d}\nu_s} \frac{F_d}{\eta_r \eta_d \eta_{FP}} = \frac{\delta\nu^2}{N_s} \frac{F_d}{\eta_r \eta_d \eta_{FP}} \quad (7.157)$$

where the integral in the denominator is simply the spectral variance, $\delta\nu^2$. Thus the efficiency scaled statistical bound (Equation 7.8) is achieved. In principle, high performance can be achieved provided the etalon efficiency, η_{FP}, can be made high while maintaining a narrow resolution $\Delta\nu$. This means both the FSR and reflecting finesse must be kept low, while keeping the analyzer resolution, $\Delta\nu$, less than the signal spectral width. Under low finesse and imperfect etalon conditions, the theory of Equation (7.157) is not valid and a more detailed analysis based upon the convolution of the receiver-plane spectrum with the etalon transmission function,[107] is required to obtain the correct $N_s(\nu_s)$ needed in Equation (7.154).

If Equation (7.156) were precise, then it would apply to the two-channel fringe-imaging frequency analyzer described in the previous section. For a shot-noise limited two-channel analyzer, Equation (7.156) can be written as

$$\text{var}[\hat{\nu}_s] = \left(\Delta\nu_c N_1(\nu_s)/\beta_{D1}^2(\nu_s) + \Delta\nu_c N_2(\nu_s)/\beta_{D2}^2\right)^{-1} \frac{F_d}{\eta_r \eta_d T_m}$$

$$(7.158)$$

where the Fabry–Perot transmission, η_{FP}, is replaced by the two-channel transmission, T_m. The two signal count levels are found by integrating the spectrum. This produces

$$N_1(\nu_s) = \eta_r \eta_d T_m N_s(1 - \text{erf}[\nu_{sn}])/2$$
$$\text{and} \quad (7.159)$$
$$N_2(\nu_s) = \eta_r \eta_d T_m N_s(1 + \text{erf}[\nu_{sn}])/2$$

where, as before, $\nu_{sn} = \nu_s/\sqrt{2}\delta\nu$ is a normalized signal frequency. The sensitivities for each channel are given by

$$\beta_{Di}(\nu_s) = \frac{1}{N_i(\nu_s)} \frac{\partial N_i(\nu_s)}{\partial \nu_s} \qquad (7.160)$$

For the two-channel receiver with a Gaussian spectrum the two sensitivities can be shown to be given by

$$\beta_{D1}(\nu_s) = -\sqrt{\frac{2}{\pi \delta \nu}} \frac{\exp[-\nu_s^2]}{(1 - \mathrm{erf}[\nu_s])} \quad \text{and} \quad \beta_{D2}(\nu_s) = \sqrt{\frac{2}{\pi \delta \nu}} \frac{\exp[-\nu_s^2]}{(1 + \mathrm{erf}[\nu_s])}$$

$$(7.161)$$

Substituting Equations (7.159) and (7.161) into Equation (7.158) produces Equation (7.152), thus the expression for the multichannel receivers indeed reduces to the two-channel receiver performance described in the previous section.

7.4.2.5. Detector Truncation or Overlap and Alignment Efficiency

In general, direct detection systems are not diffraction limited and are, therefore, relatively insensitive to small alignment errors. Nevertheless, in the near-field they usually suffer from an efficiency loss associated with the overlap between the receiver FOV and the transmitted beam for bistatic systems and defocusing losses for either monostatic or bistatic systems. However, the near-field loss is acceptable, due to the large available signal. Furthermore, these systems often employ a telescope with a central obscuration so the efficiency in the far-field is roughly the ratio of the receiver area with and without the obscuration. The losses from overlap efficiency or central obscuration are small in most systems where the obscuration area is small compared to the total collection area. This loss, which was captured in the analysis via the detector truncation efficiency, η_d, is well described in Measures' text[134] and will not be detailed here.

The defocusing loss is, essentially, the ratio of the defocused image of the target-volume transmit beam to the effective detector area, which defines the receiver's FOV. Often, the physical detector area is larger than the effective area, which is governed by the field stop, which limits the receiver field

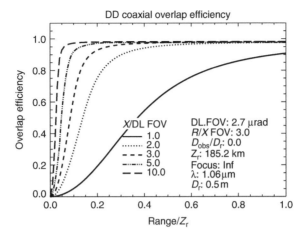

Figure 7.34 Coaxial direct detection overlap efficiency, for a 50 cm aperture parametric in the FOV of the transmitter (in times diffraction limit of the transmit aperture), with a receiver FOV equal to three times the transmit FOV.

of view. Figure 7.34 shows the combined loss for a 50 cm receiver with a coaxial 2 cm transmitter and a 2 cm obscuration diameter as a function of range corresponding to Measures' Equation (7.77).

7.4.2.6. Refractive Turbulence Effects

In general, since direct detection systems are not phase sensitive receivers they are less susceptible to the deleterious effects of refractive turbulence. Thus, apertures much larger than $2\rho_0$ can be efficiently employed. Irradiance scintillation, on the other hand, should not be ignored, since velocity is derived from the integrated irradiance. Fortunately, in the wind sensing application, high levels of spatial and temporal integration are employed, which effectively mitigates this source of signal modulation, provided the number of signal photons in a single spatio-temporal mode is less than 1, which is often the case for high-altitude molecular DDLs.

7.4.2.7. Direct Detection Wind Lidar Theory Summary

For the direct detection Doppler wind lidar, the velocity vs. range is calculated from a measurement of the centroid of the detected spectra (fringe-imaging) or from the relative imbalance between the signals in the edge or two-channel techniques. The velocity estimate variance can be derived purely from the statistical properties of the signal. Table 7.1 summarizes the velocity estimation performance of various DDL frequency analyzers. The advantage of the multichannel fringe-imaging and its two-channel variant is evidenced by the fact that the ultimate precision is limited by the signal spectral width, $\delta\nu$, not the discriminator resolution, $\Delta\nu$, enabling operation near the statistical limit. Due to this relative precision advantage, the primary techniques in use today are the fringe-imaging and two-channel receivers. Variants of the two-channel analyzer are called "double-edge" analyzers by the DDL community, since the channels operate on the edges of the spectrum. In Section 7.5 this is the nomenclature used.

Table 7.1 Summary of Key DDL Frequency Analyzer Characteristics[a,b]

	β_D	$(\eta_r\eta_d N_s / F_d)\,\mathrm{var}[\hat{\nu}_s]$	$(\eta_r\eta_d N_s / F_d)$ $\mathrm{var}[\hat{\nu}_{s|\nu_s}=0]$
Single edge	$\dfrac{T_m}{\Delta\nu}\dfrac{\eta_{bs}}{(1-\eta_{bs})}$	$\dfrac{\Delta\nu^2}{T_m}(0.5-\nu_s/\Delta\nu)^2\left(\sqrt{T_m}+\sqrt{2}\right)^2$	$5.83\Delta\nu^2/4T_m$
Double-edge	$\dfrac{2}{\Delta\nu}$	$\dfrac{\Delta\nu^2}{T_m}\left(0.25-(\nu_s/\Delta\nu)^2\right)$	$\Delta\nu^2/4T_m^c$
Two-channel FI	$\sqrt{\dfrac{2}{\pi}}\dfrac{\exp[-\nu_{sn}^2]}{\delta\nu}$	$\dfrac{\pi\delta\nu^2}{T_m}(1-\mathrm{erf}[\nu_{sn}]^2)\exp[2\nu_{sn}^2]$	$\pi\delta\nu^2/T_m^d$
MC FP FI	$\dfrac{1}{\delta\nu}$	$\delta\nu^2/\eta_{FP}$	$\delta\nu^2/\eta_{FP}$

[a]Assumes signal shot-noise limited operation, equal receiver and detector overlap efficiency and no interference between aerosol and molecular returns.
[b]Aerosol discriminator performance is degraded by a factor corresponding to the molecular to aerosol excess noise factor $(1+N_M/N_A)$.
[c]As much as half this value for designs that use a single edge for both signals (i.e., transmission for I_1 and reflection for I_2).
[d]Half this value if the 50/50 beam splitter were eliminated.

In practical terms, non-ideal receiver components result in real system performance that is degraded from the predictions above. In fringe-imaging systems, the non-zero spectral width of the etalon pass-band broadens the spectral distribution. The optimal design point is achieved by trading the signal throughput with the spectral width — the optimum for real systems being when the etalon spectral width approximately equals the signal spectral width. In two-channel systems, the filters are not as sharp, and total throughput is reduced from unity. In edge systems, the edges are not linear over the entire operating range. Models of etalon reflectance losses and spectral broadening have been treated in detail for the fringe imager and the edge technique in the cited references. Finally, many DDL designs are susceptible to sensor bias and systematic errors, requiring precise calibration and knowledge of both molecular and aerosol spectral characteristics, which are altitude and temperature dependent.

Based on these equations, it is possible to simulate the anticipated statistical component of the wind errors for a given lidar system and an assumed atmospheric model for the "fringe-imaging" and "double-edge" techniques using either, or both, the aerosol and molecular scattered signals. The received and the detected backscattered lidar signals can be computed, incorporating the spectral characteristics of the laser source and the characteristics of the lidar receiver. It is also possible to include the effect of the molecular backscattered signal as a noise source that degrades the wind error from the aerosol-scattered component (i.e., for an optimized aerosol receiver).

Sensor efficiency is a key concern for both coherent and direct detection lidar. In Table 7.2, a sample efficiency tabulation is presented, which is expected to be achievable in well-designed DDL systems, although systems in the field today are operating at 2 to 5% total efficiency. While this table is specific to a Fabry–Perot etalon imaging system, it is also representative of a well-designed edge system. The net efficiency is estimated to be ~15%, which is similar to that obtainable with coherent Doppler wind lidar (see Section 7.4.1.8).

Table 7.2   Expected Efficiency Achievable in a Well-Designed Direct Detection Doppler Lidar

Component	Efficiency (%)
Transmitter optics efficiency, η_{ox}	95
Detector truncation efficiency, η_d	100
Receiver optics efficiency (primary/secondary/imaging/routing)η_r/	90
Background-light (BGL) blocking interference filter	60
Filter etalon for additional BGL reduction	80
Optical analyzer throughput with recycling, $\eta_D = \eta_{FP}$ with other discriminator optics losses included	50
Detector quantum efficiency, η_q (CCD assumed)	75
Excess noise factor	100
Total efficiency (CCD assumed)	15

The most valuable feature of direct detection wind lidar systems, contrasted with coherent system, relates to the potential for noiseless detection, through the use of low-noise high-gain PMTs, APDs, or cooled CCDs. Background light excess noise is substantially reduced in the UV compared to the IR, whereas, for coherent receivers, LO shot-noise is always present, even in the absence of signal. This is discussed in more detail in the following section.

7.4.3.   Comparison of Coherent and Direct Detection Receivers

In this section, some of the main points that have been previously discussed in Sections 7.4.1 and 7.4.2 are summarized and some key similarities and differences between the coherent and direct detection lidar systems are highlighted. In particular, the relative velocity measurement performance of coherent and DDL systems as a function of the number of detected signal photons is compared and contrasted. In order to draw attention to the key parameter dependences and physics, some details are left out and approximations are made.

In both direct and coherent detection lidar systems, the signal collected by the receiver is a result of scattering from the atmosphere. The variance of the signal spectrum is the sum of the variances due to the pulse width and to the relative motions of the scatterers in the pulse volume. The scatterer relative motion is typically of order 0.5 m/sec for aerosol particles, due to the wind turbulence in the measurement volume, and ~300 m/sec for molecular particles, due to Brownian motion — both in standard-deviation widths. Therefore, for example, a Doppler lidar operating at a wavelength of $2\,\mu m$ and having a transform-limited 500 nsec FWHM duration Gaussian pulse would have a signal spectral width (standard deviation) that is approximately 0.6 MHz from the aerosol target and approximately 300 MHz from the molecular target. Using a lidar operating at 355 nm and 10 nsec this becomes 19 MHz for the aerosol target and 1.7 GHz from the molecular target.

Barring efficiency losses, well-designed Doppler lidar systems can achieve velocity measurement precision that is within approximately a factor of 2 of the statistical (or Gagne) lower bound, given by $\sigma_V \sim \delta V_r/\sqrt{N_s}$, where N_s is the total number of signal photons accumulated during the measurement time and $\delta V_r = \lambda \delta \nu/2$. Due to the large difference in the spectral width of the collected signals, the number of accumulated photoelectrons required to achieve a velocity precision of approximately 1 m/s range from a few, for an optimized long-pulse coherent detection lidar using aerosol scatter, to about 100,000 for a direct detection lidar using molecular scatter. Given the photon detection efficiency of 5 to 10% of typical direct detection systems, the latter number results in a requirement of 1 to 2 million photons collected by the primary lidar aperture.

If not optimally designed, the performance of Doppler lidar systems can vary significantly from the statistical bound. An approximate equation for the velocity estimate variance of a speckle-modulated signal in the presence of other noise, for either receiver type (coherent or direct), is given by the sum of the variances of noise terms due to signal shot-noise, other noise (dark or background counts for direct and one-photon zero-point noise per diversity mode for coherent), and the speckle-induced phase fluctuations of the signal,

$$\text{var}[\hat{\overline{V}}_r] = \left(\frac{\lambda}{2}\right)^2 \text{var}[\nu_s] \approx \left(\frac{\lambda}{2}\right)^2 \delta\nu^2 \left[\frac{\kappa}{N_{sd}} + \frac{\kappa N_n}{N_{sd}^2} + \frac{1}{2M_e}\right]$$

(7.162)

In this expression, $\delta\nu$ is the standard-deviation width of the signal, N_{sd} is the total number of detected photons during the measurement (i.e., $N_{sd} = \eta_r \eta_h N_s$ for either coherent or direct detection receivers, where N_s is the total number of signal photons captured by the receiver primary optic), N_n is the total number of noise counts in the measurement (for a coherent system $N_n = M_e$ as will be described later), M_e is the total signal diversity, and κ is a constant that depends on the design and efficacy of the receiver.

In the case of a coherent receiver, this equation can be derived directly from Equation (7.103) in Section 7.4.1.7 by using the relations, $M_e\text{CNR}_n = \eta_R M_e N_{sc} = N_{sd}$ and $N_n = M_e$, which are appropriate for a coherent receiver. The equations match with $\kappa = 2$.

Although the expressions for the velocity measurement variance of direct detection systems derived in Section 7.4.2.4 were developed for signal shot-noise limited performance, when finite dark and background noise counts and speckle are included the parameter dependence shown in the above equation is obtained. The shot-noise limited performance for various direct detection systems, summarized in Table 7.1, shows that κ is 1 for an ideal multichannel fringe-imaging receiver and is π for an ideal two-channel receiver. It is emphasized again that N_{sd} is the total number of *detected* signal photons, so η_D and η_h (for direct and coherent systems, respectively) are already included. It is important that these efficiencies be high so that the photons incident on the receiver have the opportunity to be detected. Below, we assume a direct detection receiver with $\kappa = 2$, which is reasonable for real direct detection receivers.

It is emphasized that the equation above is approximate only; in particular, it does not properly account for anomalies that occur at a low diversity number. A more rigorous treatment using Monte Carlo simulations (e.g., see Frehlich and Yadlowsky[190]) can be applied when detailed performance predictions at low diversity are desired.

In the following two subsections, the purpose is to examine the key similarities and differences between coherent and direct detection receivers. For this purpose, the analytic equation above allows the key characteristics of the different receiver types to be understood.

7.4.3.1. Coherent Detection Receiver

The coherent receiver has two key characteristics that govern performance.

First, the coherent receiver only detects a single spatial mode of the signal field — that mode which is matched to the quantum efficiency weighted local oscillator field. This results in unity diversity for single pulse, single range gate measurements. Independent lidar pulses, range gates, frequencies, polarizations, detectors (pixels), etc., can be utilized to increase the diversity. Incoherent averaging of the signals from the different diversity modes can improve the overall measurement precision and accuracy as long as there is an average of one or more detected photons per mode.

Second, the noise in a coherent receiver is equivalent to one detected noise photon per second per unit bandwidth, resulting in one detected noise photon for each diversity mode. Increasing the total diversity of the measurement to M_e results in the total number of effective detected noise photons for the measurement increasing to M_e.

These characteristics have a significant influence on the velocity measurement performance of a coherent receiver. Figure 7.35 illustrates the velocity measurement performance (Equation 7.162) as a function of the total number of detected signal photons parametric in the signal diversity. For any total detected photon count, N_{sd}, the optimal performance is achieved when the number of independent modes, or diversity, M_e, closely matches that count, so that only a few photons per diversity mode are detected. For each independent mode,

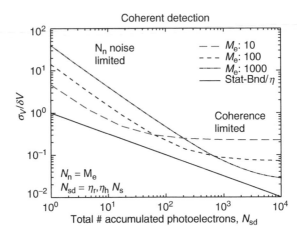

Figure 7.35 Approximate velocity measurement performance of a coherent receiver as a function of total accumulated detected signal photons and diversity, M_e.

an additional effective detected noise photon is added, resulting in rapidly degrading performance (of slope $1/N_{sd}$) as the total detected signal photon count drops below the diversity number. If the number of detected photons per mode drops significantly below 1, it is better to use a smaller number of modes. If the number of detected photons per mode is significantly greater than 1, the signal has too much degeneracy, resulting in the saturated performance shown in the figure. Degenerate photons originating from the same mode have the same amplitude and phase content — the frequency (phase) content of one being the same as all the rest. In cases where the photon count per mode is high, it is better to increase the number of modes to avoid saturation in performance. It should be noted that a coherent lidar designed to achieve an acceptable level of performance at the maximum operating range, where the photon return is lowest, will also perform at least as well (but maybe not much better) at shorter ranges (higher photon counts).

When the total detected signal photons number, N_{sd}, is significantly less than the total diversity, M_e (i.e., $CNR_n \ll 1$) the velocity precision scales as $1/(M_e^{1/2} N_{sd})$. When in this operating region, a reduction of the pulse energy by 10 times results in approximately 100 times the number of independent samples (e.g., pulses) being required to achieve the same velocity precision. The net result being that the total energy required for the measurement increases by 10 times. The best efficiency will result when the transmitted energy is set so that there is approximately one detected photon (or at most a few photons) per pulse in the coherent integration time. As shown in the figure, there is a $\sim$20 dB region near the optimal operating point ($CNR_n \sim 0$ dB) where the slope follows closely that of the statistical bound. In this region, the photons per mode and the number of modes can be traded without a major impact on performance.

7.4.3.2. Direct Detection Receiver

Unlike a coherent receiver, the noise and diversity in a direct detection receiver are not coupled. By using narrowband filters to reduce background light and low-noise photon-counting detectors to reduce the dark counts, it is possible to keep the total noise counts per pulse, per range resolution cell, smaller than 1, even when the spatial diversity is very high. This is more challenging when viewing bright sunlit objects such as clouds or snow, or especially when viewing glint reflections from the sun.

Since the etendue of a direct detection receiver can be many times diffraction limited, high levels of mode diversity are much more easily obtained than for a coherent receiver. As an example, consider a 355 nm aerosol direct detection lidar that uses a 10 nsec pulse duration and a 1 m aperture diameter to measure winds to a range resolution of 250 m. A spatial diversity of about 10,000 is achieved by setting the receiver field of view and illumination to 50 µrad full angle, which is 100 times larger than the diffraction limit of about 0.5 µrad. A range-gate (or temporal) diversity of over 1500 per pulse is obtained by averaging the returns from the 250 m range resolution cell (1.67 µs) compared to the approximately 1.5 m long independent range gates; a new independent speckle-produced intensity pattern is presented at the receiver approximately every 10 ns as the pulse propagates through the scatterers. Finally, multiple pulses are averaged adding additional diversity. The net result being a several million strong diversity number, which eliminates the saturation in performance at high signal levels observed in coherent detection lidar systems.

The decoupling of the diversity and noise is the key fundamental difference between a direct detection receiver and a coherent receiver. Direct detection systems that happen to have diversity and are noise-matched will have performance characteristics that are very similar to those of a coherent receiver.

The effect of total accumulated noise on performance, for a direct detection receiver, is illustrated in Figure 7.36. Equation (7.162), with $\kappa = 2$, was used to create the figure. As with coherent receivers, when the accumulated detected signal count falls below the accumulated noise count, the performance degrades rapidly with slope of $1/N_{sd}$, resulting in a significant penalty in total energy required to make the measurement. The lack of saturation in performance for a

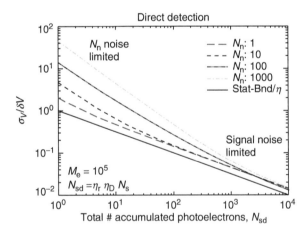

Figure 7.36 Approximate velocity measurement performance of a high-diversity direct detection receiver as a function of the total accumulated detected signal photons and the total accumulated noise equivalent photons, N_n.

high signal count (up to the diversity mode number) means that the available energy can be distributed between pulses without penalty as long as the accumulated signal exceeds the accumulated noise. Practically, this means that the energy per pulse can be reduced and the number of pulses (PRF) increased as is convenient as long as the signal count per pulse per range gate dominates the detector dark and background light count per pulse per range gate.

The performance of a limited-diversity direct detection receiver is illustrated in Figure 7.37. As with the coherent receiver, the performance saturates when more than one photon per mode is detected. For single modes, the variation from independent sample-to-sample is a real variation in the signal itself and any measurement technique will see these actual signal phase and amplitude fluctuations at high signal level. A limited-diversity DDL might be employed as a diffraction-limited DDL for imaging hard targets, or for measuring wind fields with high spatial resolution. The primary difference between Figure 7.37 and Figure 7.35 is that for the coherent

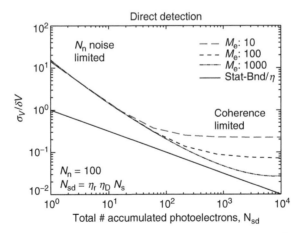

Figure 7.37 Approximate velocity measurement performance of a direct detection receiver against a distributed diffuse target when the diversity, M_e, is finite. Most direct detection Doppler wind lidar systems have very large diversity, so the saturation is not observed.

receiver the total effective noise count is necessarily equal to the diversity, whereas, in the direct detection case the noise count is decoupled from the diversity and is selected to be 100.

7.4.3.3. Example Transmitter Requirement Comparison

In this subsection, the nominal lidar requirements for an example wind measurement are presented. Assume the task is to measure atmospheric winds, to 1 m/s accuracy, 10 km ahead of an aircraft (along a horizontal path) when the aircraft is at an altitude of 10 km. The purpose is to provide information on turbulence ahead of the aircraft. The range resolution requirement is 150 m. The aircraft installation can only support a 15 cm diameter clear aperture. Wind measurements must be provided at a 0.5 Hz update rate. The task is to compare transmitter requirements for a DDL using 355 nm 10 nsec pulses and a CDL using 2.1 μm 500 nsec pulses.

As discussed above, to minimize the total power using a coherent lidar, a few signal photons should be detected per pulse range resolution cell. Given a total efficiency (optical and heterodyne, see Figure 7.24) of 15%, two detected photons per range resolution cell imply that approximately 13 photons per 500 nsec (or 2.5 pW) should be collected by the 15 cm diameter aperture from the 10 km range.

The lidar equation (Equation 7.2), along with estimates of the atmospheric attenuation and backscatter (see Figure 7.3–Figure 7.5), can be used to calculate the pulse energy required to achieve this signal level. At this altitude, the attenuation at the 2.1 μm wavelength is negligible, and the aerosol backscatter is $\sim 5 \times 10^{-10}$ m^{-1} sr^{-1} (see background mode curve of Figure 7.3). For these values the lidar equation predicts that a pulse energy of 190 mJ will produce an average of two photons per range resolution cell.

Since the range resolution of the 500 nsec pulse is about half of the required measurement resolution of 150 m, two cells and two pulses will be averaged to increase the total measurement diversity to 4, with two photons per diversity mode. This diversity and photon count significantly reduces the probability of signal "drop outs" due to speckle fades. Therefore, a 190 mJ, 500 nsec, 1-Hz-PRF coherent lidar operating at 2.1 μm is suitable for the job; an average transmitted power of only 190 mW.

Since the 190 mJ 1 Hz design is near the optimal operating point, the average power required to maintain the same performance is relatively flat as long as the energy is not lowered by more than an order of magnitude from the optimal operating point. Therefore, if laser technology limitations drive the pulse energy down by an order of magnitude to 19 mJ, the measurement can still be made, but instead of 1 Hz PRF it would require operation at little higher than 10 Hz, slightly increasing the average power. On the other hand, to do the same measurement with a pulse energy that was 100 times less, 1.9 mJ, the average power would have to be increased by, roughly, an order of magnitude to maintain high probability of detection of the signal in the presence of the noise.

For the DDL using molecular scatter, the total number of photons that must be detected to achieve the required 1 m/sec measurement accuracy is about 150,000. An excess noise factor of $k = 2$ above the statistical bound is assumed and the velocity standard deviation of the molecular target at this altitude is ~275 m/sec. Given a practical total DDL optical efficiency of 10%, the total number of photons collected by the 15 cm diameter aperture from the 10 km range over the 150 m range gate length (1 μsec) and over the 1 sec measurement period must be about 1.5 million. Using the 0.1 dB/km atmospheric attenuation, and the 2.6×10^{-6} m^{-1} sr^{-1} molecular backscatter coefficient (see Figure 7.3), the lidar equation results in a total energy requirement of 20 J (10 W average power, since 20 J is required in the update time of 0.5 sec).

With moderate care in the design of the detector and background light filters, the total noise counts during the measurement period are expected to be many orders of magnitude below the required signal count of 150,000. The pulse energy and PRF can be freely traded over a wide range to achieve the 10 W average power as long, as the dark counts are dominated by the signal counts.

The large spectral width of the molecular return drives the DDL system to significantly higher average powers than the optimally designed coherent system. The total energy required to make a given measurement increases as the square of the signal spectral width and, inversely, as the square of the required precision. If the velocity measurement precision requirement is decreased to 2 m/s instead of 1 m/s, the average power required by the DDL will drop by a factor of 4 to 2.5 W.

7.5. SYSTEM ARCHITECTURES AND EXAMPLE SYSTEMS

In this section, examples of coherent and direct detection Doppler lidar systems are presented. There have been many systems demonstrated in the four decades since the advent of the laser. Only a few examples are presented here, covering a variety of different instrument types that have been used to collect a large fraction of the wind measurement data that has been published. Coherent lidar systems are presented in Section 7.5.1 and DDL systems are presented in Section 7.5.2.

7.5.1. Coherent Detection Lidar Systems

In Section 7.5.1.1, the basic lidar system architecture that is typically used for coherent detection systems is discussed. This is followed by representative examples of specific coherent lidar systems using CO_2 and solid-state lasers in Sections 7.5.1.2 and 7.5.1.3, respectively.

7.5.1.1. Coherent Lidar System Architecture

Coherent lidar system architectures can vary significantly, but Figure 7.38 illustrates the key features of most systems. Much of the specific detail that is part of any given system is intentionally left out of the figure so that the primary function is better represented. The system starts with a stable CW single-frequency master oscillator (MO) laser. This oscillator output is split into three parts. The first part of the MO output is

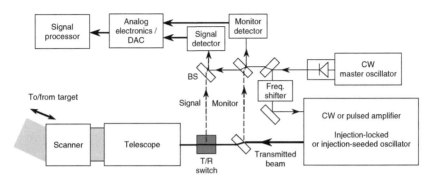

Figure 7.38 Simplified block diagram illustrating the key elements of a coherent lidar system.

first frequency shifted (or not, in the case of homodyne detection) and then used to either: (1) transmit directly to the atmosphere without additional amplification, (2) seed a CW amplifier or CW injection-locked oscillator to create an amplified CW transmit beam, or (3) seed a pulsed amplifier or Q-switched oscillator to create a higher-power pulsed transmit beam. An acousto-optic modulator (AOM) or other frequency shifter is used in many systems to allow a frequency offset of the transmitted beam from that of the MO. This removes the ambiguity between positive and negative frequency shifts, associated with positive and negative radial velocities of the target. The second part of the MO output, which may be absent in some systems, is used to mix with a small portion of the outgoing transmitted beam. This mixed signal which is detected using the monitor detector, allows for detailed knowledge of the transmitter beam frequency/phase and amplitude that can be used to improve the measurement accuracy. The third part is used to illuminate the signal detector and is known as the local oscillator (LO) laser beam. In some systems a separate laser offset-frequency-locked to the MO may be used for the LO.

The resulting transmit beam is expanded by a telescope and directed into the atmosphere using a scanner. The backscatter from the atmosphere is then collected by the telescope (or by a second telescope in the case of a bistatic system) and mixed with the local oscillator radiation on the detector. The resulting heterodyne signal is amplified and filtered using analog electronics, then digitized and processed with digital signal processing techniques to extract estimates of the aerosol backscatter coefficient, mean radial wind velocity, velocity turbulence strength or signal spectral width, and other information.

The MO must produce sufficient single-frequency power to illuminate the detectors and either transmit directly into the atmosphere or properly seed the amplifiers or oscillators used to increase the transmit power/pulse energy. The wavelength of the MO should be tuned to atmospheric extinction minima and the frequency must be of sufficient stability, over the round trip time of flight of the light to the longest measurement range and back, to allow for the desired velocity (frequency) measurement precision.

7.5.1.1.1.   *Transmitter Architectures*

The power transmitter can either be CW or pulsed. Pulsed coherent lidar systems have advantages compared to CW systems since they provide a simple means for range-resolved measurements while also allowing for measurements at longer ranges compared to equivalent-average-power CW systems. As described in Section 7.4.1, for a focused lidar in spatially uniform aerosol backscatter, the majority of the heterodyne signal is dominated by light scattered within the depth of focus around the waist of the transmitted beam.[174] Therefore, a CW system can provide range-resolved measurements by focusing, but the measurement is constrained to only one range unless the focus is varied over time. At long range, diffraction, due to the finite beam or aperture size, does not permit the beam to be effectively brought to a waist. Therefore, range-resolved measurements, in CW systems with practical aperture sizes, are limited to only a few hundred meters. Frequency or phase modulation of the CW laser and matched-filter detection of the modulated return allows CW systems to have improved range resolution. In this case, the resolution is determined by the bandwidth of the transmitted signal, $\Delta R \approx c/2B$, where ΔR is the range resolution and B is the modulation bandwidth.

A pulsed system having a FWHM pulse duration of ΔT provides a range resolution of $\Delta R \approx c\Delta T/2$ (i.e., $B \approx 1/\Delta T$). Range-resolved measurements are simply achieved in pulsed systems by time gating the return signal. For pulsed coherent lidar systems, the transmitter architecture can either be a master oscillator power amplifier (MOPA) or an injection-seeded slave oscillator, possibly followed by one or more power amplifiers (SOPA). A CW laser is typically used for the MO in both types of systems. In the MOPA architecture, pulses are formed by using amplitude modulators (and frequency or phase modulators in some cases) to create the desired pulses shape from the CW MO output. Some of the advantages and disadvantages of each of the architectures are summarized in Table 7.3.

The MOPA architecture allows for increased agility of the frequency and intensity of the transmitted waveform. In very high gain MOPA systems, gain saturation effects result in distortion of the output pulse profile and some pre-shaping of the input pulse profile is necessary to achieve the desired final output pulse shape. In very high gain MOPA systems, the isolation required between the multiple amplifier stages, or between the multiple passes of a single amplifier stage, typically results in a large size of the entire optical layout.

In the SOPA architecture, the injection-seeded slave oscillator is used to efficiently increase the energy either directly to the desired output level or to a level that allows the following amplifiers to operate efficiently at low gain. Injection seeding is a process whereby a very low power signal is injected into the high-power Q-switched, or gain-switched, slave oscillator.[83,202,203] If the injected signal is of sufficient power, and if it is sufficiently well matched to a resonator mode, it allows that mode to dominate the other resonator modes during the pulse build up, resulting in single mode output from what would otherwise have been multimode. The output frequency is at the frequency of the pulsed laser resonator mode, which does not exactly match that of the injected signal — although, typically, it is close. In order to measure the frequency variation between the transmitted pulse and the MO induced by

Table 7.3 Comparison of Master Oscillator Power Amplifier (MOPA) and Injection-Seeded Slave Oscillator Power Amplifier (SOPA) Architectures

Feature	MOPA	SOPA
Pulse profile agility	Allows for arbitrary intensity vs. time, but input pulse preshaping is required in high-gain systems.	Reduced agility, but some control is possible by using high-speed loss modulation in the oscillator during the pulse formation time
Pulse frequency agility	Allows for increased frequency agility since the output largely follows the input frequency modulation	Reduced agility, but some frequency chirp is possible by cavity length or index modulation during the pulse formation time
Beam quality	Care must be taken to avoid beam distortion due to nonuniform gain, nonuniform gain saturation, and intensity dependent nonlinear effects	Spatial mode filtering provided by the oscillator and lower gain of following amplifiers can result in very high output beam quality
Efficiency	Works most efficiently with high-gain materials	Easy to achieve high efficiency, even with low-gain materials
Size	Can be large in high-gain systems due to isolation required to prevent self-lasing or high levels of amplified spontaneous emission	Can be very compact, but achieving long duration pulses in high-gain materials requires either a long, often folded, cavity or high-speed active control of the cavity loss

the injection seeding process, a frequency monitor detector is added to SOPA systems to allow for correction of the frequency difference in the processing. In lower gain materials, such as Tm:YAG operating at 2.0 μm, the injection seeded slave oscillator, followed by low-gain power amplifiers (i.e., a SOPA), is the only practical architecture to achieve high pulse energies, because it is not possible in practical MOPA architectures to achieve sufficient amplifier gain. The SOPA architecture can be very compact and efficient with both high and low gain laser materials, but achieving long pulse durations, as required for good velocity precision, in a compact package using a high gain laser material may require pulse stretching approaches to be applied to the laser. Although it is possible to use high-speed cavity loss modulation or an intra-cavity nonlinear converter to stretch the pulses, generating high-energy 100 ns to 1 μs pulses in high-gain materials (e.g., Nd:YAG at 1.06 μm), it has proven to be difficult and complex.[204,205]

7.5.1.1.2. Optical Aberrations

As shown in Section 7.4.1, the coherent receiver only detects that portion of the signal field collected by the receive aperture that is matched with the quantum efficiency weighted LO field. Therefore, in well designed systems the CNR is maximized when the received signal is coherent over as much of the aperture as possible. For a diffuse distributed target, such as atmospheric aerosol particles, the Van Cittert–Zernike theorem (Section 7.3.8) shows that the spatial scale, over which the received signal phase and intensity are well correlated (i.e., the size of the coherent lobes of the speckle field), is maximized at the receiver by minimizing the spot size at the target range. To obtain a speckle coherence area that is matched to the aperture size the spot size at the target has to be the diffraction-limited spot size possible at that range using the focused full transmit aperture. In other words, efficient coherent detection of the received power, scattered by the aerosol particles, requires that the transmitted beam be diffraction limited and that it fill most of the available aperture.

The quantitative analysis of the loss in lidar efficiency due to different beam aberrations is given in Rye[168] and Spiers.[194] A 1 dB loss in mixing efficiency is experienced when the RMS wavefront deviation of the beam is between $\sim\lambda/10$ and $\sim\lambda/20$, the exact value depending on the type of aberration. For a 3 dB loss, the RMS wavefront deviation must be between $\sim\lambda/5$ and $\sim\lambda/10$. Similarly, the wavefront of the local oscillator and the wavefront quality of the optics in the entire receive path must be consistent with these requirements (i.e., the BPLO must also have high wavefront quality). It is easier to obtain these levels of wavefront quality at longer wavelengths, but in well-designed systems better than $\lambda/10$ RMS can be achieved even at shorter infrared wavelengths such as 1.5 or 2 μm.

7.5.1.1.3. Coherent Receiver

The transmit-receive (T/R) switch, shown in Figure 7.38, is used to spatially separate the return signal from the transmitted beam path. Many different techniques can be used to achieve this including: spatially separated (bistatic) transmit and receive paths; a polarization switch consisting of a quarter-wave retarder and polarizer combination; a Faraday isolator; or electro-optic switches for time multiplexing. The most common approach, due to its simplicity, is to use a polarization switch. The quarter-wave retarder is used to create circularly polarized light from an initially linearly polarized transmit beam. Upon aerosol or target reflection, a significant fraction of the reflected signal is not depolarized and remains circularly polarized but with opposite circular polarity, due to direction reversal. At the receiver, the signal passes through the quarter-wave retarder a second time producing a signal which is linearly polarized but orthogonal to the transmit beam. The polarizer is then used to separate the return beam from the transmit beam path.

As described in Section 7.4.1, heterodyne detection of the signal with the strong LO radiation results in a detection sensitivity of approximately one photon (neglecting other efficiency losses), even with a relatively noisy detector. In order to maximize the CNR the important features of the detector in a coherent lidar system are that it has high quantum efficiency and a sufficiently high saturation level to allow shot-noise, due to the local oscillator radiation, to dominate combined receiver noise sources, which are usually dominated by the electronic preamplifier initially following the detector. In most designs, the shot-noise power dominates the detector noise by over 10 times with only a few hundred microwatts to a few milliwatts of LO power incident on the detector.

In the 1 to 2.2 μm wavelength range, InGaAs and extended-wavelength InGaAs detectors[206] exhibit 60 to 80% quantum efficiency, dark currents less than 1 μA (which is very much smaller than the local oscillator currents in typical systems) and bandwidths beyond 1 GHz, which is sufficient for most terrestrial wind measurement applications. In the 9 to 11 μm wavelength range, the HgCdTe detector is widely used. In order to achieve shot-noise limited performance at these longer wavelengths, the detectors are typically cooled using liquid nitrogen Dewars, or other cryogenic coolers. HgCdTe detectors have lower bandwidth than InGaAs detectors, however, the Doppler shift at these longer wavelengths is also smaller, offsetting the reduced detection bandwidth.

The remainder of this section presents example coherent lidar systems that have been used for wind measurements. Comprehensive coverage of all systems is beyond the scope of this chapter; rather a few examples covering the major different types of deployed wind lidar systems are described. Additional examples can be found in previous summary publications that provide an overview of coherent lidar activity in the United States[11] and in Europe.[12] Systems employing CO_2 coherent lidar are described in Section 7.5.1.2 and systems employing solid-state lasers are described in Section 7.5.1.3. Example wind measurements using these systems are provided in Section 7.6.

7.5.1.2. Representative CO_2 Coherent Lidar Systems

In Section 7.2.1, the background and motivation for coherent lidar systems using CO_2 lasers was discussed (see References

7–35). In this section, key points relative to coherent Doppler lidar systems employing CO_2 lasers are first summarized followed by the description of several representative CO_2 systems.

The high output power and frequency stability proven in early CO_2 laser demonstrations made it a good choice for some of the first coherent lidar applications. Some of the earliest wind measurement lidar systems utilized CW CO_2 coherent transceivers. CW CO_2 ladar systems are still in use today for many short-range measurement applications. Pulsed systems are required when range-resolved measurements at longer ranges are needed. There are three primary types of pulsed CO_2 transmitters used in coherent lidar: the MOPA, the hybrid, and the gain-switched transverse-excited atmospheric (TEA) CO_2 architectures.

The first pulsed CO_2 systems demonstrated in the early 1970s were in the MOPA architecture. These early systems were used for a variety of wind measurement applications ranging from wake vortex detection and tracking to airborne air data measurements. Several groups are still using CO_2 MOPA systems for boundary layer wind measurements today.

In the hybrid transmitter[207,208] a high-pressure pulsed gain module is combined with a CW gain section in the same resonator. The CW lasing serves to self-seed the resonator and acts as a mode filter constraining the pulsed output to a single transverse and longitudinal mode, as is required for optimum coherent lidar performance.

In order to increase the lifetime of intracavity components and better extract the energy available in the gain cell, operation without the CW gain section within the high-power resonator was preferred beginning in the mid-1980s. This was made possible by the advent of injection seeding (see previous section). With injection seeding, large gain volumes could be accessed by using large-transverse-mode unstable resonators. The injection-seeded unstable-resonator TEA CO_2 laser[203,209] allowed for pulse energies up to 1 J/pulse.[29]

In the following two subsections, representative examples of CO_2 laser based coherent Doppler wind lidar systems are presented.

7.5.1.2.1. CW CO₂ Coherent Lidar

A representative example of CW coherent lidar using CO_2 lasers is the laser true airspeed system (LATAS) and its derivatives developed by the remote sensing group at QinetiQ, formerly Royal Signal and Radar Establishment (RSRE), Defense Research Agency (DRA) and Defense Evaluation and Research Agency (DERA), led initially by Michael Vaughan in Malvern, England. A photograph of the LATAS transceiver optical head is shown in Figure 7.39. This focused CW lidar and its derivatives have been used for numerous atmospheric wind and backscatter measurements.

The LATAS is a focused, CW lidar utilizing a homodyne-detection receiver. Since the LATAS is designed for forward-looking operation on a moving aircraft there is no benefit to include a frequency shifter between the transmit and LO beams — the Doppler shift is sufficient to avoid ambiguity. The waveguide CO_2 laser that the LATAS uses has an output power of 4 W, is linearly polarized, and operates on the P(20) line. The transmit/receive optics are monostatic and consist of

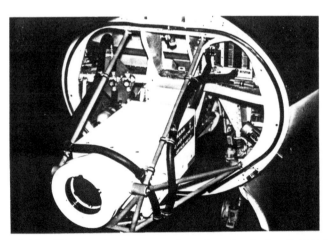

Figure 7.39 The optical transceiver head of the LATAS lidar mounted in its frame in the nose of a HS-125 aircraft with the 15 cm germanium lens at the front. The covering nose cone and the transmitting window have been removed for the photograph (see References 210 and 211).

a 15 cm germanium Galilean telescope. The telescope focus is remotely controlled to define the sensing range. A polarization switch is used to separate the received beam from the transmitted beam axis. Received radiation is mixed with the local oscillator radiation, derived by splitting off a few milliwatts of laser power, on a HgCdTe photodiode cooled to 77 K by a Joule–Thomson minicooler operating on compressed air. The spectrum of the detector output signal is obtained by use of a surface acoustic wave spectrum analyzer with a 6 MHz bandwidth processing 25 μsec blocks every 50 μsec, followed by a digital integrator to increase the signal-to-noise ratio to an acceptable level.

The LATAS transceiver is thermally insulated and a single 100 W heater mat and fan maintain uniform conditions with external temperatures down to −70°C. The weight of the 70 × 30 × 30 cm optics head is about 24 kg. A 20 cm germanium output window, with diamond-like surface coating, was designed to withstand abrasion, insects, and rain drop impact. In practice, throughout nearly 15 years of flying, the equipment proved extremely reliable during measurements around Europe, the United States, and North and South Atlantic.

The first LATAS trials were for avionic applications[19, 210–212] with measurements of free stream airspeed and wind shear; the potential for accurate pressure error calibration was also demonstrated. An example of wind shear detection using the system is provided in Section 7.6.4. In addition to wind measurements, LATAS has been used extensively to characterize the atmospheric 10.6 μm aerosol backscatter coefficient. Many of these extensive, global, airborne aerosol backscatter measurements have been well documented.[20,213–222] A summary of the body of work is provided by Vaughan.[21]

Several LATAS-like, ground-based systems have also been constructed by the Malvern group. One CW system with conical scanning is in a Land Rover vehicle and has been widely deployed across the United Kingdom and Europe in support of site studies, helicopter landing simulations, shipboard operation, comparison with balloon sondes, rocket firing trials, and airfield operation. This highly mobile and versatile coherent lidar system may be brought into operation within 5 min of arrival at a site. Example aircraft wake vortex measurements using one of these ground-based systems is provided in Section 7.6.5

7.5.1.2.2. Pulsed CO_2 Lidar

The most widely used pulsed CO_2-based lidar, to date, is one developed and used by a group at the National Oceanic and Atmospheric Administration in Boulder, CO, led originally by Huffaker and more recently by Hardesty. The genesis of the NOAA TEA CO_2-based lidar dates from a concept in the late 1970s to improve long-range weather forecasting by monitoring winds globally from satellite-borne Doppler lidar.[223] The system built to examine the capabilities of the coherent lidar for long-range wind measurements has been deployed both within a ground-based trailer and from an aircraft. The system uses an injection-seeded unstable-resonator TEA CO_2 laser to produce up to 1 J pulse energies at repetition rates up to 20 Hz. This system and example wind measurements it has made are summarized in Reference 11. An example wind measurement with this system is presented in Section 7.6.1.

Other representative examples of pulsed CO_2 coherent lidar are the Wind and infrared Doppler Lidar (WIND) and Transportable Wind Lidar (TWL) systems. The WIND system was developed as a cooperation between two research teams led by Flamant of École Polytechnique in Paris, France and Christian Werner of Deutsches Zentrum für Luft- und Raumfahrt (DLR) in Munich, Germany.[35]

The WIND system is designed around a CO_2 TEA Laser transmitter to provide a single-mode high-energy pulse at 10 Hz to achieve long-range high-accuracy wind measurements. The airborne-based WIND system was designed to easily fit into medium size jet aircraft (e.g., the DLR Falcon 20). The TWL is a ground-based lidar of similar design that was simultaneously developed at École Polytechnique and is operated from a 20-ft-long sea container. The airborne WIND instrument is equipped with a conical scanner and is primarily for measuring wind profiles from an airborne platform, while the ground-based TWL has a full hemispherical scanner allowing for increased flexibility. The availability of both instruments has facilitated cross validation during field campaigns.

There are two primary objectives of the overall WIND and TWL projects. The first is to make a significant contribution to mesoscale meteorology by investigation of phenomena like the orographic influence on atmospheric flows, the land–sea interaction, the dynamics of convective and stratiform clouds, and the transport of humidity. The second objective is to act as a precursor for spaceborne wind lidar projects.

Key characteristics of the WIND and TWL systems are presented in Table 7.4. The primary difference between the two systems is a laser average power, 1.5 to 3 W for the WIND system compared with 0.3 W for the TWL transceiver. The high performance transmitter laser for the WIND and TWL instruments was specified by the École Polytechnique group and developed at the SAT Company in France.[224]

The WIND and TWL systems utilize a transceiver design that is represented by the general lidar functional diagram of Figure 7.38, with one key exception. The pulsed TEA slave oscillator in these systems is not injection seeded but, rather, a short resonator is used to ensure single longitudinal mode operation. An offset locking servo uses the monitor detector signal to correct for the slow frequency drift maintaining a near-fixed frequency offset between the transmitter and LO lasers. Residual uncorrected frequency jitter is measured by the monitor detector and adjusted accordingly in the signal processing.

The LO is a CW RF-excited waveguide laser, which illuminates both the HgCdTe signal detector, operating at 77 K, and outgoing-pulse monitor detector. The TEA CO_2 laser produces high-energy pulses, which are transmitted into the atmosphere through an off-axis Dall Kirkham telescope. A polarization switch is used to separate the return signal from the outgoing pulse axis. The lidar returns are digitized and processing takes place in a digital signal processor.

Figure 7.40 is a photograph of the WIND instrument and Figure 7.41 is a schematic view showing the WIND system configuration in the DLR Falcon 20 jet aircraft. A lightweight

Table 7.4 Primary Characteristics of the Airborne WIND and Ground-Based TWL Lidar Systems (Courtesy Pierre Flamant)

Parameter	WIND System	TWL System
Wavelength	10.59 μm (10P20 line of $C^{16}O_2$) selected with an intracavity diffraction grating	Same
Local oscillator	CW SLM RF-excited WG CO_2 laser	Same
Slave oscillator pulse energy	100–300 mJ	150 mJ
Pulse repetition frequency	10 Hz	2 Hz
Pulse FWHM temporal width	2.5 μsec with ~200 nsec gain switched spike	Same
Resonator type	TEA, unstable resonator super-Gaussian variable reflectivity mirror	Same
Spatial mode	Near-Gaussian far-field profile ~2× diffraction-limited divergence	Same
Spectral characteristic	Injection-seeded SLM pulse with frequency chirp of ~2 MHz	Same except frequency chirp only ~1 MHz
Telescope type	Monostatic off-axis	Same
Telescope diameter	20 cm	17 cm
Platform	DLR Falcon 20 jet aircraft	20-ft-long sea container
Scanner	30° off nadir conical scan using wedged germanium optic	Full hemispheric using elevation over azimuth mirrors
Scanner rotation rate	18°/sec (nominal)	Variable rate and patterns
Detector (quantum efficiency)	HgCdTe photovoltaic at 77 K (70%)	Same

aluminum structure supports the main subunits: optical bench, laser bench, laser, telescope, and scanner. The design of the mechanical structure was selected to withstand the flight and emergency landing loads while isolating the instrument subsystems from vibration and thermal deformations.

Figure 7.40 Claude Loth, the French project engineer with the WIND instrument prior to integration in the DLR Falcon 20 aircraft. Components of the optical interferometer and mixing unit are recognizable on the upper part of the inner frame of the instrument. The pulsed CO_2 laser is below the bench. (Courtesy Pierre Flamant)

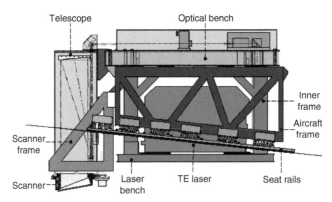

Figure 7.41 Schematic diagram illustrating how the WIND instrument is mounted into the DLR Falcon 20 aircraft. (From Werner C., P.H. Flamant, O. Reitebuch, C. Loth, F. Köpp, P. Delville,

The remaining subunits, such as, signal preprocessing, data handling and storage, electronics control, laser power supply and cooling system, are installed in three aircraft racks (not shown in the figure).

Since their first operation in the late 1990s, the TWL and WIND systems have been involved in several wind measurement programs.[225–231] In addition, the instruments during and after their development have had a role in several fundamental studies of the optimization of lidar performance including signal processing to maximize the performance of the velocity estimates given the laser transmitter frequency chirp and the nonstationary atmospheric target.[232–234] For vertical wind profiles, the wind velocity estimates are made using an algorithm that allows fitting of the data over a full conical scan to improve performance.[235] Improvement of the performance of the coherent lidar, especially at low CNR, has been demonstrated using various methods to average over speckle fluctuations.[141,142,236]

Example measurements made with the WIND and TWL systems can be found in Sections 7.6.1 and 7.6.3.

7.5.1.3. Representative Solid-State Coherent Lidar Systems

In Section 7.2.1, the background and motivation for solid-state coherent lidar systems was discussed (see References 36–73). In this section, key points relative to solid-state transceivers are summarized, followed by a description of several representative solid-state systems.

As described earlier, a significant fraction of coherent lidar remote sensing performed to date has utilized CO_2 gas lasers at wavelengths of 9 to 11 μm. In the late 1980s, single-frequency solid-state lasers in the 1 and 2 μm wavelength

G. Wildgruber, P. Drobinski, J. Streicher, B. Romand, H. Herrmann, L. Sauvage, E. Nagel, C. Boitel, M. Klier, D. Bruneau, M. Schrecker, M. Messionnier, S. Rahm, A. Dabas, R. Häring, and P.H. Salamitou, "Wind infrared Doppler lidar instrument," *Opt. Eng.*, 40, 115–125, 2001. With permission.)

regions were developed which enabled coherent Doppler lidar measurement at shorter wavelengths. Advantages of these, over the longer-wavelength CO_2 systems, include improved atmospheric transmission, increased backscatter coefficient, less beam divergence for equivalent aperture sizes, and better range resolution for equivalent wind velocity resolution. Disadvantages include tighter alignment and optical surface tolerances, and increased signal loss due to atmospheric refractive turbulence.

Because of the significant eye-safety advantage, 2 μm wavelength systems have been utilized much more frequently than the 1 μm systems. During the past 10 years, significant progress has been achieved in increased measurement capability, reduction in size, reduction in prime power requirements, and increased reliability of the 2 μm solid-state coherent lidar systems. Very recently, systems at a wavelength near 1.5 μm are being developed. Two primary advantages of operating at this wavelength are: 1) the availability of components developed for the telecommunications industry as well as 2) the potential for even more compactness and efficiency. There are four types of systems currently being explored at the 1.5 μm wavelength. These are systems using either Raman-shifted Nd:YAG, bulk Er:glass, Er:YAG, or fiber lasers. To date, only limited wind measurement data exist from these 1.5 μm systems.

Some of the earliest work at 1 and 2 μm wavelengths is covered elsewhere (see additional discussion and references in Section 7.2.1), therefore, it will not be repeated here. In the following subsections, more recent representative solid-state coherent Doppler lidar systems operating at 2 and 1.5 μm are presented.

7.5.1.3.1.   WindTracer® Lidar at 2.02 μm

The successes of the preliminary work with 2 μm wavelength solid-state lasers and lidar systems in the early 1990s[43] lead to it being the wavelength of choice by the group at Coherent Technologies for continued development into a lidar system known as the WindTracer®. The nominal operating characteristics are shown in Table 7.5. Figure 7.38 appropriately

Table 7.5 WindTracer® Nominal Operating Characteristics

Parameter	Nominal Value	Comments
Wavelength	2.022 μm	Tm:LuAG, excellent atmospheric transmission
Pulse energy	2 mJ	3 mJ demonstrated in lab
PRF	500 Hz	4 msec lifetime facilitates CW diode pumping
Pulse width	400 nsec	60 m range resolution
Beam quality, M^2	<1.1	Near-diffraction-limited Gaussian beams
Aperture size	10 cm	8 cm beam width (e^{-2} intensity)
Scanner	Full hemispheric	Elevation over azimuth
LO/SO offset frequency	105 MHz	Allows for velocity ambiguity resolution
Detection bandwidth	80–130 MHz	±25 m/sec radial wind velocity (higher possible)

describes the transceiver function with the transmitter being an injection-seeded Q-switched slave laser. The master oscillator and the pump-diodes for the slave oscillator are removed from the primary optical bench and coupled to the rest of the transceiver using, fiber optics, allowing for easy maintenance.

The WindTracer® transceiver utilizes the Tm:LuAG laser material, because: it has a long upper-state lifetime which is compatible with efficient diode-pumping; it naturally produces 0.2 to 0.5 μsec pulse durations, allowing for precision velocity measurements; and it operates at a wavelength near 2.022 μm for good atmospheric transmission and eye-safety. The Tm:LuAG slave laser is diode-pumped using fiber-coupled CW diode lasers. Typical systems require about 10 W of diode pump power to produce the 2-mJ 500-Hz pulses from the injection-seeded slave laser.

The injection seeding is maintained even in high-vibration environments by using a ramp-and-fire servo technique.[237] In short, the ramp-and-fire servo technique allows the proper resonant match between the slave and seed laser's frequencies to be found quickly on a pulse-by-pulse basis. In practice, the jitter of the frequency from the slave oscillator with respect to that of the master oscillator due to the injection-seeding process is only ∼1 to 2 MHz RMS. As described in the text associated with Figure 7.38, the signal processor corrects for this frequency jitter in estimating the wind velocity by using the output of the monitor detector.

The WindTracer® is designed to be a hands-off, turnkey lidar. A diagram and photograph of the transceiver, as well as a photograph of the shelter and scanner utilized in the Wind-Tracer® are shown in Figure 7.42. The transceiver uses a graphite bench for dimensional stability over temperature variations and for reduced weight compared to a metal bench. The use of graphite composite also alleviates the need for an expensive, heavy, and inefficient refrigerated chiller to stabilize the temperature of the optical bench and other components. Only a small and simple above-ambient-temperature liquid-to-air heat exchanger is needed to remove waste heat directly from the heat-producing components. The graphite optical bench is vibration isolated from the rest of the structure removing susceptibility to vibration of the mounting platform. The control electronics have built-in test (BIT) capability to detect and report system errors or failures and can be either remotely controlled from a computer or directly controlled from its front panel. The electronics cards are modular and cross-compatible with other Coherent Technologies transceivers, simplifying manufacturing and maintenance.

The transceiver, control electronics, heat exchanger, signal processor, and scanner, are all installed and operated from an 8 ft. × 8 ft. × 8 ft. shelter. Typical horizontal measurement ranges are 8 to 10 km and vertical ranges are 2 to 5 km.

Example WindTracer® measurements are presented in Sections 7.6.1, 7.6.5, and 7.6.6.

7.5.1.3.2. Er-Fiber Lidar at 1.54 μm

During the 1990s, several groups began to investigate lidar systems operating near 1.5 μm. The goal being to have very compact and efficient lidar systems that leverage the mass-produced components from the telecommunications industry. The alignment and assembly of the lidar is simplified and the stability improved by using fiber components. Early fiber-laser-

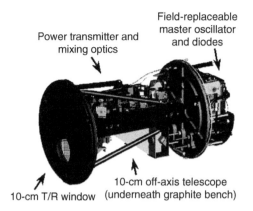

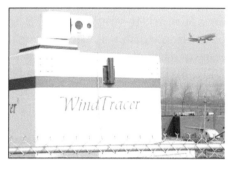

Figure 7.42 (Top) Isometric view of the transceiver used in the WindTracer® coherent lidar with protective dust covers removed (computer rendered). The slave oscillator laser and all coherent lidar mixing optics are located on top of the graphite composite bench. The 10 cm clear aperture window on the front of the unit indicates the scale. A 10 cm off-axis telescope is located beneath the bench. The fiber-coupled master oscillator and pump diodes are mounted on the back of the unit facilitating maintenance. (Bottom left) Photograph of transceiver in its enclosure mounted beneath the full-hemispheric scanner in the Windtracer® shelter. Shown are the transceiver, the field replaceable unit (FRU), and the transceiver support structure (TSS). (Bottom right) Photograph of the Wind-Tracer® shelter with the full-hemispheric scanner mounted on top.

based coherent lidar systems were demonstrated for hard target imaging and vibration detection applications.[238] CW and pulsed fiber transmitters have recently been investigated for wind measurement applications 67–73.

One example is the all-fiber lidar system developed and demonstrated in a collaboration between the QinetiQ group in Malvern, England and the group managed by Ove Steinvall at the Swedish Defense Research Establishment (FOA).[69,70] This system consisted of a CW MOPA arrangement with a low-power (∼5 mW) narrow-linewidth (∼20 kHz) semiconductor laser source, combined with a high-power erbium-doped fiber amplifier (EDFA) to give ∼1 W of lidar output. This simple system relied on the Fresnel reflection from the output fiber facet to provide the LO beam. The facet was precisely polished at the appropriate angle in order to provide the correct LO strength. Separation of the transmit and receive beams was achieved with a standard fiber-optic circulator.

A bistatic CW EDFA wind sensor has also been reported.[71] The bistatic telescope arrangement allows superior definition of the probe volume in comparison to monostatic systems, although there is a penalty in terms of overall signal strength due to the reduced overlap at short range. The all-fiber approach permits relatively straightforward alignment of the bistatic system, in comparison to free-space systems. The layout is shown in Figure 7.43. The single-frequency diode laser serves as the MO and is used to both seed the EDFA and to illuminate the signal and monitor detectors. The AOM shifts the frequency of the input to the EDFA providing the frequency offset between the signal and the MO so that the velocity ambiguity is eliminated.

The CW fiber lidar technology has also been extended to the pulsed mode of operation. The same COTS telecommunication components can be used, but the final-stage transmitter EDFA usually has a larger than normal core size to allow increased energy without optical damage. One system produced 40-μJ 40-ns pulses at a high pulse repetition rate (50 kHz) and utilized a 2.5 cm diameter monostatic aperture to give range resolved aerosol backscatter and LOS velocity data out to ranges of a few hundred meters with a temporal

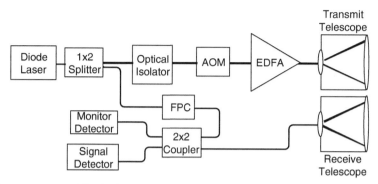

Figure 7.43 Block diagram showing the primary elements of the bistatic CW Er-doped fiber coherent lidar demonstrated at QinetiQ. Shown are components: optical isolator, acousto-optic modulator (AOM), Er-doped fiber amplifier (EDFA), and fiber polarization controller (FPC).

resolution of 0.5 sec.[239] In another system, longer (1 μsec) pulses were used to produce 10-μJ pulse energies at a pulse repetition rate of 50 kHz.[240] Using a 5 cm diameter monostatic aperture, this system was able to measure winds to approximately 1 km range by averaging the signals from 1000 pulses.

Er-doped fiber laser technology continues to advance at a rapid pace. Use of large core fibers should allow coherent fiber laser transmitters having hundreds of microjoule pulse energies at hundreds of nanoseconds pulse durations and 10 to 50 kHz pulse repetition rates in the near future. These sources will be very useful for wind lidar applications, still, it has yet to be proven whether taking advantage of the telecommunications developments at 1.5 μm will significantly reduce the overall systems cost.

7.5.2. Direct Detection Lidar

This section begins with a discussion of the DDL system architecture for wind measurement applications. This is followed by a presentation of representative DDL systems in Sections 7.5.2.4 and 7.5.2.5.

7.5.2.1.   Direct Detection Lidar Architecture

As with coherent lidar systems, the detailed implementation of DDL system architectures can vary significantly. Even so, the basic functionality of most DDL systems is common as illustrated in Figure 7.44. Note the similarity with the key elements of the coherent lidar architecture shown in Figure 7.38. The system starts with a stable CW single-frequency MO. This oscillator is used to seed an amplifier or oscillator to create a higher-power pulsed beam. Although, in principle, all of the transmitter architectures discussed in Section 7.5.1 for coherent lidar systems can be used, the short-pulse injection-seeded Q-switched Nd:YAG oscillator is used as the primary transmitter in most DDL systems, due to its availability and reliability. The single-frequency output of the Q-switched oscillator can be used directly or its wavelength can be converted using nonlinear crystals. The nonlinear conversion efficiency is aided by the high-peak-power short pulse durations.

The transmitted beam is expanded by a telescope and directed into the atmosphere. Unlike coherent lidar systems, the DDL systems do not necessarily suffer a loss in efficiency by illuminating a spot in the target plane that is larger than the diffraction limit of the receiver aperture. Therefore, many

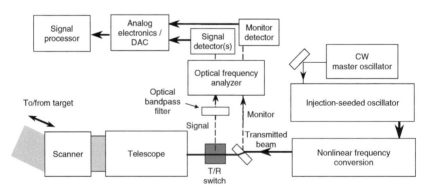

Figure 7.44 Simplified block diagram illustrating the key functionality of direct detection lidar systems. Most systems use a bistatic telescope arrangement (see text).

DDL systems use a separate and smaller-aperture telescope or a portion of the larger receiver telescope aperture for the transmit beams (i.e., bistatic). This also allows for a large number of speckles in the receiver and reduced signal fluctuation (speckle averaging). The signal backscattered from the atmosphere is collected by the receiver telescope and sent to an optical frequency analyzer. The purpose of the frequency analyzer is to convert an optical frequency change (i.e., Doppler shift) into a change in power or power spatial pattern, which is then, in turn, directly detected. Example of optical frequency analyzers utilized for DDL systems include: the transmission edge of a molecular absorption; the edge of a transmission fringe of an interferometer; or fringe pattern imaging the output of an optical interferometer. A small portion of the transmitted beam is also sent to the optical frequency analyzer allowing for the Doppler shift of the return signals to be corrected for any jitter or drift in the transmitted frequency. In most systems, this is achieved by using the monitor signal to provide feedback to the optical frequency analyzer, allowing its frequency response characteristic to be locked (or at least known) with respect to that of the transmitted light. The monitor signal can also serve to provide knowledge of the transmitted energy. The optical bandpass filter is used prior to the optical analyzer to eliminate the majority of the background light that would otherwise contaminate the measurement. This can consist of an interference filter and/or a combination of an interference filter and an etalon when narrower bandwidths are required.

The detector signals are processed using analog and digital electronics and processed with digital signal processing techniques to extract range-resolved estimates of the mean radial wind velocity. In the following subsection, the transmitter and receiver choices are described in greater detail.

7.5.2.1.1. Transmitter

Like coherent detection lidar, single-frequency transmitters with well-characterized spectra are needed in DDL, because in both approaches the goal is to measure the radial wind velocity to better than a few meters per second. Shorter pulse durations are often used in DDL, especially those using the molecular scatter, since, as described in Section 7.3.6, the spectral width of the molecular scatter, due to the Brownian motions of the molecules, is the order of a few GHz. As long as the spectral width of the transmitter is small compared to that of the scattered signal the impact on measurement precision is small. The short pulses also increase speckle diversity, if the signal processing range gate is larger than the pulse width. For example, for a 10 nsec (1.5 m) pulse with a 75 m range gate, the speckle diversity from the pulse width alone is approximately 50.

As mentioned above, DDL systems normally exploit Q-switched and injection-seeded lasers such as the Nd:YAG laser. It is relatively straightforward to operate the Nd:YAG laser in a single longitudinal mode by injection seeding the 1064 nm laser.[83] Typical pulse durations are 5 to 20 nsec resulting in transform-limited spectral bandwidths of 100 to 30 MHz FWHM at the fundamental 1064 nm wavelength. The output pulses at the very useful shorter wavelengths of 532, 355, or 266 nm are produced by frequency-doubling, -tripling, and -quadrupling the fundamental with appropriate non-linear crystals. In part, direct detection is used at these shorter wavelengths because implementing efficiently the coherent detection technique is generally more challenging than at longer, infrared, wavelengths due to the tight tolerances required on the surface quality of the optics and the alignment of the system. However, the most important feature at these shorter wavelengths is the dramatic increase of molecular scatter cross-section, making it possible to exploit the backscattered molecular signal in the upper troposphere and stratosphere, where aerosol concentration is very weak as compared to the planetary boundary layer and lower troposphere.

Until recently, the majority of DDL systems in use at major observatories for atmospheric research, have operated at the frequency-doubled wavelength of 532 nm (molecular and aerosol measurements) while some systems using aerosol scattering only have used the 1064 nm wavelength.

Increasingly, however, (as wind lidar systems are developed for applications where the user, or the general public may potentially come into close proximity to the lidar system, particularly the intense outgoing laser beam), the Nd:YAG frequency-tripled and -quadrupled wavelengths (355 and 266 nm) are being used. This is because the 355 and 266 nm wavelengths are significantly more eye-safe than at 532 or 1064 nm, since the near-UV radiation is not transmitted by the eye's cornea and lens. The strength of the molecular scattered signal is proportional to λ^{-4}, and is therefore stronger at 355 and 266 nm than at 532 nm, considerably offsetting the reduced number of transmitted photons available at the shorter wavelengths, as the result of the frequency conversion process. However, at 266 nm, ozone extinction is high, limiting it to short-range applications, like optical air data sensors for aircraft.

7.5.2.1.2. Receiver

The DDL receiver measures the mean Doppler frequency shift of the scattered light using an optical frequency analyzer. The velocity measurement accuracy and precision required for most wind measurement applications is of order 1 m/sec. Since $\delta\nu = (2/\lambda)\delta V$, the corresponding frequency-measurement precision required of the optical frequency analyzer in a DDL is, for example, 5.6 MHz at 355 nm wavelength. Written in terms of a wavelength measurement precision using $\lambda\nu = c$, this becomes $\delta\lambda = -(2\lambda/c)\delta V$, or 2.4 fm at 355 nm.

Simple prism or grating spectrometers cannot readily obtain this level of frequency or wavelength measurement precision, combined with high etendue (optical throughput — aperture solid angle product), therefore optical interferometers are typically used — or, in some cases, the edge of a narrow molecular absorption line.

In addition to the precision requirement, the velocity accuracy requirement of most applications demands that the difference frequency (i.e., the Doppler shift) between the transmitted and received light also be measured to an accuracy surpassing the precision level. The requirement for absolute frequency measurement accuracy by the DDL optical frequency analyzer is eliminated either by monitoring the transmitted laser pulse signal, as described above, or alternatively measuring the backscattered signal from a stationary or very low velocity target.

The most widely used interferometer for the optical spectrum analyzer in DDL systems is the Fabry–Perot etalon, and receivers using it are described in some detail in Section 7.5.2.2. There are alternate interferometer types under consideration for use in DDL systems, including the Fizeau and Mach–Zehnder. These later types of interferometers are also briefly discussed in the following subsections.

7.5.2.1.2.1. Signal Spectral Width and Velocity Precision: The design of the interferometer, or other optical frequency analyzer, has to be such that it properly accounts for the spectral width of the return signal. The spectral width of the signals to be analyzed is a convolution of the spectral width of the transmitted laser light and the additional broadening due to the differential motions of the scattering particles within the measurement volume. As discussed above, the lasers typically used in DDL systems have pulse durations of 5 to 20 nsec, resulting in FWHM spectral widths of 30 to 100 MHz for the transmitted light. In terms of the standard deviation of the spectrum this is 21 to 42 MHz. The velocity spread of the target is typically about 1 m/sec for the aerosol target and 250 to 300 m/sec for the N_2 molecular target (see Section 3.6, assuming T between approximately 210 and 300 K over the 20 to 0 km altitude range). Therefore, using a 5 to 10 nsec pulse at 355 nm, the spectral width (standard deviation) of the return is of order 20 to 40 MHz for aerosol scatter, limited by the transform limit of the 5 to 20 nsec pulse duration, and of order 1.4 to 1.7 GHz (3.3 to 4 GHz FWHM) for molecular scatter.

As discussed in Section 7.3.6, assuming a low noise detector and little or no background light, the statistical lower bound for the velocity precision of a quantum-limited system is $\sigma_V = \lambda \delta\nu / 2 N_s^{1/2}$. For the molecular signal, the spectral width is dominated by the Brownian motion of the molecules. Therefore, $\delta\nu \sim (2/\lambda)\delta V_T$, resulting in $\sigma_V = \delta V_T / N_s^{1/2}$, where δV_T is the thermal velocity spread of the molecules — that is, there is no wavelength dependence on the number of detected photons required to obtain a given velocity measurement precision when the target spread of velocities dominates the pulse-width spread. Due to this large velocity spread, the primary challenge of the molecular DDL technique is to measure the mean frequency of the signal to a precision and accuracy better than one part in 100 of the RMS velocity width. The statistical bound shows that order of 100,000 *detected* photons are required to achieve a velocity precision standard deviation of about 1 m/sec. Given realistic lidar system efficiencies, background noise, and detector noise, the number of signal photons required at the receiver telescope must be somewhat higher, typically 10 to 20 times higher in a well-designed system. Fortunately, the molecular backscatter coefficient from the atmosphere is sufficient, especially at UV wavelengths, to supply these numbers of signal photons with achievable transmitter energy-aperture area products. The energy referred to here is the total energy transmitted over the measurement period. The total signal can be collected over many pulses as long as the required number of photons is detected in the required measurement update time and the signal counts dominate the background light and noise counts.

7.5.2.2. Direct Detection Lidar Receiver Details

In this section, we describe in some detail the two DDL optical receiver types primarily being utilized today — the double-edge and fringe-imaging interferometers. These are illustrated schematically and compared in Figure 7.45.[241]

As mentioned previously, the Fabry–Perot etalon[197,242] has been widely used as the device of choice for providing the frequency analysis for direct detection wind lidar systems. It is regarded as an efficient user of incident light, based on its high etendue[243] (product of its acceptance solid angle and the aperture area). In particular, if the defect finesse (due to misalignment, absorption, and aberration) is much higher than the reflective finesse, the peak in-band transmission can exceed 90%, since the transmission and scattering losses within modern high-performance multilayer dielectric coatings are very small. The surface defects of state-of-the-art optics and coatings do not limit this.[244]

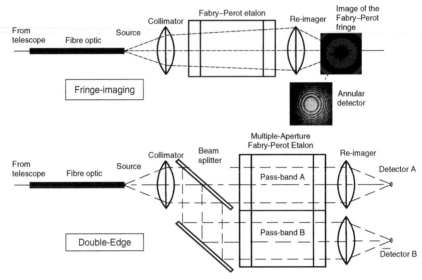

Figure 7.45 The fringe-imaging and double-edge detection methods for direct detection are shown conceptually.

With different (but appropriate) choices of the etalon performance (free spectral range, finesse, and resolution) the fringe-imaging and the double-edge techniques can be used to implement direct detection for either aerosol or the molecular backscattered signals. For either the fringe-imaging or the double-edge technique, the key objective of the detection system is the determination of the centroid of the spectrum of the backscattered signal. For the fringe-imaging technique, the centroid of the spectrum measured at a number of discrete points over the entire spectrum is determined, while for the double-edge technique, the two detection channels correspond approximately to the intensities of two spectral regions of very similar width (ideally identical) located at equal spectral intervals (ideally) about the laser line. This is described in greater detail in the following section. Note: in this section we refer to the two-channel receiver in the figure above, as a double-edge receiver to match the vernacular used by the lidar community. The two-channel theory presented in Section 7.4.2.4 is the appropriate approximate theory of operation.

7.5.2.2.1. Fringe-Imaging Receiver

In the fringe-imaging technique, the slightly divergent signal light is incident on the Fabry–Perot etalon. Specific frequencies (wavelengths) are only transmitted through the etalon at angles, θ, that satisfy the resonance condition $2d \cos \theta = m\lambda$, where d is the spacing between the etalon plates and m is an integer. Therefore, if the spectral width of the incident light is smaller than the free spectral range (FSR) of the etalon (given by $\text{FSR} = c/2d$, with c the speed of light) the output of the fully illuminated etalon imaged at infinity will be a series of concentric rings. These rings are spaced according to $\sqrt{n}$ where n is the order of interference relative to the central order.

The frequency resolution of the etalon, $d\nu$, is given by $d\nu = \text{FSR}/\mathfrak{I}$, where $\mathfrak{I}$ is the finesse of the etalon. As long as the resolution of the etalon is much smaller than the frequency spread of the incident light, the transmission fringe position and shape reflect the mean frequency and spectrum of the incident light. Detection and analysis of the transmission fringe image, that is, "fringe imaging" allows the Doppler shift of the signal to be determined.

For optimized performance using either the molecular or the aerosol signal, the fringe-imaging technique performs a full spectral analysis of the backscattered signal, placing a number of detection channels over the full width of the molecular spectrum or the aerosol signal, respectively. The appropriate characteristics of the Fabry–Perot etalon and the detector must be carefully chosen in order to optimize the performance for the specific measurement requirements. The detector has to be capable of imaging the fringe pattern and range gating. The multichannel PMT and the CCD have both been successfully exploited as appropriate detectors for fringe-imaging DDL systems. Specific approaches to allow efficient detection of the concentric ring pattern are described later.

7.5.2.2.2. Double-Edge Receiver

Figure 7.46, adapted from Gentry et al.,[87,111] shows schematically the combined molecular and aerosol spectrum. The relative amplitudes of the aerosol and the molecular signal

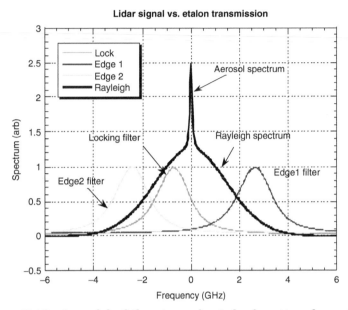

Figure 7.46 A model of the atmospheric backscattered spectrum and the three etalon pass-bands (Edge 1, Edge 2, and locking filter) of the molecular double-edge detection system. (From Gentry, B.M., Chen, H., and Li, S.X., *Opt. Lett.*, 25, 1231–1233, 2000. With permission.)

will vary by several orders of magnitude, depending on the nature of the atmosphere (dependent on altitude and meteorology). As described above, in the case of the fringe-imaging technique, the spectrum is normally sampled (and rangegated) at regular intervals of wavelength over the full extent of the aerosol or molecular-scattered spectrum (~3.3 to 4 GHz FWHM).

In the double-edge technique, the Fabry–Perot etalon(s) are used as filters to isolate two matched spectral pass-bands corresponding to the two edges of the molecular-broadened spectrum, or the aerosol spectrum, respectively. The Fabry–Perot etalon filters for the double-edge technique can be implemented as two distinct etalons. Alternatively, it is possible to build these as separate subapertures on a single etalon structure. In the latter case, the step in optical path difference is implemented as a layer of silica applied onto the etalon surface prior to deposition of the multilayer dielectric coating of the etalon itself.

Ideally, neither the "Edge 1" nor the "Edge 2" filter pass-bands should transmit the aerosol signal. The laser spectrum backscattered by the atmosphere always includes both molecular and aerosol components, ignoring for the moment the Raman scattering processes — since that involves wavelength shifts that considerably exceed the range of Doppler-broadened and Doppler-shifted wavelength changes. In the absence of the aerosol-scattered signal, the optimization of the two band-passes would be relatively straightforward. However, in practice, as shown by Garnier and Chanin,[245] (and as discussed briefly in Section 7.4.2) the increasing aerosol signal, where present, will cause the sensitivity of the differential signal observed in the two band-pass channels (to wind velocity) to degrade.

Since, in a double-edge system, the amplitude of the aerosol signal cannot be determined readily, the sensitivity to Doppler shift of a system optimized purely for sensitivity to the molecular signal[79] is not fully determined. Garnier and Chanin,[245] however, have shown that by subtly changing the FWHM and separation of the two channels (Edge 1 and Edge 2), the response to an unknown aerosol signal could be minimized. In practice, all current double-edge systems incorporate an optimized choice of interedge separation and appropriate finesse of the analysis etalons. An additional channel — the "locking filter channel" (see Figure 7.46) — has been introduced[113–115] in order to ensure the optimum balancing of the Edge 1 and Edge 2 filter channels about the zero Dopple-shifted laser signal.

For the double-edge method, several relatively simple detectors are available, for example, the PMT, APD, or CCD. The choice may depend on the wavelength, or on the criticality of detector noise or quantum efficiency for the application. These are discussed in greater detail in a later subsection.

7.5.2.2.3. *Spectral Resolution Requirements*

The spectral resolution requirements for the molecular signal are relatively similar for both fringe-imaging and double-edge techniques. For $\lambda = 355\,\text{nm}$, the spectral resolution requirements can be met by a Fabry–Perot etalon with a plate separation, d, of order 12.5 mm, and with an overall finesse of order 6 to 8. The resulting FSR (FSR $= c/2d$) of $\sim$12 GHz is large enough to ensure that there is no spectral wrapping (or aliasing) of the molecular signal through the different orders of the etalon. The finesse ($\Im$) of 7 to 8 results in a resolution, $d\nu =$ FSR/$\Im$ or approximately 1.6 GHz FWHM, which sufficiently resolves the 3.3 to 4 GHz FWHM molecular spectrum. The Fabry–Perot etalon design requirements for the double-edge technique are somewhat more critical than for the fringe-imaging technique in order to accurately meet the need to minimize the response to the highly variable aerosol signal. The process of optimization of the etalon design for the double-edge technique has been discussed in detail by Garnier and Chanin[103] and by Flesia and Korb.[113]

If the aerosol signal is to be the target of the Doppler wind lidar, then significantly higher spectral resolution is required to properly resolve the narrow aerosol signal spectral width. The resolution requirements (at $\lambda = 355$ nm) can, however, be met (again for double-edge or fringe-imaging methods) by a Fabry–Perot etalon of plate separation, d, of order 70 mm and finesse (($\Im$) of order 40 or higher. This 2.1 GHz FSR etalon provides a resolution $d\nu$ of about 50 MHz.

The fringe-imaging system depends on illumination of the etalon using a beam that has a modest angular distribution, corresponding to approximately 1 FSR. For a Fabry–Perot etalon, a change of $\delta\theta$ in the incidence angle results in a change in resonance frequency by an amount $\delta\nu = \nu \tan\theta \, \delta\theta$. At normal incidence, this expression can be used to show that the illumination solid angle, $\delta\Omega$, required to cover a spectrum of resonance frequencies of width $\delta\nu$ is given by

$$\delta\Omega = \pi\delta\theta^2 = 2\pi\delta\nu/\nu \qquad (7.163)$$

where $\delta\theta$ is the half angle of the illumination pattern. For example, if a 355-nm wavelength molecular-scatter fringe imaging system with a 12 GHz FSR etalon is used, a half angle of 5.3 mrad would use the full FSR of the etalon. Illuminating the entire FSR ensures that the signal will transmit at some angles no matter where it resides in absolute frequency. For aerosol fringe imaging systems, using a Fabry–Perot etalon with a 2.1 GHz FSR, for example, the half-angle would be approximately 2.2 mrad.

Double-edge systems, which operate on the central fringe at near normal incidence, require a higher level of beam collimation for optimal performance. In this case, the same equation applies, but $\delta\nu$ is the required etalon resolution ($\delta\nu = \text{FSR}/3$) so that the signal effectively transmits through the single etalon fringe. Due to the large spectral width of the molecular signal, the resolution required of the etalon is therefore relatively low and the angular acceptance of the interferometer is therefore relatively large. For example, the acceptance angle of the etalon described above for the molecular signal, with $\delta\nu \sim 1.6$ GHz at 355 nm, is $\delta\theta \approx 1.9$ mrad. The acceptance angle for the higher resolution etalon required for the aerosol signal, with $\delta\nu \approx 50$ MHz, is $\delta\theta \approx 0.34$ mrad.

The etendue of the beam, which is the product of the beam solid angle divergence and the beam area, $A\delta\Omega$, must be conserved in optical system. Smaller beam sizes have larger divergence angles than larger beam sizes of fixed etendue. The etalon aperture must be large enough so that the divergence angle of the received beam is contained within the acceptance angle of the etalon.

As an example, consider a DDL with a 1 m diameter aperture and a half-angle transmitted beam divergence of 50 μrad. In order to meet the acceptance angle requirements using the etalon discussed above for the aerosol edge technique at 355 nm, the beam (and the etalon aperture) must be about 15 cm in diameter. A smaller etalon aperture could be used if the divergence of the transmitted beam were lowered. For example, in order to use a 4 cm diameter etalon aperture the half-angle divergence of the beam transmitted must be reduced from 50 to 13 μrad (which is still 50 times the diffraction limit of the 1 m aperture at 355 nm). For the molecular edge etalon the beam can be decreased to about 2.5 cm in diameter using the 50 μrad beam divergence due to the significantly lower resolution required and resulting larger etalon acceptance angle.

For measurement of either the molecular or the aerosol signal, the fabrication of an optimized Fabry–Perot etalon is within current state-of-the-art. Table 7.6 summarizes the characteristics of the Fabry–Perot etalons that will accomplish the spectral analysis requirements using the fringe-imaging and double-edge techniques for the molecular signal and using the fringe-imaging technique for the aerosol signals at 355 nm. In order to provide precise control and operational flexibility, including the capability to track small changes of

Table 7.6 Summary of Representative Optical Interferometer Requirements for Aerosol (Fringe-Imaging) and Molecular (Double-Edge or Fringe-Imaging) at $\lambda = 355$ nm

Requirements	Aerosol Channel	Molecular Channel
Technique	Fringe imaging	Fringe image (FI) or double edge (DE)
Number of subpupils	1	1 (FI); 3 (DE)
Working aperture of etalon subpupil	40 mm	40 mm
Acceptance angle	± 2.2 mrad for FI	$\pm 1.9/5.3$ mrad (DE/FI)
Optical path difference	70 mm	12.5 mm
Free spectral range	2.1 GHz	12.0 GHz
Plate flatness before Coating (at $\lambda = 355$ nm)	$<\lambda/150$ rms	$<\lambda/50$ rms
Plate flatness after Coating (at $\lambda = 355$ nm)	$<\lambda/120$ rms	$<\lambda/40$ rms
Plate relative alignment (at $\lambda = 355$ nm)	<0.1 μrad	<0.5 μrad
Mirror reflectivity /reflectivity finesse	$\sim 95\%$; $F_{REF} = 61$	$\sim 75\%$; $F_{REF} = 11$
Overall finesse (including defect finesse)	41	7.7
Transmission fringe FWHM resolution	52 MHz	1.6 GHz
Overall T of etalon at transmission peaks	80%	$>87\%$
Edge 1 step height	–	0 nm
Locking step height	–	27 ± 1 nm (for DE) $\delta\nu$ from Edge 1 of 1.8 GHz
Edge 2 step height	–	77 ± 1 nm (for DE) $\delta\nu$ from Edge 1 of 5.2 GHz

the laser wavelength, etalons typically employ piezoelectric crystals and a capacitance-stabilization technique[95] to keep the plates aligned and locked to the transmitted frequency. In some systems, the receiver is tuned, rather than locked, to the mean wavelength to a small fraction of the FSR, and a "suitable calibration strategy" is utilized to ensure that the zero velocity wavelength of the laser is calibrated on an adequately regular and representative basis.

7.5.2.2.4. Comparison of Double-Edge and Fringe-Imaging Systems

Although differing in details, the fringe-imaging and the double-edge techniques share requirements for stability of the outgoing laser wavelength, the stability of all spectroscopic, optical and detection elements within the entire receiver chain, and for careful calibration of the various optical, detector and electronic elements. Details of the optimization of the fringe-imaging and double-edge detection systems are discussed, for example, by Flesia and Korb[113] and McKay.[246,247]

It has been shown by McKay[246,247] and McGill et al.[81] that for equally optimized systems, there is no inherent advantage between the double-edge and fringe-imaging systems. In purely analytical terms, the double-edge can be considered as a partly degenerate fringe-imager, where the number of spectral intervals has been reduced to two, more or less balanced about the center wavelength, as indicated in Figure 7.46. As will be discussed later, there may be practical aspects such as the possibility of conserving the available incident signal by recycling between the components or channels of the optical system that may offer advantages in practical systems.

7.5.2.2.5. Alternate Interferometers

As mentioned earlier, alternate interferometers such as the Fizeau and the Mach–Zehnder have been considered for DDL wind measurement applications. These interferometers have the advantage when used for the fringe imaging technique in that they can directly present a linear fringe pattern at the detector, which is more easily read by a linear or rectangular array detector, such as a CCD. The Fizeau etalon is essentially a Fabry–Perot etalon with a small wedge angle between the plates. The Mach–Zehnder interferometer has been investigated for DDL application by Bruneau and Pelon[125,126] It is a two-optical-path device that can be used in either the double-edge or fringe-imaging configurations. In the fringe-imaging configuration, a small angular misalignment of one of the beams from the two optical paths results in a linear fringe interference pattern. Since the Fizeau or Mach–Zehnder interferometers have not been realized in an operational DDL they will not be discussed further in this text.

7.5.2.3. Detectors and Detection Efficiency for Direct Detection Doppler Lidar

The ideal detector for any lidar would have very high quantum efficiency, very low intrinsic and read noise, and would be infinitely flexible in terms of range gating and reading the signal from the detector — permitting flexible "postprocessing" of the lidar data.

7.5.2.3.1. Detectors for Double-Edge Systems

The information presented to the detector in an edge detection system is the image of the small on-axis solid angle corresponding to the central on-axis fringe of the Fabry–Perot etalon with the necessary spectral FWHM. A suitable detector will be one that has high sensitivity (high quantum efficiency), low noise, the capability for photon counting, or analog readout, depending on the intensity of the available signal, and which can be "time-gated" to provide altitude or range-resolved information.

The conventional PMT, the APD, and the CCD are among several that have been used successfully, depending on the spectral region of the wind lidar. The PMT is a device that is essentially noise-free when used in photon-counting mode. After detection of the signal photon within the photocathode, the PMT uses either a conventional dynode structure or a microchannel plate stack to obtain very high electron gain, presenting a charge packet of order 10^6 electrons at the anode corresponding to each detected photon. With well-designed external electronic circuitry (preamplifier and discriminator) and using photocathode cooling to reduce thermionic emission, the intrinsic noise per range gate per laser pulse is negligible. Due to the negligible read-out and electronic noise, the PMT signal may be postintegrated with complete flexibility, leading to the PMT being widely used as a detector of choice, particularly at 355 and 532 nm. Its drawback is the modest quantum efficiency of the photocathode of the device, normally limited to values of order 35% or less, depending on the spectral region.

7.5.2.3.2. Detectors for Fringe-Imaging Direct Detection

For the fringe-imaging technique, by its nature, a multi-element imaging detector is required. Further, the detector must also be capable of being used in a time-gated mode in order to provide the essential range-resolved sampling of the backscattered signal. There is the additional subtle difficulty that the Fabry–Perot etalon presents its spectral information as concentric circular fringes. Presently, no off-the-shelf commercial imaging detector is available that meets all of the requirements. Therefore, fringe-imaging lidar systems have proven relatively difficult to implement. Several fringe-imaging detector implementations are described in the following.

7.5.2.3.2.1. Multichannel Detectors: The first approach follows concepts used earlier with, for example, the high-resolution Doppler interferometer (HRDI) instrument of Upper Atmospheric Research Satellite (UARS).[97] At ALOMAR, and for a number of other direct detection fringe-imaging Doppler wind lidar, an imaging photomultiplier detector (IPD), incorporating a 24-channel concentric-ring anode read-out designed to match the fringe pattern presented by the Fabry–Perot etalon, has been used.[106,248]

The device uses a stack of microchannel plates to achieve the high electronic gain (order 10^6) necessary to use photon-counting electronics and to obtain essentially noise-free signal detection. The 24-channel photon counting and data acquisition system provides 512 range gates of 100 or 200 m resolution, while a 24-channel analog system provides range resolution down to the 1 to 2 m equivalent pulse length of the laser itself.

The primary advantage of this class of photomultiplier device is that the electron gain process of a device based on micro-channel plates is essentially noise-free when the device is used in photon-counting mode. The downside is that, generally, the quantum efficiency of the photocathode is usually significantly lower than that obtainable with devices such as CCDs. However, due to the negligible read-out noise and very low thermionic emission of a properly used device, it is possible to postprocess data with complete operational flexibility. It is quite normal, with wind lidar using the PMT class of imaging detector, to integrate data obtained over a period of an hour, in order to obtain wind data to 40 km altitude or above.[106]

7.5.2.3.2.2. Circle-to-Line Imaging Optic: An alternative approach, permitting use of linear or rectangular-array detectors, uses a circle to line imaging optic (CLIO) to convert the circular fringes formed by a Fabry–Perot etalon into a linear pattern of spots. A conventional linear array detector, such as a CCD, can then be used to read the linear fringe pattern. The CCD quantum efficiency, ranging from 40 to 90%, is generally higher than that of most of the photocathode materials used in photomultiplier type devices.

The choice between a CCD detector and an imaging photomultiplier depends on a number of issues such as the quantum efficiency of either detector at the specific wavelength of a given lidar and the overall design of the optical and electronics systems required to implement range gating (and multilaser pulse data accumulation on the detector itself in the case of the CCD).

A detailed description is not included in this text, however, the CLIO is described in detail by Hays.[249,250] As shown in Figure 7.47, the linear interference pattern is produced by transforming a 90° sector of the circular Fabry–Perot fringe pattern generated by a Fabry–Perot interferometer into a series of line segments formed on the axis of a 90° full-angle internally reflecting cone.[250,251] The cone only acts to convert light from the original 90° sector onto the series of line segments, and light from outside this region is not used.

7.5.2.3.3. Improving the Efficiency of the Interferometers

When Fabry–Perot etalons are used for either the fringe-imaging or double-edge techniques, there is a significant amount of loss since only a portion of the incident light signal is transmitted through the interferometer. Several techniques that have been developed in recent years to improve the frac-

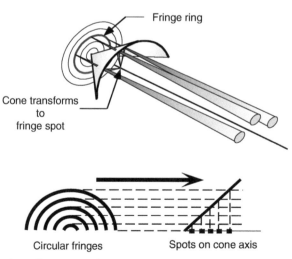

Figure 7.47 Conceptual ray trace showing the CLIO concept. (After Nardell, C., Hays, P.B., Pavlich, J., Dehring, M., and Sypitkowski, G. *Proceedings of the International Symposium on Optical Science and Technology*, SPIE 4484, pp. 36–50, San Diego, CA, July 30–31, 2001. With permission.)

tion of the incident light that contributes to the measurement are described in the following subsections.

7.5.2.3.3.1. Fringe-Imaging   Efficiency   Improvement: In the fringe-imaging technique, if light is incident on the Fabry–Perot etalon over a range of angles corresponding to one complete order (FSR), then the fraction of the light incident on the etalon that is transmitted, T_E, and the amount that is reflected, R_E, is given by

$$\eta_{FP} = T_E = \left[\frac{1-R}{1+R}\right] \quad \text{and} \quad R_E = \left[\frac{2R}{1+R}\right] \qquad (7.164)$$

where R is the reflectivity of the etalon (see Equation 7.119). This means that if $R = 0.90$, then 95% of the incident light is reflected from the etalon. In the early direct detection wind lidar systems, this very high proportion of the signal was totally lost to the subsequent signal detection stages and was thus wasted, significantly reducing system efficiency.

Three primary approaches have been developed to reduce the loss. In the first approach, known as *fractional fringe illumination*, the etalon is illuminated by a solid angle corresponding to only a fraction of the full FSR. In the second approach, known as *interferometer photon recycling*, the portion of incident light reflected by the etalon is collected and recycled with appropriate additional optics to reilluminate the etalon multiple times at angles that will be efficiently transmitted by the etalon. This approach can result in an increase of as much as a factor of three in overall transmission. In the third approach, defined as *channel photon recycling*, which is used in systems that have separate channels (etalons) for the aerosol and molecular signal analysis, the light reflected from one of the etalon channels is used as input into the second etalon channel. Some combination of all three approaches can lead to efficiency improvements of 3 to 5 over systems that use none of the approaches. Each of the approaches are described in more detail below.

Fractional Fringe Illumination: In this approach, the etalon is illuminated by a solid angle corresponding to only a fraction of the full FSR,[102] which can result in a significantly higher fraction of the signal being transmitted by the etalon. Given practical issues and realistic optical fabrication limitations — such as the need to be able to measure the Doppler shift over a representative range of line-of-sight wind velocities, a gain of as much as a factor of 2.5 could be achieved by this partial fringe illumination method.

Interferometer Photon Recycling: The interferometer photon recycling approach was taken in the development of the Michigan Aerospace Corporation (MAC) "GroundWinds" system, which is described in Section 7.5.2.5. The recycling implementation is illustrated in the figures below. In the MAC recycling approach,[251] the input fiber is placed just off the optical axis of the etalon (shown below the optical axis in Figure 7.48). In part, this is required to avoid a large loss in the CLIO imaging system, which operates with a 90° sector of the full fringe pattern. The reflected (potentially wasted) input signal is directed above the optical axis. A second fiber placed symmetrically above the optical axis can be used to collect this reflected signal. In turn, the output of this fiber can be used to illuminate the etalon from below the optical axis. The reflected (potentially wasted) signal from this second fiber is also reflected above the optical axis. A third fiber can be used to further recycle the signal. The residual light can then be injected backward through the fiber recycler so that

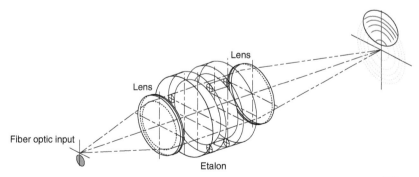

Figure 7.48 The optical configuration used by GroundWinds[251] to exploit the Fabry–Perot etalon.

illumination is achieved from above the etalon optical axis, resulting in a total of six passes (five recycles) through the etalon, as illustrated in Figure 7.49 and Figure 7.50. In addition to being a convenient way to route the light, the fibers also serve to spatially scramble the light so that the transmission fringes are re-illuminated on each pass.

Channel Photon Recycling: With the exception of DDL measurements of winds in the stratosphere, there is a strong case for using separate optimized channels to measure the aerosol and molecular signals. The molecular signal, being ever-present, ensures that the wind can always be measured, even in very clean air, while the aerosol signal, even when relatively weak, gives an improved measurement accuracy, due to the narrower signal spectrum.

Consider the analysis of the spectrum of the combined molecular and aerosol backscattered signals by two separate (and optimized) receiver channels: one for the molecular signal and a second for the aerosol signal. In a zero-order approach, the total incoming signal would be divided by a 50% beam splitter. This beam splitter would feed 50% of the available signal to each of the separate optimized molecular and aerosol detection channels. However, since the aerosol signal cannot be used efficiently by the molecular receiver channel and vice versa (for either the fringe-imaging or double-edge techniques), this approach can exploit only 50% of the available signal for either channel, and thus at least 50% of the total available signal is wasted.

An alternate approach that is more efficient is to use all the return light to illuminate the aerosol etalon, which will transmit only a small fraction of the incident light, as defined by Equation (7.164). The remaining reflected light can be collected and used to illuminate the molecular etalon. The reflected light can be separated from the incident light by either spatial separation (like that described above for the interferometer photon recycling) or polarization separation using a polarizer and a quarter-wave plate, which takes advantage of the polarization of the signal. This recycling of the reflected light from the aerosol channel into the molecular

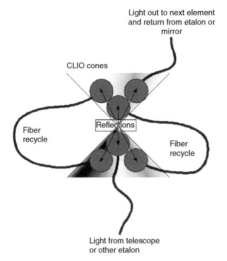

Light out to next element
and return from etalon or
mirror

CLIO cones

Fiber
recycle

Reflections

Fiber
recycle

Light from telescope
or other etalon

Figure 7.49 Schematic of the fiber optic ferule used for light recycling. In this implementation the light is recycled twice through the instrument for a total of three illuminations of the etalon from below the optical axis. By reinjection of the residual signal backwards through the fiber recycler using a mirror, additional three illuminations are achieved from above the etalon optical axis.

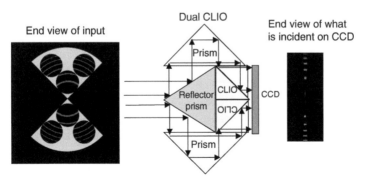

End view of input

Dual CLIO

Prism

End view of what
is incident on CCD

Reflector
prism

CLIO

CLIO

CCD

Prism

Figure 7.50 The "end view of input" shows how the Fabry–Perot ring pattern is illuminated with the fiber optic recycler when the recycler is used in both "forward" and reinjected "reverse" illumination directions. The relative intensity is not shown to scale. The "Incident on CCD" on the right shows the ring pattern after it has been optically transformed by a special dual CLIO optic on the detector.

channel has been utilized in demonstrated systems[250,251] and is proposed for others. By using a multimode fiber to collect and route the light reflected from the aerosol etalon, the spatial structure of the reflected light is rescrambled (eliminating the dark rings that coincide with the transmission rings of the aerosol etalon) prior to illumination of the molecular etalon.

It is also possible to use a third high-resolution etalon to spectrally separate the two signals, transmitting the narrow aerosol signal, with a high degree of efficiency, and reflecting the majority of the broad molecular signal. The spectral separation of the two components, prior to their spectral analysis, would not disturb the separate and efficient analysis of the molecular signal by one dedicated channel, and the aerosol signal by a separate dedicated channel. The benefit of this approach, compared to that described in the previous paragraph, is usually not significant enough to warrant adding the third high-resolution spectral separation etalon (which is physically large) to the system.

7.5.2.3.3.2. Double-Edge Detection Efficiency Improvement: In early edge detection systems, a beam splitter has been used to divide the signal between the different detection channels, which resulted in at least a 50% loss when operating at the 50% transmission point. An alternative concept, with improved efficiency, is to use a polarizing beam-splitter and a pair of quarter-wave plate, to exploit the linear polarization of the backscattered lidar signal.[252] The unused signal back-reflected from the aerosol etalon can be redirected efficiently to the two edge channels of the etalon of an optimized double-edge molecular detection system.

An even more efficient concept is shown in the right panel of Figure 7.51. In this arrangement, the polarized signal is first passed through the polarizer and is incident on the molecular Edge 1 channel. The light reflected from the Edge 1 channel will contain the entire spectrum of the Edge 2 signal as well as the entire spectrum of the aerosol signal. The reflected signal is then incident on the molecular Edge 2 channel. The reflected light from the Edge 2 channel still

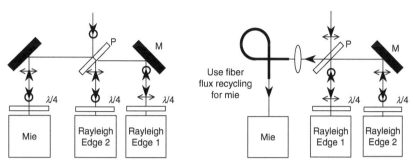

Figure 7.51 Photon recycling concepts for a combined molecular and aerosol receiver exploiting the polarization of the signal. Signal polarizations in the various propagation directions are indicated by the arrows (p polarized) and circles (s polarized). A double pass of the signal through a quarterwave plate rotates the polarization by 90°. (Left) The signal is first incident on the aerosol channel. The reflection from the aerosol (Mie) channel is then incident on the molecular (Rayleigh) Edge 1 channel. Finally, the reflection from the Edge 1 channel is incident on the Edge 2 channel. (Right) An even more efficient arrangement, whereby the signal is first incident on the molecular Edge 1 channel. The reflection from the Edge 1 channel is then incident on the Edge 2 channel. The reflection from Edge 2 is then fiber coupled to the aerosol channel whose reflection is recycled through the aerosol channel using the interferometer recycling approach described earlier.

contains the entire spectrum of the aerosol signal. Finally, the light is incident upon the aerosol channel. The signal can then be recycled in the aerosol channel using the interferometer photon recycling approach discussed earlier. Care must be taken to accurately calibrate the receiver when these recycling approaches are utilized.

There are several other methods, exploiting polarizing beam splitters and other relatively simple optical components, by which all the light usable by the aerosol channel can be directed to and used by that channel, while essentially 100% of the available signal usable by molecular Edge 1 and Edge 2 channels can be directed to those channels, with negligible loss or cross-contamination of the signals. The overall efficiency of a carefully designed system combining the analysis of the molecular and aerosol signals can be very high. In addition to the work by the University of Michigan and MAC,[250,251] a number of these concepts have been developed during the detailed studies of Doppler wind lidar systems for the European Space Agency.[253] This has resulted in ESA's adopting the AEOLUS Mission[123] (formerly called ALADIN and the Atmospheric Dynamics Mission) to develop and demonstrate the capability of a near-UV Doppler wind lidar for global wind measurements of the troposphere and low to middle stratosphere.

7.5.2.4. A Representative Double-Edge Wind Lidar System

The direct detection Doppler wind lidar system developed at the NASA Goddard Space Flight Center (Goddard Lidar Observatory for Wind — GLOW, Gentry[111]) is representative of the double-edge technique.[79,114,115,245] The GLOW lidar system is shown schematically in Figure 7.52 and is integrated into a modified van, providing mobility for field operations. The laser, a 45 cm aperture (Dall–Kirkham) receiving telescope, and the beam pointing optics are mounted on an optical bench that is bolted to the truck frame. The outgoing laser beam is coaxial with the receiving telescope. A 45 cm aperture scanner, providing full hemispherical pointing, is mounted on the roof to allow access to the atmosphere. The light collected by the telescope is coupled directly into a fiber optic cable. The input of the fiber serves as the field-stop for the system and delivers the signal to the Doppler receivers.

The laser is an injection-seeded, flash lamp-pumped Nd:YAG laser (10 Hz pulse repetition rate). The pulse length is 15 nsec and produces a spectral FWHM of 40 MHz at the fundamental wavelength of 1064 nm. The laser also includes second and third harmonic generators to produce output pulses at 532 or 355 nm. The output energy of the laser is 120 mJ/pulse at $\lambda = 1064$ nm for the aerosol measurements. For the molecular measurements, an additional optical amplifier stage provides 70 mJ/pulse at $\lambda = 355$ nm.

Two receivers are used, one for aerosol backscatter wind measurements ($\lambda = 1064$ nm) and one optimized for molecular

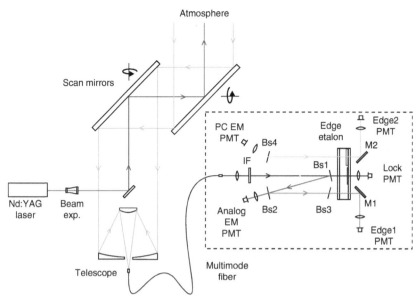

Figure 7.52 Optical layout of the GLOW lidar system. The molecular receiver is shown in the dashed box. (From Gentry, B.M., Chen, H., and Li, S.X., *Opt. Lett.*, 25, 1231–1233, 2000. With permission.)

backscatter ($\lambda = 355$ nm). The receivers are mounted on separate base-plates coupled to the receiving telescope via multimode fiber optic cables.

7.5.2.4.1. The GLOW Molecular Receiver

The GLOW molecular double-edge receiver was built at NASA Goddard Space Flight Center as an engineering prototype of Zephyr, a direct detection Doppler lidar demonstrator proposed for a Shuttle mission.[107,111] Flesia and Korb[113] have described the general principles of the design, conceptually similar to the molecular Doppler lidar system successfully operated at the Observatoire Haute Provence (OHP) in France.[79,245]

In the receiver, the fiber optic from the telescope is coupled to the collimator to produce a collimated beam of 35 mm diameter. Beam-splitters divide this beam into a total of five channels, three of which are directed along parallel paths through a Fabry–Perot etalon filter. The "edge" channels transmitted through the etalon use PMTs operated in the photon-counting mode. The third etalon channel (reference channel) uses a PMT operated in analog mode. The other two channels serve as energy monitor channels. One has a photon-counting PMT and the other has an analog mode PMT. The energy monitor channels provide intensity normalization of the respective etalon channels during calibration.

A capacatively stabilized, piezo-electrically tunable Fabry–Perot etalon (with three subapertures of 37 mm diameter) is used to create the high spectral resolution edge filters. The spectral band-passes of the three subapertures are offset from one another to provide the optimum interchannel spectral separation. The magnitude of the offset is created by small "step" coatings of silicon monoxide deposited over two of the subapertures on one of the etalon plates, prior to deposition of the reflective etalon coating. This produces the double-edge configuration described in Flesia and Korb[113] along with a third intermediate channel (sometimes called the locking channel), used to sample the outgoing laser frequency as a reference. A calibration scan of the etalon transmission, scanning the three etalon band-passes simultaneously through approximately one FSR (12 GHz), provides precise knowledge of the etalon transmission and the spectral separation of the edge channels, as is required to determine the winds from the lidar observations.

A simulated backscattered signal spectrum is shown in Figure 7.46. The characteristic spectral signature of the narrow aerosol spike is superimposed on the thermally broadened molecular backscattered spectrum. The two-edge filter channels (labeled Edge 1 and Edge 2) are located in the wings of the molecular broadened spectrum at a position which has equal sensitivity to Doppler shifts for either the molecular or the aerosol backscattered signal. This makes the wind measurement essentially insensitive to the aerosol/molecular backscatter ratio.

The reference channel (the "locking filter") is located such that the outgoing laser frequency appears on the edge of the "locking" fringe. A measurement of the outgoing laser frequency is made as a reference to remove the short-term frequency jitter and drift of the laser and also can be used to actively "lock" the etalon to the laser frequency in a servo loop in order to maintain the symmetric arrangement of the two "Edge" filters about the outgoing laser frequency.

Photon-counting PMTs provide high detection sensitivity in the upper troposphere and stratosphere where the detected signal levels are small. Analog PMT signals are sampled with a boxcar integrator and the data are stored for every 10 laser pulses. The photon-counted signals are binned in a multichannel scalar and integrated for a preselectable number of laser shots prior to storage.

A narrowband interference filter is used to restrict the broadband solar background (at $\lambda = 355\,nm$) during daytime operation. The 25 mm aperture of this filter is smaller than the receiver design beam diameter of 37 mm. This limits the effective telescope aperture for daytime measurements to about 30 cm.

7.5.2.4.2. The GLOW Aerosol Receiver for $\lambda = 1064\,nm$

The GLOW system also contains a double-edge receiver for wind measurements at $\lambda = 1064\,nm$, using the aerosol-scattered signal. Full technical details, which will not be described here, can be found in References 87 and 111. Example wind measurement data taken with the GLOW instrument are presented in Section 7.6.2 (see Figure 7.60).

7.5.2.5. A Representative Fringe-Imaging Doppler Wind Lidar System

The Doppler wind lidar system developed by Michigan Aerospace Corporation[85,251] is a representative example of a fringe-imaging DDL. The key elements of the system are illustrated in Figure 7.53, including the laser, the transmitting telescope, the receiving telescope, the interferometer, and the instrument control and data processing system.

The laser beam is transmitted via a beam expander, reducing the divergence angle of the emerging light and permitting a small field of view to be used in the receiving

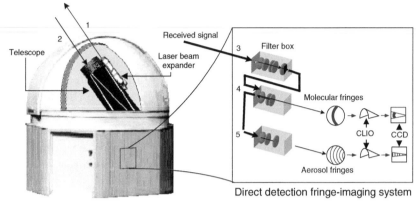

GroundWinds lidar facility

Direct detection fringe-imaging system

Figure 7.53 GroundWinds instrument concept drawing. (From Nardell, C., Hays, P.B., Pavlich, J., Dehring, M., and Sypitkowski, G. *Proceedings of the International Symposium on Optical Science and Technology*, SPIE 4484, pp. 36–50, San Diego, CA, July 30–31, 2001. With permission.)

telescope. The small divergence of the output beam minimizes the system etendue, which in turn reduces the aperture of critical elements of the detection optical system (such as the etalons) and minimizes the transmission of background light.

The backscattered signal is collected by the primary receiving telescope. A fiber optic cable transmits the signal from the focal plane of the main receiving telescope to the interferometer subsystem. A small fraction of the transmitted light is used as a wavelength reference for the detection system. The backscattered signal light and reference beam are passed through a narrowband prefilter system composed of a dielectric filter and a low-resolution Fabry–Perot etalon. After prefiltering, the signal is passed into the molecular interferometer. The resulting spectrum (circular fringes from the Fabry–Perot etalon) is converted, using a CLIO system[249] into a linear image detected by a CCD.

Since the circular fringe pattern created by the molecular etalon is present in transmitted light, the complement of the transmitted Fabry–Perot fringe pattern is reflected back from the molecular etalon.

In the MAC interferometer, this reflected light from the molecular etalon is collected and transmitted into the aerosol etalon via a fiber optic. The aerosol etalon creates a second velocity-sensitive circular fringe pattern imaged as a second linear pattern via a CLIO coupled to a second CCD camera. The reflected light from this aerosol etalon illuminates a PMT, providing a photometric intensity profile used to measure integrated energy returned, and providing correction for any misalignment of the etalons.

Three sets of information are recorded: (1) the aerosol fringe pattern — optimized for detection of the aerosol signal, (2) the molecular direct fringe pattern — optimized for detection of the molecular scattered signal, and (3) the integrated photometric return. The MAC wind lidar system[85] utilizes a unique CCD imaging technology to record the range-gated spectrum of the backscattered light. This CCD has high quantum efficiency and allows electronically programmable range resolution and spectral resolution.

Example wind measurement data taken with the GroundWinds instrument are presented in Section 7.6.2 (see Figure 7.60)

7.6. WIND MEASUREMENT APPLICATIONS

Doppler lidar sensors provide a unique capability to remotely measure atmospheric winds with high resolution. The fact that the lidar beam is quite narrow significantly reduces, if not eliminates, artifacts associated with sidelobe-induced ground clutter that are experienced using longer-wavelength radar systems. The Doppler lidar systems have advanced significantly in recent years, offering the ability to measure winds for a variety of applications from both ground-based and airborne platforms, and have made possible the ability to deploy compact and robust systems that are unattended. This section presents representative coherent and direct detection lidar measurements from prominent U.S. and European lidar groups. Representative examples of wind shear, clear-air-turbulence, aircraft wake vortex, ground and airborne wind profiling and aerosol backscatter are provided.

7.6.1. Wind Shear, Gust Front, and Turbulence Measurements

Coherent lidar is well suited to measure mesoscale variations in the wind field due to its good volume resolution and fast update rate. There have been hundreds of deployments over the past 25 years aimed at detailed wind flow measurements. These include investigations of flow around complex terrain, thunderstorm microbursts, convective outbursts, and frontal-passage gust fronts, to name a few. This section provides example coherent lidar measurements of wind shear, gusts, and turbulence from airport and other meteorological deployments.

7.6.1.1. NOAA's 10.6 μm Coherent Lidar

In October 2000, the NOAA Environmental Technology Lab (ETL) 10.6 μm CO_2 lidar[29] was used for wind measurements near Salt Lake City, Utah. The lidar was positioned near the center of the Salt Lake City basin to examine the implications for air quality of the interactions between the along-basin flow (here from the south) and the drainage flow from canyons within the range of mountains situated to the east. Figure 7.54 contains a plan-position indicator (PPI) display of data obtained on October 20, 2000. Radial wind velocities are color-coded as indicated on the color bar below the scan, with warm colors (red to yellow) describing flows towards the lidar and cool colors (green to blue) flows away from the lidar. The nominal lidar parameters during the measurement were 200 mJ, 10 Hz PRF, 28 cm aperture diameter, and 1 μsec pulse duration. The lidar operated with a circular scan at a rate of 3.3°/sec and at a constant elevation angle of 0.5°. The beam was obstructed to the west and north. Signals from three shots were averaged, and a range gate length of 300 m was used for calculating the radial velocity estimates. Some of the more interesting features of the measured wind field are the strong flow through Jordon Narrows from the south (the southerly basin flow) and the area 5 to 12 km to the northeast

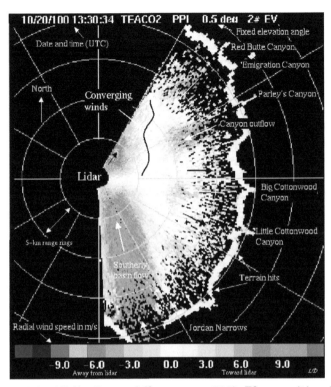

Figure 7.54 (Color figure follows page 398). Plan-position indicator (PPI) scan at 0.5° elevation of the radial Doppler wind velocity measured by the NOAA ETL TEA CO_2 coherent lidar positioned near the center of the Salt Lake City basin on October 20, 2000. The circular grid lines represent 5 km range intervals. A convergence zone is indicated to the northeast of the lidar at ranges between about 5 and 12 km (After Banta, R. M., Darby, L. S., Fast, J. D., Pinto, J. O., Whiteman, C. D., Shaw, W. J., Orr, B. D., J. Appl. Meteor., 43, 1348–1365, 2004. With permission.).

where the winds from the south meet the canyon outflow winds from the east and northeast (the convergence zone). Note that in azimuth directions where the lidar beam is not obscured, very good wind measurements are made at essentially all range gates out to ranges of 12 to 15 km. The abrupt termination of the data beyond about the 20 km range is due to hitting the local mountainous terrain. Between ~15 km range and the mountainsides, thresholding of weak signals results in range gates where the winds are not measured (black regions in the figure).

7.6.1.2.   École Polytechnique 10.6 μm Transportable Wind Lidar

In 1999, the WIND and TWL systems developed by Flamant and Werner (described in Section 7.5.1.2) were utilized to make wind measurements during the Mesoscale Alpine Program (MAP) in the Rhine valley near Bad Ragaz, Switzerland. The primary objective of the program was to improve the understanding and forecasting of the life cycle of Foehn-related phenomena, to study 3D gravity wave breaking, and associated wave drag effects in numerical models. As described in Section 7.5.1.2.2 the TWL lidar uses a 2 Hz pulsed, 10.6 μm CO_2 TEA laser with 150 mJ energy and a 17 cm diameter telescope aperture. An example 3D wind field measurement made with the TWL using a raster scan during this campaign is shown in Figure 7.55. The raster scan displays the wind flow which is generally from Chur (from the right in the figure) and going toward Walensee (Zurich) and Bodensee (Lake Constance). Several features of the wind field that are of interest include: the wind direction reversal above about 2 km and the high level of turbulent flow seen in several regions of the measurement volume. Details are described by Drobinski et al.[229]

7.6.1.3.   NOAA's 2.0 μm Coherent High-Resolution Doppler Lidar

Much of NOAA ETL work is directed towards statistical studies of the boundary layer, for which high-speed short-range measurements are required. For this application, a relatively low-energy high-PRF source is optimal. NOAA's High Resolution Doppler Lidar (HRDL)[50] was developed to meet these requirements.

The lidar is normally mounted in a 20-ft seatainer that has been operated in ground-based and shipborne field experiments; HRDL has also been deployed in aircraft. The lidar parameters include: 20 cm monostatic aperture, 2.02 μm

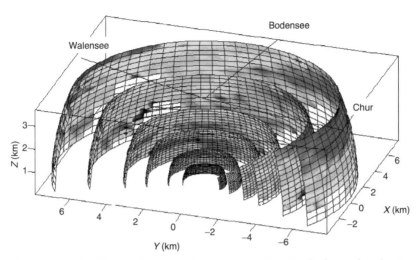

Figure 7.55 (Color figure follows page 398). Radial wind velocity measured by the TWL system during the 1999 Mesoscale Alpine Program shown on the surface of a series of concentric sections of spheres at various ranges. The radial wind velocities are in the range of ± 15 m/s. The negative velocities (toward the lidar) are in blue; the positive velocities (away from the lidar) are in red.

wavelength using Tm:LuAG lasers, 2 mJ pulse energy, and 200 Hz PRF (limited by processor speed, the laser can actually operate at up to 1 kHz). The shorter wavelength compared to the CO_2 lidar systems allows for improved range-Doppler resolution.

During the 1999 Cooperative Atmosphere/Surface Exchange Study (CASES-99) the NOAA HRDL was used to study the dynamics of the stable boundary layer (SBL). Its ability to resolve the evolution of flow structures with scales on the order of 50 m provided an excellent complement to sensitive, high-rate *in situ* turbulence sensors. Figure 7.56 shows an example of data obtained during one intensive operating period at CASES-99. This figure displays the radial wind velocity field at two different times separated by approximately 10 min during the passage of a density current in the SBL. The data displayed in each panel of the figure were collected by using a 20 sec duration RHI scan. A density current is a

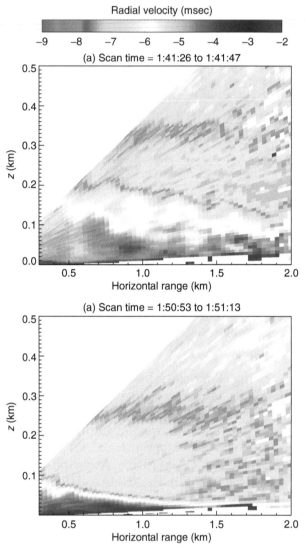

Figure 7.56 (Color figure follows page 398). NOAA HRDL RHI scan display from the CASES-99 campaign showing passage of a strong boundary layer density current. The data on the bottom were taken 10 min after the data on the top. Velocities are given by the color bar at the top of the left panel. Negative radial velocities indicate flow towards the lidar (After Sun, J., Lenschow, D. H., Burrs, S. P., Banta, R. M., Newsom, R. K., Coulter, R. L., Frasier, S., Ince, T., Nappo, C., Balsley, B., Jensen, M., Miller, D., Skelly, B., Cuxart, J., Blumen, W., Lee, X., Hu, X. Z., *Boundary Layer Meteor.* 105, 199–219, 2002. With permission.).

pool of colder, denser air that propagates along the surface. Analysis of the lidar data and *in situ* measurements indicate that eddy motions in the upper part of the density current led to periodic overturning of the stratified flow (top panel of the figure). Descending motion following the passage of the head of the density current resulted in a strongly stratified laminar flow (bottom panel of the figure). Additional details can be found in Sun's paper.[255]

7.6.1.4. Coherent Technologies' WindTracer

Coherent Technologies 2 μm WindTracer® (see Section 7.5.1.3.1 for a more more complete description of the Wind-Tracer® sensor) was deployed at the Dallas Fort Worth (DFW) airport to measure airport windshear and turbulence. The lidar operates at 2 mJ pulse energy at 500 Hz PRF, 350 nsec FWHM pulse duration and uses a 10 cm diameter off-axis telescope. Two examples of radial velocity data sets taken 10 min apart are shown in Figure 7.57. The upper portion of the figure is an RHI display at 25° azimuth (NNE) and the lower portion is a PPI display of the measured radial velocity for a 5° elevation angle. The three runways (17R, 17C, and 17L) are indicated by the boxed-in regions. The radial velocity color map range is −3 (dark purple) to +7 (white) m/sec with negative flow towards the lidar.

These data were collected during a 3 h window in the mid-to-late afternoon of a very hot and humid day. Dramatic velocity eddies of various sizes are seen to "bloom" and propagate with the general wind direction, which was out of the south. The largest positive velocity eddies in the lower kilometer of the atmosphere have a maximum radial velocity of approximately 6 m/sec and persist for longer than 10 min. The predicted 3.6 km distance traveled by the large (approximately 2 km) 6 m/sec velocity eddy during the 10-minute interval is consistent with the data. The data in each of the panels are collected in a 20 sec period determined by the scan rate. The radial velocity estimates were calculated by averaging the signals from 100 shots over a 96 m length range interval. Note that the 2 mJ pulse energy with averaging is sufficient to provide winds to 10 km horizontal range at high update rates.

7.6.2. Ground Wind Profiling

Vertical profiles of wind speed and direction are needed for many applications including numerical weather forecasting models and rocket launch vehicle safety and load alleviation. Wind profiling is typically conducted using a velocity azimuth display (VAD) scan consisting of three or more independent lines of sight, which are nominally uniformly distributed in

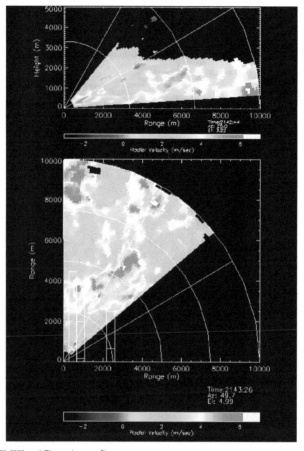

Figure 7.57   (*Continued*)

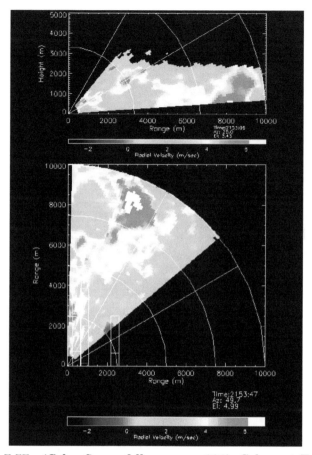

Figure 7.57 (Color figure follows page 398). Coherent Technologies' WindTracer® RHI and PPI radial velocity measured at the Dallas Fort Worth Airport in July, 2000. The horizontal range is 10 km and the velocity scale is −3 to +7 m/sec. The data in the top panel were taken 10 min after the data in the bottom panel.

azimuth (see Section 7.3.7). A series of pulse echoes are accumulated, along each LOS and a range-dependent radial velocity estimate is produced. Data from multiple LOS are combined using a least-squares or other appropriate algorithm to produce 3D vector velocity profiles.

Since DDL systems can use the molecular backscattered signal from molecules in clear air, in addition to the aerosol signal, it can be the technique of choice for wind measurements extending into the stratosphere, measuring in atmospheric regions where normally the aerosol content is variable and very low. In this section, examples of coherent and direct detection lidar data are presented.

7.6.2.1.   Coherent Technologies' Flashlamp-Pumped 2.0 μm Coherent Lidar

Figure 7.58 shows two independent successive wind profiles averaged over 5 min and one balloon sonde profile, using a six LOS VAD with a 20° cone half-angle. Researchers at Coherent Technologies collected these data in 1997, with a flashlamp-pumped, 28 mJ, 10 cm, 2.09 μm, 6 Hz coherent laser radar using six LOS and 300 pulses averaged per LOS. The two lidar measurements show strong agreement, especially above the planetary boundary layer. The balloon launch was com-

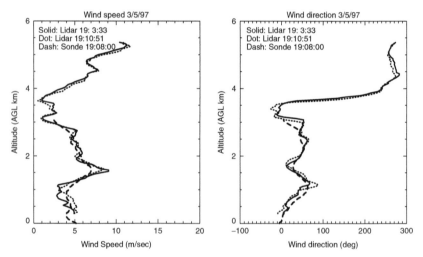

Figure 7.58   Ground wind profiles from Yuma, Arizona collected by researchers at Coherent Technologies, Inc. in 1997, with a 28 mJ, 10 cm, 2.09 μm coherent laser radar using six LOS and 300 pulses per LOS compared to a single balloon-sonde velocity profile. The two lidar measurements (solid and dotted lines) were taken 6 min apart and the balloon sonde (dashed line) was launched at the beginning of the first measurement.

mensurate with the beginning of the first lidar wind profile and its ascent rate was approximately 1000 ft/min corresponding to 12 min to ascend to 3.65 km. Excellent agreement is observed between the two consecutive lidar measurements. Agreement between the lidar data and the sonde data is good. For another example of wind profiling with this sensor see Figure 7.62 where profiles up to 15 km altitude were obtained (dotted line in Figure 7.62).

7.6.2.2. ALOMAR DWTS 532 nm "Fringe-Imaging" Direct Detection Lidar

The Doppler Wind and Temperature System (DWTS)[86,106] is a fringe-imaging system using dual 1 J, 532 nm, 30 Hz lasers and two 1.8 m diameter receiving telescopes of the ALOMAR observatory in northern Norway. This provides it with the highest energy-aperture product of any direct detection system in existence. This high system performance is, however, offset by the need to divide the backscattered signal between the many detection channels of the rather complex lidar facility (discussed in detail by von Zahn et al.[117]).

Figure 7.59 shows a cross-comparison test carried out at ALOMAR. The DWTS wind measurements extend well above the 30 km altitude limit shown in the figure, however, the radiosonde wind measurements are limited to 26.5 km. This often occurs in the Arctic winter atmosphere, due to the bursting of the balloon envelope under conditions of an extremely cold winter middle stratosphere.

7.6.2.3. Goddard's "Double-Edge" Direct Detection Lidar

The Goddard Lidar Observatory for Winds (GLOW)[87] is a mobile Doppler lidar system which uses direct detection Doppler lidar techniques to measure wind profiles from the surface into the lower stratosphere. The sensor which is described in Section 5.2.4 includes a 45 cm diameter telescope

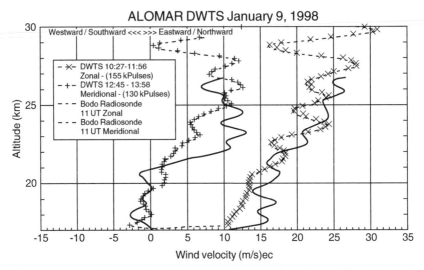

Figure 7.59 Comparison between the zonal and meridional wind measurements from the DWTS system at ALOMAR (Norway) and the nearby radiosonde station at Bodo, January 9, 1998.

and a 120 mJ, 15 nsec, 1.06 μm channel for aerosol wind measurements and a 70 mJ, 355 nm channel for molecular winds.

In September 2000 a ground-based intercomparison campaign was held at North Conway, New Hampshire, U.S.A. to compare wind measurements from several coherent and incoherent wind lidar systems. As part of the experiment, the GLOW lidar was employed to make stratospheric wind measurements using the molecular signal with the double-edge receiver. Example data from these measurements are provided in the top panel of Figure 7.60. During these measurements the lidar was pointed N-NW, the elevation angle was 45°, and the integration time for each measurement was 1 min. These data show mean and standard deviation from 50 consecutive 1-min lidar profiles. The error bars are attributed to a combination of sensor error and wind variability over the 50 min measurement period. Two comparative wind profiles (projected onto the lidar Line-of-Sight) were determined from Rawinsondes launched within a few hours of the lidar measurement time. The agreement between the lidar and radiosonde wind measurements is quite good through the entire altitude range up to 20 km.

7.6.2.4. Michigan Aerospace's GroundWinds 532 nm "Fringe-Imaging" Direct Detection Lidar

The Michigan Aerospace GroundWinds Lidar[251] utilizes the fringe-imaging technique (Section 7.5.2.5) for both the aerosol and the molecular signals. The transmitter is a frequency-doubled Nd:YAG laser operating at a PRF of 10 Hz and an energy of 400 mJ in the 8 nsec duration 532 nm wavelength pulses. The receiver telescope used during the measurements was 0.5 m in diameter with a 100 μrad field of view (half-angle). Example data collected from this sensor during the New Hampshire GroundWinds experiment are shown in the lower panel of Figure 7.60. In the figure, the squares are used for the aerosol wind data, while the asterisks are used for the molecular wind data. Radiosonde data are shown as the solid line. These data include aerosol measurements from a thin cloud layer between 6 and 9 km, which demonstrates close agreement of aerosol and molecular wind data.

7.6.3. Airborne Wind Profiling

Airborne wind profiling has commercial and military applications, which include precision airdrop, ballistic fire control, and weather prediction and modeling. Coherent lidar systems have been utilized successfully in many airborne measurement campaigns over the past 25 years. The majority of the measurements have been made using 10.6 μm CO_2 lidar systems and 2.1 μm solid-state lidar systems. This section provides several airborne coherent laser radar wind profile measurement examples.

7.6.3.1. DLR's 10.6 μm CW WIND Coherent Lidar

The first validation flight of the WIND system (see Table 7.4 in Section 7.5.1.2) integrated onto the DLR Falcon 20 was on

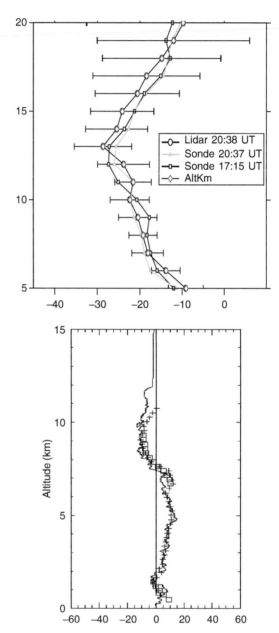

Figure 7.60 September 2000 New Hampshire Experiment: (Top) mean and error bars for 50 one-minute wind profiles along a 315° azimuth obtained with the 355 nm GLOW sensor along with two

October 12, 1999 at 13:30 UTC. The wind profiles measured by the WIND instrument were compared with a ground-based Wind Profiler Radar (WPR) of the Lindenberg Meteorological Observatory and the local model (LM) of the German Weather Service. Figure 7.61 shows the result of the wind profile measurement. The wind profile measured with the radar wind profiler WPR (averaged over 30 min) is the solid line.[256] The WIND data are profiles of the horizontal wind velocity accumulated over five conical scans (i.e., over 100 sec measurement time) from an aircraft altitude of 11 km. The LM of the German Weather Service has a horizontal resolution of 7 km. It is a nonhydrostatic model with 35 levels and it includes vertical turbulent exchange by turbulent kinetic energy. For maximum utility, airborne Doppler lidar systems have to meet at least the same specifications in horizontal resolution and accuracy as the state-of-the-art numerical models.[230,233] The data of Figure 7.61 show good agreement between the lidar, radar wind profiler measurements, and the model.

7.6.3.2. Coherent Technologies' 2 μm Coherent Lidar

Researchers at Coherent Technologies, Inc. developed a 2 μm 50 mJ, 10 Hz, 10 cm coherent lidar for airborne measurement of wind profiles below the aircraft for improved precision in cargo and troop deployment by parachute and improved ballistics accuracy.[56] Figure 7.62 shows a comparison of two wind profiles' measurements from the airborne Precision Air-Drop Lidar (PADL) sensor with two ground-based lidar systems

---

Rawinsonde profiles (September 27, 2000); (Bottom) radial wind velocity from the Michigan Aerospace 355 nm GroundWinds lidar along an azimuth angle of 350° (September 25, 2000). The plus and square symbols are measurements from the molecular and aerosol channels, respectively. The solid line is a radiosonde measurement resolved into the plane of measurement. A cloud deck at 7 to 9 km attenuates the signal, but providing high altitude aerosol (Mie) channel data for comparison with the molecular channel. The horizontal scale is wind speed in meters per second.

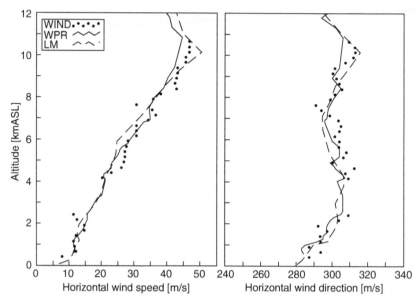

Figure 7.61 Comparison of wind speed and direction profiles form the airborne WIND lidar with the WindProfiler radar (WPR) in Lindenberg and the local model (LM) of the German Weather Service.

(the Coherent Technologies/Air Force Research Labs flash-lamp-pumped 28 mJ, 6 Hz, 10 cm aperture, 2 μm coherent lidar and the Phillips Labs 100 mJ, 20 Hz, 30 cm aperture, 10.6 μm coherent lidar), Hanscom AFB, and a Rawinsonde measurement. Each of the airborne profiles was made over a 1 min measurement time. The VAD scan pattern used for the airborne measurements consisted of a step-and-stare conical scan with nadir angle of approximately 20° and six LOS. The ground-based 2 μm wind profiles were obtained by averaging the returns from 300 pulses along each of six LOS at 30°

Figure 7.62 Intercomparison of the Coherent Technologies' airborne PADL sensor with ground-based lidar sensors and a rawinsonde: wind speed (Top) and wind direction (Bottom). The mutual agreement between the two independent lidar sensors is better than the lidar-sonde agreement, indicating superior accuracy and less bias.

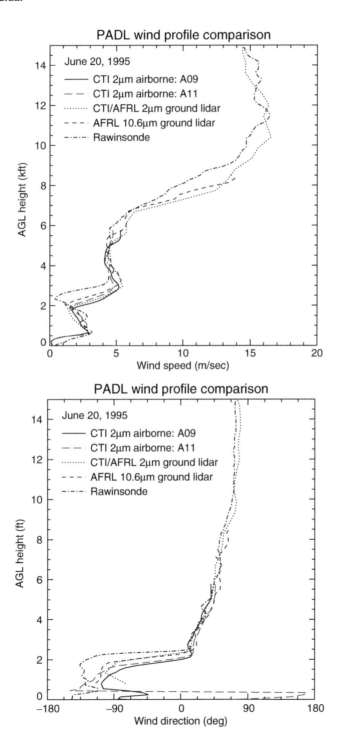

elevation angle used in the VAD scan pattern, resulting in a measurement time of about 5 min. The aircraft flight altitude was ~6 km during the measurement explaining the lack of airborne data above 6 km. Note that the ground-based 28 mJ, 10 cm aperture, 2 μm lidar was able to make measurements to 15 km altitude. Agreement between the various sensors is excellent. The variation in wind-direction near the ground is expected since the wind speed is approaching 0 m/sec and direction is meaningless. All the sensors detect the 90° wind direction change at about 2 km altitude.

7.6.4. Airborne Clear Air Turbulence Detection

The world-wide commercial aviation major accident rate has been nearly constant over the past two decades. While the rate is very low, increasing traffic over the years has resulted in an increasing absolute number of accidents. Several injuries occur each year due to encounters with turbulent winds during flight. In many instances the turbulence occurs in clear air conditions. Several studies aimed at defining sensors that can detect clear air turbulence (CAT) have been performed over the past 20 years. Conventional Doppler radar clear air sensitivity is insufficient to measure CAT. One of the more recent studies in the United States is NASA's Weather Accident Prevention (WxAP) program, which is a component of NASA's Aviation Safety Program (AvSP). One key objective of the WxAP is to provide commercial aircraft sensors with 90% probability of detection of convective and clear air turbulence 1 min before encounter. The capability of infrared Doppler lidar to measure CAT was first demonstrated in the 1970s by Huffaker and Hardesty.[11] In this section, several examples of CAT detection using lidar systems are provided.

7.6.4.1. QinetiQ's (formerly DRA) LATAS 10.6 μm CW Coherent Lidar

The Laser True Airspeed System (LATAS) developed by the remote sensing group at QinetiQ is a focused, 4 W CW 10.6 μm lidar with a 15 cm aperture (Section 7.5.1.2). In July 1982, the CW CO_2 LATAS system took part in the Joint Airport

Weather Studies (JAWS) project in Colorado. In the course of a 3-week trial the HS-125 Executive, with LATAS mounted in the nose cone, flew through many severe microbursts associated with the frequent thunderstorms prevalent at that time of year. Figure 7.63 shows a sequence of spectra in the lidar record in which the headwind changed by over 40 knots (approximately 20 m/s). These measurements contributed to the development of a descending vortex ring model for microburst behavior, in contrast to the more usual vertical jet model. The vortex ring model also explains several features observed in the JAWS flights such as dust curtains rising to over 1000 ft (300 m) around the perimeter of several microbursts. Additional detail can be found in References 12, 210 and 211.

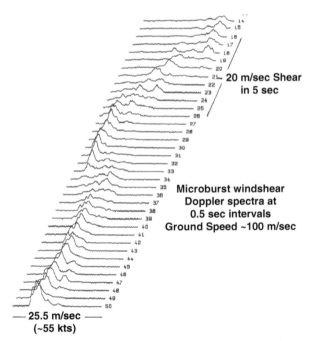

20 m/sec Shear
in 5 sec

Microburst windshear
Doppler spectra at
0.5 sec intervals
Ground Speed ~100 m/sec

—— 25.5 m/sec ——
(~55 kts)

Figure 7.63 A time sequence of lidar data spectra taken using the LATAS to measure the wind ahead of the aircraft. (From Vaughan J.M., K.O. Steinvall, C. Werner, and P.H. Flamant, "Coherent laser radar in Europe," *Proc. IEEE*, 84, 205–226, 1996. With Permission).

7.6.4.2.   Coherent Technologies' ACLAIM
2.012 μm Coherent Lidar

Under NASA funding, Coherent Technologies developed the Airborne Coherent LIDAR for Advanced In-flight Measurements (ACLAIM)[55] sensor for NASA's aviation safety program. The ACLAIM transceiver is a solid-state, diode pumped, 2 μm, 10 cm, 10 mJ, 100 Hz, 350 nsec coherent wind sensor. This sensor has flown on three different platforms for a total of about 200 flight hours. Examples of CAT data are provided in Figure 7.64. The top panel of the figure shows a 100-pulse accumulation (1 sec) range-Doppler spectrum of forward looking lidar measurements, while on board NOAA's Electra aircraft. These data show a strong, 20 m/sec, wind shear at 2 km range ahead of the aircraft. In the right panel of the figure, data taken 2 years later on board NASA's DC-8 are presented. Shown in the figure is a time series of the velocity estimates as a function of range and time (colored surface plot) along with *in situ* true airspeed (TAS) measurements (black curve). High correlation between these two sensors was achieved.

7.6.5.   Aircraft Wake Vortex Detection and Tracking

Accurate remote detection and tracking of aircraft wake vortices along approach and departure corridors have garnered significant interest in the past several years. This capability is a key to increasing capacity at airports. During Instrument Meteorological Conditions (IMC), the mandated spacing between aircraft is very often too conservative. Moreover, the number of available runways (ones with instrument landing systems, or ILS) is reduced during IMC. With knowledge of wake vortex movement and dissipation, airport operators may be able to increase the number of operations in a given time period, and therefore increase capacity.[257]

Wake vortices are induced by the lift of the aircraft and when made visible by condensation are the familiar contrails. The horizontal tornado-like rotational winds in each of the two vortices can cause a very serious flight hazard, particularly in the case of small aircraft following larger ones, and

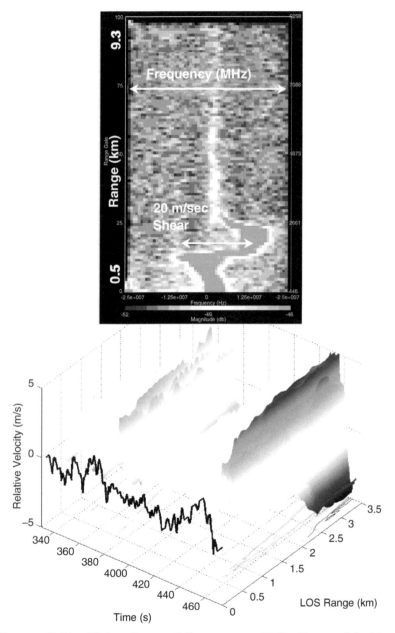

Figure 7.64 (Color figure follows page 398). Forward-looking ACLAIM CAT detection: range-resolved Doppler spectrum from the 1998 Electra flight tests (Top) and radial velocity time series onboard NASA's DC-8 from the 2001 CAMEX-IV flight tests compared to the *in situ* true air speed sensor (Bottom).

the ability to detect and track them is of high interest. Extensive studies have been made over the past 10 years of aircraft wake vortices. While pulsed lidar is most appropriate for detection and tracking of aircraft wakes, the detailed study of vortex airflow is best carried out using focused CW systems at short range.[258] Example wake vortex detection and track data from Coherent Technologies, DLR and QinetiQ are provided in this section.

7.6.5.1.   QinetiQ's (formerly DRA) 10.6 μm CW Coherent Lidar

In the late 1980s, a CO_2 CW system was developed in Malvern that was based on the LATAS design (see description in Section 7.5.1.2), but with a larger (12 in.) output aperture to improve both spatial resolution and the maximum achievable range. In later versions of the instrument, an AOM has also been added to provide a direction-sensing capability. Many campaigns have now been conducted across Europe, including at Heathrow in the period 1993 to 1995 and, subsequently, at Toulouse (1998) and Tarbes (2002) in France, and Munich (2001) in Germany. The trial at Heathrow recorded a database of vortices from nearly 2000 aircraft.[12,259,260] A number of measurement geometries were employed, including viewing the vortex pair from beneath and from the side. Scanning was achieved using a pan and tilt arrangement; the scan pattern usually involved preset 1D linear motion in a plane perpendicular to the aircraft flight path. The precise scan speed and angles were chosen for the appropriate measurement conditions (lidar location, aircraft altitude, crosswind, etc.) and the beam was manually focused at the predicted optimum range.

Figure 7.65 shows the reconstruction of an unusual (and fortunately rare) "bouncing vortex" trajectory in which the vortex, with undiminished strength, returns to the glide slope nearly 70 sec after original passage of the aircraft. The insert in the figure shows the characteristic cusp-like feature of the spectra of the lidar return as the beam is scanned across the vortex core. Sophisticated algorithms have now been developed for rapid reconstruction of vortex trajectories with

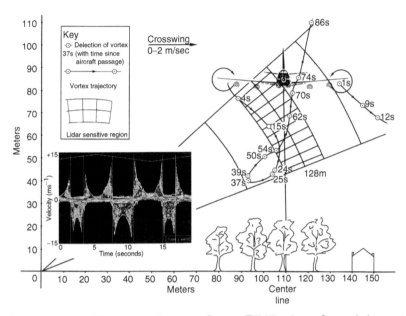

Figure 7.65 Vortex trajectory for a B747 aircraft arriving at Heathrow. The lidar was scanned $\pm 10°$ at a rate of $3°$/sec. Note the initial descent of the near-wing vortex followed by ascent close to the glideslope about 70 sec later. It must be emphasized that this is a very unusual record — in most cases the vortices are blown away and dissipated by the prevailing wind. The inset figure shows an example of the lidar signal spectra as the lidar is scanned through three times through a vortex pair. The vertical scale is line-of-sight velocity in meters per second, with positive values indicating motion away from the lidar. The horizontal axis represents time in seconds (with the aircraft overhead approximately at $t = 0$), which is converted to distance using the scan information (From Vaughan J. M., Steinvall, K. D., Werner, C., Flamant, P. H., *Proc. IEEE*, 84(2), 181–204, 1996. with permission; Inset from Hannon, S. M., Alex Thomson, J. A., Proceedings of the 9th Conference on Coherent Laser Radar, Linköping, Sweden, 202–205, June 23–27, 1997. With permission.).

a scanning lidar. The CW technique with high Doppler frequency resolution permits a good analysis of vortex velocity profile; recent work has compared wind tunnel and lidar data[261] and given evidence of an inner vortex for B747 wakes.[17]

7.6.5.2. Coherent Technologies' 2.0 μm Coherent Lidar

Pulsed coherent lidar systems are better suited to detect and track aircraft wake vortices from long range. Coherent Technologies has used the WindTracer® and its predecessor 2 μm coherent lidar systems to measure wake vortices since the early 1990s.[39,40,262] Figure 7.66 shows the wind measurement data resulting from two RHI scans of a 2 μm pulsed lidar beam through a wake vortex created by a DC 10. Each scan is separated in time by 20 sec and the first scan occurred 20 sec after the DC-10 flew through the area. The scan was perpendicular to the vortex axis and the radial velocity dipole signature, due to the two counter-rotating vortices, as is clearly evident in the data. During the 20 sec interval the core center advected approximately 50 m away from the lidar position (at range of 0 m) and sank approximately 30 m.

The location and strength of the vortices can be estimated by using a 3D maximum likelihood algorithm to detect the resulting radial velocity patterns.[263] Sample output results from a tracking filter developed by Coherent Technologies are shown in Figure 7.67 for a C-130H military transport aircraft passing at an altitude near 300 m above ground level (AGL). A shear layer is observed to halt the descent of the wake vortices at an altitude of 175 m.

7.6.5.3. DLR 2.0 μm Lidar

For long-range measurements of aircraft wake vortices, a pulsed 2 μm lidar system using the Coherent Technologies' WindTracer® transceiver (2 mJ, 500 Hz, 10 cm, 450 n) has been developed by a group from DLR in Germany.[264–266] The system is equipped with a flexible 2D scanner, consisting of an oscillating mirror and two counter-rotating prisms, and a fast data acquisition unit. The data processing algorithm includes four main stages of estimation: the Doppler spectra (power spectra of coherently detected backscatter signal), the radial velocity and velocity envelopes, the positions of vortex cores, and the vortex circulation.

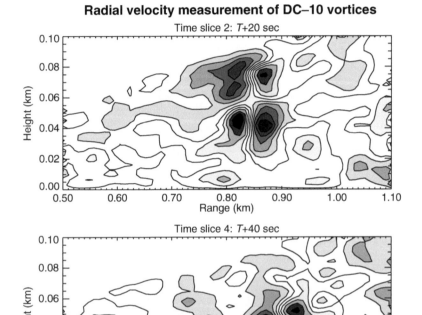

Figure 7.66 (Color figure follows page 398). Two radial velocity RHI scans, separated by 20 sec, through a pair of DC-10 wake vortices from Coherent Technologies 2 μm pulsed coherent Doppler lidar. The scan was perpendicular to vortex axis and the core's dipole signature is clearly evident in the data. Red colors represent velocities away from the lidar and blue toward the lidar. The velocity contours are separated by 0.5 m/sec.

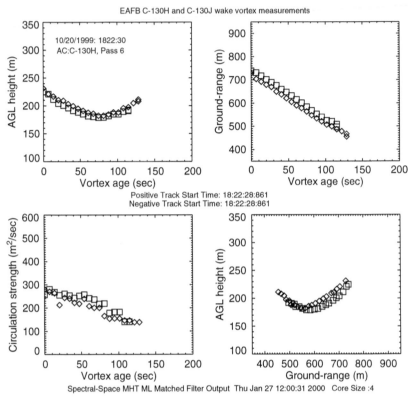

Figure 7.67 C-130H aircraft wake vortices position and circulation strength vs. time measured using the Coherent Technologies' 2 μm, 28 mJ, 6 Hz, 200 nsec, 10 cm coherent lidar on the morning of October 20, 1999. Diamonds are positive (starboard, near) vortex and squares are negative (port, far) vortex. Panel descriptions: AGL height time history (upper left); ground-range time history (upper right); circulation strength time history (lower left); height vs. ground range scatter plot (lower right). The vortices are observed to bounce at a height of 175 m, at the boundary with a strong shear layer.

During extended field experiments, the systematic characterization of wake vortices in their atmospheric environment has been demonstrated. Details of the measurements shown below are described by Köpp et al.[266] The top panel of Figure 7.68 shows an example of the velocity distribution measured by scanning through a vortex pair generated by a very large transport aircraft. From these profiles of tangential

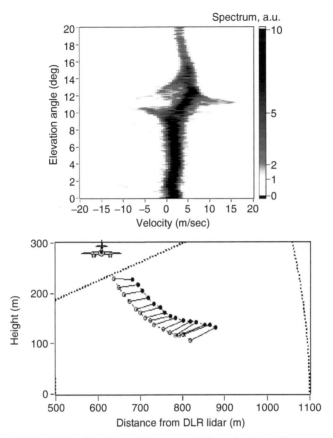

Figure 7.68 Distribution of tangential velocity of a very large transport aircraft vortex pair measured at a range of 885 m (top) and trajectories of a VLTA vortex pair observed by the pulsed Doppler lidar between 30 and 190 sec after aircraft passage (bottom). (From Keane, M., Buckton, D., Redfern, M., Bollig, C., Wedekind, C., Köpp, F., Berni, F. *J. Aircraft*, 39, 850–862, 2002. With permission.)

velocity special vortex parameters, such as trajectory (Figure 7.68, right panel), core separation, tilt angle, and circulation can be derived. Moreover, the 2 μm pulsed lidar has been successfully used for wake-vortex detection under very small aspect angles (i.e., near longitudinally), as it is the scenario for a forward-looking wake-vortex warning system.

7.6.6. Aerosol Detection and Tracking

Doppler lidar sensors provide a unique capability to generate high-resolution 3D distributions of wind and aerosol data for ground-based and airborne applications. Appropriately processed, these data can yield useful detection, tracking, and short-term prediction information relating to the extent, density, and location of potentially dangerous isolated aerosol plumes. The aerosol data are analyzed to detect isolated above-threshold inhomogeneities, and the wind and turbulence data are used to provide short-term prediction of plume propagation. Using these data in plume dispersion models can enable robust prediction of dispersion and propagation over longer time periods.

Coherent Technologies deployed and operated a WindTracer® Doppler lidar at Dugway Proving Ground (DPG), Utah, in September 2000. The primary objective of the measurement program was to evaluate the capability of the WindTracer® pulsed coherent lidar to remotely detect and track aerosol and chemical plumes while, at the same time, measuring the local wind state such that prediction of plume dispersion and advection is improved. Sample results illustrating the ability to detect and track aerosol plumes are shown in Figure 7.69. The radial velocity (left panels) and the aerosol backscatter coefficient (right panels) are shown in false color. The measurement data were updated in 1 min intervals but two measurements separated by about 20 min are shown in the figure. The arrows in the left panels represent estimates of the horizontal wind velocity vector with spatial position. The tangential component of the velocity that is used to calculate the velocity vector was estimated by tracking the cross-track motion of velocity eddies. Also shown in the right panels is the track of the centroid of the aerosol plume up to the frame time shown (red line) and an expanded view of the backscatter coefficient in the vicinity of the plume (inset in right panels). Approximately halfway through the 20 min measurement period, a gust front turned the plume track from toward the west to toward the northwest.

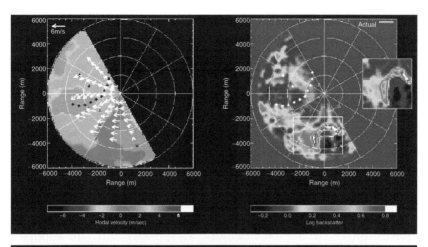

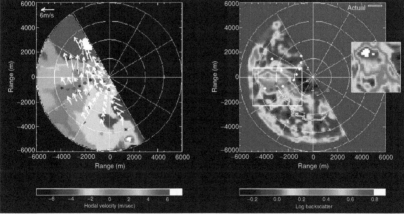

Figure 7.69 (Color figure follows page 398). WindTracer® radial velocity (left) and aerosol backscatter coefficient (right) estimates from measurements conducted at Dugway Proving Ground, Utah in September 2000, over a 12 km diameter circular area. The time difference between the data in the top and bottom panels is approximately 20 min. The ability to detect, track, and predict the aerosol plume is clearly evident in the data.

The diamond-shaped objects in the displays represent the position of point sensors. In these data, the WindTracer® Doppler lidar demonstrates its ability to detect and track aerosol plumes. These data illustrate the unique functionality offered by the pulsed coherent Doppler lidar technology.

7.7. SUMMARY AND FUTURE PROSPECTS

Doppler lidar technology is a powerful remote sensing tool for many diverse applications. Through numerous field measurements over the past three decades, the efficacy of Doppler lidar systems in measuring atmospheric wind fields has been demonstrated. There has been significant progress in the technology, signal processing, and system implementations for both direct and coherent detection Doppler lidar systems, particularly in the last decade. Essentially all the reported work on direct detection Doppler lidar systems for wind measurements to date has been directed at systems for measuring vertical wind profiles. Several systems have been built and are currently operational. Molecular backscatter based systems have demonstrated wind measurements to altitudes of approximately 50 km. Some of the improvements in system efficiency realized with direct detection Doppler lidar systems in recent years may lead to applications in a number of other areas where the capability for good range resolution can be exploited.

Using only a few watts of average laser power, pulse energies of a few to a few tens of millijoules, and apertures of only 10 to 20 cm diameter, coherent systems have shown the ability to make high resolution wind measurements at high update rates in the atmospheric boundary layer to horizontal ranges exceeding 10 km. Shorter wavelength systems offer an improved range-velocity resolution product. Examples of demonstrated capability using coherent lidar systems include:

- atmospheric vertical wind profiling from the ground to several tens of kilometers in altitude;
- windshear and gust front detection, and other high-resolution high-density wind measurements at high data rates;
- aircraft wake vortex detection and tracking in the vicinity of airports;

- airborne wind measurements for the detection of clear air turbulence ahead of aircraft and the measurement of more favorable winds above or below;
- airborne measurement of wind profiles below the aircraft for improved precision in cargo and troop deployment by parachute and improved ballistics accuracy; and
- environmental monitoring by measuring aerosol backscatter levels and detection and tracking of elevated aerosol plumes.

Historically, the primary factors limiting more widespread use of Doppler lidar systems have been the reduced performance in adverse weather, the lack of reliable and near-autonomous operation, and the high cost. In applications where the reduced performance in adverse weather (higher atmospheric transmission losses due to fog, rain, snow, or clouds) is not acceptable, the only practical solution is to combine the lidar with a longer wavelength radar system. For example, a low-cost radar could be used; one that would not be able to measure in clear air but would work well in many of the adverse conditions that limit lidar. The significant investment in $2\,\mu m$ coherent wind lidar systems in the past decade has improved the reliability of those systems to the point that complete coherent lidar systems are now commercially available (the WindTracer®). One system has now been operating autonomously and near continuously at the Hong Kong international airport since June 2002. Cost, which is still an issue for many applications, is expected to improve as applications further develop and the volume of sales increases. Although promising, it is still yet to be seen whether taking advantage of the telecommunications developments at $1.5\,\mu m$ will significantly reduce the overall systems cost.

The implementation of Doppler lidar systems for global wind measurements from space have been a goal of the Doppler lidar community since the late 1970s. Global wind profiles would primarily be used to improve the accuracy of global weather forecasting models as well as benefit our understanding of atmospheric dynamics and the climate. A satellite-borne

Doppler lidar has the potential to make wind measurements in remote areas (such as over the oceans and deserts) where little information currently exists. Several system concept studies have occurred in the past 25 years, beginning with CO_2 coherent lidar systems in the late 1970s through the early 1990s and more recently solid-state coherent and direct detection Doppler lidar systems. Currently, the European Space Agency (ESA) is moving forward with the development of a direct detection Doppler lidar that operates at 355 nm using frequency-tripled Nd:YAG lasers (the AEOLUS mission) with plans to launch the lidar in 2007.[123] Other coherent and direct detection system concepts continue to be investigated for this application by groups at NASA and NOAA in the United States[267] and by a group at the Communications Research Laboratory (CRL) in Japan.[268]

With continued improvements in laser, detector, optical, and processing hardware, Doppler lidar systems are expected to become smaller, more efficient, and lower cost and therefore will find more widespread use. There are significant investments being made in eye-safe diode-pumped solid-state laser technology by commercial groups and government agencies for numerous applications. Efficient production of high pulse energies and/or high average powers with near diffraction-limited performance in compact packages, as needed by many Doppler lidar applications, is the focus of much of the development. In many near-term applications, where more energy or processing power are not required, most of the development emphasis is being placed on reducing cost and increasing reliability. With continued improvements in the speed of signal processing hardware, more sophisticated real-time processing algorithms and data displays will be used to improve system sensitivity and ease of use without having a significant system cost impact. Multifunction lidar systems that measure other parameters in addition to the winds (e.g., measurements of temperature and water vapor profiles, the presence of other selected molecular species — for example, related to pollution, and 3D images) are possible and are expected to be developed and find application in the future.

ACKNOWLEDGMENTS

The authors would like to express our sincere thanks to a number of scientists for contributing their ideas, data, and other valuable contributions toward this work. In particular, the following individuals provided significant material and comments: Alan Brewer, Didier Bruneau, Pierre Flamant, Cristina Flesia, Rod Frehlich, Bruce Gentry, Steve Hannon, Mike Hardesty, Mike Harris, Jack McKay, Carl Nardell, Guy Pearson, Stephan Rahm, Barry Rye, Paul Suni, Michael Vaughan, Christian Werner, and Dave Willetts. The authors would like to acknowledge the patience, support, and encouragement of their wives, Paula Henderson, Jillyanne Gatt, Karen Elphick-Rees, and Neva Huffaker through the many long hours spent producing this chapter.

Glossary of Key Variable Symbols

In the following definitions, example equations are provided for quick reference, however, the assumptions are not always stated. The text in the subsections following the glossary list the appropriate assumptions.

Physical Constant Variables

c	Speed of light; $c = 2.998 \times 10^8$ m/s
h	Planck's constant; $h = 6.626 \times 10^{-34}$ J s
k_B	Boltzmann's constant; $k_B = 1.3806 \times 10^{-23}$ J/K
e	Charge of an electron; $e = 1.602 \times 10^{-19}$ C

Optical Field Variables

k	Wavenumber; $k = 2\pi/\lambda$ (m^{-1})
λ	Wavelength (m); $\lambda = c/\nu$ in a vacuum
ν	Optical frequency; $\nu = c/\lambda$ (Hz) in a vacuum
ν_x	Transmitter optical frequency (Hz)
ν_r	Received signal optical frequency (Hz)
ν_d	Doppler frequency shift (Hz); $\nu_d = 2V_r/\lambda$

U Electric field with the carrier frequency embedded (V/m); $U(t) = |U(t)|\cos[2\pi\nu_c t + \theta(t)]$, where ν_c is the carrier frequency

$\boldsymbol{U}$ Complex electric field with the carrier frequency embedded (V/m); $\boldsymbol{U}(t) = |U(t)|\exp[i2\pi\nu_c t + i\theta(t)]$, where $i = \sqrt{-1}$

$\boldsymbol{u}$ Normalized electric field complex amplitude ($\sqrt{\text{W}}$/m); $\boldsymbol{u} = \boldsymbol{U}/\sqrt{z_m}$ where z_m is the impedance of the medium (377 Ω for free space)

I Electric field irradiance (W/m^2); $I = |\boldsymbol{u}|^2 = \boldsymbol{u}\boldsymbol{u}^*$

x, y, ρ Transverse distances (m); $\rho^2 = x^2 + y^2$

Gaussian Beam Variables

F Focus distance or radius of curvature of the transmit beam.

R_r Rayleigh range of the Gaussian beam (m); $R_r = \pi\omega_0^2/\lambda$

ω e^{-2} beam irradiance radius (m)

ω_0 e^{-2} beam irradiance radius at the beam waist, used for the transmit beam radius (m)

ω_R e^{-2} beam irradiance radius at the target range (m); $\omega_R^2 = \omega_0^2 [(1 - R/F)^2 + (R/R_r)^2] + (\lambda R/\pi\rho_0)^2$

Target Variables

β Backscatter coefficient (m^{-1} sr^{-1})

β_A Aerosol backscatter coefficient (m^{-1} sr^{-1})

β_M Molecular backscatter coefficient (m^{-1} sr^{-1})

ΔR Range resolution; $\Delta R = c\Delta T/2$

P_r Reflected signal power (W)

R Target range (m)

t_r Round trip time to range R (s); $t_r = 2R/c$

V Scalar velocity or speed (m/s)

V_r Radial velocity (m/s)

Nontarget Atmospheric Variables

C_n^2 Refractive index structure constant (m$^{-2/3}$)

L_{bg} Background light radiance (W/m^2/sr); $L_{bg} = \dfrac{\varepsilon 2hc^2}{\lambda^5[\exp(h\nu/kT_{bb})-1]}$ for a blackbody radiator

P_{bg} Background light power (W); $P_{\text{bg}} = TA_{\text{r}} \, \Omega \Delta \lambda L_{\text{bg}}$

N_{bg} Number of background light photons in the integration or measurement time $N_{\text{bg}} = P_{\text{bg}} T_{\text{int}}/h\upsilon$

T One-way atmospheric transmission

T_x One-way atmospheric transmission for the transmit beam

T_{b} One-way atmospheric transmission for the BPLO beam

Transmitter Variables

E_x Transmitter energy (J)

δt Transmitted pulse power standard deviation width (s)

ΔT Transmit pulse power full-width half-maximum width (s)

I_x Transmit irradiance (W/m^2)

$I_{\text{n}x}$ Normalized transmit irradiance such that $\iint I_{nx} da = 1$ before the truncating transmit aperture (1/m^2)

$\eta_{\text{o}x}$ Transmitter optical efficiency, not including $\eta_{\text{T}x}$

$\eta_{\text{T}x}$ Transmit power truncation efficiency from the limiting aperture

P_x Transmitter power (Watts)

$\omega_{\text{r}x}$ Exp[-2] beam irradiance radius of Gaussian transmit beam (m)

Common Receiver Variables

A_{r} Receiver area (m^2)

A_{d} Detector area (m^2)

a_{r} Receiver aperture radius (m)

B Receiver bandwidth (Hz); $B = 1/T_{\text{int}}$ for direct detection or $1/2T_{\text{int}}$ for coherent detection

$\delta\theta_{\text{L}}$ Lag angle between the transmitter and receiver (rad); $\delta\theta_{\text{L}} = (2R/c)\Omega_{\text{s}}$

E_{s} Received signal energy collected by the telescope assuming no loss (W)

η_{d} Detector truncation efficiency; $\eta_{\text{d}} = \iint_{A_{\text{d}}} I_{\text{sd}} \, da / \iint_{\infty} I_{\text{sd}} \, da$

η_{q} Detector quantum efficiency without gain (electrons/photon)

η_{or} Receiver optics efficiency not including direct detection frequency discriminator efficiency or heterodyne receiver beam splitter efficiency

η_{r} Receiver efficiency; $\eta_{\text{r}} = \eta_{\text{q}}\eta_{\text{bs}}\eta_{\text{or}}$ for the coherent receiver and $\eta_{\text{r}} = \eta_{\text{q}}\eta_{\text{or}}$ for a direct detection receiver

η_{R} Receiver total efficiency, $\eta_{\text{R}} = \eta_{\text{r}}\eta_{\text{h}}$ for coherent detection receivers and $\eta_{\text{R}} = \eta_{\text{r}}\eta_{\text{D}}$ for direct detection receivers

I_{b} BPLO irradiance (W/m^2)

I_{nb} Power-normalized BPLO irradiance just before the truncating transmit aperture (1/m^2)

I_{s} Receiver-plane signal irradiance (W/m^2)

I_{sd} Detector-plane signal irradiance (W/m^2)

i Photocurrent (A); $i = \int\int \Re I \, da$

i_{bg} Background light current

i_{d} Detector dark current

i_{s} Baseband signal photocurrent in a direct detection receiver (A), note in a coherent receiver the interference or heterodyne current, i_{h} is the signal photocurrent

i_{n} Noise current (A)

i_{sn} Shot-noise current (A); $i_{\text{sn}} = 2eiB$ where i is an unmultiplied (signal, LO, background light, dark, etc.) photocurrent

N_{n} Total number of effective noise equivalent photons in a measurement or integration time; $N_{\text{n}} = M_{\text{e}}N_{\text{nc}}$

N_{nc} Number of effective integrated noise photons in a coherence time. $N_{\text{nc}} = F_{\text{h}}/\eta_{\text{r}}\eta_{\text{h}}$ for a coherent receiver

N_{s} Total number of signal photons in a measurement time; $N_{\text{s}} = M_{\text{e}}N_{\text{sc}}$ or $N_{\text{s}} = P_{\text{s}}T_{\text{int}}/h\upsilon$

N_{sc} Number of signal photons in a coherence time; $N_{\text{sc}} = P_{\text{s}}\tau_{\text{c}}/h\upsilon$

N_{sd} Total number of detected signal photons in a measurement or integration time; $N_{\text{sd}} = \eta_{\text{R}}N_{\text{s}}$

P_{s} Received signal power after telescope but before photoreceiver (W); $P_{\text{s}} = \eta_{\text{ox}}\eta_{\text{Tx}} \, E_t T^2 \, (c\beta/2) \, (A_{\text{r}}/R^2)$ or $P_{\text{s}} = \int\int A_R I_s da$

P_{sd} Detector-plane received signal power (W)

$\Re$ Detector responsivity (A/W); $\Re = \eta_{\text{q}}e/h\upsilon$

$\hat{\eta}_q$ Effective heterodyne quantum efficiency; $\hat{\eta}_q = \int\int \mid \eta_q$ $\boldsymbol{u}_{\mathrm{od}}(x,y)\mid^2 \, \mathrm{d}x\,\mathrm{d}y/\int\int \eta_q \mid \boldsymbol{u}_{\mathrm{od}}(x,y)\mid^2 \, \mathrm{d}x\,\mathrm{d}y$

ρ_T Beam truncation ratio; $\rho_T = \omega_0/a$

S_n Amplifier noise spectral density (A^2/Hz)

$\mathrm{SNR_{QL}}$ Quantum-limited SNR (power ratio); $\mathrm{SNR_{QL}} = M\langle N_s\rangle/$ $(M + \langle N_s\rangle)$

Ω Receiver antenna solid angle (sr)

Ω_s Scanner rotation rate (rad/s)

Coherent Detection Receiver Variables

α_b BPLO diffraction ratio; $\alpha_b = \omega_{Rb}/\omega_{dRb}$

a_0 Effective LO beam aperture radius where it is mixed with the signal (m)

CNR   Coherent detection wideband carrier-to-noise ratio

$\mathrm{CNR_n}$ Coherent detection narrowband or matched-filter carrier-to-noise ratio

F_h Heterodyne receiver excess noise factor (total noise power divided by LO shot-noise power)

$\delta\rho$ Radial offset between BPLO and transmit beams (m)

$\delta\theta$ Angular offset between BPLO and transmit beams (rad)

η_a Heterodyne antenna efficiency; $\eta_a = \eta_h\eta_{Tx}$

η_a^u Untruncated Gaussian beam heterodyne antenna efficiency

η_a^{um} Matched BPLO and transmit untruncated Gaussian beam antenna efficiency; $\eta_a^{um} = 1/2\alpha^2$

η_{BS} Beam splitter (combiner) signal throughput efficiency

$\eta_{\delta\rho}$ Gaussian beam offset efficiency; $\eta_{\delta\rho} = \exp\left[-\delta\rho^2/\omega_R^2\right]$

$\eta_{\delta\theta}^{FF}$ Far-field angular misalignment efficiency for diffraction limited Gaussian beams; $\eta_{\delta\theta}^{FF} = \exp[-\delta\theta^2/\theta_d^2]$, where θ_d is the diffraction limited beam angle.

η_h Heterodyne efficiency

η_{rt} Refractive turbulence efficiency

η_{Tb} BPLO power truncation loss from the limiting aperture

i_o Local oscillator photocurrent (A)

i_h Heterodyne signal photocurrent (A)

NET   Coherent receiver noise equivalent temperature, that is, the temperature of a blackbody radiator for which its signal power equals the CD receiver's noise power; $\mathrm{NET} = h\nu/k \, \ln(1 + \varepsilon\eta_r)$

P_h	Heterodyne signal power (W)
P_o	Local oscillator power before the beam splitter (W)
P_{od}	Detector-plane local oscillator power (W); $P_{od} = P_o$ $(1 - \eta_{BS})$
ω_{ob}	e^{-2} beam irradiance radius the pupil-plane Gaussian BPLO beam (m)
ω_{rb}	e^{-2} beam irradiance radius a Gaussian BPLO beam (m) at range R
ω_{dRb}	e^{-2} beam irradiance radius of a diffraction-limited Gaussian BPLO beam (m); $\omega_{dRb} = \lambda R/\pi\omega_{ob}$

Direct Detection Receiver Variables

A_{FP}	Fabry–Perot power absorption
β_D	Frequency discriminator sensitivity (Hz)
$\Delta\lambda$	Receiver background light filter wavelength width
F_{FP}	Fabry–Perot coefficient of finesse
$\mathfrak{I}$	Fabry–Perot reflectivity finesse; $\mathfrak{I} = (\pi/2)\sqrt{F_{FP}}$
FSR	Fabry–Perot free spectral range
F_d	Amplified detector excess noise factor
G	Detector linear gain (electrons/electron)
η_D	General direct detection frequency discriminator throughput efficiency
R_{FP}	Fabry–Perot power reflection
S	Frequency discriminator signal derived from two or more photocurrents
SNR	Direct detection power signal-to-noise ratio
SNR_n	Direct detection narrowband, or matched-filter power, signal-to-noise ratio
T_{FP}	Fabry–Perot power transmission
T_m	Peak power transmission through the discriminator optic (edge-detection or Fabry–Perot frequency discriminator)
T_o	Edge-detection frequency discriminator transmission at the center frequency
T_s	Frequency discriminator transmission at the signal frequency ν_s

Measurement and Signal Variables

$\langle x \rangle$ Mean of x, also denoted by $\bar{x}$

α_D Degeneracy factor, also mean signal spectrum height above the mean noise floor in a coherent receiver; $\alpha_D = \sqrt{2}CNR_n$

δt Transmitted pulse power standard deviation width (s); $\delta t = \Delta T/\sqrt{8\ln(2)}$ for a Gaussian pulse

$\delta \nu$ Signal power spectrum spectral standard deviation width (Hz); $\delta \nu = 1/\sqrt{2}\pi\tau_c$ for a Gaussian spectrum

$\delta \nu_p$ Transmitted pulse energy spectrum spectral standard deviation width (Hz); $\delta \nu_p = 1/4\pi\delta t$ for a Gaussian pulse

δV_r Effective radial velocity standard deviation width (m/sec); $\delta V_r = \lambda \delta \nu/2$

$\Delta \nu$ Power spectrum full-width half-maximum width (Hz); $\Delta \nu = \delta \nu\sqrt{8\ln(2)}$ for a Gaussian spectrum

ΔT Transmit pulse power full-width half-maximum width (s) $\Delta T = \delta t\sqrt{8\ln(2)}$ for a Gaussian pulse

F_e Velocity estimate variance excess noise factor

FOM   Coherent detection figure of merit representing the ratio of the signal spectrum height above the mean noise floor to the RMS noise variations in a frequency bin; $FOM = \sqrt{N}CNR_n$

N_p Number of independent pulse echoes averaged

N_g Number of independent range gate echoes averaged

σ_V Velocity measurement precision (m/s)

T_{int} Integration or measurement time such that $N_s = P_s T_{int}/h\nu$

τ_c e^{-1} coherence time (s)

var[x]   Measurement variance of x

$\overline{V}_r$ Mean radial velocity estimate (m/sec)

Speckle Variables

A_c Field coherence area; $A_c = \pi\rho_c^2$ for a circular complex coherence factor (CCF) (m^2)

A_{cg} Goodman's field coherence area; $A_{cg} = \pi\rho_{cg}^2$ for a circular CCF (m^2)

M Goodman's speckle diversity not counting temporal (multiple pulses and range gates) diversity, includes spatial, polarization, and spectral diversity

M_e Total effective diversity including averaging; $M_e = MN_pN_g$

μ_{12} Field normalized mutual intensity or CCF, which by the Van Cittert–Zernike theorem is proportional to the Fourier transform of the target irradiance distribution function;

$$\mu_{12}\ (\Delta x, \Delta y) = \frac{\exp(-j\psi)\iint I_R(\xi,\eta)\exp(j2\pi(\Delta x\xi + \Delta y\eta)/\lambda R)\mathrm{d}\xi\mathrm{d}\eta}{\iint I_R(\xi,\eta)\mathrm{d}\xi\mathrm{d}\eta}$$

ρ_c Field coherence radius defined as the e^{-1} point on the CCF for circular CCF; $\rho_c = \sqrt{2}\omega_0$ for an untruncated Gaussian transmitter and an extended diffuse target (m)

ρ_{cg} Goodman's field coherence radius; $\rho_{cg} = \omega_0$ for an untruncated Gaussian transmitter and an extended diffuse target (m)

ρ_o Transverse coherence radius due to refractive turbulence (m); $\rho_o \sim 3.1 r_o$, where r_o is Fried's coherence diameter; $\rho_o^{5/3} = 2.91 k^2 \int_0^R C_n^2\ (z)(1 - z/R)^{5/3}\ \mathrm{d}z$ for Kolmogorov turbulence

SNR_I Irradiance SNR; $\mathrm{SNR}_I = \langle I\rangle^2/\mathrm{var}[I] = M_e$

REFERENCES

1. Oliver B.M., "Signal-to-noise ratios in photoelectric mixing," *Proc. IRE*, 49, 1960–1961.
2. Forrester A.T., *J. Opt. Soc. Am.*, 51, 253, 1961.
3. Gould G., S.F. Jacobs, J.T. LaTourrette, M. Newstein, and P. Rabinowitz, "Coherent detection of light scattered from a diffusely reflecting surface," *Appl. Opt.*, 3, 648–649, 1964.
4. Goldstein I. and A. Chabot, "Characteristics of a traveling-wave ruby single-mode laser as a laser radar transmitter," *IEEE J. Quantum Electron.*, QE-2, 519, 1966.
5. Lucy R.F., K. Lang, C.J. Peters, and K. Duval, "Optical superheterodyne receiver," *Appl. Opt.*, 6, 1333, 1967.
6. Jelalian A.V., *Laser Radar Systems* (Artech House, Boston, MA, 1992).
7. Patel C.K.N., "Interpretation of CO_2 optical maser experiments," *Phys. Rev. Lett.*, 12, 588–590, 1964.

8. Thomson J.A.L. and M.F. Dorian, "Heterodyne detection of monochromatic light scattered from a cloud of moving particles," Tech Ret GDC-ERR-AN-1090, Convair Division of General Dynamics, San Diego, CA, 1967.

9. Menzies R.T., "Coherent and incoherent lidar — an overview," in *Tunable Solid State Lasers for Remote Sensing*, R.L. Byer, E.K. Gustafson, and R. Trebino, Eds. (Springer Verlag, Berlin, 1985).

10. Killinger D.K. and N. Menyuk, " Laser remote sensing of the atmosphere," *Science* 235, 37, 1987.

11. Huffaker R.M. and R.M Hardesty, "Remote sensing of atmospheric wind velocities using solid-state and CO_2 coherent laser systems," *Proc. IEEE*, 84(2), 181–204, 1996.

12. Vaughan J.M., K.O. Steinvall, C. Werner, and P.H. Flamant, "Coherent laser radar in Europe," *Proc. IEEE*, 84, 205–226, 1996.

13. Huffaker R.M., "Aero-astrodynamics research review," no. 3, NASA TMX-53389, Oct 1965.

14. Huffaker R.M., "Laser Doppler detection systems for gas velocity measurements," *Appl. Opt.*, 9(5), 1026, 1970.

15. Huffaker R.M., A.V, Jelalian, and J.A. Thomson, "Laser-Doppler system for detection of aircraft trailing vortices," *Proc. IEEE*, 58, 322–326, 1970.

16. Libby J.T., T.J. Dasey, and R.M. Heinrichs, "Wake vortex and hover downwash measurements of the V-22 and XV-15 tiltrotor aircraft," Proceedings of the 11th Coherent Laser Radar Conference, 113–114, Malvern, UK, 2001.

17. Vaughan J.M. and M. Harris, "Lidar measurement of B747 wakes; observation of a vortex within a vortex," *Aerosp. Sci. Technol.*, 5, 409–411, 2001.

18. Kopp F., R.L. Schwiesow, and C. Werner, " Remote measurements of boundary-layer wind profiles using a cw Doppler lidar," *J. Climate Appl. Meteorol.*, 23, 148–154, 1984.

19. Woodfield A.A. and J.M. Vaughan, "Using an airborne CO_2 laser for free stream airspeed and windshear measurements," AGARD Conference Proceedings, no. 373, 22.1–22.18, 1984.

20. Vaughan J.M., D.W. Brown, P.H. Davies, C. Nash, G. Kent, and M.P. McCorrick "Comparison of SAGE II solar extinction data with airborne measurements of atmospheric backscattering in the troposphere and lower stratsosphere," *Nature*, 332(6166), 709–711, 1988.

21. Vaughan J.M., "Scattering in the atmosphere," in *Scattering — Scattering and Inverse Scattering in Pure and Applied*

Science, E.R. Pike and P.C. Sabatier, Eds., Chapter 2.4.3 (Academic Press, San Diego, CA, 2002).

22. Bilbro J.W. and W.W. Vaughan "Wind measurements in the nonprecipitous regions surrounding severe storms by an airborne pulsed Doppler lidar system," *Bull. Am. Metorol. Soc.*, 59, 1095, 1978.

23. Bilbro J.W., et al., *Appl. Opt.*, 25, 2952, 1986.

24. Pearson G.N., B.J. Rye, and R.M. Hardesty, "Design of a high repetition rate CO_2 Doppler lidar for atmospheric monitoring," *Proc. SPIE*, 1222, 142–153, 1990.

25. Pearson G.N., "A high-pulse-repetition-frequency CO_2 Doppler lidar for atmospheric monitoring," *Rev. Sci. Instrum.*, 64(5), 1155–1157, 1993.

26. Brewer W.A., B.J. Rye, R.M. Hardesty, and W.L. Eberhard, "Performance characteristics of a compact RF-excited MOPA CO_2 Doppler lidar," Proceedings of the 9th Conference on Coherent Laser Radar, Linköping, Sweden, June 23–27, 1997.

27. Menzies R.T. and R.M. Hardesty, "Coherent Doppler lidar for measurements of wind fields," *Proc. IEEE*, 77, 449–462, 1989.

28. Post M.J. and W.D. Neff, "Doppler lidar wind measurements in a narrow mountain valley," *Bull. Am. Meteorol. Soc.*, 67, 274–281, 1985.

29. Post M.J. and R.E. Cupp, "Optimizing a pulsed Doppler lidar," *Appl. Opt.*, 29, 4145–4158, 1990.

30. Banta R.M., L.D. Olivier, P.H. Gudiksen, and R. Lange, "Implications of small-scale flow features to modeling dispersion over complex terrain," *J. Appl. Meteorol.*, 35, 330–342, 1996.

31. Rothermel J., D.R. Cutten, R.M. Hardesty, R.T. Menzies, J.N. Howell, S.C. Johnson, D.M. Tratt, L.D. Olivier, and R.M. Banta, "The multi-center airborne coherent atmospheric wind sensor," *Bull. Am. Meteorol. Soc.*, 79, 581–599, 1998.

32. Doran J.C., J.D. Fast, and J. Horel, "The VTMX campaign," *Bull. Am. Meteorol. Soc.*, 83, 537–551, 2002.

33. Mayor S.D., R.L. Schwiesow, D.H. Lenschow, and C.L. Frush, "Validation of radial velocity measurements from the NCAR Doppler lidar," Proceedings of the Coherent Laser Radar Conference, 299–272, Keystone, CO, July 23–27, 1995.

34. Kavaya M.J. and R.T. Menzies, "Lidar aerosol backscatter measurements: systematic, modeling, and calibration error considerations," *Appl. Opt.*, 244, 3444–3453, 1985.

35. Werner C., P.H. Flamant, O. Reitebuch, C. Loth, F. Köpp, P. Delville, G. Wildgruber, P. Drobinski, J. Streicher,

B. Romand, H. Herrmann, L. Sauvage, E. Nagel, C. Boitel, M. Klier, D. Bruneau, M. Schrecker, M. Messionnier, S. Rahm, A. Dabas, R. Häring, and P.H. Salamitou, "Wind infrared Doppler lidar instrument," *Opt. Eng.*, 40, 115–125, 2001.

36. Kane T.J., W.J. Kozlovsky, R.L. Byer, and C.E. Byvik, "Coherent laser radar at 1.06 μm using Nd:YAG lasers," *Opt. Lett.*, 12, 239, 1987.

37. Kavaya M.J., S.W. Henderson, J.R. Magee, C.P. Hale, and R.M. Huffaker, "Remote wind profiling with a solid-state Nd:YAG coherent lidar system," *Opt. Lett.*, 14, 776–778, 1989.

38. Hale C.P., S.W. Henderson, J.R. Magee, and S.R. Vetorino, "Compact high-energy Nd:YAG coherent laser radar transceiver," Proceedings of the 6th Conference on Coherent Laser Radar, Snowmass, CO, July 8–12, 1991.

39. Hannon S.M., J.A. Thomson, J.R. Magee, S.W. Henderson, and R.M. Huffaker, "Airport surveillance using a pulsed solid-state coherent laser radar," Proceedings of the 7th Conference on Coherent Laser Radar Applications and Technology, Paris, France, July 19–23, 1993.

40. Richmond R., S. Hannon, and J.A. Thomson, "Laser radar mapping of aircraft wake vortices," Proceedings of the 7th Conference on Coherent Laser Radar Applications and Technology, Paris, France, July 19–23, 1993.

41. Hawley J.G., R. Targ, S.W. Henderson, C.P. Hale, M.J. Kavaya, and D. Moerder, "Coherent launch-site atmospheric wind sounder (CLAWS): theory and experiment," *Appl. Opt.*, 32, 4557, 1993; also in *SPIE Milestone Series 141*, Selected Papers on Laser Applications in Remote Sensing, W.B. Grant, et al., Eds., 1997.

42. Henderson S.W., C.P. Hale, J.R. Magee, M.J. Kavaya, and A.V. Huffaker, "Eye-safe coherent laser radar systems at 2.1 μm using Tm, Ho:YAG lasers," *Opt. Lett.*, 16, 773–775, 1991.

43. Henderson S.W., P.J.M. Suni, C.P. Hale, S.M. Hannon, J.R. Magee, D.L. Bruns, and E.H. Yuen, "Coherent laser radar at 2 μm using solid-state lasers," *IEEE Trans. Geosci. Remote Sensing* 31, 4, 1993; also in *SPIE Milestone Series 133*, Selected Papers on Laser Radar, G. Kamerman, Ed., 1997.

44. Suni P.J.M. and S.W. Henderson, "1-mJ/pulse Tm:YAG laser pumped by a 3-W diode laser," *Opt. Lett.*, 16, 817–819, 1991.

45. Suni P.J.M., G. Gates, E.H. Yuen, D.L. Bruns, S.R. Vetorino, and T.J. Valle, "A diode-pumped 2-μm transceiver for ground

and airborne Doppler lidar measurements," Proceedings of the 7th Conference on Coherent Laser Radar, 206–209, Paris, France, July 19–23, 1993.

46. Targ R., B.C. Steakley, J.G. Hawley, L.L Ames, P. Forney, D. Swanson, R. Stone, R.G. Otto, V. Zarifis, P. Brockman, R.S. Calloway, S.H. Klein, and P.A. Robinson, "Coherent lidar airborne wind sensor II: flight test results at 2 and 10 μm," *Appl. Opt.*, 35, 7117–7127, 1996.

47. Wilson G.A., S.W. Henderson, C.P. Hale, and S.M Hannon, "Development and testing of an optical air turbulence sensor," Proceedings of the 9th Conference on Coherent Laser Radar, 198, Linköping, Sweden, June 23–27,1997; also in *Laser Focus World*, September, 1997, p. 55.

48. Grund C., "Coherent Doppler lidar for boundary layer wind measurement employing a diode-pumped Tm:Lu,YAG laser," Proceedings of the 8th Coherent Laser Radar Conference, 14–16, Keystone, CO, 1995.

49. Wulfmeyer V., M. Randall, A. Brewer, and R.M. Hardesty, "2 μm Doppler lidar transmitter with high frequency stability and low chirp," *Opt. Lett.*, 25, 1228–1230, 2000.

50. Grund C.J., R.M. Banta, J.L. George, J.N. Howell, M.J. Post, R.A. Richter, and A.M. Weickmann, "High resolution Doppler lidar for boundary layer and cloud research," *J. Atmos. Ocean. Technol.*, 18, 376–393, 2001.

51. Lenschow D.H., V. Wulfmeyer, and C. Senff, "Measuring second- through fourth-order moments in noisy data," *J. Atmos. Ocean. Technol.*, 17, 1330–1347, 2000.

52. Newsom R.K. and R.M. Banta, "Shear-flow instability in the stable nocturnal boundary layers observed by Doppler lidar during CASES-99," *J. Atmos. Sci.*, 60, 16–33, 2003.

53. Poulos G.S., W. Blumen, D.C. Fritts, J.K. Lunquist, J. Sun, S.P. Burns, C. Nappo, R. Banta, R. Newsom, J. Cuxart, E. Terradellas, B. Balsey, and M. Jensen, "CASES-99: a comprehensive investigation of the stable nocturnal boundary layer," *Bull. Am. Meteorol. Soc.*, 83, 555–581, 2002.

54. Suni P.J.M., M.W. Phillips, C.P. Hale, G.J. Wagner, and S.W. Henderson, "Diode-pumped 2-micron laser technology for coherent lidar applications," Proceedings of the 8th Coherent Laser Radar Conference, 306–309, Keystone, CO, 1995.

55. Hannon S.M., H.R. Bagley, D.C. Soreide, D.A. Bowdle, R.K. Bogue, and L. Jack Ehernberger, " Airborne turbulence de-

tection and warning; ACLAIM flight test results," Proceedings of the 10th Coherent Laser Radar Conference, 20–23, June 28–July 2, 1999.

56. Richmond R.D., P.D. Woodworth, R. Fetner, J.A. Overbeck, M. Salisbury, S.W. Henderson, S.M. Hannon, and S.R. Vetorino, "Eye-safe solid-state ladar for airborne wind profiling," Proceedings of the 8th Coherent Laser Radar Conference, 138–140, Keystone, CO, July 23–27, 1995.

57. Yu J., U.N. Singh, N.P. Barnes, and M. Petros, "125-mJ diode-pumped injection-seeded Ho,Tm:YLF laser," *Opt. Lett.*, 23, 780–782, 1998.

58. Phillips M.W. and S.W. Henderson, "High energy two micron laser development at CTI for airborne and space-borne coherent lidar," Proceedings of the 11th Coherent Laser Radar Conference, 188, Malvern, Worcestershire, UK, July 1–6, 2001.

59. Phillips M.W., D. Schnal, C. Colson, C.P. Hale, D. D'Epagnier, M. Gibbens, and S.W. Henderson, "SPARCLE coherent lidar transceiver," invited paper, Proceedings of the Tenth Biennial Coherent Laser Radar Technology and Applications Conference, Mount Hood, OR, 98, June 28–July 2, 1999.

60. Koch G.J., M. Petros, B.W. Barnes, J.Y. Beyon, F. Amjajerdian, J. Yu, M.J. Kavaya, U.N. Singh, "VALIDAR: a testbed for advanced 2-μm coherent Doppler wind lidar," Proceedings of the 12th Conference on Coherent Laser Radar, Bar Harbor, ME, June 15–20, 2003.

61. Frehlich R.G., S.M. Hannon, and S.W. Henderson "Performance of a 2-μm coherent Doppler lidar for wind measurements," *J. Atmos. Ocean. Technol.*, 11, 1517–1528, 1994.

62. Frehlich R.G., S.M. Hannon, and S.W. Henderson, "Coherent Doppler lidar measurements of winds in the weak signal regime," *Appl. Opt.* 36(15), 3491–3499, 1997.

63. Henderson S.W., E.H. Yuen, and S.M. Hannon, "Autonomous lidar wind field sensor: design and performance," in *Proc. SPIE*, number 3757, Applications of Lidar to Current Atmospheric Topic III, 18–27, Denver, CO, July, 1999.

64. Hannon S.M., "Automomous infrared Doppler lidar: airport surveillance applications," Proceedings of 1st European Conference on Radar Meteorology, Bologna, Italy, Sept 4–8, 2000 (to appear in *J. Phys. Chem. Earth: Part B*).

65. McKinnie I.T., T.W. Monarski, G.T. Bennett, A. Oien, D.D. Smith, S.M. Hannon, and S.W. Henderson, "Eye-safe coher-

ent laser radar at 1.56-μm using solid-state Raman lasers," Proceedings of the 11th Coherent Laser radar Conference, 159, Malvern, Worcestershire, UK, July 1–6, 2001.

66. Monarski T.W., S.M. Hannon, and P. Gatt, "Eye-safe coherent lidar detection using a 1.5-μm Raman laser," *Proc. SPIE*, 4377, 229–236, 2001.

67. Yanagisawa T., K. Asaka, K. Hamazu, and Y. Hirano, "11-mJ 15-Hz single-frequency diode-pumped Q-switched Er,Yb phosphate glass laser," *Opt. Lett.*, 26, 1262–1264, 2001.

68. Stoneman R.C. and A.I.R. Malm, "High-power Er:YAG laser for coherent laser radar," Conference on Lasers and Electro-Optics, paper CThZ6, 2004.

69. Karlsson C., F. Olsson, D. Letalick, and M. Harris, "All-fibre multifunction continuous-wave 1.55 micron coherent laser radar for range, speed, vibration and wind measurements", *Appl. Opt.*, 39, 3716–3726, 2000.

70. Harris M., G.N. Pearson, K.D. Ridley, C. Karlsson, F. Olsson, and D. Letalick, "Single-particle laser Doppler anemometry at 1.55 micron," *Appl. Opt.*, 40, 969–973, 2001.

71. Harris M., G. Constant, and C. Ward, "Continuous-wave bistatic laser Doppler wind sensor," *Appl. Opt.*, 40, 1501–1506, 2001.

72. Pearson G.N., P.J. Roberts, J.R. Eacock, and M. Harris, "Analysis of the performance of a coherent pulsed fiber-lidar for aerosol backscatter applications," *Appl. Opt.*, 41, 6442–6450, 2002.

73. Kameyama S., T. Fujisaka, K. Asaka, T. Ando, Y. Koyata, Y. Hirano, and S. Wadaka, "All-fiber coherent integration lidar," Proceedings of the 11th Coherent Laser Radar Conference, 76–80, Malvern, UK, 2001.

74. Henderson S.W., J.A.L. Thomson, S.M. Hannon, T.J. Carrig, P. Gatt, and D. Bruns, "Wide-bandwidth eye-safe coherent laser radar for high resolution hard target and wind measurements," Proceedings of the 9th Conference on Coherent Laser Radar, Linköping, Sweden, 160–163, June 23–27, 1997.

75. Hannon S.M., J.A.L. Thomson, S.W. Henderson, P. Gatt, R. Stoneman, and D. Bruns, "Agile multiple pulse coherent lidar for range and micro-Doppler measurement," Proceedings of the Laser Radar Technology and Applications Conference, Orlando, FL, SPIE 3380, 259, 1998.

76. Lombardi G., J. Butman, T. Lyons, D. Terry, and G. Piech, "Multiple-pulse coherent laser radar waveform," Proceedings of the 12th Conference on Coherent Laser Radar, Bar Harbor, ME, June 15–20, 2003.

77. von Cossart G., J. Fiedler, von Zahn U., Fricke K.H., Nussbaumer V., Nelke G., Hubner F., Hauchcorne A., Marcovici J.P., Fassina F., Nedelkovic D., Rees D., and Meredith P. J., "Modern technologies employed in the ALOMAR Rayleigh/ Mie/Raman lidar", Proceedings of the 12th European Rocket and Balloon Programmes and Related Research, ESA SP-370, 387–394, 1995.

78. McDermid I.S., J.B. Laudenslager, and D. Rees, "UV-excimer laser based incoherent Doppler lidar system," *Proceedings of the NASA Symposium on Global Wind Measurements*, W.E. Baker and R.J. Curran, Eds., 1985.

79. Chanin M.L., A. Garnier, A. Hauchecorne, and J. Porteneuve, "A Doppler lidar for measuring winds in the middle atmosphere," *Geophys. Res. Lett.*, 16, 1273–1276, 1989.

80. Rees D., M. Vyssogorets, N.P. Meredith, E. Griffin, and Y. Chaxell, "The Doppler wind and temperature system of the ALOMAR lidar," *J. Atmos. Terr. Phys.*, 58(16), 1827–1842, 1996.

81. McGill M. J., W.R. Skinner, and T.D. Irgang, "Validation of wind profiles measured using incoherent Doppler lidar," *Appl. Opt.*, 36, 1928–1939, 1997.

82. Collins S.C., T.D. Wilkerson, V.B. Wickwar, J.C. Walling, D.F. Heller, and D. Rees, "The alexandrite ring laser: a spectrally narrow lidar light source for atmospheric fluorescence and absorption observations," *Proceedings of the 18th International Laser Radar Conference* (Springer Verlag, Berlin, 1996).

83. Park Y.K., G. Guiliani, and R.L. Byer, "Single axial mode operation of a Q-switched Nd:YAG oscillator by injection seeding," *IEEE J. Quantum Electron.*, QE-20, 117, 1984.

84. Zhou B., T.J. Kane, G.J. Dixon, and R.L. Byer, "Efficient, frequency-stable laser-diode-pumped Nd:YAG laser," *Opt. Lett.*, 10, 62, 1985.

85. Hays P.B. and C.A. Nardell, "The GroundWinds New Hampshire instrument and the LIDAR-Fest 2000 Campaign," *Proc. SPIE*, 4484, 36–50, 2001.

86. Baumgarten G., D. Rees, and N.D. Lloyd, Observations of Arctic stratospheric Winds by the ALOMAR Doppler Wind and Temperature System, ESA SP-437, 331–334, 1999.

87. Gentry B. and H. Chen, "Tropospheric wind measurements obtained with the Goddard Lidar Observatory for Winds (GLOW): validation and performance," Proceedings of the

International Symposium on Optical Science and Technology, Vol. 4484, 74–81, San Diego, CA, July 30–31, 2001.

88. Benedetti-Michellangeli G., F. Congeduti, and G. Fiocco, "Measurement of aerosol motion and wind velocity in the lower troposphere by Doppler optical radar," *J. Atmos. Sci.*, 29, 906–910, 1972.

89. G.W. Grams, Atmospheric Technology, Nos. 6 and 7, National Center for Atmospheric Research, USA, 1974 and 1975.

90. Congeduti F., G. Fiocco, A. Adriani, and C. Guarrelli, "Vertical wind velocity measurements by a Doppler lidar and comparisons with a Doppler sodar," *Appl. Opt.*, 20, 2048–2054, 1981.

91. Abreu V.J., "Wind measurements from an orbital platform using a lidar system with incoherent detection: an analysis," *Appl. Opt.*, 18, 2992, 1979.

92. Hays P.B., T.L. Killeen, and B.C. Kennedy, "The Fabry–Perot interferometer on Dynamics Explorer", *Space Sci. Instrum.*, 5, 395–416, 1981.

93. Rees D., T.J. Fuller-Rowell, A. Lyons, T.L. Killeen, and P.B. Hays, "Stable and rugged etalon for the Dynamics Explorer Fabry–Perot interferometer: 1. Design and construction," *Appl. Opt.*, 21, 3896–3902, 1982.

94. Rees D., P.A. Rounce, I. McWhirter, A.F.D. Scott, A.H. Greenaway, and W. Towlson, "Observations of atmospheric absorption lines from a stabilised balloon platform and measurements of stratospheric winds," *J. Phys. E: Sci. Instrum.*, 15, 191–206, 1982.

95. Rees D., I. McWhirter, P.B. Hays, and T. Dines, "A stable, rugged capacitance-stabilised piezo-electric scanned Fabry–Perot etalon," *J. Phys. E.: Sci. Instrum.*, 14, 1320–1325, 1982.

96. McCleese D.J. and J.S. Margolis, "Remote sensing of stratospheric and mesospheric winds by gas correlation electro-optic phase-modulation spectroscopy," *Appl. Opt.*, 22, 2528, 1983.

97. Hays P.B., V.J. Abreu, M.D. Burrage, D.A. Gell, H.J. Grassl, A.R. Marshall, Y.T. Morton, D.A. Ortland, W.R. Skinner, D.L. Wu, and J.-H. Yee, "Remote sensing of mesospheric winds with the high resolution Doppler imager," *Planet. Space Sci.*, 40(12), 1599–1606, 1992.

98. Hays P.B., V.J. Abreu, J. Sroga, and A. Rosenberg, "Analysis of a 0.5 micron space-borne wind sensor," Conference on Satellite/Remote Sensing Applications, Clearwater Beach, FL, 1984.

99. Rosenberg A. and J. Sroga, "Development of a 0.5 μm incoherent doppler lidar for space application," *Proceedings of the*

NASA Symposium on Global Wind Measurements, W.E. Baker and R.J. Curran, Eds., 157–162 (Deepak Publishing, Hampton, VA, 1985).

100. Rees D., I. McWhirter, and D. Wade, "Development of a Doppler wind lidar system for atmospheric wind measurements," Proceedings of the 8th ESA Symposium on European Rocket and Balloon Programmes and Related Research, ESA SP 276, 99–106, 1987.

101. Rees D., et al., "Upper atmospheric wind and temperature measurements using imaging Fabry–Perot interferometers," *WITS Handbook*, Vol. 2, 188–223, C.H. Liu, Ed. (SCOSTEP, University of Illinois, 1989).

102. Rees D. and I.S. McDermid, "Doppler lidar atmospheric wind sensor: re-evaluation of the 355-nm incoherent Doppler lidar," *Appl. Opt.*, 29, 4133–4144, 1990.

103. Garnier A. and M.L. Chanin, "Description of a Doppler Rayleigh lidar for measuring winds in the middle atmosphere," *Appl. Phys. B*, 54, 1992.

104. Abreu V.J., J.E. Barnes, and P.B. Hays, "Observations of winds with an incoherent lidar detector," *Appl. Opt.*, 31, 4509, 1992.

105. Rees D., U. von Zahn, G. von Cossart, G. Nelke, K.H. Fricke, and N.D. Lloyd, "The Doppler wind and temperature system of the ALOMAR lidar," 13th European Rocket and Balloon Programmes and Related Research" Oland, Sweden, Published by ESA SP-397, July 1997.

106. Rees D., U. von Zahn, W. Singer, G. von Cossart, G. Nelke, K-H. Fricke, R. Rüster, W. Eriksen, and N.D. Lloyd, "Observations of winds in the arctic stratosphere by the ALOMAR Doppler wind and temperature system," *Proceedings of the 24th Annual European Meeting on Atmospheric Studies and Optical Methods, Sentraltrykkeried, Bado, Norway*, ISBN 82-994583-0-7, 122–128, 1998.

107. McGill M.J. and J.D. Spinherne, "Comparison of two direct-detection Doppler lidar techniques," *Opt. Eng.*, 37, 2675–2686, 1998.

108. Korb L.C., B.M. Gentry, and C.Y. Weng, "Edge technique: theory and application to the lidar measurement of atmospheric winds," *Appl. Opt.*, 31, 4002, 1992.

109. Gentry B.M. and C.L. Korb, "Edge technique for high accuracy Doppler velocimetry," *Appl. Opt.*, 33, 5770–5777, 1994.

110. Korb C.L., B. Gentry, and S. X. Li, "High accuracy atmospheric wind field measurements with an edge technique lidar," in *Advances in Atmospheric Remote Sensing with Lidar*, A. Ansmann, Ed., 259–262 (Springer-Verlag, Berlin, 1997).

111. Gentry B.M., Chen H., and Li S.X., "Wind measurements with a 355 nm molecular Doppler lidar," *Opt. Lett.*, 25, 1231–1233, 2000.

112. Friedman J.S., C. Tepley, P. Castleberg, and H. Roe, "Middle-atmosphere Doppler lidar using a iodine-vapor edge filter," *Opt. Lett.*, 22, 1648–1650, 1997.

113. Flesia C. and C.L. Korb, "Theory of the double-edge molecular technique for Doppler lidar wind measurement," *Appl. Opt.*, 38, 432–440, 1999.

114. Korb C.L., C. Flesia S. Lolli, and C. Hirt, "Double-edge molecular measurement of lidar wind profiles at the Observatoire de Haute Provence," 2000.

115. Flesia C., C.L. Korb, and C. Hirt, "Double-edge molecular measurement of lidar wind profiles at 355 nm," *Opt. Lett.*, 25, 1466–1468, 2000.

116. She C.Y. and J. R. Yu, "Doppler-free saturation fluorescence spectroscopy of Na atoms for atmospheric applications," *Appl. Opt.*, 34, 1063–1075, 1995.

117. von Zahn, U., G. von Cossart, J. Fiedler, K.H. Fricke, G. Nelke, G. Baumgarten, D. Rees, A. Hauchecorne, and K. Adolfsen, "The ALOMAR Rayleigh/Mie/Raman lidar: objectives, configuration, and performance," *Ann. Geophys.*, 18, 815–833, 2000.

118. She C.Y., S. Chen, Z. Hu, J. Sherman, J.D. Vance, V. Vasoli, M.A. White, J. Yu, and D.A. Krueger, "Eight-year climatology of nocturnal temperature and sodium density in the mesopause region (80 to 105 km) over Fort Collins, CO (41°N, 105°W)", *Geophys. Res. Lett.*, 27, 3289–3292, 2000.

119. She C.Y., J.D. Vance, B.P. Williams, D.A. Krueger, H. Moosuller, D. Gibson-Wilde, and D.C. Fritts, "Lidar studies of atmospheric dynamics near polar mesopause," *EOS Trans. Am. Geophys. Union*, 83(27), 289, 2002 (see also p. 293).

120. See Chapter 6 of this book.

121. Pirronen P. and E.W. Eloranta, "Demonstration of a high-spectral-resolution lidar based on an iodine absorption filter," *Opt. Lett.*, 19(3), 234, 1994.

122. Smart A.E., "Optical velocity sensor for airdata applications," *Opt. Eng.*, 31, 166, 1992.

123. ESA Official announcement of approval and funding of the AEOLUS Project (Space-based Doppler Wind Lidar), October 2002.

124. Morancais D., F. Fabre, P. Berlioz, R. Mauer, and A. Culoma, "Spaceborne wind lidar concept for the atmospheric dynamics mission ALADIN," in *Advances in Laser Remote Sensing*, 20th International Laser Radar Conference, A. Dabas, C. Loth, and J. Pelon, Eds., 15–18, Ecole Polytechnique, Palaiseau, France, 2000.

125. Bruneau D., "Fringe-imaging Mach–Zehnder interferometer as a spectral analyzer for molecular Doppler wind lidar," *Appl. Opt.*, 41, 503–510, 2002.

126. Bruneau D. and J. Pelon, "Simultaneous measurements of particle backscattering and extinction coefficients and wind velocity by lidar with a Mach–Zehnder interferometer: principle of operation and performance assessment," *Appl. Opt.*, 42(6), 1101–1114, 2003.

127. Selah B., *Photoelectron Statistics*, (Springer-Verlag, New York, 1978).

128. Gatt P. and S.W. Henderson, "Laser radar detection statistics: a comparison of coherent and direct detection receivers," in *Laser Radar Technology and Applications VI*, G.W. Kamerman, Ed., Proceedings of SPIE, Vol. 4377, 2001.

129. Xiao M., L.A. Wu, and H.J. Kimble, "Detection of amplitude modulation with squeezed light for sensitivity beyond the shot noise limit," *Opt. Lett.*, 13, 476–478, 1988.

130. Srivastava V., et al. "Wavelength dependence of backscatter by use of aerosol microphysics and lidar data sets: application to 2.1-μm wavelength for space-based and airborne lidars," *Appl. Opt.*, 40(27), 4759–4769, 2001.

131. Global TroposphericWind Sounder (GTWS) Science Definition Team (SDT), "Global Tropospheric Wind Sounder Science and Operational Wind Data Product Requirements," September 28, 2001.

132. Bowdle D.A., S.F. Williams, J. Rothermel, and J.E. Arnold, "Global Backscatter Experiment (GLOBE)," in *Coherent Laser Radar: Technology and Applications*, 12 of OSA 1991, Technical Digest Series (Optical Society of America, Washington, DC, 1991).

133. Hannon S.M., P. Gatt, and S.W. Henderson, "Continental U.S. aerosol backscatter at 2 μm: distribution and implications for ground- and space-based coherent lidars," Proceedings of the 9th Conference on Coherent Laser Radar, 176, Linköping, Sweden, June 23–27, 1997.

134. Measures R.M., *Laser Remote Sensing* (John Wiley and Sons, New York, 1984).

135. Anderson G.P. and J.H. Chetwynd, "Fascode3p User Guide," U.S. Air Force Phillips Laboratory, Hanscom Air Force Base, MA, 1992.

136. Rothman L.S., et al. "HITRAN database: 1986 edition," *Appl. Opt.*, 26(19), 4058, 1987.

137. Gagne J.-M,. J.-P. Saint-Dizier, and M. Pickard, Methode d'echantillonage des fonctions deterministes en spectroscopie: application a un spectrometre multi-canal par comptage photonique, *Appl. Opt.*, 13, 581–588, 1974.

138. Rye B.J., and R.M. Hardesty, "Discrete spectral peak estimation in incoherent back-scatter heterodyne lidar. I: Spectral accumulation and the Cramer–Rao lower bound," *IEEE Trans. Geosci. Remote Sensing*, 31, 16–27, 1993.

139. Francon, M., *Laser Speckle and Application in Optics* (Academic Press, New York, 1979).

140. Gatt, P., T.P. Costello, D.A. Heimmermann, D.C. Costellanos, and C.M. Stickley, "Coherent optical array receivers for the mitigation of speckle effects," *Appl. Opt.*, 35, 5999–6009, 1996.

141. Favreau X., A. Delaval, P.H. Flamant, A. Dabas, and P. Delville, "Four-element receiver for pulsed 10 μm-heterodyne Doppler lidar," *Appl. Opt.*, 39, 2441–2448, 2000.

142. Drobinski P., P.H. Flamant, and P. Salamitou, "Spectral diversity technique for heterodyne Doppler lidar that uses hard target returns," *Appl. Opt.*, 39, 376–385, 2000.

143. Goodman, J.W., *Statistical Optics* (John Wiley and Sons, New York, 1985).

144. MacKerrow E.P. and M.J. Schmitt, "Measurement of integrated speckle statistics for CO_2 lidar returns from a moving, nonuniform, hard target," *Appl. Opt.*, 36, 6921–6937, 1997.

145. Goodman, J.W., in *Laser Speckle and Related Phenomena*, 2nd edn, J.C. Dainty, Ed. (Springer-Verlag, New York, 1984).

146. Dereniak E.L. and D.G. Crowe, *Optical Radiation Detectors* (John Wiley and Sons, New York, 1984).

147. Siegman, A.E., "The antenna properties of optical heterodyne receivers," *Proc. IEEE*, 54(10), 1350–1356, 1966, also *Appl. Opt.*, 5, 1588–1594, 1966.

148. Van Vleit K.M., "Noise limitations in solid state detectors," *Appl. Opt.*, 6, 1145–1169, 1967.

149. Gatt, P. "Efficient Wideband Coherent Lidar," Final Report to NASA contract NAS8-99050, CTI-9913, Coherent Technologies, Inc., Lafayette, CO, June 1999.

150. Yuen H.P. and V.W.S. Chan, "Noise in homodyne and heterodyne detection," *Opt. Lett.*, 8, 177, 1983.

151. Shapiro J.H., "Quantum noise and excess noise in optical homodyne and heterodyne receivers," *IEEE J. Quantum Electron.*, QE-21, 237, 1985.

152. Schwartz M., *Information Transmission, Modulation, and Noise*, Chapter 6.6, 4th edn (McGraw Hill, New York, 1990).

153. Frehlich R.G. and M.J. Kavaya, "Coherent laser radar performance for general atmospheric refractive turbulence," *Appl. Opt.*, 30(36), 5325–5352, 1991.

154. Rye B.J. and R.G. Frehlich, "Optimal truncation and optical efficiency of an apertured coherent lidar focused on an incoherent backscatter target," *Appl. Opt.*, 31, 2891–2899, 1992.

155. Shapiro J.G., "Correlation scales of laser speckle in heterodyne detection," *Appl. Opt.*, 24(12), 1883–1888, 1985.

156. Belmonte A. and B.J. Rye, "Heterodyne lidar returns in the turbulent atmosphere: performance evaluation of simulated systems," *Appl. Opt.*, 39(15), 2401–2411, 2000.

157. Frehlich R., "Effects of refractive turbulence on ground-based verification of coherent Doppler lidar performance," *Appl. Opt.*, 39(24), 4237–4246, 2000.

158. Osche G.R., K.N. Seeber, Y.I. Lok, and D.S. Young, "Laser radar cross-section estimation from high-resolution image data," *Appl. Opt.*, 31(14), 2452, 1992.

159. Bachman C.G., *Laser Radar Systems and Techniques* (Artech House, Dedham, MA, 1979).

160. Skolnik M.I., *Radar Handbook*, Equation 2, Chapter 27 (McGraw Hill, New York, 1970).

161. Kavaya MJ., R.T. Menzies, D.A. Haner, U.P. Oppenheim, and P.H. Flamant, "Target reflectance measurements for calibration of lidar atmospheric backscatter data," *Appl. Opt.*, 22(17), 2619, 1983.

162. For incoherent signal addition, the average power of the sum is the sum of the constituent signal average powers.

163. Rye B.J., "Antenna parameters for incoherent backscatter heterodyne lidar," *Appl. Opt.*, 18, 1390–1398, 1979.

164. Frehlich R.G., "Heterodyne efficiency for a coherent laser radar with diffuse aerosol targets," *J. Mod. Opt.*, 41, 2115–2159, 1994. Also see Frehlich R.G., "Optimal local oscillator

field for a monostatic coherent laser radar with a circular aperture," *Appl. Opt.*, 32, 4569–4577, 1993.

165. Huffaker, M. Private communication; see Huffaker, R.M., H.B. Jeffrerys, E.A. Weaver, J.W. Bilbro, G.D. Craig, R.W. George, E.H. Gleason, P.J. Marerro, E.J. Reinbold, and J.E. Shirey, "Development of a laser Doppler system for the detection, tracking and measurement of aircraft wake vortices," Report FAA-RD-74-213 (FAA, Washington, DC, 1975).

166. Olaofe, G.O., "Diffraction by Gaussian apertures," *J. Opt. Soc. Am.*, 60, 1654–1657, 1970.

167. Degnan, J.J. and B.J. Klein, "Optical antenna gain 2: receiving antennas," *Appl. Opt.*, 13(10), 2397–2401, 1974.

168. Rye B.J., "Primary aberration contribution to incoherent backscatter heterodyne lidar returns," *Appl. Opt.*, 21, 839, 1982.

169. Wang, J.W., "Optimal truncation of a lidar transmitted beam," *Appl. Opt.*, 27, 4470–4474, 1988.

170. Zhao, Y., M.J. Post, and R.M. Hardesty, "Receiving efficiency of a pulsed coherent lidars: theory," *Appl. Opt.*, 29, 4111–4119, 1990.

171. Priestly J.T., NOAA Wave Propagation Laboratory, Boulder, CO, Personal communication with Barry Rye, 1980.

172. Thomson, J.A. and F.P Boynton, "Development of design procedures for coherent lidar measurements of atmospheric winds," Final Report to NOAA contract NOAA-0307-022-35106, PD-B-77-137, Physical Dynamics Corp., Berkley, CA, June 1977, revised Jan 1978.

173. Sonnenschein, C.M. and F.A. Horrigan, "Signal-to-noise relationships for coaxial systems that heterodyne backscatter from the atmosphere," *Appl. Opt.*, 10(7), 1600, 1971.

174. Kavaya M.J. and P.J.M. Suni, "Continuous wave coherent laser radar: calculation of measurement location and volume," *Appl. Opt.*, 30, 2634, 1991.

175. Su, J.Z., et al., "Conceptual solutions for optical scanning lag angle of scanning ladar," *Proc. SPIE*, 4893, 2002.

176. Bealand, R.R., "Propagation through atmospheric turbulence", in *IREO Handbook*, F.G. Smith, Ed., Vol. 2, Equation 2.179, Chapter 2 (Environmental Research Institute of Michigan and SPIE Optical Engineering Press, 1993).

177. Fried, D.L. "Atmospheric modulation noise in an optical heterodyne receiver," *IEEE J. Quantum Electron.*, QE-3(6), 213–221, 1967.

178. Yura, H.T., "Signal-to-noise ratio of heterodyne lidar systems in the presence of atmospheric turbulence," *Optica Acta*, 26(5), 627–644, 1979.

179. Clifford S.F. and S. Wandzura, "Monostatic heterodyne lidar performance: the effect of the turbulent atmosphere," *Appl. Opt.*, 20, 514–516, 1981; *Appl. Opt.*, 20, 1502(E), 1981.

180. Targ R., M.J. Kavaya, R.M. Huffaker, and R.L. Bowles, "Coherent lidar airborne windshear sensor: performance evaluation," *Appl. Opt.*, 30, 2013, 1991, using $\eta_a = 1/\text{SRF}$.

181. Jakeman E., "Enhanced backscattering through a deep random phase screen," *J. Opt. Soc. Am. A*, 5(10), 1638, 1988.

182. Andrews L.C., C.Y. Young, and W.B. Miller, "Coherence properties of a reflected optical wave in atmospheric turbulence," *J. Opt. Soc. Am. A*, 13(4), 851, 1996.

183. Rye B.J., "Refractive-turbulence contribution to incoherent backscatter heterodyne lidar returns," *J. Opt. Soc. Am.*, 71(6), 687–691, 1981.

184. Frehlich R.G., "Effects of refractive turbulence on coherent laser radar," *Appl. Opt.*, 32, 2122–2139, 1993.

185. Van Trees, *Detection, Estimation, and Modulation Theory*, Part. I, Equation 2.516 (John Wiley and Sons, New York, 1968).

186. Levin M.J., "Power spectrum parameter estimation," *IEEE Trans. Inform. Theory*, IT-11, 100–107, 1965.

187. Rye J.B., R.M. Hardesty, "Estimate optimization parameters for incoherent backscatter heterodyne lidar," *Appl. Opt.*, 36, 9425–9436, 1997.

188. Rye J.B., "Estimate optimization parameters for incoherent backscatter heterodyne lidar including unknown signal bandwidth," *Appl. Opt.*, 39, 6086–6096, 2000.

189. Frehlich R.G., "Cramer–Rao bound for Gaussian random process and application to radar processing of atmospheric signals," *IEEE Trans. Geosci. Remote Sensing*, 31, 1123–1131, 1993.

190. Frehlich R.G. and M.J. Yadlowsky, "Performance of mean frequency estimators for Doppler radar and lidar," *J. Atmos. Ocean. Technol.*, 11(5), 1217–1230, 1994.

191. Doviak R.J. and D.S. Zrnic, *Doppler Radar and Weather Observations*, Equation 6.22b (Academic Press, New York, 1984).

192. Shapiro J.W., "Performance analysis of peak-detecting laser radars," *SPIE Proc.*, 663, 38–56, 1986.

193. Dabas A., "Semiempirical model for the reliability of a narrow band frequency estimator for Doppler lidar," *J. Atmos. Ocean. Technol.*, 16, 19–28, 1999.

194. Spiers G., "The effect of optical aberrations on the performance of coherent Doppler lidar," Proceedings of 10th Coherent Laser Radar Conference, Mount Hood, OR, June 28–July 2, 1999.

195. McIntyre R.J., "Multiplication noise in uniform avalanche diodes," *IEEE Trans. Electron Devices*, ED-13(1), 164–168, 1966.

196. Saleh B.E.A. and M.C. Teich, *Fundamentals of Photonics*, Chapter 9 (John Wiley and Sons, New York, 1991).

197. Vaughan J.M., *The Fabry–Perot Interferometer: History, Theory, Practice and Applications* (Hilger, Philadelphia, PA, 1989).

198. Papoulis A., *Probability, Random Variables, and Stochastic Processes*, 2nd edn, Equation 7.21 (McGraw Hill, New York, 1984).

199. Rye B.J., "Comparative precision of distributed-backscatter Doppler lidars," *Appl. Opt.*, 34, 8341–8344, 1995.

200. McKay J.A., "A comparison of Doppler measurement techniques," presented at the Meeting of the NOAA Working Group on Space-Based Lidar Winds, Daytona, FL, January 21–13, 1997.

201. Bloom S.H., R. Kremer, P.A. Searcy, M. Rivers, J. Menders, and E. Korevaar, "Long-range, noncoherent laser Doppler velocimeter," *Opt. Lett.*, 16, 1794, 1991.

202. Henderson S.W., J.R. Magee, and C.P. Hale, "Injection seeded operation of a Q-switched Cr,Tm,Ho:YAG laser," invited paper, Proceedings of the Meeting on Advanced Solid-State Lasers 6, Salt Lake City, UT, 127, March 5–7, 1990.

203. Lachambre J.L., P. Lavigne, G. Otis, and M. Noel, "Injection locking and mode selection in TEA-CO_2 laser oscillators," *IEEE J. Quantum Electron.*, QE-12(12), 756–764, 1976.

204. Schmid W.E., "Pulse stretching in a Q-switched Nd:YAG laser," *IEEE J. Quantum Electron.*, QE-q6, 790, 1980.

205. Arsen'ev V.A., I.N. Matveev, and N.D. Ustinov, "Nanaosecond and microsecond pulse gereration in solid state lasers (review)," *Sov. J. Quantum Electron.*, 7, 321, 1977.

206. Hamamatsu Photonics KK, InGaAs pin photo-diode product literature, Hamamatsu Photonics KK, Solid State Division, Japan.

207. Willetts D.V. and M.R. Harris, "An investigation into the origin of frequency sweeping in a hybrid TEA CO_2 laser," *J. Phys. D: Appl. Phys.*, 15, 51–67, 1982.

208. Pearson G.N. and C.G. Collier, "A pulsed coherent CO_2 lidar for boundary-layer meteorology," *Q. J. R. Meteorol. Soc.*, 125, 2703–2721, 1999.

209. Tratt D.M., A.K. Karr, and R.G. Harrison, "Spectral control of gain-switched lasers by injection-seeding: applications to TEA CO_2 systems," *Prog. Quantum Electron.*, 10, 229–266, 1985.

210. Woodfield A.A. and J.M. Vaughan, "Airspeed and wind shear measurements with an airborne CO_2 CW laser," *AGARDograph*, 272, 7.1–7.17, 1983.

211. Woodfield A.A. and J.M. Vaughan, "Airspeed and windshear measurements with an airborne CO_2 CW laser," *Int. J. Aviation Safety*, 1, 207–224, 1983.

212. Keeler R.J., "An airborne laser air motion sensing system. Part I: concept and preliminary experiment," *J. Atmos. Ocean. Technol.*, 4(1), 113–127, 1987.

213. Bowdle D.A., J. Rothermel, J.M. Vaughan, and M.J. Post, "Aerosol backscatter measurements at 10.6 micrometers with airborne and ground-based CO_2 Doppler lidars over the Colorado high plains 1. Lidar intercomparison," *J. Geophys. Res.*, 96(D3), 5327–5335, 1991.

214. Bowdle D.A., J. Rothermel, J.M. Vaughan, and M.J. Post, "Aerosol backscatter measurements at 10.6 micrometers with airborne and ground-based CO_2 Doppler lidars over the Colorado high plains 2. Backscatter structure," *J. Geophys. Res.*, 96(D3), 5337–5344, 1991.

215. Bowdle D.A., J. Rothermel, J.M. Vaughan, and M.J. Post, "Evidence of a tropospheric aerosol backscatter background mode," *Appl. Opt.*, 28(6), 1040–1042, 1989.

216. S.B. Alejandro, G.G. Koenig, J.M. Vaughan, and P.H. Davies, "SABLE: a South Atlantic aerosol backscatter measurement program," *J. Am. Meteorol. Soc.*, 71(3), 281–287, 1990.

217. Alejandro S.B., "Atlantic atmosphere aerosol studies — 1. Programme overview," *J. Geophys. Res.*, 100(D1), 1035–1041, 1995.

218. Vaughan J.M., D.W. Brown, C. Nash, S.B. Alejandro, and G.G. Koenig, "Atlantic atmospheric aerosol studies — 2. Compendium of airborne backscatter measurements at $10\,\mu m$," *J. Geophys. Res.*, 100(D1), 1043–1065, 1995.

219. Vaughan J.M., N.J. Geddes, P.M. Flamant, and C. Flesia, "Establishment of a backscatter coefficient and atmospheric database," Report to ESTEC CR980139, Noordwijk, ESA, 1998.

220. Vaughan J.M., D.W. Brown, and D.V. Willetts, "The impact of atmospheric stratification on space-borne Doppler wind lidar," *J. Mod. Opt.*, 45, 1583–1599, 1998.

221. Vaughan J.M., N.J. Geddes, and R.H. Maryon, "Atmospheric backscatter variations: correlation with air parcel movement and fluctuation statistics," Report to ESTEC CR992291, February 2000.

222. Vaughan J.M., R.H. Maryon, and N.J. Geddes, "Comparison of atmospheric aerosol backscattering and air mass back trajections," *Meteorol. Atmos. Phys.*, 79, 33–46, 2002.

223. Huffaker R.M., Ed., Feasibility Study of Satellite-Borne Lidar Global Wind Monitoring System, NOAA Tech. Memo, ERL-WPL-37, 1978. Also see: R.M. Huffaker, T.R. Lawerence, M.J. Post, J.T. Priestly, F.F. Hall, Jr., R.A. Richter, and R.J. Keeler, "Feasibility studies for global wind measuring satellite system (Windsat): analysis of simulated performance," *Appl. Opt.*, 23, 2532–2536, 1984.

224. Delville P., C. Loth, D. Bruneau, P.H. Flamant, Th. Le Floch, and J.-C. Farcy, "A New TE-CO_2 laser for coherent lidar and wind applications," in Proceedings of the Conference on Coherent Laser Radar, Keystone, CO, 297–300, 1995.

225. Delaval A., P.H. Flamant, C. Loth, A. Garnier, C. Vialle, D. Bruneau, R. Vilson, and D. Rees, "VALID-2: performances validation of direct detection and heterodyne detection Doppler wind Lidars", ESA, Final Report. (The report is available on request to ESA/ESTEC.), 2000.

226. Drobinski P., R.A. Brown, P.H. Flamant, and J. Pelon, "Evidence of organized large eddies by ground-based Doppler lidar, sonic anemometer and sodar," *Bound.-Layer Meteorol.*, 88, 343–361, 1998.

227. Drobinski P., A.M. Dabas, C. Haeberli, and P.H. Flamant, "On the small-scale dynamics of flow splitting in the Rhine valley during a shallow foehn event," *Bound.-Layer Meteorol.*, 99, 277–296, 2001.

228. Drobinski P., A.M. Dabas, C. Haeberli, and P.H. Flamant, "Statistical characterization of the flow structure in the Rhine valley," *Bound.-Layer Meteorol.*, 106, 483–505, 2002.

229. Drobinski P., C. Haeberli, E. Richard, M. Lothon, A.M. Dabas, P.H. Flamant, M. Furger, and R. Steinacker, "Scale interaction processes during MAP-IOP 12 south foehn event in the Rhine valley," *Q. J. R. Meteorol. Soc.*, 128, 1–36, 2002.

230. Reitebuch O., Ch. Werner, I. Leike, P. Delville, P.H. Flamant, A. Cress, and D. Engelbart, "Experimental validation of wind profiling performed by the airborne 10 fm-heterodyne Doppler lidar WIND," *J. Atmos. Ocean. Technol.*, 18, 1331–1344, 2001.

231. Reitebuch O., H. Volkert, C. Werner, A. Dabas, P. Delville, P. Drobinski, P.H. Flamant, and E. Richard, "Determination of air flow across the alpine ridge by a combination of airborne Doppler lidar, routine radio-sounding and numerical simulation," *Q. J. R. Meteorol. Soc.*, 129, 715–727, 2003.

232. Dabas A., P. Drobinski, and P.H. Flamant, "Chirp induced bias in velocity measurements by a coherent Doppler CO2 lidar," *J. Atmos. Ocean. Technol.*, 15, 407–415, 1998.

233. Dabas A., P. Drobinski, and P.H. Flamant, "Velocity biases of adaptive filter estimates in heterodyne Doppler lidar measurements," *J. Atmos. Ocean. Technol.*, 17, 1189–1202, 2000.

234. Drobinski P., A. Dabas, and P.H. Flamant, "Remote measurement of turbulent wind spectra by heterodyne Doppler lidar technique," *J. Appl. Meteorol.*, 39, 2434–2451, 2000.

235. Smalikho I. "Technique of wind vector estimation from data measured with a scanning coherent Doppler lidar," *JTECH* 20, 276–291, 2003.

236. Guérit G., P. Drobinski, B. Augère, and P.H. Flamant, "Effectiveness of simultaneous independent realizations at low CNR to improve heterodyne Doppler lidar performance. Part 2: experimental results," *Appl. Opt.*, 41, 7510–7515, 2002.

237. Henderson S.W., E.H. Yuen, and E.S. Fry, "Fast resonance-detection technique for single-frequency operation of injection-seeded Nd:YAG lasers," *Opt. Lett.*, 11, 715, 1986.

238. Phillips M.W., S.M. Hannon, P.G. Wanninger, P.J.M. Suni, and J.A.L. Thomson, "Range-Doppler imaging with a coherent laser radar based on optical fiber amplifiers," Proceedings of the 8th Conference on Coherent Laser Radar, Keystone, CO, 250–253, July 23–27, 1995 (also see, *Laser Focus World*, May 1995).

239. Pearson G.N., J.R. Eacock, K.D. Ridley, and M. Harris, "Pulsed Doppler lidar at 1.55-μm," Proceedings of the 12th Conference on Coherent Laser Radar, Bar Harbor, ME, June 15–20, 2003.

240. Kameyama S., T. Ando, K. Asaka, Y. Hirano, and H. Inokuchi, "A compact all-fiber pulsed coherent Doppler lidar system and its performance evaluation," Proceedings of the 12th Conference on Coherent Laser Radar, Bar Harbor, ME, June 15–20, 2003.

241. McKay J.A. and D. Rees, "Design of a direct detection Doppler wind lidar for spaceflight", *SPIE Proc.* 3494, Laser Radar Techniques (Ranging and Atmospheric Lidar) II, Barcelona, Sept 1998.
242. Hernandez, G., *Fabry–Perot Interferometers* (Cambridge University Press, New York, 1986).
243. Jacquinot, P., "The luminosity of spectrometers with prisms, gratings or Fabry–Perot etalons," *J. Opt. Soc. Am.*, 44, 761–765, 1954.
244. McKay J.A. and D. Rees, "Space qualified etalon," *Opt. Eng.*, 39(1), 315–319, 2000.
245. Garnier A. and M.L. Chanin, "Description of a Doppler Rayleigh lidar for measuring winds in the middle atmosphere," *Appl. Phys. B*, 55, 35–40, 1992.
246. McKay, J.A., "Modeling of direct detection Doppler wind lidar: I — The edge technique," *Appl. Opt.*, 37, 6480–6486, 1998.
247. McKay J.A., "Modeling of direct detection Doppler wind lidar: II — Fringe Imaging," *Appl. Opt.*, 6487–6493, 1998.
248. Rees D., M. Vyssogorets, N.P. Meredith, E. Griffin, and Y. Chaxell, "The Doppler wind and temperature system of the ALOMAR lidar," *J. Atmos. Terr. Phys.*, 58(16), 1827–1842, 1996.
249. Hays P.B., "Circle to line interferometer optical system," *Appl. Opt.*, 29, 1482–1489,1990.
250. Irgang T.D., P.B. Hayes, and W.R. Skinner, "Two-channel direct-detection Doppler lidar employing a charge-coupled device as a detector," *Appl. Opt.*, 41, 1145–1155, 2002.
251. Nardell C., P.B. Hays, J. Pavlich, M. Dehring, and G. Sypitkowski, "GroundWinds New Hampshire and the Lidarfest 2000 campaign," Proceedings of the International Symposium on Optical Science and Technology, SPIE 4484, 36–50, San Diego, CA, July 30–31, 2001.
252. Pain T., Private communication, 2000.
253. Morancais D., F. Fabre, P. Berlioz, R. Mauer, and A. Culoma, "Spaceborne wind lidar concept for the atmospheric dynamics mission ALADIN", in *Advances in Laser Remote Sensing*, 20th International Laser Radar Conference, A. Dabas, C. Loth, and J. Pelon, Eds. Ecole Polytechnique, Palaiseau, France, 15–18, 2000.
254. Banta, R.M., L.S. Darby, J.D. Fast, J.O. Pinto, C.D. Whiteman, W.J. Shaw, and B.D. Orr, 2004: Nocturnal low-level jet in a mountain basin complex. I: Evolution and implications to other flow features. J. Appl. Meteor., 43, 1348–1365.

255. Sun J., D.H. Lenschow, S.P. Burns, R.M. Banta, R.K. Newsom, R.L. Coulter, S. Frasier, T. Ince, C. Nappo, B. Balsley, M. Jensen, D. Miller, B. Skelly, J. Cuxart, W. Blumen, X. Lee, and X.Z. Hu, "Intermittent turbulence in stable boundary layers and its relationship with density currents," *Bound.-Layer Meteorol.*, 105, 199–219, 2002.

256. Steinhagen H., J. Dibbern, D. Engelbart, U. Görsdorf, V. Lehmann, J. Neisserand, and J. Neuschaefer, "Performance of the first European 482 MHz wind profiler radar with RASS under operational conditions," *Meteorol. Z.* NF7, 248–261, 1998.

257. Hinton, D.A., J.K. Charnock, D.R. Bagwell, and D. Grigsby, "NASA aircraft vortex spacing system development status," AIAA 99-0753, 37th Aerospace Sciences Meeting & Exhibit, January 11–14, Reno, NV, 1999.

258. Constant G., R Foord, P.A. Forrester, and J.M. Vaughan, "Coherent laser radar and the problem of aircraft wake vortices," *J. Mod. Opt.*, 41, 2153, 1994.

259. Vaughan J.M., D.W. Brown, G. Constant, J.R. Eacock, and R. Foord, "Structure, trajectory and strength of B747 aircraft wake vortices measured by laser," *NATO AGARD*, CP-584, 10.1–10.10, 1996.

260. Greenwood J. and J.M. Vaughan, "Measurements of aircraft wake vortices at Heathrow by laser Doppler velocimetry," *Air Traffic Control Q.*, 6, 179, 1998.

261. Harris M., J.M. Vaughan, K. Huenecke, and C. Huenecke, "Aircraft wake vortices: a comparison of wind tunnel data with field-trial measurements by laser radar," *Aerosp. Sci. Technol.*, 4, 363–366, 2000.

262. Hannon, S.M. and J. Alex Thomson, "Aircraft wake vortex detection and measurement with pulsed solid-state coherent laser radar," *J. Mod. Opt.*, 41(11), 2175–2196, 1994.

263. Hannon, S.M. and J.A. Thomson, "Real time wake vortex detection, tracking and strength estimation with pulsed coherent lidar," Proceedings of the 9th Conference on Coherent Laser Radar, Linköping, Sweden, 202–205, June 23–27, 1997.

264. Harris, M., R.I. Young, F. Köpp, A. Dolfi, and J.-P. Cariou, "Wake vortex detection and monitoring," *Aerosp. Sci. Technol.*, 6, 325–331, 2002.

265. Keane, M., D. Buckton, M. Redfern, C. Bollig, C. Wedekind, F. Köpp, and F. Berni, "Axial detection of aircraft wake vortices using Doppler lidar," *J. Aircraft*, 39, 850–862, 2002.

266. Köpp, F., S. Rahm, and I. Smalikho, Characterisation of air-craft wake vortices by 2 μm pulsed Doppler lidar, *J. Atmos. Ocean. Technol.*, 21(2), 194–206, 2004.
267. Kavaya M.J., U.N. Singh, W.S. Heaps, and T. Cazeau, "NASA's new laser risk reduction program for future space lidar missions," Proceedings of the 12th Conference on Coherent Laser Radar, Bar Harbor, ME, June 15–20, 2003.
268. Mizutani K., T. Itabe, M. Ishizu, S. Ishii, and K. Asai, "JEM/CDL: A space-borne coherent Doppler lidar program," Proceedings of the 11th Coherent Laser Radar Conference, 15–18, Malvern, UK, July 1–6, 2001.

8

Airborne Lidar Systems

EDWARD V. BROWELL, WILLIAM B. GRANT,
and SYED ISMAIL

Lidar Applications Group,
NASA Langley Research Center, Hampton, VA, USA

723

8.1. INTRODUCTION

Airborne lidar systems have proven very useful for atmospheric and oceanic studies during the past three decades and more recently for surface and vegetation canopy studies. Typical applications of airborne lidar for atmospheric studies include studying the long-range transport of pollutants, taking large-scale surveys of tropospheric aerosols and ozone (O_3) over remote regions of the Earth, studying water vapor (H_2O) and the hydrologic cycle, and investigating various processes associated with biomass burning emissions, desert dust transport, stratospheric aerosol transport following volcanic eruptions, polar O_3 changes, and polar stratospheric clouds (PSCs), and metal ion concentrations in the ionosphere. Airborne lidar systems that are participating in studies of aerosols, O_3, and H_2O can also be used to make correlative measurements of space-based remote-sensing instruments and serve as test beds on the way to space-based lidar systems.

In addition to atmospheric studies, airborne lidar systems have been used for diverse hydrospheric studies, including measurements of chlorophyll, phytoplankton, dissolved organic matter, inorganic suspended material, water depth, and even fish school detection. Airborne lidar systems have been used to study surface properties such as the infrared (IR) reflectivity of desert geological features. Airborne lidar systems have also recently been applied to study the density and structure of the vegetation canopy in forests.

The main advantages of airborne lidar systems are that they expand the geographical range of studies beyond those possible by surface-based fixed or mobile lidar systems by virtue of being able to fly to high altitudes and to remote locations. Thus, they permit measurements at locations inaccessible to surface-based lidar systems. For atmospheric studies, they permit measurements over large regions in times that are short, compared with atmospheric motion, so that large-scale patterns are discernible. Another advantage of an airborne lidar is that the normal lidar technique, which uses aerosols and molecules as distributed reflectors, performs better in the nadir direction than in the zenith direction since the atmospheric density increases with range r (decreasing altitude), compensating somewhat for the $1/r^2$ falloff in lidar signal with range. For the zenith direction, the advantage is that the airborne lidar system is higher and thus closer to the atmosphere being measured. The main disadvantage of using an airborne lidar is the complexity and cost of conducting aircraft operations, a fact that limits the number of airborne missions.

This chapter will examine the specific requirements for airborne lidar, review the important application areas of airborne lidar, and then indicate the direction of future airborne lidar applications.

8.2. SPECIFIC REQUIREMENTS FOR AIRBORNE LIDAR

Operating a lidar on an aircraft places some constraints on the lidar system design and performance. The primary constraints include limitations on size, mass, power, and receiver aperture, and the ability to operate in an environment with high- and low-frequency vibrations and temperature and cabin pressure variations. For example, if prisms are used in the laser cavity to control the laser wavelength, the cavity will have to be enclosed in a pressurized container so that changes in the index-of-refraction of air do not affect the laser wavelength. There may be both 60- and 400-Hz power available, and often the power sources must be shared with other instruments. In addition, the lidar system is generally designed to stay intact under 9-G impact forces so that personnel onboard the aircraft are not hit by flying debris during a hard landing. Finally, the use of toxic and hazardous materials such as flammable solutions should be kept to a minimum.

Airborne lidar systems should be designed and developed to function with limited operator intervention during flight. An extreme example of this design and development is the Lidar Atmospheric Sensing Experiment (LASE) (Browell et al., 1997a; Moore et al., 1997), which was developed to operate in the ER-2 high-altitude aircraft. In the original configuration, it had two switches — "on/off" and "acquire data" — in part because the pilot had his hands full flying the aircraft, and in part because it was developed as a test bed for a future differential absorption lidar (DIAL) system in space.

Another requirement for airborne lidar systems is to work in an eye-safe mode when flying over populated areas or in the vicinity of other aircraft. The simplest way to meet this requirement is to operate at an eye-safe wavelength, such as at a wavelength longer than 1.5 μm. However, while this approach is useful for aerosols and clouds, it is not always possible for DIAL systems that have to be tuned to absorption features of various gases. Thus, the general way to operate in an eye-safe mode involves knowing the eye-safe operating range and reducing the power or stopping laser transmission when the altitude separation falls below the eye-safe level for observers on the ground or in aircraft flying in the path of the beam.

The engineering requirements for airborne lidar systems have been satisfied, in general, as evidenced by the large number of such systems that have been developed and used successfully in many field experiments all over the world.

8.3. SPECIFIC AIRBORNE LIDAR APPLICATION AREAS

8.3.1. Aerosols

Aerosols play important roles in the atmosphere, including affecting visibility, causing direct and indirect effects on the Earth's radiation budget, and contributing to global climate change (Charlson et al., 1992). The transport of material such as desert dust, acid deposition, and chemical reactions in the

atmosphere are also factors. In addition, aerosols play important roles in atmospheric studies by providing tracers of
atmospheric structure and dynamics and by serving as distributed reflectors for lidar measurements involving atmospheric backscatter.

Lidar systems measure backscattered radiation as a
function of time-of-flight or, equivalently, range. From this
measurement, relative aerosol scattering can be determined.
If the molecular profile is known and an aerosol-free region is
found for normalization, the aerosol scattering ratio (ASR) can
be calculated. (ASR is defined as the ratio of aerosol backscattering to molecular backscattering.) Unfortunately, finding
a nearly aerosol-free region for nadir measurements is often
difficult. However, if a zenith measurement is also made, a
calibration in the zenith can be transferred to the nadir by
comparing nadir and zenith measurements during a vertical
ascent or descent. The longer the laser wavelength from the
ultraviolet (UV) to the near-IR spectral region, the higher the
ASR tends to be. This results from the molecular (Rayleigh)
scattering varying as λ^{-4}, while the aerosol (Mie) scattering
generally varies from λ^{-0} to λ^{-2}, depending on the aerosol size
distribution and composition. From the ASR at two wave-
lengths, one can determine the wavelength dependence of
aerosol scattering, $\lambda^{-\alpha}$, where α is similar to the Ångström
coefficient for aerosol extinction:

$$\alpha = 4 - \ln((\text{ASR}_{\text{IR}}/\text{ASR}_{\text{VIS}})/(\lambda_{\text{IR}}/\lambda_{\text{VIS}})) \qquad (8.1)$$

A value near 4 indicates particles that are very small compared to the laser wavelengths, while a value near 0 indicates
particles that are very large compared to the laser wavelengths.

It would be useful to measure aerosol extinction separately from backscatter, but that is generally difficult to do
with the usual elastic backscatter lidar. However, it can be
done by using Raman lidar since the signal at the Raman
wavelength does not contain any backscattered radiation,
which would obscure the extinction. It can also be determined
by using a high-spectral-resolution lidar, which can separate

the Mie scattering (narrowband) from the Rayleigh scattering (broadband) (Shipley et al., 1983; Hair et al., 2001).

In addition, the depolarization ratio can be measured if a polarizer is included in the receiver to separate the parallel and perpendicular components of backscatter. The depolarization ratio is a measure of the departure from sphericity of the aerosols: the more irregularly shaped the aerosols are, the larger the depolarization, with ice crystals (Sassen et al., 2000), PSCs (Browell et al., 1990a,b; Toon et al., 1990), and wind-blown dust (di Sarra et al., 2001; Murayama et al., 2001) having high depolarization ratios.

The pioneering airborne lidar measurements of the atmosphere were made in the late 1970s to the early 1980s using Nd:YAG lasers. These measurements were of aerosol distributions in the planetary boundary layer (PBL) (McElroy et al., 1981; McElroy and Smith, 1986), the behavior of power plant plumes as they were transported in the boundary layer (Uthe, 1984) and the free troposphere (Shipley et al., 1984), and the transport of Saharan dust across the Atlantic Ocean (Talbot et al., 1986). These measurement programs demonstrated that airborne lidar systems could provide useful information and paved the way for other lidar systems that followed.

Airborne lidar measurements of aerosols have been made routinely during such field programs as NASA's Global Tropospheric Experiment (GTE), which have included deployments to many remote regions of the Earth, including Amazonia (Browell et al., 1988), the Canadian boreal forest (Browell et al., 1992), Africa (Anderson et al., 1996; Browell et al., 1996b), the western Pacific (Browell et al., 1996a), and the tropical and South Pacific (Fenn et al., 1999; Browell et al., 2001). One interesting topic is the study of aerosols from biomass burning (Andreae et al., 1988; Anderson et al., 1996; Browell et al., 1996a,b). The aerosols measured by airborne lidar on the GTE missions, in conjunction with the O_3 measurements, are also used to help characterize the air masses encountered (Browell et al., 1996a,b, 2002; Fenn et al., 1999). Other airborne tropospheric aerosol studies include those of Arctic haze (Radke et al., 1989; Brock et al., 1990; Khattatov et al., 1997), volcanic emissions (Hobbs et al., 1991), urban plumes (Hoff and Strawbridge, 1997), aerosol sources in the Lower Fraser Valley, British Columbia (Hoff et al., 1997), and the study of extinction by continental aerosols over the Indian Ocean during the Indian Ocean Experiment (INDOEX) (Leon et al., 2001). Aerosols were also a focus of the Tropospheric Aerosol Radiative Forcing Observational Experiment (TARFOX) conducted over the Atlantic Ocean off the eastern coast of the United States near 38°N (Ferrare et al., 2000a,b; Ismail et al., 2000).

Airborne lidar systems have been used to study sulfate aerosols formed in the stratosphere following the eruption of volcanoes. McCormick and Swissler (1983) used airborne lidar to determine the stratospheric aerosol mass and latitudinal distribution of aerosols following the eruption of El Chichón in 1982. Winker et al. (1992a,b) made similar studies following the eruption of Mt Pinatubo in June 1991. Browell et al. (1993) showed that the Artic polar vortex set up in 1991 before the aerosols from the eruption could reach the Arctic. Grant et al. (1992, 1994) reported on the effects of aerosols from the Mt Pinatubo eruption in reducing the O_3 in the tropical strato-spheric reservoir (TSR). Grant et al. (1994, 1996) also reported on the width of the TSR edge, using aerosol measurements from airborne lidar and space-based instruments. The key finding was that the edge is very narrow ($\sim 1°$) in winter, but much broader ($\sim 10°$) in summer.

Wind measurement using a space-based lidar system would generally rely on aerosols to provide the backscattered signals. There have been several airborne missions designed to study aerosol concentrations and distributions over the oceans in anticipation of placing a Doppler lidar in space. Data from a pair of missions using airborne CO_2 lidar systems at 10.6 μm to determine the latitudinal distribution of aerosol backscatter over the Atlantic Ocean from 1988 to 1990 were reported by Vaughan et al. (1995). While the earlier Doppler winds measuring systems were based on CO_2 lasers, more recent work has moved toward shorter wavelengths as the technology for solid-state lasers has matured (Henderson et al., 1993). Thus, the GLOBE program that studied the latitudinal distribution of aerosols over the Pacific Ocean used lasers operating at 1.06, 9.11, and 9.25 μm to gather information on aerosol backscatter for possible space-based Nd:YAG and CO_2 isotope laser Doppler lidar systems (Menzies and Tratt, 1997; Cutten et al., 1998). The northern hemisphere was found to have similar aerosol backscatter to that in the southern hemisphere at 9.25 μm in November, but one to two orders of magnitude more in the 3 to 12 km altitude region during May and June systems (Menzies and Tratt, 1997). An increased backscatter was found in the western Pacific northern mid-latitudes compared to that in the eastern Pacific above 5 km in May and June.

Several airborne Nd:YAG laser lidar systems measured aerosols and clouds along the ground track of the Lidar In-space Technology Experiment (LITE) during the September 1994 mission (Winker et al., 1996). One set of measurements helped establish that multiple scattering is important even in the PBL for optical depths of a few tenths (Grant et al., 1997). Another set confirmed that LITE could see aerosols in urban plumes (Strawbridge and Hoff, 1996). In addition, it was shown that aircraft measurements of aerosol size distributions and backscatter-to-extinction ratios from the literature can be used to estimate sulfate emission rates from urban/industrial regions (Hoff and Strawbridge, 1997).

Desert dust can be advantageously studied by using airborne lidar systems. The NASA Goddard Space Flight Center's Large Aperture Scanning Airborne Lidar (LASAL) (Palm et al., 1994) participated in the LITE validation program and encountered Saharan dust plumes off the west coast of Africa (Karyampudi et al., 1999). Data from both lidar systems were used, along with a model, to show the influence of meteorological conditions on the plume transport and the effect of the plume to generate a middle-level jet near its southern boundary that affected the transport. Chazette et al. (2001), using an airborne lidar system, later demonstrated that Saharan dust is transported in multiple thin layers of a few hundred meter thicknesses.

The PBL is a region of interest that can be studied easily using airborne lidar systems. The development of the PBL during the diurnal cycle was studied in Amazonia (Browell et al., 1988). Schwiesow et al. (1990) studied the intersection of an aerosol layer with the marine BL. Dupont et al. (1994) compared lidar measurements in the PBL with Mie scattering calculations to show that the extinction coefficient increased with relative humidity (RH). Airborne lidar aerosol measurements can be used with advanced analytical techniques such as wavelet analysis to determine sharp aerosol boundaries at the top of the PBL or for the retrieval of multiple layers in complex, stably stratified regions of the lower troposphere (Davis et al., 2000). Flamant et al. (2001) studied the PBL structure during a cold air outbreak.

The NASA Langley airborne UV DIAL system has been used to develop several techniques for acquiring and processing aerosol data. For one, the dynamic range normally available, which uses one analog-to-digital (A/D) converter, was extended from 12 to 16 bits by employing two A/D converters in series.

8.3.2. Clouds

Clouds, along with aerosols, are the key uncertain factors in the Earth's radiation budget and global climate change (Wielicki et al., 1995; Hansen et al., 1997). They are highly variable in time and space, yet have properties that, on a time-averaged basis, may be fairly regular but have statistical fluctuations. Lidar systems measure clouds in a manner similar to that for aerosols, with the primary differences being that the cloud particles are generally much larger than non-cloud aerosols and have higher extinction; consequently, it is more difficult to probe deeply into many clouds.

Airborne lidar systems are particularly well suited for the study of thin cirrus clouds since the full vertical extent of the cloud can be probed (e.g., Sassen et al., 2000; Pfister et al., 2001). Early airborne lidar studies of cirrus clouds were reported by Spinhirne and Hart (1990). Tropical cirrus clouds were found to have an average depth of 4.5 km and to occur within ±2 km of the tropopause using the multiwavelength lidar system developed by Spinhirne et al. (1997) (Sassen et al., 2000). Pfister et al. (2001), relying on data from the

NASA Langley airborne UV DIAL system, found that tropical cirrus clouds were of two types, those formed by convective outflow and those formed *in situ*, often aided by gravity waves that elevated the air mass to where it cooled and formed thin clouds. Airborne lidars have also been used to follow the microphysics of contrail evolution in which it was determined that the small ice crystals generated by the contrail form the nuclei on which ambient H_2O condenses (Spinhirne et al., 1998). Airborne lidars, by virtue of high sensitivity to aerosols, are able to observe subvisible cirrus clouds. In a summary of observations from the 1970s and the 1980s, it was determined that the majority of thin cirrus clouds is subvisible and has a non-negligible impact on the radiation budget of the tropics (McFarquhar et al., 2000). These papers are examples of how airborne lidar can lead to new insight into such phenomena as the detailed properties of cirrus clouds and their formation.

Optically thick clouds are difficult to characterize because of the large attenuation and multiple scattering associated with them. However, it is possible to analyze cloud properties well out to cloud optical depths (CODs) of 5 to 10, and there is the potential to do so for CODs beyond 10 (Pelon et al., 2000).

Airborne lidar systems can easily determine cloud tops (Spinhirne et al., 1982). The NASA Langley airborne UV DIAL system often senses the tops of marine stratus clouds during the oceanic portions of the GTE missions (e.g., Browell et al., 1996a,b, 2001; Fenn et al., 1999). Another airborne lidar system demonstrated that LITE was able to determine the marine stratus cloud top height to within 50 m (Strawbridge and Hoff, 1996).

8.3.3. Polar Stratospheric Clouds

PSCs are liquid or solid particles that form in the polar stratosphere during cold winters. Type 2 PSCs are composed of large water ice crystals. Type 1 PSCs are smaller and are composed of a combination of other compounds. Type 1a particles are solids formed from HNO_3 and H_2O and are referred to as nitric acid trihydrate or dihydrate. Type 1b PSCs can be liquid and are assumed to be composed of ternary solutions of $H_2SO_4/HNO_3/H_2$ (Browell et al., 1990a,b; Toon et al., 1990, 2000). There is an additional category, Type 1c, which has small, water-rich, nitric acid solid particles (Toon et al., 2000).

Airborne lidar systems are ideal instruments for studying PSCs for several reasons: (1) they can fly long distances under the PSCs, permitting the rapid measurement of their spatial distribution; (2) they can measure backscatter at several wavelengths, providing an indication of the aerosol size distribution; and (3) they can measure the depolarization ratio, permitting a determination of the liquid or solid phase of the particles. Thus, airborne lidar measurements led to the classification of Type 1a and 1b (Browell et al., 1990a,b; Toon et al., 1990) and Type 1c PSCs (Toon et al., 2000).

Airborne lidar systems have also made important contributions to the understanding of the mechanisms of PSC formation. Godin et al. (1994) studied the PSC formation in the lee wave of mountains. Carslaw et al. (1998) analyzed airborne lidar data of PSCs that were formed in the lee wave of Norwegian mountains. Lee waves are due to orographically lifted air that is cooled due to the lower pressure. Thin (100 m) layers were clearly seen, and the progress of formation was plainly evident. Such measurements are made uniquely using lidar systems. Tsias et al. (1999) studied enhanced type Ia PSCs.

8.3.4. Wind Fields

Generally, wind fields are measured with lidar systems by looking at the Doppler shift of aerosol backscatter along the line of sight. These small frequency shifts are generally measured by using heterodyne detection techniques originally developed for radar. To measure vector wind fields, measurements have to be made in at least two directions so that the components along orthogonal directions can be determined. This two-directional measurement can be made from airborne platforms by scanning forward and backward.

The ultimate goal of airborne winds' measurements is the measurement of winds from a space-based Doppler lidar (Baker et al., 1995). Currently, global wind fields are determined primarily from twice-daily radiosonde measurements at numerous sites as well as from limited commercial aircraft reports in flight corridors. Thus, there are vast regions of the Earth where wind fields are known in only a limited fashion with considerable uncertainty, so a space-based winds-measuring lidar system could provide the missing data if sufficient aerosols are present and if the lidar technology can be developed. While CO_2 lasers were originally proposed for space-based wind measurements, solid-state lasers are now the leading candidates due to technological advances and the intended applications (Frehlich, 1995).

The first airborne measurements of winds were made by using a pulsed-CO_2-laser-based Doppler lidar system (Bilbro and Vaughan, 1978; Bilbro, 1984). The lidar system was pointed in the fore and aft directions in order to obtain the horizontal vector wind field. These winds' measurements provided the initial proof of principle of this approach from an aircraft.

Another airborne wind lidar system called wind infrared Doppler lidar (WIND) has been developed through French–German cooperation (Reitebuch et al., 2001; Werner et al., 2001). The system is based on a pulsed 10.6-μm laser transmitter, a heterodyne receiver, and a conical scanning device. The derived wind vector is displayed using a velocity-azimuth display (VAD) technique with a vertical sampling of 250 m. It has been shown that the airborne WIND measurements of the horizontal wind vector are within 1.5 m/s and 5° of other wind measurements and model estimates. These results show the excellent capability of conical scanning Doppler lidars to provide unique insights into mesoscale dynamic processes and the progress that has been made toward a future space-based wind lidar system.

The atmospheric lidar groups of the National Oceanic and Atmospheric Administration (NOAA) Environmental Technology Laboratory, NASA Marshall Space Flight Center, and the Jet Propulsion Laboratory have developed and flown a scanning, 1 J/pulse, CO_2 coherent Doppler lidar capable of mapping three-dimensional volume of atmospheric winds and aerosol backscatter in the PBL and free troposphere. Applications have included the study of severe and nonsevere atmospheric dynamics, intercomparisons with other sensors, and the simulation of prospective satellite Doppler lidar wind profilers. Wind measurements were made for the marine boundary layer (MBL) and near the coastline of the western United States (Rothermel et al., 1998a,b).

Another application area for airborne winds-measuring lidar systems is in wake vortex, microburst, and turbulence detection. The use of an airborne laser radar (lidar) to measure wind velocities and to detect turbulence in front of an aircraft in real time can significantly increase fuel efficiency, flight safety, and terminal area capacity. Two coherent lidar airborne shear sensor (CLASS) systems have been developed: the 10.6-μm CO_2 system (CLASS-10), which is a flying brassboard lidar system, and the 2.02-μm Tm:YAG solid-state system (CLASS-2). Both lidars have shown a wind measurement accuracy of better than 1 m/s (Targ et al., 1996). One problem experienced with CLASS-10 was strong attenuation by H_2O during tests in Florida. Wake vortex turbulence is also an area of interest, especially in terminal areas (Harris et al., 2000; Vaughan and Harris, 2001).

A noncoherent edge technique has also been proposed as a way to measure winds from space (Korb et al., 1997; Flesia and Korb, 1999). In the edge technique, the Doppler-shifted backscattered signal is attenuated by different amounts as it passes through an interferometer in the receiver, and the amplitude of the resulting signal is a measure of frequency shift. While the edge technique has not been used on an airborne platform, it will be tested on such a platform if it is deemed a likely candidate for space-based lidar wind measurements.

The airborne Doppler lidar systems have provided much of the information required to evaluate the prospects for future aircraft- and space-based wind measuring lidar systems. The emphasis has now shifted to developing the appropriate lasers and other technology that will operate reliably from space.

8.4. Differential Absorption Lidar

8.4.1. Global O_3 Measurements

The first airborne DIAL system was flown for O_3 and aerosol investigations in conjunction with the Environmental Protection Agency's Persistent Elevated Pollution Episodes (PEPE) field experiment conducted over the east coast of the United States in the summer of 1980 (Browell, 1983; Browell et al., 1983). This initial system has evolved into the advanced UV DIAL system that will be described in the next section. Airborne O_3 DIAL systems have been developed by several other groups as well (Uthe et al., 1992; Moosmueller et al., 1994, Wirth and Renger, 1996; Ancellet and Ravetta, 1997, 1998).

The current version of NASA Langley's airborne UV DIAL system has been described in detail in two recent publications (Richter et al., 1997; Browell et al., 1998). This system uses two 30-Hz, frequency-doubled Nd:YAG lasers to sequentially pump two dye lasers that are frequency-doubled into the UV to produce the on-line (288.2 nm/301 nm) and off-line (299.6 nm/310 nm) wavelengths for DIAL O_3 measurements during tropospheric and stratospheric missions, respectively. The residual 1064- and 590- to 620-nm beams from the frequency doubling processes of the Nd:YAG and dye lasers, respectively, are also transmitted for aerosol and cloud measurements. This system has a demonstrated absolute accuracy for O_3 measurements of better than 10% or 2 parts per billion by volume (ppbv), whichever is larger, and a measurement precision of 5% or 1 ppbv with a vertical resolution of 300 m and averaging times of 5 min (about 70-km horizontal resolution at typical DC-8 ground speeds) (Browell, 1983, 1989; Browell et al., 1983, 1985; Grant et al., 1998). Table 8.1 gives the parameters of the NASA LaRC airborne UV DIAL system.

The NASA Langley airborne UV DIAL system has been used as a test bed for developing advanced techniques for processing O_3, aerosol, and cloud data sets. Real-time data analysis is conducted onboard the aircraft during missions. The processed O_3 and aerosol data (as false color images and line profiles) are distributed throughout the aircraft for use in

Table 8.1 Parameters of the NASA LaRC Airborne UV DIAL System

Lasers		Nd:YAG-pumped dye lasers, frequency doubled into the UV	
Pulse repetition frequency (Hz)		30	
Pulse length (nsec)		8–12	
Pulse energy (mJ)			
At 1.06 µm		250–300	
At 600 nm		50–70	
For troposphere, at 288/300 nm		20	
For stratosphere, at 301/310 nm		20	
Dimensions (l × w × h) (cm)		594 × 102 × 109	
Mass (kg)		1735	
Power requirement (kW)		30	
Receiver			
Wavelength region (nm)	289–311	572–622	1064
Area (m^2)	0.086	0.086	0.864
Receiver optical efficiency (%)	30	40	30
Detector quantum efficiency (%)	26 (PMT)	8 (PMT)	40 (APD)
Field-of-view (mrad)	≤1.5	≤1.5	≤1.5
System performance			
Measurement range (km)	Up to 10–15 (nadir and zenith)		
Vertical resolution (m)	300–1500, depending on range		
Horizontal resolution (km)	≤70		
Absolute measurement accuracy	≤10% or 2 ppbv, whichever is greater		
Measurement precision	≤5% or 1 ppvb, whichever is greater		

possible flight changes and for interpretation of *in situ* measurements. For final processing, O_3 data are submitted to the campaign archive within 3 to 6 months after the end of a field mission. In addition, the O_3 data are compared to *in situ* O_3

measurements from instruments onboard the aircraft or ozonesonde data to double check that the absorption coefficients used in the DIAL calculation are correct. One of the early improvements in data processing was the application of the Bernoulli technique to reduce the O_3 DIAL error in the presence of inhomogeneous aerosol distributions (Browell et al., 1985). In the Bernoulli technique, the lidar equation is solved to derive aerosol scattering and extinction profiles by assuming that a relationship exists between backscattering and extinction, and that the value of scattering (or extinction) is known at one location on the profile. This approach helps to correct for erroneous O_3 features that occur in regions of rapidly changing aerosol scattering along the DIAL profile. Another advancement is the use of *in situ* O_3 data measured at the aircraft altitude to provide data to use for interpolating the O_3 values across the region where the UV DIAL system does not measure, which extends from about 750 m above to 750 m below the aircraft. Since the aircraft is a moving platform, it often rolls during turns and changes altitude. Both maneuvers affect the data and are compensated for by indexing the lidar data to the vertical component of range and by referencing the data to altitude instead of range from the aircraft. Both corrections are made automatically by using the platform data stored during the flight.

The NASA Langley airborne UV DIAL systems have made significant contributions to the understanding of both tropospheric and stratospheric O_3. They have been used in 23, mainly international, field experiments since 1980, including 17 tropospheric missions, five stratospheric missions, and one upper-tropospheric, lower-stratospheric mission.

These field experiments have ranged over, or near, most of the Earth's oceans and all of the Earth's continents. A few examples of the scientific contributions made by these airborne UV DIAL systems are given in Table 8.2 (see also the data sets obtained by these systems at our web site, http://asd-www.larc.nasa.gov/lidar/lidar.html and the review by Browell et al., 1998) as well as in Figure 8.1 to Figure 8.5.

One of the standard products of the UV DIAL system measurements on the GTE field missions is a characterization

Table 8.2 Examples of Significant Contributions of Airborne O_3 DIAL Systems to the Understanding of Tropospheric and Stratospheric O_3

Measurements	References
Troposphere	
Tropopause fold event	Browell et al. (1987)
Air mass characterizations	Browell et al. (1996a,b, 2001) and Fenn et al. (1997, 1999)
Biomass burn plumes	Browell et al. (1988, 1994, 1996b)
Continental pollution plumes	Browell et al. (1994, 1996a) and Alvarez et al. (1998)
Convective outflow	Browell et al. (1996a)
Stratospheric intrusions and stratospherically influenced air	Browell et al. (1987, 1992, 1996a,b, 2001) and Ancellet (2001)
Cirrus cloud investigations	Newell et al. (1996) and Pfister et al. (2001)
Effect of biomass burning on tropospheric O_3 production	Browell et al. (1996b)
Absorbing aerosol interference with TOMS O_3 measurements	Fishman et al. (1996)
Power plant plume studies	Banta et al. (1998), Senff et al. (1998), and Valente et al. (1998)
Warm conveyor belt transport	Grant et al. (2000)
Decay of a cutoff low	Ravetta and Ancellet (2000)
Pollution capping by stratospheric intrusion	Cho et al. (2001)
Springtime Arctic O_3 maximum	Browell et al. (2003)
Stratosphere	
Chemical explanation of behavior of Antarctic O_3	Ko et al. (1989)
Intercomparison of O_3 measurements over Antarctica	Margitan et al. (1989)
Quantification of O_3 depletion in the Arctic	Browell et al. (1990a, 1993)
Polar stratospheric cloud particle characterizations	Browell et al. (1990b), Toon et al. (2000), and Butler et al. (2001)
Cross-vortex boundary transport	Flentje et al. (2000)
O_3 reduction in TSR after Mt Pinatubo and TSR edge extent	Grant et al. (1994, 1996)
Intercomparison with ground- and space-based instruments	Grant et al. (1998)

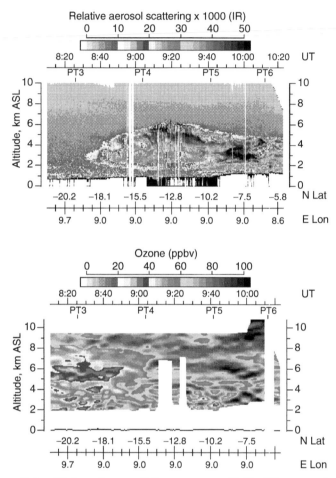

Figure 8.1 (Color figure follows page 398). Biomass burning plume over Atlantic Ocean arising from biomass burning in central part of western Africa observed on October 14, 1992, during the TRACE-A mission.

of the observed air masses (Browell et al., 1996a,b, 2001; Fenn et al., 1999). To classify the air masses, DIAL-measured O_3 and aerosol profiles are used along with the meteorological analysis of potential vorticity (PV). A reference O_3 profile is developed for the mission that maximizes the sensitivity to discrimination of air with elevated O_3 and of air with low O_3. For example, the reference O_3 profile for the South Pacific

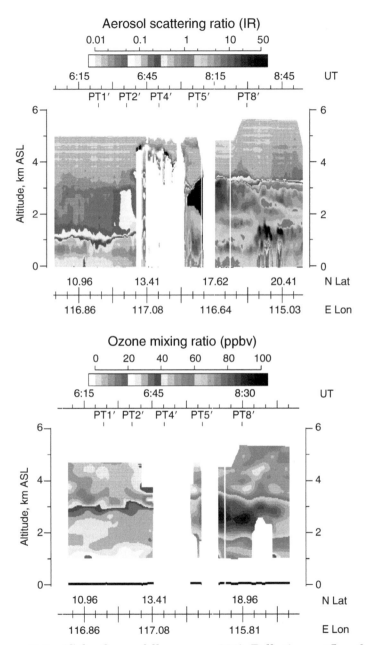

Figure 8.2 (Color figure follows page 398). Pollution outflow from China over South China Sea (right side of figure) with clean tropical air on south side of front (left side of figure) observed on February 21, 1994, during PEM West B.

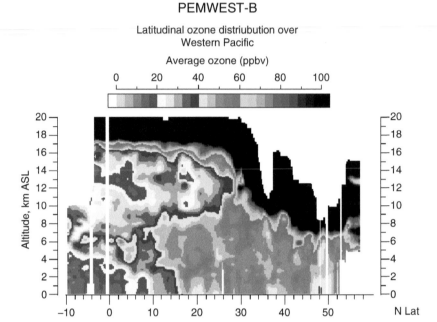

PEMWEST-B

Latitudinal ozone distriubution over
Western Pacific

Figure 8.3 (Color figure follows page 398). Latitudinal distribution of ozone over western Pacific Ocean obtained during Pacific Exploratory Mission (PEM West B) in February to March 1994.

during September to October 1996 started at 20 ppbv at the surface, increased to 42 ppbv at 5 km, and then to 65 ppbv at 18 km (Fenn et al., 1999). The amount of PV in the air was also used to indicate the contribution of stratosphere–troposphere exchange to the air mass characteristics. A total of 8 to 11 air mass types have been classified using this approach. Among the elevated-O_3 air masses are high plume air masses with high O_3 and enhanced aerosols. This air mass type is further subdivided according to the amount of PV or stratospherically influenced air contained in it. Following the classification procedure, we can use the *in situ* data to generate tables that contain the atmospheric properties and the average aerosol and gas values for each air mass type.

The NASA Langley UV DIAL system made some interesting observations of the TSR in the months and years following the eruption of Mt Pinatubo in the Philippines in June

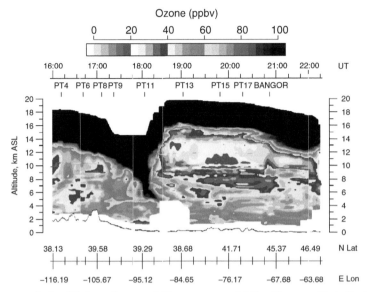

Figure 8.4 (Color figure follows page 398). Ozone distribution observed on October 13, 1997, on a flight across the United States during the SASS (Subsonic Assessment) Ozone and Nitrogen Experiment (SONEX). Stratospheric intrusion clearly evident on the left side of the figure, and low ozone air from tropics transported to midlatitudes can be seen in the upper troposphere on the right.

1991, a time when, due to the extremely heavy aerosol loading during the first year after the eruption, some of the instruments, such as the Stratospheric Aerosol and Gas Experiment II (SAGE II), a solar occultation instrument, were unable to measure in the TSR. The UV DIAL system under flew the TSR in January and February 1992 and was able to clearly see the relationship between the aerosol loading and the O_3 loss as well as the sharp edge of the TSR on the winter side (Grant et al., 1992, 1994).

Airborne O_3 DIAL systems can also be used to study fine structure such as filaments and laminae in the stratosphere. Heese et al. (2001) used an airborne O_3 DIAL system to study stratospheric O_3 filaments extruded from the polar vortex and compared them with a stratospheric transport model. In an NASA mission conducted in 1995 to 1996, the UV DIAL system under flew the TSR and observed laminae of midlatitude

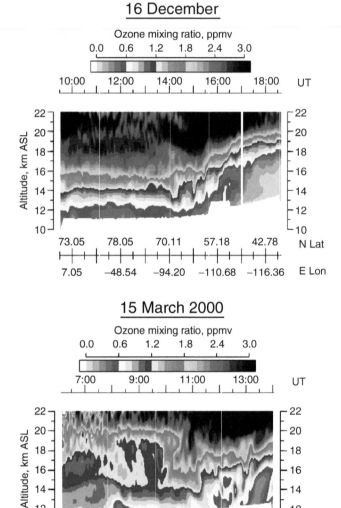

Figure 8.5 (Color figure follows page 398). Ozone cross-sections in stratosphere measured in winter of 1999/2000 during SOLVE mission. The change in ozone number density in the Arctic polar vortex due to chemical loss during winter is clearly evident at latitudes north of 72°N.

air masses with reduced aerosol loading and elevated O_3 intruding into the TSR, with an amplitude that decreased toward the equator (Grant et al., in preparation).

An interesting case study was that of a low O_3 air mass encountered between 8 and 14 km over most of the eastern United States in October 1997. The NASA Langley UV DIAL system was on the first leg of a mission, flying from Edwards Air Force Base, California, to Bangor, Maine. A stratospheric intrusion was also observed at the western edge of this air mass. The O_3 column measurements by the Total Ozone Mapping Spectrometer (TOMS) also showed low O_3 over the eastern United States, with higher O_3 over the Rocky Mountains. Backward trajectories showed that the air mass originated in the tropics, an example of a warm conveyor belt transport of tropical air north and upward, with return flow from the Arctic south and downward, which occurs fairly frequently over the United States from September through March.

The UV DIAL system has often observed stratospheric intrusions into the troposphere in the midlatitudes starting with the first airborne lidar observation of a tropospheric fold event over the southwest United States in 1984 (Browell et al., 1987). The fold occurred shortly after the eruption of El Chichón, and elevated sulfate aerosols arising from that eruption were also clearly seen along with the elevated O_3 in the fold. Examples are given in Browell et al. (1996a,b; 1998; 2003), Fenn et al. (1999), and Cho et al. (2001).

A number of key issues could be addressed by a space-based O_3 DIAL system. Space-based DIAL systems have been considered since the early 1970s (Wright et al., 1975) and include a number of issues relating to O_3, such as photochemical O_3 production/destruction and transport in the troposphere; location of the tropopause; and stratospheric O_3 depletion and dynamics. High-resolution airborne O_3 DIAL and other aircraft measurements show that to study tropospheric processes associated with biomass burning, transport of anthropogenic pollutants, tropospheric O_3 chemistry and dynamics, and stratosphere–troposphere exchange, a vertical profiling capability with a resolution of 2 to 3 km is needed (e.g., Browell et al., 1996b; Fenn et al., 1999; Newell et al.,

1999), and this capability is not currently available from passive remote-sensing satellite observations.

A space-based O_3 DIAL system would be able to make critically needed global measurements of O_3 across the troposphere, which would yield a much better understanding of tropospheric O_3 production and transport, especially in remote regions. An example of the type of latitudinal O_3 cross-section that could be provided by a space-based O_3 DIAL system is shown in Figure 8.3 (Browell et al., 1998). This figure shows many different aspects of O_3 loss and production and vertical and horizontal transport and stratosphere–troposphere exchange that occur from the tropics to high latitudes. This type of data would be available from just one pass from a space-based O_3 DIAL system. In addition, a space-based O_3 DIAL system optimized for tropospheric O_3 measurements would also permit high-resolution O_3 measurements in the stratosphere (1-km vertical, 100-km horizontal), along with high-resolution aerosol measurements (100-m vertical, 10-km horizontal). In addition, these DIAL measurements will be useful to assist in the interpretation of passive remote-sensing measurements and to help improve their data processing algorithms (e.g., Browell et al., 1994; Grant et al., 1998). The development of a space-based O_3 DIAL system has been recently described by Browell et al. (1997b, 1998).

8.4.2. Global Water Vapor Measurements

Water vapor (H_2O) and the hydrologic cycle are very important factors in weather and climate (Renno et al., 1994). H_2O also enhances the greenhouse effect through feedback (Ramanathan and Vogelmann, 1997; Held and Soden, 2000). H_2O contributes much of the energy to hurricanes, so that improved measurements of H_2O around hurricanes would improve the ability to forecast their strength and direction (Dabberdt and Schlatter, 1996; Kamineni et al., 2003; Kamineni et al., 2005). H_2O and O_3 are important to the formation of OH in the troposphere, and OH is at the center of most of the chemical reactions in the lower atmosphere. In addition to influencing the production of OH, H_2O is an excellent tracer of vertical and horizontal transport of air masses in the troposphere, and it can be used as a tracer of stratosphere–troposphere exchange. Increased aerosol sizes due to high relative humidities can also affect heterogeneous chemical processes in the boundary layer and in cloud layers. Thus, knowledge of H_2O distributions can be used in many different ways to better understand chemical and transport processes that influence the composition of the global troposphere.

More accurate knowledge of H_2O distributions in the upper troposphere is needed for more accurate calculations of radiation budgets and a better understanding of the impact of H_2O on global warming (Hansen et al., 1984; Arking, 1990; Rind et al., 1991; Shine and Sinha, 1991). Strong H_2O absorption in the center of the pure rotation band (100 to $400\,\mathrm{cm}^{-1}$) is the principal contribution to upper tropospheric cooling rate (Clough et al., 1992). In addition to this direct effect on radiative fluxes, upper tropospheric humidity has a strong influence on the development and dissipation of cirrus clouds, which, in turn, influence the atmospheric radiative balance. Upper tropospheric H_2O measurements are also required to study atmospheric transport, convection, and cirrus clouds (Newell et al., 1996), stratosphere–troposphere exchange (Langford and Reid, 1998), and polar stratospheric studies (Browell et al., 1990a,b).

The first H_2O DIAL system was flown in 1982 (Browell et al., 1981, 1984; Browell, 1983), and several new airborne H_2O DIAL systems have been developed and deployed by this same group over the last 16 years (Browell et al., 1991, 1997a; Higdon et al., 1994; Browell and Ismail, 1995). Other groups have also developed airborne H_2O DIAL systems (see, e.g., Ehret et al., 1993, 1996, 1998, 1999; Kiemle et al., 1997; Quaglia et al., 1997; Fix et al., 1998; Wulfmeyer and Feingold, 2000; Bruneau et al., 2001a,b; Poberaj et al., 2001; Wulfmeyer and Walther, 2001a,b).

In the initial step toward the development of a space-based H_2O DIAL system, the first airborne H_2O DIAL system was developed and demonstrated in 1982 (Browell, 1983). This system was based on Nd:YAG-pumped dye laser technology, and it was used in the first airborne H_2O DIAL investigation of the MBL over the Gulf Stream (Browell et al., 1984). This laser was later replaced with a flashlamp-pumped solid-state alexandrite laser, which had high spectral purity. Spectral purity is defined as the ratio of energy contained within a narrowly defined spectral region to the energy outside that region. This system was used to make accurate H_2O profile measurements across the lower troposphere (Higdon et al., 1994). The first fully autonomously operating DIAL system called LASE (Lidar Atmospheric Sensing Experiment) was developed as a prototype for a space-based H_2O DIAL system (Browell and Ismail, 1995; Browell et al., 1997a; Moore et al., 1997). While the LASE system was initially designed for operation from the NASA high-altitude ER-2 aircraft, it was later modified to fly on conventional medium-altitude aircraft as well (Ferrare et al., 2000a,b; 2001; Ismail et al., 2001).

LASE uses a Ti:sapphire laser that is pumped by a double-pulsed, frequency-doubled Nd:YAG to produce laser pulses in the 815-nm absorption band of H_2O. The wavelength of the Ti:sapphire laser is controlled by injection seeding with a diode laser that is frequency locked to an H_2O line by using an absorption cell. Each pulse pair consists of on-line and off-line wavelengths for the H_2O DIAL measurements. Operation of LASE on the side of the H_2O absorption line permits the use of the same H_2O line for different absorption cross-sections (Sachse et al., 1993, 1995), permitting the measurement of H_2O concentrations across four orders of magnitude of dynamic range over a 20-km altitude range. The accuracy of LASE H_2O profile measurements was determined to be better than 6% or 0.01 g/kg, whichever is larger, over the full dynamic range of H_2O concentrations in the troposphere (Browell et al., 1997a). LASE has participated in over eight major field experiments since 1995 (e.g., Browell et al., 1997a, 2000a,b, 2001; Ferrare et al., 1999, 2000a,b, 2001; Ismail et al., 2000, 2001; Clayton et al., 2001). The use of LASE data in Florida State University forecast models from two NASA field experiments over the Atlantic in 1998 and 2001 showed improvements in moisture, track and intensity forecasts consistently (Kamineni et al., 2003 and 2005). See

Table 8.3 for the parameters of the LASE H_2O DIAL system and Table 8.4 for a listing of topics that have been studied using airborne H_2O DIAL measurements, as well as Figure 8.6 to Figure 8.8.

The technology for a space-based H_2O DIAL system is rapidly maturing in the areas of high-efficiency, high-energy, high-spectral-purity, long-life lasers with tunability in the 815- and 940-nm regions (see, e.g., Barnes, 1998); low-weight, large-area, high-throughput, high-background-rejection receivers; and high-quantum-efficiency, low-noise, photon-counting detectors. With the expected advancements in lidar technologies, a space-based H_2O DIAL system could be flown on a long-duration space mission in the near future.

Space-based DIAL measurements can provide global H_2O profiling capability, which when combined with passive remote sensing with limited vertical resolution, can lead to three-dimensional measurements of global H_2O distributions. High vertical resolution H_2O ($\leq 1\,km$), aerosol ($\leq 100\,m$), and cloud top ($\leq 50\,m$) measurements from the lidar along the satellite ground track can be combined with the horizontally contiguous data from nadir passive sounders to generate a more complete high-resolution, three-dimensional H_2O, aerosol, and cloud fields for use in the various studies indicated above (Smith, 1991). In addition, the combination of active and passive measurements can provide significant synergistic benefits leading to improved temperature and relative humidity measurements. There is also strong synergism with aerosol and cloud imaging instruments and with future passive instruments that are being planned or proposed for missions addressing atmospheric chemistry, radiation, hydrology, natural hazards, and meteorology.

8.5. RESONANCE FLUORESCENCE LIDAR

Resonance fluorescence lidar systems are used to measure metal ions in the ionosphere to study various processes in that atmospheric region such as dynamics and the origin and chemical evolution of the ions. Ground-based fluorescence

Table 8.3 Parameters of the LASE H_2O DIAL System

Laser — Ti:sapphire	
Wavelength (nm)	813–818
Pulse energy (mJ)	100
Pulse-pair repetition frequency	5 (on- and off-line pulses separated by 300 μsec)
Linewidth (pm)	<0.25
Stability (pm)	<0.35
Spectral purity (%)	>99
Beam divergence (mrad)	<0.6
Pulse width (nsec)	35
Receiver	
Area (m^2)	0.11
Receiver optical efficiency (%)	50 (night), 35 (day)
Avalanche phododiode (APD) detector quantum efficiency (%)	80
Field-of-view (mrad)	1.0
Noise equivalence power (W $Hz^{-0.5}$)	2×10^{-14}
Excess noise factor	3
System performance	
Measurement range (altitude) (km)	15
Range resolution (m)	300–500
Measurement accuracy (%)	5–10

Table 8.4 Examples of Significant Contributions of Airborne H_2O DIAL Systems to the Understanding of H_2O Distributions

Measurements	References
Marine boundary layer over Gulf Stream	Browell et al. (1984)
H_2O transport at a land/sea edge	Higdon et al. (1994)
Large-scale H_2O distributions across troposphere	Browell and Ismail (1995)
Correlative *in situ* and remote measurements	Browell et al. (1997a)
Boundary layer humidity fluxes	Kiemle et al. (1997)
Boundary layer development	Ismail et al. (1998)
Cirrus cloud measurements	Browell et al. (1998)
Hurricane studies	Ferrare et al. (1999), Browell et al. (2000a), and Ismail et al. (2001)
Lower-stratospheric H_2O studies	Ehret et al. (1999)

Table 8.4 Examples of Significant Contributions of Airborne H₂O DIAL Systems to the Understanding of H₂O Distributions—(Continued)

Measurements	References
Relative humidity effects on aerosol sizes	Ferrare et al. (2000)
Ice supersaturation in the upper troposphere	Ferrare et al. (2001a)
Stratospheric intrusions	Clayton et al. (2001)
H₂O distributions over remote Pacific Ocean	Browell et al. (2000b, 2001)
Characterization of upper tropospheric water vapor	Ferrare et al. (2004)
4-D water vapor forecast model development and to improve Quantitative Precipitation Forecast	Wulfmeyer et al. (2004)
Convection initiation studies	Wakimoto et al. (2005)

lidar systems have been used to study the ionosphere since the 1960s, but aircraft fluorescence lidar systems were not deployed until 1990. They have made important contributions to the study of both the chemistry and dynamics of metal ions in the ionosphere.

For example, gravity waves that can propagate upward from convective activity, as well as horizontal winds that are advected upwards by orographic features such as mountain ranges, can be detected using resonance fluorescence of Na (Hostetler and Gardner, 1994; Manson et al., 1998). For these measurements, an excimer laser-pumped dye laser was employed. Gravity waves excite sporadic Na layer formation because the gravity waves move air masses up or down to different temperature regions and because the existence of Na as a free ion increases rapidly with temperature (Gardner et al., 1995).

The measurements of the ionosphere during the Airborne Lidar and Observations of Hawaiian Airglow-90 (ALOHA-90) campaign are presented in the July 1991 issue of *Geophysical Research Letters*, and the important contributions to the understanding of the ionosphere resulting from ALOHA-93

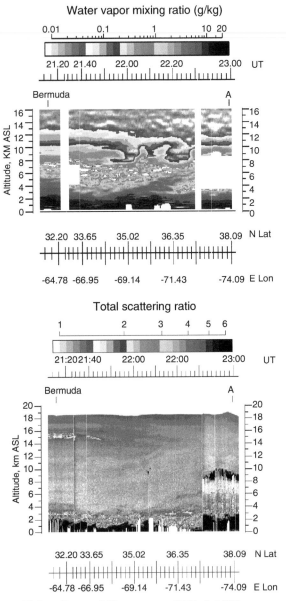

Figure 8.6 (Color figure follows page 398). LASE measurements of water vapor (top) and aerosols and clouds (bottom) across troposphere on an ER-2 flight from Bermuda to Wallops on July 26, 1996, during Tropospheric Aerosol Radiative Forcing Experiment (TARFOX).

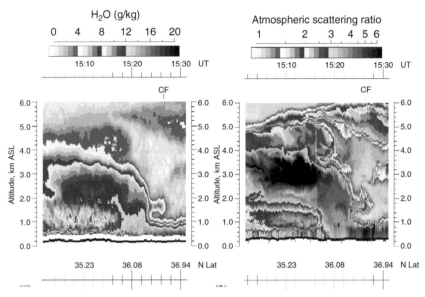

Figure 8.7 (Color figure follows page 398). Water vapor and aerosol cross-section obtained on July 14, 1997, on a flight across a cold front during Southern Great Plains (SGP) field experiment conducted over Oklahoma.

campaign are summarized in the March 27, 1998, issue of the *Journal of Geophysical Research*. The University of Illinois Na lidar was flown in these campaigns. Qian et al. (1998) have presented results from studies of the sporadic Na layers, and Swenson et al. (1998) have described the dynamical and chemical aspects of a mesospheric Na "wall" event.

Another topic of interest is the source and sinks of metal ions. Meteor ablation is the primary source in this region. Thus, studying meteor Fe ablation trails helps in this effort (Chu et al., 2000). A frequency-doubled alexandrite laser was used, based on an idea presented by Gelbwachs (1994). The main advantage of an airborne fluorescence lidar over a similar ground-based lidar is the ability to cover more sky since the aircraft speed is three to five times that of the ionospheric winds, and this provides the opportunity to generate a "snapshot" of the region. This advantage is especially important for short-lived phenomena such as meteor showers

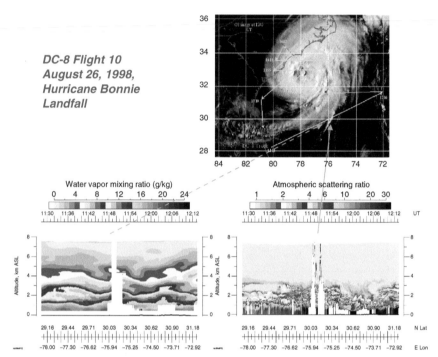

Figure 8.8 Measurements of water vapor, aerosols, and clouds in the inflow region of Hurricane Bonnie on August 26, 1998, during the Convection and Moisture Experiment (CAMEX-3). A rain band can be clearly seen at the middle of Leg-AB on satellite and LASE cross-sections.

where it is desired to observe as many events as possible in order to obtain better statistical confidence in the results. In addition, the aircraft has the ability to go to regions of the Earth where the features of interest can be better studied.

8.6. RAMAN LIDAR

A Raman lidar was developed for methane (CH_4) measurements so that changes in polar stratospheric O_3 could be related to a conservative tracer (Heaps and Burris, 1996). This system was applied in the NASA Tropical Ozone Transport Experiment/Vortex Ozone Transport Experiment (TOTE/

VOTE), which was conducted primarily from Hawaii and Fairbanks, Alaska, between December 1995 and February 1996. The uncertainties in the CH_4 measurements prevented them from being used in conjunction with the O_3 DIAL measurements to quantify the chemical O_3 loss in the Arctic vortex. This system also demonstrated a capability for making Raman measurements of H_2O distributions in the lower stratosphere.

8.7. ATMOSPHERIC TEMPERATURE, DENSITY

The Raman lidar approach has been used for an airborne lidar system that can be used for profiling stratospheric temperature (Heaps et al., 1997). The capability was initially demonstrated during the TOTE/VOTE field mission, and this system was used to make temperature measurements from ~ 4 km above the aircraft to over 70 km above sea level (ASL) (Burris et al., 1998). When compared with microwave temperature profiles, the lidar measurements were an average of 2.3 K warmer. This approach was also used by the NASA Goddard Space Flight Center and Langley Research Center's Airborne Raman Ozone, Aerosol, and Temperature Lidar (AROTEL) in the SOLVE field experiment conducted in the Arctic from December 1999 to March 2000 (Burris, J., et al.). In addition to the Raman temperature measurement across the lower stratosphere, the AROTEL system used the Rayleigh backscatter from the aerosol-free atmosphere above about 30 km to derive the atmospheric density and the atmospheric temperature. The combination of these two techniques allowed the AROTEL system to provide continuous temperature profiles from a few kilometers above the aircraft to above 40 km altitude. In addition, AROTEL also provides multiple wavelength lidar backscatter measurements at 532 and 1064 nm for studies of stratospheric aerosols and PSC and DIAL measurements of O_3 profiles using 308 and 355 nm laser wavelengths from a XeCl excimer laser and tripled Nd:YAG laser, respectively. The O_3 measurements from AROTEL and the NASA Langley UV DIAL system, which was also in SOLVE, were used in the estimation of polar O_3 loss during the winter of 1999/2000 (Schoeberl et al., 2002; Grant et al., 2002).

8.8. HYDROSPHERIC LIDAR

Just as lidars have proven invaluable in studying the atmosphere, they have also been applied to the study of oceans and other bodies of water. Some of the same techniques used in atmospheric studies are also used in hydrospheric studies: time-of-flight is used to determine depth of the bottom; fluorescence is used to measure organic material, to study phytoplankton distribution, and to detect oil films on the surface; Raman scattering by water is used to normalize the lidar returns (Bristow et al., 1981); multiple scattering considerations are used to help determine the turbidity of the water. There are a number of reviews of the field: Measures (1984), Guenther (1989), and Lutomirski (1994).

Water has its minimum attenuation between about 480 and 580 nm, depending on what is in the water. Thus, the blue–green spectral region is best suited to bathymetry (depth-sounding) lidar. Thus, the frequency-doubled Nd:YAG laser is well suited for this purpose (Northam et al., 1981), especially when a high laser PRF is used, such as the 168-Hz laser discussed by Penny et al. (1989).

Hydrospheric lidar can be used to study subsurface scattering layers in an analogous manner, as in studying aerosol layers with atmospheric lidar systems. Hoge and Swift (1993a) discussed using the NASA Airborne Oceanographic Lidar (AOL) to demonstrate the capabilities of their lidar system for this application.

Much of the work with lidar systems in hydrospheric studies has been done either to demonstrate the potential for space-based lidar systems (Bartsch et al., 1992) or to help characterize and validate passive optical remote-sensing instrument measurements of the characteristics of bodies of water (Hoge, 1988). An example of using a lidar system to help validate the use of a passive remote-sensing system for eventual use in space is given in the study of the upwelling spectral radiances from chlorophyll pigment (Hoge et al., 1993a). They used the NASA AOL, which uses a frequency-doubled Nd:YAG laser (150 mJ/pulse at 10 Hz), a spectroradiometer system in the lidar receiver, as well as a passive solar-induced spectroradiometer (Hoge and Swift, 1986). The spectrometer is normally used with 32 contiguous channels, each 11.25 nm in width, between 410 and 770 nm. For this mission over the North Atlantic Ocean, the spectrometer was adjusted to a range from 381 to 740 nm. The system is mounted in a P-3A Orion aircraft, which is ideally suited for flying slowly at moderately low altitudes over the ocean. The water-Raman-normalized lidar signals at 685 nm were compared with signals at various pairs of wavelengths from the passive radiometer chosen to simulate a space-based instrument to determine the best pair to use.

One of the more interesting experiments in which the AOL was involved was to test the iron hypothesis for phytoplankton; that is, that iron is the limiting factor controlling the growth of phytoplankton in the open oceans. It was found that iron was indeed a limiting factor (Martin et al., 1994). This finding might open an avenue for sequestering carbon in the oceans. Additional studies were reported later (Hoge et al., 1998a).

Recent studies with this fluorescent lidar showed that the fluorescence or chromophoric (colored) dissolved organic matter (CDOM) is very nearly linearly related to the absorption of the CDOM (Hoge et al., 1993b). Further studies of DOM were made off the coasts of the North Pacific and North Atlantic Oceans to determine, among other things, the correlation with phytoplankton (Hoge et al., 1993c; Vodacek et al., 1995). The airborne fluoresensor has been used to study the enhancement in phytoplankton arising from iron enrichment of the ocean's surface (Hoge et al., 1998a) and the distribution of phytoplankton around the Galapagos Islands, with the findings linked to the dynamics of the ocean (Hoge et al., 1998b). More recently, Hoge et al. (2002) reported measurements of the spatial variability of the phytoplankton in the Middle Atlantic Bight and confirmed the observation of Sea-viewing Wide Field-of-view Sensor (SeaWiFS) that CDOM was being exported from the Cape Hatteras coastal area to the Gulf Stream, where it could then travel a long distance.

While not as large a field of study as atmospheric probing using lidar, hydrospheric studies using lidar are turning up interesting findings and may become more important in the future for such uses as detecting oil spills, studying organic matter in the water, studying the structure of subsurface water, and continuing uses in validating space-based instrument measurements while paving the way for future use on orbiting platforms.

8.9. LASER ALTIMETERS

The time-of-flight information from a lidar system can be used for laser altimetry from air- or space-based platforms. Laser altimeters were first flown in the 1960s and used in the 1960s and 1970s to measure terrain features and sea ice roughness. In the 1980s and 1990s, more detailed measurements were made (Ritchie, 1996).

Typical airborne laser altimeters are generally based on the Nd:YAG laser, have PRFs of 50 to 500 Hz, pulse energies of up to 5 mJ, and divergences of 1 to 2 mrad (Ridgway et al., 1997; Blair et al., 1999). If based on a pulsed gallium–arsenide diode laser at 904 nm, it can have a PRF of 4 kHz (Weltz et al., 1994). The aircraft are generally flown at altitudes 500 m to 1 km above the surface. The Global Positioning System (GPS) is used as a position reference. The accuracy of airborne laser altimeters is in the few centimeter range. Ridgway et al. (1997) report repeated measurements of a dome in California, finding that there was a mean difference of 2.6 cm and a standard deviation of 11 cm. A 10-cm accuracy was reported by Bamber et al. (1998, 2001). Airborne laser altimeters have been used to check the accuracy of space-based radar altimeters (Bamber et al., 1998).

Lidar can also be used to study the forest canopy. A short-pulsed lidar system with a fast digitization rate can yield profiles of backscatter from leaves and branches that can be used to estimate forest biomass and volume, aerodynamic roughness, and leaf area indices (Ritchie, 1996). The eventual goal is to place a vegetation canopy lidar into space (Drake et al., 2002). Several airborne lidar demonstrations of the approach have been conducted.

Forest and range vegetation studies in Arizona and Texas were reported by Ritchie et al. (1995). A study of vegetation and land surface changes caused by erosion and channel development in Niger was reported by Ritchie et al. (1996). Harding et al. (2001) studied four deciduous forest stands in eastern Maryland. They were able to demonstrate the ability of the their lidar system, Scanning Lidar Imager of Canopies by Echo Recovery (SLICER), to document such interesting features as the successional sequence of closed-canopy, broadleaf forest stands. Drake et al. (2002) flew their lidar system, Laser Vegetation Imaging Sensor (LVIS), over the La Selva Biological Station in Costa Rica. They confirmed the ability of large-footprint lidar to estimate important structural attributes, such as mean stem diameter and above ground biomass. Lefsky et al. (2002) reported the use of their airborne lidar system for measuring leaf area index.

The study of ice sheets is critical because thinning of the ice sheets is an important indicator of climate change. Based on flights over the Greenland ice sheet in 1993/1994 and 1998/1999, thinning at lower elevations of 1 m/year led to an estimated net loss of 51 km^3/year, sufficient to raise sea level by 0.13 mm/year (1995 Krabill et al., 2000). This portion is 7% of the observed loss for the total Greenland ice sheet. A more recent paper presented reports that showed variations in thickening and thinning across the ice sheet (Abdalati et al., 2001).

There are also applications to land surface features. Since lidar can penetrate the vegetation canopy, it can be used to map the ground surface elevation. Jansma et al. (2001) reported measurements in Puerto Rico where a residual bias of -0.5 m was found for the laser altimeter measurements relative to postprocessed kinematic GPS elevations. Another application is monitoring coastal beach erosion (1995 Krabill et al., 2000; Brock et al., 2002).

8.10. FUTURE DEVELOPMENTS EXPECTED

Advances in lidar technology, including miniaturization and further advances in autonomous operation, should lead to the untended deployment of lidar systems on either piloted or unpiloted aeronautical vehicles (UAVs). *In situ* O_3 measuring instruments are already carried on commercial aircraft to make continuous measurements along the flight track as done in the MOZAIC program (Morgenstern and Carver, 2001). A likely next step is the placement of such a lidar system on a dedicated NASA aircraft for validation of tropospheric O_3 measuring instruments on the Aura Earth Observing satellite due to be launched in early 2004. Placement of H_2O DIAL systems on UAVs for hurricane studies and monitoring is also a possibility. Since UAVs can fly for long periods (24+ h), to high altitudes (20 km), and at times and places where piloted aircraft would not fly, there may also be applications for O_3 DIAL systems on UAVs. Such applications might include studying the global changes in atmospheric composition and chemistry, studying polar O_3 depletion, global pollution transport, and so forth.

Lidar technology development using the aircraft platform is a valuable intermediate step toward the evolution of lidar technology to space. The aircraft platform can be used as a surrogate to space-based system and it permits testing in many ways including: operation in the nadir mode, making measurements over spatially varying atmospheric conditions including the background, studying impact of high-altitude clouds, and some limited testing of the mechanical stability of the lidar systems. The lidar systems can also be tested for autonomous operation under the above-mentioned conditions to enhance lidar technology toward space-based deployment (Browell et al., 1997a). The airborne lidar systems can be used as space simulators for developing and testing data collection and retrieval algorithms. Projection of lidar performance can be made by a choice of optimum lidar parameters by scaling and from the knowledge gained from airborne measurements. Such a performance prediction was conducted during the development of LITE (McCormick et al., 1993). Airborne lidar systems are also valuable for the validation of space-borne lidar products.

8.11. SUMMARY AND CONCLUSION

Airborne lidar systems have been developed and used for a large variety of applications in studying the atmosphere from the surface to 100 km, the near-surface waters, solid Earth surfaces, and the vegetation canopy. These studies have added significantly to the store of knowledge of the Earth and will continue to do so. They have also helped pave the way to placing lidar systems in space, which further extends the spatial and temporal reach of lidar applications.

REFERENCES

Abdalati, W., et al., Outlet glacier and margin elevation changes: near-coastal thinning of the Greenland ice sheet, *J. Geophys. Res.* — *Atmos.*, 106, 33729–33741, 2001.

Alvarez, R. J. II, C. J. Senff, R. M. Hardesty, D. D. Parrish, W. T. Luke, T. B. Watson, P. H. Daum, and N. Gillani, Comparisons of airborne lidar measurements of ozone with airborne in situ measurements during the 1995 Southern Oxidants Study, *J. Geophys. Res.* — *Atmos.*, 103, 31155–31171, 1998.

Ancellet, G., and F. Ravetta, The airborne lidar for tropospheric ozone (ALTO), in A. Ansmann, R. Neuber, P. Rairoux, and U. Wandinger (eds.), *Advances in Atmospheric Remote Sensing with Lidar*, Springer-Verlag, Berlin, pp. 399–402, 1997.

Ancellet, G., and F. Ravetta, Compact airborne lidar for tropospheric ozone: description and field measurements, *Appl. Opt.*, 37, 5509–5521, 1998.

Ancellet, G., Airborne ozone lidar measurements for studying stratosphere-troposphere exchanges and transport in the polluted planetary boundary layer, in A. Dabas. C. Loth, and J. Pelon (eds.), *Advances in Laser Remote Sensing*, Selected papers presented at the 20th International Laser Radar Conference (ILRC), Vichy, France, July 10–14, 2000, Edition d'Ecole Polytechnique, Palaiseau Cedex, France, pp. 357–360, 2001.

Anderson, B. E., et al., Aerosols from biomass burning over the tropical South Atlantic region: distributions and impacts, *J. Geophys. Res. — Atmos.*, 101, 24117–24137, 1996.

Andreae, M. O., et al., Biomass-burning emissions and associated haze layers over Amazonia, *Geophys. Res. — Atmos.*, 93, 1509–1527, 1988.

Arking, A., Feedback processes and climate response, Proceedings of the Conference on Climate Impacts of Solar Variability, NASA Conf. Pub. CP-3086, 1990.

Baker, W. E., et al., Lidar-measured winds from space — a key component for weather and climate prediction, *Bull. Am. Meteorol. Soc.*, 76, 869–888, 1995.

Bamber, J. L., S. Ekholm, and W. Krabill, The accuracy of satellite radar altimeter data over the Greenland ice sheet determined from airborne laser data, *Geophys. Res. Lett.*, 25, 3177–3180, 1998.

Bamber, J. L., S. Ekholm, and W. Krabill, A new, high-resolution digital elevation model of Greenland fully validated with airborne laser altimeter data, *J. Geophys. Res. — Solid Earth*, 106, 6733–6745, 2001.

Banta, R. M., et al., Daytime buildup and nighttime transport of urban ozone in the boundary layer during a stagnation episode, *J. Geophys. Res. — Atmos.*, 103, 22519–22544, 1998.

Barnes, N. P., Remote sensing of planet Earth — challenges for solid-state lasers, *Laser Phys.*, 8, 25–28, 1998.

Bartsch, B., T. Braeske, and R. Reuter, Oceanic lidar operating at high altitudes: a computer model study, *EARSeL Adv. Remote Sens.*, 1, 79–84, 1992.

Bilbro, J. W., and W. W. Vaughan, Wind field measurement in the nonprecipitous regions surrounding severe storms by an airborne pulsed Doppler lidar system, *Bull. Am. Meteorol. Soc.*, 59, 1095–1100, 1978.

Bilbro, J., Airborne Doppler lidar wind field measurements, *Bull. Am. Meteorol. Soc.*, 65, 348–359, 1984.

Blair, J. B., D. L. Rabine, and M. A. Hofton, The laser vegetation imaging sensor: a medium-altitude, digitisation-only, airborne laser altimeter for mapping vegetation and topography, *ISPRS J. Photogram. Remote Sens.*, 54, 115–122, 1999.

Bristow, M., D. Nielsen, D. Bundy, and F. Furtek, Use of water-Raman emission to correct airborne laser fluorosensor data for effects of water optical attenuation, *Appl. Opt.*, 20, 2889–2906, 1981.

Brock, C. A., L. F. Radke, and P. V. Hobbs, Sulfur in particles in Arctic hazes derived from airborne in situ and lidar measurements, *J. Geophys. Res. — Atmos.*, 95, 22369–22387, 1990.

Brock, J. C., C. W. Wright, A. H. Sallenger, W. B. Krabill, and R. N. Swift, Basis and methods of NASA airborne topographic mapper lidar surveys for coastal studies, *J. Coast. Res.*, 18, 1–13, 2002.

Browell, E. V., Remote sensing of tropospheric gases and aerosols with an airborne DIAL system, in D. K. Killinger and A. Mooradian (eds.), *Optical Laser Remote Sensing*, Springer-Verlag, New York, pp. 138–147, 1983.

Browell, E. V., A. K. Goroch, T. D. Wilkerson, S. Ismail, and R. Markson, Airborne DIAL water vapor measurements over the Gulf Stream, Abstracts, 12th International Laser Radar Conference, Aix en Provence, France, August 13–17, 1984, pp. 151–155.

Browell, E. V., E. E. Danielsen, S. Ismail, et al., Tropopause fold structure determined from airborne lidar and in situ measurements, *J. Geophys. Res. — Atmos.*, 92, 2112–2120, 1987.

Browell, E. V., Differential absorption lidar sensing of ozone, *Proc. IEEE*, 77, 419–432, 1989.

Browell, E. V., A. Carter, and T. Wilkerson, An airborne differential absorption lidar system for water vapor investigations, *Opt. Eng.*, 20, 84–90, 1981.

Browell, E. V., et al., NASA multipurpose airborne DIAL system and measurements of ozone and aerosol profiles, *Appl. Opt.*, 22, 522–534, 1983.

Browell, E. V., S. Ismail, and S. T. Shipley, Ultraviolet DIAL measurements of O_3 profiles in regions of spatially inhomogeneous aerosols, *Appl. Opt.*, 24, 2827–2836, 1985.

Browell, E. V., E. Danielsen, S. Ismail, G. Gregory, and S. Beck, Tropopause fold structure determined from airborne lidar and in situ measurements, *Geophys. Res. — Atmos.*, 92, 2112–2120, 1987.

Browell, E. V., G. L. Gregory, R. Harriss, and V. Kirchhoff, Tropospheric ozone and aerosol distributions across the Amazon Basin, *J. Geophys. Res. — Atmos.*, 93, 1431–1451, 1988.

Browell, E. V., et al., Airborne lidar observations in the wintertime Arctic stratosphere: ozone, *Geophys. Res. Lett.*, 17, 325–328, 1990a.

Browell, E. V., et al., Airborne lidar observations in the wintertime Arctic stratosphere: polar stratospheric clouds, *Geophys. Res. Lett.*, 17, 385–388, 1990b.

Browell, E. V., N. S. Higdon, C. F. Butler, M. A. Fenn, B. E. Grossmann, P. Ponsardin, W. B. Grant, and A. S. Bachmeier,

Tropospheric water vapor measurements with an airborne lidar system, Preprints, Seventh AMS Symposium on Meteorological Observations and Instrumentation, New Orleans, Louisiana, January 14–18, 1991.

Browell, E. V., C. F. Butler, S. A. Kooi, M. A. Fenn, R. C. Harriss, and G. L. Gregory, Large-scale variability of ozone and aerosols in the summertime Arctic and subarctic troposphere, *J. Geophys. Res. — Atmos.*, 97, 16433–16450, 1992.

Browell, E. V., et al., Ozone and aerosol changes observed during the 1992 Airborne Arctic Stratospheric Expedition, *Science*, 261, 1155–1158, 1993.

Browell, E. V., M. A. Fenn, C. F. Butler, W. B. Grant, R. C. Harriss, and M. C. Shipham, Ozone and aerosol distributions in the summertime troposphere over Canada, *J. Geophys. Res. —Atmos.*, 99, 1739–1755, 1994.

Browell, E., and S. Ismail, First lidar measurements of water vapor and aerosols from a high-altitude aircraft, Proc. OSA Optical Remote Sensing of the Atmosphere, Salt Lake City, Utah, February 5–9, OSA Technical Digest 2, pp. 212–214, 1995.

Browell, E. V., et al., Large-scale air mass characteristics observed over the Western Pacific during summertime, *J. Geophys. Res. — Atmos.*, 101, 1691–1712, 1996a.

Browell, E. V., et al., Ozone and aerosol distributions and air mass characteristics over the South Atlantic Basin during the burning season, *J. Geophys. Res. — Atmos.*, 101, 24043–24068, 1996b.

Browell, E. V., et al., LASE validation experiment, in A. Ansmann, R. Neuber, P. Rairoux, and U. Wandinger (eds.), *Advances in Atmospheric Remote Sensing with Lidar*, Springer-Verlag, Berlin, pp. 289–295, 1997a.

Browell, E. V., S. Ismail, T. C. McElroy, R. M. Hoff, and A. Dudelzak, Global Measurements of ozone and aerosol distributions with a space lidar system, *EOS*, 78, F89, 1997b.

Browell, E. V., S. Ismail, and W. B. Grant, Differential absorption lidar (DIAL) measurements from air and space, *Appl. Phys. B*, 67, 399–410, 1998.

Browell, E. V., S. Ismail, W. B. Grant, and R. Ferrare, Airborne Lidar Water Vapor, Ozone, and Aerosol Measurements, Proceedings, AMS Symposium on Lidar Atmospheric Monitoring, Long Beach, CA, January 9–14, 2000a.

Browell, E. V., S. Ismail, and R. Ferrare, Hurricane Water Vapor, Aerosol, and Cloud Distributions Determined from Airborne

Lidar Measurements, Proceedings, AMS Symposium on Lidar Atmospheric Monitoring, Long Beach, CA, January 9–14, 2000b.

Browell, E. V., et al., Large-scale air mass characteristics observed over the remote tropical Pacific Ocean during March–April 1999: results from PEM Tropics B Field Experiment, *J. Geophys. Res. — Atmos.*, 106, 32481–32501, 2001.

Browell, E. V., et al., Ozone, aerosol, potential vorticity, and trace gas trends observed at high latitudes over North America from February to May 2000, *J. Geophys. Res.*, 108 (D4), art. no. 8369, 2003.

Bruneau, D., P. Quaglia, C. Flamant, M. Meissonnier, and J. Pelon, Airborne lidar LEANDRE II for water-vapor profiling in the troposphere. I. System description, *Appl. Opt.*, 40, 3450–3461, 2001a.

Bruneau, D., P. Quaglia, C. Flamant, M. Meissonnier, and J. Pelon, Airborne lidar LEANDRE II for water-vapor profiling in the troposphere. II. First results, *Appl. Opt.*, 40, 3462–3475, 2001b.

Burris, J, W. Heaps, B. Gary, W. Hoegy, L. Lait, T. McGee, M. Gross, and U. Singh, Lidar temperature measurements during the Tropical Ozone Transport Experiment (TOTE)/Vortex Ozone Transport Experiment (VOTE) mission, *J. Geophys. Res. — Atmos.*, 103, 3505–3510, 1998.

Burris, J., et al., Validation of temperature measurements from the airborne Raman ozone temperature and aerosol lidar during SOLVE, *J. Geophys. Res.*, 107(D20), 8286, 2002, doi:10.1029/2001JD001028.

Butler, C. F., E. V. Browell, W. B. Grant, V. G. Brackett, O. B. Toon, J. Burris, T. McGee, M. Schoeberl, and M. J. Mahoney, Polar stratospheric cloud characteristics observed with airborne lidar during the SOLVE Campaign, in A. Dabas. C. Loth, and J. Pelon (eds.), *Advances in Laser Remote Sensing*, Selected papers presented at the 20th International Laser Radar Conference (ILRC), Vichy, France, July 10–14, 2000, Edition d'Ecole Polytechnique, Palaiseau Cedex, France, pp. 397–400, 2001.

Carslaw, K. S., et al., Particle microphysics and chemistry in remotely observed mountain polar stratospheric clouds, *J. Geophys. Res. — Atmos.*, 103, 5785–5796, 1998.

Charlson, R. J., S. E. Schwartz, J. M. Hales, R. D. Cess, J. A. Coakley, J. E. Hansen, and D. J. Hofmann, Climate forcing by anthropogenic aerosols, *Science*, 255, 423–430, 1992.

Chazette, P., J. Pelon, C. Moulin, F. Dulac, I. Carrasco, W. Guelle, P. Bousquet, and P-H. Flamant, Lidar and satellite retrieval of dust aerosols over the Azores during SOFIA/ASTEX, *Atmos. Environ.*, 35, 4297–4304, 2001.

Cho, J. Y. N., R. E. Newell, E. V. Browell, W. B. Grant, C. F. Butler, and M. A. Fenn, Observation of pollution plume capping by a tropopause fold, *Geophys. Res. Lett.*, 28, 3243–3246, 2001.

Chu, X.-Z., W. L. Papen, G. Papen, C. S. Gardner, G. Swenson, and P. Jenniskens, Characteristics of Fe ablation trails observed during the 1998 Leonid meteor shower, *Geophys. Res. Lett.*, 27, 1807–1810, 2000.

Clayton, M. B., et al., Stratosphere–troposphere exchange events observed by LASE in tropics, mid- and high-latitude regions, in A. Dabas. C. Loth, and J. Pelon (eds.), *Advances in Laser Remote Sensing*, Selected papers presented at the 20th International Laser Radar Conference (ILRC), Vichy, France, July 10–14, 2000, Edition d'Ecole Polytechnique, Palaiseau Cedex, France, pp. 361–364, 2001.

Clough, S. A., M. I. Iacono, J. L. Moncet: Line-by-line calculations of atmospheric fluxes and cooling rates: application to water vapor, *J. Geophys. Res. — Atmos.*, 97 (D14), 15761–15785, 1992.

Cutten, D. R., et al., Intercomparison of pulsed lidar data with flight level CW lidar data and modeled backscatter from measured aerosol microphysics near Japan and Hawaii, *J. Geophys. Res. — Atmos.*, 103, 19649–19661, 1998.

Dabberdt, W. F., and T. W. Schlatter, Research opportunities from emerging atmospheric observing and modeling capabilities, *Bull. Am. Meteorol. Soc.*, 77, 305–323, 1996.

Davis, K. J., N. Gamage, C. R. Hagelberg, C. Kiemle, D. H. Lenschow, and P. P. Sullivan, An objective method for deriving atmospheric structure from airborne lidar observations, *J. Atmos. Ocean. Technol.*, 17, 1455–1468, 2000.

di Sarra, A., T. Di Iorio, M. Cacciani, G. Fiocco, and D. Fua, Saharan dust profiles measured by lidar at Lampedusa, *J. Geophys. Res. — Atmos.*, 106, 10335–10347, 2001.

Drake, J. B., R. O. Dubayah, D. B. Clark, R. G. Knox, J. B. Blair, M. A. Hofton, R. L. Chazdon, J. F. Weishampel, and S. D. Prince, Estimation of tropical forest structural characteristics using large-footprint lidar, *Remote Sens. Environ.*, 79, 305–319, 2002.

Dupont, E., J. Pelon, and C. Flamant, Study of the moist convective boundary-layer structure by backscattering lidar, *Boundary-Layer Meteorol.*, 69, 1–25, 1994.

Ehret, G., C. Kiemle, W. Renger, and G. Simmet, Airborne remote sensing of tropospheric water vapor with a near-infrared differential absorption lidar system, *Appl. Opt.*, 32, 4534–4551, 1993.

Ehret, G., A. Giez, C. Kiemle, K. J. Davis, D. H. Lenschow, S. P. Oncley, and R. D. Kelly, Airborne water vapor DIAL and in situ observations of a sea–land interface, *Contrib. Atmos. Phys.*, 69, 215–228, 1996.

Ehret, G., A. Fix, V. Weiss, G. Poberaj, and T. Baumert, Diode-laser-seeded optical parametric oscillator for airborne water vapor DIAL application in the upper troposphere and lower stratosphere, *Appl. Phys. B*, 67, 427–431, 1998.

Ehret, G., K. P. Hoinka, J. Stein, A. Fix, C. Kiemle, and G. Poberaj, Low stratospheric water vapor measured by an airborne DIAL, *J. Geophys. Res. — Atmos.*, 104, 31351–31359, 1999.

Fenn, M. A., E. V. Browell, and C. F. Butler, Airborne lidar measurements of ozone and aerosols during PEM-West A and PEM-West B, in A. Ansmann, R. Neuber, P. Rairoux, and U. Wandinger (eds.), *Advances in Atmospheric Remote Sensing with Lidar*, Springer-Verlag, Berlin, pp. 355–358, 1997.

Fenn, M. A., et al., Ozone and aerosol distributions and air mass characteristics over the South Pacific during the burning season, *J. Geophys. Res. — Atmos.*, 104, 16197–16212, 1999.

Ferrare, R., et al., LASE measurements of water vapor, aerosols, and clouds during CAMEX-3, Proc. 1999 OSA Symposium on Optical Remote Sensing of the Atmosphere, pp. 114–116, 1999.

Ferrare, R., et al., Comparison of aerosol optical properties and water vapor among ground and airborne lidars and sun photometers during TARFOX, *J. Geophys. Res. — Atmos.*, 105, 9917–9933, 2000a.

Ferrare, R., et al., Comparisons of LASE, aircraft, and satellite measurements of aerosol optical properties and water vapor during TARFOX, *J. Geophys. Res. — Atmos.*, 105, 9935–9947, 2000b.

Ferrare, R. A., et al., Lidar measurements of relative humidity and ice supersaturation in the upper troposphere, in A. Dabas. C. Loth, and J. Pelon (eds.), *Advances in Laser Remote Sensing*, Selected papers presented at the 20th International Laser Radar Conference (ILRC), Vichy, France, July 10–14, 2000, Edition d'Ecole Polytechnique, Palaiseau Cedex, France, pp. 317–320, 2001.

Ferrare, R.A., E.V. Browell, S. Ismail, S. Kooi, L.H. Brasseur, V.G. Brackett, M. Clayton, J. Barrick, H. Linné, A. Lammert, G. Diskin, J. Goldsmith, B. Lesht, J. Podolske, G. Sachse, F.J. Schmidlin, D. Turner, D. Whiteman, D. Tobin, H. Revercomb, Characterization of upper troposphere water vapor measurements during AFWEX using LASE, *J. Atmos. Oceanic Tech.*, 21, 1790-1808, 2004.

Fishman, J., V. G. Brackett, E. V. Browell, and W. B. Grant, Tropospheric ozone derived from TOMS/SBUV measurements during TRACE A, *J. Geophys. Res. — Atmos.*, 101, 24069–24082, 1996.

Fix, A., V. Weiss, and G. Ehret, Injection-seeded optical parametric oscillator for airborne water vapor DIAL, *Pure Appl. Opt.*, 7, 837–852, 1998.

Flamant, C., M. Georgelin, L. Menut, J. Pelon, and P. Bougeault, The atmospheric boundary-layer structure within a cold air outbreak: comparison of in situ, lidar and satellite measurements with three-dimensional simulations, *Boundary-Layer Meteorol.*, 99, 85–103, 2001.

Flentje, H., W. Renger, M. Wirth, Martin, and W. A. Lahoz, Validation of contour advection simulations with airborne lidar measurements of filaments during the Second European Stratospheric Arctic and Midlatitude Experiment (SESAME), *J. Geophys. Res. — Atmos.*, 105, 15417–15437, 2000.

Flesia, C., and C. L. Korb, Theory of the double-edge molecular technique for Doppler lidar wind measurement, *Appl. Opt.*, 38, 432–440, 1999.

Frehlich, R., Comparison of 2-μm and 10-μm coherent Doppler lidar performance, *J. Atmos. Ocean. Technol.*, 12, 415–420, 1995.

Gardner, C. S., X. Tao, and G. C. Papen, Observations of strong wind shears and temperature enhancements during several sporadic Na layer events above Haleakala, *Geophys. Res. Lett.*, 22, 2809–2812, 1995.

Gelbwachs, J. A., Iron Boltzmann factor lidar: proposed new remote-sensing technique for mesospheric temperature, *Appl. Opt.*, 33, 7151–7156, 1994.

Godin, S., G. Megie, D. C. Haner, C. Flesia, and Y. Emery, Airborne lidar observation of mountain-wave-induced polar stratospheric clouds during EASOE, *Geophys. Res. Lett.*, 21, 1335–1338, 1994.

Grant, W. B., et al., Observations of reduced ozone concentrations in the tropical stratosphere after the eruption of Mt. Pinatubo, *Geophys. Res. Lett.*, 19, 1109–1112, 1992.

Grant, W. B., et al., Aerosol-associated changes in tropical stratospheric ozone following the eruption of Mount Pinatubo, *J. Geophys. Res. — Atmos.*, 99, 8197–8211, 1994.

Grant, W. B., et al., Use of volcanic aerosols to study the tropical stratospheric reservoir, *J. Geophys. Res. — Atmos.*, 101, 3973–3988, 1996.

Grant, W. B., E. V. Browell, C. F. Butler, and G. D. Nowicki, LITE measurements of biomass burning aerosols and comparisons with correlative airborne lidar measurements of multiple scattering in the planetary boundary layer, in A. Ansmann, R. Neuber, P. Rairoux, and U. Wandinger (eds.), *Advances in Atmospheric Remote Sensing with Lidar*, Springer-Verlag, Berlin, pp. 153–156, 1997.

Grant, W. B., et al., Correlative ozone measurements with the airborne UV DIAL system during TOTE/VOTE, *Geophys. Res. Lett.*, 25, 623–626, 1998.

Grant, W. B., et al., A case study of transport of tropical marine boundary layer and lower-tropospheric air masses to the northern mid-latitude upper troposphere, *J. Geophys. Res. — Atmos.*, 105, 3757–3769, 2000.

Grant, W. B., E. V. Browell, C. F. Butler, S. C. Gibson, S. A. Kooi, and P. von der Gathen, Estimation of Arctic polar vortex ozone loss during the winter of 1999/2000 using vortex-averaged airborne differential absorption lidar ozone measurements referenced to N_2O isopleths, J. Geophys. Res.-Atmos. 108, 4309, doi:10.1029/2002JD002668, 2003

Guenther, G. C., Airborne laser hydrography to chart shallow coastal waters, *Sea Technol.*, March, pp. 55, 57–59, 1989.

Hair, J. W., L. M. Caldwell, D. A. Krueger, and C. Y. She, High-spectral-resolution lidar with iodine-vapor filters: measurement of atmospheric-state and aerosol profiles, *Appl. Opt.*, 40, 5280–5294, 2001.

Hansen, J., A. Lacis, D. Rind, G. Russell, P. Stone, et al., Climate sensitivity: Analysis of feedback mechanisms, in J. E. Hansen and T. Takahashi (eds.), *Climate Processes and Climate Sensitivity*, Geophys. Monogr. 29, AGU, Washington, D.C., pp. 130–163, 1984.

Hansen, J., Sato, M., Lacis, A., and R. Ruedy, The missing climate forcing, *Phil Trans. Roy. Soc. London Ser. B — Bio. Sci.*, 352, 231–240, 1997.

Harding, D. J., M. A. Lefsky, G. G. Parker, and J. B. Blair, Laser altimeter canopy height profiles — methods and validation for closed-canopy, broadleaf forests, *Remote Sens. Environ.*, 76, 283–297, 2001.

Harris, M., J. M. Vaughan, K. Huenecke, and C. Huenecke, Aircraft wake vortices: a comparison of wind-tunnel data with field trial measurements by laser radar, *Aerosp. Sci. Technol.*, 4, 363–370, 2000.

Heaps, W. S., and J. Burris, Airborne Raman lidar, *Appl. Opt.*, 35, 7128–7137, 1996.

Heaps, W. S., J. Burris, and J. A. French, Lidar technique for remote measurement of temperature by use for a vibrational–rotational Raman spectroscopy, *Appl. Opt.*, 36, 9402–9405, 1997.

Heese, B., S. Godin, and A Hauchecorne, Forecast and simulation of stratospheric ozone filaments: a validation of a high-resolution potential vorticity advection model by airborne ozone lidar measurements in winter 1998/1999, *J. Geophys. Res.*, 106, 20011–20024, 2001.

Held, I. M., and B. J. Soden, Water vapor feedback and global warming, *Ann. Rev. Energy Environ.*, 25, 441–475, 2000.

Henderson, S. S., P. J. M. Suni, C. P. Hale, S. M. Hannon, J. R. Magee, D. L. Bruns, and E. H. Yuen, Coherent laser-radar at 2 μm using solid-state lasers, *IEEE Trans. Geosci. Remote Sens.*, 31, 4–15, 1993.

Higdon, N. S., et al., Airborne differential absorption lidar system for measurements of atmospheric water vapor and aerosols, *Appl. Opt.*, 33, 6422–6438, 1994.

Hobbs, P. V., L. F. Radke, J. H. Lyons, R. J. Ferek, D. J. Coffman, and T. J. Casadevall, Airborne measurements of particle and gas emissions from the 1990 volcanic eruptions of Mount Redoubt, *J. Geophys. Res. — Atmos.*, 26, 18735–18752, 1991.

Hoff, R. M., M. Harwood, A. Sheppard, F. Froude, J. B. Martin, and W. Strapp, Use of airborne lidar to determine aerosol sources and movement in the Lower Fraser Valley (LFV), BC, *Atmos. Environ.*, 31, 2123–2134, 1997.

Hoff, R. M., and K. B. Strawbridge, LITE observations of anthropogenically produced aerosols, in A. Ansmann, R. Neuber, P. Rairoux, and U. Wandinger (eds.), *Advances in Atmospheric Remote Sensing with Lidar*, Springer-Verlag, Berlin, pp. 145–148, 1997.

Hoge, F. E., and R. N. Swift, Active–passive correlation spectroscopy: a new technique for identifying ocean color algorithm spectral regions, *Appl. Opt.*, 25, 2571–2583, 1986.

Hoge, F. E., Oceanic and terrestrial lidar measurements, in R. M. Measures (ed.), *Laser Remote Chemical Analysis*, John Wiley and Sons, New York, pp. 409–503, 1988.

Hoge, F. E., and R. N. Swift, The influence of chlorophyll pigment upon upwelling spectral radiances from the North Atlantic Ocean: an active–passive correlation spectroscopy study, *Deep-Sea Res. II*, 40, 265–277, 1993a.

Hoge, F. E., A. Vodacek, and N. V. Blough, Inherent optical properties of the ocean: retrieval of the absorption coefficient of chromophoric dissolved organic matter from fluorescence measurements, *Limnol. Oceanogr.*, 38, 1394–1402, 1993b.

Hoge, F. E., R. N. Swift, J. K. Yungel, and A. Vodacek, Fluorescence of dissolved organic matter: A comparison of North Pacific and North Atlantic Oceans during April 1991, *J. Geophys. Res. — Oceans*, 98, 22779–22787, 1993c.

Hoge, F. E, C. W. Wright, R. N. Swift, J. K. Yungel, R. E. Berry, and R. Mitchell, Fluorescence signatures of an iron-enriched phytoplankton community in the eastern equatorial Pacific Ocean, *Deep-Sea Res. Part II: Top. Studies Oceanogr.*, 45, 1073–1082, 1998a.

Hoge, F. E., C. W. Wright, T. M. Kana, R. N. Swift, and J. K. Yungel, Spatial variability of oceanic phycoerythrin spectral types derived from airborne laser-induced fluorescence emissions, *Appl. Opt.*, 37, 4744–4749, 1998b.

Hoge, F. E., C. W. Wright, P. E. Lyon, R. N. Swift, and Y. K. Yungel, Inherent optical properties imagery of the western North Atlantic Ocean: horizontal spatial variability of the upper mixed layer, *J. Geophys. Res. — Oceans*, 106, 31129–31140, 2002.

Hostetler, C. A, and C. S. Gardner, Observations of horizontal and vertical wave number spectra of gravity wave motions in the stratosphere and mesosphere over the mid-Pacific, *J. Geophys. Res. — Atmos.*, 99, 1283–1302, 1994.

Ismail, S., E. V. Browell, R. A. Ferrare, S. A. Kooi, M. B. Clayton, V. G. Brackett, and P. B. Russell, LASE measurements of aerosol and water vapor profiles during TARFOX, *J. Geophys. Res. — Atmos.*, 105, 9903–9916, 2000.

Ismail, S., E. V. Browell, R. Ferrare, S. Kooi, M. Clayton, V. Brackett, W. Edwards, and F. J. Schmidlin, LASE measurements during CAMEX-3 to characterize the hurricane environment, in A. Dabas. C. Loth, and J. Pelon (eds.), *Advances in Laser Remote Sensing*, Selected papers presented at the 20th International Laser Radar Conference (ILRC), Vichy, France, July 10–14, 2000,, Edition d'Ecole Polytechnique, Palaiseau Cedex, France, pp. 341–344, 2001.

Ismail S., E. V. Browell, R. A. Ferrare, C. Senff, K. J. Davis, D. H. Lenschow, S. Kooi, V. Brackett, and M. Clayton, LASE measurements of atmospheric boundary layer development during SGP97, Extended Abstracts, 19th International Laser Radar Conf., spons. by NASA and Am. Meteorol. Soc., Annapolis, MD, July 6–10, 1998.

Jansma, P., G. Mattioli, and A. Matias, SLICER laser altimetry in the eastern Caribbean, *Surv. Geophys.*, 22, 561–579, 2001.

Kamineni, Rupa and T. N. Krishnamurti, Richard A. Ferrare, Syed Ismail, and Edward V. Browell, *Geophysical Research Letters*, 30, doi:10.1029/2002GL016741, 2003.

Kamineni, R., T. N. Krishnamurti, S. Pattnaik, E. V. Browell, S. Ismail, and R. A. Ferrare, Impact of CAMEX-4 data sets for hurricane forecasts using a global model, *J. Atmos. Sci.*, in press, 2005.

Karyampudi, V. M., et al., Validation of the Saharan dust plume conceptual model using lidar, Meteosat and ECMWF data, *Bull. Am. Meteorol. Soc.*, 80, 1045–1075, 1999.

Khattatov, V. U., A. E. Tyabotov, A. P. Alekseyev, A. A. Postnov, and E. A. Stulov, Aircraft lidar studies of the Arctic haze and their meteorological interpretation, *Atmos. Res.*, 44, 99–111, 1997.

Kiemle, C. G. Ehret, A. Giez, K. J. Davis, D. H. Lenschow, and S. P. Oncley, Estimation of boundary layer humidity fluxes and statistics from airborne differential absorption lidar (DIAL), *J. Geophys. Res. — Atmos.*, 102, 29189–29203, 1997.

Ko, M. K. W., J. M. Rodriguez, N. D. Sze, M. H. Proffitt, W. L. Starr, A. Krueger, E. V. Browell, and M. P. McCormick, Implications of AAOE observations for proposed chemical explanations of the seasonal and interannual behavior of Antarctic ozone, *J. Geophys. Res. — Atmos.*, 94, 16705–16715, 1989.

Korb, C. L., B. M. Gentry, and S. X. F. Li, Edge technique Doppler lidar wind measurements with high vertical resolution, *Appl. Opt.*, 36, 5976–5983, 1997.

Krabill, W. B., R. H. Thomas, C. F. Martin, R. N. Swift, and E. B. Frederick, Accuracy of airborne laser altimetry over the Greenland ice-sheet, *Int. J. Remote Sens.*, 16, 1211–1222, 1995.

Krabill, W., et al., Greenland ice sheet: high-elevation balance and peripheral thinning, *Science*, 289, 428–430, 2000.

Langford, A. O., and S. J. Reid, Dissipation and mixing of a small-scale stratospheric intrusion in the upper troposphere, *J. Geophys. Res. — Atmos.*, 103, 31265–31276, 1998.

Lefsky, M. A., W. B. Cohen, G. G. Parker, and D. J. Harding, Lidar remote sensing for ecosystem studies, *Bioscience*, 52, 19–30, 2002.

Leon, J. F., et al., Large-scale advection of continental aerosols during INDOEX, *J. Geophys. Res. — Atmos.*, 106, 28427–28439, 2001.

Lutomirski, R. F., Lidar remote sensing of ocean waters, *Proc. SPIE*, 2222, 12–19, 1994.

Manson, A. H., C. E. Meek, J. Qian, and C. S. Gardner, Spectra of gravity wave density and wind perturbations observed during

Arctic Noctilucent Cloud (ANLC-93) campaign over the Canadian Prairies: synergistic airborne Na lidar and MF radar observations, *J. Geophys. Res. — Atmos.*, 103, 6455–6465, 1998.

Margitan, J. J., et al., Intercomparison of ozone measurements over Antarctica, *J. Geophys. Res. — Atmos.*, 94, 16557–16569, 1989.

Martin, J. H., et al., Testing the iron hypothesis in ecosystems of the equatorial Pacific Ocean, *Nature*, 371, 123–129, 1994.

McCormick, M. P., and T. J. Swissler, Stratospheric aerosol mass and latitudinal distribution of the El Chichón eruption cloud for October 1982, *Geophys. Res. Lett.*, 10, 877–880, 1983.

McCormick, M. P., et al., Scientific investigations planned for the Lidar In-space Technology Experiment (LITE), *Bull. Am. Meteorol. Soc.*, 74, 205–214, 1993.

McElroy, J. L., J. A. Eckert, and C. J. Hager, Airborne down-looking lidar measurements during STATE 78, *Atmos. Environ.*, 15, 2223–2230, 1981.

McElroy, J. L., and T. B. Smith, Vertical pollutant distributions and boundary layer structure observed by airborne lidar near the complex southern California coastline, *Atmos. Environ.*, 20, 1555–1566, 1986.

McFarquhar, G. M., A. J. Heymsfield, J. Spinhirne, and B. Hart, Thin and subvisual tropopause tropical cirrus: observations and radiative impacts, *J. Atmos. Sci.*, 57, 1841–1853, 2000.

Measures, R. M., *Laser Remote Sensing, Fundamentals and Applications*, John Wiley and Sons, New York, 1984.

Menzies, R. T, and D. M. Tratt, Airborne lidar observations of tropospheric aerosols during the Global Backscatter Experiment (GLOBE) Pacific, *J. Geophys. Res. — Atmos.*, 102, 3701–3714, 1997.

Moore, A. S., et al., Development of the Lidar Atmospheric Sensing Experiment (LASE) — an advanced airborne DIAL instrument, in A. Ansmann, R. Neuber, P. Rairoux, and U. Wandinger (eds.), *Advances in Atmospheric Remote Sensing with Lidar*, Springer Verlag, New York, pp. 281–288, 1997.

Moosmueller, H., R. J. Alvarez II, R. M. Jorgensen, et al., An airborne UV-DIAL system for ozone measurements: field use and verification, Optical Sensing for Environmental Monitoring, SP-89, Air & Waste Management Assoc., pp. 413–422, 1994.

Morgenstern, O., and G. D. Carver, Comparison of cross-tropopause transport and ozone in the upper troposphere and lower stratosphere region, *J. Geophys. Res. — Atmos.*, 106, 10205–10221, 2001.

Murayama, T., et al., Ground-based network observation of Asian dust events of April 1998 in east Asia, *J. Geophys. Res. — Atmos.*, 106, 18345–18359, 2001.

Newell, R. E., Y. Zhu, E. V. Browell, S. Ismail, W. G. Read, J. W. Waters, et al., Upper tropospheric water vapor and cirrus: Comparison of DC-8 observations, preliminary UARS Microwave Limb Sounder measurements and meteorological analyses, *J. Geophys. Res. — Atmos.*, 101, 1931–1941, 1996.

Newell, R. E., V. Thouret, J. Y. N. Cho, P. Stoller, A. Marenco, and H. G. Smit, Ubiquity of quasi-horizontal layers in the troposphere, *Nature*, 398, 316–319, 1999.

Northam, G. B., M. A. Guerra, M. E. Mack, et al., High repetition rate frequency-doubled Nd:YAG laser for airborne bathymetry, *Appl. Opt.*, 20, 968–971, 1981.

Palm, S. P., S. H. Melfi, and D. L. Carter, New airborne scanning lidar system: Applications for atmospheric remote sensing, *Appl. Opt.*, 33, 5674–5681, 1994.

Pelon, J., C. Flamant, V. Trouillet, and P. H. Flamant, Optical and microphysical parameters of dense stratocumulus clouds during mission 206 of EUCREX '94 as retrieved from measurements made with the airborne lidar LEANDRE 1, *Atmos. Res.*, 55, 47–64, 2000.

Penny, M. F., B. Billard, and R. H. Abbot, LADS — the Australian laser airborne depth sounder, *Int. J. Remote Sens.*, 10, 1463–1479, 1989.

Pfister, L., et al., Aircraft observations of thin cirrus clouds near the tropical tropopause, *J. Geophys. Res. — Atmos.*, 106, 9765–9786, 2001.

Poberaj, G., A. Assion, A. Fix, C. Kiemle, M. Wirth, and G. Ehret, Airborne all-solid-state DIAL for water vapor measurements in the tropopause region, in A. Dabas. C. Loth, and J. Pelon (eds.), *Advances in Laser Remote Sensing*, Selected papers presented at the 20th International Laser Radar Conference (ILRC), Vichy, France, July 10–14, July, Edition d'Ecole Polytechnique, Palaiseau Cedex, France, pp. 325–328, 2001.

Qian, J., Y. Y. Gu, and C. S. Gardner, Characteristics of the sporadic Na layers observed during the airborne lidar and observations of Hawaiian airglow airborne noctilucent cloud (ALOHA/ANLC-93) campaigns, *Geophys. Res. — Atmos.*, 103, 6333–6347, 1998.

Quaglia, P., D. Bruneau, A. Abchiche, M. Lopez, F. Fassina, J. P. Marcovici, et al., The airborne water-vapor lidar LEANDRE II: design, realization, tests and first validations, in A. Ansmann,

R. Neuber, P. Rairoux, U. Wandinger (eds.), *Advances in Atmospheric Remote Sensing with Lidar*, Springer-Verlag, Berlin, pp. 297–300, 1997.

Radke, L. F., C. A. Brock, J. H. Lyons, P. V. Hobbs, and R. C. Schnell, Aerosol and lidar measurements of hazes in midlatitude and polar air masses, *Atmos. Environ. Part A: Gen. Top.*, 23, 11, 2417–2430, 1989.

Ramanathan, V., and A. M. Vogelmann, Greenhouse effect, atmospheric solar absorption and the Earth's radiation budget: from the Arrhenius–Langley era to the 1990s, *Ambio*, 26, 38–46, 1997.

Ravetta F. and G. Ancellet, Identification of Dynamical Processes at the Tropopause during the Decay of a Cutoff Low Using High-Resolution Airborne Lidar Ozone Measurements, *Monthly Weather Review*: Vol. 128, No. 9, pp. 3252–3267.

Reitebuch, O., C. Werner, I. Leike, P. Delville, P. H. Flamant, A. Cress, and D. Engelbart, Experimental validation of wind profiling performed by the airborne 10-mu m heterodyne Doppler lidar WIND, *J. Atmos. Ocean. Technol.*, 18, 1331–1344, 2001.

Renno, N. O., K. A. Emanuel, and P. H. Stone, Radiative–convective model with an explicit hydrologic-cycle. 1. Formulation and sensitivity to model parameters, *J. Geophys. Res. — Atmos.*, 99, 14429–14441, 1994.

Richter, D. A., E. V. Browell, C. F. Butler, and N. S. Higdon, Advanced airborne UV DIAL system for stratospheric and tropospheric ozone and aerosol measurements, in A. Ansmann, R. Neuber, P. Rairoux, U. Wandinger (eds.), *Advances in Atmospheric Remote Sensing with Lidar*, Springer-Verlag, Berlin, pp.395–398, 1997.

Ridgway, J. R., J. B. Minster, N. Williams, J. L. Bufton, and W. B. Krabill, Airborne laser altimeter survey of Long Valley, California, *Geophys. J. Int.*, 131, 267–280, 1997.

Rind, D., E. W. Chiou, W. Chu, S. Oltmans, J. Lerner, et al., Satellite validation of water vapor feedback in GCM climate change experiments, *Nature*, 349, 500–503, 1991.

Ritchie, J. C., Airborne laser altimeter measurements of landscape topography, *Remote Sens. Environ.*, 53, 91–96, 1995.

Ritchie, J. C., Remote sensing applications to hydrology: airborne laser altimeters, *Hydrol. Sci. J.*, 41, 625–636, 1996.

Ritchie, J. C., M. Menenti, and M. A. Weltz, Measurements of land surface features using an airborne laser altimeter: the HAPEX-Sahel experiment, *Int. J. Remote Sens.*, 17, 3705–3724, 1996.

Rothermel, J., D. R. Cutten, R. M. Hardesty, R. T. Menzies, J. N. Howell, S. C. Johnson, D. M. Tratt, L. D. Olivier, and B. M. Banta, The multi-center airborne coherent atmospheric wind sensor, *Bull. Am. Meteorol. Soc.*, 79, 581–599, 1998a.

Rothermel, J., L. D. Olivier, R. M. Banta, R. M. Hardesty, J. N. Howell, D. R. Cutten, S. C. Johnson, R. T. Menzies, and D. M. Tratt, Remote sensing of multi-level wind fields with high-energy airborne scanning coherent Doppler lidar, *Opt. Express*, 2, 2, 40–50, 1998b.

Sachse, G. W., L.-G. Wang, S. Ismail, E. V. Browell, and C. Banziger, Multi-wavelength sequential seeding method for water vapor DIAL measurements, Proc. OSA Optical Remote Sensing of the Atmosphere, Salt Lake City, Utah, March 8–12, 1993, OSA Tech. Dig. Ser. 5, p. 162, 1993.

Sachse, G., L.-G. Wang, C. Antill, Jr., S. Ismail, and E. Browell, Line-center/side-line diode laser seeding for DIAL measurements of the atmosphere, Proc. OSA Optical Remote Sensing of the Atmosphere, Salt Lake City, Utah, February 5–9, 1995, OSA Tech. Dig. Ser. 2, p. 121, 1995.

Sassen, K., R. P. Benson, and J. D. Spinhirne, Tropical cirrus cloud properties derived from TOGA/COARE airborne polarization, *Geophys. Res. Lett.*, 27, 673–676, 2000.

Schoeberl, M. R., P. A. Newman, L. R. Lait, T. J. McGee, J. F. Burris, E. V. Browell, W. B. Grant, E. C. Richard, P. von der Gathen, R. Bevilacqua, I. S. Mikkelsen, and M. J. Molyneux, An assessment of the ozone loss during the 1999–2000 SOLVE/THESEO 2000 Arctic campaign, *J. Geophys. Res.*, 107 (D20), art. no. 8261, 2002.

Schwiesow, R. L., S. D. Mayor, V. M. Glover, and D. H. Lenschow, Intersection of a sloping aerosol layer observed by airborne lidar with a cloud-capped marine boundary layer, *J. Appl. Meteorol.*, 29, 1111–1119, 1990.

Senff, C. J., R. M. Hardesty, R. J. Alvarez II, and S. D. Mayor, Airborne lidar characterization of power plant plumes during the 1995 Southern Oxidants Study, *J. Geophys. Res. — Atmos.*, 103, 31173–31189, 1998.

Shine, K. P., and A. Sinha, Sensitivity of the Earth's climate to height-dependent changes in the water vapor mixing ratio, *Nature*, 345, 382–384, 1991.

Shipley, S. T., D. H. Tracy, E. W. Eloranta, et al., High spectral resolution lidar to measure optical scattering properties of atmospheric aerosols. 1: Theory and instrumentation, *Appl. Opt.*, 22, 3716–3724, 1983.

Shipley, S., E. Browell, D. McDougal, B. Orndorff, and P. Haagenson, Airborne lidar observations of long-range transport in the free troposphere, *Environ. Sci. Technol.*, 18, 749–756, 1984.

Smith, W. L., Atmospheric soundings from satellites: false expectation or the key to improved weather prediction? *Q. J. R. Meteorol. Soc.*, 117, 267–297, 1991.

Spinhirne, J. D., M. Z. Hansen, and L. O. Caudill, Cloud top remote sensing by airborne lidar, *Appl. Opt.*, 21, 1564–1571, 1982.

Spinhirne, J. D., and W. D. Hart, Cirrus structure and radiative parameters from airborne lidar and spectral radiometer observations: the 28 October 1986 FIRE study, *Mon. Wea. Rev.*, 118, 2329–2343, 1990.

Spinhirne, J. D., S. Chudamani, J. F. Cavanaugh, and J. L. Bufton, Aerosol and cloud backscatter at 1.06, 1.54 and 0.53 μm by hard-target-calibrated airborne Nd:YAG/methane Raman lidar, *Appl. Opt.*, 36, 3475–3490, 1997.

Spinhirne, J. D., W. D. Hart, and D. P. Duda, Evolution of the morphology and microphysics of contrail cirrus from airborne remote sensing, *Geophys. Res. Lett.*, 25, 1153–1156, 1998.

Strawbridge, K. B., and R. M. Hoff, LITE validation experiment along California's coast: preliminary results, *Geophys. Res. Lett.*, 23, 73–76, 1996.

Swenson, G. R., J. Qian, M. M. C. Plane, P. J. Espy, M. J. Taylor, D. N. Turnbull, and R. P. Lowe, Dynamical and chemical aspects of the mesospheric Na "wall" event on October 9, 1993 during the Airborne Lidar and Observations of Hawaiian Airglow (ALOHA) campaign, *Geophys. Res. — Atmos.*, 103, 6361–6380, 1998.

Talbot, R., R. Harriss, E. Browell, et al., Distribution and geochemistry of aerosols in tropical North Atlantic troposphere: relationship to Saharan dust, *J. Geophys. Res.*, 91, 5173–5171, 1986.

Targ, R., et al., Coherent lidar airborne wind sensor. 2. Flight-test results at 2 and 10 μm, *Appl. Opt.*, 35, 7117–7127, 1996.

Toon, O. B., E. V. Browell, S. Kinne, and J. Jordan, An analysis of lidar observations of polar stratospheric clouds, *Geophys. Res. Lett.*, 17, 393–396, 1990.

Toon, O. B., A. Tabazadeh, E. V. Browell, and J. Jordan, Analysis of lidar observations of Arctic polar stratospheric clouds during January 1989, *J. Geophys. Res. — Atmos.*, 105, 20589–20615, 2000.

Tsias A, et al., Aircraft lidar observations of an enhanced type Ia polar stratospheric clouds during APE-POLECAT, *Geophys. Res. — Atmos.*, 104, 23961–23969, 1999.

Uthe, E. E., Cooling tower plume rise analyses by airborne lidar, *Atmos. Environ.*, 18, 107–119, 1984.

Uthe E. E., J. M. Livingston, and N. B. Nielsen, Airborne lidar mapping of ozone concentrations during the Lake Michigan ozone study, *J. Air Waste Manage. Assoc.*, 42, 1313–1318, 1992.

Valente, R. J., et al., Ozone production during an urban air stagnation episode over Nashville, Tennessee, *J. Geophys. Res. — Atmos.*, 103, 22555–22568, 1998.

Vaughan, J. M., D. W. Brown, C. Nash, S. B. Alejandro, and G. G. Koenig, Atlantic atmospheric aerosol studies 2. Compendium of airborne backscatter measurements at 10.6 μm, *J. Geophys. Res. — Atmos.*, 100, 1043–1065, 1995.

Vaughan, J. M., and M. Harris, Lidar measurement of B747 wakes: observation of a vortex within a vortex, *Aerosp. Sci. Technol.*, 5, 409–411, 2001.

Vodacek, A., F. E. Hoge, R. N. Swift, J. K. Yungel, E. T. Peltzer, and N. V. Blough, The use of in-situ and airborne fluorescence measurements to determine BTV absorption coefficients and DOC concentrations in surface waters, *Limnol. Oceanogr.*, 40, 311–415, 1995.

Wakimoto, R. M., H. V. Murphey, E. V. Browell, and S. Ismail, The "Triple Point" on 24 May 2002 during IHOP. PartI: Airborne dopplet and LASE analyses of the frontal boundaries and convection initiation, Mon. Wea. Rev., in press, 2005.

Weltz, M. A., J. C. Ritchie, and H. D. Fox, Comparison of laser and field-measurements of vegetation height and canopy cover, *Water Res. Res.*, 30, 1311–1319, 1994.

Werner, C., et al., Wind infrared Doppler lidar instrument, *Opt. Eng.*, 40, 115–125, 2001.

Wielicki, B. A., R. D. Cess, M. D. King, D. A. Randall, and E. F. Harrison, Mission to Planet Earth — role of clouds and radiation in climate, *Bull. Am. Meteorol. Soc.*, 76, 2125–2153, 1995.

Winker, D. M., and M. T. Osborn, Airborne lidar observations of the Pinatubo volcanic plume, *Geophys. Res. Lett.*, 19, 167–170, 1992a.

Winker, D. M., and M. T. Osborn, Preliminary analysis of observations of the Pinatubo volcanic plume with a polarization-sensitive lidar, *Geophys. Res. Lett.*, 19, 171–174, 1992b.

Winker, D. M., R. H. Couch, and M. P. McCormick, An overview of LITE: NASA's Lidar In-space Technology Experiment, *Proc. IEEE*, 84, 164–180, 1996.

Wirth, M., and W. Renger, Evidence of large scale ozone depletion within the Arctic polar vortex 94/95 based on airborne lidar measurements, *Geophys. Res. Lett.*, 23, 813–816, 1996.

Wright, M. L., E. K. Proctor, L. S. Gasiorek, and E. M. Liston, A Preliminary Study of Air Pollution by Active Remote Sensing Techniques, Final Report, NAS1-11657, Stanford Research Institute, Menlo Park, CA, NASA CR-132724, 1975.

Wulfmeyer, V., and G. Feingold, On the relationship between relative humidity and particle backscattering coefficient in the marine boundary layer determined with differential absorption lidar, *J. Geophys. Res. — Atmos.*, 105, 4729–4741, 2000.

Wulfmeyer, V., and C. Walther, Future performance of ground-based and airborne water-vapor differential absorption lidar. I. Overview and theory, *Appl. Opt.*, 40, 5304–5320, 2001a.

Wulfmeyer, V., and C. Walther, Future performance of ground-based and airborne water-vapor differential absorption lidar. II. Simulations of the precision of a near-infrared, high-power system, *Appl. Opt.*, 40, 5321–5336, 2001b.

Wulfmeyer, V., H.-S. Bauer, M. Grzeschik, A. Behrendt, F. Vandenberghe, E. V. Browell, S. Ismail, and R. A. Ferrare, 4-dimentional variational assimilation of water vapor differential absorption lidar data: The first case study within IHOP 2002, submitted to Mon. Wea. Rev., October 2004.

9

Space-Based Lidar

UPENDRA N. SINGH, SYED ISMAIL, MICHAEL J. KAVAYA,
DAVID M. WINKER, FARZIN AMZAJERDIAN

NASA Langley Research Center, Hampton, VA, USA

9.1. INTRODUCTION

This chapter discusses several atmospheric measurements that can be made by Earth-orbiting, Earth-looking lidar (light detection and ranging, laser radar, optical radar, ladar) systems. We have seen that lidar systems of many different types operated from a number of different platforms (laboratories, vehicles, airplanes, and balloons) have been developed to fill a variety of measurement needs. However, many application that require understanding of global scale phenomena such as climate, atmospheric chemistry, water and carbon cycles require observations from satellite platforms.' Even if the cost of building a large network of ground-based instruments is not excessive, the oceans and many other areas of the globe are largely inaccessible. Traditionally, these global measurements have been made using passive sensors operating in the optical or microwave regions of the electromagnetic spectrum. Space-based lidar technology is being developing to fit applications where passive sensors cannot meet current measurement requirements. Lidar remote sensing enjoys the advantages of excellent vertical and horizontal resolution; easy aiming; independence from natural light for the signal and from background noise; and control and knowledge of transmitted wavelength, pulse shape, and polarization and received polarization.

Starting in the early 1960s, passive satellite sensors have been used to observe aerosols and clouds, gases such as water vapor and ozone, to profile atmospheric temperature, and perform many other measurements. Lidar offers capabilities that complement passive techniques. Passive techniques offer limited vertical resolution, a few range bins within the atmosphere, at best, and often are limited to measuring total column amounts. The primary advantage of lidar is very high vertical resolution. While lidar signals are blocked by dense clouds that are most often found in the lower troposphere, lidar can often penetrate thin cirrus and profile down to the tops of dense clouds.

Many passive sensing techniques rely at least in part on reflected solar radiation. Lidar systems are not dependent on solar illumination, so are as capable at night as during the day, if not more so due to reduced solar background levels. The range-resolved nature of the lidar technique means the surface reflectance can be subtracted from the atmospheric signal, allowing the sensing of weak atmospheric signals in the presence of strong or variable background signals. Thus, measurements that are difficult for passive systems, such as sensing weak aerosol scattering above a bright land surface, can be performed readily with lidar.

Technical requirements for space lidar are more demanding than for ground-based or airborne systems. First, the range to a target within the atmosphere is much larger, with orbital altitudes generally falling in the range of 250 to 700 km, so that R^2 losses are much larger. For example, a system at an orbital altitude of 400 km has a signal which is reduced by a factor of 40^2, or 1600, relative to the same system operating on an aircraft at a range of 10 km (Appendix A). This means that system sensitivity is a major concern. The space-based lidar system sensitivity may be increased through higher laser pulse energy and pulse repetition frequency (PRF), larger receiver collection area, and increased lidar system photon sensitivity. However, satellite allocations for electrical power, instrument mass, telemetry, thermal control, etc. are limited. Design of a lidar to be operated on a satellite involves careful balancing of many different constraints.

Perhaps, the most distinguishing characteristics of space lidars are the environmental requirements. Space lidar systems must be specially designed to survive the mechanical vibration loads of launch and operate in the vacuum of space, where exposure to ionizing radiation limits the electronic components available. Finally, space lidars must be designed to be highly reliable because they must operate without the need for repair or adjustment. Lifetime requirements tend to be important drivers of the overall system design. The maturity of the required technologies is a key to the development of any space lidar system.

Although space lidar systems have initial high development and launch costs, the operational years of the lidar mission enjoy relatively low cost because there is no large network of ground-based sensors to access and monitor and maintain (Table 9.1).

Lidar technology is evolving rapidly and lidar systems have been used in a number of applications to study the atmosphere, land surface, and ocean (Chapter 8). Because of the enormous potential that lidar measurements from space hold for these applications, a large number of feasibility studies have been conducted over the last three decades. The earliest study was conducted in the late 1960s (Evans, 1968), and was followed by a number of studies in the U.S. (e.g., Huffaker, 1978; Browell et al., 1998; Kavaya et al., 2002), in Europe (Endeman, 1983; Ehret et al., 2002; Flamant, 2002), and in Japan (Uchino, 1992). The pace of space-borne lidar studies has continued during the past decade and resources have been devoted for technology development that is applicable to space; however, only a few of these studies have

Table 9.1 Advantages and Disadvantages of Space-Based Lidars

Advantages	Disadvantages
Global coverage	Long range to target
No physical or political inaccessibility	Large launch forces
Land, ocean, and ice coverage	Limited power source and heat removal
Low operational cost	Lifetime requirement
No obscuration from thin or low-level clouds	High altitude and thick cloud blockage
Atmospheric backscattering increases with range, yielding lower dynamic range	High radiation levels

borne fruit due to the lack of enabling technologies or re-sources needed for implementing these missions. The Lidar In-space Technology Experiment (LITE; Winker et al., 1996) and the Mars Orbiter Laser Altimeter (MOLA; Abshire et al., 2000a,b) have been two highly successful lidar missions in space. Although LITE was flown on the space shuttle as a technology demonstration experiment in 1994, it was one of the most successful lidar experiments to retrieve global atmospheric aerosol and cloud properties (Osborn et al., 1998; Winker and Trepte, 1998; Karyampudi et al., 1999; Platt et al., 1999; Stephens et al., 2001). The MOLA ranging altimeter was launched in 1996 to investigate the topography of Mars and it provided valuable data to evaluate the geophysical, geographic, and atmospheric properties of Mars (Smith et al., 2001; Neumann et al., 2003).

Lidar technology that is applicable to space remains the pacing element in the development of future space-based lidar missions. A discussion of technology developments at NASA to enable future lidar missions is included in this chapter. Also presented is a description of four classes of lidar missions, viz., aerosols and clouds, lidar altimetry, Doppler wind lidar (DWL), and DIfferential Absorption Lidar (DIAL) systems. These missions are at various stages of development, deployment, and operation. These typical space missions and the discussion do not necessarily cover all of the lidar systems being pursued.

9.2. TECHNOLOGY DEVELOPMENT FOR SPACE-BASED LIDAR MISSIONS

9.2.1. Introduction

Lasers were flown in space less than 10 years after invention of the laser (Mims, 1975). The first lidar in space was the Apollo Laser Altitmeter that was flown successfully in lunar orbit on Apollo missions 15–17, starting in 1971 (Jelalien, 1992). Since then, various laser systems have been deployed in space programs. A successful example is a high-energy flash lamp-pumped Nd:YAG laser in the space shuttle-based LITE, used for stratospheric and tropospheric aerosols and clouds measurement in 1994 (Couch et al., 1991). The development of diode-pumped solid-state lasers, however, opens a new era for space laser missions since this technology offers the key features that space laser requires such as high-power efficiency, long lifetime, compactness and light weight, and reliable and safe operation. The first diode-pumped solid-state laser to operate in space was the laser in MOLA (Afzal, 1994). The follow-on mission, Mars Global Surveyor (MOLA; Afzal, 2000), which was in space for over 5 years, helped revolutionize our understanding of the planet Mars. The Geoscience Laser Altimeter System (GLAS), launched in 2003, has a goal of operation lifetime of 5 years (Abshire et al., 2000b). All of these space lasers were based on diode-pumped Nd:YAG laser technology, which is the most mature technology with special consideration of space laser requirements. In spite of these successful missions, lidar technology for other applications remains at low technology readiness levels.

9.2.2. The Laser Risk Reduction and Active Optical Remote Sensing Programs

An Independent Laser Review Panel set up to examine NASA's space-based lidar missions and the technology readiness of lasers appropriate for space-based lidars indicated a critical need for an integrated research and development strategy to move laser transmitter technology from low technical readiness levels to the higher levels required for space missions. Based on the review, a multiyear Laser Risk Reduction Program (LRRP) was initiated by NASA in 2002 to develop technologies that ensure the successful development of the broad range of lidar missions envisioned by NASA (Singh and Heaps, 2002). Three key recommendations have been made by the LRRP team to advance the space laser technology. First, to establish space-hardened laser transmitter test beds; second, to develop space-qualified laser diode arrays (LDAs); and third, to advance wavelength conversion technologies for multiple lidar applications.

The first recommendation involves developing two laser transmitters to provide the primary wavelengths for carrying out six priority measurements for Earth sciences, and chemical and biological agent detection for civilian and warfare applications. These are the Nd:YAG/YLF laser at 1.06 μm and Ho:Tm:YLF/LuLiF laser at 2.05 μm. A 2-μm laser can be used to measure carbon dioxide using the DIAL technique, and wind in lower atmosphere by using the coherent Doppler technique. A third harmonic generation of 1-μm laser at 0.355 μm can be used for wind in the upper atmosphere by using the noncoherent Doppler technique. Second harmonic generation at 0.532 μm is the wavelength of choice for surface mapping and oceanography. One- and 2-μm lasers can further be down- or up-wavelength shifted by utilizing nonlinear wavelength conversion techniques for generating wavelength regions, such as 0.30 to 0.32 μm for ozone measurement, 0.28 to 0.35 μm for biological agent detection, and 3 to 12 μm for chemical agent detection.

The second recommendation involves high-power LDAs. LDA technologies are critical to space-based laser systems. Currently, no vendor can supply a space-qualified, conductively cooled, long-life, high-power LDA for use in space-based lidar systems. NASA has initiated and established a multiyear program with several LDA vendors for research, development, testing, and characterization of LDAs for space applications.

The third recommendation is to develop eye-safe space-based solid-state laser transmitters for multiple lidar applications. The concern of eye safety is promoting the development of solid-state eye-safe laser transmitters in the ultraviolet (UV) and mid-infrared (IR). Using nonlinear optical devices, such as the optical parametric oscillator (OPO), optical parametric amplifier (OPA), and wave-mixing technologies, the tunable wavelength range can be extended to target the different atmospheric constituents, such as ozone, carbon monoxide, ammonia, methane, etc.

In addition to the LRRP, NASA has considered developing a multienterprize Active Optical Remote Sensing (AORS) program as the next logical step in developing this important, nascent remote-sensing capability. The objective of this initiative is to advance the required technologies to a state of readiness appropriate for routine research and operational status within the desired 10-year timeframe. Under the new AORS initiative, NASA will engage with other U.S. Government entities, as well as industry and academia, to leverage common AORS interests and complementary skills/expertise to formulate an integrated technology development plan which will address each member agency's requirements in a synergistic and cost-effective manner.

9.2.3. Lidar Technologies for Space Missions

9.2.3.1. Lasers

The most challenging tasks involving the space laser design and qualification are laser system ruggedness and long lifetime operation. All the optical components in the laser resonator, including laser crystals, Q-switch, harmonic generator crystals, wave plates, mirrors, and other optical elements, must survive and keep alignment after experiencing the launch-induced random vibration levels of greater than $10\,g_{rms}$ (rms, root mean square). The mechanical and acoustic vibration environment of a spacecraft launch can be particularly detrimental to the performance of the optical system. Crossed-Porro laser resonator designs are popular for building rugged lasers and are often considered for space-based instruments. Another discipline in the laser design is to use as few optical components as possible. For example, the LITE laser transmitter oscillator used only eight optical components. Misalignment may also occur between the transmitter and receiver due to vibration during launch and due to thermal induced structure deformations in space. Extensive structural and thermal analyses along with component and system level testing need to be performed to validate the laser transmitter design. Good mechanical and thermal design is required to maximize the mission success probability.

Another serious challenge is the long operational lifetime required by the space laser. The lasers are designed to continuously operate in the orbiting mission for years without degradation in performance. Two disciplines are applied to this consideration. The first is to operate all optical components at appropriately derated levels. The pump diodes should never operate at full power. The laser itself will operate at the fluence level that does not threaten the optical components. The second is to design the space laser with enough redundancy. The LITE laser transmitter deployed two identical lasers and the GLAS laser transmitter uses three identical lasers, although there is only one laser working at any particular time in both cases.

The Cloud–Aerosol Lidar and Infrared Pathfinder Satellite Observations (CALIPSO) mission is a space operational clouds and aerosol lidar planned to launch in 2005. A robust space-based laser was developed utilizing NASA Langley Research Center (LaRC), Ball Aerospace Technologies Corp. and Fibertek, Inc., expertise (Hovis et al., 2003). The mission requirements for the laser are summarized in Table 9.2.

The global wind, carbon dioxide, and water vapor vertical profile measurements can be made with a 2-μm lidar system. In 1998, researchers at LaRC developed a prototype 2-μm laser transmitter breadboard for NASA's Earth Orbitor-2 Shuttle Mission SPARCLE (SPAce Readiness Coherent Lidar Experiment) (Kavaya and Emmitt, 1998; Yu et al., 1998a). The laser resonator consisted of eight mirrors forming a ring configuration. The resonator design was compact and rigid to fit in a 19-in.-diameter can. This laser delivered single-frequency Q-switched pulse of 109 mJ at 6 Hz. The perform-

Table 9.2 CALIPSO Laser Requirements

Energy 1064 nm	110 mJ
Energy 532 nm	100–125 mJ
Pulse width	>15 nsec, <50 nsec
Repetition frequency	20 Hz
Beam quality	<10 mm mrad (both λ)
Line width 1064 nm	<70 pm
Line width 532 nm	<35 pm
Polarization 532 nm	>100:1 Linear
Beam colinearity	Output divergence <10%
Beam jitter	Output divergence <10%
Lifetime	2×10^9 Shots
Power	102 W
Cooling	Conductive

ance of the laser met or exceeded the requirements for the SPARCLE transmitter (Yu et al., 1998b). The mission requirements for the laser are summarized in Table 9.3.

Development of a UV laser transmitter capable of operating from a space platform is a critical step in enabling global Earth observations of aerosols and ozone at resolutions greater than current passive instrument capabilities. The specifications for the UV laser transmitter are listed in Table 9.4. The "on-line" wavelength must be between 305 and 308 nm while the "off-line" wavelength must be between 315

Table 9.3 SPARCLE Laser Requirements

Wavelength (nm)	2051
Energy/pulse (mJ)	>100
Pulse width (nsec)	>150, <600
Repetition frequency (Hz)	6
Beam quality M^2	<1.2
System bandwidth (GHz)	15
Pulse spectral bandwidth (FWHM)	$1.2 \times$ FTL
Spectral purity	SLM,[a] TEMoo
Polarization	>100:1 Linear
Drift (MHz) w.r.t. master oscillator	<10
Pulse energy fluctuation (%)	Less than ± 10
Pump design margin (%)	25
Lifetime	5×10^6 Shots

[a]Single longitudinal mode.

Table 9.4 UV Transmitter Specifications

Wavelength (nm)	308, 320
Energy/pulse (mJ)	500
Repetition frequency (Hz)	10, Doublets
Doublet delay (µs)	300–400
Pulse width (nsec)	<20
Line width (pm)	<50
Spectral purity (%)	>99
Beam divergence (µrad)	150
Efficiency (%)	>2
Mass (kg)	150
Power (W)	500

and 320 nm and there must be a 10-nm separation between the "on line" and "off line."

The proposed laser system is based on Nd laser technology and high-energy OPO and sum-frequency-mixing technology. A good beam quality, seeded (for line narrowing) Nd-based pump laser (YAG or YLF), determines the timing for the laser system. This Nd laser must possess good frequency control, good beam quality, and high electrical-to-optical efficiency. The Nd laser is frequency doubled to pump a beta barium borate (BBO)-based OPO and BBO sum frequency mixer. If the conversion efficiency is high in the OPO, no amplification stages will be necessary. However, if the OPO conversion efficiency is lower than expected, high-efficiency amplification will be necessary. Either an OPA or Ti:Sapphire amplifier can be used. The design results in a minimal number of nonlinear conversions, has a low quantum defect, and minimizes the number of UV optical elements (Barnes, 1998; Stadler et al., 1998). In fact, the last nonlinear process in our design produces the UV wavelength needed for the ozone measurement. In addition, the design can be frequency controlled so that multiple laser systems can be locked to the same wavelength. An OPO/mixer design using no amplification stages has already been demonstrated at 130 mJ by researchers at NASA Langley at efficiency levels of 5% (optical-to-optical) from 1064 to 320 nm using a 10-Hz Nd:YAG laser.

9.2.3.2. Lidar Telescope

The optical performance required by most space-based lidar instruments, envisioned for future deployment, coupled with their demanding physical and environmental constraints, creates the need for advanced telescope concepts with novel optical and opto-mechanical designs. The lidar telescope must endure launch forces, the transition to zero gravity, and occasional orbit reboosts, and must maintain its alignment in the harsh thermal environment of space. The most challenging trades for a lidar telescope are high rigidity and long-term stability vs. fabrication cost and mass. For most lidar instruments, the telescope accounts for most of the weight and size budget of the instrument. Therefore, any reduction in the telescope mass will have a major impact on the total mission cost. The telescope mass is even more critical for the lidar instruments, such as DWLs (Baker et al., 1995), that require scanning the laser beam. The main approach for scanning the laser beam is to rotate the telescope. For this approach to succeed, the weight of the telescope has to be kept to a minimum to permit the rapid step and stare with its quick starting and stopping of the rotation to occur without unduly taxing the rotation motor. However, a rotating telescope would need to have a primary mirror of significant size near the diameter of the desired outgoing beam, which is in the 0.5- to 1.5-m range for many lidar concepts under consideration. Therefore, the mass of the telescope has to be reduced for three reasons. The telescope weight needs to be minimized to permit the scanning of such a large primary mirror with its significant inertia to allow step–stare operation. A reduction in weight would reduce the load on the scanning motor and permit lower power requirements, less thermal load (a smaller scanner motor), and greater procurement options. Also, a lighter but structurally stiff telescope would permit faster motion with less deceleration time to come to a stop without ringing.

There are also lidar concepts, like ozone DIAL (Browell et al., 1998), that require a large photon collection aperture area with diameter of the order of 2 m or greater in order to overcome the limitation of available laser power. Limited by the launch vehicles, such lidars will have to resort to segmented, deployable mirror panels to provide the required aperture area for collection of return photons. The mirror panels must be lightweight not just for meeting the launch constraints, but also for avoiding the distortions of mirror panels and their structure after deployment under 0-g conditions (Lake et al., 1999).

In general, most space-based lidar systems require the telescope to have low surface scatter from the mirrors and minimal backscatter toward the detector, highly reflective coatings on the mirrors for the IR laser, a compact optical path to minimize misalignments, and limited reflections to disrupt the beam polarization. To maintain operation on orbit, the telescope should be athermal (uniform coefficient of thermal expansion) (Ahmad et al., 1996). To maintain optical alignment between the time it is assembled (under a 1-*g* environment) and when it reaches orbit (under a *μ-g* environment), the telescope material needs to have high stiffness and minimal self-weight deflection.

Based on these requirements and the need for low mass, conventional telescope designs with glass optics and metal structure are not suitable for space-based lidars. There are several approaches for achieving meter-class lightweight telescopes suitable for space-based lidar applications. In the past, lightweight beryllium mirrors and structure (Hsu and Johnson, 1995) were successfully used for space-based lidar systems such as shuttle laser altimeter and MOLA (Ramos-Izquierdo *et al.*, 1994). The disadvantages associated with beryllium are high thermal expansion coefficient and fabrication limitations due to its toxic nature.

Telescope weight reduction is also achievable by using composite materials and a specialized manufacturing technique (Anapol et al., 1995; Chen et al., 1998). Fabrication of carbon/silicon carbide (C/SiC) mirrors is continuing to advance, demonstrating great potential for producing high-quality meter-class mirrors with both spherical and parabolic figures. The carbon fiber materials are attractive because of their low thermal expansion and nontoxicity, allowing the use of conventional machines and tools. The latter directly translates to a much lower manufacturing cost compared to beryllium.

Another novel approach being considered is based on using metal alloy shells. This technique utilizes recent progress on the development of thin mirrors for the next generation space telescope. For this approach, the shells and optical surfaces are formed by electrochemical deposition, typically nickel, to coat a near-final-shaped mandrel, often aluminum, to create as continuous and homogenous a shell structure over the mandrel as possible (Hibbard, 1995). Metal plating processes over aluminum mandrels are well matured, and nickel's relatively high hardness makes it suitable to diamond-turn and easy to polish and it has excellent micro-yield properties. Nickel-plated aluminum is popular among mirror designers since the nickel alloy can be tailored to the requirements of the part to help control strength, rigidity, ease of machining, and hardness. The object is to balance the proportions of the various elements in the nickel alloy (phosphor and others) such that the metal achieves a glass-like behavior with high microyield strength but without becoming so hard that machining and finishing become a problem. Using the nickel shell concept, the lidar telescope can be made completely athermal using the same materials for the optics and structure (Peters et al., 2000). The mass of a nickel shell telescope could be compatible with a SiC telescope. Its most important advantage over a C/SiC telescope may be the potential of fabricating mirrors with more complex optical figures such as a very fast off-axis mirror.

9.2.3.3. Detector

Another critical component of the lidar receiver is its photo-detector with associated signal coupling optics and preprocessing electronics. Fortunately, ongoing advances in areas such as material growth, thin-film coating, packaging, and device integration are continuing to yield more sophisticated and cost-effective products. Thus, it is likely that these robust and efficient devices will continue to create new opportunities for lidar applications. Even though today's detection devices cover much of the spectral regions of interest, there are still several lidar applications that require considerably higher sensitivity than is currently available. For these ambitious lidar instrument concepts, the transmitter power and the telescope size have been pushed to their practical limits, leaving its detector devices to increase their sensitivity in order to permit cost-effective science products. Examples of such lidar concepts include DIAL instruments for measuring CO_2 and water vapor (Browell et al., 1998; Abedin, 2002). The DIAL measurements of CO_2 at 2-μm wavelength may demand about two orders of magnitude increase in the detector's specific detectivity (D^*) compared with current state-of-the-art. For other lidars for which the required detectors are readily available, their design optimization and careful matching of the signal coupling optics and postdetection electronics are important for minimizing waste of valuable collected photons from the atmosphere.

Lidar's optical receivers are often categorized either as direct detection or heterodyne detection. A direct detection receiver measures the power of the return signal collected by the telescope and incident on the detector active area, while heterodyne detection measures the amplitude and phase of the optical field of the return signal. Both optical detection schemes are well understood and their theories well established (Kingston, 1995). Therefore, the detailed analyses of their performance are not discussed here, but rather some of the key performance factors are overviewed.

9.2.3.4. Direct Detection

For ideal direct detection, the signal-to-noise ratio (SNR) would be limited by the noise generated by the current induced by incident optical power. This is referred to as signal-noise or shot-noise-limited detection. Operating in weak signal regime, a direct detection lidar is almost never shot-noise limited and the other noise sources dominate the detector shot noise. The SNR of a direct detection lidar can be expressed in its simplest form as

$$\frac{S}{N} = \frac{(\rho P_s)^2}{(2e\rho P_B + 2e\rho P_s + 2eI_d + (4KT_e/R_{in}) + (i^2_{N_{amp}}))B}$$

$$(9.1)$$

where the first term in the denominator is the background noise resulting from solar or blackbody radiation incident on the detector. The second term is the shot noise followed by the detector dark current noise induced by the applied bias voltage. The next term is the amplifier thermal (Johnson) noise determined by its effective input impedance. The last term is the amplifier total noise power that is the combination of the amplifier input current and input voltage noise powers. The selection of the detector, at the given operating wavelength and bandwidth, must focus on the efficiency of the detector quantum conversion of photons to electrons. A wide range of detectors operating at wavelengths of interest to lidar applications are currently available. Many of these detectors are already operating close to their quantum efficiency limits at spectral bands of interest from UV to mid-IR. The lidar designers must also pay careful attention to the signal processing electronics to avoid or at least to minimize the excessive noise resulting from signal amplification. This is obvious from the simple SNR expression above, where the amplifier Johnson noise and intrinsic noise can easily dominate the detector-induced noise terms particularly from lidars requiring a bandwidth greater than 15 MHz. Amplification of the detector signal may also be possible through internal gain by using photomultiplier tubes (PMTs) or avalanche photodiode detectors (APDs) (McIntyre, 1966; Kingston, 1995).

APDs may be thought of as a solid-state version of PMTs for which each generated photoelectron emits additional electrons in a successive multiplicative manner. In an APD, the photon-generated electrons are accelerated in an electric field generated by a large reverse bias voltage. As these electrons collide with the atomic lattice of the APD structure, additional electrons are released via secondary emission. The secondary carriers are also accelerated resulting in a "multiplication avalanche." The popularity of APDs has been on the rise in recent years despite the fact that their gains are usually several orders of magnitude less than those of PMTs. This can be partly attributed to the insensitivity of APDs to magnetic fields and their size and robustness compared with the PMT's bulky vacuum tube. APDs offer several other advantages including higher quantum efficiency, better linearity, and efficient operation in the near IR where PMTs cannot operate. For both types of detectors with intrinsic gain, the SNR equation can be written as

$$\frac{S}{N} = \frac{(\rho P_s)^2}{((2e\rho P_B + 2e\rho P_s + 2eI_{dg})F + (1/M^2)(4KT_e/R_{in}) + (1/M^2)(i^2_{N_{amp}}))B}$$

(9.2)

where M is the detector internal gain and F is the detector excess-noise factor. This expression shows that by increasing the detector gain M, the excessive amplifier noise and the Johnson noise become insignificant and the SNR becomes detector noise limited. By comparing Equation (9.2) with Equation (9.1), one can see that the APDs or PMTs are viable alternatives to PIN a photodiode having a large intrinsic layer sand-wiched between p-type and n-type layers. photodiodes only when the amplifier noise terms (Johnson noise and the amplifier excess noise) cannot be overcome by the combined detector noise power.

In view of the practical issues associated with PMTs, let us only consider APDs. The intrinsic gain of an APD is limited by the fabrication imperfections and nonuniformities, and the thermally induced lattice vibrations resulting in excessive collisions of electrons. But for most practical cases, an APD is best operated at a gain lower than its maximum value. The optimum APD gain is a function of the noise power generated by the detector's amplifier and the APD's excess-noise factor, which itself increases with the gain. The APD excess-noise factor (Kingston, 1995) can be expressed as

$$F = kM + \left(2 - \frac{1}{M}\right)(1 - k) \qquad (9.3)$$

where k is the ionization ratio of the photodetector, which is a measure of electron, and hole-induced ionization due to the application of an electric field. Therefore, there is an optimum gain value, which is a function of the photodetector dark current, signal and background optical powers, and the amplifier noise power (Figure 9.1). The latter in turn is determined by the required electrical bandwidth and the photodetector junction capacitance. Currently, APDs cover a wide spectral range from UV to near IR and the developmental effort for further expanding this range is under way (Abedin, 2002). Silicon APDs are the most popular because of their wide spectral range, high gain, and low excess-noise factor. Silicon APDs cover from 400 to 1100 nm, and can operate at a gain of the order of 200 with an excess-noise factor of less than

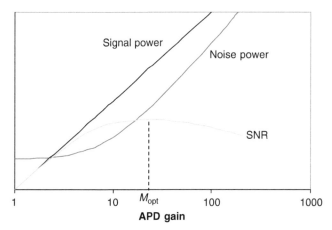

Figure 9.1 APD performance vs. its multiplication gain.

3.0. The InGaAs APDs are also attractive because of their relatively wide-frequency bandwidth and ability to operate at longer wavelengths. Current InGaAs APDs operate from 900 to 1700 nm with a gain of around 10 to 15 and an excess-noise factor of about 5.0 to 8.0.

9.2.3.5. Coherent Detection

The main advantages of coherent or heterodyne detection are high sensitivity and ability to directly provide the signal's phase and frequency. The price for these benefits is the addition of a local oscillator (LO) laser, higher-quality optics preserving the wavefront of the incoming signal (Rye, 1982), and strict optical alignment (Wang, 1984). Another important factor that must be considered, when employing heterodyne detection, is the effect of the atmospheric refractive turbulence (Fried, 1967). The refractive turbulence combined with stringent relative alignment between the LO and the return signal usually restricts coherent lidars to operation at longer wavelengths in the near- and mid-IR regions and limit the size of the receiver aperture. The theoretical prediction of coherent detection performance has been well documented for many years (Yura, 1974; Churnside and Yura, 1983; Shapiro, 1985; Amzajerdian and Holmes, 1991; Frehlich and Kavaya, 1991) and will not be addressed here. Rather, the attempt is to discuss the key design and operational parameters of a het-erodyne detection photoreceiver.

We start with a coherent detection SNR equation given by

$$\frac{S}{N} = \frac{2\rho^2(1 - 2\alpha P_{\mathrm{LO}})^2 P_{\mathrm{LO}} P_{\mathrm{s}} F_0}{(2e\rho(1 - \alpha P_{\mathrm{LO}})P_{\mathrm{LO}} + 2eI_{\mathrm{d}} + (4KT_{\mathrm{e}}/R_{\mathrm{in}}) + (i^2_{N_{\mathrm{amp}}}))B}$$

$$(9.4)$$

where P_{LO} is the applied LO power, α is the detector non-linearity coefficient, and F_0 accounts for the signal power reduction due to speckle, turbulence, atmospheric transmission, misalignment, and other systematic losses (Amzajerdian and Holmes, 1991). The effect of detector nonlinearity can be significant and is the limiting factor for the LO power. Ideally, one would apply more LO power until the detector shot noise (first term in the denominator) dominates the other noise terms. In this case, the coherent detection approaches the shot-noise-limited operation for which the SNR is independent of LO power, given by

$$\frac{S}{N} = \frac{\rho P_{\mathrm{s}} F_0}{eB}$$

$$(9.5)$$

However, in most practical cases, the detector linearity is limited and it starts to saturate before reaching the shot-noise-limited level. In fact, there is an optimum LO power level beyond which any additional increase in LO power results in reduction of the SNR (Holmes and Rask, 1995). Figure 9.2 gives a typical responsivity curve, showing the detector's nonlinear behavior with increasing incident optical power, for an InGaAs detector with an active area diameter of 75 μm operating at 2-μm wavelength (Amzajerdian, 1996).

If the detector responsivity and nonlinearity parameters and the preamplifier noise characteristics are known, the optimum LO power can be obtained from Equation (9.6) (Amzajerdian, 2001):

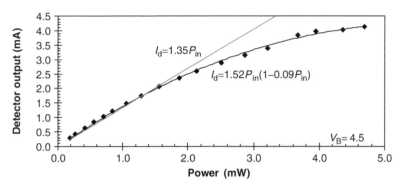

Figure 9.2 Responsivity of an InGaAs detector at 2-μm wavelength.

$$P_{LO}|_{opt} = \frac{1}{2\alpha} - \frac{1}{2\alpha}(1+\beta)^{1/2}$$
$$\times \left[\cos\left(\frac{1}{3}atan\sqrt{\beta}\right) - \sqrt{3}\sin\left(\frac{1}{3}atan\sqrt{\beta}\right)\right] \quad (9.6)$$

where

$$\beta = \frac{2\alpha N}{e\rho} \quad \text{and} \quad N = 2eI_d + \frac{4KT_e}{R_f} + (i^2_{N_{amp}})$$

Figure 9.3 illustrates the dependence of the optimum LO power on the detector nonlinearity and the preamplifier noise power for the detector of Figure 9.2. As can be seen from this plot, the dependence of the optimum LO power level on the detector nonlinearity increases with the preamplifier noise power.

Figure 9.4 shows the optimum LO power of a heterodyne photoreceiver using the 2-μm InGaAs depicted in Figure 9.2, as a function of operating bandwidth. Figure 9.4 shows the strong dependence of the optimum LO power on the operating bandwidth. It also indicates that excessive LO power, particularly for narrower-bandwidth applications, can significantly reduce the receiver's sensitivity.

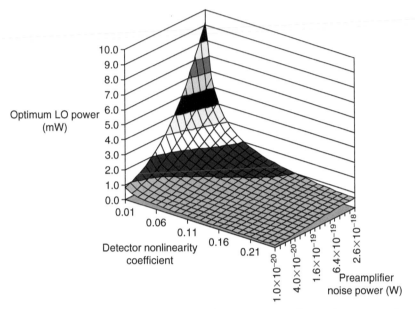

Figure 9.3 Optimum local oscillator power dependence on detector nonlinearity and preamplifier noise power.

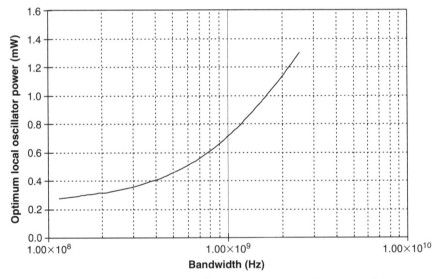

Figure 9.4 Optimum local oscillator power as a function of operating bandwidth.

9.3. SPACE-BASED LIDAR FOR OBSERVATION OF AEROSOLS AND CLOUDS

9.3.1. Introduction

The worldwide scientific community is now engaged in an effort to improve our understanding of the Earth's climate and the potential for climate change due to increasing levels of greenhouse gases in the atmosphere and other consequences of human activities (IPCC, 2001). Aerosols and clouds play important, but poorly understood, roles in the climate system. Unlike greenhouse gases, tropospheric aerosols are highly variable in space and time due to variable sources and short atmospheric residence times. Our understanding of many cloud processes, including the interaction of clouds with solar and thermal radiation, is limited by the inability of current passive satellite instruments to observe the detailed vertical structure of cloud systems. To understand the global characteristics of clouds and aerosols, satellite monitoring is required. The LITE flew on the Space Shuttle STS-64 mission in September 1994 as a proof-of-concept demonstration (Winker et al., 1996). Space agencies around the world are now engaged in developing satellite-based lidar instruments to provide the vertically resolved measurements of aerosols and clouds needed by the scientific community.

Figure 9.5 illustrates the ability of lidar to observe complex distributions of aerosols and clouds from space. A dust layer lies on top of a marine boundary layer (BL), which is heavily loaded with aerosol and embedded broken clouds. To the left, the dust layer has become detached from the BL. Note that aerosols can be observed below an attenuating cirrus layer located at 10- to 15-km altitude at the right side of the figure.

Although about half of all cloud systems are multilayered, current cloud retrievals from passive satellite sensors detect only an upper cloud layer or retrieve one equivalent cloud layer. Figure 9.6 indicates the capabilities of lidar to observe the vertical structure of a multilayer cloud. Although the lidar beam is blocked by dense BL clouds and deep convective clouds, LITE demonstrated that space lidar has a significant ability to profile multilayer clouds.

The first generation of satellite lidars has been designed around solid-state lasers such as Nd:YAG and Nd:YLF, which represent relatively mature technologies. Wavelength selection is not critical for this application and tunability is not required, as aerosols and clouds do not have fine spectral features. Wavelength selection is driven more by the availability of laser sources and sensitive, low-noise detectors. Lasers based on Nd:YAG and Nd:YLF operate in the near IR near $1\,\mu m$, and they can be efficiently doubled and tripled. While the exact wavelength of operation is not critical for cloud and aerosol studies, the availability of several wavelengths provides useful capabilities. LITE was intended to explore and demonstrate the potential applications of backscatter lidar from space and so was designed around a Nd:YAG laser which was doubled and tripled to provide output at 1064, 532, and 355 nm. While the lidar backscatter signal from clouds is independent of wavelength in this spectral range, the signal from aerosols is weakly dependent on wavelength and molecular backscatter is strongly dependent on wavelength. Thus, the use of multiple wavelengths allows the discrimination of cloud and aerosol layers based on the spectral behavior of the signals. Molecular return signals at 532 and 355 nm are large enough so that molecular backscatter at high altitudes can be used to calibrate the instrument responsivity, to an accuracy of a few percent (Reagan et al., 2002). Because molecular scattering can act to mask weak aerosol scattering, aerosol signals have higher contrast with respect to the background molecular signal at longer wavelengths. Data from LITE were used to demonstrate the capabilities of satellite lidar to detect and profile clouds (Winker and Trepte, 1998; Platt et al., 1999), observe aerosols in the stratosphere and the troposphere (Kent et al., 1998; Osborn et al., 1998), and even measure temperature profiles by using the 355-nm channel to measure the profile of molecular density in the midstratosphere (Gu et al., 1997). In the remainder of this section, we will discuss the Cloud–Aerosol LIdar with Orthogonal Polarization (CALIOP) instrument, which will be the first polar-orbiting lidar dedicated to atmospheric studies.

9.3.2. CALIPSO and the CALIOP Instrument

CALIOP is a two-wavelength polarization lidar that will fly on the CALIPSO satellite. The CALIPSO mission is being developed within the framework of collaboration between the U.S. and French space agencies, NASA and CNES (Winker et al., 2003). The CALIPSO payload also includes two passive instruments: the imaging IR radiometer and the wide field camera. The two passive instruments are nadir viewing and coaligned with the lidar. CALIPSO is planned for launch in the summer of 2005, followed by a 3-year on-orbit mission.

The CALIOP instrument builds on the experience of LITE and is designed to provide global observations of aerosol and cloud vertical structure and properties. CALIOP is built around a doubled Nd:YAG laser operating at 1064 and 532 nm. The wavelength dependence of the backscatter signals at 1064 and 532 nm allows a qualitative classification of aerosol size and aids discrimination of cloud and aerosol layers. Unlike LITE, CALIOP uses two 532-nm receiver channels to measure components of the backscatter return signal polarized parallel and perpendicular to the linearly polarized transmitted laser pulse. The backscatter from spherical particles retains the polarization state of the incident laser radiation, whereas the backscatter from irregular ice crystals or aerosol particles is depolarized. Thus, the lidar depolarization signal can be used to discriminate between ice and water clouds (Sassen, 1991) or between hydrated aerosol droplets and dry aerosol particles. Data from CALIOP and the two passive instruments will be used together to measure the radiative and physical properties of cirrus clouds.

The design of CALIOP is shown schematically in Figure 9.7. A diode-pumped Nd:YAG laser produces linearly polarized pulses of radiation at 1064 and 532 nm. The atmospheric return is collected by a 1-m telescope that feeds a three-channel receiver, measuring the backscattered intensity at 1064 nm and the two orthogonal polarization components at 532 nm (parallel and perpendicular to the polarization plane of the transmitted beam). The receiver subsystem consists of the telescope, relay optics, detectors, preamps, and line

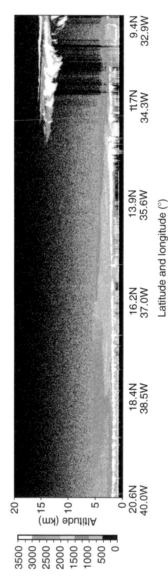

Figure 9.5 (Color figure follows page 398). LITE 532-nm nighttime raw data showing Saharan dust over the Atlantic Ocean on Sepember 17, 1994. A layer of dust (below 5 km) is located above a cloud-capped marine boundary layer (≈1 km). At the right, a layer of cirrus is seen at an altitude of 15 km.

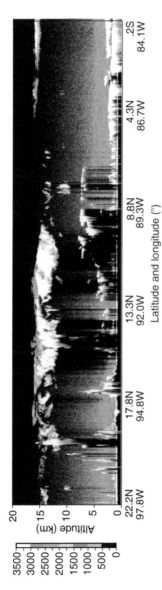

Figure 9.6 (Color figure follows page 398). LITE 532-nm nighttime raw data showing multiple cloud layers observed over the tropical Pacific Ocean on Sepember 14, 1994.

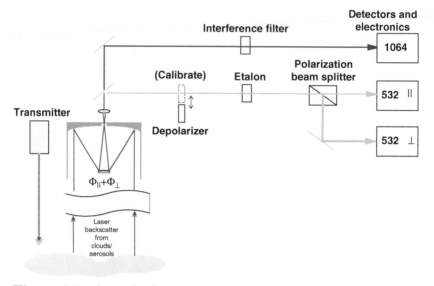

Figure 9.7   Optical schematic of CALIOP.

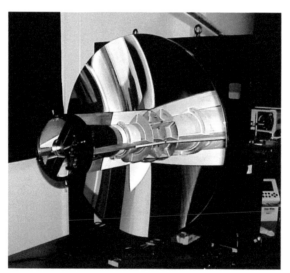

Figure 9.8   Flight telescope in final alignment.

drivers mounted on a stable optical bench. Signal processing and control electronics are contained in boxes mounted on the payload housing. In a satellite lidar, a large receiver collecting area as well as low mass are necessary. Therefore, the receiver telescope (Figure 9.8) is an all-beryllium 1-m diameter design. The telescope primary mirror, secondary mirror, metering structure, and inner baffle are all made of beryllium to avoid differential thermal expansion between the different components of the telescope. A carbon composite sunshade prevents direct solar illumination of the mirrors. A fixed field stop is located at the telescope focus. A mechanism located in the collimated portion of the beam contains a shutter and a depolarizer used in calibrating the 532-nm perpendicular channel.

Given the long range between the instrument and the atmosphere, satellite lidars must be capable of detecting very low signal levels. In CALIOP, PMTs, of a design similar to those used in LITE, are used for detection of the 532-nm return. PMTs offer a very low dark current, a large dynamic range, and reasonable responsivity. An APD is used for the 1064-nm channel, as available PMTs have very low quantum efficiency at that wavelength. The 532-nm channel uses a narrowband etalon to reduce the solar background illumination. A dielectric interference filter provides sufficient solar rejection for the 1064-nm channel. Dual 14-bit digitizers on each channel operating at 10 MHZ provide the effective 22-bit dynamic range needed to measure both strong cloud and weak molecular backscatter signals. An active beam-steering system, which allows adjustment on-orbit, is used to ensure alignment between the transmitter and the receiver.

The laser transmitter subsystem consists of two redundant Nd:YAG lasers (Figure 9.9), each with a beam expander, and the beam-steering system. The lasers are Q-switched and frequency doubled to produce simultaneous pulses at 1064 and 532 nm. Although only one laser is operated at a time, each laser produces 110 mJ of energy at each of the two wavelengths at a PRF of 20.2 Hz. Beam expanders reduce the angular divergence of the laser beam to produce a beam diameter of 70 m at the Earth's surface. The laser system is required to have good electrical efficiency given the limited

Figure 9.9   Flight laser in pressurized housing.

electrical power available from a satellite platform. Therefore, the lasers are diode pumped and passively cooled, using a dedicated radiator panel, avoiding the use of pumps and coolant loops. Each laser is housed in its own sealed canister filled with dry air at standard atmospheric pressure so that the operating environment of the laser on-orbit is similar to that of a laboratory environment on the ground.

The fundamental sampling resolution of the lidar is 30 m vertical and 333 m horizontal, determined by the receiver electrical bandwidth and the laser PRF. Backscatter data will be acquired from the surface to 40 km with 30-m vertical resolution. Low-altitude data will be downlinked at full resolution. To reduce the required telemetry bandwidth, data from higher altitudes will be averaged on-board the satellite, both vertically and horizontally. The lidar profiles are averaged to a resolution of 60 m vertical and 1 km horizontal in the upper troposphere and to 180 m vertical and 1.7 km horizontal in the lower stratosphere. The lidar calibration is obtained by normalizing the 532-nm return signal to the predicted return in the altitude range of 30 to 35 km where the return signal is mostly due to molecular scattering. The response of the perpendicular channel relative to the parallel channel can be calibrated on-orbit by inserting a depolarizer into the 532-nm beam path. The 1064-nm channel is calibrated relative to the 532-nm total backscatter signal using cirrus clouds as targets (Reagan et al., 2002). Design parameters of the lidar are summarized in Table 9.5.

9.3.3. Detection Sensitivity

CALIOP is required to accurately measure signal returns from the aerosol-free region between 30 and 35 km as well as the strongest cloud returns. Therefore, it has been designed so that the linear dynamic range encompasses the full range of molecular, aerosol, and cloud backscattering encountered in the atmosphere, which spans many orders of magnitude. Strongly scattering targets can be detected using single-shot profiles, but signal averaging to increase the SNR is required to detect weaker targets. The sensitivity for detection of cloud and aerosol layers can be computed using standard target detection theory (Kingston, 1979). The problem is one of detecting a cloud or aerosol layer which produces a signal s_c, which must be distinguished from the signal

Table 9.5 Key Lidar Parameters

Characteristic	Value
Laser	Diode-pumped Nd:YAG
Pulse energy (mJ)	110 (532 nm); 110 (1064 nm)
PRF (Hz)	20.2
Polarization purity (%)	>99 (532 nm)
Cooling	Passive
Boresight adjustment (°)	± 1
Beam divergence (μrad)	100
Receiver	
Telescope diameter (m)	1
FOV (μrad)	130
Detector/passband	
532 nm	PMT/35 pm
1064 nm	APD/400 pm
Spatial resolution	
Lower troposphere	333 m Horizontal, 30 m vertical
Upper troposphere	1 km Horizontal, 60 m vertical
Stratosphere	1.7–5 km Horizontal, 180–300 m vertical

from the molecular background, s_m. Noise contributions to the return signals may come from detector dark current, solar background illumination, and the statistical fluctuations of the lidar backscatter signal itself. If we define a threshold signal level, s_t, where $s_m < s_t < s_c$, we can compute the probability of detection and the false alarm probability for a given threshold level. Detectability is parameterized in terms of $x_d = (s_c - s_t)/\sigma_c$ and $x_{fa} = (s_t - s_m)/\sigma_c$, where σ_c^2 is the variance of the total noise on the signal.

The general solution for the minimum detectable enhancement of the signal above the background is rather complicated, but there are simple solutions for two cases of interest. When detector noise and solar background are negligible so that the noise is dominated by the statistical fluctuations of the signal itself and for $x_d = x_{fa} = x$, the minimum detectable enhancement above the molecular background, β_{min}, is

$$\beta_{min} = \beta_m[(x/\mathrm{SNR}_m + 1)^2 - 1]$$

where β_m is the backscatter coefficient of the molecular atmosphere surrounding the layer and SNR_m is the SNR of the lidar return signal outside the layer. When the measurement noise is dominated by either dark current or the solar background, so that the noise is essentially the same inside and outside the layer, the minimum detectable enhancement is given by

$$\beta_{min} = \beta_m[(x_d + x_{fa})/\mathrm{SNR}_m]$$

Figure 9.10 shows the layer detection sensitivity of the CALIOP 532-nm channel for different amounts of horizontal signal averaging, computed using the complete, general expression for detectability. The results shown assume vertical averaging of 60 m. The horizontal bars in the figure show representative magnitudes of aerosol and cirrus backscatter. The range of aerosol backscattering shown corresponds to the 5th and 95th percentiles of the lower troposphere values observed by LITE. Detectability improves with altitude because the magnitude of the molecular scattering decreases with altitude. The altitude dependence is different for day and night conditions.

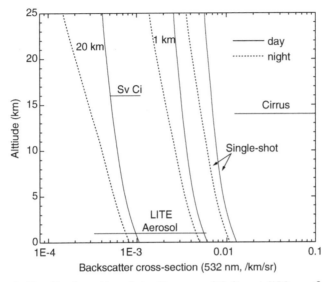

Figure 9.10 Backscatter detection sensitivity at 532 nm for vertical averaging of 60 m and (1) no horizontal averaging, (2) horizontal averaging over 1 km, or (3) horizontal averaging over 20 km. Solid and dashed lines indicate daytime and nighttime performance, respectively.

CALIOP has sufficient sensitivity to detect most cloud layers at the highest resolution of the downlinked data, using single shots below an altitude of 8 km, and averaged horizontally over 1 km above 8 km. Averaging is required to detect most aerosol layers and the weakest clouds. The results shown in Figure 9.10 indicate that most aerosol layers will be detected with 20-km horizontal averaging, but additional averaging can be performed to provide even greater detection sensitivity. In the production CALIOP detection algorithm, the signal averaging is increased in several steps to obtain the optimum balance between spatial resolution and detection sensitivity (Vaughan and Winker, 2002).

9.3.4. Aerosol and Cloud Measurements

In Figure 9.11 and Figure 9.12, LITE profiles over southern Africa are used to simulate 532-nm raw data from CALIOP.

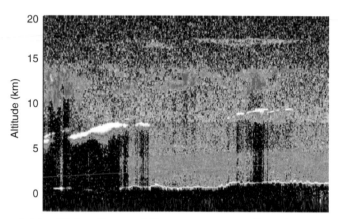

Figure 9.11 (Color figure follows page 398) Simulated CALIOP 532-nm observations (parallel channel).

Figure 9.12 (Color figure follows page 398) Simulated CALIOP 532-nm observations (perpendicular channel).

This data segment shows several layers of cloud located above a layer of biomass smoke that is capped by an inversion at an altitude of about 4 km. Clouds vary from subvisible cirrus, near the tropopause at 17 km, to optically thin cirrus between altitudes of 10 and 15 km, to highly attenuating cloud, probably supercooled water droplets, located below 10 km. Because the LITE instrument did not measure depolarization, clouds above ca. 10 km were prescribed to be ice in this simulation, with

Table 9.6 CALIOP Data Products and Uncertainties

Data product	Measurement capabilities and uncertainties
Aerosols	
Height, thickness	Layers with $\tau > 0.005$
Extinction profile (%)	40
Clouds	
Height	Layers with $\tau > 0.01$
Thickness	Layers with $\tau < 5$
Extinction profile	Within a factor of 2 for $\tau < 5$
Ice/water phase	Layer by layer

particle depolarization of 30%. Clouds below ca. 10 km, aerosols, and the Earth's surface are prescribed to be nondepolarizing.

The primary science data products to be retrieved from CALIOP are listed in Table 9.6. Although the retrieval algorithms used to produce these products include some features which are uniquely adapted to a satellite instrument, they are based on standard techniques which have been used in ground-based and airborne lidars to detect aerosol and cloud layers (Winker and Vaughan, 1994) and to retrieve cloud and aerosol extinction (Young, 1995).

9.4. LIDAR ALTIMETRY

9.4.1. Introduction

Airborne and space-based lidars have demonstrated precise measurements of surface heights and atmospheric backscatter (Bufton, 1989; Bufton et al., 1991; Gardner, 1992; Geoscience Laser Altimeter System, 1995). The history of laser altimeters in space starts with the Apollo Laser Altimeter systems that were successfully flown in lunar orbit starting in 1971 (Jelalien, 1992). The Clementine lidar was developed to map the topography of the Moon (Smith et al., 1997), the NEAR rangefinder was developed to conduct a topologic study of Asteroid 433 Eros (Cole, 1998), and the Shuttle Laser Altimeter was used to conduct Earth's topography (Garvin et al., 1998). Most of the work in hardening lasers for severe environments has been done by the military for tank mounted and

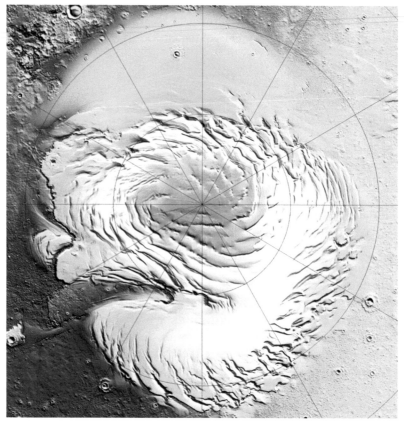

Figure 9.13 Mapping of the Martian North Polar Icecaps by MOLA, Altitude variations are shown. (courtesy of NASA/MGS project/MOLA investigation)

airborne target designators (Jelalian, 1992). This also applies to some of the optics and detectors. The laser design used for the MOLA was started by McDonnell Douglas as a potential diode laser-based replacement for a flash-lamp pumped laser in an airborne target designator. The MOLA space-based laser altimeter (Zuber, 1992) has already provided information about Martian surface and atmosphere (Abshire et al., 2000b; Smith et al., 2001). Figure 9.13 shows a planetary profile of Mars from MOLA data (Smith et al., 1999). The GLAS is a space-based lidar designed for NASA's Earth Science Enterprise's (ESE) ICESat Mission. The primary mission for GLAS is to measure the seasonal and annual changes in the heights of the ice sheets that cover Greenland and Antarctica. To do this, GLAS will measure the distance to the ice sheet from orbit at nadir with 1064-nm laser pulses. Each laser pulse results in a single range determination, and the single shot radial ranging accuracy is 10 cm for ice surfaces sloped <3°. Seasonal and annual fluctuations in regional ice sheet heights will be determined by comparing successive GLAS measurement sets. These should permit assessment of changes in the polar ice sheet topography on 3- to 6-month time scales. The information gained from ICESat will dramatically improve our knowledge of the short- and long-term changes of the Earth's major ice sheets and their possible contribution to the rise in global sea levels.

GLAS will also determine the vertical distributions of clouds and aerosols below its flight path by measuring atmospheric backscatter profiles at both 1064 and 532 nm. The 1064-nm measurements will use an analog detector and profile of the height and vertical structure of thicker clouds. Measurements at 532 nm are more sensitive since they use photon-counting detectors, and they will be used to measure the height distributions of very thin clouds and aerosol layers. With averaging, these can be used to determine the height of the planetary BL. The lidar measurements of the vertical aerosol distribution over a global scale will help improve understanding of aerosol climate effects. The measurement performances of the altimeter and lidar have been estimated by using analysis and simulations. The mission, instrument, and subsystem designs are briefly discussed in this segment. Since the launch of GLAS it has conducted global observation of aerosol, clouds and surface topography of the Earth's surface over the past two years (Sun et al., 2004; Mahesh et al., 2004).

9.4.2. Mission Overview

GLAS was launched into a 598-km circular polar orbit in January 2003. The orbit and mission parameters are summarized in Table 9.7. The orbit's 94° inclination was selected to optimize

Table 9.7   GLAS Orbit and Measurements

Orbit altitude (km)	598
Orbit inclination (°)	94
Orbit repeat tracks (km)	1 every 183 days
Ground track spacing (km)	15 at equator; 2.5 at 80° latitude
Postprocessed pointing knowledge (arcsec)	0.2 (all axes)
Position requirements	Postprocessing
Radial orbit height (cm)	<5
Along-track (cm)	<20
Laser measurement	Nadir viewing
Direction	(Nominal)
Laser pulse repetition frequency (Hz)	40
Measurement wavelengths	
Surface and cloud tops (nm)	1064
Atmospheric aerosols (nm)	532
Spot diameter on surface (m)	66 (e-2 points)
Along-track laser spot separation (m)	170 (center to center)

the crossing ground track patterns over Greenland and Antarctica and to enable data comparison with other NASA ESE instruments. The spacecraft was selected as a commercial derivative. It will determine the orbit altitude, position, and time from an on-board GPS receiver. The GLAS instrument utilizes the GPS receiver's orbit altitude for initial surface height estimates and the GPS 1-sec time markers as a long-term time standard. GLAS is being designed to operate continuously for 3 years with a goal of 5 years. NASA also plans for two follow-on ICESat missions. Since the launch of GLAS it has conducted global observation of aerosol, clouds and surface topography of the Earth's surface over the past two years (Sun et al., 2004; Mahesh et al., 2004)

9.4.3. Instrument Description

The GLAS instrument specifications are summarized in Table 9.8. The instrument incorporates three diode-pumped Q-switched Nd:YAG lasers, a 100-cm-diameter beryllium receiver telescope, a narrow ($\approx$25 pm) optical filter at 532 nm, Si APD detectors for 1064 and 532 nm, and a subsystem to measure the pointing angles of each laser firing to the

Table 9.8 GLAS Instrument Specifications

Laser type	ND:YAG slab, three-stage Q-switched, diode pumped
Number of lasers	Three each, one operated at any time
Laser PRF (Hz)	40
Laser pulse width (nsec)	4–6
Laser divergence angle (μrad)	110
Telescope diameter (cm)	100
1064-nm Detector	Si APD — analog (two each)
532-nm Detector	Si APD — Geiger (eight each)
Mass (kg)	300
Power (W)	300 average
Instrument duty cycle (%)	100
Data rate (kbps)	$\approx$500
Physical size	$\approx$110 cm $\times$ 140 cm $\times$ 110 cm
Thermal control	Radiators with variable conductance heat pipes

arcsecond level. The instrument components are mounted on an L-shaped optical bench, with one side of the L carrying the lasers and stellar reference system (SRS) and the other side serving as the telescope interface and carrying the aft-optics and detector assemblies. In the present configuration, the instrument radiator is mounted in a separate plate, which is parallel to the laser bench.

9.4.3.1. Laser Design

In order to meet the mission lifetime requirements, GLAS uses three identical laser transmitters. One laser is used at any one time, the second is needed to meet the lifetime goal, and the third is available as a spare. There is one transmitter optical path, and the second and third lasers are optically selected as needed via flip mirror assemblies. The laser specifications are summarized in Table 9.9. The Nd:YAG lasers are passively Q-switched, diode pumped, conductively cooled, and emit $\approx$5-nsec-wide pulses in a transverse electromagnetic (TEM) beam (Zuber et al., 1992; Afzal and Selker, 1994). The nominal transmitted pulse energy is 75 mJ at 1064 nm and 30 mJ at 532 nm and the laser pulse frequency is 40 Hz. The laser design uses an $\approx$2-mJ energy Q-switched master oscillator to establish

Table 9.9 Laser Specifications

Number of flight lasers	Three
Energy/pulse (mJ)	
1064 nm	75
532 nm	35
Pulse repetition frequency (Hz)	40
Pulse shape/width stability (%)	± 5
Pulse width (FWHM) (nsec)	4–6
Beam divergence (μrad)	110
Spatial beam profile	Nominal Gaussian
Beam-pointing stability (μrad)	± 10
Mass (each laser) (kg)	12.7
Size (each laser)	<18 cm $\times$ 30 cm $\times$ 15 cm
Power (W)	<105 @ 28 V
Wall plug efficiency (%)	−6
Lifetime goal (three lasers)	5 years continuous
Total laser shots in 5 years	6.3×10^9
Target lifetime per laser	3.15×10^9 Shots

the pulse length and spatial beam quality, followed by two double-pass zigzag slab laser amplifiers to increase the pulse energy, and a nonlinear crystal to produce the 532 nm.

9.4.3.2. Receiver Design

The laser backscatter from clouds, aerosols, and the surface are collected by the receiver telescope. The telescope is an all-beryllium Cassegrain design, with a diameter of 100 cm and a field of view (FOV) of 350 μrad. The 1064-nm detectors are silicon avalanche photodiodes with ≈200-MHz bandwidths. One serves as the prime detector and the backup unit can be switched into the path via a flip mirror assembly. The 1064-nm optical receiver is a larger, higher-performance version of the MOLA (Ramos-Izquierdo *et al.*, 1994) receiver.

The 1064-nm surface echo is spread in time due to the slope and roughness of the terrain surface (Abshire et al., 1991; McGarry et al., 1991). A fast electronics-timing unit is used to measure the time of flight of the laser pulse. The GLAS 1064-nm electronics receiver uses an all-digital approach, which digitizes and records both the transmitted and the echo signals at a 1-GHz rate. This permits postdetection algorithms to search the digitized window, find the surface echo signal, and calculate the transmitted and echo pulse energies and centroid occurrence times. This receiver approach permits a close approximation of a maximum likelihood estimator for range, and results in better performance and more flexible operation than the threshold, pulse width, and energy measurements used on MOLA (Abshire et al., 1991; McGarry et al., 1991). The 1064-nm cloud lidar electronics utilize a filter that preserves the waveform with 2-MHz (75 m) resolution, for the lowest 30 km in range. The atmospheric profiles at 1064 nm are reported for every laser firing.

To reject the daytime background light, the 532-nm lidar receiver uses a 170-μrad FOV. A movable mirror assembly is used to keep this aligned with the laser signal. A narrowband 532-nm filter and etalon are also used to reject background light. The etalon has a bandwidth of ≈25 pm, and its center wavelength is thermally tuned to track the average laser wavelength. The received signals that pass through the filter are distributed via beam splitters to eight photon-counting detectors. These operate in the Geiger mode and have >50% photon-counting efficiency. Using eight detectors permits >80-MHz photon-counting rates from the 532-nm receiver, which allows photon-counting even with sunlit clouds in the telescope's FOV. At 532 nm, the sum of 40 individual lidar profiles measured by the photon-counting detectors is reported once per second.

9.4.4. Laser Pointing Angle Determination

Near their edges, the ice sheets have sloped surfaces, and pointing of the laser beam away from nadir can bias the GLAS surface height measurements. For example, over surfaces with 2° slopes, knowledge to ≈8 μrad is required to achieve 10-cm height accuracy. As a consequence, accurate knowledge of the pointing angle for each laser firing is critical for the ICESat mission. GLAS uses an SRS to accurately determine the pointing angle of each laser firing relative to inertial space (Afzal et al., 1998). The GLAS SRS approach uses a high-precision star camera oriented toward local zenith

(Millar and Sirota, 1998). Its measurements are coupled with a gyroscope to measure the inertial orientation of the SRS optical bench. The far-field pattern of the laser beam is measured on every laser firing with a laser reference system (LRS). It incorporates a special camera that views zenith and digitizes images with a higher resolution and a narrower FOV than the star camera. The LRS folds a small fraction of the outgoing laser beam into the laser reference camera with two highly stable cube corner assemblies. The first cube corner, which has a transparent entry face, folds a small fraction of the laser beam angle into the second cube corner, which directs it into the laser reference camera. The laser reference camera digitizes each laser far-field pattern along with several alignment markers. These are optical signals from the alignment reference surfaces of the star camera and gyroscope, which are folded via cube corners into the laser reference camera.

Optically measuring each laser far-field pattern relative to the star camera and gyroscope alignment points permits the angular offsets of each laser pulse to be determined relative to inertial space in the ground-based data processing. The laser reference camera can also view stars that pass through its FOV. When these events occur roughly once every 8 min, they permit the alignment biases between the star camera and laser reference camera to be determined with subarcsecond precision.

9.4.5. Surface Measurements

The GLAS 1064-nm pulses will be used to measure the range to the surface, and from orbit will illuminate the surface with footprints of 66-m diameter and 170-m center-to-center spacing. The error budget for the ice surface measurements is summarized in Table 9.10. Over ice surfaces with slopes $<2°$, every laser measurement of range should have <10-cm resolution. When combined with 10-cm orbit uncertainty and up to 18-cm vertical uncertainty due to laser pointing biases over sloped surfaces, this results in measurement of the ice sheet vertical height with 20-cm accuracy. Subsequent data processing and averaging will grid measurements over

Table 9.10 Ice Altimetry Error Budget

Source	Error type	Magnitude (cm)
Instrument	Single-shot accuracy	
	(3° surface features)	<10
	Range bias	<5
	Laser beam pointing angle uncertainty	
	(1 arcsec, 2° surface)	18
	Radial orbit uncertainty	5
	Clock synchronization (1 μsec)	1
Spacecraft	Distance uncertainty from S/C POD*	
	to GLAS zero reference point	0.5
Environment	Atmospheric error (10-mbar error,	
	0.23 cm/mbar)	2
	RSS error	0.20

*Spacecraft precise orbit determination

surface areas of 150 km × 150 km to allow a determination of 1.5-cm height changes per year in these areas.

When over land, GLAS will profile the heights of the topography and vegetation. The GLAS measurements will allow the Earth's land surface to be referenced, for the first time, to a common global grid with 1-m-level accuracy. For its surface measurements, GLAS utilizes a single-channel receiver with a 1064-nm detector. The detector signal is sampled by an all-digital receiver that records each surface echo waveform with 1-nsec resolution and a length of 200, 400, or 600 samples. Analysis of the echo waveforms permits discrimination between cloud and surface returns, measurements of the roughness or slopes of the surface and the vertical distributions of vegetation or trees illuminated by the laser beam.

9.4.6. Atmospheric Measurements

GLAS will also measure atmospheric backscatter profiles at 1064 and 532 nm. These measurements share the transmitter, the receiver telescope, and the 1064-nm detector with the surface lidar. The 1064-nm atmospheric measurements will be used to profile the heights and vertical distribution of clouds and dense aerosols every shot, resulting in 75-m vertical and 175-m horizontal resolution.

The 532-nm atmospheric backscatter measurements are used to measure the vertical distribution of optically thin aerosols during both day and night. They utilize a frequency doubler in the laser, and a dichroic beam splitter, narrowband filter, and photon-counting detectors in the receiver. The daytime background count rate is reduced by spatial and spectral filtering. The 532-nm FOV is 170 μrad, and is kept aligned to the laser and accommodates the echo's lag angle by using a movable mirror assembly in the 532-nm receiver channel. A two-stage optical band-pass filter is used to restrict the receiver's spectral band-pass to ≈25 pm. In it a thermally tunable etalon is used to track the slow variations in the laser's 532-nm wavelength. After the filter, the optical signal is split equally to eight identical Geiger-mode Si APDs used as the 532-nm photon counting detectors. At mid altitudes, GLAS will accumulate 20 to 40 consecutive 532-nm atmospheric measurements, producing an average 532-nm lidar profile at a 1-Hz rate with 75-m vertical and 3.5- to 7-km along-track resolution.

9.5. WIND MEASUREMENT FROM SPACE

9.5.1. Introduction

An earlier chapter described the ability of lidar systems to measure wind velocity. This section deals with the as yet unfulfilled desire of scientists and technologists to place a DWL system in earth orbit, aim the laser pulses at the atmosphere, and obtain vertical profiles of the wind velocity over the whole earth. The benefits of such global wind profiling would be profound, especially if the measurements were highly accurate, had good vertical and horizontal spatial resolution, and occurred often enough in the right places (Halem and Dlouhy, 1984; Baker and Curran, 1985; Curran et al., 1987; Baker, 1991; Keller and Johnson, 1992; Baker et al., 1995; Cordes, 1995; Canavan and Butterworth, 1999; ESA, 1999; NASA, 2000; Clausen et al., 2002). The foreseen applications of the space-based wind measurements include: (1) direct use of the lidar wind profiles when and where they occur, and (2) infusion (assimilation) of the wind profiles into climate and weather computer models producing more accurate predictions of several meteorological parameters, including wind, over regions both near and far from the lidar measurements. Applications in the first category include use of data in the vicinity of hurricanes for improved track and intensity prediction; use of data to calculate the source or predict the motion of pollutants and other airborne objects; and regional and battlefield prediction of fog, dust, smoke, etc., migration. The second category, assimilation into models, has received much more attention from customers like NASA and NOAA. Applications include improved climate understanding, weather prediction, understanding of atmospheric processes, and weather hazard prediction and loss mitigation.

9.5.2. Wind Measurement Requirements

Since at least 1978 (Huffaker, 1978), there have been many attempts to list requirements for the wind measurements from space-borne DWL systems. Although sometimes unclear, these requirements have mostly been focused on the assimilation of the wind measurements into climate and weather computer models, rather than the "where-is, when-is" use of the measurements. The early requirement for the DWL to provide horizontal wind velocities transitioned to a requirement for single-component-only wind velocities, but taken in synergistic patterns to still provide the computer models with horizontal wind "information" (Lorenc et al., 1992). Most of these requirement lists have consisted of only wind speed accuracy, and horizontal and vertical resolution. As lidar system researchers attempted to match lidar hardware and "point designs" to the requirements, it was noted that an unambiguous set of requirements would need a great deal more definition of the atmosphere and desired wind measurements. Recently, NASA and NOAA brought together climate and numerical weather prediction (NWP) scientists and lidar researchers to create a more defined and usable and up-to-date requirement specification. Inputs from lidar technologists were used to ensure a complete, unambiguous set of requirements for reasonable DWL systems while avoiding any "technologist push" of the requirements themselves.

The joint NASA ESE/NOAA National Environmental Satellite Data and Information Service (NESDIS) "Draft Global Tropospheric Winds Sounder (GTWS) Science and Operational Data Specification" was released in October 2001. Two complete sets of numbers representing "threshold" and "objective" wind missions were listed. The threshold requirements were deemed to represent the minimum data requirements that would result in a meaningful impact on science and operational weather prediction. A brief synopsis of the threshold requirements is presented in Table 9.11, but note that the entire NASA/NOAA document is needed for a full, unambiguous specification. Some errata in the original release have been corrected here. Table 9.11, although an abbreviation of the full NASA/NOAA specification, is considerably more complex than earlier statements of wind requirements. This depth of specification is needed for the lidar designer to have an unambiguous task. It reveals the complicated interaction of different requirements with each other, with the atmospheric environment, with the orbit, and with the DWL technology limitations. Previous short-version requirement lists led to lidar point designs with possibly tens of decibels of ambiguity.

Table 9.11 acknowledges the fact that foreseeable DWL systems will have to employ shot accumulation to make a single line-of-sight (LOS) wind measurement, contrary to earlier emphasis on single-shot DWL winds. The DWLs will have to employ a lidar scanner to meet the requirements on cross-track locations and collocated biperspective pairs of LOS winds. (The locations directly under the orbiting satellite on the earth's surface are referred to as the ground track.) The biperspective LOS wind pairs that provide horizontal wind "information" must not extend more than 100 km each, and must repeat with an along-track spacing of no more than 350 km. There must be at least four cross-track locations where this occurs, covering at least 800 km, centered on the ground track. Two cloud layers are specified, one with defined laser extinction, and one with defined laser blockage statistics.

Table 9.11 Draft NASA/NOAA Threshold Wind Data Product
Requirements

Vertical depth of regard (DOR) of wind measurements	km	0–20
Vertical resolution		
Tropopause to top of DOR	km	Not required
Top of BL to tropopause	km	1
Surface to top of BL (BL top specified at 2 km)	km	0.5
Vertical location accuracy of LOS wind measurements	km	0.1
Horizontal location accuracy of LOS wind measurements	km	0.5
Number of collocated LOS wind measurements for horizontal[a] wind calculation	—	2 = pair
Allowed angular separation of LOS wind pair, projected to a horizontal plane	°	30–150
Maximum allowed horizontal separation of LOS wind pair	km	35
Maximum horizontal extent of each horizontal[a] wind measurement	km	100
Maximum along-track horizontal spacing of horizontal[a] wind measurements	km	350
Minimum horizontal cross-track width of regard of wind measurements	km	±400
Minimum number of cross-track locations for horizontal[a] wind measurements[b]	—	4
Maximum cross-track spacing of adjacent cross-track locations	km	350
Design maximum horizontal wind speed		
Above BL	m/sec	75
Within BL	m/sec	50
Maximum 1σ LOS wind random error, projected to a horizontal plane; from all lidar, geometry, pointing, atmosphere, signal processing, and sampling effects[c]	m/sec	3 (2 in BL)
Design 1σ wind turbulence level	m/sec	1.2 (1 in BL)
Maximum LOS wind unknown bias error, projected to a horizontal plane	m/sec	0.1
Minimum design a priori velocity knowledge window, projected to a horizontal plane (using nearby wind measurements and contextual information)	m/sec	26.6

Table 9.11 Draft NASA/NOAA Threshold Wind Data Product
Requirements—Continued

Design cloud field		
Layer from 9 to 10 km, extinction coefficient	km^{-1}	0.14
Layer from 2 to 3 km, 50% of lidar shots untouched, 50% blocked	%	50, random
Aerosol backscatter coefficient: two vertical profiles provided	$m^{-1} sr^{-1}$	Provided
Aerosol backscatter Probability density function (PDF)	m sr	Lognormal
PDF width	$m^{-1} sr^{-1}$	Provided
Atmospheric extinction coefficient: two vertical profiles provided	km^{-1}	Provided
Minimum wind measurement success rate, referenced to these requirements, including two specified cloud layers	%	50
Orbit latitude coverage	°	80N–80S
Downlinked data	—	All raw data
Mission life	year	2

[a]Horizontal winds are not actually calculated; rather two LOS winds with appropriate angle spacing and collocation are measured for an "effective" horizontal wind measurement. The two LOS winds are reported to the user.
[b]The four cross-track measurements do not have to occur at the same along-track coordinate; staggering is permitted.
[c]The true wind is defined as the linear average, over a 100-km × 100-km box centered on the LOS wind location, of the true three-dimensional wind projected onto the lidar beam direction provided with the data.

The LOS wind error, projected into the horizontal plane, and including all lidar, geometry, spacecraft, software, and sampling errors, must be less than 2 m/sec in the BL, and less than 3 m/sec elsewhere (1 and 2 m/sec for the objective requirements, not shown). The overall wind measurement success rate, given all other requirements including the two cloud layers, must be greater than 50%. This last requirement allows for clouds, for otherwise failed wind measurements including measurements with large errors, and statistically infrequent events involving convergence of several error sources. Although there is still some ambiguity or "wiggle room" in these requirements, it is far less than what existed with prior requirement statements.

9.5.3. Space-Based DWL Measurement Technique

The space-based DWL measurement technique consists of transmitting laser pulses at the atmosphere and detecting light that is scattered by the atmosphere in the reverse direction back into the lidar receiver. This "monostatic" lidar architecture has both the transmitter and the receiver at the same location. The geometry of lidar remote sensing from space is described in Appendix A. The targets in the atmosphere that backscatter the light are air molecules or aerosol particles suspended in the air. In both cases the scattering objects move with the wind velocity and encode the velocity information onto the backscattered light in the form of a Doppler shift of the original transmitted laser light frequency. By detecting the backscattered light, determining the total Doppler shift from the original frequency, and using knowledge of the geometry of the measurement, the wind velocity along the LOS direction of the laser beam may be calculated.

Consider, for simplicity, the case where the only motion of the transmitting lidar system and the reflecting aerosols/molecules is along the LOS direction. Let the total relative velocity be V_T, which is positive if the relative motion is "approaching" or "closing"; that is, the separation distance is decreasing with time. Then, Einstein (1905) and Van Bladel (1984, section 5.11) have shown that

$$f_R = f_T \frac{1 + (2V_T/c) + (V_T^2/c^2)}{1 - (V_T^2/c^2)}$$

where f_T is the transmitted laser frequency, f_R is the detected backscattered frequency, and c is the speed of light.

The above equation may be simplified by ignoring the terms with higher powers of V_T/c, leading to

$$f_R = f_T(1 + (2V_T/c))$$

The detected light frequency is larger than the transmitted frequency when there is an overall "approaching" geometry. A "departing" or "receding" geometry would cause a smaller detected frequency, which is the familiar red shift of astronomy. A factor of two is present in the equation since both the light emitter and the detector are moving relative to the target. At this time, the two leading contenders for a space-based DWL use laser wavelengths of 2.05 and 0.355 μm. The coherent-detection DWL (CDWL) technique currently favors 2.05 μm, and the noncoherent-detection (direct detection) DWL (NDWL) technique currently favors 0.355 μm. From the above equation, and using $c = f\lambda$, we see that the Doppler shifts will be 0.976 and 5.63 MHz/m/sec, respectively.

The actual situation is that the laser beam's LOS direction makes one angle with the forward velocity vector of the spacecraft/lidar, and makes another angle with the total motion of the aerosols/molecules. The total air motion, in turn, consists of the earth rotation and wind velocity vectors. The fully relativistic equation in terms of these two angles and with higher powers of V_T/c does not appear to be published. Nevertheless, the error in calculating wind velocity from using the simplified equation vs. the more rigorous equation may be explored by replacing the velocities in the above two equations with the projected velocities, that is, $V \cos(\theta)$, and calculating f_R using both equations. Consider a 2.05-μm lidar orbiting at 400 km with a nadir angle of 45° and an azimuth angle of 0° (looking straight ahead along the flight path) (Appendix A). Let both the earth rotation and the wind velocities equal zero. The simplified equation yields a value of f_R that is approximately 140 kHz below the result from the rigorous equation. This causes a wind velocity error of approximately 0.15 m/sec, which is significant when trying to achieve an overall error budget of 1 m/sec. This error is a function of orbit height, but is independent of the lidar's optical wavelength or pointing direction. There are several additional pitfalls to avoid for wind measurement from space (Gudimetla and Kavaya, 1999; Kavaya et al., 2001).

The total Doppler shift of the backscattered light is proportional to the total relative motion of the lidar system and atmospheric target projected along the LOS direction. In addition to the wind velocity, the spacecraft motion and the rotation of the earth also contribute to the total Doppler shift.

These must both be subtracted from the total in order to obtain the wind velocity. Figure A.3 (in Appendix A) shows that the projection of the spacecraft velocity into the laser beam LOS direction will be very much larger than the required wind velocity error. The same is typically true of earth rotation velocity. This subtraction step leads to strict requirements on final pointing knowledge for each laser shot.

After each laser pulse is transmitted, and after a time delay as the pulse reaches the atmosphere and the backscattered light returns to the lidar, an optical signal is received that is continuous in time, representing increasing range from the lidar and decreasing altitude in the atmosphere. The direction of the lidar receiver's optical axis must be pointed in the direction of each transmitted laser pulse, within some tolerance, until the atmospheric signal is completely received. This is a short-duration (a few milliseconds) pointing control (stability) requirement on the lidar and spacecraft, and it depends on the type of DWL and its optics size (Kavaya et al., 2001).

The continuous atmospheric signal from each pulse is divided into N_{ALT} time intervals for separate processing that will yield an altitude profile. Today's DWL technology is unable to meet the NASA/NOAA measurement requirements with a single laser pulse. Depending on the orbit parameters, the lidar's size and hence its sensitivity, and the atmospheric parameters, a consecutive number N_{ACC} of laser pulses is transmitted and combined (accumulated) (Frehlich, 1996; Rye and Hardesty, 1993a,b). The primary benefit of laser shot accumulation is a reduced velocity error for NDWL and reduced threshold aerosol particle concentration for CDWL. Reduced threshold aerosol concentration translates into greater coverage of the overall atmosphere. During the accumulation time, the lidar scanner is stationary and ideally the laser pulses are all parallel to each other. The N_{ACC} pulses are spaced by a distance proportional to the forward velocity of the laser beam in the atmosphere, V_T, and inversely proportional to the PRF. The accumulated signals are segmented into time intervals and then processed to yield total relative LOS velocity. The measurement volume may be pictured as the parallel strings of a harp that are tilted from the vertical and are slanting through the atmosphere. Two horizontal planes bound any particular time or altitude interval. The intersection of the measurement volume with a horizontal plane is a series of N_{ACC} circles in a straight line parallel to the ground track, ignoring earth rotation and unwanted pointing variations. Subtraction of the spacecraft and earth velocities results in a single profile of LOS wind velocity vs. altitude. The N_{ALT} velocities in a single profile are each assigned a time, location, altitude, and pointing direction. For example, the time and location may be chosen as corresponding to the middle shot of the N_{ACC} laser shots, when none of the shots are blocked by clouds, and adjusted correspondingly when clouds block some of the shots. The above sequence is then repeated for the remainder of the $N_{CT} = 4$ cross-track locations by adjusting the azimuth angle of the assumed fixed-nadir-angle lidar scanner. As the spacecraft travels forward and the geometry is just right, the above sequence is repeated while viewing each LOS wind location from a different (aft) perspective. The NASA/NOAA requirements call for $N_{PER} = 2$ perspectives or different measurement angles in each location. ($N_{PER} = 2$ is the maximum number possible with one satellite, a constant nadir angle scanner, and strict perspective collocation requirements. The two perspectives may be called fore and aft.) The angle between these two perspectives, projected into a horizontal plane, must lie in the 30 to 150° range in order to contain sufficient horizontal wind "information." The coincidence of the two LOS wind lines, of N_{ACC} shots each, must be within 35 km horizontally. The total number of laser shots fired is $N_{ACC}N_{CT}N_{PER}$, and the total number of LOS wind velocities measured is $N_{ALT}N_{CT}N_{PER}$. The entire sequence described above is repeated at an along-track spacing of no more than 350 km.

Most types of DWL systems allow both vertical resolution and N_{ALT} to be changed if desired during later data processing, if sufficient raw data are downlinked. Depending on the type of DWL, N_{ACC} may also be changed during data processing. Then, the data may be processed with multiple choices for these three parameters, yielding, for example, better vertical and horizontal resolution when the SNR is large.

9.5.4. DWL Scan Pattern

In 2001 to 2002, NASA lidar technologists created space-based DWL point designs that were intended to meet the NASA/NOAA draft threshold requirements. A point design for a CDWL system was performed at NASA LaRC (Kavaya et al., 2002), and one for an NDWL system was performed at NASA Goddard Space Flight Center (Gentry, 2002). Mandatory conditions for the DWL point designs included orbit height $= 400$ km, $N_{PER} = 2$, and $N_{CT} = 4$. The combination of the NASA/NOAA requirements, the orbit height, and geometry greatly limited the number of possible operating points.

The spacecraft/lidar orbital tangential velocity was 7676 m/sec, the ground track's advance velocity at the surface was 7222 m/sec, and the orbit period was 92.4 min. Figure A.2 (Appendix A) shows that a laser beam nadir angle at the spacecraft of at least $45°$ was needed to cover the required 400 km on both sides of the ground track. Both the CDWL and the NDWL point designs chose a nadir angle of $45°$. The slant range to the earth surface was 585 km, and the round trip time of light from the surface was 3.9 msec. The nadir angle of the laser beam as it intercepted the earth surface was $48.7°$. Any velocity error in the LOS direction was multiplied by a geometric factor of 1.33 when projected into a horizontal plane. Both point designs also had $N_{AZ} = 8$ lidar scanner azimuth angles; which is calculated from $N_{AZ} = N_{CT}N_{PER}$. The eight scanner azimuth angles were required to make four pairs of LOS wind measurements at the required $N_{CT} = 4$ cross-track positions. Figure A.4 (Appendix A) shows that the time taken to advance 100 km in the along-track direction, with an orbit height of 400 km, is 13.8 sec, and therefore a 350-km advance takes 48.3 sec. This time must be apportioned between scanner movement time and data collection time. Approximately 1 sec was budgeted for each change in the scanner azimuth angle. The eight azimuth angle changes used 8.3 sec of the available 48.3 sec. The remaining 40 sec were used for the eight LOS wind measurements, performing each one in 5 sec. Each 5-sec-duration LOS "line" of wind was 36 km long on the earth's surface, and was as wide as the laser

beam, assuming excellent pointing stability. The laser beam
width is a function of the laser wavelength, transmitted beam
size, beam quality that may cause excess diffraction spread-
ing, and slant range to the target. For the CDWL point design,
it was approximately 3 to 6 m. The 2-μm pulsed laser devel-
opers at NASA LaRC believed that it was best to limit the
CDWL design laser PRF to 12 Hz from laser physics and
design considerations (Yu et al., 2000b). With 5 sec available
to perform each LOS measurement, the maximum possible
value of $N_{ACC} = 60$ was selected for the CDWL design. The
earth surface intercept locations of the 12-Hz pulsed laser
shots were spaced by 600 m.

The exact values of the eight lidar scanner azimuth
angles and their sequence must be chosen to produce sufficient collocation and adequate cross-track positions of the four
pairs of biperspective LOS wind measurements. A secondary
consideration is to minimize the angular change that the
scanner must undergo in the allotted 1-sec intervals. Figure
9.14 shows one example of a scan pattern that meets the
primary criteria. In the figure, the 400-km high lidar system
is crossing the equator and is moving northwest with a sun-
synchronous orbit inclination angle of 97° (7° east of due north
at the equator). The eight scanner azimuth angles, in sequential order, are 114, 73.8, −18.4, −155, −109, −64.2, 25.3, and
152°. The azimuth angle is defined as 0° in the forward along-
track direction, and positive in a counterclockwise direction

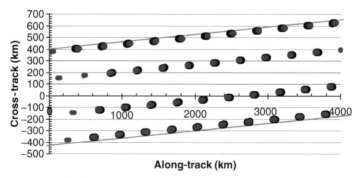

Figure 9.14   Lidar scan pattern.

when looking downward. The cross-track coordinate is positive to the left of the ground track when looking forward. The along-track and cross-track axes are defined for the beginning of the shot sequence, and fixed to the earth, so that the effect of earth rotation can be seen. Since the earth rotates to the east, the shot sequence slants to the west, which is left or positive for a north moving satellite. The slant angle in the figure reflects the combined effect of the ground-track speed, 7222 m/sec, and the equatorial earth rotation speed, 463 m/sec. The two gray lines show the maximum possible field of regard (FOR) of the assumed lidar scanner, reaching up to 414 km from the ground track. The sequence begins with the lidar at coordinate (AT, CT) = (200 km, 0) (AT = along track, CT = cross track). The arbitrary along-track offset moves all shots to positive along-track coordinates. The first lidar shot strikes the earth at coordinates (32 km, 378 km). Since this first shot at azimuth 114° is looking in the back direction, it is plotted as gray. Likewise, the forward-looking shots are colored black. The coincidence of the two desired perspectives (fore and aft, black and gray) can be seen in the figure. Comparison of Figure 9.14 with Table 9.11 reveals that the attempted wind measurement locations meet the required positions. An algorithm to find an optimum scan pattern has not yet been developed, and Figure 9.14 is almost certainly not the optimum. The maximum angle change for the scanner in this pattern is 136° degrees, with 180° being the worst case. The desired coincidence of the two perspectives fails at other latitudes for this scan pattern, and it may be necessary to adjust the azimuth angles for different latitudes during a mission.

9.5.5. DWL Velocity Error

There are numerous publications regarding the velocity error or accuracy of each type of DWL sensor. However, for space-based operation in earth orbit, the overall wind error of the measurement has many more elements than the basic sensor error. It is important to understand the overall error budget and how the DWL sensor contributes to it. A simple portrayal of the velocity error budget is shown in Figure 9.15. The

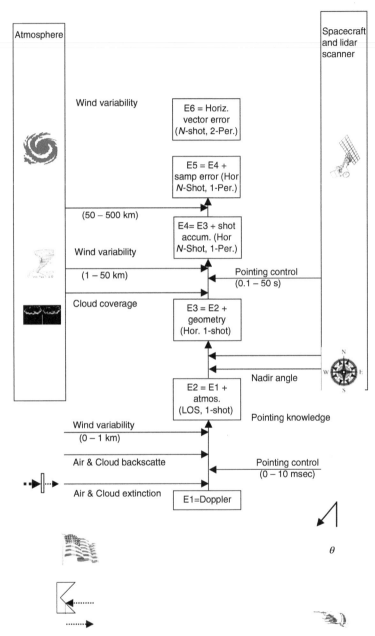

Figure 9.15 Space-based DWL measurement error. Hor. = hori-horizontal; Per. = perspective; Samp. = sampling; LOS = line of sight; E = wind error.

column to the left portrays the several interactions and contributions of the atmosphere to the velocity error. The column to the right similarly portrays how the lidar spacecraft, lidar-to-spacecraft mechanical attachment, and lidar scanner contribute to velocity error. The center of the figure, from bottom to top, conveys how the velocity error is built up from various contributions. Six levels of wind error are shown. Level 1 velocity error (i.e., E1) represents the theoretical DWL error most frequently discussed in the lidar technology literature. Besides the DWL hardware, it includes the signal processing and velocity estimation hardware and software algorithms.

Closely connected to the basic lidar velocity error are four atmospheric qualities: the extinction it imposes on the laser beam, the backscatter efficiency of the molecules or aerosol particles in sending photons back to the lidar, the variability of the LOS wind velocity over the volume used to make one LOS wind estimate, and the innate velocity width of the molecular or aerosol target from Brownian motion. The first two of these four qualities affects the velocity error through the DWL SNR:

$$\mathrm{SNR} = \frac{S}{N} \propto \frac{ED^2 T^2_{\mathrm{ATM}}[\beta_{\mathrm{AER}} + \beta_{\mathrm{MOL}}]\eta_{\mathrm{OPT}}\eta_{\mathrm{DET}}}{R^2}$$

where S is the signal power, N is the noise power, E is the laser pulse energy (J), D is the lidar receiver telescope diameter (m), $T_{\mathrm{ATM}}(\lambda)$ is the one-way beam transmission through the atmosphere to range R,

$$T^2_{\mathrm{ATM}} = \exp\left[-2\int_0^R \alpha(x)\mathrm{d}x\right]$$

$\alpha(x)$ is the atmospheric extinction coefficient (m^{-1}), β_{AER} and β_{MOL} are the atmospheric aerosol and molecular backscatter coefficients $(\mathrm{m}^{-1}\,\mathrm{sr}^{-1})$, respectively, η_{OPT} is the lidar's photon efficiency including transmission, reflection, and diffraction, and η_{DET} is the optical detector's efficiency at converting photons into electrons. The atmospheric parameters all depend on the exact laser wavelength, the altitude, and many other factors. The CDWL and NDWL noise terms differ significantly from each other (Menzies, 1985).

In the case of DWL, velocity error is much more important than SNR as a figure of merit. In general, the DWL LOS wind velocity error is a function of the SNR, the spectral width of the received signal σ_{SIG}, and other conditions,

$$\sigma_{\mathrm{VEL}} = f(\mathrm{SNR}, \sigma_{\mathrm{SIG}}, \ldots)$$

for both CDWL and NDWL. The spectral width of the signal is due to the varying LOS velocity over the single-lidar-shot measurement volume (wind variability), as depicted in the figure, the innate spectral width of the atmospheric target, and the spectral width of the illuminating laser light. The innate spectral width of the target is a function of its mass as well as the temperature. Considering the approximate masses of an air molecule and aerosol particle to be 5×10^{-26} kg and 1.3×10^{-16} kg, respectively, the full width at half maximum (FWHM) values of the velocity distribution of molecules and aerosol particles are approximately 300 and 0.006 m/sec, respectively, in the troposphere. The molecular target velocity width is much larger than the desired velocity error leading to a requirement for large SNR.

For space-based operation with long round trip times of light, the SNR reduction due to misalignment of the receiver, at the time of signal reception, with the original transmitted pulse direction must also be included at this basic level. This is a short-term (3.9 msec in the NASA point designs) pointing control (stability) requirement. The lidar's internal alignment, the mounting of the lidar to the spacecraft, and spacecraft vibration and rotation may contribute to the overall misalignment. The latter two contributors of the three may or may not be considered part of the lidar system photon efficiency η_{OPT}, but in either case the loss in SNR must be considered. Adding these four effects produces level 2 velocity error; which is referenced to one lidar shot in the LOS direction.

Level 2 wind error includes some atmospheric effects. Next, we must add geometric effects. Imperfect pointing knowledge will result in imperfect subtraction of the spacecraft and earth rotation velocities, and also imperfect labeling of the direction of the wind data when delivered to the data user. Also, the error in the LOS direction is magnified when projected into the horizontal plane by the lidar scanner nadir angle. These two effects bring us to level 3 velocity error, which is referenced to one lidar shot in the horizontal plane.

We now consider the shot accumulation discussed above. The atmospheric LOS wind variability over the longer distances of shot accumulation (e.g., 36 km) and the possibility of cloud extinction or blockage of some of the shots are added. Any imperfect pointing control over this time interval (e.g., 5 sec) will cause additional velocity error. Wind variability on scales corresponding to the set of accumulated lidar shots (1 to 50 km) will cause additional error. Level 4 velocity error is referenced to a single LOS shot-accumulated wind velocity estimate projected into the horizontal plane.

The data user will most likely use each pair of LOS wind velocity profiles to represent the wind velocity in a volume of air that is much larger than the laser-illuminated measurement volume, especially in the cross-track direction. For example, the scan pattern discussed above will ideally produce two LOS wind estimates that are collocated in a long rectangle of length 36 km and width ideally as narrow as 5 m (top view). The requirements state that the wind information will be used as representative of square patches of atmosphere with sides of length 100 km. Depending on the true wind field during the measurement time, the imperfect spatial sampling by the DWL will produce various sampling or representative errors. This is depicted as wind variability over 50 to 500 km in the figure. Level 5 velocity error includes this error source. The NASA/NOAA requirements did not ask for a calculation of horizontal wind vector from each pair of two collocated LOS perspectives. However, the assimilation of the wind data into computer models may involve such a calculation, explicit or implicit. Level 6 velocity error is referenced to a final horizontal velocity vector that uses two LOS perspectives.

The NASA/NOAA requirement of total LOS wind error less than 2 to 3 m/sec refers to level 5 velocity error. Since each increasing level of velocity error includes more sources of error, the error increases with level. The basic lidar error from the standard lidar equation, which usually refers to level 2 wind error, must be less than the final 2- to 3-m/sec requirement, fitting into a comprehensive error budget. The complete draft NASA/NOAA requirements specified many of the atmospheric parameters necessary to calculate level 5 velocity error. Atmospheric extinction, backscatter, and cloud effects were specified. The NASA/NOAA GTWS Science Definition Team provided a sampling error value of 0.7 m/sec to use for this scan pattern and the 100-km box. These conditions were integrated with the basic DWL performance calculation models to provide a tool for the point designs.

9.5.6. Types of DWL Systems

There are three DWL techniques being primarily considered for the space measurement of wind. They differ in the nature of their receiver section that must decode the Doppler shift information in the backscattered light. These three techniques are the CDWL, the NDWL — fringe imager (NDWL-FI), and the NDWL — double edge (NDWL-DE). (Other types of noncoherent receivers, such as the iodine molecular filter, Mach-Zehnder interferometer, and multibeam Fizeau wedge interferometer, have also been proposed, but will not be discussed here.)

Space does not permit the listing of all the publications on these techniques, and only a small sampling is given. Theoretical discussions divide into: (1) the SNR obtained from a given combination of lidar hardware, measurement geometry, and atmospheric conditions, and (2) the wind measurement performance given an SNR and particular lidar hardware and software algorithms. Discussions of CDWL theory (Zhao et al., 1990; Targ et al., 1991; Frehlich and Kavaya, 1991; Rye and Hardesty, 1993a,b; Frehlich and Yadlowski, 1994) and wind measurement results (Bilbro, 1980; Kavaya et al., 1989a; Henderson et al., 1991; Huffaker and Hardesty, 1996) have been published. Articles on NDWL-FI theory (Abreu, 1979; Chanin et al., 1989; McKay, 1998b) and wind measurement performance (Chanin et al., 1989; Abreu et al., 1992; Souprayen et al., 1999), and on NDWL-DE theory (Korb et al., 1992, 1998; McKay, 1998a; Fleisa and Korb, 1999) and wind measurement performance (Korb et al., 1997; Fleisa et al., 2000; Gentry et al., 2000) have also been written.

Comparisons of coherent and noncoherent measurement of wind have also been done (Menzies, 1986; Rye, 1995), as well as comparisons of the NDWL-FI with the NDWL-DE (McGill and Spinhirne, 1998; McGill et al., 1999; McKay, 1999; McKay and Rees, 2000). The latter comparisons showed that the two noncoherent techniques have similar theoretical performance, with NDWL-FI having some practical advantages.

The favored optical wavelengths are 2 and 0.355 µm for CDWL and NDWL, respectively. The NDWL wavelength is the third harmonic of the Nd:YAG laser. The monostatic lidar Doppler shifts are 0.976 and 5.63 MHz/m/sec, respectively. The CDWL reflecting targets are aerosol particles (including cloud particles) that move with the wind. The NDWL reflecting targets are the air molecules. The intrinsic spectral widths of the two targets are 0.006 and 300 m/sec, respectively, because of their different masses. All three techniques utilize an optical detector to convert backscattered photons into electrons.

By adding an LO laser optical field onto the optical detector, CDWL uses the optical detector additionally as a mixer, translating the optical signal spectrum from near 1.5×10^{14} Hz down to the Megahertz region (Menzies, 1976; Kingston, 1979). The Megahertz signal is then digitized and processed in a computer using signal processing and frequency estimation algorithms. The computer has access to the preserved spectrum of the optical signal. The frequency estimation for CDWL may be thought of as tone detection in noise. When the correct tone frequency is identified, the frequency/velocity error is naturally small, giving a highly accurate wind measurement. As the SNR decreases for whatever reason, the probability of misidentifying the proper tone frequency increases. When misidentification occurs, there is no valid frequency/velocity measurement. However, even under low SNR, a fraction of the frequency/velocity measurements will be good with high accuracy. The figures of merit for CDWL are the percentage of velocity estimates that are good, and the velocity error of the good estimates (Frehlich and Yadlowsky, 1994).

Both the NDWL-FI and the NDWL-DE techniques detect the Doppler shift of the backscattered light by using optical components to discriminate between different Doppler shifts. Optical detectors are used after the optical components to measure only intensities. The Doppler shift is encoded in those measured intensities. The NDWL-FI technique sends the backscattered light through a Fabry–Perot etalon. Transmission at different angles through the etalon favors slightly different wavelengths of light (i.e., Doppler shifts). An array of optical detectors in a nested bull's eye pattern, or an equivalent of that, are used to make the measurement. The NDWL-DE technique sends both the transmitted and the backscattered light through an optical filter with a sharply varying transmission vs. wavelength. Optical detectors monitor the intensity of light entering and leaving the "edge" filter. Since the NDWL techniques use optical intensity measurements to determine wind velocity, their single figure of merit is the SNR. As the SNR decreases, the velocity error rises (Table 9.12).

Figure 9.16 is an illustrative example of how the CDWL and NDWL techniques differ as the signal level changes. Signal levels will fall for CDWL when aerosol concentrations are lower, such as in the mid troposphere. Signal levels for NDWL will fall when molecular concentrations fall, such as at higher altitudes, and when extinction becomes large, such as at very low altitudes. Note that the relative positions of the curves, both horizontally and vertically, depend strongly on the assumptions about the atmosphere and lidar system sizes, and that this figure should be used only to understand the concepts. The upper panel shows velocity error of a "good" wind estimate. CDWL "good" wind estimates have very low velocity error over the entire range of signal level. Both NDWL techniques have a velocity error that is proportional to the inverse square root of the signal level, which becomes

Table 9.12 Comparison of the Three DWL Techniques

	CDWL	NDWL-FI	NDWL-DE
Anticipated wavelength (μm)	Near 2	0.355	0.355
Doppler shift (MHz/m/sec)	0.976	5.63	5.63
Atmospheric target	Aerosols and clouds	Molecules	Molecules
Target spectral width (m/sec)	0.006	300	300
Doppler shift determination	Heterodyne translation of optical spectrum to lower frequency; digitization; computer/software spectral estimation	Transmission angle through Fabry–Perot etalon vs. Doppler shift; optical intensity measurements	Differential transmission of transmitted and received light by Fabry-Perot etalon; optical intensity measurements
Low signal disadvantage	Declining probability of wind measurement	Increasing velocity error	Increasing velocity error
Low signal advantage	Remaining wind measurements are very accurate	No dropouts; always get a measurement	No dropouts; always get a measurement

large at low signal levels. The lower panel shows the difference in atmospheric coverage vs. signal level. At high signal levels, all three techniques will achieve the same atmospheric coverage due to cloud blockage. As the CDWL signal level falls, the probability of a good wind estimate falls, reducing atmospheric coverage. The coverage of the NDWL techniques does not fall with signal level. They continue to make a wind measurement; albeit with increasing velocity error. These plots only refer to the level 1 wind velocity error discussed above, and actual velocity error and coverage numbers will be combined with all other contributors to velocity error to obtain the overall performance.

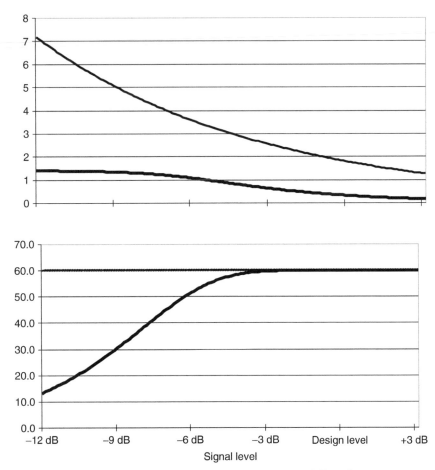

Figure 9.16 Qualitative behavior of DWL LOS velocity error (upper panel) and atmospheric coverage (lower panel, %) vs. signal level referenced to an arbitrary mission design level. Cloud coverage assumed equal to 40%. (Heavy curves, coherent detection; light curves, direct detection.)

9.5.7. Planned Space Missions

As of this writing, there is a space-based Doppler wind lidar mission under development, the European Space Agency's Atmospheric Dynamics Mission (ADM) (ESA, 1999). The ADM, scheduled to be launched in 2007 for a 3-year mission, will use the NDWL technique with both double-edge and fringe imager receivers. The NDWL-DE receiver will be used for the molecular backscatter. The orbit height will be 400 km, and the nadir angle will be 35 deg. A diode-pumped Nd:YAG pulsed laser, tripled to 355 nm, will emit 150-mJ pulses at 100 Hz. Due to electrical power constraints, the laser will operate at a 25% duty cycle with a 28 sec period. A 1.5 m diameter telescope will collect the backscattered photons. The telescope will not scan, but will stare sideways at the night side of the earth to minimize spacecraft motion Doppler shift and background light. This will produce LOS, but not "horizontal vector", wind profiles. ADM is considered a demonstration mission, paving the way to future operational missions.

9.6. DIFFERENTIAL ABSORPTION LIDAR

9.6.1. Introduction

The DIAL technique was first used to measure lower tropospheric water vapor (H_2O) using a coincidence of a H_2O absorption line in the 690-nm band of H_2O by Schotland (1966). In the DIAL method, two spectrally close laser pulses are transmitted near simultaneously with one pulse that is called the on-line, at a strongly absorbing spectral location due to the presence of an absorbing gas, and another, called the off-line, at a less absorbing spectral location. In principle, concentration profiles of the absorbing gas can be retrieved using the on- and off-line lidar signals and the knowledge of the differential absorption cross-section ($\sigma_{on} - \sigma_{off}$) (Measures, 1984). The advantages of the DIAL method are ease of inversion of gas concentration profiles from lidar data, absolute concentration profiling without a need for calibration, day and night measurements employing suitable background rejection, and simultaneous species and aerosol backscatter profiling. The DIAL technique has been used successfully to measure a large number of trace gases in the atmosphere from the laboratory, ground, and aircraft (Browell et al., 1998). The DIAL technique has been employed most successfully for atmospheric ozone (O_3) and H_2O measurements in field campaigns and monitoring (Browell et al., 1997b; Leblanc and McDermid,

2000, Chapter 8). In recent years a great deal of interest has arisen for profiling atmospheric CO_2 from space because of its applications to climate and the global carbon cycle studies. While the principle of DIAL for O_3, H_2O, and CO_2 is the same, the application of the DIAL methodology and technologies are vastly different. For example, the space-based O_3 measurements require careful selection of wavelengths to penetrate through the high O_3 stratospheric layer to reach the troposphere. In the case of water vapor, the use of narrowband absorption features and the ability to measure H_2O concentrations that vary considerably over altitude, latitude, and the season are critical. And in the case of CO_2, high-precision DIAL measurements are needed to understand and evaluate the sources and sinks of CO_2 with measurement precision of $<0.5\%$. In addition, the laser, receiver, and detector technologies are vastly different because of operation in different regions of the spectrum from UV to IR for optimum DIAL measurement capability. Over the last decade NASA has invested a significant amount of resources to develop technology for space-based measurement of O_3, H_2O, and CO_2. Descriptions of the concepts of space-based DIAL O_3, H_2O, and CO_2 systems are presented in the following sections.

9.6.2. Space-Based O_3 DIAL

9.6.2.1. Rationale

Tropospheric chemistry is considered to be the "next frontier" of atmospheric chemistry, and understanding and predicting the global influence of natural and human-induced effects on tropospheric chemistry will be the next challenge for atmospheric research in the foreseeable future. In particular, obtaining the global distribution of tropospheric O_3 with high vertical resolution (1 to 3 km) would greatly enhance the understanding of the atmospheric processes related to transport, dynamics, O_3 production and loss, atmospheric radiation balance, and photochemistry. The simultaneous high-vertical resolution (100 to 200 m) measurements of aerosol and cloud distributions along with the O_3 measurements provide important complementary information about air mass types, and their origin, evolution, chemistry, and transport as has been demonstrated in many NASA Global Tropospheric Experiment field missions.

A space-based O_3 lidar system will be able to provide the high-resolution measurements of O_3, aerosols, and clouds that are needed to complement other passive satellite measurements. The total ozone mapping spectrometer, in conjunction with the solar backscatter UV and Stratospheric Aerosol and Gas Experiment (SAGE II), has provided global information on the total column of O_3 across the troposphere (Fishman et al., 1990, 1991), and this has contributed significantly to our general understanding of tropospheric processes. It is expected that the Tropospheric Emission Spectrometer (TES) will provide low-vertical resolution (≈ 5 km) O_3 profile measurements in the middle and upper portion of the troposphere. This will make an improvement in our ability to understand tropospheric O_3 and chemistry, but it is recognized that even higher-vertical resolution measurements (<2.5 km) are needed to identify and resolve most of the complex tropospheric structure, chemistry, and transport. A space-based O_3 lidar system can provide these measurements, and it can also provide accurate information concerning the altitude location of atmospheric layers containing enhanced or depleted levels of O_3, aerosols, and clouds. This information can be directly assimilated into three-dimensional chemical-transport models, and it can be used to aid in the interpretation of the passive measurements. Lidar systems can also make measurements day or night to address diurnally dependent BL, transport, and photochemical processes.

9.6.2.2. Measurement Objective

The capability of aircraft-based DIAL systems for measuring O_3 to conduct process studies over regional scale has been fully demonstrated from data obtained during major field campaigns (Chapter 8); the technology needed for tropospheric O_3 measurements from space is still being developed. Studies have been conducted to evaluate the feasibility and technology needed for future space-based O_3 DIAL systems that have led to the identification of critical lidar component technologies (Browell et al., 1997b; Stadler et al., 1998). As a result of these studies the 305- to 320-nm spectral region was identified as being optimum for the tropospheric O_3 profiling of a space-based DIAL system. Spectral regions at shorter wavelengths than this would lead to a strong absorption in the stratosphere and longer-wavelength regions will have a very low sensitivity for O_3 measurements in the troposphere. A wavelength separation (dλ [delta lambda]) of 10 to 12 nm between the on-line and off-line has been found to be optimum in order to achieve measurement sensitivity in the troposphere and minimize errors cased by aerosol inhomogeneities in the atmosphere (Browell et al., 1997a). High-power lasers, large collection area receivers, high-throughput low-noise detection system, and larger data averaging are needed to overcome the lower SNR expected due to the long range of space-based measurements. The objective of a space-based O_3 system would be to measure tropospheric O_3 profiles with a vertical resolution of 1 to 2 km over horizontal scales of 200 km with an accuracy of 10% over background tropospheric O_3 conditions.

9.6.2.3. Measurement Simulations

The projected performances of space-based O_3 DIAL measurements using lidar parameters in Table 9.13 are shown in Figure 9.16. A U.S. standard (1976) model atmosphere and O_3 profile have been assumed. A 3-m class deployable telescope (Lake and Peterson, 1998) has been used that provides an effective area that is about 6.4 times that of a 1-m fixed aperture telescope and about 64 times the area of a 15-in.-size telescope that is normally used on aircraft lidar systems. Two different laser systems have been used in this calculation — a high-pulse energy (0.5 J) and low-repetition rate (10 Hz) laser (HELR) and a low-pulse energy (0.010 J) and high-repetition rate (1000 Hz) laser (LEHR). The high-energy laser approach has been described by Barnes (1998) and Stadler et al. (1998) and alternate approaches to obtaining tunable solid-state lasers in the UV that have the potential for LEHR have been

Table 9.13 Space-Based Ozone DIAL Parameters

Transmitter	
Energy (mJ)	500 (on-line and off-line)
Repetition frequency (Hz)	10 (double pulsed — 20-Hz effective frequency)
Wavelength (nm)	308/320 (on and off)
Beam divergence (mrad)	0.15
Pulse width (nsec)	<100
Spacecraft altitude (km)	400
Receiver	
Telescope diameter	3-m class deployable
FOV (mrad)	0.2
Filter bandwidth (nm)	0.2 (day)
Optical transmittance	0.65
Detector efficiency (%)	31 (PMT)
Noise equivalent power (W/Hz$^{1/2}$)	1×10^{-15}

discussed by Fix et al. (1998) and Richter et al. (2000). HELR is more desirable than LEHR in order to minimize the influence of day background. However, generation of high-energy pulses in the UV remains a technological challenge. A dawn-to-dusk orbit and the associated twilight background in the UV have been used in these simulations. It can be seen from Figure 9.17 that, under the twilight background conditions, the HELR performance is better than that of the LEHR system. In fact, the performance of the HELR system degrades only slightly during moon-lit night background conditions. The HELR system would provide measurements even under full daylight condition but with vertical resolutions of about 3 km in the troposphere. The LEHR system's performance would be still poorer; however, if performance at nighttime alone is desired, then the LEHR system would provide good troposphere measurements as does the HELR. Therefore, assuming steady progress in laser technology, the potential exits for lidar systems to enable high-vertical resolution O_3 measurements for global tropospheric measurements to address environmental, chemical, dynamic, and climatic issues. It is also seen that in spite of having twice as much

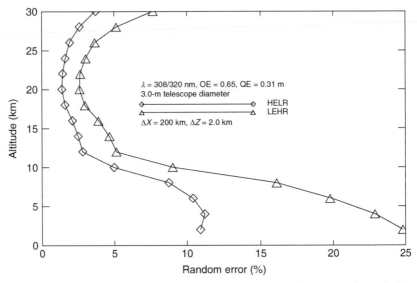

Figure 9.17 Profiles of measurement errors of a space-based O_3 DIAL systems under twilight background conditions.

total power (energy times pulse repetition rate) in the LEHR compared to the HELR system, the latter performs better under twilight background condition.

9.6.3.   Space-Based H_2O DIAL

9.6.3.1.   Rationale

A space-based H_2O DIAL system would provide needed global H_2O distributions that would enable improved prediction of weather and climate, and improve our knowledge of the global hydrological cycle. Water vapor is the primary greenhouse gas in the atmosphere and it is the principal carrier of energy fluxes in the atmosphere. Consequently, knowledge of the distribution of H_2O is very important in the study of climate and short- and long-term weather. Upper tropospheric H_2O and its variations have significant climatic implications (Del Genio *et al.* 1994). Even though upper tropospheric H_2O number densities are three to four orders of magnitude less than near the surface, model calculations have shown that small percentage changes in upper tropospheric H_2O can affect the climate as much as similar percentage changes in the BL (Arking, 1990). Calculations of upper tropospheric cooling rate require the knowledge of upper tropospheric H_2O (Clough et al., 1992). Upper tropospheric H_2O distributions can be used to study transport, dynamics, convection, and cirrus formation (Newell et al., 1996). Knowledge of H_2O distributions is required in initiating NWP models, weather and hurricane forecast improvements (Krishnamurti et al., 1994; Kamineni et al., 2003). Atmospheric H_2O is the main source of OH that controls many atmospheric chemical distributions. Water vapor varies considerably with altitude, location, and season. Knowledge of the global distribution of tropospheric H_2O is not adequate and that of the upper troposphere is poor.

9.6.3.2. Measurement Approach and Objective

Atmospheric H_2O was the first trace gas species that was detected by the DIAL technique in 1966. Since then, considerable progress has been made in the development of tunable lasers and operation of DIAL systems on aircraft (Chapter 8, and the references therein). The development and operation of an autonomous H_2O DIAL system called LASE (Lidar Atmospheric Sensing Experiment) has demonstrated some of the technology needed for space-based DIAL and the ability to measure H_2O distributions over varying atmospheric conditions and over a large concentration range (Ismail and Browell, 1995; Browell et al., 1997a). The magnitude of absorption cross-section of H_2O absorption lines in the near IR varies considerably such that H_2O optimum DIAL measurements can be made over the full range of atmospheric concentrations. The optimum altitude is the location where a minimum in the random errors profile occurs. This point of minimum error is caused by a combination of increase of errors with range due to the loss of the signal from absorption by the species and an increase in sensitivity with range due to stronger absorption at higher ranges as it occurs due to distribution of H_2O in the atmosphere. As a rule-of-thumb the optimum error point occurs at a location where the one-way optical depth τ (optical depth is given by the expression $\Sigma n d\sigma dz$, where n is the H_2O

number density, σ is the on-line cross-section, and z is the range) is ≈ 1 (e.g., Remsberg and Gordley, 1978). A table containing a list of the strongest H_2O lines that are suitable for DIAL measurements along with the altitudes for optimum measurements is given in Table 9.14. Generally, depending on how sharply peaked the optimum altitude region is, measurements with good accuracy can be made over several kilometers above and below the optimum altitude region. The optical depth calculations were made using the U.S. 1976 Standard Model atmosphere H_2O profile, assuming the HITRAN H_2O line parameters (Rothman et al., 1998), and the "Voigt" profile to represent the altitude variation of the H_2O line absorption profile. Table 9.14 shows that for upper tropospheric H_2O measurements, lines at least as strong as those in the 940- or 1130-nm band would be needed. Water vapor measurement over the entire altitude region below the optimum altitude region can be made using the sideline technique employed in LASE (Sachse et al., 1995).

Excellent Si:APD detectors in the 940-nm region with high quantum efficiency ($>40\%$), low noise, and ability to detect lidar signals in the analog or photon-counting mode are commercially available. Also technology developments to

Table 9.14 Strong Absorption Lines Suitable for H_2O DIAL Measurements

H_2O band wavelength (nm)	Wavelength of strong line (nm)	Cross-section ($\times 10^{-22}$ cm^2)	Lower-energy state E'' (cm^{-1})	Optimum altitude (km)
590	594.421	1.0×10^{-1}	285.419	
690	694.572	2.03×10^{-1}	224.838	0
720	724.571	1.30	285.419	4.5
820	817.223	1.60	224.838	5.0
940	934.805	1.8×10	224.838	9
1130	1123.825	5.62×10	224.838	11.5
1400	1382.721	6.20×10^2	285.419	30
1850	1842.052	8.43×10^2	224.838	35
2600	2596.015	6.89×10^3	224.838	52
6000	5896.466	9.73×10^3	224.838	58

obtain high-energy tunable lasers in the 940-nm region are in progress (Barnes et al., 2002). These technology developments favor the development of future space-based H_2O DIAL systems at 940 nm. NASA has conducted some initial evaluation of such a system (Browell et al., 1998) and ESA is conducting phase A studies for a Water Vapor Lidar Experiment in Space (WALES; Gerard *et al.*, 2004) that would operate in the 940-nm region. The highest resolutions are generally needed in the lower troposphere because of the high gradients in the atmospheric BL and to study processes on mesoscale sizes. The vertical and horizontal resolutions in the BL needed are 0.5 and 50 km, respectively, with an accuracy of 10%. In the mid and upper troposphere, vertical resolutions of about 1 km would suffice.

9.6.3.3. Measurement Simulations

The projected performance, based on the methodology developed by Ismail and Browell (1989) of a space-based H_2O DIAL system using the lidar parameters in Table 9.15 is shown in Figure 9.18. A midlatitude summer atmospheric model has been used for the temperature and molecular number density. Also, U.S. standard model H_2O profiles along with a midday background level have been used in these calculations. The on-line cross-sections for one of the strongest lines ($\sigma = 1.8 \times 10^{-21}$ cm^2) in the 940-nm band along with two sidelines have been used in the night background calculations (solid lines). The advantage of lidar measurements is the ability to smooth the data vertically and average the number of shots needed to achieve the desired resolution and precision. In the case of H_2O DIAL it is estimated that systematic effects due to uncompensated instrumental and atmospheric effects cause an additional 3 to 4% errors (Ismail and Browell, 1989). Figure 9.18 also shows the DIAL performance in the low and mid troposphere over day background conditions (dashed lines). The measurement degrades with altitude and significantly more data averaging is needed to improve the measurement accuracy at the cost of resolution. However, good performance over tropospheric altitudes would be possible during night and low day background conditions like over the oceans and

Table 9.15   Space-Based H_2O DIAL Parameters

Transmitter	
Energy (mJ)	500 (on-line and off-line)
Linewidth (pm)	0.25
Stability (pm)	0.25
Spectral purity (%)	>99.0
Repetition frequency (Hz)	10 (double pulsed — 20 Hz effective frequency)
Wavelength (nm)	944
Beam divergence (mrad)	0.1
Pulse width (nsec)	<100
Spacecraft altitude (km)	400
Receiver	
Area (effective) (m^2)	2.5 (2-m-diameter telescope)
FOV (mrad)	0.25
Filter bandwidth (nm)	0.4 (day), 1.0 (night)
Optical transmittance (%)	40 (day), 65 (night)
Detector efficiency (%)	40 (APD)
Noise equivalent power (W/Hz$^{1/2}$)	1×10^{-15}
Excess noise factor	3.0

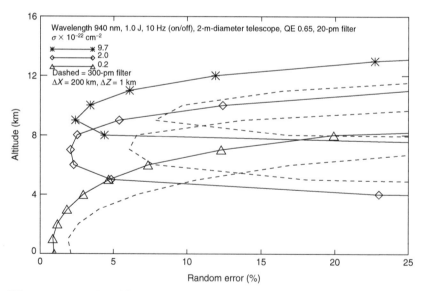

Figure 9.18   Profiles of measurement errors of a space-based H_2O DIAL system (solid lines indicate night background conditions, and dashed lines indicate day background conditions).

during morning and evening conditions when the solar background is not as intense.

9.6.4. Space-Based CO_2 DIAL

9.6.4.1. Rationale

Atmospheric CO_2 concentrations have increased rapidly as a by-product of economic growth and industrialization during the last 100 years. The recent rate of increase of CO_2 is about 100 times higher than at any other time during the past 1000 years. Historical data records have indicated a direct correlation between increase in the atmospheric CO_2 levels and atmospheric temperatures. While the long-term trend of CO_2 in the atmosphere and its seasonal variations are well documented (Keeling, 1976), the sources and sinks of CO_2 on continental and regional scales are not well understood. A sink for CO_2 comprising more than 25% of the 7.1 petagrams (10^{15} g) of CO_2 that is released annually into the atmosphere remains unaccounted for (Kaiser and Schmidt, 1998). It is also known that both land and ocean sinks fluctuate with space and time. Knowledge of the spatial and temporal distribution of CO_2 and an understanding of the causes of these variations are essential for predicting future levels of CO_2 and the impact of CO_2 on future climate change and, consequently, for formulating sound social and economic policy decisions.

9.6.4.2. Measurement Requirements

To fully understand the global atmospheric CO_2 cycle requires measurements of atmospheric CO_2 profiles on continental and global scales that are not currently available. The difficulty in making profile measurements arises from CO_2 exhibiting small variations in concentration over large spatial scales. Atmospheric gradients of CO_2 over continents are typically 2 to 10 parts per million by volume (ppmv). Typical horizontal scales are approximately 1000 km except near coastlines, where they decrease rapidly to as low as 1 km. Vertical scales, as deduced from airborne measurements over Amazonia, are typically 1 to 2 km. Contrasts of this magnitude develop daily between the planetary BL and the free troposphere as a direct consequence of the activity of the biosphere below. A network of 70 ground stations involving *in situ* observations from towers located over land has been developed only recently to monitor long-range variations of CO_2 near the surface. Although the quality of these measurements is high, these *in situ* tower measurements only represent the small footprint area of the tower, provide no coverage over the oceans, and are limited in height. Such observations cannot provide representative sampling of large geographical regions and may be subject to systematic effects due to localized influences of terrain, wind patterns, and meteorological factors, etc. High-precision global measurements of atmospheric CO_2 profiles from space will enhance the present understanding of sources and sinks of CO_2 in the atmosphere. High-precision global measurements of atmospheric CO_2 profiles from space are required to improve our understanding of the sources and sinks of CO_2. To assist in understanding many carbon cycle processes, weekly averages of CO_2 measurements with accuracy of 1 ppmv near the surface and 1 to 2 ppmv in the free troposphere are needed on horizontal scales of 200 to 500 km over 1- to 2-km vertical range intervals in the troposphere. In addition, column CO_2 measurements with 1- to 2-ppmv precision with high horizontal resolution ($\approx$100 km) are needed. This represents a CO_2 DIAL measurement accuracy of 0.3 to 0.5% for both range-resolved and column measurements. These requirements will allow the instrument to measure the influence of the land surface on CO_2 concentrations in the atmosphere, even where this influence is relatively weak. Observations made over time and space provide the basic data for determining the magnitude of atmosphere–biosphere fluxes, a principal objective of carbon cycle science. The strongest sources and sinks will be readily identified and quantified.

9.6.4.3. Measurement Approach and Technology Developments

One of the important issues in the DIAL measurement of CO_2 profiles is the selection of suitable CO_2 lines. Spectral scans of CO_2 and H_2O absorption lines in the 2.05-μm region are shown in Figure 9.19. These spectra were derived from the HITRAN database (Rothman et al., 1998) and show the presence of weak H_2O absorption lines. Analysis in a recent publication shows that these are the most suitable CO_2 absorption lines for range resolved DIAL measurements in the near- and mid-IR regions on the basis of the strength of the lines (Rothman et al., 1998), noninterference from water vapor, and temperature insensitivity of the cross-section (Ambrico et al., 2000 and Tratt, 2003). There are a number of strong lines in this spectrum with atmospheric optical depths due to CO_2 absorption exceeding 10. These lines have a range of ground-state energy levels that are suitable for temperature-insensitive mixing ratio measurements (Ismail and Browell, 1989). Operation in the sideline mode would enable the selection of optical depth values needed for optimum measurements, and this method will be relatively insensitive to known atmospheric effects.

Solid-state lasers, operating in the 2-μm region, based on Ho:Tm:YLF, that are being developed at NASA Langley offer several advantages as a transmitter for DIAL measurements of CO_2. First, they are tunable in a spectral region containing strong CO_2 absorption lines that are suitable for high-sensitivity CO_2 measurements. Second, there has been

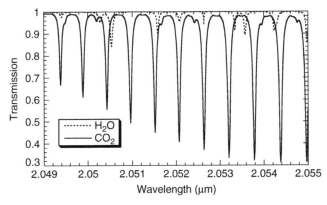

Figure 9.19 Atmospheric transmission over a 1-km vertical path near 2.05 μm.

considerable development in this laser technology for application in coherent lidar measurements of wind. As much as 600 mJ per pulse with a single-frequency spectrum has been demonstrated with diode-pumped Ho:Tm:YLF (Singh et al., 1998). This capability for high pulse energy, and parallel NASA research efforts for further increase in energy, provides a framework for evolution of the lidar application described here toward aircraft and, ultimately, space platforms. Third, the 2-μm wavelength offers a high level of eye safety — a maximum permissible exposure of 100 mJ/cm^2 (American National Standard Institute, 2000). Three specific laser developments have already been demonstrated to enable DIAL measurements at 2 μm: (1) tunability across a number of temperature-insensitive lines (Koch et al., 2000, 2002), (2) line locking on selected absorption using an absorption cell (Koch et al., 2000, 2002), and (3) double-pulsed operation for on and off wavelength generation (Yu et al., 2000a). The unique feature of this operation is that it provides an on-and-off wavelength pulse pair with a single pump pulse.

An InGaAsSb APD with a peak spectral response near 2 μm has been demonstrated (Shellenbarger et al., 1997). This and other promising detectors in the 2-μm region are being acquired, tested, and characterized to develop a high-efficiency and low-noise detector system needed for space-based DIAL applications. Technology developments are under way to reduce the noise associated with some of these detectors.

CO_2 in the atmosphere has been already detected using differential absorption techniques (Taczak and Killinger, 1998; Koch, 2001) and heterodyne detection. The accuracy of these initial measurements was limited due to the quality of the transmitter and the collection efficiency of the receiver. With an improved laser system, more accurate measurements have been made and a measurement precision of about 2% has been demonstrated (Koch et al., 2004). Heterodyne detection requires averaging of a large number of lidar pulses to remove speckle noise. Consequently, a direct detection lidar employing a large collection area receiver would be more suitable for operation from space. Lidar parameters needed for a space-

Table 9.16 Space-based CO_2 DIAL Parameters

Transmitter	
Energy (mJ)	1000–2000 (on-line and off-line)
Linewidth (pm)	0.05
Stability (pm)	0.05
Spectral purity (%)	>99.0
Repetition frequency (Hz)	10 (double pulsed — 20 Hz effective frequency)
Wavelength (nm)	2005
Beam divergence (mrad)	0.1
Pulse width (nsec)	<100
Spacecraft altitude (km)	400
Receiver	
Area (effective) (m^2)	6.4 (3-m-diameter telescope)
FOV (mrad)	0.25
Filter bandwidth (nm)	0.4 (day), 1.0 (night)
Optical transmittance (%)	40 (day), 65 (night)
Detector efficiency (%)	40 (APD)
Noise equivalent power (W/Hz$^{1/2}$)	1×10^{-15}
Excess noise factor	3.0

based CO_2 DIAL demonstration are given in Table 9.16. Significant progress in the laser development at 2 μm has already been achieved under various NASA technology programs. Progress toward the demonstration of technology for a deployable telescope from space is under way at the University of Colorado under NASA funding.

ACKNOWLEDGMENTS

The authors would like to acknowledge numerous contributions over several years from Robert Atlas, Wayman Baker, C. Belsches, James Bilbro, David Bowdle, Edward Browell, David Emmitt, Rod Frehlich, Bruce Gentry, William Grant, Johnathan Hair, Mike Hardesty, Sammy Henderson, Milton Huffaker, Steve Johnson, Susan Kooi, Robert Menzies, Ken Miller, Tim Miller, Barry Rye, Gary Spiers, and Jirong Yu. The authors are grateful to James Abshire for many contributions to the sections on Lidar Altimetry.

APPENDIX A

Geometry of Space-Based Lidar Remote Sensing

The signal level captured by the lidar is proportional to the inverse of the square of the distance from the lidar to the target. The one-way range R_T from the lidar to the target is given by

$$R_T = \text{SQRT}[(R_E + Z_L)^2 + (R_E + Z_T) - 2(R_E + Z_L)$$
$$\times (R_E + Z_T)\cos(\theta_T - \theta_L)]$$

where R_E is the earth's radius, Z_L is the height of the lidar system above the earth (orbit height), Z_T is the height of the target atmosphere above the earth, θ_L is the laser beam nadir angle at the lidar system, and θ_T is the laser beam nadir angle at the target given by

$$\theta_T = \text{arcsin}[(R_E + Z_L)\sin\theta_L/(R_E + Z_T)]$$

(The earth's radius varies by 21.4 km from the equator to the poles. We ignore that here and use an average radius, 6371.315 km. The variation is approximately 0.3%.) Figure A.1 shows how the one-way range to the earth's surface (let $Z_T = 0$) varies with lidar orbit height and laser beam

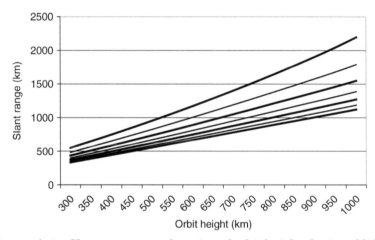

Figure A.1 Slant range as a function of orbit height (km) and lidar nadir angle (°) in 5° steps from 25 (bottom curve) to 55.

nadir angle at the lidar. For the parameters listed, the one-way distance varies by a factor of 6.6, which means the lidar signal level will vary by a factor of 43 or 16 dB. To maintain a constant signal level over this range of mission parameters, the sensitivity of the various lidar systems would have to vary by 16 dB. This sensitivity derives from parameters like the laser pulse energy, pulse rate, receiver area, and photon efficiency of the other lidar components. From a lidar technology point of view, the lower orbit heights and nadir angles present much less of a challenge. A secondary effect of the large distance is the delay time between firing a laser pulse and receiving the backscattered signal. This delay time is found by multiplying the slant range in Figure A.1 by 2 and dividing by the velocity of light. It also varies by a factor of 6.6, from 2.2 msec at 300-km orbit height and $25°$ nadir angle to 14.6 msec at 1000-km orbit height and 55 nadir angle. Each lidar measurement and measurement technique will have a different requirement for keeping the receiver aimed in the transmitted pulse's direction while waiting for the return signal. This is a pointing control (stability) requirement that favors shorter round trip times of light.

The lidar orbit height and laser beam nadir angle also determine the distance of the laser beam in the atmosphere from the point directly below the lidar satellite. The "horizontal" great circle distance, following the curvature of the earth, is given by

$$R_{FOR} = (R_E + Z_T)(\theta_T - \theta_L)$$

This distance varies slightly with the target height Z_T. At any instant the FOR of the lidar is a circle of radius R_{FOR} centered on the subsatellite point. The azimuth angle setting of the lidar scanner determines the exact position of the laser beam on this circle. (We assume here that the lidar scanner has a constant nadir angle θ_L.) The terms "swath width" and "coverage" have come into common use because the field of space-based remote sensing began with passive imaging sensors, followed later by active radar sensors. Both of these sensors have a large instantaneous FOV compatible with the terms swath width and coverage. The instantaneous FOV of a lidar system is much smaller, probably in the range of 5 to 100 m across. The closest analogy to the common meaning of FOV would be the lidar's FOR. As the lidar travels forward, the FOR covers a swath below the lidar of width $D_{FOR} = 2R_{FOR}$. Figure A.2 shows that D_{FOR}, when $Z_T = 0$, varies by a factor of 13 over the entries plotted. Since the laser beam is not always on, but rather is pulsed, and since the illuminated area when it pulses has a diameter in the 5- to 100-m range, the term coverage must be understood to have a different meaning than in the passive imager or active microwave radar sense. Nevertheless, it is interesting to compare D_{FOR} to the displacement of the earth by the time the lidar flies overhead on its next orbit. The equatorial displacement from the rotation of the earth during one orbit time of the satellite is shown in Figure A.3. Comparing the one-orbit equatorial displacements to the widths D_{FOR} in Figure A.2 reveals that in most cases there will be a gap between the FORs of successive orbits. This may or may not be a disadvantage, depending on the science mission. The FOR gap between successive orbits decreases as the latitude increases from the equator toward the poles. It decreases until it reaches 0 at some value of

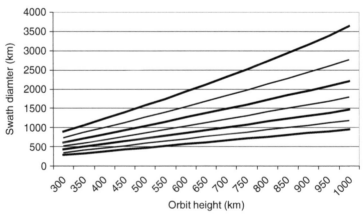

Figure A.2 Swath diameter (km) (FOR) as a function of orbit height (km) and lidar nadir angle (°) in 5° steps from 25 (bottom curve) to 55.

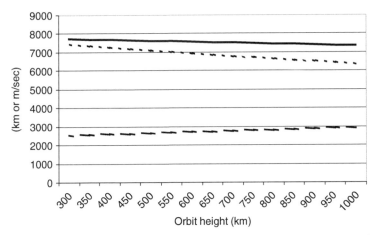

y-axis label: (km or m/sec)

x-axis label: Orbit height (km)

Figure A.3 The earth's equatorial displacement in one orbit (km) as a function of orbit height (km) (long dash); velocity of subspacecraft point/laser spot (m/sec) (short dash); tangential velocity of spacecraft (m/sec) (solid).

latitude that depends on orbit height and nadir angle. These "no-gap" latitude values are plotted in Figure A.4.

Another important geometry effect is the limited amount of time available to make the desired measurement since the lidar system is moving very fast in its orbit around the earth. Assuming the lidar scanner is in a fixed position, and imagining for the moment that the laser beam is always on, the laser beam position in the atmosphere moves in the orbit direction with a speed that varies slightly with target height Z_T. The forward velocity of the laser beam position is given by

$$V_T = (GM_E)^{1/2}(R_E + Z_T)/(R_E + Z_L)^{3/2}$$

where M_E is the mass of the earth, and the gravitation constant $G = 6.67259 \times 10^{-11}$ m^3/sec^2/kg. Once again letting $Z_T = 0$, Figure A.3 shows the laser spot's forward velocity and Figure A.5 shows the time available for the lidar measurement if the desired horizontal resolution of the measurement is 100 km. The time available for other horizontal resolutions can be easily calculated. This time varies very little with orbit height, and ranges from 13 to 16 sec. It is independent of nadir angle. The available time must be apportioned between lidar data collection, and scanner movement and settling time.

The above equation yields the tangential velocity of the lidar satellite when $Z_T = Z_L$. These are large velocities in the neighborhood of 7.5 km/sec, and are also plotted in Figure A.3.

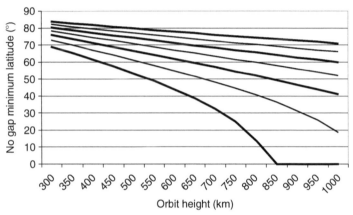

Figure A.4 Latitude (°) above which the FOR covers the earth on successive orbits as a function of orbit height (km) and nadir angle (°) in 5° steps from 25 (top curve) to 55.

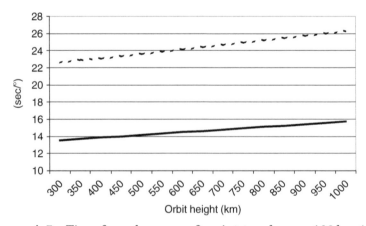

Figure A.5 Time for subspacecraft point to advance 100 km (sec) (solid line); earth rotation in one orbit of spacecraft (°) (dashed line).

Two effects of these large velocities are a change in the absorption of the laser beam by the atmosphere, and a large Doppler shift of the light backscattered to the lidar system. The normal atmospheric absorption models used to predict the absorption vs. height of laser beams assume that the laser and atmosphere have no relative motion. For space missions, one must adjust the laser wavelength used in the models by the amount of the Doppler shift due to the relative motion in order to properly predict absorption. Since most lidar missions to date have only aimed in the nadir direction, this phenomenon has not mattered. For future DIAL missions, the atmospheric absorption at two laser wavelengths must be well known for all heights. If nonnadir angle pointing is employed, the Doppler shift, and hence the atmospheric absorption, will change as the lidar scanner changes the azimuth angle. The second effect of the large velocities particularly affects Doppler lidar measurement of wind. The DWL receiver is designed to be sensitive to the Doppler shift of the return light. The large Doppler shifts from the satellite motion, as well as from the earth's rotation, must accurately be removed. Note that the Doppler shifts for atmospheric absorption and for the backscattered light frequency differ by a factor of 2 (Kavaya et al., 2001).

The equation for the laser beam nadir angle at the atmospheric target θ_T has been given above. It is a function of orbit height, nadir angle at the lidar, and target height. One of the parameters determined by this angle is the ratio between a length along the laser beam in the LOS direction and the corresponding length projected to the vertical direction. This is important when converting the lidar's fundamental range resolution in the LOS direction to the final vertical resolution of the measurement. Figure A.6 gives the ratios of the LOS to vertical distances. In order to obtain enough signal strength, or possibly because of the laser's pulse duration, a lidar measurement cannot go below some threshold LOS range resolution. Therefore, larger LOS to vertical ratio values indicate the possibility of achieving better (or smaller) vertical resolutions for a given LOS resolution. For the entries in Figure A.6, the ratio varies by a factor of 2.8.

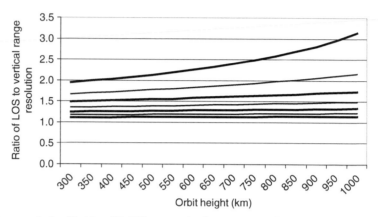

Figure A.6 Ratio of LOS to vertical range resolution as a function of orbit height (km) and lidar nadir angle (°) in 5° steps from 25 (bottom curve) to 55.

Another parameter determined by the laser beam nadir angle at the target is the ratio of a horizontal length to the same length projected in the LOS direction of the laser beam. This ratio is particularly important for wind measurement since the wind velocity error made by the lidar system in the fundamental LOS direction will be geometrically amplified into a larger error in the horizontal direction. Figure A.7 shows the horizontal to LOS ratios. These numbers represent the amplifying factors of the LOS wind velocity error into the error of the projection of the LOS wind into a horizontal plane. This amplification factor ranges from 1.1 to 2.3.

It would be convenient if certain combinations of orbit height and nadir angle simultaneously benefited all aspects of the lidar measurement from space. Unfortunately, the results are mixed. The slant range and light round trip time benefit from small orbit heights and nadir angles, while the FOR, measurement time, vertical resolution, and horizontal wind error all benefit from larger orbit heights and nadir angles. Similarly, the time to make the measurement favors larger orbit heights while the one-orbit rotation of the earth favors smaller orbit heights. There are also many other spacecraft and mission constraints that must be considered in addition to these factors to obtain an overall optimum selection (Kavaya et al., 1989b, 1994, 2001).

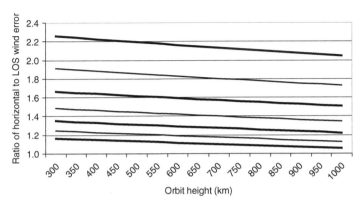

Figure A.7 Ratio of horizontal to LOS wind error as a function of orbit height (km) and lidar nadir angle (°) in 5° steps from 25 (top curve) to 55.

APPENDIX B

List of Acronyms

AORS	Active Optical Remote Sensing
APD	avalanche photodiode detector
BBO	beta barium borate
BL	boundary layer
CALIOP	Cloud–Aerosol LIdar with Orthogonal Polarization
CALIPSO	Cloud–Aerosol Lidar and Infrared Pathfinder Satellite Observations
CDWL	coherent-detection Doppler wind lidar
DIAL	DIfferential Absorption Lidar
DOR	depth of regard
ESE	Earth Science Enterprise
DWL	Doppler wind lidar
FOR	field of regard
FOV	field of view
FWHM	full width at half maximum
GLAS	Geoscience Laser Altimeter System
GTWS	Global Tropospheric Winds Sounder
HELR	high pulse energy, low repetition rate
IR	infrared
LaRC	Langley Research Center (NASA)

LASE	Lidar Atmospheric Sensing Experiment
LASER	light amplification through stimulated emission of radiation
LDA	laser diode array
LEHR	low pulse energy, high repetition rate
lidar	light detection and ranging
LITE	Lidar In-space Technology Experiment
LO	Local oscillator
LOS	line of sight
LRS	laser reference system
MOLA	Mars Orbiter Laser Altimeter
NDWL	noncoherent-detection (direct) Doppler wind lidar
NDWL-DE	NDWL — double edge
NDWL-FI	NDWL — fringe imager
NWP	numerical weather prediction
OPA	optical parametric amplifier
OPO	optical parametric oscillator
PDF	probability density function
PMT	photomultiplier tube
ppmv	parts per million by volume
PRF	pulse repetition frequency
RMS	root mean square
SLM	single longitudinal mode
SNR	signal-to-noise ratio
SPARCLE	SPAce Readiness Coherent Lidar Experiment
SRS	stellar reference system
TEM	transverse electromagnetic
UV	ultraviolet
WALES	Water Vapor Lidar Experiment in Space
YAG	yttrium aluminum garnet
YLF	yttrium lithium fluoride

REFERENCES

Abedin, M. N., Noise measurement of III-V compounds detectors for 2-μm LIDAR/DIAL remote sensing applications, *Int. J. High Speed Electron. Syst.*, 12(2), 531–540, 2002.

Abreu, V. J., Wind measurements from an orbital platform using a lidar system with incoherent detection: an analysis, *Appl. Opt.*, 18(17), 2992, 1979.

Abreu, V. J., J. E. Barnes, and P. B. Hays, Observations of winds with an incoherent lidar detector, *Appl. Opt.*, 31(22), 4509, 1992.

Abshire, J. B., S. S. Manizade, W. H. Schaefer, R. K. Zimmerman, S. Chitwood, and J. C. Caldwell, Design and Performance of the Receiver for the Mars Observer Laser Altimeter, Technical Digest — CLE091, Paper CF14, Optical Society of America, Washington, DC, 1991.

Abshire, J. B., X. Sun, and R. Afzal, Mars Orbiter Laser Altimeter: receiver model and performance analysis, *Appl. Opt.*, 39, 2449–2460, 2000a.

Abshire, J. B., X. Sun, E. A. Ketchum, R. S. Afzal, and P. S. Millar, The Geoscience Laser Altimeter System (GLAS) for the ICES at mission, OSA Conference on Laser and Electro Optics, San Francisco, CA, May 2000b.

Afzal, R. S., Mars observer laser altimeter: laser transmitter, *Appl. Opt.*, 33, 3184–3188, 1994.

Afzal, R. S., *Space Operation of the MOLA Laser, Advanced Solid-State Lasers*, TOPS, Vol. XXXIX, edited by H. Injeyan, U. Kellerand, and C. Marshall, 66–69, 2000.

Afzal, R. S. and M. D. Selker, Design considerations for the GLAS laser transmitter, Proceedings, 1994 International Laser Radar Conference (ILRC), Sendai, Japan, 1994.

Afzal, R. S., et al., Laser transmitter design for the Geoscience Laser Altimeter System, 19th International Laser Radar Conference (ILRC) (NASA/CP-1998-207671/PT2), edited by U. N. Singh, S. Ismail, and G. Schwemmer, Annapolis, MD, 613–616, 1998.

Ahmad, A., F. Amzajerdian, C. Feng, and Y. Li, Design and fabrication of a compact lidar telescope, *Proc. SPIE*, 2832, 34–42, 1996.

Ambrico, P. F., A. Amodeo, P. D. Girolamo, and N. Spinelli, Sensitivity analysis of differential absorption lidar measurements in the mid-infrared region, *Appl. Opt.*, 39, 6847–6865, 2000.

American National Standard Institute, Z136.1, 2000.

Amzajerdian, F., Experimental evaluation of InGaAs photodetectors for 2-micron coherent lidars, Conference on Lasers and Electro-Optics, Anaheim, CA, June 2–7, 1996.

Amzajerdian, F., Improved analytical formulations for optical heterodyne receivers, 11th Coherent Laser Radar Technology and Applications Conference, Great Malvern, UK, July 1–6, 2001.

Amzajerdian, F. and J. F. Holmes, Time-delayed statistics for a bistatic coherent lidar operating in atmospheric turbulence, *Appl. Opt.*, 30, 3029–3033, 1991.

Anapol, M., L. Gardner, T. Tucker, and R. Koczor, Lightweight 0.5-m silicon carbide telescope for a geo-stationary Earth observatory mission, *Proc. SPIE*, 2543, 164–172, 1995.

Arking, A., Feedback processes and climate response, Proceedings of the Conference on Climate Impacts of Solar Variability, NASA Conference Publication CP-3086, 1990.

Baker, W. E., Utilization of satellite winds for climate and global change studies, *Palaeogeogr. Palaeoclimatol. Palaeoecol. (Global Planetary Change Sect.)* 90, 157, 1991.

Baker, W. E. and R. J. Curran, editors, Report of the NASA Symposium on Global Wind Measurements, STC-2081, A. Deepak Publishing, Hampton VA 1985. (Also companion Proceedings and Air Force volumes.)

Baker, W. E., G. D. Emmitt, F. Robertson, R. M. Atlas, J. E. Molinari, D. A. Bowdle, J. Paegle, R. M. Hardesty, R. T. Menzies, T. N. Krishnamurti, R. A. Brown, M. J. Post, J. R. Anderson, A. C. Lorenc, and J. McElroy, Lidar-measured winds from space: a key component for weather and climate prediction, *Bull. Am. Meteorol. Soc.*, 76, 869–888, 1995.

Barnes, J. C., Solid state laser technology and atmospheric sensing applications, 19th International Laser Radar Conference (ILRC) (NASA/CP-1998-207671/PT2), edited by U. N. Singh, S. Ismail, and G. Schwemmer, Annapolis, MD, 619–622, 1998.

Barnes, N. P., B. M. Walsh, E. D. Filer, R. L. Hutcheson, and R. W. Equall, Compositional wavelength tuning predictions for garnets: comparison with experimental results, *Advanced Solid State Lasers*, edited by Martin Freeman and Larry Marshall. Vol. 68, OSA Trends in Optics and Photonics Series, Optical Society of America, Washington, DC, 280–283, 2002.

Bilbro, J. W., Atmospheric laser Doppler velocimetry: an overview, *Opt. Eng.*, 19(4), 533–542, 1980.

Browell, E. V., S. Ismail, and W. B. Grant, Differential absorption lidar (DIAL) measurements from air and space, *Appl. Phys. B*, 67, 399–410, 1998.

Browell, E. V., S. Ismail, W. M. Hall, A. S. Moore, S. A. Kooi, V. G. Brackett, M. B. Clayton, J. D. W. Barrick, F. J. Schmidlin, N. S. Higdon, S. H. Melfi, and D. Whiteman, *LASE Validation Experiment, in Advances in Atmospheric Remote Sensing with Lidar,* edited by A. Ansmann, R. Neuber, P. Rairoux, and U. Wandinger, Springer-Verlag, Berlin, 289–295, 1997a.

Browell, E. V., S. Ismail, T. C. McElroy, R. M. Hoff, and A. E. Dudelzak, Global measurements of ozone and aerosol distributions with a space-based lidar system, AGY Fall Meeting, December 8–12, 1997b.

Bufton, J. L., Laser altimetry measurements from aircraft and spacecraft, Proceedings IEEE, Vol. 77, 463, 1989.

Bufton, J. L, J. B. Garvin, J. F. Cavanaugh, L. Ramos-Izquierdo, T. D. Clem, and W. B. Krabill, Airborne lidar for profiling of surface topography, *Opt. Eng.*, 30, 72, 1991.

Canavan, G. H. and J. E. Butterworth, Economic Value of Global Weather Measurements, LA-UR-99-0903, Los Alamos National Laboratory, 1999.

Chanin, M. L., A. Garnier, A. Hauchecorne, and J. Porteneuve, A Doppler lidar for measuring winds in the middle atmosphere, *Geophys. Res. Lett.*, 16(11), 1273–1276, 1989.

Chen, P. C., T. T. Saha, A. M. Smith, and R. Romero, Progress in very lightweight optics using graphite fiber composite materials, *Opt. Eng.*, 37, 666–676, 1998.

Churnside, J. H. and H. T. Yura, Speckle statistics of atmospherically backscattered laser light, *Appl. Opt.*, 22, 2559–2565, 1983.

Clausen, M., M. Kalb, G. McConaughy, R. Muller, S. Neeck, M. Seablom, and M. Steiner, Advanced Weather Prediction Technologies: NASA's Contribution to the Operational Agencies, Report to NASA's Earth Science Technology Office, May 31, 2002.

Clough, S. A., M. I. Iacono, J. L. Moncet, Line-by-line calculations of atmospheric fluxes and cooling rates: application to water vapor, *J. Geophys. Res.*, 97(D14), 15761–15785, 1992.

T. D. Cole, NEAR Laser Rangefinder: a tool for the mapping and topologic study of asteroid 433 Eros, Johns Hopkins Appl. Technical Digest, 19, no. 2, 1998.

Cordes, J. J., Economic Benefits and Costs of Developing and Deploying a Space-Based Wind Lidar, Economics Discussion Paper D-9502, Department of Economics, George Washington University, 1995.

Couch, R. H., et al., Lidar In-space Technology Experiment (LITE): NASA's first in-space lidar system for atmospheric research, *Opt. Eng.*, 30, 88–95, 1991.

Curran, R. J., et al, LAWS — Laser Atmospheric Wind Sounder, Vol. IIg, NASA EOS Instrument Panel Report, 1987.

Del Genio, A. D., W. Kovari, M.-S. Yao, Climatic implications of the seasonal variation of upper troposphere water vapor, *Geophys. Res. Lett.*, 21, 2701–2704, 1994.

Ehret, G., C. Kiemle, B. Mayer, and M. Wirth, *WALES (Water Vapor Lidar Experiment in Space): A Sensitivity Analysis, in Lidar Remote Sensing in Atmospheric and Earth Sciences*, edited by L. Bissonnette, G. Roy, and G. Vallee, Defence R&D Canada, Valcartier, 759–762, July 2002.

Einstein, A., On the electrodynamics of moving bodies, *Ann. Phys.*, 17, 891, 1905.

Endemann, M., Application of Lasers for Climatology and Atmospheric Research, Final Report for ESTEC Contract N. 4868/81/ NL/HP (SC), 1983.

ESA, Atmospheric Dynamics Mission, SP-1233 (4), European Space Agency, 1999.

Evans, W. E., Remote Probing of High Cloud Cover via Satellite-Borne Lidar, NASA CR-96893, 1968.

Fishman, J., K. Fakhruzzaman, B. Cros, and D. Nganga, Identification of widespread pollution in the southern-hemisphere deduced from satellite analyses, *Science*, 252, 1693–1696, 1991.

Fishman, J., C. E Watson, J. C. Larsen, and J. A. Logan, Distribution of tropospheric ozone determined from satellite data, *J. Geophys. Res. Atmos.*, 95, 3599–3617, 1990.

Fix, A., V. Weiss, and G. Ehret, Injection seeded optical parametric oscillator by radiation injection, *Pure Appl. Opt.*, 7, 837–852, 1998.

Flamant, P. H., *ESA <AEOLUS-ADM> to Probe Atmospheric Winds, in Lidar Remote Sensing in Atmospheric and Earth Sciences*, edited by L. Bissonnette, G. Roy, and G. Vallee, Defence R&D Canada – Valcartier, pp. 815-826, July 2002.

Fleisa, C. and C. L. Korb, Theory of the double-edge molecular technique for Doppler lidar wind measurement, *Appl. Opt.*, 38(3), 432, 1999.

Fleisa, C., C. L. Korb, and C. Hirt, Double-edge molecular measurement of lidar wind profiles at 355 nm, *Opt. Lett.*, 25(19), 1466, 2000.

Frehlich, R., Simulation of coherent Doppler lidar performance in the weak-signal regime, *J. Atmos. Ocean. Technol.*, 13, 646, 1996.

Frehlich, R. G. and M. J. Kavaya, Coherent laser radar performance for general atmospheric refractive turbulence, *Appl. Opt.*, 30, 5325–5352, 1991.

Frehlich, R. G. and M. J. Yadlowsky, Performance of mean-frequency estimators for Doppler radar and lidar, *J. Atmos. Ocean. Technol.*, 11(5), 1217–1230, 1994.

Fried, D. L., Atmospheric modulation noise in an optical heterodyne receiver, *IEEE J. Quant. Electron.*, QE-3, 213–221, 1967.

Gardner, C. S., Ranging performance of satellite laser altimeters, *IEEE Trans. Geosci. Remote Sensing*, 30, 1061, 1992.

J. Garvin, J. Bufton, J. Blair, D. Harding, S. Luthcke, J. Frawley, D. Rowlands, Observations of the Earth's Topography from the Shuttle Laser Altimeter (SLA): Laser-pulse Echo-recovery Measurements of Terrestrial Surfaces, Physics and Chemistry of the Earth, 23, no. 9, 1053–1068, 1998.

Gentry, B., Technology assessment of direct detection in support of the GTWS, 17th Meeting of the Working Group on Space-Based Lidar Winds, 2002.

Gentry, B. M., H. Chen, and S. X. Li, Wind measurements with 355-nm molecular Doppler lidar, *Opt. Lett.*, 25(17), 1231, 2000.

Geoscience Laser Altimeter System, *MTPE/EOS Reference Handbook*, NASA, Washington, DC, 132, 1995.

Gerard, E., et al., Major advances foreseen in humidity profiling from the Water Vapour Lidar Experiment in Space (WALES), *Bull. Amer. Metor. Soc.*, 85, 237–252, 2004.

Gu, Y. Y. Y., C. S. Gardner, P. A. Castelberg, et al., Validation of the Lidar In-space Technology Experiment: stratospheric temperature and aerosol measurements, *Appl. Opt.*, 36, 5148–5157, 1997.

Gudimetla, V. S. R. and M. J. Kavaya, Special relativity corrections for space-based lidars, *Appl. Opt.*, 38, 6374, 1999.

Halem, M. and R. Dlouhy, Observing system simulation experiments related to space-borne lidar wind profiling. part I: forecast impacts of highly idealized observing systems, Proceedings of the AMS Conference on Satellite/Remote Sensing and Application, 272, 1984.

Henderson, S. W., C. P. Hale, J. R. Magee, M. J. Kavaya, and A. V. Huffaker, Eye-safe coherent laser radar system at 2.1 μm using Tm,Ho:YAG lasers, *Opt. Lett.*, 16(10), 773–775, 1991.

Hibbard, D. L., Electrochemically deposited nickel alloys with controlled thermal expansion for optical applications, *Proc. SPIE*, 2543, 236–243, 1995.

Holmes, J. F. and B. J. Rask, Optimum optical local-oscillator power levels for coherent detection with photodiodes, *Appl. Opt.*, 34, 927–933, 1995.

Hovis, F., G. Witt, E. Sullivan, K. Andes, B. Suliga, K. Le, E. Fakhoury, and M. Fromm, Space-based laser design, *Advanced Solid-State Photonics*, edited by John J. Zayhowski OSA, Technical Digest, 93–96, 2003.

Hsu, Y. W. and R. A. Johnson, Design and analysis of one meter beryllium space telescope, *Proc. SPIE*, 2543, 244–257, 1995.

Huffaker, R. M., editor, Feasibility Study of Satellite-Borne Lidar Global Wind Monitoring System, NOAA Technical Memorandum ERL WPL-37, 1978.

Huffaker, R. M. and R. M. Hardesty, Remote sensing of atmospheric wind velocities using solid-state and CO2 coherent laser systems, *Proc. IEEE*, 84(2), 181–204, 1996.

IPCC, 2001: Climate Change 2001 — The Scientific Basis, The Intergovernmental Panel on Climate Change Third Assessment Report, Cambridge University Press. (Available at www.ipcc.ch.)

Ismail, S. and E. V. Browell, Airborne and spaceborne lidar measurements of water vapor profiles: a sensitivity analysis, *Appl. Opt.*, 28, 3603–3615, 1989.

Ismail, S. and E. V. Browell, NASA LASE water vapor differential absorption lidar measurements and performance evaluation, IEEE Co-Meas Meeting, Atlanta, GA, April 6–10, 1995.

A. V. Jelalien, *Laser Radar Systems*, Artech House, Boston, pp. 139–140, 1992.

Kaiser, J. and K. Schmidt, Coming to grips with the world's greenhouse gases, *Science*, 281, 504–506, 1998.

Kamineni, R., T. N. Krishnamurti, R. A. Ferrare, S. Ismail, and E. V. Browell, Impact of high resolution water vapor cross-sectional data on hurricane forecasting, *Geophys. Res. Lett.*, 130, 1234–1239, March 2003.

Karyampudi, V. M., et al., Validation of the Saharan dust plume conceptual model using lidar, meteosat, and ECMWF data, *Bull. Am. Meteorol. Soc.*, 80, 1045–1075, 1999.

Kavaya, M. J. and G. D. Emmitt, The space readiness coherent lidar experiment (SPARCLE) space shuttle mission, *Proc. SPIE*, 3380, 2–11, 1998.

Kavaya, M. J., G. D. Emmitt, R. G. Frehlich, F. Amzajerdian, and U. N. Singh, A space-based point design for global coherent Doppler wind lidar profiling matched to the recent NASA/NOAA draft science requirements, Digest of the 21st International Laser Radar Conference, Quebec City, Quebec, Canada, 817, July 8–12, 2002.

Kavaya, M. J., S. W Henderson, J. R. Magee, C. P. Hale, and R. M. Huffaker, Remote wind profiling with a solid-state Nd:YAG coherent lidar system, *Opt. Lett.*, 14(15), 776–778, 1989a.

Kavaya, M. J., S. W. Henderson, E. C. Russell, R. M. Huffaker, and R. G. Frehlich, Monte Carlo computer simulations of ground-based and space-based coherent DIAL water vapor profiling, *Appl. Opt.*, 28, 840, 1989b.

Kavaya, M. J., G. D. Spiers, E. S. Lobl, J. Rothermel, and V. W. Keller, Direct global measurements of tropospheric winds employing a simplified coherent laser radar using fully scalable technology and technique, *Proc. SPIE*, 2214, 237, 1994.

Kavaya, M. J., G. D. Spiers, and, R. G. Frehlich, Potential pitfalls related to space-based lidar remote sensing of the earth with an emphasis on wind measurement, *Proc. SPIE*, 4153, 385, 2001.

Keeling, C. D, Atmospheric Carbon Dioxide Variations at Mauna Loa Observatory, Hawaii, Tellus, Stockholm, Vol. 26, No. 6, 538–551, 1976.

Keller, L. M. and D. R. Johnson, An atmospheric energy analysis of the impact of satellite lidar winds and TIROS temperatures in global simulations, *Monthly Weather Rev.*, 120, 2831, 1992.

Kent, G. S., C. R. Trepte, K. M. Skeens, and D. M. Winker, LITE and SAGE II measurements of aerosols in the southern hemisphere upper troposphere. *J. Geophys. Res.*, 103, 19111–19127, 1998.

Kingston, R. H., *Detection of Optical and Infrared Radiation*, Springer-Verlag, Berlin, 1979.

Kingston, R. H., *Optical Sources, Detectors, and Systems Fundamental and Applications*, Academic Press, San Diego, CA, 1995.

Koch, G. J., Coherent Differential Absorption Lidar for Combined Measurement of Wind and Trace Atmospheric Gases, Ph.D. Dissertation, Old Dominion University, May 2001.

Koch, G. J., A. N. Dharamsi., C. M. Fitzgerald, and J. C. McCarthy, Frequency stabilization of a Ho:Tm:YLF laser to absorption lines of carbon dioxide, *Appl. Opt.*, 39, 3664–3669, 2000.

Koch, G. J., M. Petros, J. Yu, and U. N. Singh, Precise wavelength control of a single-frequency pulsed Ho:Tm:YLF laser, *Appl. Opt.*, 41, 1718–1721, 2002.

Koch, G. J., et al., Coherent differential absorption lidar measurements of CO_2, *Appl. Opt.*, 43, 50–92–5099, 2004.

Korb, C. L., B. M., Gentry, and S. X. Li., Edge technique Doppler lidar wind measurements with high vertical resolution, *Appl. Opt.*, 36(24), 5976, 1997.

Korb, C. L., B. M. Gentry, S. X. Li, and C. Fleisa, Theory of the double-edge technique for Doppler lidar wind measurement, *Appl. Opt.*, 37(15), 3097, 1998.

Korb, C. L., B. M. Gentry, and C. Y. Weng, Edge technique: theory and application to the lidar measurement of atmospheric wind, *Appl. Opt.*, 31(21), 4202, 1992.

Krishnamurti, T. N., G. Rohaly, and H. S. Bedi, On the improvement of precipitation forecast skill from physical initialization, *Tellus*, 46A, 598–614, 1994.

Lake, M. S. and L. D. Peterson, Research on the problem of high-precision deployment for large-aperture space-based science instruments, Space Technology and Applications International Forum 1998, Part I, January 1998.

Lake, M. S., J. E. Phelps, J. E. Dyer, D. A. Caudle, A. Tam, J. Escobedo, and E. P. Kasl, A deployable primary mirror for space telescopes, *Proc. SPIE*, 3785, 14–25, 1999.

Leblanc, T. and I. S. McDermid, Stratospheric ozone climatology from lidar measurements at Table Mountain and Mauna Loa, *J. Geophys. Res.*, 105, 14613–14623, 2000.

Lorenc, A. C., R. J Graham, I. Dharssi, B. Macpherson, N. B. Ingleby, and R. W. Lunnon, Preparation for the Use of Doppler Wind Lidar Information in Meteorological Data Assimilation Systems, UK Meteorological Office, Final Report on ESA Contract No. 9063/90/HE-I, 1992.

A. Mahesh, M. A. Grey, S. P. Palm, W. D. Hart and J. D. Spinhirne, Passive and active detection of clouds: Comparisons between MODIS and GLAS observations, *J. Geophys. Res.*, 108, L04108–L04108, 2004.

McGarry, J. F., L. K. Pacini, J. B. Abshire, and J. B. Blair, Design and Performance of an Autonomous Tracking System for the Mars Observer Laser Altimeter Receiver, Technical Digest — CLE091, Paper CThR27, Optical. Society of America, Washington, DC, 1991.

McGill, M. J., W. D. Hart, J. A. McKay, and J. D. Spinhirne, Modeling the performance of direct-detection Doppler lidar systems including cloud and solar background variability, *Appl. Opt.*, 38(30), 6388, 1999.

McGill, M. J. and J. D. Spinhirne, Comparison of two direct-detection Doppler lidar techniques, *Opt. Eng.*, 37(10), 2675–2686, 1998.

McIntyre, R. J., Multiplication noise in uniform avalanche photodiodes, *IEEE Trans. Electron. Devices*, ED-13, 164–168, 1966.

McKay, J. A., Modeling of direct detection Doppler wind lidar I. The edge technique, *Appl. Opt.*, 37(27), 6480, 1998a.

McKay, J. A., Modeling of direct detection Doppler wind lidar II. The fringe imaging technique, *Appl. Opt.*, 37(27), 6487, 1998b.

McKay, J. A., Fabry-perot etalon aperture requirements for direct detection Doppler wind lidar from earth orbit, *Appl. Opt.*, 38(27), 5859, 1999.

McKay, J. A. and D. Rees, Space-based Doppler wind lidar: modeling of edge detection and fringe imaging Doppler analyzers, *Adv. Space Res.*, 26(6), 883–891, 2000.

Measures, R. M., *Laser Remote Sensing Fundamental and Applications*, Krieger Publications Co., Malabar, FL, 1984.

Menzies, R. T., Laser heterodyne detection techniques, *Laser Monitoring of the Atmosphere*, edited by E. D. Hinkley, Springer-Verlag, Berlin, Ch. 7, 1976.

Menzies, R. T., Coherent and incoherent lidar — an overview, *Tunable Solid State Lasers for Remote Sensing*, edited by R. L. Byer, E. K. Gustafson, and R. Trebino, Springer-Verlag, Berlin, 17, 1985.

Menzies, R. T., Doppler lidar atmospheric wind sensors: a comparative performance evaluation for global measurement applications from earth orbit, *Appl. Opt.*, 25(15), 2546–2553, 1986.

Menzies, R. T. and D. M. Tratt, Differential laser absorption spectrometry for global profiling of tropospheric carbon dioxide: selection of optimum sounding frequencies for high precision measurements, *Appl. Opt.*, 42, 6569–6577, 2003.

Millar, P. S. and J. M. Sirota, The Geoscience Laser Altimeter System stellar reference system, 19th International Laser Radar Conference (ILRC) (NASA/CP-1998-207671/PT2), edited by U. N. Singh, S. Ismail, and G. Schwemmer, Annapolis, MD, 215–218, 1998.

Mims, F. M. III, *Light-Beam Communications*, Howard W. Sams & Co, 151–152, Indianapolis, 1975.

NASA, Understanding Earth System Change, NASA's Earth Science Enterprise Research Strategy for 2000–2010, December 2000.

Neumann, G. A., D. E. Smith, and M. T. Zuber, Two Mars years of clouds detected by the Mars Orbiter Laser Altimeter, *J. Geophys. Res.*, 108(E4), 5023, 2003.

Newell, R. E., Y. Zhu, E. V. Browell, S. Ismail, W. G. Read, et al., Upper tropospheric water vapor and cirrus: DC-8 observations, UARS MLS measurements and meteorological analysis intercompared, *J. Geophys. Res.*, 101, 1931–1941, 1996.

Osborn, M. T., G. S. Kent, and C. R. Trepte, Stratospheric aerosol measurements by the lidar in space technology experiment, *J. Geophys. Res. — Atmos.*, 103, 11447–11453, 1998.

Peters, B. R., P. J. Reardon, F. Amzajerdian, and T. S. Blackwell, Lightweight lidar telescopes for space applications, SPIE's 2nd International Asia-Pacific Symposium on Remote Sensing of the Atmosphere, Environment, and Space, Sendai, Japan, October 2000.

Platt, C. M. R., D. M. Winker, M. A. Vaughan, and S. D. Miller, Backscatter-to-extinction ratios in the top layers of tropical mesoscale convective systems and in isolated cirrus from LITE observations, *J. Appl. Meteorol.*, 38, 1330–1345, 1999.

Ramos-Izquierdo, L., J. L. Bufton, and P. Hayes, Optical system design and integration of the Mars observer laser altimeter, *Appl. Opt.*, 33, 307, 1994.

Reagan, J. A., et al., Spaceborne lidar calibration from cirrus and molecular backscatter returns. *Trans. Geosci. Remote Sensing*, 40, 2285–2290, 2002.

Remsberg, E. E. and L. L. Gordley, Analysis of differential absorption lidar, *Appl. Opt.*, 17, 624–630, 1978.

Richter, D. A., W. D. Marsh, N. S. Higdon, T. H. Chyba, and T. Zenker, Tunable UV generation with a frequency-mixed type II OPO, Aerospace Conference Proceedings, 2000 IEEE, Piscataway, NJ, Vol. 3, 35–41, 2000.

Rothman, L. S., C. P. Rinsland, A. Goldman, et al., The HITRAN Molecular Spectroscopic Database and HAWKS (HITRAN Atmospheric WorkStation), 1966 edition, JQSRT, 60, 665–710, 1998.

Rye, B. J., Primary aberration contribution to incoherent backscatter heterodyne lidar returns, *Appl. Opt.*, 21, 839–844, 1982.

Rye, B. J., Comparative precision of distributed-backscatter Doppler lidars, *Appl. Opt.*, 34(36), 8341–8344, 1995.

Rye, B. J. and R. M. Hardesty, Discrete spectral peak estimation in incoherent backscatter heterodyne lidar. I: spectral accumulation and the cramer-rao lower bound, *IEEE Trans. Geosci. Remote Sensing*, 31, 16, 1993a.

Rye, B. J. and R. M. Hardesty, Discrete spectral peak estimation in incoherent backscatter heterodyne lidar. II: correlogram accumulation, *IEEE Trans. Geosci. Remote Sensing*, 31, 28, 1993b.

Sachse, G. W., L.-G. Wang, S. Ismail, and E. V. Browell, Line center/ side line diode laser seeding for DIAL measurements of the atmosphere, Proceedings of Optical Remote Sensing of the Atmosphere, OSA, Salt Lake City, UT, 121–123, February 5–9, 1995.

Sassen, K., The polarization lidar technique for cloud research: a review and current assessment. *Bull. Am. Meteorol. Soc.*, 72, 1848–1866, 1991.

Schotland, R. M., Some observations of the vertical profile of water vapor by means of ground based optical radar, Proceedings of the Fourth Symposium on Remote Sensing of the Environment, Ann Arbor, MI, 12–24, April 1966.

Shapiro, J. H., Correlation scales of laser speckle in heterodyne detection, *Appl. Opt.*, 24, 1883–1888, 1985.

Shellenbarger, J. A., M. G. Mauk, M. I. Gottfried, J. D. Lesko, L. C. Dienetta, GaInAsSb and InAsSbP photo detectors for mid-infrared wavelengths, *Proc. SPIE*, 2999, 25–33, February 12–14, 1997.

Singh, U. N. and W. S. Heaps, Integrated NASA lidar system strategy for space-based remote sensing, *Laser Remote Sensing in Atmosphere and Earth Sciences*, edited by L. R. Bissonnette, G. Roy, and G. Vallee, 11–13, Defence R&D Canada, Valcartier 2002.

Singh, U. N., J. Yu, M. Petros, N. P. Barnes, J. A. Williams-Byrd, G. E. Lockhard, and E. A. Modlin, Injection-seeded, room-temperature, diode-pumped Ho:Tm:YLF laser with output energy of 600 mJ at 10 Hz, Advanced Solid-State Laser Conference, *OSA Trends in Optics and Photonics Series*, Vol. 19, 194, 1998.

Smith, D. E., M. A. Zuber, H. V. Frey, J. B. Gervin, J. W. Head, et al., Mars Orbiter Laser Altimeter: experiment summary after the first year of global mapping of Mars, *J. Geophys. Res.*, 106, 23689–23722, 2001.

Smith, D. E., et al., The global topography of Mars and implications for surface evolution, *Sci. Mag.*, 284, 1495–1503, 1999.

D. E. Smith, M. T. Zuber, G. A. Neumann, and F. G. Lemoine, Topography of the Moon from the Clementine lidar, *J. Geophys. Res.*, 102, 1591–1611, 1997.

Souprayen, C., A. Garnier, A. Hertzog, A. Hauchecorne, and J. Porteneuve, Rayleigh-Mie Doppler wind lidar for atmospheric measurements, I. Instrumental setup, validation, and first climatological results, *Appl. Opt.*, 38(12), 2410, 1999.

Stadler, J. H., E. V. Browell, S. Ismail, A. E. Dudelzak, and D. J. Ball, Ozone research with advanced cooperative lidar experiment (OR-ACLE) implementation study, Nineteenth International Laser Radar Conference, edited by U. N. Singh, S. Ismail, and G. Schwemmer, NASA CP-1998-207671/PT2, 845–847, 1998.

Stephens, G. L., R. J. Engelen, M. Vaughan, and T. L. Anderson, Toward retrieving properties of the tenuous atmosphere using space-based lidar measurements, *J. Geophys. Res. — Atmos.*, 106, 28143–28157, 2001.

Sun, X., J. B. Abshire, H. Riris, J. McGarry, and M. Sirota, Geoscience Laser Altimeter System (GLAS) on the ICES at Mission: Sscience measurement performance since launch, AGU Fall Meeting, Paper C21B-03, December 14, 2004.

Taczak, T. M. and D. K. Killinger, Development of a tunable, narrow-linewidth, cw 2.066 μm Ho:YLF laser for remote sensing of atmospheric CO_2 and H_2O, *Appl. Opt.*, 37, 8460–8476, 1998.

Targ, R., M. J. Kavaya, R. M. Huffaker, and R. L Bowles, Coherent lidar airborne windshear sensor: performance evaluation, *Appl. Opt.*, 30, 2013, 1991.

Uchino, O., Japanese backscatter and DIAL lidar studies in the J-POP perspective, GEWEX Symposium Abstracts, Center National d'Etudes Spatiales, Paris, France, 1–3, June 15–18, 1992.

Van Bladel, J., *Relativity and Engineering*, Springer-Verlag, Berlin, 1984.

Vaughan, M. A. and D. M. Winker, SIBYL: a selective iterated boundary location algorithm for finding cloud and aerosol layers in CALIPSO lidar data. Proceedings of the 21st International Laser Radar Conference, Quebec City, Quebec, Canada, 791–794, 2002.

Wang, J. Y., Detection efficiency of coherent optical radar, *Appl. Opt.*, 23, 3421–3427, 1984.

Winker, D. M., R. H. Couch, and M. P. McCormick, An overview of LITE: NASA's Lidar In-space Technology Experiment, Proc. IEEE, 84, 164–180, 1996.

Winker, D. M., J. Pelon, and M. P. McCormick, The CALIPSO mission: spaceborne lidar for observation of aerosols and clouds, *Proc. SPIE*, 4893, 1–11, 2003.

Winker, D. M. and C. R. Trepte, Laminar cirrus observed near the tropical tropopause by LITE, *Geophys. Res. Lett.*, 25, 3351–3354, 1998.

Winker, D. M. and M. A. Vaughan, Vertical distribution of clouds over Hampton, Virginia observed by lidar under the ECLIPS and FIRE ETO programs. *Atmos. Res.*, 34, 117–133, 1994.

Young, S. A., M. A. Vaughan, and D. M. Winker, Adaptive methods for retrieving extinction profiles from space applied to CALIPSO lidar data, Proceedings of the 21st International Laser Radar Conference, Quebec City, Quebec, Canada, 743–746, 2002.

Yu, J., U. N. Singh, J. C. Barnes, N. P. Barnes, and M. Petros, An efficient double pulsed 2-micron laser for DIAL applications, *Advances in Laser Remote Sensing*, edited by A. Dabas, C. Loth, and J. Pelon, 53–55, 2000a.

Yu, J., U. N. Singh, N. P. Barnes, J. C. Barnes, M. Petros, and M. W. Phillips, An eye safe all solid state laser for coherent wind lidar in space, *Proc. SPIE*, 3504, 152–158, 1998a.

Yu, J., U. N. Singh, N. P. Barnes, J. C. Barnes, M. Petros, and M. W. Phillips, An all solid-state 2-μm laser system for space coherent wind lidar, Proceedings of IEEE 2000 Aerospace Conference, Vol. 3, 27–33, 2000b.

Yu, J., U. N. Singh, N. P. Barnes, and M. Petros, 125-mJ diode-pumped injection-seeded Ho;Tm:YLF laser, *Opt. Lett.*, 23, 780–782, 1998b.

Yura, H. T., Optical heterodyne signal power obtained from finite sized sources of radiation, *Appl. Opt.*, 13, 150–157, 1974.

Zhao, Y., M. J. Post, and R. M. Hardesty, Receiving efficiency of monostatic pulsed coherent lidars. 1: Theory, *Appl. Opt.*, 29(28), 4111–4119, 1990.

Zuber, M. T., D. E. Smith, S. C. Solomon, D. O. Muhleman, J. W. Head, J. B. Garvin, J. B. Abshire, and J. L. Bufton, The Mars observer laser altimeter investigation, *J. Geosci. Res.*, 97, 7781, 1992.

T - #0286 - 071024 - C924 - 229/152/40 - PB - 9780367392529 - Gloss Lamination